Recent Trends and Advances in Model Based Systems Engineering

Azad M. Madni • Barry Boehm • Daniel Erwin
Mahta Moghaddam • Michael Sievers
Marilee Wheaton
Editors

Recent Trends and Advances in Model Based Systems Engineering

Springer

Editors
Azad M. Madni
Department of Astronautical Engineering
University of Southern California
Los Angeles, CA, USA

Daniel Erwin
Department of Astronautical Engineering
University of Southern California
Los Angeles, CA, USA

Michael Sievers
Department of Astronautical Engineering
University of Southern California
Los Angeles, CA, USA

Barry Boehm
Department of Industrial and Systems Engineering
University of Southern California
Los Angeles, CA, USA

Mahta Moghaddam
Department of Electrical and Computer Engineering
University of Southern California
Los Angeles, CA, USA

Marilee Wheaton
Department of Astronautical Engineering
University of Southern California
Los Angeles, CA, USA

ISBN 978-3-030-82082-4 ISBN 978-3-030-82083-1 (eBook)
https://doi.org/10.1007/978-3-030-82083-1

This Springer imprint is published by the registered company Springer Nature Switzerland AG
The registered company address is: Gewerbestrasse 11, 6330 Cham, Switzerland

Preface

Systems engineering is undergoing an exciting transformation that is motivated by mission and system complexity and paced by advances in model-based systems engineering (MBSE) and digital engineering (DE). This transformation is enabled by Industry 4.0, the Internet of Things (IoT), and the ongoing convergence of systems engineering with other disciplines. The central theme of the 2020 Conference on Systems Engineering Research (CSER) was motivated by these developments. Specifically, this conference was focused on exploring recent trends and advances in model-based systems engineering (MBSE) and the synergy of MBSE with other disciplines (e.g., digital engineering) and technologies (e.g., simulation, AI, machine learning).

Systems engineering today has two major thrusts: traditional methods that work well for relatively mature and not overly complicated or complex systems, and new and innovative methods that are specifically driven by the increasing complexity of sociotechnical systems and advances in engineering, materials, computation, and convergence. The latter has become increasingly important for addressing problems in the twenty-first century. MBSE is rapidly becoming a vital system engineering advance to address such problems.

Researchers from academia, industry, and government submitted papers on a variety of MBSE topics for this conference. These include ontologies and MBSE, MBSE processes, model-based methods in systems architecting, modeling approaches in MBSE, MBSE standards, MBSE languages, synergy between MBSE and digital engineering (DE), economic analysis of MBSE, MBSE application areas (e.g., manufacturing, aerospace, defense), and the future of MBSE.

This volume is a compendium of peer-reviewed research papers from university, government, and industry researchers who participated in 2020 CSER. It brings together diverse domains and technical competencies of model-based systems engineering (MBSE) in a single, comprehensive volume. To help the reader conveniently navigate this volume, the chapters are organized into seven parts. Each part represents a key MBSE research area.

It is our hope that this volume will get the readers interested in pursuing MBSE research beyond the traditional application areas and take on complex scientific and societal problems of national and global significance. On behalf of the editors, I want to thank all who contributed to this volume. We hope that you find the contents of this volume inspiring and potentially useful building blocks for future research.

University of Southern California,
Los Angeles, CA, USA

Azad M. Madni

Contents

About the Editors

Azad M. Madni is a member of the National Academy of Engineering and the University Professor of Astronautics, Aerospace and Mechanical Engineering. He also has a joint appointment in Civil and Environmental Engineering at the University of Southern California's Viterbi School of Engineering. He is the holder of the Northrop Grumman Foundation Fred O'Green Chair in Engineering. He is the executive director of USC's systems architecting and engineering program and the founding director of the Distributed Autonomy and Intelligent Systems Laboratory. He is the founder and CEO of Intelligent Systems Technology, Inc., a hi-tech company that conducts research and offers educational courses in intelligent systems for education and training. He is the chief systems engineering advisor to The Aerospace Corporation. He received his Ph.D., M.S., and B.S. degrees in Engineering from UCLA. His recent awards include *2021 INCOSE Benefactor Award, 2021 IEEE AESS Judith A. Resnik Space Award, 2020 IEEE SMC Norbert Wiener Award, 2020 NDIA's Ferguson Award* for Excellence in Systems Engineering, *2020 IEEE-USA Entrepreneur Achievement Award, 2019 IEEE AESS Pioneer Award, 2019 INCOSE Founders Award, 2019 AIAA/ASEE Leland Atwood Award, 2019 ASME CIE Leadership Award, 2019 Society for Modeling and Simulation International Presidential Award, and 2011 INCOSE Pioneer Award.* He is a Life Fellow/Fellow of IEEE, INCOSE, AIAA, AAAS, SDPS, IETE, AAIA, and WAS. He is the co-founder and current chair of IEEE SMC Technical Committee on Model Based Systems Engineering. He is the author of *Transdisciplinary Systems Engineering: Exploiting Convergence in a Hyper-Connected World* (Springer 2018). He is the co-author of *Tradeoff Decisions in System Design* (Springer, 2016).

Barry Boehm is a member of the National Academy of Engineering and Distinguished Professor of Computer Science in the Viterbi School of Engineering at the University of Southern California. His honors and awards include INCOSE Pioneer, guest lecturer at the USSR Academy of Sciences, the AIAA Information Systems Award, the J.D. Warnier Prize for Excellence in Information Sciences, the ISPA Freiman Award for Parametric Analysis, the NSIA Grace Murray Hopper Award, the Office of the Secretary of Defense Award for Excellence, the ASQC Lifetime Achievement Award, and the ACM Distinguished Research Award in Software Engineering . He is an AIAA Fellow, an ACM Fellow, an IEEE Fellow. He received his B.A. degree from Harvard University, and his M.S. and Ph.D. degrees from UCLA, all in mathematics.

Daniel Erwin is a professor and chair of astronautical engineering, and an associate fellow of AIAA. He is also the faculty member who oversaw the all-student research team that set the student altitude record for rockets to pass the Karman line into outer space. He is the co-director of USC's Distributed Autonomy and Intelligent Systems Laboratory. His awards include the 2017 Engineers' Council Distinguished Engineering Project Achievement Award, 2016 Engineers' Council Distinguished Engineering Educator Award, 2006 USC-LDS Student Association Outstanding Teaching Award, 1995 USC School of Engineering Outstanding Teaching Award, and 1993 TRW, Inc. TRW Excellence in Teaching Award. He received his B.S. in applied physics from California Institute of Technology and his M.S. and Ph.D. in electrical engineering from the University of Southern California.

Mahta Moghaddam is Distinguished Professor of Electrical and Computer Engineering and a member of the National Academy of Engineering. She is an IEEE Fellow and a past president of IEEE Antennas and Propagation Society. She is the editor-in-chief of the *IEEE Antennas and Propagation* journal. She has received several awards and honors including NASA Honor Award, Outstanding Public Service Leadership Medal, and numerous group achievement awards. She is the president and co-founder of Maxwell Medical Corporation, a USC startup that is developing medical diagnostic therapy and interoperative monitoring devices using microwave technology. She received her B.S. in electrical engineering from the University of Kansas with highest distinction, and her M.S. and Ph.D. degrees in electrical engineering from the University of Illinois at Urbana-Champaign.

Marilee Wheaton is a systems engineering fellow of The Aerospace Corporation and president of the International Council on Systems Engineering. She is a fellow of AIAA and INCOSE, and a senior member of IEEE. Previously, she was a general manager at The Aerospace Corporation. She has also served as an adjunct associate professor of systems architecting and engineering program at the University of Southern California for 11 years. Marilee is a life member and fellow of the Society of Women Engineers. She is an advisory board member of California State University, Northridge College of Engineering and Computer Science. She received her B.A. degree in mathematics from California Lutheran University and her M.S. degree in systems engineering from University of Southern California. Marilee has been involved in the Conference on Systems Engineering Research in a leadership role for nearly a decade. She is currently pursuing her doctorate at the University of Southern California in astronautical engineering with a specialization in systems engineering.

Michael Sievers is a senior systems engineer at Caltech's Jet Propulsion Laboratory in Pasadena, California, and is responsible for developing and analyzing spacecraft and ground systems architectures. He is also an adjunct lecturer at the University of Southern California where he teaches classes in systems and system of systems architectures, engineered resilience, and model-based systems engineering. Dr. Sievers earned his Ph.D. and master's degrees in computer science and a Bachelors' degree in electrical engineering all from UCLA. He has published over 50 journal and conference papers and is an INCOSE Fellow, AIAA Associate Fellow, and IEEE Senior Member.

Part I
MBSE and Digital Engineering

Toward a Reference Architecture for Digital and Model-Based Engineering Information Systems

Hayden C. Daly and Paul T. Grogan

Abstract Digital and model-based engineering envisions a future where software systems are intricately involved in systems engineering and engineering design efforts. Recent advances in the field of software engineering have the potential to enable more flexible, reconfigurable, and updateable systems for engineering applications. This paper introduces an information system reference architecture for digital and model-based engineering activities based on modern web-based architectural styles. An application case explains how the reference architecture shaped the implementation of the Tradespace Analysis Tool for Constellations (TAT-C) Knowledge Base, a software component for space systems engineering that maintains a resource library conforming to common object schemas. Database, back-end, and front-end software components serve as architectural layers connected by simple information protocols based on semantic linked data models for improved interoperability.

Keywords Digital engineering · Layered architecture · Model-based engineering · Model interoperability · Semantic web technology · Software architecture

1 Introduction

Digital and model-based engineering (DMbE) envisions the widespread use of digital artifacts, digital environments, and digital tools to support engineering activities (Hale et al. 2017). It encompasses recent efforts in model-based systems engineering to adopt semantic frameworks and graphical modeling languages as a means to represent systems models in a common, interoperable format (Bone et al. 2018, 2019). However, developing an infrastructure platform for information technology that is "flexible, reconfigurable, and updateable" remains a significant

H. C. Daly · P. T. Grogan (✉)
Castle Point on Hudson, Stevens Institute of Technology, Hoboken, NJ, USA

A. M. Madni et al. (eds.), *Recent Trends and Advances in Model Based Systems Engineering*, https://doi.org/10.1007/978-3-030-82083-1_1

challenge in DMbE (Hale et al. 2017). This challenge sits at the intersection between two fields that increasingly overlap: systems engineering and software engineering.

The field of software engineering has experienced significant growth, innovation, and change in recent decades. Indeed, many of the systems modeling languages including SysML, OPM, and IDEF0 evolved from software engineering practice established in the 1990s to standardize software design and use object-oriented programming styles to accommodate greater levels of product complexity (Dori 2016). More recent trends in software engineering focus on service-orientation, web-based application programming interfaces (APIs), and containerization as further systems-level techniques to accommodate increased product complexity with more distributed software architectures.

Drawing from the state-of-the-art in software engineering practice, this paper advances a reference software architecture to support DMbE practices. The proposed reference architecture has been successfully implemented for the Tradespace Analysis Tool for Constellations Knowledge Base (TAT-C KB), a systems engineering software tool in the domain of space systems. Based on insights and experience from this application case, the reference architecture has the potential to serve as the foundation for future DMbE information systems.

2 Background

The term *reference architecture* originated from the field of software engineering but, over the past decade, has been adopted in systems engineering to describe "the essence of existing architectures, and the vision of future needs and evolution to provide guidance to assist in developing new system architectures" (Cloutier et al. 2010). From a system design perspective, a reference architecture encodes patterns or rules and "constrains the instantiations of multiple architectures and solutions" to provide a foundation and comparison point for individual solutions (Office of the Assistant Secretary of Defense 2010).

Large-scale software systems such as the Internet follow a layered architecture as a form of modularity enabling robustness (disturbances do not easily propagate across layers) while also preserving changeability (component layers can be updated in relative isolation) (Doyle et al. 2011). In the case of the Internet, various layers encapsulate the data link, network, transport, session, and presentation components. Each component layer is constrained by a set of protocols; however, shared constraints across layers deconstrain the overall system by allowing piece-wise substitution and evolution.

Software development practice increasingly emphasizes layers at the individual application level to achieve similar lifecycle objectives. Architectural styles such as representational state transfer (REST) constrain protocols to stateless resource operations to minimize latency and maximize independence and scalability (Fielding and Taylor 2002). Improved access to customize server-side components using common scripting languages such as Python (Flask) and JavaScript (Node.js/Express) and availability of real-time, bidirectional, client–server communication libraries (WebSockets and Socket.IO) support layered architectures even for small-scale applications.

Despite recent advances in software engineering, new architectural styles are slow to translate to DMbE environments which still emphasize centralized platforms for aggregating graphical models and proprietary software without exposed APIs. A vision for a future DMbE information system resembles that of the Internet where component models are orchestrated in layers with well-defined constraining protocols. Drawing from recent work in the space systems engineering domain, this paper outlines a reference architecture to explain how a modern web-based layered architecture can support DMbE activities. The reference architecture highlights the key components (layers) and protocols to exchange information.

3 Proposed Reference Architecture

3.1 System Components

The proposed software architecture consists of a back-end, front-end, and database. The database component can be any data storage solution and could be implemented in different ways. It could be implemented as a file system, relational database, non-relational database, and in-memory store in this architecture. The database solution can be respective to the data used in the application, and all that really matters is that it will be able to communicate directly with the back-end.

The back-end connects the database and front-end/client. The primary purpose of the back-end is to convert the database command line interface into an easily accessible HTTP (HyperText Transfer Protocol) service. The back-end acts as an API that communicates with the front-end and user through the same communication strategy. This API is a wrapper for the underlying technical models and provides analysis services. This back-end can take the form of various technologies such as Flask, Node.js/Express, Apache, Nginx, and more as long as the solution allows for HTTP accessible services.

The front-end can be implemented by choice and is application specific. The main function of this component is to allow to directly interact with the service. The front-end can take many forms such as a browser-based GUI (Graphical User Interface), a mobile application, or direct communication with the client. The proposed communication strategy between these components is HTTP requests as they are a common and straightforward communication method utilized within web technology. The benefits of this communication strategy will be stressed in the following section.

As shown in Fig. 1, the proposed architecture consists of three simple components with communication only done through HTTP. The back-end can communicate with the database in whatever method applicable to the database strategy/usage. Many applications of this architecture could have three entirely isolated components

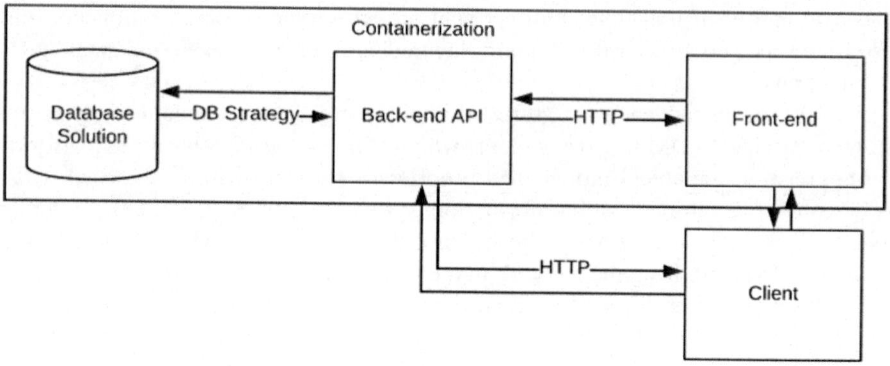

Fig. 1 UML diagram of reference architecture

or just two if the back-end and the database are stored in the same component. The entire application can be containerized using recently popular technologies like Docker and Kubernetes.

3.2 System Interface

Communication through HTTP requests is beneficial for numerous reasons, but the primary three are to reduce development work, reduce integration challenges, and improve overall operation. HTTP is the Internet standard for communication and involves the transfer of data over a URL (Uniform Resource Locator) request. These requests can contain data in the form of JSON (JavaScript Object Notation), XML (Extensible Markup Language), or other types. There are a few different fundamental forms of HTTP requests utilized in this software application which are GET, POST, and DELETE. A GET request has a specified URL and retrieves information from the server. A POST request sends information to the server. A DELETE request is meant to delete specified information from the server.

By utilizing HTTP for communication, the API has direct communication with not only the front-end but the user as well leading to greater interoperability. This reduces the development work by allowing the developer to fixate their responsibility solely on building a functional API and not the logistics of how it would communicate with the front-end. This standardized communication reduces integration difficulties for the same reason. The engineers can work with a standard interface of the HTTP requests rather than worry about developing an API and communication strategy. This leads to an understanding of a standard model for communication with support of numerous web technologies already. Lastly, it improves the overall operation because the front-end and back-end run independent of one another so their individual performances will not have an effect on one another.

3.3 Data Encoding

By utilizing communication through HTTP, the model also has to choose what syntax to format data in. JSON is a popular format for HTTP data transfer that uses a key-value (dictionary) structure to encode data. This type of data is very flexible for engineers to use as it does not require typing for any of the object properties and allows for appending of extra fields. The JSON format is also very easy for engineers to use as it is human readable and does not require much additional training.

Despite JSON's flexibility, it can still follow schema specifications. This can be achieved through the use of JSON-LD which provides standard guidelines for JSON communication (Sporny et al. 2019). Using the JSON-LD format requires the implementation of standardized schema of the data. A schema includes simple documentation about what should be sent and what variables fields represent including units, reference, and other parameters. This stresses the concept of the semantic web where everything can be in communication in a way that is easily accessible and has consistent semantics. Another major benefit of the schema is that enables interoperability based on common understanding. With the recent applications of machine learning in the field of systems engineering to discover insights on data, the usage of a standardized data will lead to much more ease on the application of processing.

For usage of standardized schema, resources such as Schema.org can be utilized (Guha et al. 2016). Schema.org is a database of schema for structured data on the Internet with the overarching goal of facilitating the semantic web.

4 Example Application Case

4.1 Tradespace Analysis Tool for Constellations (TAT-C)

The Tradespace Analysis Tool for Constellations (TAT-C) is a software modeling tool to support pre-Phase A conceptual design of Earth-observing spacecraft constellations (Le Moigne et al. 2017). Based on a mission concept and constraints on available constellation geometries, spacecraft buses, and instruments, TAT-C enumerates and searches a combinatorial tradespace to identify desirable mission architectures. Software modules in TAT-C perform specific functions such as orbital propagation, launch vehicle selection, instrument performance analysis, cost analysis, and search execution.

The TAT-C Knowledge Base (KB) module documents schema definitions for TAT-C objects and maintains a library of conforming object models gathered from historical missions. Designed as a layered architecture with database, back-end, and front-end components, other TAT-C components including a browser-based GUI access KB data resources using standard web-based protocols. A publicly accessible version of the KB application is available at https://tatckb.org.

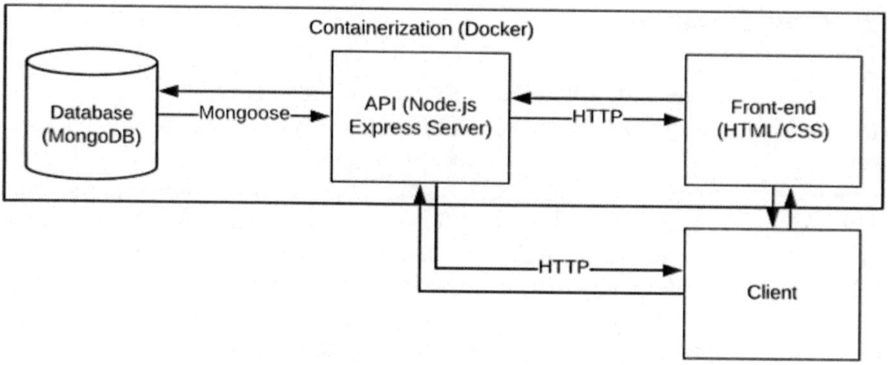

Fig. 2 UML diagram of architecture for TAT-C

Although small in scale, TAT-C exhibits many of the challenges of DMbE information systems. The KB implementation details in the following sections address the implementation of individual components, protocols exchanging information between layers, and the data encoding system selected to improve model interoperability between software components.

4.2 System Components

The TAT-C KB software architecture consists of a front-end website, a back-end API, and a database. The decision was made to isolate the front-end code from the back-end code and came into fruition as a front-end GUI and a back-end server only in communication through HTTP requests. Figure 2 shows the KB architecture.

The front-end component uses common web technology and takes the form of an HTML/CSS/JavaScript website to act as an interface for the API. The front-end interface can be seen in Fig. 3. The back-end component uses a Node.js/Express server to act as wrapper for the database and provide technical data analysis. The back-end uses Mongoose to make queries from the MongoDB database.

Containerization was also utilized on this project as it reduced friction with integration and will be expanded upon in Sect. 4.5.

4.3 System Interface

Communication between the components relies on HTTP requests. As shown in Fig. 4, the client has the choice of either interacting with the front-end GUI or making a request from the API directly via HTTP. For this function specifically, the only parameter the API requires is the collection being requested. The API

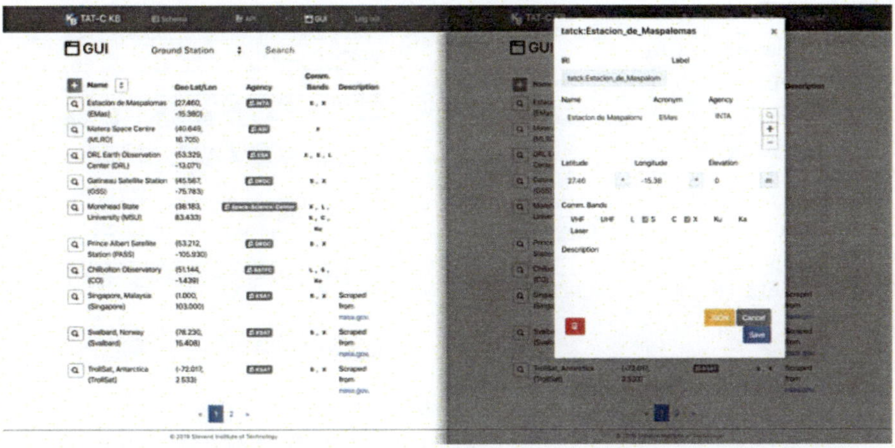

Fig. 3 The front-end interface of TAT-C KB

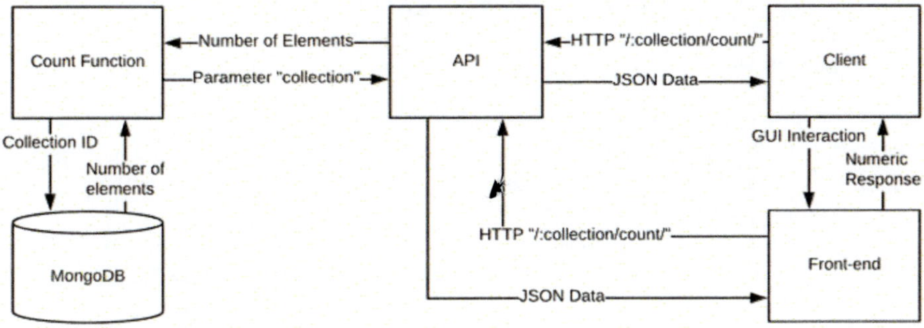

Fig. 4 Communication diagram for count HTTP requests

then redirects their request to the internal count function with the parameter of the collection and returns the number of elements in the specified collection. All the count function does is query the database for the number in that collection.

The communication with the API allowed for a few different kinds of HTTP requests which were: `count`, `list`, `get`, `add`, and `delete`. All of these have specific parameters and queries. The parameters are required fields by the API for the function. The arguments are additional fields the user can use when constructing API requests to get certain responses. Table 1 documents all of the parameters and queries which summarizes the API documentation.

The `count` function is pretty straightforward and provides the number of a specified type within the database. This allows for the additional query `search` where you are able to find the number of objects that match a specified string within the type. The `list` function provides a list of all the objects within a specified type and allows for the following queries: search, sort, offset, populate, and limit. The `get` function is the simplest and just allows you to get an object of a specified type

Table 1 All HTTP requests

Request	Method	URL	Parameters	Arguments
Count	GET	`/:type:/count/`	`type`	`search`
List	GET	`/:type/list/`	`type`	`search`, `sort`, `offset`, `populate`, `limit`
Get	GET	`/:type/:id/` `/:type/instance/:id`	`type`, `id`	`populate`
Add	POST	`/:type/add/` `/add/`	`type`	`token`
Delete	DELETE	`/:type/delete/:id/`	`type`, `id`	`token`

```
{
    @context: {
      owl: "http://www.w3.org/2002/07/owl#",
      rdf: "http://www.w3.org/1999/02/22-rdf-syntax-ns#",
      rdfs: "http://www.w3.org/2000/01/rdf-schema#",
      schema: "http://schema.org/",
      tatckb: "http://tatckb.org/schema/2.0/"
    },
    @graph: [
        {
          @id: "tatckb:ASI",
          @type: "tatckb:Agency",
          tatckb:name: "Agenzia Spaziale Italiana",
          tatckb:acronym: "ASI",
          tatckb:agencyType: "GOVERNMENT",
          @lastUpdated: "2019-07-24T20:10:06.558Z"
        }
    ]
}
```

Fig. 5 Output of "/api/agency/list?search=ASI"

and `@id` field. This request only allows the query `populate` which will populate all sub-objects by the `@id` field within the object. The add function takes the input of an object and allows you to add it to a specified type. The delete function allows you to delete an object within a specified type by the `@id` field within the object and accepts no additional queries. An example of the JSON response from the API is in Fig. 5 and will be explained in the following section.

4.4 Data Encoding

The overall goal of the KB is to set a standard for tradespace analysis data and help organize it. To increase interoperability, specific JSON-LD schemas were created for

Table 2 Schema documentation for data type `GroundStation`

Property	Expected type	Description
`name`	`schema:Text`	The full name of an entity
`latitude`	`schema:Number`	Latitude (decimal degrees) with respect to the WGS 84 geodetic model. Ranges between −90° (south) and 90° (north) where 0 degrees represents the equator
`longitude`	`schema:Number`	Longitude (decimal degrees) with respect to the WGS 84 geodetic model. Ranges between −180° (west) and 180° (east) where 0 degrees represents the prime meridian
`elevation`	`schema:Number`	Elevation (m) above mean sea level with respect to the WGS 84 geodetic model
`agency`	`tatckb:Agency`	Designer, provider, or operator, of this object

each of the 22 different data types. Some of the data fields allow for more variation than others, accomplished by storing the data in the form of JSON objects. All of the data types have three base fields allowing for better organization purposes, these are `@id`, `@type`, and `@lastUpdated`. The `@id` field is assigned to the object when created and is used for a unique identifier. The `@type` field specifies the collection type the object is meant to fit into which is later used in the HTTP requests. The `@lastUpdated` field was added for a timestamp of the last time the data was changed/updated which makes finding recently manipulated data easier.

Each type has its own specified documentation for its fields, and Table 2 shows the properties, expected types, and descriptions for the `GroundStation` type of data.

Utilizing standard schema for each of the data types in the KB improves interoperability across multiple projects within the field of tradespace analysis.

Figure 5 shows the result of the request of a list of all objects in the `Agency` collection that include the regular expression "ASI." The response contains two portions: the context and the graph. The context provides guidelines for the JSON-LD schema and datatype including the specific documentation for the TAT-C KB. The graph contains the list of objects matching the request. The object contained has the fields as specified in the schema documentation for the data type `Agency`.

4.5 Containerization Configuration

Containerization refers to the ability to virtualize a development/deployment environment and isolate it from others. Containerization has recently become very large in the software industry as it allows for replicating the development environment exactly leading to less issues in deployment. Containerization allows for the separation of the back-end/database and the front-end entirely by isolating them into two separate environments.

For the containerization, Docker was the project's selected solution. The KB project includes two separate containers–one for the database and one for the web combining both the front and back-end components. The database container uses a lightweight environment specifically designed to hold a MongoDB database which was hosted on the port 27017. The web container runs a lightweight Node.js image hosted on port 80. One deployment challenge encountered here is that the web container needed to wait the MongoDB container to be fully initialized before attempting connection or it would fail generating an error. To ensure the startup sequence functioned properly, a script was written in a Dockerfile to delay the web server startup until after the database initialization.

The primary downside of using Docker on an application is that whenever changes are made to a container, it has to be rebuilt. Usually the rebuilding time is relatively quick but depending on the amount of dependencies and libraries the environment uses, it can take more time.

5 Conclusion

Architecting DMbE information systems pursues a goal of providing a flexible, reconfigurable, and updateable platform for systems engineering and design activities. This paper adopts and transitions practices from modern web-based software engineering as a reference architecture that promotes robustness while preserving changeability. Specifically, layered architectures interconnected with well-defined and constrained protocols based on web technologies such as HTTP support DMbE using principles that enabled large-scale software systems such as the Internet.

As demonstrated in the TAT-C KB application case, this paper identifies three key layers and their functionality: the database (data persistence), back-end (technical services), and front-end (user interface) components. Interfaces based on the HTTP request–response protocol provide a simple approach to access resources. Supporting data encoding standards such as JSON-LD provide enhanced semantic interoperability while preserving simple, human-readable formats. This architectural pattern can serve as the foundation for other DMbE projects that adopt alternative component implementations, interfaces, and encoding styles.

Acknowledgments This work was supported in part by NASA collaborative agreement/grant 80NSSC17K0586 titled "Knowledge Representation for Distributed Space Mission Design using TAT-C with Machine Learning" as a part of an Advanced Information Systems Technology (AIST) 2016 project.

References

Bone, M., M. Blackburn, B. Kruse, J. Dzielski, T. Hagedorn, and I. Grosse. 2018. Toward an interoperability and integration framework to enable digital thread. *Systems* 6(4). https://doi.org/10.3390/systems6040046.

Bone, M.A., M.R. Blackburn, D.H. Rhodes, D.N. Cohen, and J.A. Guerrero. 2019. Transforming systems engineering through digital engineering. *The Journal of Defense Modeling and Simulation: Applications, Methodology, Technology* 16(4): 339–355.https://doi.org/10.1177/1548512917751873.

Cloutier, R., G. Muller, D. Verma, R. Nilchiani, E. Hole, and M. Bone. 2010. The concept of reference architectures. *Systems Engineering* 13(1): 14–27. https://doi.org/10.1002/sys.20129.

Dori, D. 2016. *Model-Based Systems Engineering with OPM and SysML*. New York: Springer.

Doyle, J.C., and M. Csete. 2011. Architecture, constraints, and behavior. *Proceedings of the National Academy of Sciences of the United States of America* 108(Supplement 3): 15624–15630. https://doi.org/10.1073/pnas.1103557108.

Fielding, R.T., and R.N. Taylor. 2002. Principled design of the modern web architecture. *ACM Transactions on Internet Technology* 62(2): 115–150. https://doi.org/10.1145/514183.514185.

Guha, R., D. Brickley, and S. Macbeth. 2016. Schema.org: Evolution of structured data on the web. *Communications of the ACM* 59(2): 44–51. https://doi.org/10.1145/2844544.

Hale, J.P., P. Simmerman, G. Kukkala, J. Guerrero, P. Kobryn, B. Puchek, M. Misconti, C. Baldwin, and M. Mulpuri. 2017. Digital model-based engineering: Expectations, prerequisites, and challenges of infusion, Technical Report NASA/TM-2017-219633, National Aeronautics and Space Administration.

Le Moigne, J., P. Dabney, O. de Weck, V. Foreman, P. Grogan, M. Holland, S. Hughes, and S. Nag. 2017. Tradespace analysis tool for designing constellations (TAT-C). In *2017 IEEE International Geoscience and Remote Sensing Symposium (IGARSS)*, Fort Worth. https://doi.org/10.1109/IGARSS.2017.8127168.

Office of the Assistant Secretary of Defense. 2010. Reference architecture description.

Sporny, M., D. Longley, G. Kellogg, M. Lanthaler, and N. Lindstr. 2019. JSON-LD 1.1, Standard, W3C. https://www.w3.org/TR/json-ld11/.

Digital Engineering Ecosystem for Future Nuclear Power Plants: Innovation of Ontologies, Tools, and Data Exchange

Christopher Ritter, Jeren Browning, Lee Nelson, Tammie Borders, John Bumgardner, and Mitchell Kerman

Abstract The construction of megaprojects has consistently demonstrated challenges for project managers in regard to meeting cost, schedule, and performance requirements. Megaproject construction challenges are commonplace within the nuclear industry with many active projects in the United States failing to meet cost and schedule efforts by significant margins. Currently, nuclear engineering teams operate in siloed tools and disparate teams where connections across design, procurement, and construction systems are translated manually or over brittle point-to-point integrations. The manual nature of data exchange increases the risk of silent errors in the reactor design, with each silent error cascading across the design. These cascading errors lead to uncontrollable risk during construction, resulting in significant delays and cost overruns. Additionally, due to the desire to reduce schedule and avoid escalation, construction is often begun prior to full design maturity. Digital engineering (DE) embodies a deliberate transformational approach to the manner in which systems are designed, engineered, constructed, operated, maintained, and retired. DoD defines DE as "an integrated digital approach that uses authoritative sources of system data and models as a continuum across disciplines to support lifecycle activities from concept through disposal" (U.S. Department of Defense, Digital Engineering Strategy, Washington, DC, June 2018). This paper describes the ontologies (data model), tool architectures, data exchange, and process to transform engineering teams to a new digital engineering ecosystem.

Keywords Digital engineering · Enterprise transformation · Systems engineering · Model based systems engineering · MBSE

C. Ritter (✉) · J. Browning · L. Nelson · T. Borders · J. Bumgardner · M. Kerman
Idaho National Laboratory, Idaho Falls, ID, USA
e-mail: Christopher.Ritter@inl.gov

A. M. Madni et al. (eds.), *Recent Trends and Advances in Model Based Systems Engineering*, https://doi.org/10.1007/978-3-030-82083-1_2

1 Introduction

New nuclear construction represents projects with high upfront capital costs, which have continued to increase over time. In a recent study, a team from MIT analyzed the nuclear industry's primary costs. The study compared current reactor programs over an established baseline to assess overall industry competitiveness. In the United States, V.C. Summer 2&3 in South Carolina and Vogtle 3&4 were analyzed. Neither project is predicated to meet the 2009 benchmark (https://energy.mit.edu/wp-content/uploads/2018/09/The-Future-of-Nuclear-Energy-in-a-Carbon-Constrained-World.pdf). V.C. Summer was recently canceled at a cost of over $4.9 billion to rate payers in South Carolina (https://www.chooseenergy.com/news/article/failed-v-c-summer-nuclear-project-timeline/). In Georgia, Vogtle 2&3 represent greater than $10 billion cost overrun which contributed to the bankruptcy of Westinghouse (https://www.utilitydive.com/news/southern-increases-vogtle-nuke-pricetag-by-11-billion/529682/).

Construction delays and cost overruns are not unique to the nuclear domain. The European Aeronautic Defense and Space (EADS) Airbus 380 program suffered approximately $6.5 billion in losses. Electrical wiring of airframes is a complex effort involving 530,000 meters of cables, 100,000 wires, and 40,300 connectors. During this installation, electrical teams found a critical issue – the wires were cut too short. Engineers in Germany and Spain used Dassault CATIA v4, while engineers in Britain and France had upgraded to Dassault CATIA v5. This resulted in German design teams being unable to update changes to the electrical design automatically. This interoperability issue cost Airbus 20 months of delays and a loss in program confidence (What Grounded the Airbus A380?) (Fig. 1).

Complex system issues continue to affect the aerospace, defense, and nuclear industries. Recognizing this, the Department of Defense (DoD) released a Digital Engineering Strategy (U.S. Department of Defense, Digital Engineering Strategy 2018). This strategy promotes the use of digital artifacts comprising the digital representations of systems, subsystems, and components to design and sustain national defense systems. The DoD's five strategic goals for digital engineering are to:

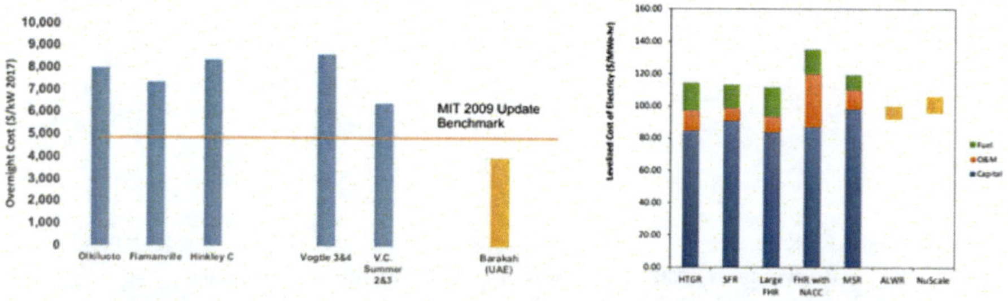

Fig. 1 (**a**) Projected LCOE for different advanced reactor concepts. (**b**) Overnight cost of recent Gen-III+ builds versus benchmark (https://energy.mit.edu/wp-content/uploads/2018/09/The-Future-of-Nuclear-Energy-in-a-Carbon-Constrained-World.pdf)

- Formalize the development, integration, and use of models to inform enterprise and program decision-making.
- Provide an enduring, authoritative source of truth.
- Incorporate technological innovation to improve the engineering practice.
- Establish a supporting infrastructure and environment to perform activities and collaborate and communicate across stakeholders.
- Transform the culture and workforce to adopt and support digital engineering across the lifecycle.

The DoD Digital Engineering Strategy recognizes the significant role digital tools have played in successful complex engineering projects. For example, Mortenson Construction recently realized a 600-day schedule savings and 25% greater productivity increase utilizing virtual design and construction (VDC) technologies across 416 VDC programs (Virtual Design and Construction). In the aerospace industry, Boeing has applied integrated digital tools since the design of the 777 and has seen a 40% improvement in first-time quality through use of digital twins on the Boeing 777X program ((https://www.aviationtoday.com/2018/09/14/boeing-ceo-talks-digital-twin-era-aviation/).

2 Three Laws of Systems Engineering

The failures of complex industry projects each share common attributes. Further context is provided through an examination of some critical lessons in engineering history. Every project shares *the three laws of systems engineering*:

1. *All systems are created equal.* The cost and complexity of today's systems require extensive engineering rigor to be applied throughout the entire lifecycle – from conceptual design through disposal. Today, even toasters can be complex systems containing both hardware and software and requiring a digital design process. The need for an extensive process is recognized through the continuing evolving list of standards: MIL-STD-499B, ANSI/EIA 632, IEEE 1220, ISO/IEC 15288, the systems engineering "Vee" model (Fig. 2), the waterfall model, the spiral development model, agile software engineering methods, and many others. Engineering teams must follow a process, or projects will become lost and result with a design that either does not meet the required intent or just cannot be completed (typically due to cost and/or schedule overruns).
2. *All systems are flawed.* All systems are conceived, designed, and built by humans, and humans are not infallible. Therefore, though not intentional, any system that humans create will be flawed. Errors, critical failures, and catastrophic failures may occur depending upon the severity of these latent, inadvertent problems. This is statistically observed through a 65% failure rate of megaprojects with failure defined as 25% over budget, 25% behind schedule, or simply unsatisfied business objectives (Merrow 2011).

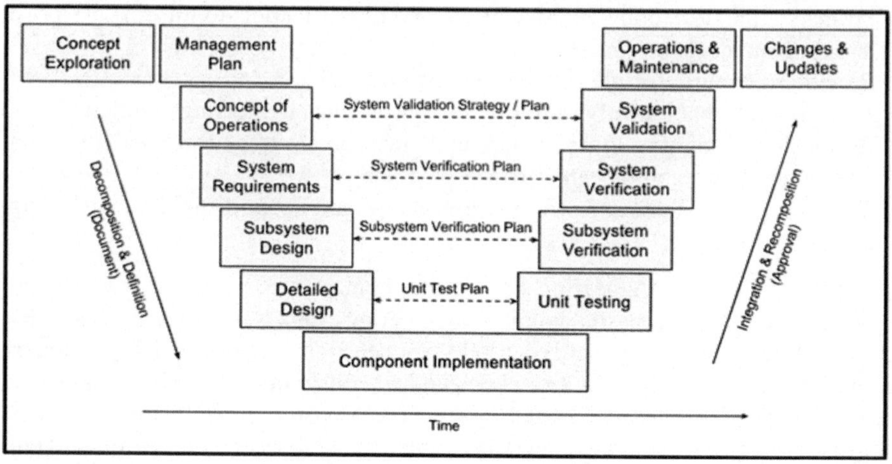

Fig. 2 The systems engineering "Vee" model (https://commons.wikimedia.org/wiki/File:Systems_ Engineering_V_diagram.jpg)

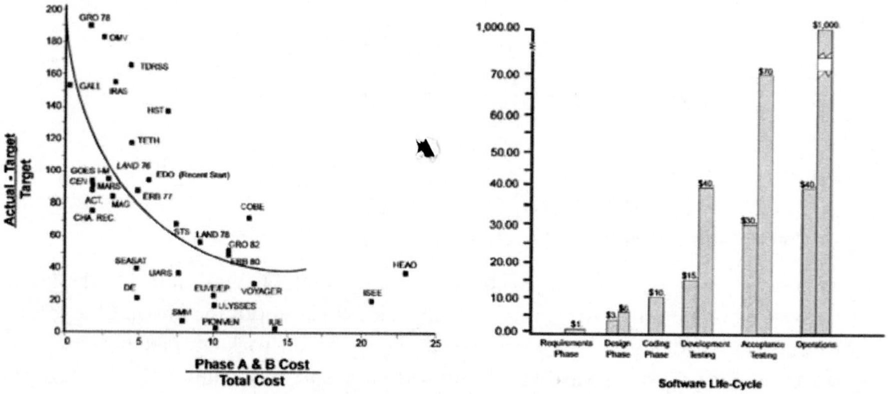

Fig. 3 Cost overruns compared to upfront design spend (https://reqexperts.com/wp-content/uploads/2015/07/Managing_Requirements.pdf; Gruhl). (**b**) Relative cost of assumption errors in software (https://reqexperts.com/wp-content/uploads/2015/07/Managing_Requirements.pdf; Extra Time Saves Money 1990)

3. *Early intervention can save the program (and the company).* Early project management studies at NASA found a direct correlation between upfront investment in systems engineering and cost overruns (Fig. 3). The cause of this correlation is the assumed error during the requirements phase in design. The cost to extract a defect early, during the conceptual phase, is up to 1000× less than errors found during operations. A classic example is the Hubble Space Telescope (HST) program where a major flaw was found in the telescope's optics system after launching the telescope in orbit. This flaw led to a challenge in space servicing

mission (SM-1) to resolve the telescope's optics issues and install a new camera system (https://www.spacetelescope.org/about/history/servicing_mission_1/).

3 Development of Digital Engineering Ecosystem for Nuclear Power Design

Given the proven cost, schedule, and risk reduction benefits in the aerospace, defense, and construction industries, the Idaho National Laboratory began a digital engineering program in 2018 to support the Versatile Test Reactor (VTR) program. The VTR is a 300-megawatt thermal fast neutron plant under development to support reliable open core testing under closely controlled environmental conditions, enabling the development and qualification of new fuels for future advanced reactor designs.

3.1 Nuclear Design Ontology

The development of an integrated ecosystem requires interoperability with precise semantics to connect tools and information. Multiple industries, such as defense (Defense Architecture Framework Meta Model 2 (DoDAF Meta-Model (DM2))) and energy (ISO 15926 (https://www.iso.org/standard/70694.html)), have domain ontologies; however a standard does not exist in the nuclear domain. Diagramming languages including Business Process Modeling Notation (BPMN), Systems Modeling Language (SysML) (OMG SysML Specifications), and Unified Modeling Language (UML) (About the Unified Modeling Language Specification Version 2.2) only specify diagram "look and feel" and fail to capture risk, cost, and other programmatic information.

To enable a standard ontology for industry, INL led development of a nuclear design ontology. Many early efforts to develop ontologies in industry necessitated the development of a top-level metamodel to explain entities within a domain ontology. Today, many top-level ontologies exist which allow for extension to support domain-specific ontologies. The Basic Formal Ontology (BFO) was selected to provide compatibility with over 300 domain-specific ontologies which already exist in the scientific community.

BFO uses ontological naming conventions (e.g., SpecificallyDependentContinuant) which caused usability concerns early in development. To mitigate usability issues, BFO was extended with the Lifecycle Modeling Language (LML) to provide simpler terminology more familiar to the systems engineering community. Accordingly, the SpecificallyDependentContinuant has been aliased with the Characteristic class name. The combination of two open industry ontologies maximizes the reach of the nuclear design ontology (Fig. 4).

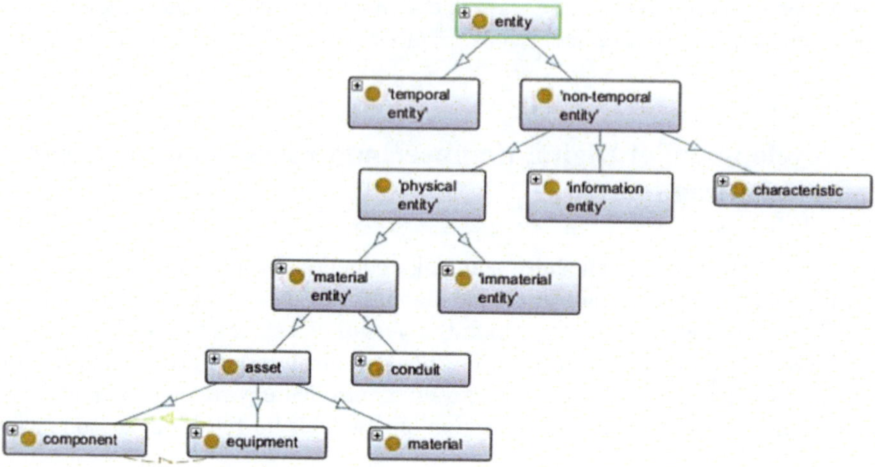

Fig. 4 Screenshot of nuclear ontology in Protégé

The BFO- and LML-derived ontology includes extensions for physical assets, for example, components (e.g., coolant), equipment (e.g., valve), and structures (e.g., turbine hall). Additionally, a fully executable extension was developed to facilitate the inclusion of behavior models. BFO and LML do not include execution specifications within their ontologies, which leads vendors to develop proprietary tree structures to store this information. To enable executable behavior models, the ontology includes a "step construct type" attribute on Action class, Planned Action class (step template), and Transition class which are used to describe the missing execution layer. This enables integration software to fully include behavior model sequencing and execution logic.

This design ontology has been extended to additionally support operations and maintenance activities of nuclear power plants. The combined design and operations ontology will be released to the open-source community as Data Integration Aggregated Model and Ontology for Nuclear Deployment (DIAMOND).

3.2 *Deployment of Digital Tools*

The development of innovative nuclear reactor facilities includes requirements, risks, schedule, cost, 2D piping and instrumentation (P&ID) diagrams, 3D plant models, material parts lists, and various engineering/procurement metadata throughout design. These requirements are similar to the DoD Digital Engineering Ecosystem (Fig. 5). Traditionally, textual data is developed in office tools (e.g., Microsoft Office), and engineering data is developed in disconnected modeling tools (e.g.,

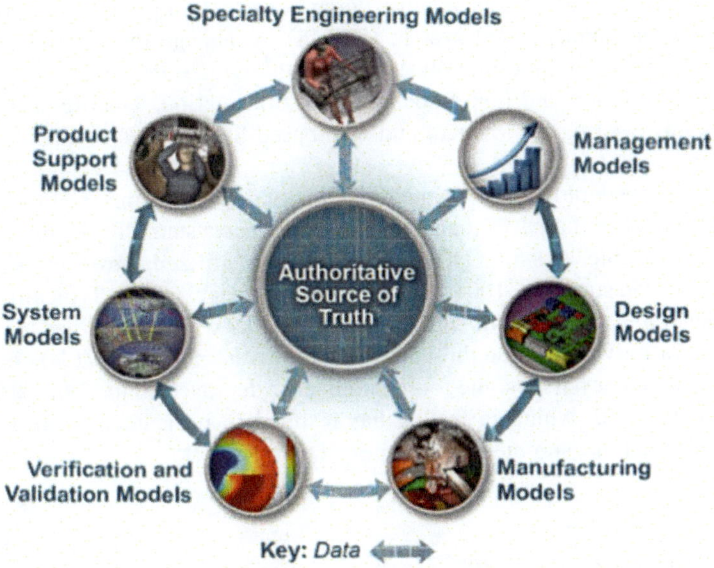

Fig. 5 DoD authoritative source of truth (U.S. Department of Defense, Digital Engineering Strategy, Washington, DC, June 2018)

Autodesk Inventor). To fully realize a digital engineering "authoritative source of truth," model-based tools have been deployed across engineering teams.

The construction industry utilized model-based digital tools, commonly referred to as building information management (BIM), in the development of complex civil engineering programs. At INL, the VTR program has deployed the AVEVA solution to support BIM which includes several capabilities including the capture of 2D piping and instrumentation diagrams (P&ID) and 3D mechanical, civil, and structural diagrams. AVEVA's suite of software includes a data-driven approach, integrations across their tools, and a web-based collaboration platform (AVEVA Net). Notably AVEVA's toolset does not natively integrate with common systems engineering tools, analytic codes (seismic, pipe stress, etc.), nor open REST APIs to connect additional tools.

INL has deployed the IBM Jazz Engineering Lifecycle Management software for the development of systems engineering artifacts. The Jazz platform includes a suite of integrated tools to manage requirements, define test cases (verification/validation), model physical/logical architectures, and manage system changes. IBM's Jazz platform includes an open standard, Open Services for Lifecycle Collaboration (OSLC), to enable typed links across tools within their ecosystem. These links enable a named association (e.g., Satisfied by) between systems engineering artifacts and exposed over web-based REST APIs. Notably, Jazz does not include native integrations for BIM tools, traceability prediction, or analytic codes.

The Department of Energy requires large construction programs to implement an earned value management system (EVMS) to manage the program schedule. Accordingly, INL has traditionally leveraged the Oracle Primavera PPM P6 tool to track the schedule and link associated cost information. The digital ecosystem leverages this existing deployment and web-based SOAP APIs available as part of the P6 deployment.

As tools were deployed, across geographically disperse teams, significant issues were discovered in the performance of the initial ecosystem. To deploy tooling as quickly as possible, existing DOE networks were utilized; however peak traffic latency was measured as 100× greater than nonpeak times. Typically, nuclear reactor development programs utilize local networks which are either inaccessible to partners or only accessible through a virtual private network (VPN) connection. To overcome latency issues, the tool suite is now deployed to the Microsoft Azure Government Cloud, which in testing has resulted in peak latency measurements equal to nominal measurements of the prior DOE network.

3.3 *Integration of Digital Tools*

The current state of the art is that each reactor vendor develops their own taxonomy and development plan. Once established, a toolset is deployed to enable the design of a new plant. Typically, these tools are disconnected and include no common interchange language to communicate outside their domain, offering varying data schemas for each stage of the lifecycle. While some vendors include integrations within their suite, most programs result in utilized comma-separated values (CSV) files to exchange information. These CSVs use specific domain languages which require manual data manipulation for exchange, thus increasing the risk of human error and creating an environment where any change requires extensive rework.

To rectify this issue, a digital engineering integration framework is under development with a team of laboratories, universities, and contractors to enable seamless connection of data across the digital engineering ecosystem. The current system, under development, stores information in the Microsoft Azure CosmosDB. CosmosDB allows for global data distribution and persistence of data in a graph database format (Gremlin). Graph databases are optimized for storage and retrieval of highly connected data, allowing for rapid queries across digital links.

The integration framework utilizes web connections through Transmission Control Protocol (TCP) Representational State Transfer (REST) interfaces which transfer data live through standard commercial networking technologies. This allows for a "digital thread" of connected data to be updated live as corresponding engineering, procurement, and construction tools are modified. Through the capture of live data, versioning, conflict resolution, and accurate analysis codes can be further developed.

Tools which expose an existing REST interface with a standard authentication mechanism (e.g., OAuth 2.0, Basic Authentication) can be natively integrated

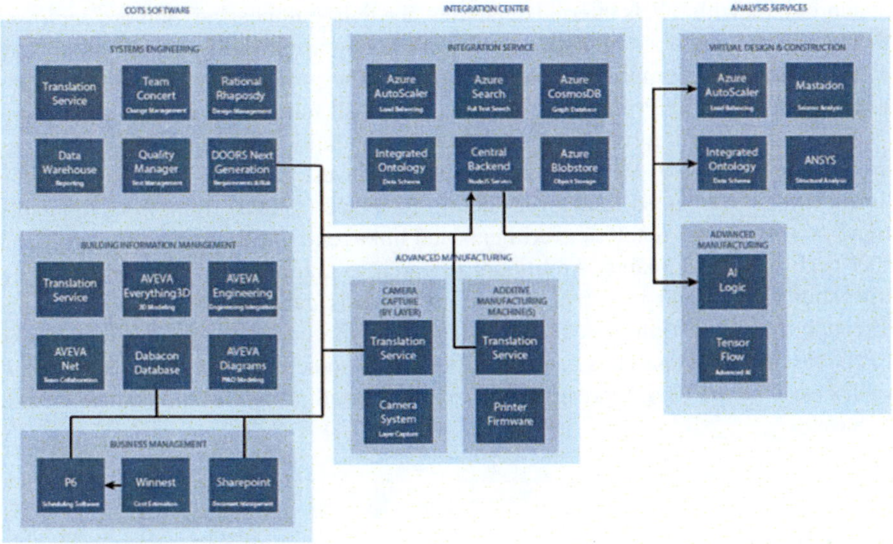

Fig. 6 Integration framework architecture

into the framework. Conversely, tools which do not expose a standard REST interface or authentication mechanism are integrated through custom adapters which expose a web-accessible interface. The AVEVA BIM provides a desktop/local server API only and thus a custom adapter has been developed. The IBM Jazz system provides an open REST interface which utilized standard authentication mechanisms (OAuth) and is straightforward to adapt into the integrated system.

The connection of engineering design data into a common, accessible database allows for the automation of analysis code integration. As part of this ecosystem, research at North Carolina State University has led to the development of automated mesh generation for finite element analysis (FEA) codes. This FEA mesh generation currently utilizes standard BIM industry foundation class (IFC) files to enable meshing through a set of Python scrips. Future iterations of this analysis integration will explore direct linkage to the integrated data store which will allow automate analysis reporting (Fig. 6).

4 Summary and Path Forward

The development of a digital engineering ecosystem is expected to provide significant improvements to cost and schedule while dramatically reducing overall program risk. These tools are currently deployed on the Versatile Test Reactor program with teams at varying stages of implementation. The BIM tools have been fully rolled out to the virtual design and construction contractor, Bechtel, and synchronized with INL's internal databases. Systems engineering tools are deployed through high-level documentation, and work is ongoing to fully deploy the toolset at the system and component levels of design. The project management team has already deployed the Primavera P6 toolset across all scheduling operations.

Looking forward, advanced modeling and simulation and analytics integrations are planned in 2020. Research is ongoing to investigate the integration of the TerraPower's Advanced Reactor Modeling Interface (ARMI) framework to enable rapid prototyping of reactor experiments and linkage to the neutronics, thermal, and fuel performance domains. Additional research to autonomously identify missing traceability links and incorrect traceability links is currently within 2020 scope. Through the connection of data sources and analytics codes, the project team looks to enable new data analytics capabilities to predict reactor performance and design issues early in the design process, minimizing cascading risk in the nuclear design process.

References

About the Unified Modeling Language Specification Version 2.2. *About the Common Object Request Broker Architecture Specification Version 3.3*, www.omg.org/spec/UML/2.2/

DoDAF Meta-Model (DM2). *Chief Information Officer*, dodcio.defense.gov/Library/DoD-Architecture-Framework/dodaf20_dm2/

Extra Time Saves Money. 1990. Warren Kuffel Computer Language, December 1990.

Gruhl, Werner M.. Chief Cost & Economics Analysis Branch. NASA Headquarters.

Lifecycle Modeling Language. *Lifecycle Modeling Language*, www.lifecyclemodeling.org/.

Merrow, E.W. 2011. *Industrial megaprojects*. Hoboken, New Jersey: Wiley.

OMG SysML Specifications. *What Is SysML? | OMG SysML*, www.omgsysml.org/specifications.htm.

U.S. Department of Defense, Digital Engineering Strategy, Washington, DC, June 2018.

Virtual Design and Construction. *Mortenson Construction*, www.mortenson.com/approach/virtual-design-construction/vdc-report

What Grounded the Airbus A380?. *Cadalyst*, www.cadalyst.com/cad/product-design/what-grounded-airbus-a380-10903

Introducing Digital Doppelgängers for Healthcare Policy Analysis

Jennifer Legaspi and Shamsnaz Virani Bhada

Abstract Healthcare policy evaluation is a time-consuming, challenging process due to the complexity of the US healthcare system which is comprised of both public and private payers; a variety of healthcare suppliers including doctors, medical device companies, and pharmacies; and patients from different insurance coverages and socioeconomic backgrounds. Systems engineering processes are intended for complex systems and are ideal for addressing healthcare policy. Specifically, model-based systems engineering (MBSE) is used to increase traceability with its model-centric approach and can be used to increase understanding of the healthcare system. In this paper, we attempt to exploit digital twin philosophy of MBSE to understand a US healthcare system as a complex system. We focus our efforts in building a digital doppelgänger which reflects most aspects of the healthcare systems digitally, but is not an exact digital twin. The doppelgänger helps navigate around the medical privacy laws of the US healthcare system and runs some analysis on healthcare policy.

Keywords Model-based systems engineering · Policy development · Model-based design · Policy analysis · Healthcare · Digital twin · Virtual prototype · Emergent properties · Digital doppelgänger

1 Introduction

Healthcare policy implementation in the USA can take years from the identification of a need for a policy change to the implementation of the policy. Legislators must work with the current system to make changes that will not detrimentally affect the actors in the system. Most recently, healthcare reform has occurred in 2010 with the Affordable Care Act (ACA) yielding new pathways of insurance coverage (Mulligan

J. Legaspi (✉) · S. V. Bhada
Worcester Polytechnic Institute, Worcester, MA, USA
e-mail: jlegaspi@wpi.edu

A. M. Madni et al. (eds.), *Recent Trends and Advances in Model Based Systems Engineering*, https://doi.org/10.1007/978-3-030-82083-1_3

2017). The implementation of the ACA in 2010 was met by strong opposition (Wei and Jarlenski 2014), but during its cycle of utilization, it garnered support. After the ACA penalty for the uninsured was repealed in 2019, only one quarter chose not to get coverage because of the repeal (Collins and Gunja 2019). Further calls for healthcare reform have been called for by various political candidates in the USA, but a 2019 survey shows that 40% of adults do not know enough about public and private insurance to have an opinion (Collins and Gunja 2019) which may be an indicator of the level of complexity of the system.

The healthcare system is a complex system that can be analyzed using systems engineering methods and approaches to be presented to an audience in a way that is easier to understand. Model-based systems engineering (MBSE) can be used to make sense of the current rules and regulations due to its inherent benefit of traceability attributed to its model-centric approach rather than a document-centric approach (Krishnan et al. 2018). Digital twins in systems engineering can also be used to address the issue of healthcare reform by accurately representing the complexity of the health records of an individual.

Typically, digital twins are implemented as a representation of a specific instance that are continually updated with respect to its twin (Madni et al. 2019). In the case of healthcare, there are rules for medical record privacy and security in the USA under the Health Insurance Portability and Accountability Act (HIPAA) (Madni et al. 2019).

Security risks that exist with a digital twin of healthcare records should be considered when using a digital twin for policy evaluation. This research addresses these security risks in healthcare reform analysis by using a digital doppelgänger instead of a digital twin.

In this case we provide a distinction between digital twin and digital doppelgänger. A digital twin of a health record would be a complete and digital version of a specific person's health records. The Meriam- Webster defines a doppelgänger as "something or someone that strongly resembles another." From this definition, we describe a digital doppelgänger in terms of health records as a synthetic version of health records that could bear strong resemblance to the health records of a person that exists in the world, but the digital doppelgänger is not related to the person's health records, nor does a change in that person's health have an effect on the synthetic health records of the digital doppelgänger. Figure 1 illustrates the differences between a digital twin and digital doppelgänger with a snapshot over multiple years of life. The living person has matched records held in the digital realm represented by a digital twin. A digital doppelgänger has similar patterns of doctor visits to the living person and may have diagnoses and treatments at age ranges considered similar with respect to health statistics. The digital doppelgänger can be viewed as a hybrid of a virtual prototype and a digital twin.

Using MBSE and digital doppelgängers or digital twins, healthcare reform can be evaluated with respect to the system requirements modeled in SysML. Digital doppelgängers as well as digital twins can enable the simulation of emergent properties in a complex system such as healthcare in the USA.

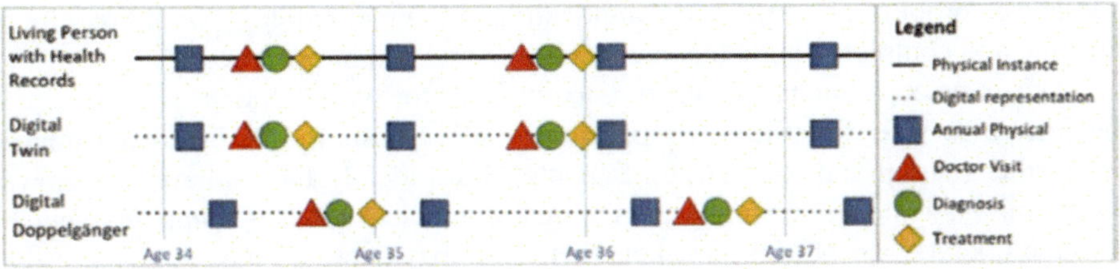

Fig. 1 Sample timeline of living person represented by a digital twin and a digital doppelgänger

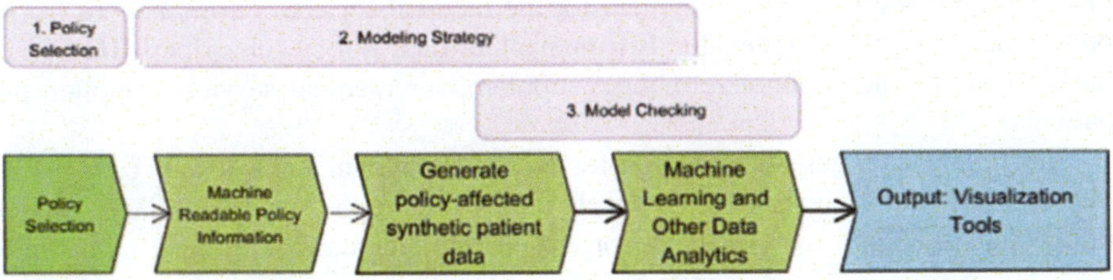

Fig. 2 Proposed methodology

2 Methodology

This section describes the proposed methodology behind the application of a digital doppelgänger with MBSE for healthcare policy evaluation. The methodology in Virani et al. is built upon to generate a methodology applicable to healthcare reform. The proposed change to the methodology follows the initial three steps of policy selection, modeling strategy, and model checking (U.S. Department of Health and Human Services 2013) and adds model distribution as a key step in the methodology as shown in Fig. 2. In previous research, models are intended to be used by researchers familiar with the tools (Krishnan et al. 2018). In this research, the intention is to provide a tool that can be used by patients, lawmakers, suppliers, and payers to make educated decisions regarding healthcare reform.

2.1 Policy Selection

The first step in the methodology is policy selection. Full applications of the methodology proposed by Virani et al. will be more rigorous for the policy selection step, but because the end goal of the methodology is to create a tool for nonexperts in the healthcare domain, policies that can potentially affect a wide range of actors in the system are prioritized over policies that may only affect a small group for the time being. Taking this into consideration, the research team selected basic, yet commonly misunderstood, aspects of health insurance such as copays and premiums for a basic proof of concept. A full application of the methodology would require a specific and complete policy that has been implemented or proposed.

Copays or copayments are fixed amounts paid for by the patient for a covered healthcare service. Premiums are paid on monthly basis by the health insurance subscriber (Virani and Rust 2016). Patients would ideally like to have no copays or premiums while still having access to quality medical services. Suppliers, which include doctors and pharmacies, benefit from copayments because it reduces the amount of upfront costs they must pay to provide a medical service to a patient. If a supplier is not reimbursed by the payer and the patient does not pay, they will only get the amount of the copayment for the medical service. Payers benefit from premiums which go into their funds from each of their policy subscribers. If payers have no funds, they cannot reimburse suppliers for medical services supplied to patients.

The policy changes selected for focus are change in copays and change in monthly premiums. Because the example policy selected is a simple policy change rather than a complex and specific proposed implementation, the use of natural language processing is not required (U.S. Department of Health and Human Services 2013). When applying this research to a specific proposition, natural language processing will likely be required to automate the policy modeling process.

2.2 *Modeling Strategy*

The modeling strategy makes use of both SysML and MATLAB with digital doppelgängers. MBSE methods emphasize the importance of traceability between SysML and MATLAB. Synthea, an open-source tool created by MITRE Corporation, provides access to digital doppelgängers which are utilized to help uncover emergent properties of the healthcare system. The digital doppelgängers generated by Synthea are based on real patient datasets and medical data to generate synthetic patient data. This synthetic electronic health record (EHR) is not connected to a human system, so it does not have the privacy and security concerns that real electronic health records have (MITRE 2019).

Using SysML, we model the effects of changes in copays and premiums on the relationships in the healthcare system, further document Synthea in model form, and document any additional work completed in MATLAB. The sequence diagram represents the effects of decreased copays on healthcare system through a patient visit to the doctor. Copays and premiums have direct effects on the actors in the relationships, but they can also indirectly affect the relationships and interactions between actors in the system. Lower copays, for example, may encourage more frequent doctor visits causing longer wait times and shorter amounts of time spent with the doctor. The lower copays may also increase the likelihood of the patient being able to afford any prescriptions provided by the doctor. According to the American Medical Association (AMA), medical nonadherence – where patients do not take prescriptions as prescribed by their doctor – is common. The AMA states that 25% of new prescriptions are not filled and 50% are not used as directed (U.S. Centers for Medicare and Medicaid Services 2019). These interactions in groups of relationships happen hundreds of millions of times annually which has a significant effect on the healthcare system overall.

Due to the broadness and complexity of the healthcare system in the USA, the SysML modeling strategy must be kept in focus by the scope of the system. The healthcare system is comprised of a range of payers, providers, patients, payer policies, as well as policies in the form of government rules and regulations. Rules and regulations can vary from state to state with some states requiring more benefits than others. By limiting the scope, the models can be better defined for the payer policies that most resemble the effects of changes in copays and premiums. In addition to limiting the scope to changes in copays and premiums, we also limit the scope by focusing on a population that resembles the state of Massachusetts.

The sequence diagram in SysML is also color coded to show how changes may either positively affect or negatively affect relationships in the system. Shown in Fig. 3, positive effects are indicated by green comment boxes. Negative effects are indicated by red comment boxes.

In addition to SysML, MATLAB is used to mathematically represent the relationships in the healthcare system. Many of the relationships are already modeled in Synthea, but aspects of the relationships that have not yet been implemented in Synthea are modeled in MATLAB. Anything modeled in MATLAB is also modeled in SysML to document functions and ensure traceability. Figures 4 and 5 show the SysML representation of a MATLAB implementation of the payers, respectively. Changes in copays and premiums were initially modeled in MATLAB until they were added to Synthea. The results in MATLAB can also be used to visualize the effects of copay and premium changes on the system. The relationships represented in MATLAB and Synthea are based on real-world data. Data from the US government as well as from Commonwealth Fund was utilized in the MATLAB models. Each function in MATLAB was modeled in SysML in order to document the purpose of the function and to provide traceability. Payers were modeled in SysML in connection to the MATLAB implementation.

At the time, payer copay data had not been implemented in Synthea, so the copays were implemented in MATLAB. Synthea models were created by MITRE Corporation and are being continually improved. While the team did not make changes to the models that were created in Synthea, changes could be requested if needed. Copay and premium data are examples of changes that were made by MITRE in parallel to this research. Recognizing the limitations of the Synthea models was important to ensuring the accuracy of the models when integrated with Synthea.

The data produced by Synthea is both a benefit and a limitation of the Synthea models by functioning a hybrid of a digital twin and a virtual prototype. We refer to this hybrid as a digital doppelgänger which we leverage to work with US medical privacy laws. In the USA, medical privacy laws such as HIPAA impose restrictions on access to medical records (Madni et al. 2019). For the purpose of analyzing health policy, restrictions on access to medical records can be prohibitive to the

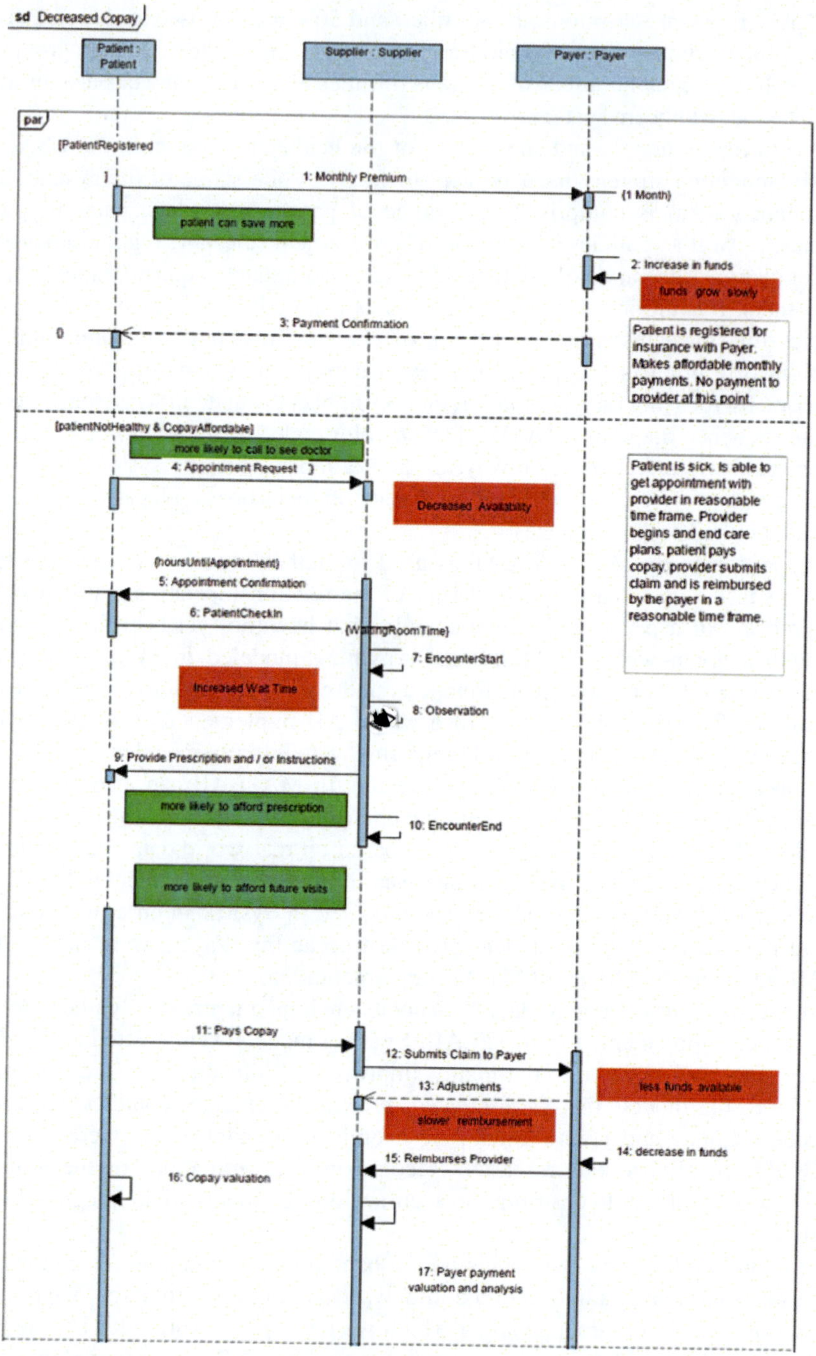

Fig. 3 Sequence diagram of patient, supplier, payer relationship

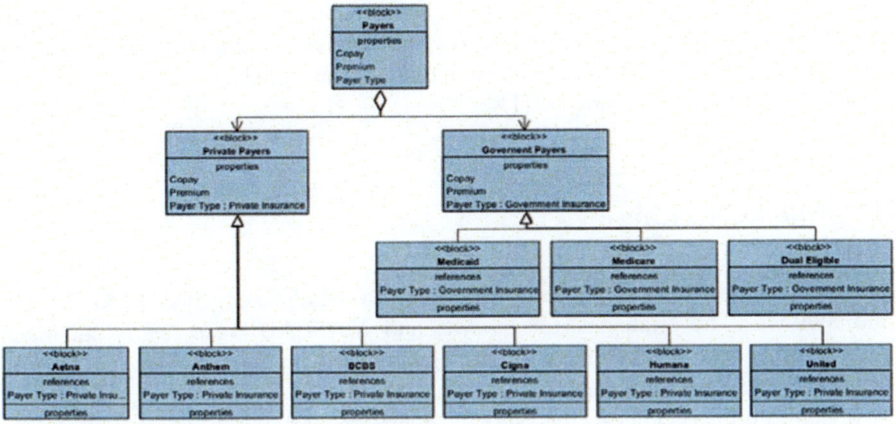

Fig. 4 Block diagram of payers

```
function copayPatient = copay(payerID)

[s, z] = size(payerID);
copayPatient = zeros(s,z);
for n = 1:s
if payerID(n)=='b3221cfc-24fb-339e-823d-bc4136cbc4ed';   %Dual Eligible
    copayPatient(n) = 100;
elseif payerID(n)=='7caa7254-5050-3b5e-9eae-bd5ea30e809c';   %Medicare
    copayPatient(n) = 40;
elseif payerID(n)=='7c4411ce-02f1-39b5-b9ec-dfbea9ad3c1a';   %Medicaid
    copayPatient(n) = 60;
elseif payerID(n)=='d47b3510-2895-3b70-9897-342d681c769d';   %Humana
    copayPatient(n) = 80;
elseif payerID(n)=='6e2f1a2d-27bd-3701-8d08-dae202c58632';   %BCBS
    copayPatient(n) = 75;
elseif payerID(n)=='5059a55e-5d6e-34d1-b6cb-d83d16e57bcf';   %United
    copayPatient(n) = 70;
elseif payerID(n)=='0e582013-fcdf-3f4b-9801-c8835acf347c';   %Aetna
    copayPatient(n) = 65;
elseif payerID(n)=='047f6ec3-6215-35eb-9608-f9dda363a44c';   %Cigna
copayPatient(n) = 60;
elseif payerID(n)=='42c4fca7-f8a9-3cd1-982a-dd9751bf3e2a';   %Anthem
    copayPatient(n) = 55;
elseif payerID(n)=='b1c428d6-4f07-31e0-90f0-68ffa6ff8c76';   %NO_INSURANCE
    copayPatient(n) = 100;
else
    copayPatient(n) = 20;
end
end
end
```

Fig. 5 MATLAB implementation of payers

analysis. Synthea takes the benefits of a digital twin such as a high level of detail and instantiations as well as the benefits of a virtual prototype but does not represent a real person to provide digital doppelgängers. Instantiations could have similar medical records to a real person but are not actually connected to a real person. Character strings are appended to automatically generated names to show that the instantiations are generated by Synthea and are not real people to prevent misuse of data and to prevent confusion (MITRE 2019).

3 Model Checking

The models have three major parts which need to be checked: SysML, MATLAB, and Synthea. The strategies for SysML and MATLAB were nearly identical. With Synthea, the model checking strategy focused on Synthea integration with MATLAB and SysML models.

In a preliminary check, the SysML diagrams and the MATLAB functions were reviewed by subject matter experts at MITRE Corporation during the model checking phase. During a complete implementation of the methodology with a specific policy instance, functions may need to be checked line by line. Modeling was completed in an iterative approach, so functionality was added incrementally and checked by experts throughout the process. With SysML, the models should be confirmed by experts who compare it to current policies. MATLAB models are also compared to policies, but these models have the added benefit of results that could be displayed. Variance from expectation could indicate the need to adjust MATLAB models. Once models are thoroughly vetted, they can be used to indicate emergent properties of the system. Accurate and sufficiently fitted models checked by experts could demonstrate unexpected effects of the complex healthcare system.

Synthea is checked by subject matter experts at MITRE Corporation. With respect to this research, SysML and MATLAB models needed to be checked for proper integration with Synthea to ensure that functionality would not be doubly implemented and that any assumptions about features in Synthea were confirmed with the MITRE team.

4 Conclusion

This paper discussed the use of digital twins and digital doppelgängers with MBSE constructs and methodologies to model and analyze healthcare policy. The approach builds on existing approaches but modifies them for the healthcare system. Digital doppelgängers are used to help simulate emergent properties of the healthcare system, but as electronic health records become increasingly prevalent and the security of those health records is addressed, digital twins can be used to even more accurately model the emergent properties of the healthcare system.

Acknowledgments We thank our colleagues at MITRE Corporation for their expertise in healthcare and Synthea, especially to Jason Walonoski and Rob Lieberthal. Additional thanks to Robi Scalfani for generating datasets from unreleased branches of Synthea.

References

Collins, S.R., and M.Z. Gunja. 2019. What Do Americans Think About Their Health Coverage Ahead of the 2020 Election? *Commonwealth Fund*, September 26 2019. [Online]. Available: https://www.commonwealthfund.org/publications/issue-briefs/2019/sep/what-do-americans-think-health-coverage-2020-election. Accessed 17 Oct 2019.

Krishnan, R., S. Virani, and R. Gasoto. 2018. Discovering toxic policies using MSBE constructs. In *Disciplinary Convergence in Systems Engineering*, Redondo Beach.

Madni, A.M., C.C. Madni, and S.D. Lucero. 2019. Leveraging Digital Twin Technology in Model-Based Systems Engineering. *Systems* 7: 7.

MITRE. *Synthea Empowers Data Driven Health IT*, MITRE, 2019. [Online]. Available: https://synthetichealth.github.io/synthea/#about-landing. Accessed Sept 2019.

Mulligan, J. 2017. The Problem of Choice: From the Voluntary Way to Affordable Care Act Health Insurance Exchanges. *Social Science and Medicine* 181: 34–42.

U.S. Centers for Medicare & Medicaid Services. *Glossary* [Online]. Available: https://www.commonwealthfund.org/publications/issue-briefs/2019/sep/what-do-americans-think-health-coverage-2020-election. Accessed 17 Oct 2019.

U.S. Department of Health and Human Services. 2013. Summary of the HIPAA Security Rule, U.S. Department of Health and Human Services, 26 July 2013. [Online]. Available: https://www.hhs.gov/hipaa/for-professionals/security/laws-regulations/index.html. Accessed 17 Oct 2019.

Virani, S., and T. Rust. 2016. Using Model Based Systems Engineering in Policy Development: A Thought Paper. In *Conference on Systems Engineering Research*.

Wei, Z., and M. Jarlenski. 2014. The Politics of Opposition to the Enactment of the Patient Protection and Affordable Care Act in the United States. *International Critical Thought* 4 (2): 208–220.

Employing Digital Twins Within MBSE: Preliminary Results and Findings

Shatad Purohit and Azad M. Madni

Abstract Model-based systems engineering (MBSE) requires greater investment than traditional systems engineering in the early phases of the system life cycle. Program management justifies this additional investment by arguing that such investments can be expected to produce continuous gains across later phases of system life cycle resulting from early detection of defects, risk reduction, improved communication, superior integration of the supply chain, product line definition, and enhanced traceability. Since systems evolve over the system life cycle, system models need to be updated continually to reflect the state of the system and realize value. However, in systems engineering organizations today, there is a tendency to reallocate modeling resources to other projects once initial modeling is completed on a particular project. This practice results in a resource gap which impedes the continuous update of system models through later phases of the system life cycle. This paper presents how digital twin technology can be exploited to address model updates throughout the MBSE life cycle. This paper also presents preliminary results from prototyping and experiments with a digital twin including data collection from a physical system operating in the real world to update the digital twin model. This paper shows how operational analysis and modeling can be enhanced by leveraging the digital twin construct.

Keywords Model-based systems engineering · Digital twin · Virtual prototype · Simulation

1 Introduction

Model-based systems engineering (MBSE) today tends to be primarily focused on the front end of the system life cycle (Madni et al. 2019). MBSE typically results in greater investment in the early phases of system life cycle when compared to early-

---

S. Purohit (✉) · A. M. Madni
University of Southern California, Los Angeles, CA, USA
e-mail: shatadkp@usc.edu

A. M. Madni et al. (eds.), *Recent Trends and Advances in Model Based Systems Engineering*, https://doi.org/10.1007/978-3-030-82083-1_4

stage investments in traditional systems engineering. Program management justifies this added expense by arguing that they expect to recover the added investment through continuous gains in the later phases of the system life cycle through early detection of defects, risk reduction, improved communication, superior integration of the supply chain, product line definition, and traceability. Furthermore, systems tend to evolve over their life cycle, generating a need for continual model update to continue to realize value. Unfortunately, in most systems engineering organizations today, modeling resources are quickly reallocated to other projects after initial modeling on a particular project is complete. This practice results in resource gaps in the ongoing project which impedes the ability to continuously update the system model through later phases of the system life cycle.

At the present time, hardware associated with a physical system and operational analysis are not an integral part of MBSE methodologies, but they need to be. Integrating hardware associated with a physical system into MBSE provides opportunities to learn from real-world data while continuously updating the system model, thereby improving model accuracy.

During the initial phases of the system life cycle, system model fidelity is limited by the available knowledge and information about the structure and behavior of the system. Therefore, system architecture and design decisions are initially based on limited data which implies decision-making under uncertainty. Additionally, the model does not extend to later stages such as manufacturing, integration, operations, and maintenance. In other words, in today's system development projects, initial phase system models rarely cover factors pertaining to later phases of the system life cycle (Madni and Purohit 2019). This omission invariably leads to complications during system (model) validation and operation. This recognition has led to the development of a digital twin-enabled closed-loop modeling approach (Madni and Erwin 2018; Madni et al. 2020). This paper describes experiments conducted using this approach.

With increasing system complexity, it is important to monitor system behavior throughout the system life cycle. This capability is essential to deal with unknown unknowns and collect data on system behavior (Datta 2016, 2017; Bone et al. 2018). The latter can increase the accuracy and fidelity of system models which in turn can improve decision quality and provide useful data for future projects (Folds & McDermott 2019; Ghosh et al. 2019). As important, model updates from a previous system can be used in defining a product line. This paper describes experiments conducted using this approach. This approach employs a combination supervised, unsupervised, and reinforcement learning for inference analytics and model upgrade.

The remainder of this paper is organized as follows. Section 2 presents the overall methodology. Section 3 describes the experiments performed with physical system and digital twin. Section 4 presents the technical risk related to digital twin-enabled model-based systems engineering, and Section 5 presents conclusions and potential for future work in this area.

2 Methodology

In MBSE, models are based on consistent assumptions, semantics, and principles. Such models can potentially answer questions of interest to stakeholders. Thereafter, experimentation using simulations can be used to verify dynamic system behaviors under a variety of assumptions.

In complex systems, hidden interactions can give rise to unexpected system behaviors. Therefore, models of such systems need to account for such interactions because the quality of predictions and decisions depends on the credibility, completeness, and accuracy of the models.

As new capabilities are being added to the twenty-first-century systems to sustain them in highly competitive markets, they are producing an increase in the interdependencies between the various components within and outside the system. Increasing dependencies increases the system's overall complexity, which results in unintended consequences including failures (Grieves and Vickers 2017; Hoffenson et al. 2019). In today's complex systems, it has become essential to embrace failures. Also, it is not enough to continue to perform adequately in the face of disruption. Today's systems need the capability to learn from responding to disruptions and failures and use that knowledge to improve system's performance and resilience.

Since detailed data on the system and its operational environment is not available in the early phases of the system life cycle, the assumptions made initially in building the models could be wrong, and the models may not have the requisite fidelity. As new data becomes available, learning can help with revision of initial assumptions and heuristics.

Failure to revisit the initial assumption is one of the most significant issues that needs to be addressed in MBSE. The initial models are created with limited data; the assumptions employed in the beginning may not be valid because they tend to be based on inadequate understanding and limited evidence (West and Pyster 2015).

Digital twin technology (from digital engineering) provides a promising means to update assumptions and increase model fidelity (Kinard 2010; Kraft 2015; Morton et al. 2009). New evidence from the physical system can be captured and stored for analysis and subsequently used to update system behavior (Chen et al. 2008; Glaessgen and Stargel 2012). The data captured in digital twins enhances traceability between system requirements and real-world behaviors.

With digital twin-enabled MBSE, the fidelity of the model can increase with new data, which in turn improves the quality of architectural and design decisions made. As shown in Fig. 1, inference and data analytics can complement MBSE. Insights and data thus produced can be used to systematically increase the fidelity and scope of the model leading to superior decisions and predictions. Also, as shown in Fig. 1, data from the physical system in the real world can be used to update system models. Subsequently, analysis and inference can lead to new insights that can complement the knowledge contained in the centralized digital models even during later phases of the system life cycle. Figure 2 depicts the MBSE life cycle phases on the horizontal axis. The MBSE, traditional SE, and digital twin-enabled

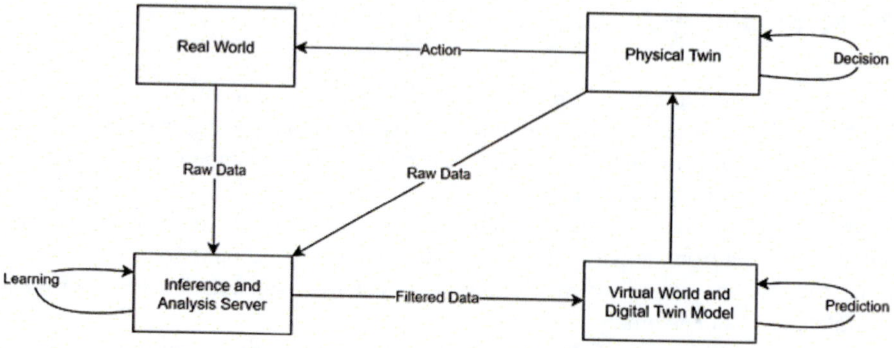

Fig. 1 Digital twin-enabled MBSE process overview

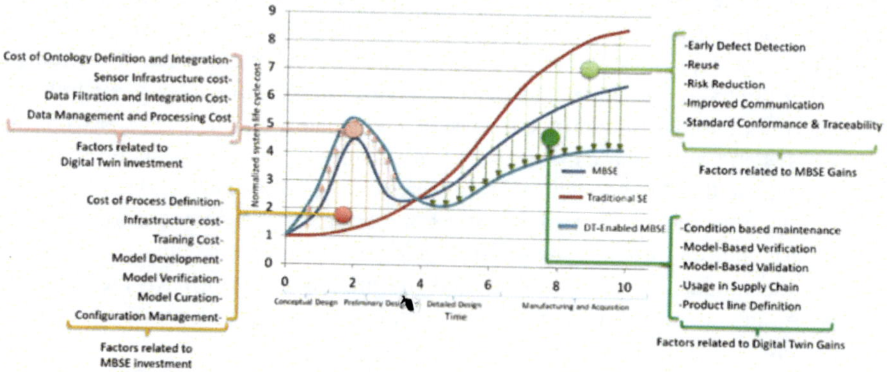

Fig. 2 Application of digital twin across MBSE life cycle with associated investments and gains (Madni et al. 2019)

MBSE cost curves are presented in Fig. 2. Color codes are used to convey traditional systems engineering, MBSE, and digital twin-enabled MBSE curves. The costs and gains associated with digital twin-enabled MBSE implementation are presented.

For digital twin-enabled MBSE, at the initial phase of the system life cycle, there are additional costs related to ontology definition and integration, sensor infrastructure implementation, data processing, data management, and configuration management. Substantial gains are expected during the later phases of the system life cycle against incurred costs considering the time value of money. Consolidating and continuously updating information from enterprise, system, and organization level facilitates superior decision-making and planning across the system life cycle.

Figure 3 presents an ontology for digital twin-enabled model update. The ontology view depicts relevant aspect of closed-loop model-based systems engineering. The ontology answers questions such as What are the different components of the digital twin? What are the different aspects of the digital twin that are updated during the system life cycle? What kind of data is collected related to the system's temporality?

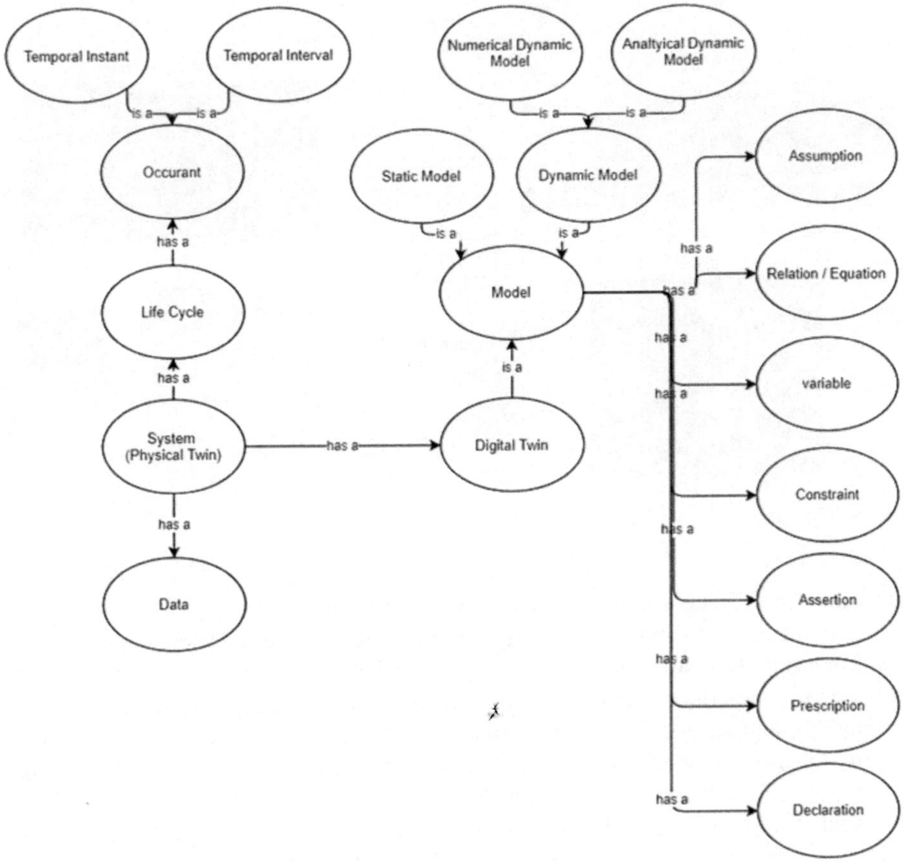

Fig. 3 Digital twin-enabled model update ontology

3 Preliminary Experiments

We conducted preliminary experiments to explore the creation of the digital twin and its update using data from real-world operation of its physical twin. In this experiment, the contextual model of the system is used to demonstrate the value of digital twin technology (i.e., modifying the model during system operation in real world). The contextual model captures the external entities present in the system's environment. Physical properties such as size, shape, location, and velocity are captured in the contextual model.

The experimentation setup consists of two UAVs (Fig. 4). Each UAV system consists of sensing subsystem, communication subsystem, battery subsystem, propulsion subsystem, and control subsystem. UAVs operate in an indoor maze-structured environment comprising static and dynamic entities and subject to

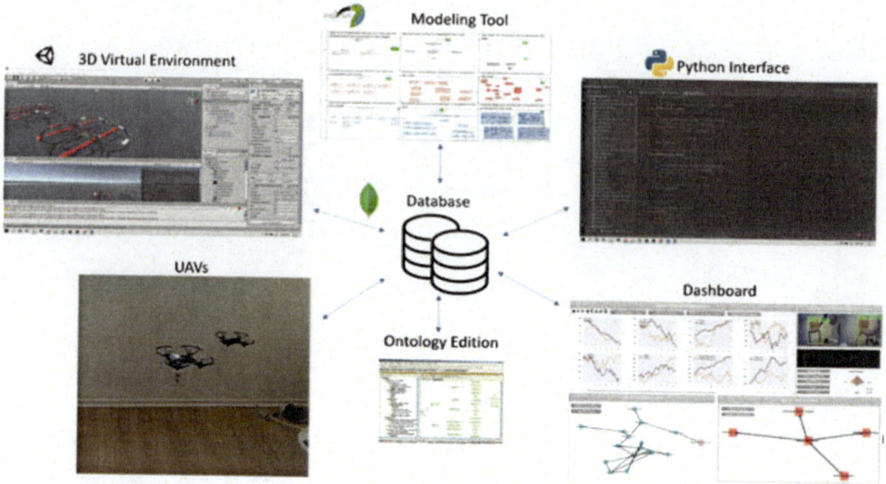

Fig. 4 Experimentation setup

external forces. The mission of the UAV system is to search for a predefined object in the environment. Both UAVs communicate the external visual and acoustic data, internal subsystem information related to battery subsystem health, internal subsystem temperature, external temperature altitude, location, communication bandwidth, and speeds. The data is transferred to the data storage system through communication network for preprocessing. The inference and analytics server performs data collection, integration, filtering, feature extraction, transformation, and analysis. For example, the video stream coming from the UAV is converted into frames of fixed size, and color threshold and contrast values are adjusted. Frames are calibrated for distortion and passed through trained Recurrent Convolutional Neural Network model, where the objects in the video feed are identified and tagged. Also, velocity vectors of objects are calculated based on the movement of objects in the frames and UAV velocity vectors. Objects with similar velocity are clustered in the same group. In this case, the contextual model of the system is continuously updated based on results from analytics.

The planning and decision-making problem involves multiple levels of decision-making, with varying levels of difficulty. For example, rule-based models, heuristic models, and probabilistic models provide increasing levels of information about the system. In the experiment, the physical system continually acquires contextual information.

The experimentation scenarios span multiple UAV operational phases. The first phase of UAVs search mission has only static objects in the environment. The system follows a rule-based model to navigate in the environment. In the next phase of the experiment, a dynamic object is introduced in the external environment of the UAVs. Thereafter, the object is identified and clustered in a new group due to its dynamic nature. The generation of new clusters induces the system to switch from a rule-based model to a state machine model. In the next phase, an external force acts on one of the UAVs, changing the stability, speed, and location of the UAV. The inference/analytics server then creates a correlation function between the sudden change in internal system attributes and the appearance of the external dynamic object. This, in turn, forces the system to shift to a probabilistic system model. In the experimentation setup, the human-system interface (Fig. 5) allows the human to intervene and add inputs. This interface enables the creation of tags, editing of the correlation function, control of the system, and updates to the model. In the experiment, operational phase data of the system is used to update the model.

4 Implementation

During simulated or real-world system operation, making model updates is challenging. Specifically, system representation needs to be sufficiently flexible (and extensible) to incorporate new information and changes. Furthermore, observations require the capability to capture and store data in digital repositories to support decision-making. This requirement, in turn, creates a requirement on the system architecture to represent the perception subsystem. Making models "closed loop" (Madni et al. 2019) requires an investment in (a) physical or virtual sensors during system prototyping, development, integration, testing, operation, and maintenance and (b) a rudimentary testbed instrumentation for data collection.

Virtual simulations capable of interacting with physical systems require communication infrastructure and computing architecture that enables the creation and automatic update of online models. An initial modest investment in the infrastructure that includes a perception subsystem (i.e., sensors) is required. However, significant benefits can be expected to accrue with repeated use of the infrastructure during simulation-based experiments. A primary technical risk in implementing a state-based model is the combinatorial explosion inherent in such models. Since data is collected continuously and the model once constructed is updated continually, dealing with a massive influx of data requires prior data filtering and processing. Our approach to containing the combinatorial explosion is to employ heuristics and contextual filtering (Madni 2018; Madni and Erwin 2018; Madni and Sievers 2018). This approach can potentially ameliorate this risk by separating data analysis and filtering from model building and communication.

5 Summary

This paper has presented a methodology, preliminary experiments, and implementation for a digital twin-enabled MBSE process. The paper provides proof of feasibility of creating a digital twin that is able to access data from the

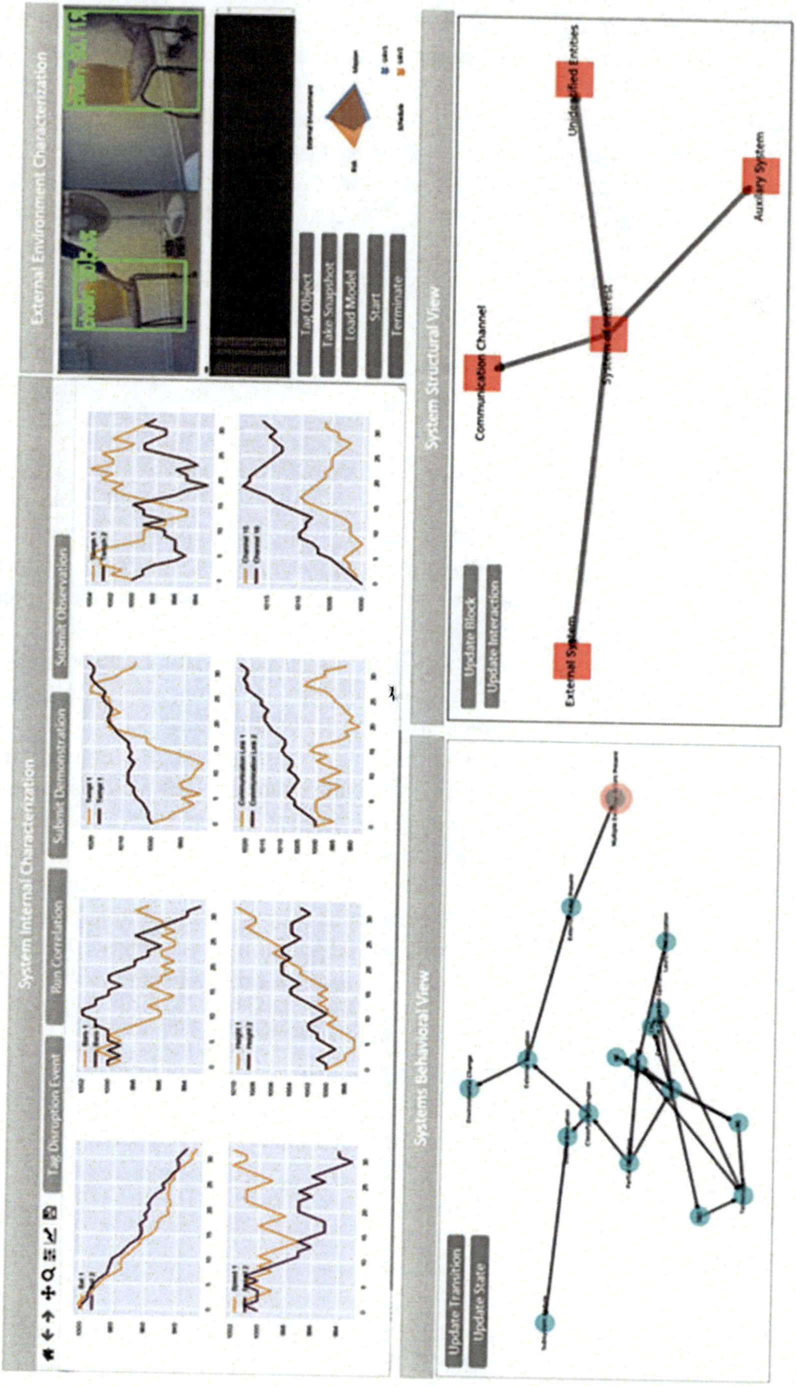

Fig. 5 Human-system interface of experimentation setup

physical twin operating in the real world to stay up-to-date with respect to the operation of the physical system. This approach essentially makes modeling into a closed-loop process (Madni et al. 2019). A multi-UAV scenario is used in the preliminary experiments to demonstrate the overall approach. This paper has also presented the overarching ontology used to integrate the major subsystems needed for experimentation and support digital twin model update based on data from the physical twin.

References

Bone, M., M. Blackburn, B. Kruse, J. Dzielski, T. Hagedorn, and I. Grosse. 2018. Toward an Interoperability and Integration Framework to Enable Digital Thread. *Systems* 6 (4): 46.

Chen, P.C., D.H. Baldelli, and J. Zeng 2008. Dynamic Flight Simulation (DFS) Tool for Nonlinear Flight Dynamic Simulation Including Aeroelastic Effects. In *Proceedings of the AIAA Atmospheric Flight Mechanics Conference and Exhibit*, Honolulu, Hawaii, USA, AIAA 2008-6376.

Datta, S.P.A. 2016. Emergence of Digital Twins, arXiv e-print. (arXiv:1610.06467).

———. 2017. Emergence of Digital Twins – Is This the March of Reason? *Journal of Innovation Management* 5 (3): 14–33. https://doi.org/10.24840/2183-0606_005.003_0003.

Folds D.J., and T.A. McDermott. 2019. The Digital (Mission) Twin: An Integrating Concept for Future Adaptive cyber-Physical-Human Systems. In *IEEE International Conference on Systems, Man and Cybernetics (SMC),* Bari, Italy, 2019, pp. 748-754, https://doi.org/10.1109/SMC.2019.8914324.

Ghosh, A.K., S. Ullah, and A. Kubo. 2019. Hidden Markov Model-Based Digital Twin Construction for Futuristic Manufacturing Systems. *Artificial Intelligence for Engineering Design, Analysis and Manufacturing*. 1–15. https://doi.org/10.1017/S089006041900012X.

Glaessgen, E.H., and D.S. Stargel 2012. The Digital Twin Paradigm for Future NASA and U.S. Air Force Vehicles.

Grieves, M., and J. Vickers. 2017. Digital Twin: Mitigating Unpredictable, Undesirable Emergent Behavior in Complex Systems. In *Transdisciplinary Perspectives on Complex Systems*, ed. F.-J. Kahlen et al. Germany: Springer.

Hoffenson, S., P. Brouse, D.S. Gelosh, M. Pafford, L.D. Strawser, J. Wade, and A. Sofer. 2019. Grand Challenges in Systems Engineering Education. In *Systems Engineering in Context, Proceedings of the 16th Annual Conference on Systems Engineering Research.*, ed. S.C. Adams et al.

Kinard, D. 2010. The Digital Thread–Key to F-35 Joint Strike Fighter Affordability, Aerospace Manufacturing and Design http://www.onlineamd.com/amd-080910-f-35-joint-strikefighter-digital-thread.aspx

Kraft, E.M. 2015. HPCMP CREATE-AV and the Air Force Digital Threat, 53rd AIAA Aerospace Sciences Meeting, Kissimmee, FL.

Madni, A.M. 2018. *Transdisciplinary Systems Engineering: Exploiting Convergence in a Hyper-connected World*. New York: Springer.

Madni, A.M., and D. Erwin, 2018. Next Generation Adaptive Cyber Physical Human Systems, Year 1 Technical Report, Systems Engineering Research Center, September.

Madni, A.M., C.C. Madni, and D.S. Lucero. 2019. Leveraging Digital Twin Technology in Model-Based Systems Engineering, MDPI Systems, special issue on Model-Based Systems Engineering, Publication, March.

Madni, A.M., and S. Purohit. 2019. Economic Analysis of Model-Based Systems Engineering. MDPI Systems, special issue on Model-Based Systems Engineering, Publication, February.

Madni, A.M., S. Purohit, and A. Madni. 2020. Digital Twin Technology-Enabled Research Testbed for Game-Based Learning and Assessment: Theoretical Issues of Using Simulations and Games in Educational Assessment, ed. O'Neil, H, Taylor & Francis, Spring.

Madni, A.M., and M. Sievers. 2018. Model-Based Systems Engineering: Motivation, Current Status, and Research Opportunities, *Systems Engineering*, Special 20th Anniversary Issue, 21 (3).

Morton, S.A., D.R. McDaniel, D.R. Sears, B. Tillman, and T.R. Tuckey. 2009. Kestrel—a Fixed Wing Virtual Aircraft Product of the CREATE Program. In *Proceedings of the 47th AIAA Aerospace Sciences Meeting*, Orlando, Fla, USA, January, AIAA 2009-338.

West, T.D., and A. Pyster. 2015. Untangling the Digital Thread: The Challenge and Promise of Model-Based Engineering in Defense Acquisition. *INSIGHT* 18 (2): 45–55.

A Review of Set-Based Design Research Opportunities

Nicholas J. Shallcross, Gregory S. Parnell, Edward Pohl, and Eric Specking

Abstract Increasing system complexity has been a driving force for the development of systems engineering methodologies. One of these methodologies is a concurrent engineering process known as set-based design (SBD). In support of ongoing SBD research, our team conducted a structured literature review to ascertain the current state of SBD research. This paper provides the results and analysis from the review of relevant SBD methodologies attempting to identify methodologies combining two or more systems analysis methodologies with SBD. The purpose of this research is to identify potential research opportunities by identifying underrepresented SBD methodologies in the literature. We specifically focus on applications combining SBD and model-based systems engineering (MBSE) with requirements development, decision analysis, risk analysis, affordability analysis, resilience, and complexity analysis applications. Our findings identified several potential research opportunities in these application areas, as well as additional opportunities for combining SBD and MBSE with systems architecting, uncertainty analysis, and intelligent adversary analysis.

Keywords Set-based design · Model-based systems engineering · Literature review · Systems engineering research opportunities

1 Introduction

The increasing complexity of engineered systems drove the development of new systems engineering (SE) methodologies for the past three decades (Bhatia and Mesmer 2019). This complexity encompasses the design of individual components, their subsequent interaction as part of a system, and the system's interaction and function with other systems in its operational environment. It logically follows

N. J. Shallcross (✉) · G. S. Parnell · E. Pohl · E. Specking
University of Arkansas, Fayetteville, AR, USA
e-mail: njshallc@uark.edu

A. M. Madni et al. (eds.), *Recent Trends and Advances in Model Based Systems Engineering*, https://doi.org/10.1007/978-3-030-82083-1_5

that engineered systems are now subject to greater uncertainty, both epistemic and aleatory, thus increasing a design program's overall risk. The need to better understand system requirements, reduce uncertainty, enhance design resiliency, and control lifecycle costs makes set-based design (SBD) an attractive product design alternative in both government and industry. SBD is a concurrent engineering methodology that facilitates enhanced product development through delaying design decisions, robust alternative development, and uncertainty reduction (Singer et al. 2009). As a methodology, SBD has the flexibility to adapt to organizational and program requirements and is well suited for complimentary implementation in conjunction with other system design processes (Singer et al. 2009). The purpose of this paper is to provide a synopsis of SBD use cases and to identify SBD application areas requiring further research and development. We specifically focus on applications combining SBD and model-based systems engineering (MBSE) with requirements development, decision analysis, risk analysis, affordability analysis, resilience, and complexity analysis applications. To accomplish this, we provide a brief description of SBD and then discuss the results and analysis of our structured literature review. Finally, we identify knowledge gaps, specifically highlighting potential research opportunities combining SBD and MBSE with modeling and simulation, system architecting, uncertainty analysis, and intelligent adversary analysis.

2 Set-Based Design

2.1 Description

Set-based design is a product design method first described by Ward et al. in their study of the Toyota Motor Corporation's design and production process (Ward et al. 1995). Ward describes SBD as a concurrent and iterative engineering process that develops sets of design alternatives, at the system and sub-system levels, and subsequently removes infeasible and sub-optimal alternatives from consideration. These set reduction activities continue until a final design solution is selected. More succinctly, SBD is a sequential product development decision-making process well suited for systems engineering programs dealing with complexity and uncertainty.

SBD develops and explores a large number of system and sub-system design alternatives, organized as sets, within the design space. Sets are groups of design alternatives sharing at least one specified design choice (Specking et al. 2018, Specking and Buchanan 2018). Set design spaces can be either discrete or continuous, and it is common for system-level design spaces to initially contain thousands and possibly millions of prospective design alternatives (Shallcross et al. 2019). It is this feature that enables SBD to explore and analyze potentially superior design alternatives compared to those developed using traditional design approaches. While there are several SBD methodologies described in the literature, these processes generally follow three principles: (1) map the design space to define feasible regions, develop multiple alternatives, and explore trade-offs, (2) identify intersections of feasible sets and develop conceptual robustness, and (3) establish feasibility before commitment by controlling uncertainty at decision points and remain within sets once committed (Raudberget 2010).

A critical component of any SBD methodology is the delaying of critical design decisions. Premature design decisions, often made in the absence of sufficient information, reduce design and program flexibility, thus increasing cost and schedule risk. Decision delay enables system requirements and design uncertainty resolution prior to commitment to specific sub-system or system alternatives (Singer et al. 2009). In this regard, decision delay is a key risk mitigation strategy for complex engineered systems and is the primary mechanism enabling many of the potential benefits of SBD.

2.2 Benefits

Numerous SBD benefits, identified from the literature, are grouped into two complimentary categories: (1) value improvement benefits and (2) risk mitigation benefits. Value improvement benefits result from the development of multiple robust design alternatives that are Pareto-optimal in regard to stakeholder value. Value improvement benefits can take many forms but are generally the result of superior alternative identification and development (Small et al. 2018). A simple thought exercise regarding continuous spectrums supports this point. Mathematically, the probability of occurrence for any single discrete point over a continuous range is zero. Many complex system design spaces can also be thought of as continuous spectrums and thus contain a nearly infinite number of design options. Traditional approaches generally consider only a few design alternatives; however, the probability that one of these complex designs is Pareto-optimal is essentially zero. By considering a multitude of design alternatives, SBD increases the identification likelihood of an optimal design or set of designs, in terms of value and some other metric. Thus by considering a significantly greater number of alternatives, we are able to identify alternatives with greater value than those developed using traditional design approaches. This is especially true during the early design phase. Figure 1 provides a simple depiction of SBD value improvement benefits when compared to traditional system design approaches (Wade et al. 2018).

Using uncertainty resolution methods can yield risk mitigation benefits. In this regard, we reduce epistemic or knowledge uncertainty by delaying design decisions. By definition, a decision is an irrevocable allocation of resources; thus decision delay forestalls premature commitment of finite resources (Parnell et al. 2011). This results in increased design and program flexibility in the face of changing and uncertain requirements. Additionally, delayed decisions enable technology maturation helping to ensure pre-commitment design feasibility (Ghosh and Seering 2014). In the end, these factors help to mitigate program schedule and cost risks by

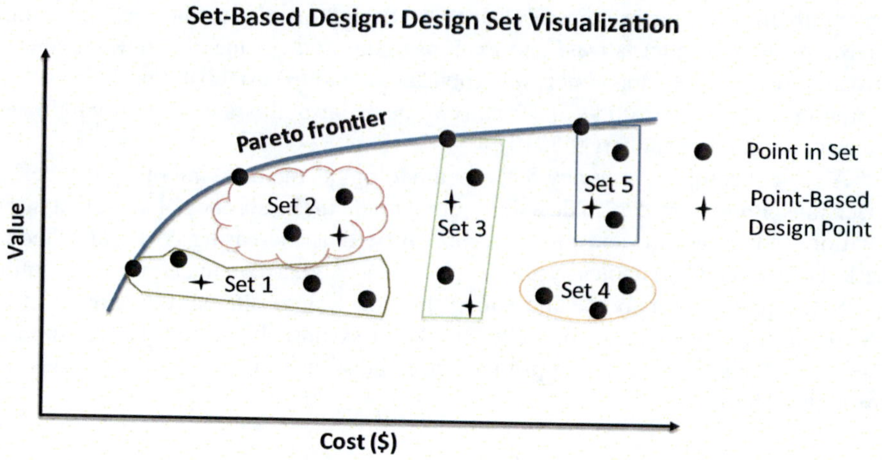

Fig. 1 Point-based design and set-based design comparison

retaining the capability to efficiently update or change design options with minimal rework, a leading driver of cost and schedule overruns (Shallcross et al. 2019).

3 A Review of the Current State of Set-Based Design

This section presents a brief summary of our literature review methodology and an overview of key findings. We undertook a structured literature review of 158 documents, primarily consisting of theses, books, journal papers, conference papers, and technical reports to identify the state of SBD application and research in government, industry, and academia. We focused specifically on identifying novel SBD approaches or applications where SBD was used in conjunction with other system engineering design and process methodologies. It is not the purpose of this paper to provide an in-depth analysis of each contribution included within the review. Instead, we provide a high-level analysis of the general contributions from the body of knowledge. As a result, we highlight documents that are either seminal contributions, provide novel approaches, or support general statements about the body of knowledge.

3.1 Literature Review Methodology

This research conducted a methodical search of systems engineering design papers whose main topics included either "set-based design," "model-based systems engineering," or "systems engineering," plus one or more of the following topics:

affordability, complexity, decision analysis, requirements development, resilience applications, risk analysis, and uncertainty analysis. Forty-one database searches, combining these keywords, yielded 572 documents which we screened for study relevance, resulting in the inclusion of 158 documents in the final review. We used Qiqqa, a reference management software, to facilitate document management, enabling efficient analysis and review. Publication dates for these documents range from 1995 to 2019, with 143 having publication dates of 2005 or later. We categorized papers using the author-provided keywords, allowing us to catalog documents under multiple keyword categories. Additionally, we undertook a thorough review of document content to ascertain the level and type of contribution to our above referenced research areas. This review, along with the keyword categorization, formed the basis of the results and analysis provided in the following sections.

3.2 Description of Author-Provided Keywords

A review of author-provided keywords identified 309 unique keywords and phrases from the 158 documents included in this survey. Table 1 provides a listing of the top 10 keywords by observation. Overall, keywords similar to those used in our search and screening process are displayed prominently in this list, as well as related topics such as decisions analysis, optimization, and design and tradespace exploration.

We binned each keyword into 24 categories enabling keyword aggregation into a set of manageable topics. For example, the category "set-based methods" captured all keywords associated with known SBD methodologies, such as set-based design, set reduction, or set-based thinking. We screened each document to ensure that keyword categories were listed only once against a specific publication, preventing double counting of categories. Table 2 provides a listing of the top 15 categories by observation and occurrence frequency.

Set-based methods appear most frequently within the literature (68 contributions), followed closely by decision analysis applications (66 contributions). Several

Table 1 Top 10 author-provided keywords by observation from the 158 reviewed articles

Rank	Author-provided keywords	Observations
1	Set-based design	63
2	Model-based systems engineering	20
3	Decision analysis	17
4	Risk analysis	16
5	Systems engineering	16
6	Uncertainty	15
7	Optimization	14
8	Multi-objective decision analysis	11
9	Trade-off analysis	11
10	Resilience	10

Table 2 Top 15 keyword categories by observation from the 158 reviewed articles

Rank	Keyword categories	# observations
1	Set-based methods	68
2	Decision analysis applications	66
3	General systems engineering	46
4	Risk applications	30
5	MBSE applications	28
6	Design and tradespace analysis	26
7	Uncertainty analysis	25
8	Resilience	23
9	Complexity	18
10	Optimization and heuristics	17
11	Defense applications	16
12	Modeling and simulation	15
13	Early design applications	14
14	Requirements development	12
15	AoA	12

categories germane to our research, such as risk and MBSE applications, as well as resilience, complexity, and requirements development occur less frequently within the literature. This helps identify potential emergent SBD research areas.

3.3 Analysis of Literature Review Research Areas

Following keyword categorization, we analyzed the research activity within the body of literature, by identifying studies contributing to three or more research categories of interest. As our future research looks to develop complex system design methodologies combining SBD with MBSE, we focused our analysis on documents containing significant contributions to both areas. Therefore, we identified all documents containing "set-based methods" and "MBSE applications." We then screen those categories against topics such as decision analysis applications, risk applications, resilience, complexity, and requirements analysis. Figure 2a–d provides a visual depiction of notable joint research by category, specifically highlighting instances of research contributing to three research categories of interest.

While the body of literature contains numerous studies contributing to any 2 of the 24 keyword categories, a limited number of publications contribute to 3 categories. For example, there are 68 publications contributing to set-based methods, 28 contributing to MBSE, and 66 contributing to decision analysis applications in our literature. However, as seen in Fig. 2a, only five documents contribute significantly to all three.

We observed similar results when we analyzed contributions pertaining to design and tradespace applications, risk applications, modeling and simulation,

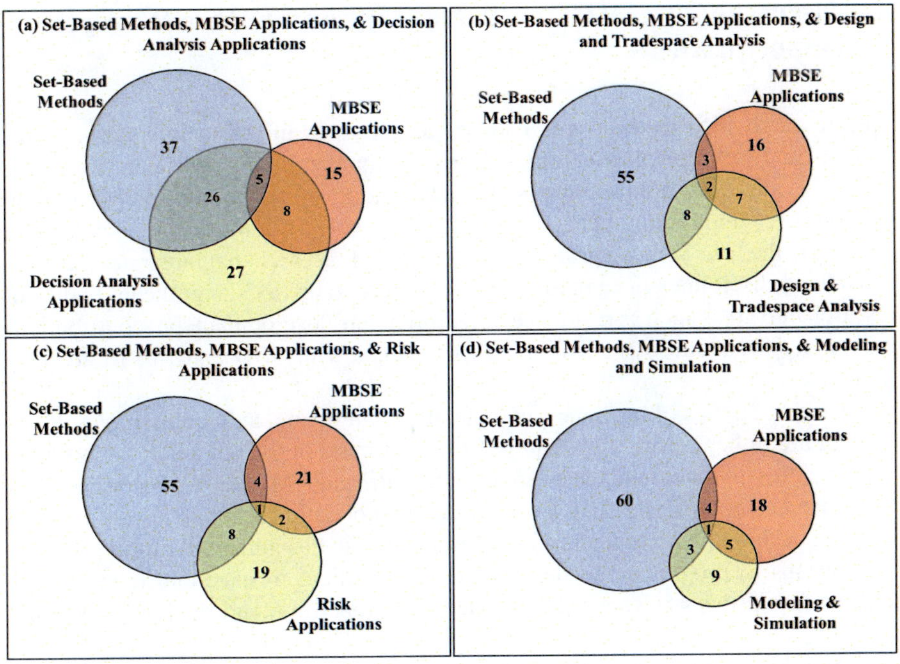

Fig. 2 Joint research activity including set-based methods, MBSE applications and (**a**) decision analysis applications, (**b**) design and tradespace analysis, (**c**) risk applications, and (**d**) modeling and simulation

affordability, and uncertainty analysis. While there exists significant and relevant research in these areas, there is a dearth of research combining these disciplines with SBD and MBSE. In this regard, it is also interesting to note that only five studies provide significant contributions to both SBD and MBSE: Blackburn (2014), Garner et al. (2015), Miller (2017), Specking et al. (2018), and Yukish et al. (2018). Several authors, such as Nahm and Ishikawa, identified the efficacy of conducting SBD in a model-based environment as early as 2005 (Nahm and Ishikawa 2005). Similarly, other researchers, such as Wong, described methodologies containing elements of SBD and MBSE (Wong et al. 2009). However, it is only recently that methodologies, such as those listed above, purposefully combining both SBD and MBSE in systems engineering and product design appear in the literature.

Ultimately, the purpose of this study was to identify knowledge gaps with the current set of SBD and MBSE literature to guide future research efforts. Figure 2a–d identifies joint contribution areas with limited research, highlighting the potential for new approaches that integrate modeling and simulation, risk, or tradespace analysis with model-centric SBD. However, this research also identified application areas completely lacking significant contributions, which we define as SBD knowledge gaps.

4 Discussion on SBD Knowledge Gaps and Future Research Opportunities

This research identified several knowledge gaps pertaining to integrating SBD, MBSE, and several of the topics listed at the beginning of Sect. 3. Figure 3a–d provides four contribution areas highlighting opportunities for integration: resilience, system requirements development, complexity, and system architecting. As seen in the four Venn diagrams, there is a lack of research that integrates the three considered disciplines in each diagram. As before, there exist significant bodies of research into areas such as resilience or requirements development, but contributions integrating these and other similar research areas with SBD and MBSE are very limited.

As the categories of resilience, complexity, architecting, and requirements development are rather broad, they require further discussion to fully elicit the research opportunities to integrate these areas with SBD and MBSE. Concepts such as resilience and complexity span a wide array of applications. In this research we are mainly concerned with resilience as it pertains to the design and platform use case. This speaks to programs and designs that are resilient to uncertain requirements and platforms that are robust in the face of competitive and evolving operating

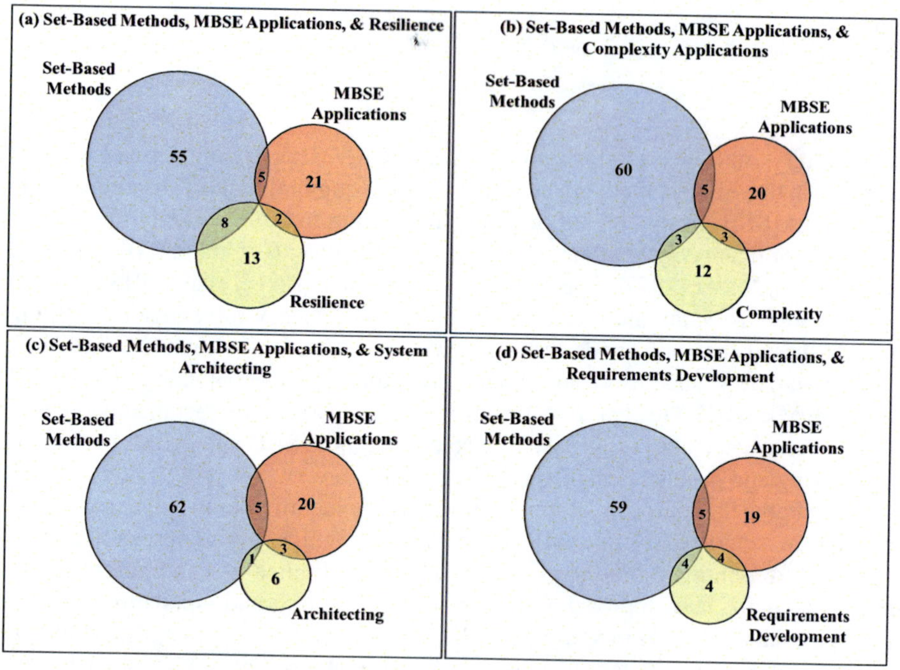

Fig. 3 Identified knowledge gaps for set-based methods, MBSE applications, and (**a**) resilience, (**b**) complexity applications, (**c**) system architecting, and (**d**) requirements development

environments. Similarly, complexity not only refers to complex systems and system of systems (SoS) but also complex environments. Complexity, and specifically system complexity, leads naturally into system architecting and how to effectively design an adaptive and affordable SoS under uncertainty. Finally, uncertainty reduction is enabled, in part, by thorough and robust requirements development. Requirements development and analysis can take on many forms, but in this context, requirements development specifically pertains to business and mission analysis. In addition to these findings, we also validated a research opportunity identified separately by Hartman and Specking regarding the use of intelligent adversary analysis to inform system design and risk analysis (Hartman 2018; Specking et al. 2018). A review of the literature found no applications combining intelligent adversary analysis with SBD or any other systems engineering application to inform design requirements or reduce epistemic uncertainty. Based on our review of the literature, these areas along with uncertainty and affordability applications create potential research opportunities for significant meaningful contributions within the engineering and design communities.

5 Conclusion

This research evaluated current SBD and MBSE literature to identify methodology development opportunities. Our research reviewed 158 journal articles, conference papers, technical reports, theses, and books to identify SBD research applications integrating additional methods such as MBSE, decision analysis, risk analysis, or requirements development. The review identified a rich source of SBD applications, generally conducted in concert with another systems engineering approach. However, the review also identified a lack of applications combining SBD with two or more specific methodologies, resulting in SBD knowledge gaps. We identified applications with limited research combining SBD and MBSE with decision analysis, tradespace analysis, risk analysis, or modeling and simulation methodologies. Additionally, we identified gaps lacking any research or methodologies incorporating SBD and MBSE with resilience, complexity, architecting, or requirements development applications. Finally, we validated a relevant research opportunity identified in the literature that combines SBD and intelligent adversary analysis to inform system design and programmatic risk. These results will inform our future research focus as we continue to develop and improve upon SBD methodologies suitable for complex and adaptive system development. The full list of reviewed sources is available to interested parties upon request.

References

Bhatia, G., and B. Mesmer. 2019. Trends in Occurrences of Systems Engineering Topics in Literature. *Systems* 7 (2): 28.

Blackburn, M. 2014. Introducing Model Based Systems Engineering Transforming System Engineering through Model-Based Systems Engineering: Technical Report for 2014, Systems Engineering Research Center, Stevens Institute of Technology.

Garner, M., N. Doerry, A. MacKenna, F. Pearce, C. Bassler, S. Hannapel, and P. McCauley. 2015. Concept Exploration Methods for the Small Surface Combatant. In *World Maritime Technology Conference*.

Ghosh, S., and W. Seering. 2014. Set-Based Thinking in the Engineering Design Community and Beyond. In *ASME 2014 International Design Engineering Technical Conferences and Computers and Information in Engineering Conference*.

Hartman, G. 2018. Enhancing Set-Based Design to Engineer Resilience for Long-Lived Systems. Dissertation, Wayne State University, Detroit, Michigan, United States.

Miller, S. 2017. Design as a Sequential Decision Process: A Method for Reducing Design Set Space Using Models to Bound Objectives, Dissertation, Pennsylvania State University, State College, Pennsylvania, United States.

Nahm, Y., and H. Ishikawa. 2005. Representing and Aggregating Engineering Quantities with Preference Structure for Set-Based Concurrent Engineering. *Concurrent Engineering* 13 (2): 123.

Parnell, G., P. Driscoll, and D. Henderson. 2011. *Decision making in systems engineering and management*. Wiley.

Raudberget, D. 2010. The Decision Process in Set-Based Concurrent Engineering-An Industrial Case Study. In *DS 60: Proceedings of DESIGN 2010, the 11th International Design Conference*, Dubrovnik, Croatia.

Shallcross, N., G. Parnell, E. Pohl, and D. Buede. 2019. Integrating Set-Based Design into the Department of Defense Acquisition System to Inform Programmatic Decisions. In *Proceedings of the International Annual Conference of the American Society for Engineering Management.*

Singer, D., N. Doerry, and E. Buckley. 2009. What Is Set-Based Design? *Naval Engineers Journal* 121 (4): 31.

Small, C., R. Buchanan, E. Pohl, G. Parnell, M. Cilli, S. Goerger, and Z. Wade. 2018. A UAV Case Study with Set-Based Design. *INCOSE International Symposium* 28: 1578.

Specking, E., G. Parnell, E. Pohl, and R. Buchanan. 2018. Early Design Space Exploration with Model-Based System Engineering and Set-Based Design. *Systems* 6 (4): 45.

Specking, E., and R. Buchanan. 2018. A Foundation for System Set-Based Design Trade-off Analytics. In *Proceedings of the International Annual Conference of the American Society for Engineering Management.*

Wade, Z., G. Parnell, S. Goerger, E. Pohl, and E. Specking. 2018. Designing Engineered Resilient Systems Using Set-Based Design. *Systems Engineering in Context*, p. 111.

Ward, A., J. Liker, J. Cristiano, and D. Sobek. 1995. The Second Toyota Paradox: How Delaying Decisions Can Make Better Cars Faster. *Sloan Management Review* 36 (3): 43.

Wong, J., K. Parrish, I. Tommelein, and B. Stojadinovic. 2009. SetPlan: A Computer Tool to Aid in Set-Based Design. In *17th Annual Conference of the International Group for Lean Construction, IGLC17.*

Yukish, M., S. Miller, J. Martin, L. Bennett, and M. Hoskins. 2018. Set-Based Design, Model-Based Systems Engineering, and Sequential Decision Processes. *Naval Engineers Journal* 130 (4): 93.

Digital Modernization for Systems Engineering

Jorge Buenfil, Ross Arnold, Benjamin Abruzzo, and Scott Lucero

Abstract Digital modernization is a relatively new concept that the Department of Defense (DoD) has embraced and recommended for implementation across the services. We explain the concepts behind digital modernization using a systems approach to avoid confusion and promote clarity. Many other definitions of these concepts include unnecessary combination of descriptions, goals, and methods. We describe digital modernization in terms of its purpose, elements, and interconnections. Additionally, we offer new simpler definitions of the digital thread, digital system model, and digital twin concepts. To conclude we propose ways in which digital modernization could be implemented in systems engineering practice.

Keywords Digital modernization · Digital system model · Digital twin · Digital thread

1 Introduction

The Department of Defense (DoD) released its Digital Modernization Strategy on July 12, 2019; this strategy will guide DoD's information technology transformation. There are four strategic initiatives in the strategy:

- Innovation for advantage
- Optimization
- Resilient cybersecurity
- Cultivation of talent

J. Buenfil (✉)
Agile Defense, Inc., Reston, VA, USA
e-mail: jorge.buenfil@alumni.stevens.edu

R. Arnold · B. Abruzzo
US Army CCDC, Picatinny Arsenal, NJ, USA

S. Lucero
DoD, Washington, DC, USA

A. M. Madni et al. (eds.), *Recent Trends and Advances in Model Based Systems Engineering*, https://doi.org/10.1007/978-3-030-82083-1_6

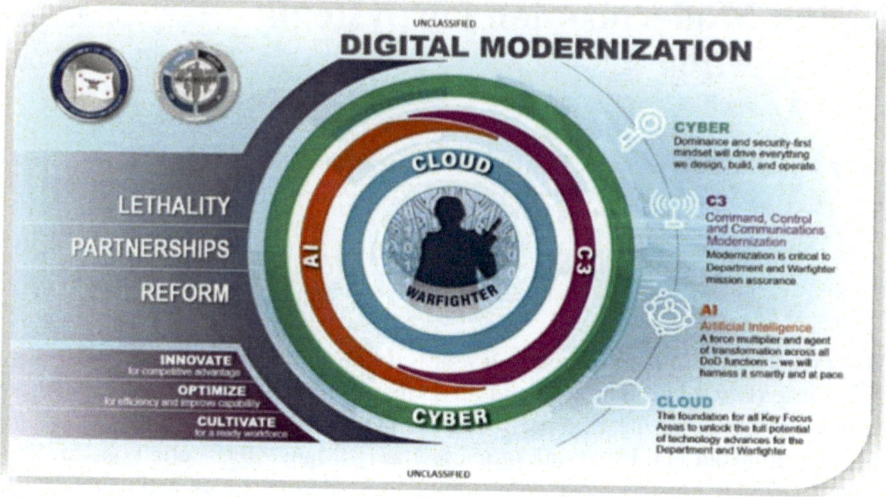

Fig. 1 Digital modernization. (US Dept. of Defense 2019)

This strategy supports the National Defense Strategy (NDS) lines of effort through the lens of cloud, artificial intelligence, command, control, communications, and cybersecurity, as shown in Fig. 1. In particular, the National Defense Strategy indicates: "The security environment is also affected by rapid technological advancements and the changing character of war."

Rapid technological advances mentioned above imply a departure from traditional developments, which used to be designed with a combination of technical drawings and small-scale prototypes. Most of the time, those prototypes were made of cheaper materials, such as wood, but still some physical properties could be directly observed (i.e., coefficient of drag in a wind tunnel or aerodynamic stability) (Rich and Janos 1994). In this framework, DoD relied on a linear process to develop complex US government work not protected by US copyright systems, typically with a design-test-evaluate iterative process hoping that the result would serve a range of missions and users. The policy-regulated acquisition process used to govern this process is document-intensive and often stove-piped, which extended the time for each cycle with systems that are cumbersome to change and sustain (DepSecDef Systems Engineering 2018).

In the now distant past, a computer's cost was very high and its processing capacity very limited, confining the use of modeling and simulation (M&S) only to those with the means and for which there was practically no alternative, as in the case of thermonuclear weapon design in the 1950s (Bird and Sherwin 2005). As digital computers became more powerful and readily available, many more of the stages of tests and analyses could be performed virtually.

We see digital modernization as the natural evolution of the movements toward "virtualization" applied to engineering efforts. By virtualization we mean the abstraction of relevant information from the real world. Virtualization has been evolving since the early days of computer simulations, through software-defined radios, to fly-by-wire technologies, digital photography, electronic documents, etc.

Now that computers are ubiquitous, powerful, and relatively inexpensive, widespread use of high-fidelity M&S is not only a possibility but also a necessity. Especially for systems engineering, this has resulted in a movement away from document-centric practices to model-centric practices. Yet, replacing traditional paper documents with electronic equivalents is not enough to modernize the practice in the terms outlined by the DoD's Digital Modernization Strategy. What is missing is the ability to close the loop: to keep virtual equivalents of every element of the system so that not only its parts but also its behaviors and interdependencies could be reproduced and visualized in high correlation with the equivalent physical implementation.

The DoD seized these developments to enact its Digital Engineering Strategy as follows:

> Incorporating the use of digital computing, analytical capabilities, and new technologies to conduct engineering in more integrated virtual environments to increase customer and vendor engagement, improve threat response timelines, foster infusion of technology, reduce cost of documentation, and impact sustainment affordability. These comprehensive engineering environments will allow DoD and its industry partners to evolve designs at the conceptual phase, reducing the need for expensive mock-ups, premature design lock, and physical testing (DepSecDef Systems Engineering 2018).

In short, DoD's goals are to reduce the time it takes to design new armament systems, while also lowering the cost of development and maintaining high availability, reliability, and adaptability to change. These goals promote the use of digital representations of systems and components and the use of digital artefacts as technical means of communication across a diverse set of stakeholders. Digital representations enable virtualization, which in turn allows us the ability to study the new system with computer simulations in order to fix any observable problems before even building prototypes of such system.

The remainder of this article is organized as follows. In Sect. 2 we discuss some of the previous definitions of digital thread, digital system model, and the digital twin. Sect. 3 covers our proposed definitions for the terms, with some discussion of open research challenges in Sect. 4. Finally we share our conclusions and discuss the future of digital modernization in Sect. 5.

2 Background

Figure 2 shows a graphical representation of the goals of digital engineering. It is noteworthy that the goals start with the use of models, since models are the artifacts that can be simulated; however, equally if not more important is that the last goal is the transformation of the enterprise culture and workforce. There are concrete technical solutions to the other goals but not the last one, which in our experience

Fig. 2 Digital engineering goals (DepSecDef Systems Engineering 2018)

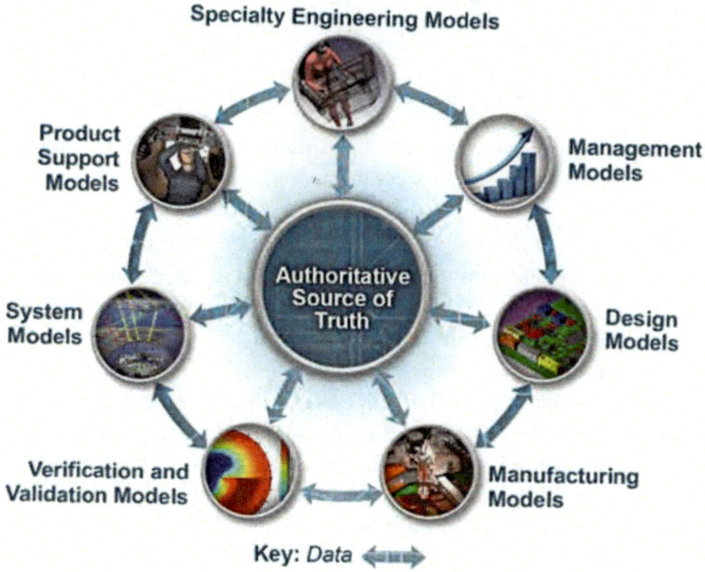

Fig. 3 Virtual system example (DepSecDef Systems Engineering 2018)

ends up being the hardest challenge that can make or break the efforts to achieve digital modernization. Figure 3 depicts an example of how an actual system is virtualized by connecting the different models that describe the system with high fidelity for simulation to the point where it becomes the authoritative source of truth for the development, manufacture, and sustainment phases of its life cycle.

In this context, the authoritative source of truth means the original data elements that represent the best data available. Other data elements may copy the original data, or create links to it, but updates to the data should be carefully designed to update the original and not the copies to prevent falling out of sync and loss of version control. The authoritative source of truth used to be called single source of truth to indicate that this was the original single source, but it led to confusion by implying that there was a single repository for the data. In reality the concept of authoritative source of truth does not require a physical single storage implementation; it may well be distributed over a wide area network as long as there is a way to tell which one is the original and where it is located.

The elements behind digital modernization are:

- Digital thread
- Digital system model
- Digital twin

The concept of digital thread was originally created and disseminated by the US Air Force (USAF), which originally defined it as "the combination of advanced modeling and simulation tools that link materials-design-processing-manufacturing" (Chief Scientist USAF, 2013).

Another well-known definition is:

> "Incorporating the use of digital computing, analytical capabilities, and new technologies to conduct engineering in more integrated virtual environments to increase customer and vendor engagement, improve threat response timelines, foster infusion of technology, reduce cost of documentation, and impact sustainment affordability. These comprehensive engineering environments will allow DoD and its industry partners to evolve designs at the conceptual phase, reducing the need for expensive mock-ups, premature design lock, and physical testing" (DepSecDef Systems Engineering 2018). "The Digital Thread refers to the communication framework that allows a connected data flow and integrated view of the asset's data throughout its lifecycle across traditionally siloed functional perspectives. The digital thread concept raises the bar for delivering the right information to the right place at the right time" (Leiva 2016). Yet another common definition, proposed by Dr. Kraft, US Air Force (USAF), expands the concept to add that:
>
> > Digital Thread is the creation and use of cross domain, common digital surrogates of a materiel system to allow dynamic, contemporaneous assessment of the system's current and future capabilities to inform decisions in the Capability Planning and Analysis, Preliminary Design, Detailed Design, Manufacturing, Testing, and Sustainment acquisition phases. (Kraft 2013)

Those definitions are not incorrect; however, they mix in ways the digital thread works, along with some justification of why it is important and how to use it. If we wanted a simple definition to say what it is exactly, we would not find it. Dr. Kraft described the digital surrogate as a "physics-based technical description of the weapon system resulting from the generation, management, and application of data, models, and information from authoritative sources across the system's life cycle." Currently, the digital surrogate concept is more widely known as a digital system model (DSM) by DoD. A DSM is the result of applying digital engineering techniques using a digital thread.

The last element to define is the digital twin, which has been defined as:

virtual instance of a physical system (twin) that is continually updated with the latter's performance, maintenance, and health status data throughout the physical system's life cycle. (Madni et al. 2019)

Another definition of digital twin describes it as:

a virtual representation of the system as an integrated system of data, models, and analysis tools applied over the entire life cycle on a tail-number unique and operator-by-name basis. (Chief Scientist USAF, 2013)

NASA and USAF researchers describe a digital twin as "an integrated multi-physics, multi-scale, probabilistic simulation of an as-built vehicle or system that uses the best available physical models, sensor updates, fleet history, etc., to mirror the life of its corresponding flying twin" (Glaessgen and Stargel 2012). Those are good definitions in general, but also mix in some of the usage together with the what is the digital twin. Particularly the last two definitions assume that digital twins apply only to aircraft.

We researched current literature in the field for other definitions, but most of them combine the definitions with extraneous concepts such as purpose, method of creation, etc., instead of focusing on what the thing they are defining is. Additionally, most definitions fail to show the connection between digital twin, digital system model, and digital thread. Therefore, we present our own definitions using their level of abstraction as the differentiating criteria, which also facilitates the description of their interconnections to each other.

3 Proposal

Systems thinking is a set of synergistic analytic skills used to improve the capability of identifying and understanding systems, predicting their behaviors, and devising modifications to them in order to produce desired effects. These skills work together as a system. (Arnold and Wade 2015)

We propose applying a systems approach (Arnold and Wade 2015; von Bertalanffy 1969; Arnold and Wade 2017) to explain the concepts behind DoD's Digital Modernization Strategy. By a systems approach, we mean observing the parts (elements), behaviors (purpose), and interfaces (interconnections) that make digital modernization a system in its own merit, as depicted in Fig. 4.

These concepts can be better analyzed by describing them at different levels of abstraction. We work with three levels: conceptual, logic, and physical. In very general terms, the conceptual level of abstraction deals with the main ideas, the what to do. This is the level of the digital thread. The logical level of abstraction deals with how to do things. This level corresponds roughly to the digital system model. The physical level of abstraction deals with what to do things. This is the level of the digital twin. To better understand those concepts, we will synthesize simple definitions from multiple definitions that can be found online. Our aim is to

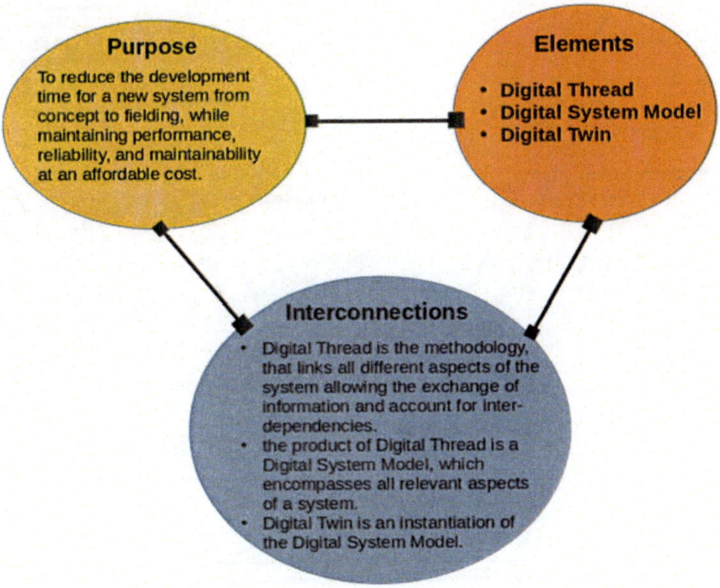

Fig. 4 Digital modernization

explain each term as a specialization of another term that is better known. We will indicate what is different with the new term that makes it unique.

In descending order by level of abstraction, we have:

Digital Thread: abstract representation of a class of physical systems as a collection of links that connect the system elements to enable computer modeling and simulation of data flows corresponding to the performance of the system's functionality

Digital System Model: an instance of a digital thread that corresponds to a particular system. For example, a laptop computer or a car

Digital Twin: an instance of a digital system model that corresponds to a unique physical entity. For example, a specific laptop computer with a unique serial number or car with a unique vehicle identification number (VIN)

The layers of abstraction can also be mapped to phases in the life cycle of the system, as shown in Fig. 5. The digital thread is a conceptual vision of a system, with its essential elements, their interconnections, and behavior/purpose of each part. This corresponds to the concept of operations and system-level requirements of systems engineering in Fig. 5. At the logical layer, the DSM is a model of the system; this corresponds to the design phases in Fig. 5. The physical level corresponds to the implementation phase in Fig. 5, where instance of the system implementation is made along with its corresponding digital twin. A common confusion happens between the concept of DSM and digital twin. One way to clear that confusion is to consider that the DSM represents a system as designed, while the

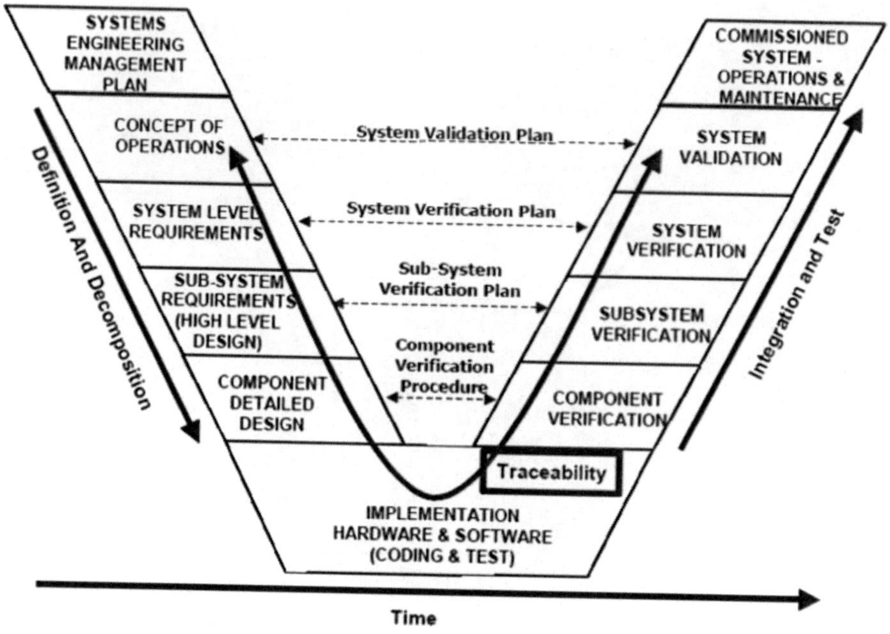

Fig. 5 Systems Engineering "V"

digital twin represents an instance of the system as built/as maintained; every digital twin is unique. One DSM typically results in several physical system instances, each of them assigned to its own digital twin.

In summary, the digital thread is the process of creation of a system in a virtual environment (a computer system). The product of the digital thread process is a digital system model that contains all the essential elements of an actual system, such as requirements, architecture, design, test cases, behavioral description, etc. that allow the system to be simulated by a computer, observed, and analyzed. Once the virtual system is perfected to meet its requirements and is produced, the physical instance of the system can have a Fig. 5 Systems Engineering V (Wasserman 2014) digital representation containing the exact configuration and maintenance record of this particular system at a particular point in time. This digital representation is the digital twin.

Planning and building a virtual system provides a virtual copy that can be studied and shared at low cost as compared with a physical prototype of the same system. These advantages are synonymous with the advantages afforded by software, which itself is a virtualization of data processing and data management. Software systems and associated digital data are easier to search, update, and share than physical counterparts such as large volumes of paper files, reports, tables with numbers, or calculations performed on mechanical devices. Similar efficiencies can be obtained by using the digital thread to create and manage digital system models and subsequently digital twins. The Tesla company is one example of a company that has utilized digital modernization concepts to its advantage. Tesla has revolutionized the automotive industry in no small part by taking the concept of virtualization to new heights. Tesla cars are designed with digital engineering; their DSMs are simulated and analyzed extensively before prototypes are made and tested on actual roads. Digital engineering is not only used for design. The DSM is used to drive the more than 1,000 factory robots that build a substantial portion of Tesla's cars (Campbell 2018); additionally, once the cars are delivered, a digital twin is maintained with all relevant data collected for that particular vehicle identification number (VIN). The digital twin includes the maintenance record, software versions, mileage, location, and many parameters that describe how a particular vehicle has been operated. The virtual world meets the physical world when software updates are delivered wirelessly to enhance the physical operation of the car's systems.

Boeing has taken a similar approach with its fleet of newer airplanes, all of which have a digital twin containing all the information necessary to remotely diagnose problems with the plane and to identify which kind of services or repairs may be required to keep an aircraft safely operational (High 2018).

4 Discussion and Open Research Challenges

The US DoD realizes that to overmatch near-peer competition challenges, it is necessary to streamline the ways new defense systems are acquired. New systems need to be fielded faster without sacrificing performance and cost. Given the complexity of new, large-scale defense systems, which often incorporate sophisticated electronics, new materials, high energy, and increasing artificial intelligence, meeting DoD goals will be extremely challenging using only today's systems engineering practices. A possible approach to this challenge is to use digital engineering via the digital thread to create and manage digital system models to develop new systems and then use digital twins to manage their maintenance operations once the systems are produced.

Adopting this paradigm implies the necessity to change the way the DoD develops and maintains new defense systems. However, there are several obstacles that must be overcome to successfully implement the Digital Modernization Strategy within the DoD. Taking a systems approach (Madni et al. 2019) to the problem, it rapidly becomes evident that it is not sufficient to simply agree to do things differently. It is necessary to adjust external factors such as the environment to foster and protect the first steps in the direction of change. The main obstacles as we see them are the following:

- Cultural inertia
- Fear of the unknown
- Lack of incentive to change

4.1 Cultural Inertia

One of the most difficult challenges that stands in the way of implementing digital modernization is to convince the workforce across the enterprise to embrace the new paradigm. Digital modernization reaches beyond new tool environments and training; it requires a change of philosophy of how systems are built and maintained. In a way, it is similar to gamification, defined as a set of activities and processes to solve problems by using or applying the characteristics of game elements (Deterding et al. 2011). Gamification commonly employs game design elements to improve user engagement. Sadly, until cultural inertia is overcome, a real digital transformation of the enterprise is unlikely. However, once cultural inertia is overcome, digital transformation will likely happen even without ideal tools or infrastructure in place.

4.2 Fear of the Unknown

Closely related to cultural inertia is the fear of the unknown, the not created here syndrome that makes people say things like "that's not how we do things" and "that will never work here." Even if the enterprise embraces digital transformation, members of the workforce may still fear some aspects of it either because they do not understand the new paradigm or because it has the potential to affect rooted interests in the old system. For example, if a department has the mission to verify engineering diagrams using human expertise, the introduction of an automatic verification system could trigger fierce resistance to those whose jobs would be replaced. The affected parties might argue that the new system could not be trusted, even in the face of factual evidence that the system performs at an equivalent or even more efficient level than that of humans on a particular task.

4.3 Lack of Incentives to Change

Even if we manage to motivate the workforce to embrace change, overcome cultural inertia, and dispel the fears of the unknown and our workforce is ready for the digital modernization, a lack of incentives may still pose a problem. If incentives to pursue the change are not established, the migration toward digital modernization will happen slowly at best or perhaps not at all. The result is a missed opportunity to transform the enterprise and align it better with the needs of future systems. The main reason to incentivize people to change is to provide the spark that ignites the explosion of innovation and creativity necessary to transform the enterprise as a whole, not just the engineering operations. Digital modernization is a holistic approach to organizational improvement and therefore requires buy-in from the entire organization.

5 Conclusion

We applied a systems approach to explain the concepts behind DoD's digital modernization in terms of its purpose, elements, and interconnections. The purpose is to reduce the development time for a new weapon system from concept to fielding, while maintaining performance, reliability, and maintainability at an affordable cost to the government. The elements of digital modernization are the digital thread, digital system model, and digital twin, which we defined in simple terms in Sect. 3 of this paper.

A summary of the interconnections of those concepts is that the digital thread is the methodology, or engineering paradigm, that links all different aspects of the system, represented as models, and allows the exchange of information among those models to account for interdependencies and second- and third-order effects of actions and events that affect the system. The product with this methodology of system development is the digital system model, which encompasses all relevant aspects of a particular system, such as a drone. Finally, the digital twin is an instantiation of the digital system model which keeps track of a particular physical system's state "as maintained." There is a one-to-one relationship between a produced copy of the system and its corresponding digital twin.

Given the absolute need to stay competitive both commercially and in the geopolitical and strategic military sense, we are left with little choice but to change how we develop engineering systems throughout their entire life cycles from conceptualization to decommissioning. The push for shorter development cycles, lower development costs, and high adaptability to changing circumstances demands that we create a digital world where new systems could exist and show us how they behave under a variety of inputs and environments. The knowledge gained by simulating complex engineering systems will be critical to the realization of correct physical implementations requiring less rework. This in turn will translate to lower costs, shorter development cycles, and greater adaptability as compared with traditional methods. Digital modernization will usher in a paradigm shift away from traditional complex system development in which a system gradually evolves by implementing a few features at a time, testing those features, updating the design to include new knowledge, and then repeatedly iterating through this cycle until the system meets all requirements and passes all operational tests. However, digital modernization will not happen by itself.

People will make the change. In particular, the systems engineers are in charge of leading this transformation. They must drive the systems engineering practice to higher level of excellence to allow their organizations not only to succeed but also to compete and win in highly contested environments. The cost of failure in this transformation is unacceptably high.

References

Arnold, R.D., and J.P. Wade. 2015. A Definition of Systems Thinking: A Systems Approach. *Procedia Computer Science* 44: 669–678. [Online]. Available: http://www.sciencedirect.com/science/article/pii/S1877050915002860.

———. 2017. A Complete Set of Systems Thinking Skills. *INCOSE International Symposium*, 27(1), Adelaide, South Australia.

Bird, K., and M.J. Sherwin. 2005. *American Prometheus. The Triumph and Tragedy of J. Robert Oppenheimer*. Random House, Inc.

Campbell, P. 2018. Here's What's Really Going on in Tesla's Factory, August 2018. [Online]. Available: https://ftalphaville.ft.com/2018/08/16/1534435108000/Here-s-what-s-really-going-on-in-Tesla-s-factory/.

Chief Scientist of the United States Air Force, Global Horizons. 2013. Final Report. United States Air Force. Global Science and Technology Vision. United States Air Force, June 2013. [Online]. Available: https://www.dodmantech.com/ManTechPrograms/Files/AirForce/ClearedDTforWebsite.pdf.

Deterding, S., D. Dixon, R. Khaled, and L. Nacke. 2011. From Game Design Elements to Gamefulness: Defining "gamification". In *Proceedings of the 15th International Academic MindTrek Conference: Envisioning Future Media Environments, ser. MindTrek '11*. New York: ACM, 2011, pp. 9–15. [Online]. Available: http://doi.acm.org/10.1145/2181037.2181040.

Glaessgen, E., and D. Stargel. 2012. The Digital Twin Paradigm for Future NASA and U.S. Air Force Vehicles. In *53rd AIAA/ASME/ASCE/AHS/ASC Structures, Structural Dynamics and Materials Conference – Special Session on the Digital Twin*; April 23, 2012 – April 26, 2012; Honolulu, HI; United States. NASA Langley Research Center; Hampton, VA, United States, 2012.

High, P. 2013. Boeing's CIO wins Forbes CIO Innovation Award by Developing the Digital Flight Deck. April c. [Online]. Available: https://www.forbes.com/sites/peterhigh/2018/04/16/boeings-cio-wins-forbes-cio-innovation-award-by-developing-the-digital-flight-deck/#452fd2bb7e6a.

Kraft, D.E. 2013. Expanding the Digital Thread to Impact Total Ownership Cost. In *201 NIST MBE Summit, 2013*.

Leiva, C. 2016. What is the digital thread? 2016. [Online]. Available: https://www.ibaset.com/blog/what-is-the-digital-thread/.

Madni, A.M., C.C. Madni, and S.D. Lucero. 2019. Leveraging Digital Twin Technology in Model-Based Systems Engineering. 2019. [Online]. Available: https://www.mdpi.com/2079-8954/7/1/7/htm.

Office of the Deputy Assistant Secretary of Defense for Systems Engineering, Department of Defense Digital Engineering Strategy. U.S. Department of Defense, June 2018. [Online]. Available: https://sercuarc.org/wp-content/uploads/2018/06/Digital-Engineering-StrategyApproved.pdf.

Rich, B.R. and L. Janos. 1994. *Skunk Works. A Personal Memoir of My Years at Lockheed*. Little, Brown and Company.

Summary of the 2018 National Defense Strategy of The United States of America. Sharpening the American Military's Competitive Edge. Office of Prepublication and Security Review, 2018. [Online]. Available: https://dod.defense.gov/Portals/1/Documents/pubs/2018-National-Defense-Strategy-Summary.pdf.

US Department of Defense, DoD Digital Modernization Strategy. DoD Information Resource Management Strategic Plan FY19-23. Office of Prepublication and Security Review, July 2019. [Online]. Available: https://media.defense.gov/2019/Jul/12/2002156622/-1/-1/1/DOD-DIGITAL-MODERNIZATION-STRATEGY-2019.PDF.

von Bertalanffy, L. 1969. *General System Theory*. New York: George Braziller, Inc.

Wasserman, S. 2014. Model-Based System Engineering – Beyond Spreadsheets. [Online]. Available: https://www.engineering.com/DesignSoftware/DesignSoftwareArticles/ArticleID/7352/Model-Based-System-Engineering%2D%2DBeyond-Spreadsheets.aspx.

Investigating Model Credibility Within a Model Curation Context

Donna H. Rhodes

Abstract Model curation can be thought of as a hallmark of digital maturity, signifying that models are highly valuable assets of the enterprise used as a trusted basis for engineering decisions. As enterprises begin to develop large model repositories, model credibility becomes a central concern. Recent studies show the decision to use, reuse, and repurpose models is contingent on the model consumer's perception of validity and trustworthiness of the model. This paper discusses an investigation of selected foundational works on credibility of models, simulations, and websites as part of a larger research effort on model curation. The objective is to leverage findings and strategies from prior work and identify useful heuristics that can inform model credibility within the context of model curation.

Keywords Model curation · Model credibility · Model discovery · Model consumer · Trust

1 Introduction

The transformation to digital engineering has brought model curation to the forefront of systems engineering research. Model curation can be defined as the *lifecycle management, control, preservation and active enhancement of models and associated information to ensure value for current and future use, as well as repurposing beyond initial purpose and context.* Curation activities include model governance, acquisition, accession, valuation, preservation, active enhancement, model discovery, and archiving. Curation practices promote formalism and provide for the strategic management and control of models and associated digital artifacts, particularly when managed as a collection at the enterprise level. Model curation infrastructure will better enable an enterprise to establish and actively enhance

D. H. Rhodes (✉)
Massachusetts Institute of Technology, Cambridge, MA, USA
e-mail: rhodes@mit.edu

A. M. Madni et al. (eds.), *Recent Trends and Advances in Model Based Systems Engineering*, https://doi.org/10.1007/978-3-030-82083-1_7

the collection of models that are of value to the enterprise (Rhodes 2018a, b). As evidenced by curation practice in institutional collections (e.g., museum, historical society, libraries), dedicated leadership and support roles are needed to carry out curation processes. Moving into the future, model curation can be expected to involve unique responsibilities at the enterprise level, motivating a new leadership role of model curator (Rhodes 2019). The lack of access to models, mistrust of models, and perception of legitimacy of models are all barriers to model reuse and longevity, which can potentially be mitigated by model curation. The terms *model* and *simulation* are used in this paper in a broad manner; useful definitions are specified by NASA (Fogg et al. 2001).

Model: A description or representation of a system, entity, phenomenon, or process (adapted from Bankes (1993)). **Note:** A model may be constructed from multiple sub-models; the sub-models and the integrated sub-models are all considered models. Likewise, any data that goes into a model are considered part of the model.

Simulation: The imitation of the behavioral characteristics of a system, entity, phenomenon, or process.

The most recently published version of the NASA Standard for Models and Simulations (2016) illustrates the life cycle of M&S as shown in Fig. 1. As completed models are selected for curation, model curation can be thought of as overlapping the latter two phases, model use/operations and model archiving. Additionally, an enterprise model collection would come into play at the start of the life cycle, where reuse of a model is an option that replaces the development of a new model.

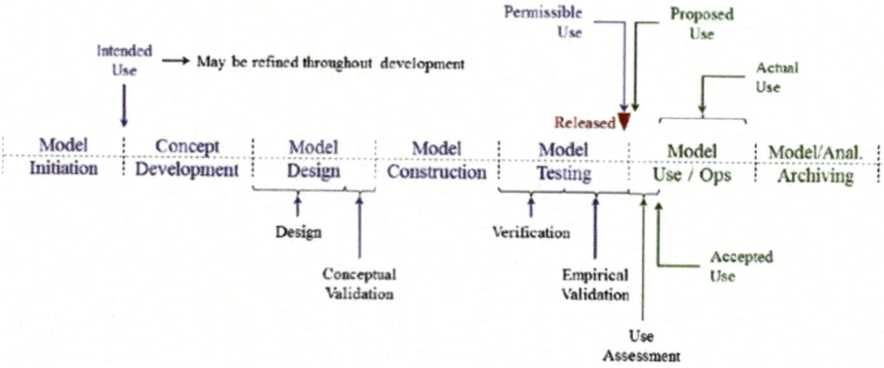

Fig. 1 M&S life cycle from NASA-STD-7009A W/CHANGE 1 (Ahn and de Weck 2007, p. 70)

2 Model Curation

Not all models are suitable for curation. Curation applies to longer duration models, rather than those developed for a quick study or to simply work out a problem. Two broad categories of longer-duration models that necessitate model curation are (1) models used throughout the lifespan of a program and (2) models to be intentionally reused for other purposes/contexts. The former includes, for instance, the set of models comprising a digital twin. The latter includes reference models and "platform" models, which enable enterprise to reuse and repurpose models (Rhodes 2018b).

Model curation use cases need to be generated through investigating the myriad situations under which curation is applicable and valuable to the enterprise and its stakeholders. This will depend uniquely on the enterprise, program characteristics, and specialized circumstances, including its strategic technology and business roadmaps. Model credibility is a key consideration because there is minimal value in placing a model under curation if the credibility is in question. This investigation is motivated by the belief that the foundational ideas and approaches for model credibility are worth re-examining respective to a model curation context. This context has many facets, including curation practice, requisite leadership, supporting infrastructure, required competencies, curating for model consumer needs, and application of innovations to enhance model discovery (Rhodes 2019). Model credibility is a construct that underpins all of these facets.

3 Model Credibility

Model credibility and its associated constructs (model confidence, model trust, model validation, model value, etc.) have been investigated and discussed in the literature for more than four decades. The earliest works come from the operations research and simulation communities (Rhodes 2018a, b, 2019). Our ongoing research is re-examining the prior work on credibility within the context of model curation. As revealed through past investigation of model credibility, there are actions that can be taken in curation practice and the supporting infrastructure to increase the likelihood that a model consumer will perceive a model as trustworthy and valid.

The following sections review selected works on model and simulation credibility. In addition to model credibility, a body of work on website credibility is discussed. It is worth noting that websites at the time of the studies were dense and content-rich static information, as contrasted with today's highly visual and more interactive experience. As such, this is an informative body of work for digital engineering in that the websites in the early website credibility studies are more akin to digital artifacts.

3.1 Credibility of Models and Simulations

Kahne (1976) proposed a new approach for examining model credibility for large-scale systems, asserting "model credibility is separated into two distinct, if not independent, issues: validity and value" (verification is assumed in his paper). Kahne states, "To be a bit more specific, we are contrasting verisimilitude (having the appearance of truth) with worth (value). For convenience we will use the words validity and value." He notes that credibility of a model will depend, among other things, upon "the quality of the match between the model and the model user," noting that the model reflects the biases and outlook of the modeler. A novelty of his approach is to take the viewpoint of buyer/seller, with a subjective approach to credibility-type questions where credibility is defined as *capable of being believed* (Kahne 1976).

In 1979, the SCS Technical Committee issued a report on *terminology for model credibility*. This was motivated by the desire to develop a standard set of terminology to facilitate effective communication between the builder of a simulation model and its potential users, believed to be the "cornerstone for establishing the credibility of a computer simulation" (SCS 1979). This committee provided a framework to review credibility of a simulation (Fig. 2). This framework divides the simulation environment into three basic elements. Inner arrows describe processes that relate the elements to each other. The outer arrows refer to procedures that are used to evaluate credibility of the processes.

Gass and Joel (1981) investigated the concepts of *model confidence*, showing model confidence to be not an attribute of a model, but of the model user (Gass and Joel 1981). Seven confidence criteria the authors proposed are model definition, model structure, model data, computer model verification, model validation, model usability, and model pedigree. The latter (originally called model demographics) is especially pertinent to perception of credibility, given its subjectivity. They state pedigree "should enable the decision maker to determine the model's status with respect to past achievements, theoretical and methodological state-of-the-art, and the expert advice that went into its development" (Gass and Joel 1981). Gass (1993) states that critical to use of a model is "the credibility or confidence that the decision maker has in the model and its ability to produce information that would be of value to the decision makers" (Gass 1993).

Balci (1986) proposed comprehensive guidelines for assessing credibility of simulation results. He characterizes a life cycle of simulation study as richly characterized with 10 phases, 10 processes, and 13 credibility assessment stages (Fig. 2).

Balci's important work demonstrates that credibility assessment is complex and involves staged assessment through the lifespan of a model or simulation (Steele 2008). His work demonstrates that during development, the acceptance of the model is a result of the model consumer's cumulative perception of validation efforts. This suggests the importance of giving a model consumer transparency into the series of validation activities that went into the original development, not only the end result.

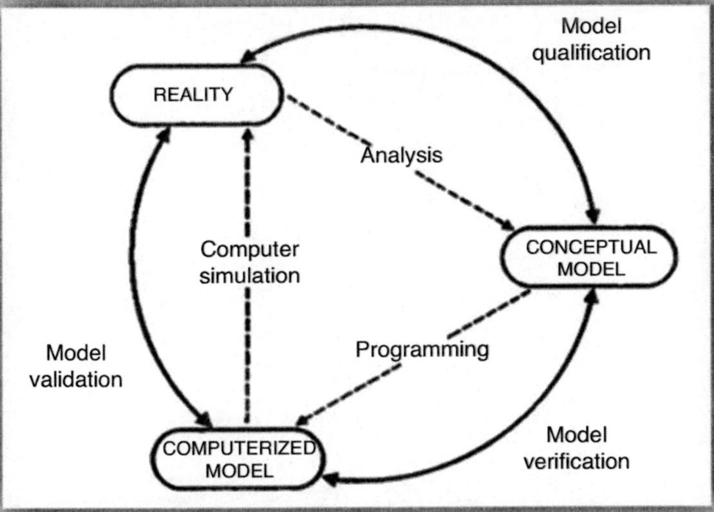

Fig. 2 SCS framework to review credibility of a simulation (SCS 1979)

Assessment of Credibility of Models and Simulations

Steele (2008) reveals the insights and thinking behind the NASA's Standard for Models and Simulations (M&S) (Fig. 3). Eight relevant factors of credibility were identified during the development of this standard, which defines credibility as *the quality to elicit belief or trust in M&S results (Steele* 2008*)*. The evolution of the NASA standard surfaced various dimensions of credibility and more recently an assessment approach. State of the practice on model credibility assessment in the systems field has emerged as part of the NASA efforts over more than a decade. A method for M&S credibility assessment is described in Appendix E of the 2016 update of *NASA Standard for Models and Simulations* (2016) [4, pp. 55–72]. Ahn et al. (2014) propose a formal procedure based on the NASA standard to assess the credibility of an M&S in an objective way using the opinions of an expert group for credibility assessment and a Delphi approach (Ahn et al. 2014), initially piloted on an M&S platform called SpaceNet by Ahn and de Weck in 2007 (Ahn and de Weck 2007).

3.2 *Website Credibility*

The Stanford University Persuasive Technology Lab, founded and directed by BJ Fogg, has investigated captology (computers as persuasive technologies) over the

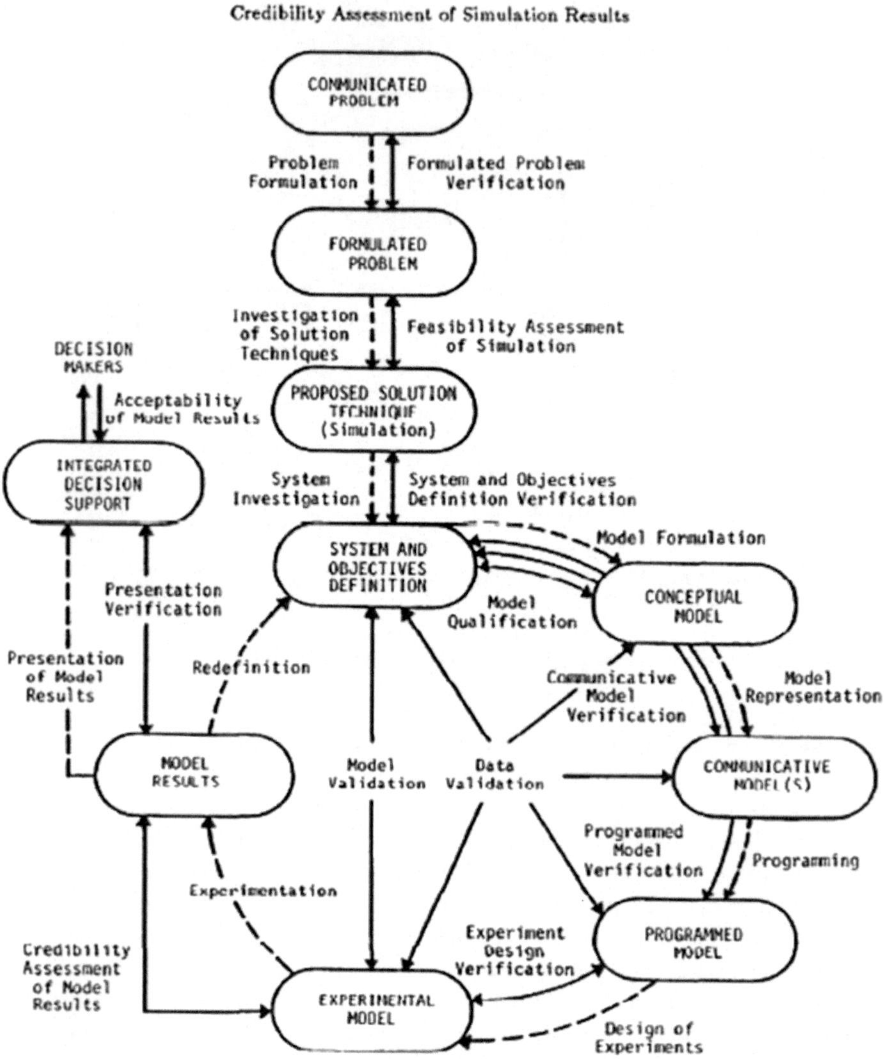

Fig. 3 Balci's life cycle of a simulation study (Steele 2008)

past two decades, with early studies on website credibility. The larger body of research of the lab seeks to create insights into how computing products can be designed to change people's beliefs and behaviors. An early study (Fogg et al. 2000) aimed to assess a broad range of elements that impact varying aesthetic, context, and technical factors on credibility of websites (Fogg et al. 2000).

Fogg et al. (2001) state "simply put, credibility can be defined as believability" and is a perception based on two factors: trustworthiness + expertise (Fig. 4). Additional clarifying points are the following: (1) credibility is a perceived quality;

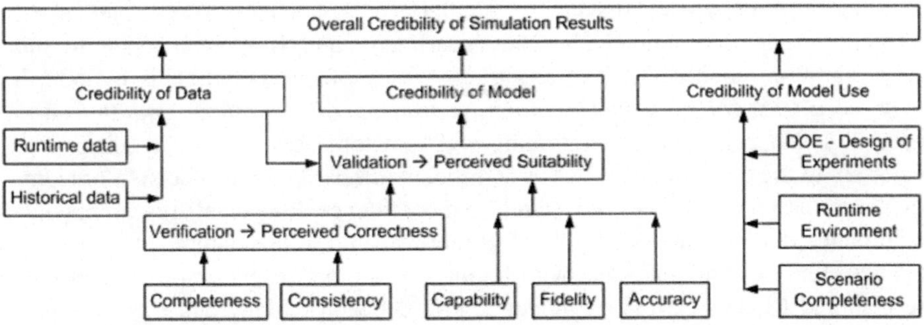

Fig. 4 De Vin's schematic of factors influencing the overall credibility of simulation results (De Vin 2015, p 154)

it does not reside in an object, person, or piece of information; and (2) when discussing credibility of a computer product, one is always discussing a perception of credibility (Fogg et al. 2001).

Several findings of this work are insightful for model curation. First, web credibility was found to increase when users perceive a real-world organization and real people behind a website. Second, small errors had a large negative impact on credibility of a website. Third, the users view websites as less credible if they experience technical problems (e.g., delays in download of information). Fogg states, "if users think a site lacks credibility – that the information and services cannot be trusted – they will abandon the site and seek to fill their needs in other ways" (Fogg et al. 2001). This appears to suggest that a poorly designed model repository would be a significant deterrent to the success of model curation in an enterprise. The research by Fogg et al. led to a collection of website design guidelines; several of these appear to be applicable for design of model curation infrastructure and enablers.

3.3 *Recent Research on Model Confidence and Trust*

Research performed by Flanagan (2012) uses case studies and a web-based experiment to investigate key challenges to model-based design: distinguishing model confidence from model validation (Flanagan 2012). The objective of her research is to understand factors that cause perception of model quality to differ from actual quality. She proposes eight factors as the key variables to misaligned model confidence and tests hypotheses for six of these in the experiment to illustrate the effect of the factors on perception of model credibility. According to Flanagan, these factors can potentially help explain behavior of decision-makers, especially in the situation where "the model would be a good tool to help solve a problem; however, the decision-maker does not agree and continues without input from the model, effectively dismissing its predictions" (Flanagan 2012). One of the hypotheses that was validated in the experiment, and is most relevant to this model curation, concerns source and transparency of the model. This hypothesis is: *A more trustworthy model author and transparent governing equations will improve model perception*. Her finding was that for cases where the source was important to the decision, there was a significant difference in the decision outcome where untrustworthy sources caused reduced confidence. Flanagan's research, while preliminary, demonstrates the value of further research of this nature.

These findings are consistent with findings of a more recent empirical study by German and Rhodes (2017) on model-centric decision-making and trust (German and Rhodes 2017). Model credibility was found to be a perceived quality, positively impacted by tailorable transparency and available model pedigree (detailing who originated the model, who subsequently enhanced the model, assumptions made, expertise of modelers, etc.). They also found that while not always needed, model consumers must have available model transparency when determining if a model should be trusted in making a specific decision.

3.4 Recent Research on Overall Credibility

Recent work by De Vin (2015) provides a significant discussion on credibility of simulation, stating it "...is thus influenced by three factors: Credibility of the model, credibility of the data, and credibility of the model use" (De Vin 2015, p. 152). He notes that without credible data (also called data pedigree as discussed in (NASA 2016)), it will not be possible to carry out trustworthy validation of the model. De Vin's paper "uses the NASA CAS model for credibility assessment of simulations to arrive at a schematic representation of how overall credibility as composed of aspect related to the model, the data, and the model's use" (De Vin 2015).

4 Toward Design Guidelines for Model Curation

Model curation requires both initial and ongoing investment, which must be outweighed by the payoffs (e.g., life cycle model usability, repurposing of models for new contexts, reuse of models at the enterprise level, etc.). This would be a poor investment if model credibility is neglected. In the spirit of Kahne's buyer/seller approach, it may be useful to think about model curation from a model acquirer (consumer)/model curator approach. The model acquirer's perception of the expertise and authority of the model curator will have influence on perception of credibility.

Model curation practices and infrastructure must be designed to support quality of individual model consumer experience and the capacity to foster perceived trustworthiness of the model. Model curation infrastructure (model repositories, interfaces for repository access, etc.) needs to be designed for cost effectiveness and security, as well as for the quality of the experience of human interaction. As can be inferred by Fogg (2000, 2001), a poorly designed model repository would negatively impact perceived credibility, as would a poorly designed user experience with access and interaction with the repository. The design implications resulting from the website credibility studies of Fogg et al. offer practical guidance for model curation. For example, including markers of expertise and markers of trustworthiness could be implemented through model pedigree information.

4.1 Heuristics for Model Curation

The technological challenges for creating model collection repositories are quite significant. The social, cognitive, and perceptual challenges are equally – if not more – challenging, yet are more difficult to comprehend and address. These challenges must all be addressed for the future success of digital engineering.

Model credibility is now generally accepted as a property of the model perceiver, and there are observed behaviors and useful strategies that the systems community can adapt in support of digital engineering goals. The foundational papers and on credibility suggest a derived set of heuristics toward establishing design guidelines that can be used by the systems community in evolving model curation practice and enabling technology. Continuing investigation is expected to contribute to expanding and refining these heuristics and formulating respective design guidelines. An initial set of nine heuristics are proposed below:

1. Model credibility is an attribute of the model consumer, not the model.
2. Model credibility is positively influenced by communication between modeler and model consumer, both active and passive.
3. Credibility of digital artifacts is influenced by both trustworthiness and expertise of a model consumer.
4. Acceptance of a model for (re)use is influenced by a model consumer's belief that the model has the ability to produce information of value to them.
5. Credibility of models in a collection influences a model consumer's trust in the enabling infrastructure.
6. A model consumer's experience in discovering and retrieving models from a repository influences perceived credibility of the model.
7. Model credibility is influenced by a model consumer's trust in the expertise of the model originator, as well as modelers who subsequently enhance and maintain the model over time.
8. Model credibility is influenced by a model consumer's capacity for transparency into the validation activities throughout its development and enhancement.

9. Credibility of the model collection is influenced by a model consumer's perception of expertise of the governance authority that accepted the model into the collection.

5 Conclusion and Further Research

Significant research is needed to realize the promise of model curation for the systems field (Rhodes 2018a, b, 2019). The ongoing larger research project is investigating four facets of model curation. The first is model curation implementation practices for performing model curation activities at the enterprise level. These must be harmonized with existing model management practices and supporting practices such as configuration management and data management. A second facet of the research investigates precursors, enablers, and barriers to model curation practice, infrastructure, and approaches to the curating of models. The third facet concerns innovations through myriad newer methods and technologies that could be applied in model curation, especially in model discovery and curating for consumer needs. The fourth concerns competencies, roles, and responsibilities. Model credibility is an overarching consideration across these four areas of research. The formulation of heuristics that draw from prior work and other fields can be used to inform the strategies and enablers for model curation such as model pedigree, accession records, and model valuation.

This paper considers selected prior research on model credibility, simulation credibility, and website credibility. Data credibility is not explored in this paper but is also of central importance to model curation. The findings suggest that prior work on credibility has significant relevance for model curation and useful strategies and practices from other fields can be adapted for digital engineering. Further investigation into the substantial body of work on credibility within the model curation context is ongoing. The linkage between model credibility and data credibility needs further examination within the model curation context. Additionally, other areas remain to be explored, including credibility in regard to information retrieval, data curation, human-computer interaction, and augmented intelligence.

Acknowledgments This material is based upon work supported, in whole or in part, by the US Department of Defense through the Systems Engineering Research Center (SERC) under Contract HQ0034-13-D-0004. SERC is a federally funded University Affiliated Research Center managed by Stevens Institute of Technology. Any opinions, findings, and conclusions or recommendations expressed in this material are those of the authors and do not necessarily reflect the views of the US Department of Defense.

References

Ahn, J., and O.L. de Weck. 2007. Pilot Study: Credibility Assessment of SpaceNet 1.3 with NASA-STD-(I)-7009, Tech Rep. MIT, Cambridge, MA.

Ahn, J., O. de Weck, and M. Steele. 2014. Credibility Assessment of Models and Simulations Based on NASA's Models and Simulation Standard Using the Delphi Method. *Systems Engineering* 17 (2): 237–248.

Balci, O. 1986. Credibility Assessment of Simulation Results. Proceedings of the 1986 Winter Simulation Conference.

Bankes, S., 1993. Exploratory modeling for policy analysis. *Operations Research* 41 (3), 435–449.

De Vin, L. 2015. Simulation, Models, and Results: Reflections on their Nature and Credibility. In *Proceedings of FAIM 2015*, 148–155.

Flanagan, G. 2012. Key Challenges to Model-Based Design: Distinguishing Model Confidence from Model Validation. Master's Thesis. Massachusetts Institute of Technology.

Fogg, B.J., J. Marshall, A. Osipovich, C. Varma, O. Laraki, N. Fang, J. Paul, A. Rangnekar, J. Shon, P. Swani, and M. Treinen. 2000. Elements That Affect Web Credibility: Early Results from a Self-Report Study. In *CHI'00 Extended Abstracts On Human Factors in Computing Systems*, 287–288. ACM.

Fogg, B.J., J. Marshall, O. Laraki, A. Osipovich, C. Varma, N. Fang, J. Paul, A. Rangnekar, J. Shon, P. Swani, and M. Treinen. 2001. What Makes Web Sites Credible?: A Report on a Large Quantitative Study. In *Proceedings of the SIGCHI Conference on Human Factors in Computing Systems*, 61–68. ACM.

Gass, S. 1993. Model Accreditation: A Rationale and Process for Determining A Numerical Rating. *European Journal of Operational Research* 66 (2): 250–258.

Gass, S.I., and L. Joel. 1981. Concepts of Model Confidence. *Computers and Operations Research.* 8 (4): 341–346.

German, E.S., and D.H. Rhodes. 2017. Model-Centric Decision-Making: Exploring Decision-Maker Trust and Perception of Models. In *15th Conf. on Systems Engineering Research.* Los Angeles, CA.

Kahne, S. 1976. Model Credibility for Large-Scale Systems. *IEEE Transactions on Systems, Man and Cybernetics.* Aug. 587–590.

NASA. 2016. NASA-STD-7009A w/CHANGE 1, Standard for Models and Simulations https://standards.nasa.gov/standard/nasa/nasa-std-7009. Accessed 30 Oct 2019.

Rhodes D.H. 2018a. Interactive Model-Centric Systems Engineering Technical Report, Phase 5. (No. SERC-2018-TR 104). Systems Engineering Research Center Hoboken NJ.

———. 2018b. Interactive Model-Centric Systems Engineering Technical Report, Phase 6. (No. SERC-2019-TR 003). Systems Engineering Research Center Hoboken NJ.

Rhodes, D.H. 2019. Model Curation: Requisite Leadership and Practice in Digital Engineering Enterprises. 17th Conference on Systems Engineering Research. Washington, DC.

SCS. 1979. Terminology for Model Credibility. *Reports of the SCS Technical Committees.*

Steele, M.J. 2008, June. Dimensions of Credibility in Models and Simulations. In *Proceedings of the 2008 Summer Computer Simulation Conference* (57). Society for Modeling & Simulation International.

Part II
Modeling in MBSE

Automated Detection of Architecture Patterns in MBSE Models

Matthew Cotter, Michael Hadjimichael, Aleksandra Markina-Khusid, and Brian York

Abstract The evaluation of a system's architecture is an essential process within the systems engineering lifecycle. Commercially available model-based systems engineering (MBSE) tools, when combined with standards-based architecture modeling languages, provide a means through which architecture information can be expressed graphically and formally in a machine-readable format; this format can be leveraged in order to improve the system architecture evaluation process. The authors propose an automated, repeatable method for detecting patterns of interest embedded within an MBSE model. This novel method uses a heuristically guided set of similarity measures that depend on textual and graphical content of a model. The proposed method has been implemented for architectures developed in IBM's Rational Rhapsody, and No Magic Inc.'s MagicDraw, and has proven to be able to identify six well-established patterns: Adapter, Bridge, Composite, Façade, Observer, and Proxy. This automation has the potential to produce cost and time savings for the evaluation process and to add an additional degree of rigor and completeness to an architecture evaluation.

Keywords Model-based systems engineering · MBSE · Architecture patterns · Design patterns · Unified modeling language · Systems modeling language · Pattern detection · Architecture analysis · Architecture evaluation

1 Introduction

Though no single accepted definition exists, a systems architecture is often described as a conceptual model that defines the structure and behavior of a real-world system; this model must be based on principles, concepts, and properties that logically relate to, and are consistent with, one another (INCOSE 2019).

M. Cotter (✉) · M. Hadjimichael · A. Markina-Khusid · B. York
The MITRE Corporation, Bedford, MA, USA
e-mail: mjcotter@mitre.org

A. M. Madni et al. (eds.), *Recent Trends and Advances in Model Based Systems Engineering*, https://doi.org/10.1007/978-3-030-82083-1_8

Systems architectures are utilized for different purposes, by different stakeholders, at different times throughout the system lifecycle. For example, system architectures are often evaluated by systems engineers during the conceptualization phase of the system lifecycle in order to quantify "the potential of said architecture to deliver a system capable of fulfilling stakeholder requirements, and to identify any potential risks associated with fulfilling said requirements" (Lassing et al. 1999; Maranzano et al. 2005).

System architecture evaluation may be informal or formal. Even if architecture evaluation is conducted informally, there exist many principles, processes, and guidelines that suggest how a systems architecture evaluation should be conducted. Most of these evaluation methods have their origins in software design and development; Babar et al. (2013) summarize popular and existing architecture evaluation methods, noting that all of the methods they summarize typically "use scenarios to characterize quality attributes....[these scenarios] are mapped onto architectural components to assess the architecture's capability to support those scenarios, or identify the changes required to handle those scenarios" (Babar et al. 2013).

Though these methods are commonly used, they are often entirely human-driven processes that rely heavily on expert observation and qualitative assessment. Thus, these evaluation techniques can at times be resource- and time-intensive if the complexity of the system is large or if the number of alternative architectures to compare with one another is large. *Architecture patterns*, also referred to as design patterns, may provide a means through which the architecture evaluation process can be made more automated and rigorous. Generally, a pattern may be defined as a reusable solution to one or more common problems (Taylow et al. 2009).

The usefulness of patterns, and more specifically architecture patterns, is twofold. First, the incorporation of patterns into a systems architecture can make said system more efficient and effective (Champin and Solnon 2003). Second, architecture patterns provide an evaluation team with a concrete set of architectural features that align with a specific capability or requirement.

This paper describes a method to automatically detect patterns of interest within a systems architecture captured in UML or SysML using either Rational Rhapsody or No Magic Inc.'s MagicDraw, two widely used system architecture development tools. The paper demonstrates that once a pattern is formalized and encoded into a UML/SysML model, there is enough contextual information in a systems architecture to have an algorithm intelligently search the graphical representation of the architecture and identify strong candidates for architecture pattern matches. This automation potentially facilitates cost/time savings for evaluation and has the potential to add an additional rigor and completeness to any evaluation method.

Though patterns are applied across multiple unique fields (e.g., software engineering, requirements engineering, mechanical engineering), it is important to note that their application in the domain of systems engineering has not yet become common practice through the community. Cloutier (2005) summarizes recent work and research on patterns within the systems engineering domain and proposes a framework through which future research can assist in the growth, and formal application, of patterns within the systems engineering (Cloutier 2005). The authors note that successful execution of research along Cloutier's framework would strengthen the methodology described in this paper; as patterns become more prevalent in practice, methods for their detection become more valuable.

Section 2 provides an overview of the technical approach that provides context to our research, before discussing complimentary prototypes that were created as part of this research in more detail. Section 3 summarizes the results of applying the prototypes to system architecture models. Section 4 concludes the paper, summarizing potential impacts and limitations to automated pattern detection.

2 Technical Approach

The authors investigated the feasibility of developing a product that guides a systems engineer during the process of system architecture evaluation by automatically detecting architecture patterns embedded within a systems architecture. In order to realize this goal, the authors made the following observations that helped scope and motivate the technical approach:

1. Commercially available model-based systems engineering (MBSE) tools provide a means to capture architecture data that is formally structured and thus easily manipulated and analyzed by external applications. This is primarily because MBSE tools force users to conform to semiformal graphical design languages such as the Unified Modeling Language (UML) and/or Systems Modeling Language (SysML). Both UML and SysML are managed and maintained by the Object Management Group (OMG) (Tsantalis et al. 2006).
2. The architecture under evaluation and architectural patterns to be detected can both be defined in UML/SysML models.
3. Architecture pattern detection is analogous to finding a labeled, directed subgraph within a larger labeled and directed graph, a complex category of problems that has been widely researched. Champin and Solnon (2003) measure the similarity of labeled graphs in order to compare cases in case-based reasoning and modeling structured objects (Bernardi et al. 2014). Tsantalis et al. (2006) used similarity scoring to detect patterns in open-source software, exploiting the inheritance hierarchies of classes and interfaces in open-source software (Weilkiens 2008). Bernardi et al. (2014) search for architecture patterns in a software system, modeling both in a design-specific language, relying on hierarchical models (Gamma et al. 1995).
4. As of the time of publication, there exists no widely accepted method of automatically detecting patterns within non-software architectures, using UML and SysML elements and constructs.

In summary, the authors recognized that a combination of MBSE tools, UML/SysML, architectural patterns, and graph theory can be utilized to improve the architecture evaluation process. For example, Fig. 1 illustrates how an architecture

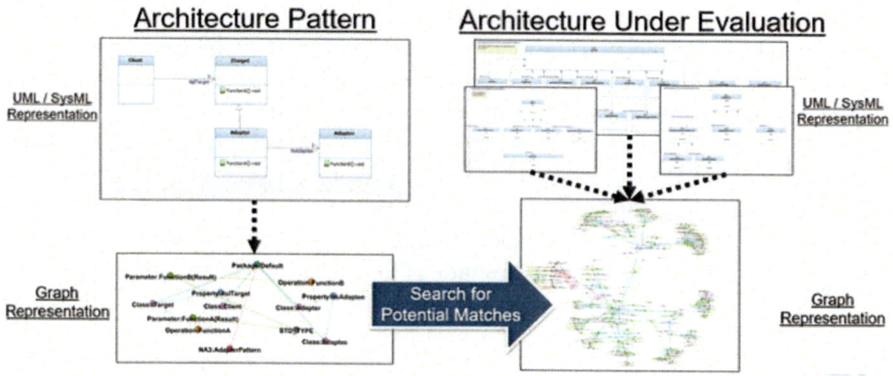

Fig. 1 Technical approach to architectural pattern detection

pattern and the system architecture under evaluation can both be parsed in an equivalent graph-based format, before matches of the pattern can be found in the architecture. The architectural diagrams in Fig. 1 are created using Rhapsody.

The authors defined the overall technical approach to MBSE-centric-based pattern detection in four sequential steps:

1. *Architecture parsing*: The process of translating text-based architecture data that has been generated from an architectural diagram in an MBSE tool into a graph-based data structure whose nodes and edges correspond to a subset of elements contained within the original MBSE model. The text-based architecture data complies with the XML Metadata Interchange (XMI) standard, as defined by the Object Management Group (OMG). The export of architecture data into XMI format is supported by most commercial MBSE tools. This parsing process has been implemented using Python 3.
2. *Architecture pattern detection*: The process of comparing two graph-based data structures generated from step #1 above. The comparison process requires that one data structure represents the system architecture to be evaluated and the other represents a potential pattern to be located. A similarity score is calculated using both natural language processing and structural graph comparisons. This process has been implemented using Python 3.
3. *Architecture (and pattern) visualization*: The process of visualizing the architecture data and pattern matching results in a web-based application. This process has been implemented using both JavaScript and Java.
4. *Pattern completion recommendation*: Partial architecture pattern matches lead naturally to a recommender system which identifies those pattern components necessary to complete the pattern and ideally improve the architectural design.

Figure 2 summarizes the information presented above into a sequential process that utilizes a UML/SysML defined system as input and provides architectural

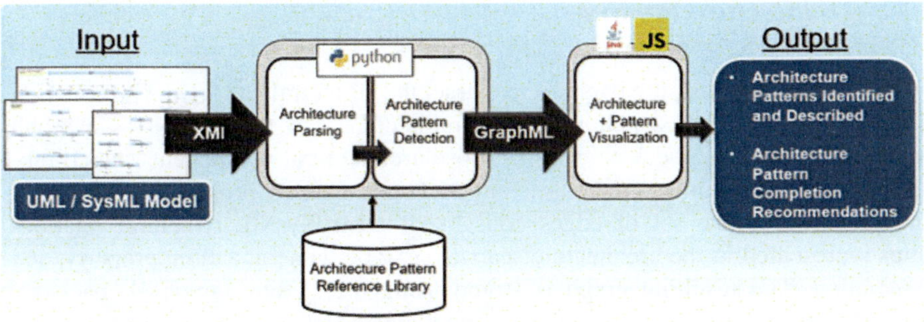

Fig. 2 Technical approach to architectural pattern detection

features/patterns as output. The following subsections describe the formation of architecture patterns and each stage of the previously described technical approach in more detail.

2.1 Architectural Pattern Definitions

Each architecture pattern used in this research can be traced back to software-specific object-oriented patterns that were originally defined in 1994 by a group of authors referred to as the "Gang of Four" (Ingram et al. 2015). Their work and these patterns have been influential in the field of software engineering and are critical to object-oriented design theory and practice. Additionally, it has been noted by some that many of these patterns can be used to describe non-software systems as well (Tupper 2011; Duell 1997). However, this statement is not as widely accepted, nor are these patterns consciously applied as regularly outside the software domain.

Note that UML and SysML are object-oriented languages; therefore, the Gang of Four design patterns can be represented using UML/SysML constructs. The following six architecture patterns were chosen for this research: Adapter, Bridge, Composite, Façade, Observer, and Proxy. These were selected as they are universally accepted and their usefulness in application is well-documented (Larman 2013; The GraphML Team 2019). Additionally, all of these patterns are referred to as *structural patterns*, in that they were originally intended to "provide a manner to define relationships between classes and objects" (Weilkiens 2008). A structural pattern can be described within a single diagram type in a UML (i.e., class diagram) or SysML (i.e., block definition diagram) model. Future work may extend pattern identification to other diagram types within UML and SysML, for example, describing system behavior such as state machines.

2.2 Architecture Parsing

The goal of the architecture parser is to extract the structural and lexical information from the SysML representation of the architecture and represent it as a graph structure. The set of nodes consists of those architectural components which represent SysML structural elements such as packages, classes, properties, operations, and parameters. The set of edges consists of UML/SysML structural relationships represented in the architecture: part-of, is-type, generalization, property, and association. The resulting graph is stored using the Python GraphML package, where GraphML is an XML-based file format created specifically for storing graph structures (Bastian et al. 2009). Any metadata and non-graph relevant material in the XMI files, such as behavioral UML/SysML elements or applied profiles, are ignored and are not included in the resulting graph structure. Future versions of the parsing capability could incorporate this sort of information into the graph structure if needed for the detection of additional architecture patterns.

2.3 Architecture Pattern Detection

The goal of the pattern detection prototype is to find all instances of an architecture pattern embedded within a system architecture model. More specifically, given two graphs, one parsed from the system architecture to be evaluated and the other parsed from the pattern definition, the objective is to search for candidate matches of a design pattern graph to subgraphs within the system architecture graph. Nodes and edges in pattern graph are compared against one or more elements of the architecture graph. The same pattern may be found in multiple places in a large architecture; therefore, multiple candidate matches, complete or partial, may be identified.

The proposed approach relies on lexical clues as a starting point for seeking out and matching architecture patterns. For example, in seeking an Adapter architecture pattern, the algorithm can trigger from any of the set of keywords as a substring of, or a similar string to, an architecture graph node label: "adapt," "client," "wrapper," or "target." Without such an assumption, every architecture node must be tested as a possible match. From each possible starting point, the matching algorithm uses a maximal set of contextual information to identify the most likely possible matching between design pattern components and architectural components.

2.4 Architecture Visualization

A key challenge in evaluating architectures is the ability to visualize features of interest while maintaining a clear view of the system from an architectural perspective. Graph-based architecture visualization tools have the potential to strike

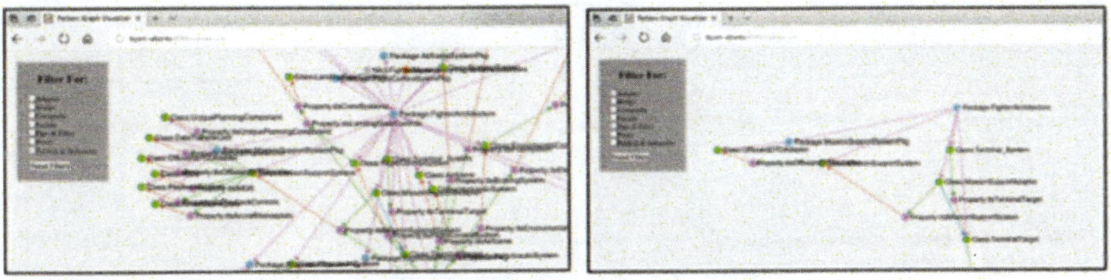

Fig. 3 (**a**) An architecture uploaded to the web application. (**b**) Filtered to show a single matched pattern

a balance between these concepts if information can be selectively displayed or suppressed by a user. The value of such architecture visualization applications is twofold. First, visualization can act as a "debugging tool" for the investigation and validation of an architecture's definition. Discrepancies in an architecture's definition will often be readily apparent to a viewer when the architecture is viewed from a new perspective. Second, a visualization tool may help the viewer understanding how components of system interact with each other.

A web-based application was developed that visualizes GraphML data created as output from the pattern detection algorithm. The web application allows a user to upload and examine any architecture that has been evaluated by the pattern matching software. The web application leverages existing functions and libraries provided by the publicly available tool, Gephi Toolkit, in order to process the input file into a user-friendly visualization (Maioriello 2002). Figure 3a, b shows a snapshot of the web application, having ingested a sample architecture and filtering said architecture for detected Adapter pattern.

3 Results and Discussion

The authors created functional prototypes for all steps of the technical approach. The prototypes were tested using systems architectures and patterns generated in both Rhapsody and MagicDraw, yielding successful pattern detection results.

Our prototypes were able to successfully detect patterns that were intentionally embedded within both small-scale and large-scale models created in either Rhapsody or MagicDraw, with limited false positives. These embedded patterns were constructed to mimic both the semantic structure of the design pattern and the natural language used to name each of the elements within it. For example, Fig. 4 presents a systems architecture test model constructed to mimic a realistic logical decomposition of a generic military aircraft, with a small portion of the architecture intentionally created to resemble an *Adapter* pattern.

The pattern matching software was able to focus its search of the systems architecture around keywords, or partial keywords, associated with any portion

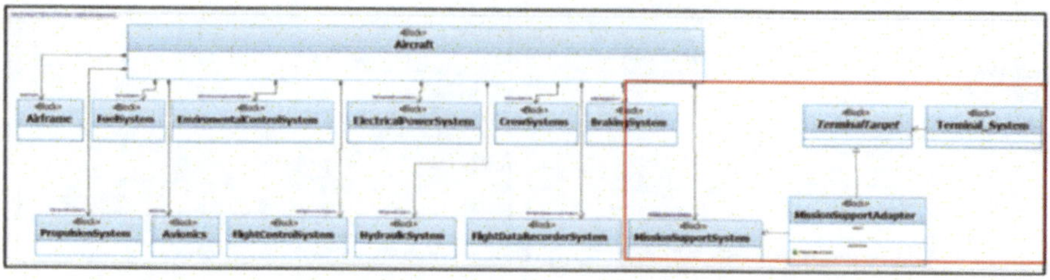

Fig. 4 Adapter pattern variant, embedded within a systems architecture model

of the anticipated pattern(s). It then successfully matched all components of the Adapter pattern to analogous representations within the systems architecture model. This behavior was consistent across all test cases and patterns; the software was able to reliably match the appropriate set of elements from the architecture under evaluation, to the appropriate element of the pattern definition. Additionally, the program was tested on a real architecture of approximately 7000 nodes and 14000 links and was able to quickly identify several design patterns in a trivial amount of time.

Several challenges to pattern detection are worth highlighting. First, there are no enforced UML/SysML standards for labeling of architectural components, either of graph nodes or edges. This can increase the number of misidentified components or missed patterns: if an implemented pattern does not match *any* keywords, it will not be detected. Therefore, it will always be necessary for a skilled systems engineer or subject matter expert to be available to verify the pattern implementations detected by the software. Second, architectural design patterns are not always clearly or consistently defined in the literature, so a specific implementation of a design pattern is chosen as search criteria; the authors recognize that there are other valid variations, and a more refined prototype must take this into consideration. Third, UML/SysML notation is not expressive enough to convey all the requirements necessary to express an architecture pattern. In particular, the authors had to create a mechanism within an MBSE tool to indicate data associated with one or more pattern components, such as the concept "zero or more of" or "one or more of." These mechanisms will allow more flexibility in architecture pattern specifications and implementations. Finally, although both Rhapsody and MagicDraw allow for the generation of XMI data as discussed previously, there remain structural representation differences that cause the export of models to differ between tools. Fortunately, the semantic content remains identical between tools. For example, architecture patterns exported from each software package differ enough that the technical approach requires the use of Rhapsody pattern templates when seeking patterns in Rhapsody-defined system architectures and MagicDraw templates for MagicDraw-defined architectures.

4 Conclusion

The authors were able to successfully demonstrate detection of patterns of interest in system architecture models built in MBSE tools. This demonstration required:

- The development of a joint MagicDraw-Rhapsody UML/SysML architecture parser
- The creation of a library of standard systems architecture patterns, expressed in UML/SysML
- The design, implementation, and testing of a unique pattern detection approach which locates potential architectural patterns in a systems architecture that potentially consists of thousands of elements and relationships
- The development of a graph-based visualization application to examine the pattern matching results

There are a number of potential next steps for this research. First, alternative patterns that utilize a wider set of UML/SysML elements, e.g., behavioral elements or profile extensions, could be incorporated into the prototypes. Second, the pattern detection capability could be directly integrated into Rhapsody or MagicDraw using one or more JAVA-based plugins that utilize the tool's native application programming interface (API) in order to collect and manipulate elements of a systems architecture.

Once a pattern is formalized and encoded into a UML/SysML model, there is enough contextual information in a systems architecture to have an algorithm intelligently search the graphical representation of the architecture and identify strong candidates for architectural pattern matches. Our approach is also capable of finding *partial* pattern matches, providing the evaluating systems engineer with an indication of how to potentially improve the architecture. The pattern matching prototype takes advantage of known keywords in order to promote computational efficiency. In some cases, this may also lead to unintentional false matches, especially in cases where a systems architecture model has dozens or hundreds of similar label names. However, even given this potential limitation, the proposed technique is likely to provide time and cost savings during architectural review, while also leading to a more objective, complete evaluation of the overall architecture and consequent system modifications and improvements.

Acknowledgments \The authors would like to acknowledge Sanith Wijesinghe and Tom Wheeler for the continuous support, advocacy, and guidance throughout the course of this research. The authors would also like to thank our collaborators, Huy Tran and Karl Thomson (University of Illinois), without whom this research, and the complimentary technical products, would not have been successful. Finally, the authors thank all reviewers of this publication for their thoughtful, thorough, and detailed feedback.

References

Babar, M., A. Brown, and I. Mistrik. 2013. *Agile Software Architecture: Aligning Agile Processes and Software Architectures*. San Francisco: Morgan Kaufmann Publishers Inc.

Bastian, M., S. Heymann, and M. Jacomy. 2009. Gephi: An open source software for exploring and manipulating networks. In *International AAAI Conference on Weblogs and Social Media.*

Bernardi, M., M. Cimitile, and G. Di Lucca. 2014. Design Pattern Detection Using a DSL-Driven Graph Matching Approach. *Journal of Software: Evolution and Process* 26 (12): 1233–1266.

Champin, P., and C. Solnon. 2003. Measuring the Similarity of Labeled Graphs. In *5th International Conference on Case-Based Reasoning: Research and Development*. Berlin: Springer.

Cloutier, R.J. 2005. Toward the Application of Patterns to Systems Engineering. In *Conference on Systems Engineering Research (CSER) 2005*.

Duell, M. 1997. Non-software Examples of Software Design Patterns. *Conference on Object Oriented Programming, Systems, Languages, and Application*, New York.

Gamma, E., R. Helm, R. Johnson, and J. Vlissides. 1995. *Design Patterns: Elements of Reusable object Oriented Software*. Boston: Addison-Wesley Longman Publishing Co.

INCOSE, Stevens Institute of Technology, IEEE Computer Society, Systems Engineering Body Of Knowledge (SEBoK). 11 September 2019. [Online]. Available: https://www.sebokwiki.org/wiki/System_Architecture.

Ingram, C., R. Payne, and J. Fitzgerald. 2015. Architecture Modeling Patterns for Systems of Systems. *INCOSE International Symposium* 25 (1): 1177–1192.

Larman, C. 2013. *Applying UML And Patterns*. 3rd ed.

Lassing, N., D. Rijesenbrij and H.V. Vliet. 1999. The Goal of Software Architecture Analysis: Confidence Building or Risk Assessment. *Proceedings of First BeNeLux Conference on Software Architecture*.

Maioriello, J. 2002. What are Design Patterns and Do I Need Them. Developer.com, 2002. [Online].

Maranzano, J., S. Rozsypal, G. Zimmerman, G. Warnken, P. Wirth, and D. Weiss. 2005. Architecture Review: Practice and Experience. *IEEE Software* 22 (2): 34–43.

Taylow, R.N., N. Medvidovic, and E.M. Dashofy. 2009. *Software Architecture: Foundations Theory and Practice*. Wiley.

The GraphML Team. The GraphML Format, 24 January 2019. [Online]. Available: http://graphml.graphdrawing.org/.

Tsantalis, N., A. Chatzigeorgiou, G. Stephanides, and S. Halkidis. 2006. Design Pattern Detection Using Similarity Scoring. *IEEE Transactions on Software Engineering* 32 (11): 896–909.

Tupper, C. 2011. Enterprise Architecture Frameworks and Methodologies. In *Data Architecture: From Zen to Reality*, 23–55. Elsevier Inc.

Weilkiens, T. 2008. Systems Engineering with SysML/UML: Modeling, Analysis, Design, Morgan Kaufmann/The OMG Press.

A Survey of Super-Resolution Techniques for a Potential CubeSat Imagery System Architecture

William Symolon and Cihan Dagli

Abstract CubeSats have the demonstrated potential to contribute to commercial, scientific, and government applications in remote sensing, communications, navigation, and research. Despite significant research into improving CubeSat operational efficiency, there remains one fundamental limitation of CubeSats for EO imaging applications: the small lenses and short focal lengths result in imagery with low spatial resolution. This paper reviews the previous research on super-resolution techniques and proposes potential applications of super-resolution to CubeSat EO imagery.

Keywords Resolution enhancement · Super-resolution · Electro-optical imagery · CubeSat · Architecture

1 Introduction

CubeSats have the demonstrated potential to contribute to commercial, scientific, and government applications in remote sensing, communications, navigation, and research at a fraction of the size, development costs, and launch costs of the large, exquisite, multifunction satellites designed to support Cold War military requirements. Poghosyan et al. (Poghosyan and Golkar 2017) and Selva and Krejci (2012) conducted reviews of the recent history of CubeSat missions and surveyed CubeSat contributions to the scientific and experimental communities with the goal of determining the applications for which CubeSats are best suited. Missions such as Earth science, astrophysics, in situ laboratory applications, and technology demonstration have already benefitted from CubeSat contributions (Poghosyan and Golkar 2017; Selva and Krejci 2012). CubeSats offer significant advantages in terms of reduced development timelines and development costs. The small size and weight

W. Symolon (✉) · C. Dagli
Engineering Management and Systems Engineering Department, Missouri University of Science and Technology, Rolla, MO, USA
e-mail: west42@umsystem.edu

A. M. Madni et al. (eds.), *Recent Trends and Advances in Model Based Systems Engineering*, https://doi.org/10.1007/978-3-030-82083-1_9

of CubeSats allow multiple satellites to be launched on the same rocket, thus greatly reducing launch costs (Selva and Krejci 2012). However, the reduced size, weight, and power margins inherent in CubeSats also have disadvantages. Smaller satellites are typically limited to single payloads or functions. The reduced functionality of CubeSats requires larger numbers of satellites to achieve the same performance. The mitigation of these disadvantages has been the subject of a significant body of research, such as Altinok et al. use of decision forests to conduct image analysis onboard a CubeSat (Altinok et al. 2016), Chang et al. and Pu et al. exploration of super-resolution through neighbor embedding (Chang et al. 2004; Pu et al. 2009), Denby's and Lucia's use of on-orbit edge computing to increase CubeSat efficiency (Denby and Lucia 2019), and Lüdenmann et al. employing sub-pixel image registration on a nanosatellite (Lüdenmann et al. 2019).

For traditional electro-optical (EO) imagery applications, high resolution (HR) requires large lenses and long focal lengths, which in turn require large satellites to support them (Buzzi et al. 2019). Past research has demonstrated that on-board image processing techniques can make more efficient use of limited satellite resources (Altinok et al. 2016; Blaschke et al. 2014; Denby and Lucia 2019; Lüdenmann et al. 2019). The continued miniaturization of electronics makes it increasingly possible to apply these preprocessing algorithms to CubeSat missions (Denby and Lucia 2019; Edeler et al. 2011; Lüdenmann et al. 2019). Work in pixel registration (Lüdenmann et al. 2019), feature classification (Chia et al. 2015), parallel computing (Denby and Lucia 2019), and radar interferometry (Hacker and Sedwick 1999) has laid the groundwork for the collection of EO imagery using multiple CubeSats flying in close formation.

Despite the data handling improvements, there remains one fundamental limitation of CubeSats for EO imaging applications: the small lenses and short focal lengths result in imagery with low spatial resolution. These low resolutions (LR) are sufficient for scientific applications such as weather forecasting and agricultural assessments (Poghosyan and Golkar 2017; Selva and Krejci 2012), but are insufficient for defense mission planning and intelligence operations. There are two primary methods for improving spatial image resolution: hardware solutions that focus on improved camera capabilities and analytical methods that focus on software solutions (Khattab et al. 2018). Hardware improvements are often restricted by cost, large size, or technology readiness limitations – all three of which we've identified as being incompatible with the CubeSat concept. Additionally, optical imaging hardware is subject to the Rayleigh criterion in which light diffraction limits the best possible resolution (Lee and Ashok 2019; Sprigg et al. 2016). Thus, a computational algorithm solution is required to improve EO spatial resolution of CubeSat images.

For reference, CubeSats are manufactured in a variety of form factors that are all based on a 1U form, which is a 10-centimeter cube (Fig. 1a) (Space Flight Now https://spaceflightnow.com/2015/10/16/nasa-to-fly-cubesats-on-three-new-commercial-launchers/). Other CubeSat form factors are based on scaling that 1U form factor, the most common variations of which are 2U, 3U, and 6U form factors. A 3U CubeSat is a 10 cm $\times$ 10 cm $\times$ 30 cm satellite, roughly the

(a) (b) (c)

Fig. 1 (**a**) NASA file photo of a 1U form factor CubeSat (Space Flight Now https://spaceflightnow.com/2015/10/16/nasa-to-fly-cubesats-on-three-new-commercial-launchers/); (**b**) SatRevolution (REC) CubeSat, 2U and 6U form factor, Earth imaging 50cm resolution; (**c**) Plant Labs (Dove) CubeSat, 3U form factor, Earth imaging 3.7 m resolution (NanoSats Database https://www.nanosats.eu/tables#constellations)

size of a loaf of bread. Figure 1b, c is 2U and 3U CubeSats, respectively. Figure 1b, c images are excerpted from a Table of Commercial CubeSat Constellations (NanoSats Database https://www.nanosats.eu/tables#constellations).

This paper reviews the previous research on super-resolution (SR) and proposes potential applications of super-resolution to CubeSat EO imagery. Sections 2 and 3 discuss the two main categories of super-resolution: single-image super-resolution and multi-image super-resolution, respectively. Section 4 proposes potential applications of super-resolution to a CubeSat imagery system architecture, followed by concluding remarks in Sect. 5.

2 Single-Image Super-Resolution

Single-image super-resolution (SISR) is a relatively recent field of research and concerns the estimation of an HR image from a single LR image (Qureshi et al. 2012). SISR requires a training database of LR and HR pairs with specific features and segments common to both and annotated for machine learning algorithms. There are three main categories of SISR algorithms: interpolation-based algorithms reconstruct HR images using existing pixels to interpolate probable missing pixels; reconstruction-based algorithms use a priori knowledge (down-sampling, blurring, and warping) to recover the HR image; learning-based algorithms use dictionary pairs of training and testing images to estimate HR images (Yao et al. 2020).

Guo et al. (2019) developed a generalized image restoration neural network called the deep likelihood network (DL-Net). This research is focused on image restoration tasks that aren't limited to narrow applications based on the original neural network training data set. Typically, these training data are generated by intentionally degrading a high-resolution image; these training sets tend to result in poor generalization by the network. The authors build upon single-image super-resolution (SISR) through neighbor embedding as developed by (Chang et al. 2004; Pu et al. 2009; Timofte et al. 2013) to design an image interpolation algorithm capable of generalizing from a variety of image degradation types. This process makes use of the local geometry within neighboring image segments to extrapolate from low-resolution to high-resolution images.

Fan and Yeung (2007) expand on the SISR application by studying the distribution of small image patches to determine the types of image structures (edges, corners, outlines, etc.) that are likely to occur in the image. The authors assume similar local geometries between the image patches to neighborhood relationships between the low-resolution and corresponding high-resolution training images. By retaining image patch geometric relationships and inter-patch relationships with neighbors, the authors are able to generate high-resolution images that are both accurate and smooth.

Ismail et al. (2020) explore applications of super-resolution where there is an insufficient quantity or quality of neural network training data. Propose the use of adaptive network-based fuzzy inference system (ANFIS) to interpolate effective mappings from low-resolution to high-resolution images, given sparse training data.

Al-Mansoori and Kunhu (2013) evaluate three well-known interpolation techniques: nearest neighbor, bilinear, and bicubic interpolation. In all three cases, the authors use a single low-resolution image and perform the interpolation techniques to compare the results. The initial experiment limited the magnification factors as a proof of concept. In all example-based super-resolution results, the bicubic interpolation yielded smoother edges and more detailed high-resolution images.

Since SISR methods by definition only use one LR input image, the SISR algorithms tend to be computationally faster because they don't require motion estimation and pixel registration between input images (Bätz et al. 2015). However, SISR algorithms have a fundamental limitation in that the training set must be similar to the desired HR image in order for the reconstruction algorithms to be effective (Qureshi et al. 2012). Additionally, the possible resolution enhancement is limited compared to multi-image super-resolution techniques (Bätz et al. 2015).

SISR techniques are fast, less computationally intensive, and capable of producing sharp HR images for specific applications. However, SISR methods do not generalize well to large-scale problems and require large databases of LR/HR image pairs in order to estimate and reconstruct HR images. Additionally, as discussed in the following section, SISR methods cannot take advantage of relative motion between a series of LR images to achieve better resolution improvements. These limitations mean that SISR techniques cannot leverage all the advantages of satellite imagery and therefore are not ideal for CubeSat SR applications.

3 Multi-image Super-Resolution

Multi-image super-resolution (MISR) is a well-studied problem which typically consists of three stages: registration estimates the shifts between LR images, relative to a reference image, with sub-pixel accuracy; interpolation obtains a uniform HR image from a nonuniform composite of LR images; and restoration removes the image blur and noise. MISR can be further subdivided into frequency domain techniques and spatial domain techniques (Qureshi et al. 2012). The relative motion between LR input images produces the sub-pixel shifts necessary to achieving higher-resolution enhancement by accounting for information from adjacent image frames (Bätz et al. 2015).

Sub-pixel motion must be present in the input sequence frames in order to realize the best possible resolution enhancement using MISR techniques. However, this sub-pixel motion also requires highly accurate motion estimation in order to avoid introducing artifacts from erroneous motion vectors (Bätz et al. 2016a). Bätz and colleagues have published a series of papers (Bätz et al. 2015, 2016a, b, 2017) proposing various methods to minimize the introduction of these artifacts and to improve the overall image enhancement results. Their proposed methods include locally adaptive denoising (Bätz et al. 2017) which introduced a step between interpolation and restoration, dual weighting (Bätz et al. 2016a) which employs both a motion confidence weight and a distance weight to resolve motion estimation errors, and hybrid SISR/MISR (Bätz et al. 2015) approach that employs both SR techniques but weights SISR more heavily in the case of static targets and MISR more heavily in the case of dynamic targets. All of these techniques showed significantly improved peak signal-to-noise ratio (PSNR) compared to more traditional SR techniques.

Mandanici et al. (2019) applied an MISR algorithm to terrestrial thermal images using a novel registration technique that computes the sum of normalized distances (SND) to a given reference image. A higher SND denotes less accurate registration. These images are then excluded from the interpolation stage, based on an SND threshold value. This methodology has the added benefit of coherence analysis to identify reconstructed pixels that are less reliable which, when combined with the image frame rejection criteria, resulted in improved thermal image resolution.

Some researchers (Cohen et al. 2019; Zhang et al. 2009) are exploring the use of super-resolution in microscopy applications to obtain image resolutions beyond Rayleigh criterion diffraction-limited resolution. Cohen et al. (2019) investigate the resolution limit of image analysis algorithms. Zhang and colleagues propose a method to capture random micro-displacement offsets of multiple images without the need for a high-cost, precision mechanical device (Zhang et al. 2009). These precision offsets allow superior HR image reconstruction compared to the more expensive fixed micro-offset technique. While this research focused on microscopic image enhancement, it would be interesting to research whether their techniques may have applicability to CubeSat image resolution enhancement.

MISR techniques are able to produce superior resolution enhancement by taking advantage of the sub-pixel motion between consecutive LR images by accounting for information from adjacent pixels, given a sufficiently accurate motion estimation algorithm. However, MISR approaches have a tendency to present ill-posed problems, either due to an inadequate number of LR images or poor estimation of image capture artifacts, such as blur (Khattab et al. 2018). Despite this limitation, past research has demonstrated that regularization techniques (Irani and Peleg 1991) help to invert an ill-posed problem to a well-posed problem (Khattab et al. 2018). Overall, MISR techniques offer better potential to take advantage of CubeSat capabilities.

4 Applications to CubeSats

Researchers have developed a number of algorithms designed to improve image quality. Three common algorithms are pixel averaging, super-resolution, and mosaicking (Lüdenmann et al. 2019). In addition to lens size and focal length, CubeSat downlink data rate and onboard storage capacity are two other limiting factors in electro-optical imaging (Altinok et al. 2016). In order to apply averaging, super-resolution, or mosaicking, the CubeSats must either downlink large image files to be processed terrestrially or the CubeSats must have real-time access to a memory-intensive catalog of georectified reference images (Altinok et al. 2016; Lüdenmann et al. 2019). In either case, the requirements are impractical for use on-board a CubeSat. Lüdenmann et al. (2019) developed a method to use a combination of correlation and regression algorithms to identify the geometric transformations between consecutive images on-board the CubeSat while keeping the data downlink requirements within the size, weight, and power restrictions imposed by the CubeSat standard.

MISR techniques are most effective when sub-pixel motion is present in the input sequence frames. This attribute of MISR makes it particularly useful in CubeSat EO imagery applications since satellites in Earth orbit are in constant motion. A formation of CubeSats can capture multiple images of the same target area on Earth. The varying locations of the CubeSats within the formation combined with the orbital velocity of the CubeSats inherently provide the input image offset required for successful MISR application.

Figure 2 depicts a high-level operational view (OV-1) of one possible system architecture for CubeSat EO SR applications. This architecture assumes a pre-determined CubeSat formation, optimized (Buzzi et al. 2019; Chia et al. 2015)

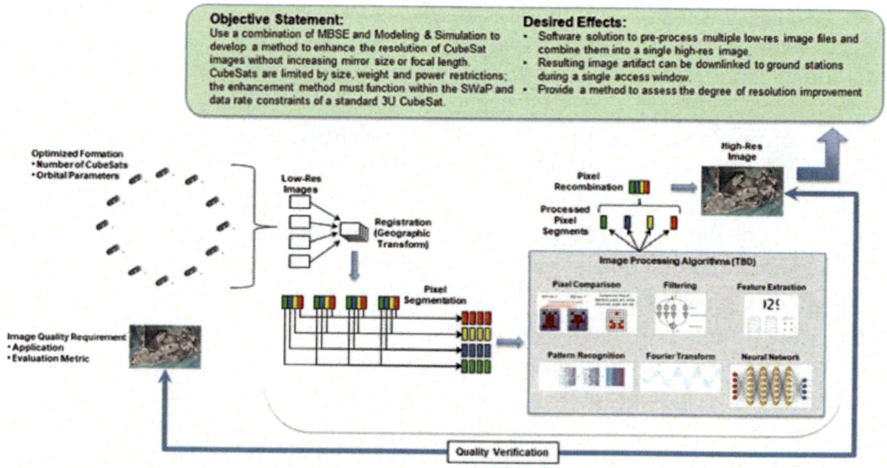

Fig. 2 OV-1 of potential system architecture for CubeSat super-resolution

for the number of CubeSats and orbital parameters necessary to collect the LR images. The LR images are then preprocessed for motion estimation and pixel registration (Bätz et al. 2015; Khattab et al. 2018; Lüdenmann et al. 2019) before being segmented (Blaschke et al. 2014; Denby and Lucia 2019) and input to a MISR computational algorithm for HR pixel interpolation. The processed image segments are then recombined, the HR image is restored, and a quality verification process certifies the resulting HR image. One possible method of quality verification is to compare the resulting resolution improvements, from the proposed architecture, against a theoretically perfect Rayleigh-limited image from current, state-of-the-art, CubeSat imaging hardware.

Since traditional resolution enhancement algorithms are impractical for CubeSat applications and techniques exist to efficiently register LR images prior to downlink, MISR becomes an attractive technique to improve the spatial resolution of CubeSat images.

5 Conclusion and Future Research

This survey paper reviews recent research published regarding SR and discusses the advantages and disadvantages of the two primary SR techniques: SISR and MISR.

SISR is a relatively new research field and consists of three main categories of algorithms: interpolation-based algorithms, reconstruction-based algorithms, and learning-based algorithms. SISR algorithms tend to be computationally faster and provide good resolution enhancement for specific applications; however, they do not generalize well and cannot take advantage of information from adjacent pixels in sequential image frames. These limitations mean that SISR is not the best choice for CubeSat SR applications.

MISR is a well-studied problem that consists of geometric registration, interpolation, and restoration to derive a single HR image from multiple LR images. The relative sub-pixel motion between input image frames is the key to achieving high-quality resolution enhancement. However, the estimation of that motion, during the pixel registration process, must be highly accurate to avoid introducing error artifacts into the HR image. An additional challenge with MISR techniques is that an inadequate number of LR input images or poor estimation of image capture artifacts can contribute to making the MISR approach an ill-posed problem. Care must be taken to understand the constraints and limitations of the imaging hardware and to apply regularization techniques to define a well-posed MISR problem.

The inherent limitations of CubeSats and the nature of satellite orbits make MISR an attractive technique for improving CubeSat EO spatial resolution. Additional research is required to develop a resolution enhancement model that can enhance image resolution sufficiently enough to extend the utility of CubeSat images to defense mission planning and intelligence operations. CubeSats have already demonstrated their potential to contribute to scientific discovery; extending that potential to include that satisfaction of national defense requirements will provide intelligence value at a fraction of the current costs of large satellites.

References

Al-Mansoori, Saeed, and Alavi Kunhu. 2013. Enhancing DubaiSat-1 Satellite Imagery Using a Single-Image Super-Resolution. In *Proceedings of SPIE – The International Society for Optical Engineering*.

Altinok, Alphan, David R. Thompson, Benjamin Bornstein, Steve A. Chien, Joshua Doubleday, and John Bellardo. 2016. Real-Time Orbital Image Analysis Using Decision Forests, with a Deployment Onboard the IPEX Spacecraft. *Journal of Field Robotics* 33 (2): 187–204.

Bätz, Michel, Andrea Eichenseer, Jürgen Seiler, Markus Jonscher, and André Kaup. 2015. Hybrid Super-Resolution Combining Example-Based Single-Image and Interpolation-Based Multi-Image Reconstruction Approaches. In *Proceedings – International Conference on Image Processing, ICIP*, 58–62.

Bätz, Michel, Andrea Eichenseer, and André Kaup. 2016a. Multi-Image Super-Resolution Using a Dual Weighting Scheme Based on Voronoi Tessellation. In *Proceedings – International Conference on Image Processing, ICIP*, 2822–2826.

———. 2016b. Multi-Image Super-Resolution for Fisheye Video Sequences Using Subpixel Motion Estimation Based on Calibrated Re-Projection. In *European Signal Processing Conference*, 1872–1876.

Bätz, Michel, Ján Koloda, Andrea Eichenseer, and André Kaup. 2017. Multi-Image Super-Resolution Using a Locally Adaptive Denoising-Based Refinement. In *IEEE 18th International Workshop on Multimedia Signal Processing, MMSP*.

Blaschke, Thomas, Geoffrey J. Hay, Stefan Lang, K. Maggi, Peter Hofmann, Elisabeth Addink, Raul Q. Feitosa, Freek van der Meer, Harald van der Werff, Frieke van Coillie, and Dirk Tiede. 2014. Geographic Object-Based Image Analysis – Towards a New Paradigm. *ISPRS Journal of Photogrammetry and Remote Sensing* 87: 180–191.

Buzzi, Pau Garcia, Daniel Selva, Nozomi Hitomi, and William J. Blackwell. 2019. Assessment of Constellation Designs for Earth Observation: Application to the TROPICS mission. *Acta Astronautica* 161: 166–182.

CubeSat Design Specification. (CDS) Rev. 13, The CubeSat Program, Cal Poly SLO. Retrieved from https://static1.squarespace.com/static/5418c831e4b0fa4ecac1bacd/t/56e9b62337013b6c063a655a/1458157095454/cds_rev13_final2.pdf. Accessed 14 June 2019.

Chang, Hong, Dit-Yan Yeung, and Yimin Xiong. 2004. Super-Resolution Through Neighbor Embedding. *Proceedings of the IEEE Computer Society Conference on Computer Vision and Pattern Recognition 1*: 275–282.

Chia, W.C., L.S. Yeong, S.I. Ch'Ng, and Y.L. Kam. 2015. The Effect Of Using Super-Resolution To Improve Feature Extraction And Registration Of Low Resolution Images In Sensor Networks. In *Proceedings of the 7th International Conference of Soft Computing and Pattern Recognition, SoCPaR*, 340–345.

Cohen, Edward, Anish Abraham, Sreevidhya Ramakrishnan, and Raimund Ober. 2019. Resolution Limits of Image Analysis Algorithms. *Nature Communications* 10 (1): 793–804.

Denby, Bradley, and Brandon Lucia. 2019. Orbital Edge Computing: Machine Inference in Space. *IEEE Computer Architecture Letters* 18 (1): 59–62.

Edeler, Torsten, Kevin Ohliger, Stephan Hussmann, and Alfred Mertins. 2011. Multi Image Super Resolution Using Compressed Sensing. In *ICASSP, IEEE International Conference on Acoustics, Speech and Signal Processing – Proceedings*, 2868–2871.

Fan, Wei, and Dit-Yan Yeung. 2007. Image Hallucination Using Neighbor Embedding over Visual Primitive Manifolds. In *Proceedings of the IEEE Computer Society Conference on Computer Vision and Pattern Recognition*.

Guo, Yiwen, Wangmeng Zuo, Changshui Zhang, and Yurong Chen. 2019. Deep Likelihood Network for Image Restoration with Multiple Degradations. *arXiv:1904.09105v1 [cs.CV]*. 19 April 2019.

Hacker, T.L, and R.J. Sedwick. 1999. Space-Based GMTI Radar Using Separated Spacecraft Interferometry. *AIAA Space Technology Conference & Exposition*, 566–579. *AIAA 99-4634*. 28–30 September 1999. Albuquerque, New Mexico.

Irani, M., and S. Peleg. 1991. Improving Resolution by Image Registration. *CVGIP: Graphical Models and Image Processing* 53: 231–239.

Ismail, Muhammad, Jing Yang, Changjing Shang, and Qiang Shen. 2020. Single Frame Image Super Resolution Using ANFIS Interpolation: An Initial Experiment-Based Approach. *Advances in Intelligent Systems and Computing* 1043: 27–40.

Khattab, M.M., A.M. Zeki, A.A. Alwan, A.S. Badawy, and L.S. Thota. 2018. Multi-Frame Super-Resolution: A Survey. In *IEEE International Conference on Computational Intelligence and Computing Research, ICCIC*.

Lee, Kwan Kit, and Amit Ashok. 2019. Surpassing Rayleigh Limit: Fisher Information Analysis of Partially Coherent Source(s). *Optics and Photonics for Information Processing XIII*: 11136.

Lüdenmann, Jürgen, Arno Barnard, and Daniël F. Malan. 2019. Sub-pixel Image Registration on an Embedded Nanosatellite Platform. *Acta Astronautica* 161: 293–303.

Mandanici, Emanuele, Luca Tavasci, Francesco Corsini, and Stefano Gandolfi. 2019. A Multi-image Super-Resolution Algorithm Applied to Thermal Imagery. *Applied Geomatics* 11 (3): 215–228.

NanoSats Database. CubeSat Tables. Retrieved from https://www.nanosats.eu/tables#constellations. Accessed 30 June 2019.

Poghosyan, Armen, and Alessandro Golkar. 2017. CubeSat Evolution: Analyzing CubeSat Capabilities for Conducting Science Missions. *Progress in Aerospace Sciences* 88: 59–83.

Pu, Jian, Junping Zhang, Peihong Guo, and Xiaoru Yuan. 2009. Interactive Super-Resolution Through Neighbor Embedding. In *9th Asian Conference on Computer Vision, Revised Selected Papers, Part III, LNCS 5996*, 496–505. September, 2009.

Qureshi, S.S., X.M. Li, and T. Ahmad. 2012. Investigating Image Super Resolution Techniques: What to Choose? In *International Conference on Advanced Communication Technology, ICACT*, 642–647.

Selva, Daniel, and David Krejci. 2012. A Survey and Assessment of the Capabilities of CubeSats for Earth Observation. *Acta Astronautica* 74: 50–68.

Space Flight Now. NASA to fly CubeSats on Three New Commercial Launchers. Retrieved from https://spaceflightnow.com/2015/10/16/nasa-to-fly-cubesats-on-three-new-commercial-launchers/. Accessed 1 Nov 2019.

Sprigg, Jane, Tao Peng, and Yanhua Shih. 2016. Super-Resolution Imaging Using the Spatial-Frequency Filtered Intensity Fluctuation Correlation. *Scientific Reports* 6: 38077.

Timofte, Radu, De Smet, Vincent, and Luc Van Gool. 2013. Anchored Neighborhood Regression for Fast-Example-Based Super-Resolution. *Proceedings of the IEEE International Conference on Computer Vision*: 1920–1927.

Yao, Tingting, Yu Luo, Yantong Chen, Dongqiao Yang, and Lei Zhao. 2020. Single-Image Super-Resolution: A Survey. *Lecture Notes in Electrical Engineering* 516: 119–125.

Zhang, Jin, Zhong Wang, Guang H. Zhou, and Sheng H. Ye. 2009. Research of Super-Resolution Reconstruction Based on Multi-Images of Random Micro-Offset. In *Proceedings of the 2nd International Congress on Image and Signal Processing, CISP'09*.

Data Analytics of a Honeypot System Based on a Markov Decision Process Model

Lidong Wang, Randy Jones, and Terril C. Falls

Abstract A honeypot system can play a significant role in exposing cybercrimes and maintaining reliable cybersecurity. Markov decision process (MDP) is an important method in systems engineering research and machine learning. The data analytics of a honeypot system based on an MDP model is conducted using *R* language and its functions in this paper. Specifically, data analytics over a finite planning horizon (for an undiscounted MDP and a discounted MDP) and an infinite planning horizon (for a discounted MDP) is performed, respectively. Results obtained using four kinds of algorithms (value iteration, policy iteration, linear programming, and Q-learning) are compared to check the validity of the MDP model. The simulation of expected total rewards for various states is implemented using various transition probability parameters and various transition reward parameters.

Keywords Data analytics · Honeypot · Cybersecurity · Markov decision process (MDP) · Q-learning · Machine learning

1 Introduction

The interaction between an attacker and a defender in an intrusion detection system (IDS) possibly includes the attacker entering the system, detecting the attack, reentering the system, re-detecting the attack, and so on. During the process, both the defender and the attacker learn about each other—their intentions, vulnerabilities, and methods. This problem can be formulated as a Markov decision process (MDP), and an optimal decision depends on state variables and model parameters that characterize the attacker's persistence and the IDS's detection rate. There are potential benefits of gathering the attacker's intelligence and considering the learning effect in a cyber-defense scenario (Bao and Musacchio 2009).

L. Wang (✉) · R. Jones · T. C. Falls
Institute for Systems Engineering Research, Mississippi State University, Vicksburg, MS, USA
e-mail: lidong@iser.msstate.edu

A. M. Madni et al. (eds.), *Recent Trends and Advances in Model Based Systems Engineering*, https://doi.org/10.1007/978-3-030-82083-1_10

An IDS and a host system controlled by a Markov decision process can serve as an efficient resiliency solution. A Markov adversary model was created with real data for four various types of cyberattacks, and how the model captures intrinsic features of malicious behaviors was investigated (Magalhaes and Lewis 2016). A two-player (defender and attacker) model of MDP was proposed to evaluate results of cyber intrusions to one or multiple substations. Various factors that influence the results of the competition between defense and intrusion were incorporated in the MDP; these factors include the skill and knowledge of an attacker, the performance indexes of proactive and reactive defense measures, etc. (Chen et al. 2018). A partially observable Markov decision process (POMDP) for a resilient design was studied, and the use of flexible contracts for multi-UAV (unmanned aerial vehicle) swarm control was illustrated (Madni et al. 2018).

A honeypot-based approach for an intrusion detection and prevention system (IDPS) was presented, and a honeypot-based IDPS was developed (Baykara and Das 2018). A honeypot is designed to attract attackers by providing attackers with misleading or false information. An attacker that compromises a honeypot system leaves valuable tracks about its attack methods (Bar et al. 2016). An SSH self-adaptive honeypot system was developed in Python to decide how to interact with an external attacker. A deep Q-learning algorithm that makes use of neural network (NN) was used in the honeypot system (Pauna et al. 2018).

Honeypots have been tools in disclosing cybercrimes. Botmasters who operate the commands and control of botnets can exploit the fact that a honeypot should not be involved in illegal actions through commanding a compromised machine to take malicious actions against targets. A honeypot operator needs to choose an optimal response that has a balance between being liable for participating in any illicit action and being disclosed (Hayatle et al. 2013).

The purpose of this paper is to perform data analytics of a honeypot system based on an MDP model, compare the results of various algorithms, and investigate results obtained using various transition probability parameters and various transition reward parameters.

2 Methods

2.1 Markov Decision Process

An MDP is a sequential decision process in which decisions lead to a sequence of Markov chains with rewards. MDP has been an important method in systems engineering research and artificial intelligence. An MDP can be defined by a tuple $<S, A, P, R, \gamma>$ (Alsheikh et al. 2015; Chen et al. 2018; Mohri et al. 2012):

- S is a set of states.
- A is a set of actions.

- P is a transition probability matrix that describes the transition from state s to state s' ($s \in S$, $s' \in S$) after the action a ($a \in A$) is taken.
- R is the immediate reward after the action a is taken.
- γ ($0 < \gamma < 1$) is a discounted factor of rewards.

A "policy" is defined as a mapping from a state to an action. The objective of an MDP is to find an optimal policy that maximizes the expected average reward per period or the expected total reward. The set of equally spaced and consecutive time period for the analysis of an MDP is called a planning horizon. An epoch is the end of a time period. The optimal policy over a finite planning horizon maximizes the vector of expected total rewards till the end of the horizon. For an infinite planning horizon, the expected total reward (discounted) is used to measure the earnings of a discounted MDP (with a discount factor γ).

2.2 *Algorithms for Solving the Markov Decision Process*

Value iteration (VI), policy iteration (PI), linear programming (LP), and Q-learning can be used to compute the optimal policy of a recurrent MDP. Data analytics results of these algorithms can be substantially different, or there are possible convergence problems of iterations during the analytics process if an MDP model is not reasonable due to an incorrect structure or inappropriate parameters of the MDP model. Therefore, the four kinds of algorithms are used in this paper and their results are compared to check the validity of the MDP model.

Value iteration: If a planning horizon is finite, the optimal policy of an MDP can be obtained using value iteration. The above four kinds of algorithms can, in principle, be employed in computing the optimal policy of an MDP over an infinite planning horizon. For each state, VI uses the following value iteration equation (Sutton and Barto 2018; van Otterlo 2009; van Otterlo and Wiering 2012) to calculate the expected total reward:

$$V(s) := \max_a \sum_{s'} P\left(s, a, s'\right)\left(R\left(s, a, s'\right) + \gamma V\left(s'\right)\right) \quad (1)$$

where $R(s, a, s')$ is the immediate reward of the transition from the state s to the state s' due to the action a. $P(s, a, s')$ is the transition probability. $V(s)$ and $V(s')$ are the expected total reward in the state s and the state s', respectively. A stopping criterion is used to evaluate the convergence during the iterative process. The criterion is that the value difference between two consecutive iterative steps is less than a given tolerance.

Policy iteration: It aims to find a better policy compared with the previous policy. It starts with an arbitrary initial policy. A greedy policy is performed through choosing the best action for each state based on the current value function. An iterative procedure of policy evaluation and policy improvement is terminated until two successive policy iterations lead to the same policy that indicates an optimal policy is obtained (Sutton and Barto 2018; van Otterlo and Wiering 2012).

Linear programming: Unlike VI and PI, the LP method attempts to get a static policy by solving a linear program because an MDP can be formulated and expressed as a linear program (Sigaud and Buffet 2013).

Q-learning: It is a kind of reinforcement learning to find the best policy for the most reward. It enables an agent to learn the optimal action-value function, i.e., Q-value function. It can converge to an optimal policy in both deterministic and nondeterministic MDPs as well as allows learning an optimal policy when a model is unknown. It can also be applied to non-MDP domains (Liu et al. 2017; Majeed and Hutter 2018; Zanini 2014).

3 A Markov Decision Process Model of a Honeypot System

3.1 The MDP Model Structure

A honeypot system is set to lure attackers. A botnet (network-based devices) is used to send spam, steal data, etc. A botmaster keeps a bot (autonomous program) online. A honeypot can be in one of three possible states under one of three actions (Hayatle et al. 2013). The three states are described as follows:

- State 1: The honeypot is not targeted by an attacker yet; it is waiting for attacks to join a botnet.
- State 2: The honeypot has been compromised; it has been a member of a botnet.
- State 3: This is a disclosed state. The honeypot is not a member of the botnet anymore because the real identity of the honeypot has been discovered or it has lost interactions with a botmaster for a long time.

At each state, the honeypot chooses one of the following three actions:

- Action 1: The honeypot permits a botmaster to compromise the system and execute commands.
- Action 2: The honeypot does not permit a botmaster to compromise the system.
- Action 3: The honeypot is reinitialized as a new honeypot and is reset to its initial state.

An MDP-based model of the honeypot system is created. State transitions of three states (State 1, State 2, and State 3) due to three various actions in the model of the honeypot system are shown in Fig. 1.

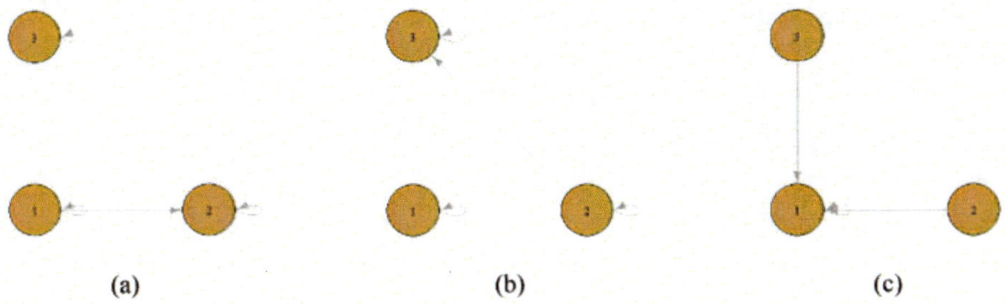

Fig. 1 State transitions under various actions: (**a**) Action 1, (**b**) Action 2, and (**c**) Action 3

3.2 The State Transition Matrix and the Reward Matrix

Transitions between various states in the MDP model of the honeypot system depend on the actions as well as two main probabilities that indicate the transitions from State 1 to State 2 and from State 2 to State 3 (Pauna et al. 2018). There is no transition from State 1 to State 3 directly and no transition from State 3 to State 2. The probability of the transition from State 3 to State 1 is 0 (under Action 1 and under Action 2) or 1 (under Action 3). The two main probabilities are further explained as follows:

1. P_{wc}: the probability of the transition from waiting (State 1) to being compromised (State 2)
2. P_{cd}: the probability of the transition from State 2 to the state of being disclosed (State 3)

State transitions (including self-transitions) lead to the following expenses or benefits (Hayatle et al. 2013):

1. E_o: the expense of operation on deploying, running, and controlling the honeypot
2. E_r: the expense in reinitializing the honeypot and resetting to its initial state
3. E_l: the expense of liability when the honeypot operator becomes liable due to illicit actions
4. B_i: the benefit of information when the honeypot gets the attacker's information such as techniques

The above parameters are selected because they are necessary and main parameters. The state transition probability matrix and the reward matrix under various actions can be computed using the parameters.

4 Data Analytics of the Honeypot System and Results

4.1 Data Analytics of the Honeypot System over a Finite Planning Horizon

Let $P_{wc} = 0.6$, $P_{cd} = 0.7$, $E_o = 1.5$, $E_r = 3$, $B_i = 30$, and $E_l = 20$. The four kinds of algorithms can be implemented. Expected total rewards for the three states are computed by value iteration over a seven-step planning horizon (seven steps are chosen as an example). The rewards at the end of the planning horizon are set equal to 0 for all the three states to begin the backward recursion of value iteration. Table 1 and Table 2 show the results of the MDP with and without a discount, respectively.

V1(n), *V2(n)*, and *V3(n)* represent the expected total rewards at step *n* for State 1, State 2, and State 3, respectively. The results for the situation without a discount can be obtained by setting $\gamma = 1.0$. The data analytics in this section and subsequent sections is performed using *R* language and its functions.

4.2 Data Analytics of the Honeypot System over an Infinite Planning Horizon

The above data (i.e., probabilities and expenses or benefits) are also used in the data analytics of the honeypot system with a discount $\gamma = 0.85$ over an infinite planning horizon. Policy evaluation is performed and the result for various policies is shown in Table 3. A policy is a mapping from a state to an action. For example, the policy c (1, 1, 3) indicates that Action 1, Action 1, and Action 3 are taken on State 1, State 2, and State 3, respectively. *V1*, *V2*, and *V3* are the expected total reward of State 1, State 2, and State 3, respectively.

Table 1 The expected total rewards of states computed by value iteration over a seven-step planning horizon (without a discount)

Epoch *n*	0	1	2	3	4	5	6	7
V1(n)	72.81	64.28	55.70	46.99	37.98	28.20	16.50	0
V2(n)	59.50	51.00	42.50	34.00	25.50	17.00	8.50	0
V3(n)	61.28	52.70	43.99	34.98	25.20	13.50	–1.50	0

Table 2 The expected total rewards of states computed by value iteration over a seven-step planning horizon (with a discount, $\gamma = 0.85$)

Epoch *n*	0	1	2	3	4	5	6	7
V1(n)	50.62	47.40	43.59	39.05	33.51	26.45	16.50	0
V2(n)	38.50	35.29	31.52	27.09	21.87	15.73	8.50	0
V3(n)	37.29	34.05	30.19	25.48	19.48	11.03	–1.50	0

Table 3 The expected total rewards of states with various policies ($\gamma = 0.85$)

Policies	c (1, 1, 1)	c (1, 1, 2)	c (1, 1, 3)	c (1, 2, 3)
V1	68.7879	68.7879	68.7879	19.2377
V2	56.6667	56.6667	56.6667	−7.4571
V3	−10.0000	−10.0000	55.4697	13.3520

Table 4 Data analytics of the honeypot system over an infinite planning horizon using various algorithms ($\gamma = 0.85$)

Methods and algorithms	*V1*	*V2*	*V3*
Value iteration (Jacob's algorithm)	66.9210	54.7998	53.6028
Value iteration (Gauss-Seidel's algorithm)	68.7879	56.6667	55.4697
Policy iteration	68.7879	56.6667	55.4697
Linear programming algorithm	68.7879	56.6667	55.4697
Q-learning	68.8000	56.6667	55.4160

As mentioned before, value iteration, policy iteration, linear programming, and Q-learning are used to check the validity of the MDP model. Table 4 shows results over an infinite planning horizon based on a discounted MDP with γ=0.85. All optimal policies obtained using various algorithms are *c* (1, 1, 3). Jacob's algorithm and Gauss-Seidel's algorithm are used in value iteration, respectively. The accuracy of Gauss-Seidel's algorithm is better than that of Jacob's algorithm. The results of Gauss-Seidel's algorithm, policy iteration, and linear programming are the same and very close to the result of Q-learning (when the number of iterations *n* is equal to 100,000), indicating that the MDP model is valid and the parameters in the model are reasonable.

4.3 Data Analytics of the Honeypot System with Various Transition Probability Parameters

Policy iteration is used in the data analytics over an infinite planning horizon with various transition probability parameters. The following data are used in the data analytics: P_{cd} =0.5, $E_o = 1.5$, $E_r = 3$, $B_i = 30$, $E_l = 20$, and γ =0.85. Expected total reward *V* = (*V1, V2, V3*) for State 1, State 2, and State 3 at various P_{wc} (P_{wc} = 0.1, 0.3, 0.5, 0.7, 0.9) is analyzed; results are shown in Fig. 2. *V2* does not change with the increase of P_{wc}, while *V1* and *V3* increase with the increase of P_{wc}. *V1* is higher than *V3* (see Fig. 2).

Let P_{wc} =0.5, $E_o = 1.5$, $E_r = 3$, $B_i = 30$, $E_l = 20$, and γ =0.85. Expected total reward *V* = (*V1, V2, V3*) at various P_{cd} (P_{cd} = 0.1, 0.3, 0.5, 0.7, 0.9) is analyzed; results are shown in Fig. 3. The optimal policy *c* (a_1, a_2, a_3) at various P_{cd} (P_{wc} = 0.5) is shown in Table 5. The optimal policy is changed from *c* (1, 2, 3) to *c* (1, 1, 3) when P_{cd} increases. The optimal policy is dependent on system

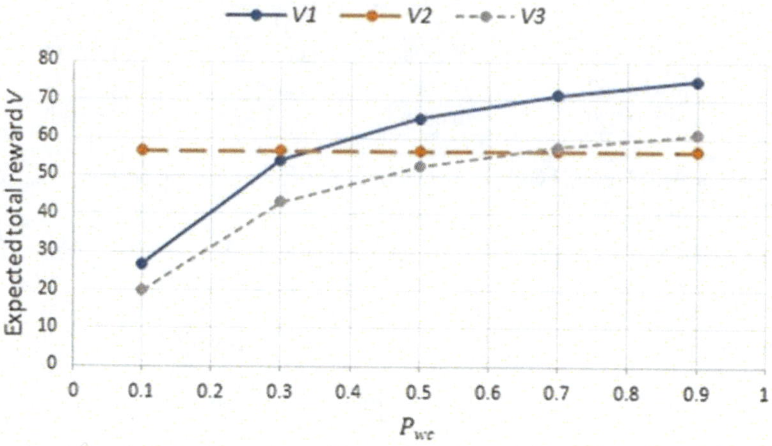

Fig. 2 Expected total reward V ($V1$, $V2$, and $V3$) at various P_{wc}

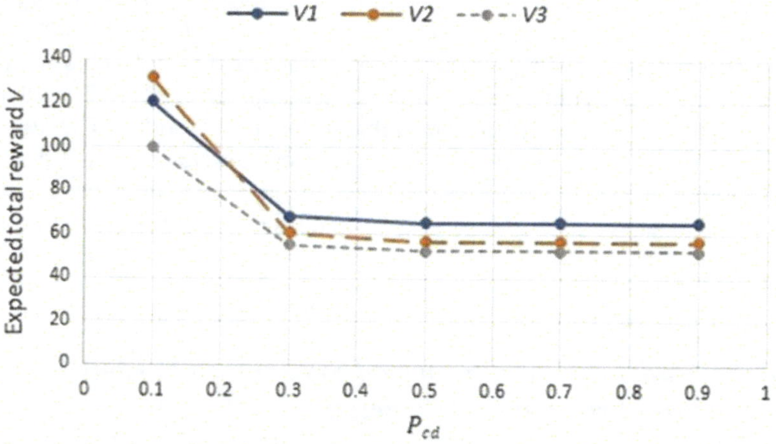

Fig. 3 Expected total reward V ($V1$, $V2$, and $V3$) at various P_{cd}

parameters, and optimal policy changes are obtained based on the computation at changed parameter(s), for example, increased P_{cd}. $V1$, $V2$, and $V3$ are decreased when P_{cd} is increased from 0.1 to 0.3; they are slightly decreased when P_{cd} is from 0.3 to 0.5; and they do not change when P_{cd} is from 0.5 to 0.9. It is shown that $V1$ has higher values than $V3$ in Fig. 3.

Table 5 The optimal policy $c\ (a_1, a_2, a_3)$ at various P_{cd} ($P_{wc} = 0.5$)

P_{cd}	0.1	0.3	0.5	0.7	0.9
a_1	1	1	1	1	1
a_2	2	2	1	1	1
a_3	3	3	3	3	3

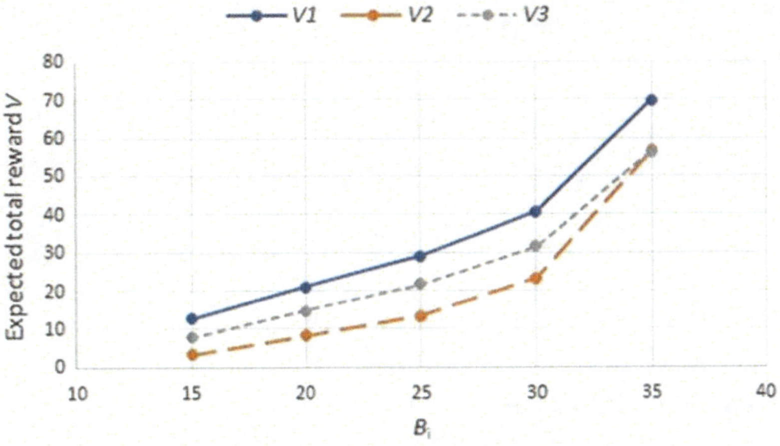

Fig. 4 Expected total reward V ($V1$, $V2$, and $V3$) at various B_i

4.4 Data Analytics of the Honeypot System with Various Transition Reward Parameters

Similarly, the policy iteration algorithm is used in the data analytics over an infinite planning horizon with various transition reward parameters. The following data are used: P_{wc} =0.5, P_{cd} =0.5, $E_o = 1.5$, $E_r = 3$, $E_l = 25$, and $\gamma = 0.85$. Expected total reward $V = (V1, V2, V3)$ at various B_i ($B_i = 15, 20, 25, 30, 35$) is analyzed; results are shown in Fig. 4. *V1*, *V2*, and *V3* are increased when B_i is increased. *V1* is higher than both *V2* and *V3*.

Let P_{wc} =0.5, P_{cd} =0.5, $E_o = 1.5$, $E_r = 3$, $B_i = 25$, and $\gamma = 0.85$. Expected total reward $V = (V1, V2, V3)$ at various E_l ($E_l = 15, 20, 25, 30, 35$) is analyzed; results are shown in Fig. 5. *V1*, *V2*, and *V3* are decreased when E_l is increased from 15 to 25; they do not change when E_l is increased from 25 to 35. It is shown that *V1* has higher values than both *V2* and *V3*.

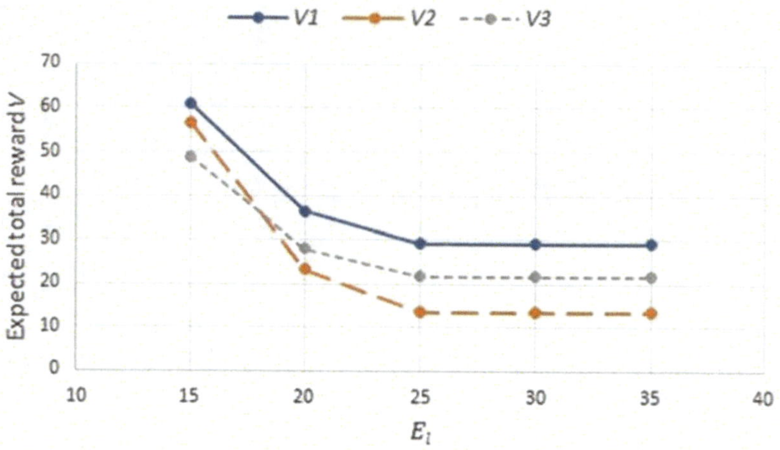

Fig. 5 Expected total reward V ($V1$, $V2$, and $V3$) at various E_l

5 Conclusion and Future Work

The data analytics of a honeypot system based on an MDP model has demonstrated that the methods in this paper are effective in finding optimal policies to maximize the expected total rewards at various transition probability parameters and transition reward parameters. The methods are efficient in the data analytics over a finite planning horizon as well as an infinite planning horizon. Value iteration (Gauss-Seidel's algorithm), policy iteration, and linear programming achieve the same result that is very close to the result of Q-learning, which demonstrates that the MDP model is valid and parameters in the model are reasonable. The integration of the four kinds of algorithms based on MDP with *R* and its functions can build up a powerful modeling approach and an efficient simulation technology that enhance the functions of honeypot system and help practice good cybersecurity. The future work is (1) the ability to add possible new states when the MDP solution does not fit with the observations as extreme situations can happen when attacks fall outside a trained pattern, (2) the big data analytics of a honeypot system based on POMDP and the real-time simulation of the POMDP, and (3) the resilient design of a honeypot system based on POMDP.

Acknowledgments This paper is based upon work performed under Contract No. W912HZ-17-C-0015 with the US Army Engineer Research and Development Center (ERDC). Any opinions, findings, and conclusions or recommendations expressed in this paper are those of the author(s) and do not reflect the views of the ERDC.

References

Alsheikh, M.A., D.T. Hoang, D. Niyato, H.P. Tan, and S. Lin. 2015. Markov Decision Processes with Applications in Wireless Sensor Networks: A Survey. *IEEE Communications Surveys & Tutorials* 17 (3): 1239–1267.

Bao, N., and J. Musacchio. 2009. Optimizing the Decision to Expel Attackers from an Information System. In *47th Annual Allerton Conference on Communication, Control, and Computing (Allerton)*, IEEE, pp 644–651.

Bar, A., B. Shapira, L. Rokach, and M. Unger. 2016. Scalable Attack Propagation Model and Algorithms for Honeypot Systems. In *2016 IEEE International Conference on Big Data (Big Data)*, IEEE, pp. 1130–1135.

Baykara, M., and R. Das. 2018. A Novel Honeypot-Based Security Approach for Real-Time Intrusion Detection and Prevention Systems. *Journal of Information Security and Applications* 41: 103–116.

Chen, Y., J. Hong, and C.C. Liu. 2018. Modeling of Intrusion and Defense for Assessment of Cyber Security at Power Substations. *IEEE Transactions on Smart Grid* 9 (4): 2541–2552.

Hayatle, O., H. Otrok, and A. Youssef. 2013. A Markov Decision Process Model for High Interaction Honeypots. *Information Security Journal: A Global Perspective* 22 (4): 159–170.

Liu, Y., H. Liu, and B. Wang. 2017. Autonomous Exploration for Mobile Robot Using Q-Learning. In *2017 2nd International Conference on Advanced Robotics and Mechatronics (ICARM)*, IEEE, pp. 614–619.

Madni, A.M., M. Sievers, A. Madni, E. Ordoukhanian, and P. Pouya. 2018. Extending Formal Modeling for Resilient Systems Design. *INSIGHT* 21 (3): 34–41.

Magalhaes, A., & G. Lewis. 2016. Modeling Malicious Network Packets with Generative Probabilistic Graphical Models, pp. 1–5.

Majeed, S.J., and M. Hutter. 2018. On Q-learning Convergence for Non-Markov Decision Processes. In *IJCAI*, pp. 2546–2552.

Mohri, M., A. Rostamizadeh, and A. Talwalkar. 2012. *Foundations of Machine Learning. Adaptive Computation and Machine Learning*. MIT Press, 31, p. 32.

Pauna, A., A.C. Iacob, and I. Bica. 2018. QRASSH-A Self-Adaptive SSH Honeypot Driven by Q-learning. In *2018 International Conference on Communications (COMM)*, IEEE, pp. 441–446.

Sigaud, O., and O. Buffet, eds. 2013. *Markov Decision Processes in Artificial Intelligence*. Wiley.

Sutton, R.S., and A.G. Barto. 2018. *Reinforcement Learning: An Introduction*. MIT Press.

van Otterlo, M. 2009. *Markov Decision Processes: Concepts and Algorithms. Course on 'Learning and Reasoning'*.

van Otterlo, M., and M. Wiering. 2012. Reinforcement Learning and Markov Decision Processes. In *Reinforcement Learning*, 3–42. Berlin\Heidelberg: Springer.

Zanini, E. 2014. *Markov Decision Processes*. [Online]. https://www.lancaster.ac.uk/pg/zaninie.MDP.pdf

Probabilistic System Modeling for Complex Systems Operating in Uncertain Environments

Parisa Pouya and Azad M. Madni

Abstract Complex systems that continuously interact with dynamic uncertain environments need the ability to adapt their decision-making based on observed outcomes of their decisions and actions. Traditional deterministic modeling approaches are invariably inadequate for modeling systems whose models are not fully known initially. For such systems, we need the ability to start with an incomplete model and then progressively complete the model with observations made along the way. To address this problem type, we propose an extendable-partially observable Markov decision process (extendable-POMDP) to model the system's state space and decision-making in the presence of uncertainties. The extendable-POMDP model is able to account for unknown-unknowns by incorporating "new hidden states" that result in expanding the model state space which in turn extends the associated probability distributions. This paper provides an online algorithm for solving a POMDP model by employing a heuristic search algorithm that estimates long-term rewards in a finite-horizon look-ahead in a sense-plan-act cycle. Heuristics are employed in model definition, expansion, and online look-ahead search to contain the otherwise inevitable computational complexity arising from state-space explosion.

Keywords Decision-making and planning · Probabilistic modeling · Partially observable Markov decision processes · Complex systems · Model-based systems engineering (MBSE) · Heuristic search

P. Pouya (✉) · A. M. Madni
University of Southern California, Los Angeles, USA
e-mail: pouya@usc.edu

A. M. Madni et al. (eds.), *Recent Trends and Advances in Model Based Systems Engineering*, https://doi.org/10.1007/978-3-030-82083-1_11

1 Introduction

Understanding the dynamic behavior of complex systems that undergo state changes as a result of ongoing interaction with the environment is a challenging system modeling problem. Existing techniques that are employed for these purposes are either deterministic or stochastic. Deterministic models fully determine a model based on setting up initial assumptions and conditions, while stochastic models exhibit required randomness needed to characterize stochastic properties. Recent studies on modeling a system's dynamic behavior can be classified as the following: (1) inferring the underlying model and parameters for classifying and making predictions about the future and (2) identifying the conditional dependencies, relationships between variables, correlations, and changes in variables over time (Robinson and Hartemink 2009). The main objective of the former class of problems is to find patterns and parameters from the behavioral data and make predictions about the future. Examples of this class of problems are speech recognition, activity or behavior detection, and anomaly detection. Markovian models, such as Markov chains and hidden Markov models (HMMs), are usually employed in these applications. The latter class of problems focuses on identifying the underlying (system or environment) states and correlations resulting from a system's interaction with its environment. The main goal in these problems is to design an abstract representation (model) of the real system-environment interaction and use it for understanding and reasoning about the system (Madni and Sievers 2018). This paper focuses on adapting existing modeling techniques for the latter class of problems that require ongoing decision-making in complex systems operating in dynamic uncertain environments.

Various techniques, such as time-varying Gaussian graphical models (Talih and Hengartner 2005; Xuan and Murphy 2007), dynamic Bayesian networks (DBNs) (Robinson and Hartemink 2009), and different Markovian models, are employed for designing models based on state variables and correlations. In this paper, we focus on Markov model family because they offer a strong mathematical framework and probabilistic structure capable of modeling systems for a wide range of applications. These models perform well in practice if applied in the right way when modeling complex systems (Rabiner 1989). For instance, HMMs (also viewed as stochastic generalization of finite-state automata) are examples of Markovian models in which state variables and dependencies are modeled as states and probability distributions, respectively. Markovian models are also employed for developing modeling tools, such as state diagrams that are mainly used in MBSE and control system design (Madni and Sievers 2018; Wray and Zilberstein 2019).

To make decisions and plan actions with respect to state variables and their correlations, decisions and their influences on state variables should be embedded within the system model. Markov decision processes (MDPs) are Markov models that capture the transitions and correlations between various state variables during system-environment interactions that occur during system's decision-making. Basically, MDPs can be viewed as Markov chains that include decisions within the model that allows for making decisions over time (Alagoz et al. 2010). POMDPs are generalization of MDP models extended to a probabilistic domain in which uncertainty regarding the state of the system model is allowed, and state variable information is completely hidden or only partially known. This implies that an observation (signal) from the environment (1) can identify more than one state at a time or (2) cannot be explained based on the existing state variables and correlations in the model. The former implies that the most probable states should be considered, instead of one unique state, while the latter means that a new "hidden state" is required in the model to represent the new observation. POMDPs are widely used in modeling sequential planning in various applications in which systems interface noisy and uncertain environments (Cassandra 1998). For instance, Hubmann et al. (2017), Song et al. (2016), and Ulbrich and Maurer (2013) employ POMDPs for decision-making in autonomous vehicles (AVs) where there exist uncertainties in observed information and intensions of passengers and drivers. Machine maintenance, structural inspection, machine vision, search and rescue, and target identification are other examples of complex systems with uncertain environments where POMDPs are successfully employed for planning and decision-making (Cassandra 1998). Generally, with POMDP applications, the former definition of hidden information is widely addressed through the probability distributions within the model. However, the latter definition of hidden information, i.e., unknown-unknowns, should also be considered in the model design to ensure accurate and correct response when faced with unknown-unknowns.

In this paper, we propose an extended version of the standard POMDP models that accounts for unknown-unknowns and unexplainable signals, in addition to uncertainties, by incorporating new hidden states, expanding the model, and retuning the probability distributions when needed (Madni et al. 2018a; b; Sievers et al. 2019a, b). Also, an online look-ahead heuristic search algorithm is provided that solves the extended POMDP model by estimating the possible future decision paths for each available decision and calculating the expected long-term reward associated with that decision and path. An example of model setup and decision-making using the new POMDP model and online algorithm for a simulated autonomous vehicle (AV) in a specific scenario is also provided.

2 Review of Markov Models and Decision Processes

2.1 Markov Models and Hidden Markov Models (HMMs)

A Markov model (chain) is defined as a triplet $<S, T, \pi>$ in which S denotes a finite set of states that are directly observable from the system-environment interaction, $T : S \times S \rightarrow [0, 1]$ is the transition function (matrix) that includes the probabilities associated with transitioning from a state to another, and $\pi : p(s_i) \rightarrow [0, 1]$ where $s_i \in S$ is the initial state probability distribution. In contrast with the Markov models

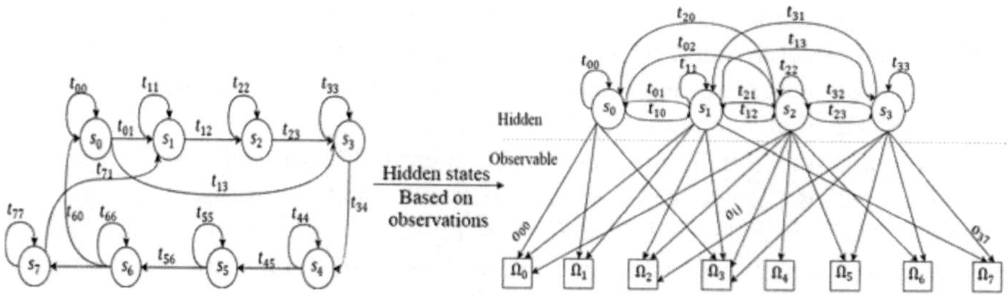

Fig. 1 (**a**) A Markov chain with eight states. (**b**) An HMM with four hidden states and eight observations. $t_{i,j}$ and $o_{i,k}$ transition and emission probabilities, respectively

that assume state variables and their changes over time are directly observable, HMMs assume an underlying "hidden" process associated with state variables that is modeled as a Markov chain and that process is obtained from noisy observations. In other words, an HMM is a statistical Markov model. Basically, an HMM can be defined as a tuple <*S*, Ω, *T*, *O*, *π*> in which *S* denotes the state space in the model; $\Omega = \{o_0, \ldots, o_n\}$ is the observation space in which distinct observations associated with states are defined; $T : S \times S \rightarrow [0, 1]$ shows the transition matrix; $O : S \times \Omega \rightarrow [0, 1]$ represents the observation probabilities in the emission matrix, where $O_i(o_j)$ is the probability associated with observing o_j at state s_i at time t; and $\pi : p(s_i) \rightarrow [0, 1]$ defines the probabilities associated with being at a state at time $t = 0$ (Rabiner 1989). Figure 1a shows a Markov model where various observations from the system-environment interaction are modeled as individual states, whereas Fig. 1b shows the same example with hidden states identified based on observations.

2.2 *Markov Decision Processes (MDPs)*

To keep track of the impacts of transitions in between states, values can be embedded within the model definition. These values can be defined with respect to an objective or goal defined for the model under a specific scenario. Moreover, the capability of making decisions or reacting to changes in state variables can also be integrated with a Markov model. Adding the ability of making decisions based on observed changes in state variables to a Markov model and storing the impacts of changes and transitions defines an MDP model. In general, an MDP model is defined as a tuple <*S*, *A*, *T*, *R*> where *S* identifies a set of finite states (state space), *A* identifies an action space, $T : S \times A \times S \rightarrow [0, 1]$ represents the transition function that identifies the transition probabilities between states based on actions, and $R : S \times A \times S \rightarrow \Re$ shows the reward function which identifies the rewards or penalties associated with being in a state and making a decision. The overall objective in MDPs is to find the most optimal mapping between the actions and states, so-called optimal policy that maximizes the sum of long-term rewards by achieving the goal of the MDP using minimum possible number of decisions or in the shortest time. A commonly applied approach for finding an optimal policy associated with an MDP model is using value iteration (Eq. 2) that employs dynamic programming for solving the Bellman's equation in an iterative process until the optimal value is achieved. For each state at time t, the action corresponding to the maximum value is considered the optimal mapping between that state and available actions (Eq. 3):

$$V_t^*(s) = \max_{a \in A} \sum_{s'} p\left(s|s',a\right)\left[R\left(s|s',a\right) + \gamma V_{t-1}^*\left(s'\right)\right] \tag{2}$$

$$\pi_t^*(s) = argmax_{a \in A} \sum_{s'} p\left(s|s',a\right)\left[R\left(s|s',a\right) + \gamma V_{t-1}^*\left(s'\right)\right] \tag{3}$$

2.3 *Partially Observable Markov Decision Processes (POMDPs)*

POMDPs are generalization of MDPs to uncertain environments where partially available data could potentially result in incomplete information about the state space. Uncertainty may appear as (1) uncertainty in actuation, whether an action is carried out successfully; (2) uncertainty in sensor and data interpretation due to sensor noise and limited sensor capabilities; (3) uncertainty about the environment; and (4) uncertainty about intensions of other systems in the environment (Koenig and Simmons 1998; Bai et al. 2015). In contrast with the MDP models that assume full access to state space, partial observability implies that the system only receives an indication of its current state that only allows for probabilistic identification of the state. A POMDP model can be defined as a tuple <*S*, *A*, Ω, *T*, *O*, *R*> in which *S* determines a finite state space which is hidden; *A* identifies a finite set of actions; $\Omega = \{o_0, \ldots, o_n\}$ is a finite set of observations; $T : S \times A \times S \rightarrow [0, 1]$ is the transition function that identifies the probabilities associated with transitions in between states; $O : S \times A \times \Omega \rightarrow [0, 1]$ defines the emission function (or matrix), which provides the probabilities associated with performing an action in a state and observing an observation from the observation space; and finally $R : S \times A \times S \rightarrow \Re$ is the reward function that provides rewards/penalties associated with performing an action in a state and transitioning to another state (Spaan 2012). Figure 2 shows the differences between a problem modeled using both an MDP and a POMDP model with four states $S = \{s_0, s_1, s_2, s_3\}$ and three actions $A = \{a_0, a_1, a_2\}$. In this figure, $[t_0, t_1, t_2]_{i,j}$ and $[r_0, r_1, r_2]_{i,j}$ represent the transition probabilities and rewards/penalties of performing actions $[a_0, a_1, a_2] \in A$ at state s_i and transitioning to state s_j in both MDP and POMDP models, respectively. In addition, $[o_0, o_1, o_2]_{i,k}$ are the emission probabilities of observing $o_k \in \Omega$ after performing $[a_0, a_1, a_2] \in A$ at state s_i in the POMDP model.

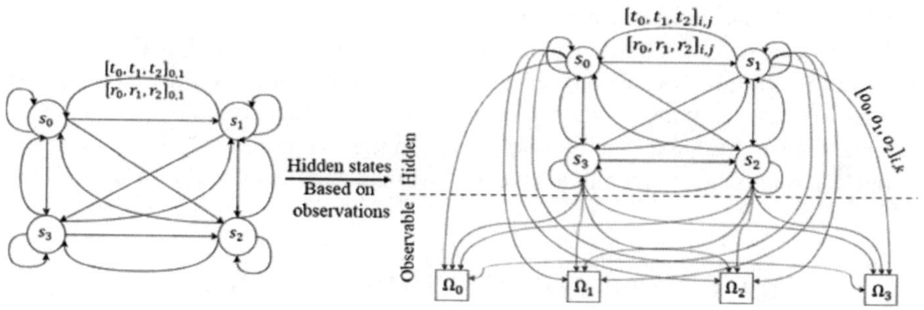

Fig. 2 A problem modeled with an MDP model if the states are observable (left) and a POMDP model where the full-observability assumption is relaxed

Since the state space is only partially observable, POMDPs employ a probabilistic distribution, so-called belief, over the state space to determine the most recent state based on received observations. At time $t = 0$, with no available observation, the belief can be initialized as a uniform distribution over all the states. Later, as the system interacts with the environment and receives feedback (i.e., observations and rewards), the belief vector gets updated based on Bayes' rule as follows:

$$b^{t+1}(s_i) = p\left(s_i | b^t, a, o\right) = \frac{p(o|s_i, a) \sum_{s \in S} p(s_i|s, a) b^t(s)}{\sum_{s' \in S} p(o|s', a) \sum_{s \in S} p(s'|s, a) b^t(s)} \tag{4}$$

where $p(o|\, s_i, a)$ and $p(s_i|\, s, a)$ show the emission probability of performing a at s_i and observing o and the probability of transitioning to s_i after performing action a at state s. A POMDP can be formulated as an MDP to find the optimal policies associated with all possible belief states by solving Bellman's equation using techniques such as dynamic programming (Eq. 5):

$$V_t^*\left(b^t\right) = \max_{a \in A}\left[\sum_{s \in S} b^t(s) R(s, a) + \gamma \sum_{o \in \Omega} p\left(o|b^t, a\right) V_{t-1}^*\left(b|b^t, a, o\right)\right] \tag{5}$$

The techniques, such as dynamic programming, that evaluate every imaginable belief and action pair and provide an optimal policy prior to execution are known as "offline algorithms." Offline algorithms assume that the initial model setup and the environment are fixed. While the offline algorithms can achieve very good performance, they often take significant amount of time, e.g., hours, to solve slightly large problems in which there exist numerous possible situations to consider (Ross et al. 2008). On the other hand, online algorithms circumvent the complexity of computing a policy by only considering the current belief and a small horizon to search for contingency plans. Since online algorithms evaluate the actual belief achieved from real interactions between a system and its environment, they can handle changes in the environment (e.g., changes in goals) without recomputing the full policy for the whole model (Sunberg and Kochenderfer 2018; Ye et al. 2017).

3 Proposed POMDP and Solution

In dynamic and uncertain environments, there is no guarantee that all information is initially known and considered in the model. This means that there may be observations that cannot be explained using the existing states in the model, which require expanding the current state space to include "new hidden states," when discovered during the execution phase. The current definition of the POMDPs and offline solutions are not able to accommodate the issues associated with unknown-unknowns.

To accommodate this issue, we further extended the standard definition of the POMDP models to allow for expanding the model and incorporating new hidden states. Thus, the definition of the POMDP updates to a tuple $<S^+, A, \Omega^+, T^+, O^+, R^+>$ in which $S^+ : S \cup H$ is the extended finite set of states including the hidden state(s), H, and $\Omega^+ : \Omega \cup \Theta$ is the extended finite set of observations. Initially, H and Θ are empty sets and the model is a tuple of $<S, A, \Omega, T, O, R>$. As the model discovers unknown-unknowns during its interaction with its environment, the sets are accumulated by the new observations and hidden states. $T^+ : S^+ \times A \times S^+ \rightarrow [0, 1]$ is the extended transition function that includes the probabilities of transitioning to and from the hidden states, $R^+ : S^+ \times A \times S^+ \rightarrow \mathfrak{R}$ determines the extended reward function, and $O^+ : S^+ \times A \times \Omega^+ \rightarrow [0, 1]$ identifies the extended observation function that contains the probabilities of observing both $o \in \Omega$ and $o' \in \Theta$ (Fig. 3). On the other hand, definition of states, observations, and reward function in the proposed POMDP model are slightly different from a standard POMDP model. **State space** is defined based on various, high-level events that generally describe different conditions in the system-environment interaction. Generalization of the state space to high-level events helps with reducing the number of state space and decreasing the related computational complexities.

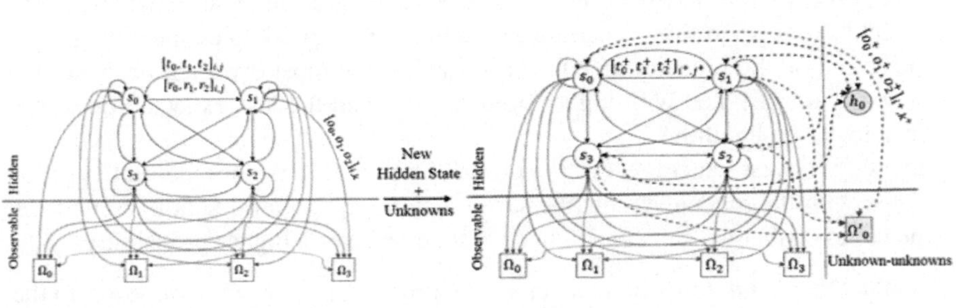

Fig. 3 Example of how a POMDP expands as a new hidden state is added to incorporate an unknown-unknown

Moreover, variables that identify the goal and failure events are also modeled as states in the model. Since the goal and failures are also embedded as states within a model, the **reward function** assigns rewards/penalties ($r > 0$ or $r < 0$) to the states depending on whether a state is identified as goal, failure, or transient, which changes the reward matrix to a reward vector $R^+ : 1 \times S^+ \rightarrow \mathfrak{R}$. **Optimal policy** in the proposed POMDP is identified as an action that updates the belief so that, considering all possible observations in the model, the goal state has a higher probability (failure has less probability) in the future belief. In other words, the optimal policy is the mapping between the belief and actions that maximizes the long-term sum of rewards by constantly moving along a plan trajectory that improves the belief state (Eq. 6):

$$\pi_t^*\left(b^t\right) = argmax_{a\in A}\left[\sum_{o\in\Omega^+}\sum_{s\in S^+} p\left(o|s,a\right)\left(b^{t+1}(s)|b^t(s),a,o\right)R^+(s)\right] \tag{6}$$

Thus, since the belief represents the probabilities assigned to most probable states and optimal policy improves the belief probability distribution by assigning a higher probability to the goal state, the combination can be used for explaining the decisions made and reasoning about the model.

3.1 *Initializing New Hidden States*

Hidden states are associated with the events that cannot be interpreted from the initialized state in the model state space (Sievers et al. 2019a, b). When a new hidden state is initialized in the proposed POMDP model, the transition and emission probabilities of the hidden state are empirically initialized using the following heuristics depending on the model and scenario:

Heuristic 1: "Expected Outcome: Assign higher probabilities to the most expected states/observations based on how the action changes the state variables of the model." After performing a certain action (e.g., a), depending on the influence of the action on the predetermined state variables, the most expected observations or states are assigned with higher probabilities, and the less expected ones will have lower probabilities.

Heuristic 2: "Safest Outcome: Assign probabilities so that the optimal policy associated with the hidden state is the safest action (e.g., neutral action) or updates the belief so that the safest/most neutral state receives a higher probability."

On the other hand, transition and emission probabilities *from known states* to the hidden state(s) are initially assigned with small probabilities (e.g., 0.01), because this transition (or emission) is not repeated enough compared to the known states.

3.2 N-Step Look-Ahead (Online Policy Estimation Algorithm)

Since the hidden states and unexplainable observations are discovered during the execution phase, there exists no prior information, such as pre-planned policies, associated with them. Thus, even if an offline solution can efficiently calculate and estimate optimal policies for a model prior to execution, the solution still lacks the optimal policies associated with the newly added hidden state. In other words, offline algorithms also lack the ability of updating a pre-estimated solution when a model or environment changes that results in changing the model objective or goal (Ross et al. 2008). This implies that the offline algorithms are not applicable to the problems with highly dynamic environments and objectives, because they require to recompute the optimal policies after any changes. On the other hand, online algorithms that rely on combining offline calculations in estimating optimal policies during the look-ahead search in estimated future beliefs are not sufficient, since there is no prior information that exists for newly initialized hidden states. To this end, we implemented the online, "N-Step Look-Ahead," policy estimation algorithm that (1) defines a belief tree with the current belief state as its top node, "root"; (2) recursively, explores the possible plan/decision paths by traversing the expected beliefs located on the lower levels of the tree; and (3) calculates the expected long-term rewards for available possible plans in that tree to select the plan with the highest long-term reward.

The algorithm recursively expands the belief states at each level ($l \leq N$) until it reaches the deepest level (N) or a termination condition for a belief is met. The tree is explored bottom to top and left to right, meaning that initially the values associated with the leftmost branches are calculated starting from the bottom of the tree and moving to the top node, then the next branch is explored, and value is calculated until all branches are traversed (Eq. (7)). A learning rate $0 < \gamma < 1$ is also considered, so that the largest rewards are collected as early as possible. The depth of the tree determines the finite horizon for the look-ahead search. Basically, at each level of the tree, the expected beliefs at level $l < N$ are calculated based on the beliefs at level $l - 1$, available actions, and possible observations:

$$V^N\left(b^t, a\right) = \sum_{o \in \Omega^{+'}} \Pr\left(o \mid b^t, a\right) * \gamma \sum_{a' \in A} V_{l+1}^N\left(b_{a,o}^{t+1}, a'\right) \tag{7}$$

Since this algorithm is designed to provide a policy anytime it's required, the execution time of the algorithm (time complexity) is very important. As N increases, the search algorithm explores deeper levels and the estimated long-term reward becomes more accurate. On the other hand, larger N requires more computation time, so there is a trade-off between the accuracy and execution time. Various techniques, such as sampling and heuristic search, are employed to reduce the computation time of a search algorithm (Ye et al. 2017; Kurniawati and Yadav 2016; Etminan and Moghaddam 2018a; b). We employ a heuristic search that reduces the time complexity of the search by only considering the exploration of the belief nodes that satisfy the conditions defined in our heuristics. Our heuristic summarizes as follows:

Search Heuristic1: "Expand non-terminal/non-failure belief nodes." Using this heuristic, the search algorithm only expands and explores the belief nodes that have low probabilities assigned to failure states.

Search Heuristic2: "Expand the belief nodes using possible actions and associated reachable observations only." Based on this heuristic, the belief nodes in the lower levels (*l-1*) of the tree are calculated based on observations with high probabilities from a parent belief node at level *l* and an action *a*. Reachable observations are identified using the criteria represented in Eq. 8:

$$o \in \Omega^{+'}\left(b^t, a\right) \; iff \; \sum_{s \in S^+} b^t(s) p\left(o|s, a\right) \geq L \tag{8}$$

Where $\Omega^{+'} \subseteq \Omega^{+}$ denotes the reachable observation and $0 \leq L \leq 1$ is the minimum reachability probability defined empirically based on the size or the problem and emission matrix. Figure 4 shows how applying the heuristic search reduces the computation time in the N-Step Look-Ahead search from an exponential growth rate to a linear growth rate for a given model that includes four states and three actions.

4 An Exemplar POMDP Model

In an exemplar scenario, we simulated an AV in a multilane freeway using PythonVTK. As shown in Fig. 5a, the AV (green) is surrounded by traffic in different lanes (different relative distances and velocity/speed). The AV has two main objectives with respect to its surrounding environment (freeway + traffic). These are (1) safely drive within one lane and (2) safely change lanes when it becomes necessary (Pouya and Madni 2020a).

For the purpose of this paper, we define a POMDP model, including the states and probabilities associated with the former objective, and test the model in the simulated multilane freeway using the N-Step Look-Ahead function for constant decision-making. Later, we tune the parameters and demonstrate model expansion for including hidden states. The initial step in POMDP model definition is identification of candidate states with respect to various high-level general conditions. However, due to partial observability (e.g., sensor noise and hidden driver intensions), the states can be identified based on observations. Figure 5b demonstrates various observation classes (candidate states) defined based on vehicle speed and relative distances (difference between distance in front and rear, $dF - dR$) using the simulated traffic data around the AV with two different traffic setups.

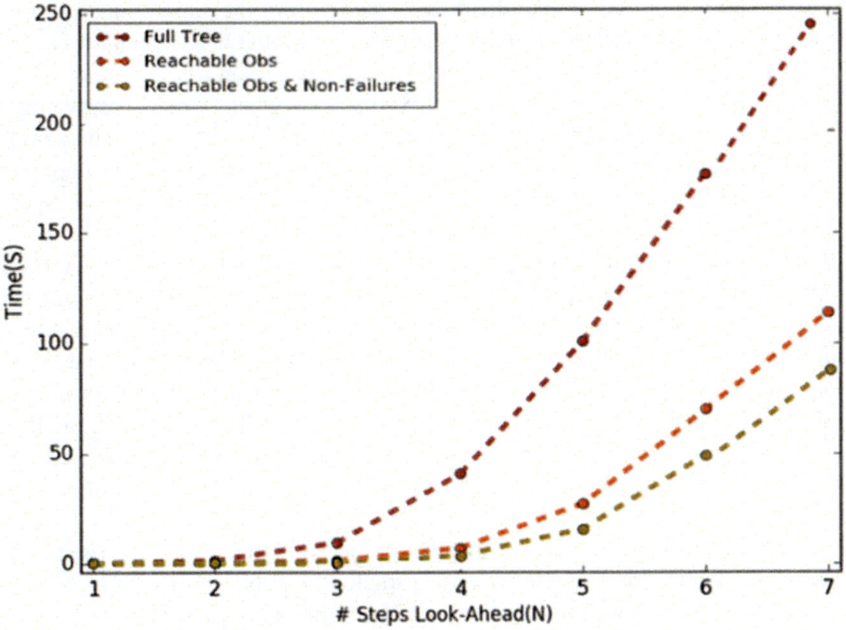

Fig. 4 Look-ahead policy estimator computation time as different heuristics are applied

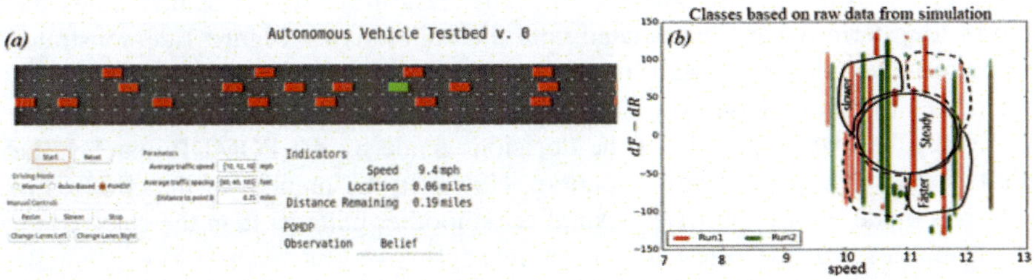

Fig. 5 (**a**) Traffic simulation dashboard. (**b**) High-level traffic conditions (patterns) based on data

The datapoints that fall into the dashed classes are observations that cannot fully determine a state, which we refer to as noisy and partially available observations. In addition, according to pre-determined speed and distance limits in the simulation, the datapoints that exceed these limits are identified as failures or crashes. The next step after identifying the state candidates is defining the probabilities and reward values. Since the objective of the model is to drive safely within a lane, the goal state is s_2 : *safe and steady* with $R(s_2) = +10$, the failure state is s_3 : *crashed/failure* with $R(s_3) = -20$, and s_0 : *slower*, s_1 : *faster* are transient states with $R(s_0) = R(s_1) = +1$ reward values. In addition, actions associated with this model are a_0 : *maintain status quo*, a_1 : *speed up*, and a_2 : *slow down*. The transition and emission matrices can be learned from the simulation data or can be initialized based on expert's judgment and tuned within the simulation. Later, the probability and reward matrices are expanded as the model receives an observation that cannot be explained using the current state space.

To expand and initialize the probability matrices when a new observation is realized, we have applied the "most expected outcome" heuristic. As an example, the transition probability associated with h_0 and a_1 has the highest probability assigned to s_1, because the vehicle expects to drive faster as it applies a_1 and speeds up.

In this simulation and for the purpose of this exemplar scenario, hidden observations are generated as outputs of a random function invoked in random times in the simulation. After tuning the probabilities and defining the extension technique within the POMDP model, the model is tested in the simulation by having the N-Step Look-Ahead function evaluate the possible decisions at every time step based on the most recent belief. N equal to 2 is selected for the depth of look-ahead search with a sampling rate of 0.1s, and the POMDP decisions and performance are compared to a rule-based algorithm designed based on time-to-collision measurements.

Figure 6 (left) shows a series of changes in the AV's belief, and the right figure represents the values for possible actions estimated for each belief with N = 2.

As a new hidden state is identified and the belief is expanded at $t = 15$, the look-ahead value estimation decides to maintain status quo (a_0) as long as the belief probability assigned to the hidden state is high but changes its decision as soon as the belief in the known states goes higher. The dashed line demonstrates the sum of long-term rewards associated with the belief series. Figure 7 demonstrates the performance of the POMDP model in comparison with a rule-based algorithm that makes decisions based on TTC criteria with full observability. As shown in the figure, the overall pattern of the decisions made by the POMDP matches the rule-based with full-observability pattern. However, the number of the changes in decisions made by the POMDP is smaller (smoother pattern) than the rule-based, which implies that the rule-based is more aggressive and reacts to every single

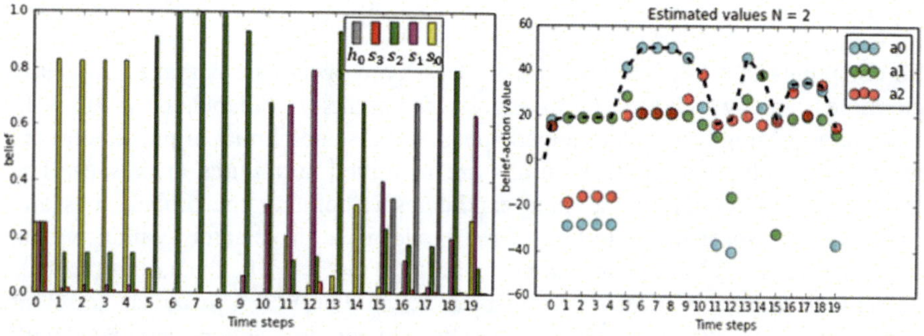

Fig. 6 (left) Belief updates, expanded at t = 15 to include a new hidden state; (right) estimated long-term reward (dashed line)

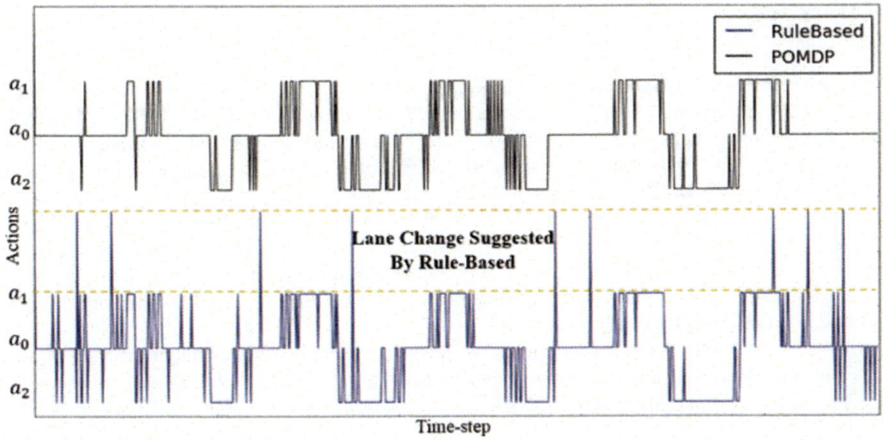

Fig. 7 POMDP performance evaluation and comparison with a TTC rule-based. (Pouya and Madni 2020b)

observation. In contrast, the POMDP ensures about the consistency of the received observation and reacts when its belief with respect to the received observation is high.

5 Summary and Future Work

In this paper, we emphasized the importance of understanding the behavior of a complex system from its interactions with its environment to design accurate models for resilient decision-making with partially available data. We presented an extendable-POMDP model that is initialized using available information, and then adapts to new information by incorporating new hidden states, and thereby extends the related probability distributions using heuristics, so they can be learned incrementally. The flexibility introduced by incorporating new hidden states results in risk associated with evaluating the accuracy of decisions made for the hidden states with less or no prior information about the state, which we manage by employing heuristics in model expansion. To address the risk associated with computational complexity, the N-Step Look-Ahead online value estimation algorithm is employed. This algorithm uses heuristic search, to solve the extendable-POMDPs in an any-time fashion. We intend to extend the work presented in this paper to realize a probabilistic modeling paradigm that can be used for decision-making and planning of complex systems and system of systems that operate in highly dynamic, uncertain environments. For model testing and verification purposes, we currently compare with rule-based algorithms and full observability, but in the future, we intend to employ machine learning (e.g., Q-learning (Pouya and Madni 2020c)) and formal reasoning methods to create a formal verification technique for POMDP models.

References

Alagoz, O., H. Hsu, A.J. Schaefer, and M.S. Roberts. 2010. Markov Decision Processes: A Tool for Sequential Decision Making Under Uncertainty. *Medical Decision Making 30* (4): 474–483.

Bai, H., S. Cai, N. Ye, D. Hsu, and W.S. Lee. 2015. Intention-aware Online POMDP Planning for Autonomous Driving in a Crowd. In *2015 IEEE International Conference on Robotics and Automation (ICRA)*, IEEE, pp. 454–460.

Cassandra, A.R. 1998. A Survey of POMDP Applications. In *Working Notes of AAAI 1998 Fall Symposium on Planning with Partially Observable Markov Decision Processes*, Vol. 1724.

Etminan, A., and M. Moghaddam. 2018a. Electromagnetic Imaging of Dielectric Objects Using a Multidirectional-Search-Based Simulated Annealing. *IEEE Journal on Multiscale and Multiphysics Computational Techniques 3*: 167–175.

——— 2018b. A Novel Global Optimization Technique for Microwave Imaging Based on the Simulated Annealing and Multi-Directional Search. In *2018 IEEE International Symposium on Antennas and Propagation & USNC/URSI National Radio Science Meeting,* IEEE, pp. 1791–1792.

Hubmann, C., M. Becker, D. Althoff, D. Lenz, and C. Stiller. 2017. Decision Making for Autonomous Driving Considering Interaction and Uncertain Prediction of Surrounding Vehicles. In *2017 IEEE Intelligent Vehicles Symposium (IV)*, 1671. 1678: IEEE.

Koenig, S., and R. Simmons. 1998. Xavier: A Robot Navigation Architecture Based on Partially Observable Markov Decision Process Models. *Artificial Intelligence Based Mobile Robotics: Case Studies of Successful Robot Systems*, (Partially), pp. 91–122.

Kurniawati, H., and V. Yadav. 2016. An Online POMDP Solver for Uncertainty Planning In Dynamic Environment. In *Robotics Research*, 611–629. Cham: Springer.

Madni, A.M., and M. Sievers. 2018. Model-Based Systems Engineering: Motivation, Current Status, and Needed Advances. In *Disciplinary Convergence in Systems Engineering Research*, 311–325. Cham: Springer.

Madni, A.M., M. Sievers, A. Madni, E. Ordoukhanian, and P. Pouya. 2018a. Extending Formal Modeling for Resilient Systems Design. *INSIGHT 21* (3): 34–41.

Madni, A., D. Erwin, A. Madni, E. Ordoukhanian, and P. Pouya. 2018b. *Formal Methods in Resilient Systems Design using a Flexible Contract Approach* (No. SERC-2018-TR-119). SYSTEMS ENGINEERING RESEARCH CENTER HOBOKEN NJ HOBOKEN United States.

Pouya, P., and A.M. Madni. 2020a. Expandable-Partially Observable Markov Decision-Process Framework for Modeling and Analysis of Autonomous Vehicle Behavior. *IEEE Systems Journal*. https://doi.org/10.1109/JSYST.2020.30.

———. 2020b. Leveraging Probabilistic Modeling and Machine Learning in Engineering Complex Systems and System-of-Systems. In *AIAA Scitech 2020 Forum*, p. 2117.

———. 2020c. A Probabilistic Online Policy Estimation for Autonomous Systems Planning and Decision Making. *In 2020 IEEE International Conference on Systems, Man, and Cybernetics (SMC),* IEEE.

Rabiner, L.R. 1989. A Tutorial on Hidden Markov Models and Selected Applications in Speech Recognition. *Proceedings of the IEEE 77* (2): 257–286.

Robinson, J.W., and A.J. Hartemink. 2009. Non-stationary Dynamic Bayesian Networks. *Advances in Neural Information Processing Systems*: 1369–1376.

Ross, S., J. Pineau, S. Paquet, and B. Chaib-Draa. 2008. Online Planning Algorithms for POMDPs. *Journal of Artificial Intelligence Research 32*: 663–704.

Sievers, S., A.M. Madni, and P. Pouya. 2019a. Trust and Reputation in Multi-agent Resilient Systems. In *2019 International Conference on Systems, Man, Cybernetics (SMC)*, 741–747. IEEE.

Sievers, M.M., A.M. Madni, and P. Pouya 2019b. Assuring Spacecraft Swarm Byzantine Resilience. In *AIAA Scitech 2019 Forum*, p. 0224.

Song, W., G. Xiong, and H. Chen. 2016. Intention-Aware Autonomous Driving Decision-Making in an Uncontrolled Intersection. In *Mathematical Problems in Engineering, 2016*.

Spaan, M.T. 2012. Partially Observable Markov Decision Processes. In *Reinforcement Learning*, 387–414. Berlin/Heidelberg: Springer.

Sunberg, Z.N., and M.J. Kochenderfer. 2018. Online Algorithms for POMDPs with Continuous State, Action, and Observation Spaces. In *Twenty-Eighth International Conference on Automated Planning and Scheduling*.

Talih, M., and N. Hengartner. 2005. Structural Learning with Time-Varying Components: Tracking the Cross-Section of Financial Time Series. *Journal of the Royal Statistical Society: Series B (Statistical Methodology) 67* (3): 321–341.

Ulbrich, S., and M. Maurer. 2013. Probabilistic Online POMDP Decision Making for Lane Changes in Fully Automated Driving. In *16th International IEEE Conference on Intelligent Transportation Systems (ITSC 2013)*, 2063–2067. IEEE.

Wray, K.H., and S. Zilberstein. 2019. Generalized Controllers in POMDP Decision-Making. In *2019 International Conference on Robotics and Automation (ICRA)*, 7166–7172. IEEE.

Xuan, X., and K. Murphy. 2007. Modeling Changing Dependency Structure in Multivariate Time Series. In *Proceedings of the 24th International Conference on Machine Learning*, 1055–1062. ACM.

Ye, N., A. Somani, D. Hsu, and W.S. Lee. 2017. Despot: Online Pomdp Planning with Regularization. *Journal of Artificial Intelligence Research 58*.

Identification of Adverse Operational Conditions in Sociotechnical Systems: A Data Analytics Approach

Taylan G. Topcu, Konstantinos Triantis, and Bart Roets

Abstract Sociotechnical systems (STSs) such as infrastructure management systems operate under highly dynamic, contextual, and environmental conditions; therefore they depend on specifically trained group of individuals known as Controllers for their safety-critical decision-making activities. The dependency of STS on human decision-makers introduces additional complexity to the system due to the intertwined social and technical factors that influence the operational decision-making process. While the role allocated to autonomous decision-making units in STSs is rapidly increasing, hard-to-estimate and the inherently unique nature of safety-critical situations render high levels of automation infeasible and require manual control in many instances. In this paper, we investigate real-world operational data from INFRABEL (Belgian Railways) and utilize data analytics techniques to understand how Controllers behave during adverse operational conditions. The identification and evaluation of adverse operational instances can also support the design of future decision support or automation tools. To achieve this, we first provide a brief discussion of social and technical factors that influence the Controllers and their decision-making process. We then introduce robust principal component analysis that is a rigorous data analytics technique to identify influential observations (leverage points and outliers). We finally provide a brief discussion of how operations during adverse conditions differ from nominal conditions. We observe that adverse operational conditions in our application typically occur in 5% of the observations. The proposed approach will be implemented at INFRABEL, the Belgian Railways.

Keywords Sociotechnical systems · Data analytics · System safety · Human-machine teaming · Systems engineering

---

T. G. Topcu (✉) · K. Triantis
Virginia Tech, Virginia, USA
e-mail: ttopcu@vt.edu

B. Roets
INFRABEL and Ghent University, Brussels, Belgium

A. M. Madni et al. (eds.), *Recent Trends and Advances in Model Based Systems Engineering*, https://doi.org/10.1007/978-3-030-82083-1_12

1 Introduction

Sociotechnical systems are complex interdependent systems that rely on the successful collaboration between human decision-makers and autonomous systems to deliver critical services that allow for a community to sustain its economy (O'Sullivan and Sheffrin 2007). We define the term *sociotechnical systems* (STSs) as systems that rely on the collaboration of humans with engineered systems to fulfil their mission. To elaborate, management of air traffic, railroad traffic, metro, nuclear power plants, and the international space station are all examples of STSs. These systems are usually large in terms of size, operate under highly dynamic conditions, are managed by hierarchical organizations that consist of multiple decision-makers with sometimes conflicting objectives, and are tasked with a safety-critical mission. The multifaceted nature of STS and the severe consequences of their failures (National Transportation Safety Board 2016; Salmon et al. 2016) require a holistic systems thinking approach to address these interrelated mechanisms.

In STSs such as infrastructure management systems, safety-critical decision-making activities are allocated to specifically trained individuals denoted as *Controllers*. Controllers work in shifts 24/7 to make decisions regarding the daily operation and service delivery. In this paper, we adopt a descriptive research approach and investigate Traffic Controllers (TCs) at INFRABEL, the Belgian national railroad company that owns and runs the Belgian railway infrastructure, within their Traffic Control Centers (TCCs). INFRABEL manages one of the world's most complex and dense railroads; therefore it constitutes an excellent observation case. Each INFRABEL TC manages a workstation that has a built-in automated decision aid system (ADAS, i.e., the automatic setting of a train route when a train approaches a signal). ADAS can be activated within the TC's dedicated control area for a given time period, a train, or a certain node within the railroad network. Moreover, each TC is an individual; thus it has strictly subjective preferences (Keeney and Raiffa 1993) and varying experience levels and could be subject to different cognitive and work environment-related influences. Consequently, it is difficult to assess which combinations of sociotechnical factors constitute nominal or adverse operational conditions.

We live in the information age (Jin et al. 2015). Our society is currently experiencing drastic changes in many aspects such as the stock exchange (Dugast and Foucault 2018), the workforce (Lohr 2012), and even election security (Russell Neuman et al. 2014). Arguably, the current abundance of data is even transforming the way research is conducted, changing the impetus from theory-driven approaches to empirical data-based approaches (Kitchin 2014). We consider that rigorous research could benefit from the strengths of both theoretical and data-driven empirical approaches. In line with this thinking, the purpose of this paper is threefold. First, by using data analytics techniques, we statistically identify deviations from nominal or regular operational conditions given sociotechnical complexity. To achieve this, we utilize the infrastructure management system at INFRABEL that provides us with a unique sociotechnical dataset that is composed of disaggregate measurements of a large-scale infrastructure management system. Second, based on the insights obtained from the data, we articulate the necessary steps to improve our understanding of safe management of a complex sociotechnical system. Third, we pave the way for a real-world implementation at Belgian Railways, which will allow for further expert feedback and face validation. The rest of the paper is organized as follows: Sect. 2 provides a concise literature review, Sect. 3 describes the methodology along with the description of the data, Sect. 4 presents the results, and Sect. 5 concludes.

2 A Concise Literature Review

We start our coverage of the literature by differentiating between the terms engineered systems and STSs. In engineered systems (such as missiles, satellites, cars, etc.), the performance of the system is purposefully decoupled from its end user (Blanchard and Fabrycky 2011). Stakeholder preferences that are considered important by the designer are reflected into the system design through the systems engineering approach (Topcu and Mesmer 2018). In the case of STSs, usually the most context-dependent, complex, and safety-critical function of the system is allocated to a group of humans (Wilson 2000). Thus, the performance of STSs is coupled with the performance of its Controllers. The issue with that is human performance is dependent on physiological and social considerations such as fatigue (Ferguson et al. 2008; Roets and Christiaens 2019), mental stress (Beehr 2014), situational awareness (Salmon et al. 2009), prospective memory (Grundgeiger et al. 2015), management style (Barling et al. 2002), and organizational culture (Reiman and Oedewald 2007), among others. While these factors appear to be only influencing Controllers and not the rest of the system, Controller errors are one of the leading contributors of system failures (Rasmussen 1997; Cook and Rasmussen 2005; Roets and Christiaens 2019). In addition, TC tasks with a highly variable workload are more prone to human error (Roets et al. 2018). A recent study documented that disregarding the influence the contextual factors on the safety-critical decision-makers may lead up to 50% underestimation of system failure risk (Topcu et al. 2019). To summarize, human performance and thus the performance of STSs is of interdisciplinary nature (Leveson 2011) and emerges from the interaction of social and technical elements (Kroes et al. 2006).

We mentioned that in STS, the operational conditions can vary drastically; therefore it is difficult to fully automate (Balfe et al. 2015). Thus, it is extremely important to learn from accident-free operational performance during adversity. Sometimes adversity factors originate from the Controllers themselves, which intuitively leads some engineers to assume that increasing levels of automation is a dominant accident prevention strategy. For example, in 2013, a train derailment in Spain was caused by excess speed around a sharp curve that killed 79 passengers, while the train driver was distracted by a telephone call (Dawber 2015). Clearly, this accident could have been prevented with an autonomous controller. However, as manifested by some of the recent tragedies, increased automation could also introduce hard-to-react failure modes. The Boeing 737 MAX anti-stall system has contributed to two fatal crashes that resulted in 346 fatalities when pilots could not retake control from the system when erroneous information detected a nonexistent stall (The Aircraft Accident Investigation Bureau of Ethiopia 2019; Tjahjono 2018). Against this kind of hard-to-estimate safety-critical circumstances, reliance on humans offers considerable benefits. This is because the Controllers have the ability to *adjust* the STS performance based on the changing needs of the environment (Osorio et al. 2011). Going back to the case of Boeing 737 MAX crashes, the same aircraft had a similar problem on the previous day, but an extra pilot correctly diagnosed and disabled the malfunctioning system, preventing an accident from occurring (Levin and Suhartono 2019). Clearly, there is no linear pattern here as even the culture and subjective personal factors play an important role (Helmreich 2000). Some in the literature proposed using STS risk measures instead of traditional engineering risk measurement approaches (Battles and Kanki 2004). However, concerns regarding completeness and holism of such approaches remain valid because of the complexity of the phenomena and the importance of lower-level contributions (Hulme et al. 2019).

We consider that it is necessary to focus on successful Controller behavior during instances of adverse operational conditions so that we could learn from best practices. This raises a fundamental question: Given the multidimensional and intertwined relationship between social and technical factors, how could one objectively differentiate between nominal and adverse operational conditions? One potential way of achieving this is through machine learning. One machine learning technique that had success in identifying and labeling outliers has been the local outlier probability approach (Kriegel et al. 2009). This approach provides a score for the degree of being an outlier, which simply labels how much of an observation is an outlier from a scale of zero to unity. However, it requires predetermining the number of nearest neighbors (Wong and Lane 1983). Another alternative, among many others (Maronna et al. 2019), is the use the robust principal component analysis (ROBPCA) (Hubert et al. 2005). This approach does not require any predetermined value, performs exceptionally well with high-dimensional data (Filzmoser et al. 2008), and subsets the data in four segments based on their degree of difference (Herrera-Restrepo et al. 2016). Due to framing concerns, in this study we only focus on the ROBPCA algorithm.

3 Methodology

3.1 The Data

The operational data used in this study is from one of INFRABEL's busiest TCCs, for an anonymized month in 2018. Our dataset consists of 1914 observations with 9 dimensions that correspond to 1 work hour of the individual TCs. Table 1 provides variable definitions and descriptive statistics.

Table 1 Data description and its descriptive statistics

Variable name	Description	Mean	Range
Manual movement decisions	Number of signals that are manually opened by TCs	397.49	[0;1355]
Auto movement decisions	Number of signals that are automatically opened by the ADAS	344.77	[0;1580]
Adaptation decisions	Number of decisions that change the state of the railroad such as merging or splitting trains, rerouting of trains, or special procedures at single-track lines, performed manually by TCs	317.49	[0;1272]
Anticipation	Measure (in seconds) of Controller time spent using the forecast tool that anticipates the future state of the network	13.65	[0;840]
Responded phone calls	Number of phone calls addressed by Controllers. Controllers receive phone calls from other INFRABEL personnel about decisions that require further information	0.43	[0;14]
Traffic complexity	Measure of traffic complexity of control area. Estimated by using the number of control signal passes and performed adaptation decisions	1022.95	[0;24000]
Traffic density	Measure of traffic density in control area. Calculated by dividing the number of train movements with the number of large traffic control signals controlled by that Controller	899.91	[0;6087.91]
Fatigue level	This variable represents the mental fatigue of TCs. It is calculated by INFRABEL's predictive tool that is conceptually based on the fatigue risk index (Roets and Christiaens 2019; Folkard, Robertson, and Spencer 2007)	0.87	[0.67;1.37]
Delay	Average train delays within the control area. Measured from the scheduled time in seconds. If negative, it indicates that the train is early. Large positive or negative delays are due to freight trains	208.74	[−7470;5563]

3.2 The Robust Principal Component Analysis

Before we proceed to the details of the ROBPCA, we would like to emphasize the importance of identifying *influential observations* that contain the most amount of information in a dataset. Given the safety-critical mission of STS, the adverse events and their consideration in the design and management of STSs could have dire consequences. Therefore, we would like to identify and learn from these instances. We will not go into the details of the ROBPCA method as it is described elsewhere in detail (Hubert et al. 2005). However, we will provide

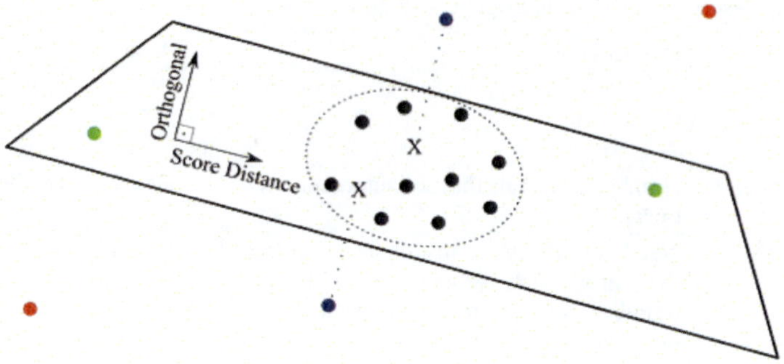

Fig. 1 Visualization of ROBPCA adopted from Hubert et al. (2005)

a summary. Essentially, ROBPCA is a robust multivariate method to classify statistically different subgroups in high-dimensional complex datasets (Rousseeuw and Hubert 2018). To elaborate, ROBPCA reduces the dimensionality of data by computing the principal component hyperplanes (PCs) that represent the majority of the observations in the dataset. In traditional principal component analysis (Wold et al. 1987), position of these PCs is computed by using variance measures; hence the definition of the hyperplanes is sensitive to outliers. Instead, ROBPCA computes the PCs by incorporating principals from projection pursuit (Huber 1985) and minimum covariance determinant estimators (Rousseeuw 1984). In short, these are median least squared bases estimators that are capable of capturing the variation of regular observations without being influenced by the outlying or extreme observations. Once the PCs that describe the majority of the observations are computed, we classify the remaining observations based on their respective Euclidian distance from these hyperplanes. Two distance measures are used for identifying the data points: the first is the orthogonal (vertical) distance and the second is the score (horizontal) distance. We provide a visualization of how these measures are used in Fig. 1.

The first subset is denoted as *regular* observations. These observations constitute the general body of the dataset. These data points are located on the PCs, meaning that their respective distance is below the threshold values on both orthogonal and score axes. We consider that these points represent *nominal operational* conditions and represent with color black in Fig. 1. The second group is denoted as *good leverage points*, and this group is composed of observations that are parallel to the PCs, but they are horizontally located away from the cluster of regular observations (only exceed the threshold for score distance). We represent these observations with color green in Fig. 1. The third group is denoted as *orthogonal outliers*. These observations are located vertically away from the PCs, meaning that they only exceed the orthogonal distance measure. They correspond to the blue dots in Fig. 1. The projection of these data points on the hyperplanes is aligned with the regular observations. The final group is represented with the color red in Fig. 1 and is denoted as *bad leverage points*. This group is composed of observations that are furthest away from the PCs both in terms of horizontal and vertical distances. In other words, these are the observations that significantly influence the distribution because they represent the most information. For the sake of this study, we will consider the bad leverage points as a proxy for *extreme operational conditions*.

4 Results and Discussion

There are 1914 observations in our dataset each corresponding to 1 hour of traffic control activities at a TC workstation. We use the "rrcov" package in R to conduct our analysis. When we optimize for the ideal number of PCs that would describe the data, we observe that two hyperplanes explain most of the variance, with PC 1 explaining %83.31 and PC 2 explaining %16.69. The cutoff distances that represent the categories the observations belong to are calculated as 2.71 for the score (horizontal) distance and 838.17 for orthogonal distance. By using these cutoff distances, we calculate that 88 observations are classified as *bad leverage points*; these are the observations that we consider as adverse operational conditions. In other words adverse operational conditions occur on %4.59 of all hours in this TCC. The number of orthogonal outliers is calculated as 104, the number of good leverage points is calculated as 81, and finally, the remaining 1641 observations are identified as regular or nominal observations. Figure 2 presents the breakdown of these points on the PCs. In Fig. 2, black dots represent nominal observations, green dots represent good leverage points, blue points represent the orthogonal outliers, and red points represent the bad leverage points.

Now that we have identified the adverse operational conditions, we could take an in-depth look into how they differ from regular data points that represent nominal operational conditions. In Table 2, we present the descriptive statistics for the regular observations that statistically represent nominal operational conditions ($n = 1641$) and the bad leverage points that statistically represent the extreme operational conditions ($n = 88$).

Table 2 indicates interesting insights regarding the operational trends. The first takeaway is that during adverse conditions, as expected, the range of the variables and their standard deviations are much larger, specifically for delays, traffic density, complexity, and anticipation usage. The second insight we observe is that during adverse conditions, the use of the ADAS is at a bare minimum with a median value of zero where far less manual movement decisions are being made. This could be attributed to the inability of the ADAS to perform adaptation decisions. Regardless, further research is necessary to understand the relationship between automation usage and extreme conditions. We don't observe a drastic difference between the amount of phone calls and mental fatigue levels. There is a strong contrast between nominal and extreme cases in terms of traffic complexity and density, which is in line with the observations of INFRABEL's domain experts. We note that the range of these variables are far larger than the median and mean values, indicating that there are outliers among the adverse conditions during which TCs have to handle excessive amounts of traffic complexity.

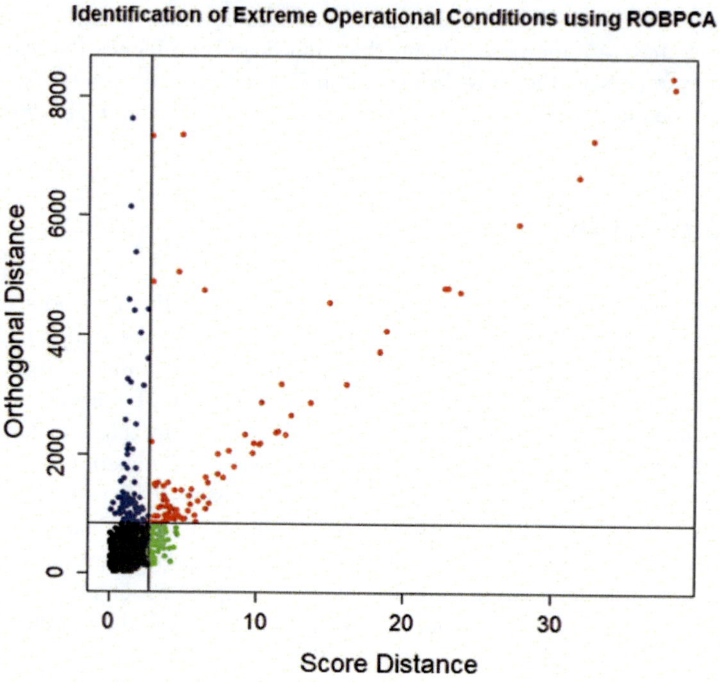

Fig. 2 Identification of extreme operational conditions through ROBPCA

Table 2 Descriptive statistics of nominal and extreme operational conditions

Variable Name	Mean		Median		Range		Standard Dev	
	Nominal	Extreme	Nominal	Extreme	Nominal	Extreme	Nominal	Extreme
Manual Move Decisions	329.50	263.69	295.00	55.00	[0;1150]	[0;1485]	245.95	400.75
Auto Move Decisions	434.20	45.40	435.00	0.00	[0;1355]	[0;780]	278.07	115.06
Adapt Decisions	310.30	341.18	270.00	165.50	[0;1,243]	[10;1,272]	259.49	359.79
Anticipation	16.42	15.11	0.00	0.00	[0;430]	[0;840]	44.13	92.17
Phone Calls	0.43	0.50	0.00	0.00	[0;14]	[0;5]	0.96	0.87
Traffic Complexity	780.84	5,215.63	757.89	3,152.78	[0;2,470.59]	[1,329.27;24,000]	570.79	4,840.46
Traffic Density	808.24	1,814.72	625.00	1,000.00	[0;3,175.82]	[0;6,087.91]	779.08	1,837.84
Fatigue Level	0.86	0.91	0.81	0.79	[0.67;1.37]	[0.67;1.35]	0.15	0.21
Delay	210.13	232.46	185.90	210.44	[-539.67;999.84]	[-7,122.75;5,173.00]	169.13	1,606.63

5 Conclusion and Future Work

As documented in this exploratory research paper, highly intertwined social and technical relationships influence the STS performance. Additionally, operational conditions can vary drastically almost instantaneously. Nevertheless, our approach demonstrated the power of interdisciplinary research in terms of bringing together theoretical and data-driven empirical approaches to pursue insights regarding complex sociotechnical phenomena. Findings of this research support our reasoning that the identification of adverse operational circumstances will lead to an increased better understanding of Controller behavior during these situations and can support the design of future automation tools. Given that we are in the age of data, we believe that there is great potential for systems research to be rooted in reality rather than crude approximations or rough mathematical models. While we haven't focused on use of other data-intensive methods such as deep learning or predictive modeling, availability of rich and large datasets offers great potential for data-driven research. That being said, data-intensive methods should be approached with caution given the complexity of the issues that we are interested in. A robust strategy to avoid misconceptions that could originate from the masking effects in data is to establish collaborations with domain experts and seek verification (at least face validity).

Acknowledgments All intellectual materials discussed in this paper are protected under the nondisclosure agreement between Virginia Tech and INFRABEL. We would like to thank Dr. Renaat van de Kerkhove, Dr. Alex Fletcher, Leslie Steen, and Kristof van der Strieckt from INFRABEL for preparing the data and assisting our research with their feedback. The views expressed in this paper are those of the authors and do not necessarily reflect the opinions of INFRABEL.

References

Balfe, Nora, Sarah Sharples, and John R. Wilson. 2015. Impact of Automation: Measurement of Performance, Workload and Behaviour in a Complex Control Environment. *Applied Ergonomics* 47 (March): 52–64. https://doi.org/10.1016/j.apergo.2014.08.002.

Barling, Julian, Catherine Loughlin, and E. Kevin Kelloway. 2002. Development and Test of a Model Linking Safety-Specific Transformational Leadership and Occupational Safety. *Journal of Applied Psychology* 87 (3): 488.

Battles, James B., and Barbara G. Kanki. 2004. The Use of Socio-technical Probabilistic Risk Assessment at AHRQ and NASA. In *Probabilistic Safety Assessment and Management*, ed. Cornelia Spitzer, Ulrich Schmocker, and Vinh N. Dang, 2212–2217. London: Springer. https://doi.org/10.1007/978-0-85729-410-4_356.

Beehr, Terry A. 2014. *Psychological Stress in the Workplace (Psychology Revivals)*. Routledge.

Blanchard, Benjamin S., and Wolter J. Fabrycky. 2011. *Systems Engineering and Analysis*. 5th ed. Upper Saddle River: Pearson Education.

Cook, R., and J. Rasmussen. 2005. 'Going Solid': A Model of System Dynamics and Consequences for Patient Safety. *BMJ Quality & Safety* 14 (2): 130–134. https://doi.org/10.1136/qshc.2003.009530.

Dawber, Alistair. 2015. Driver Facing 80 Homicide Charges over Spanish Train Crash. *The Independent, October* 8: 2015. http://www.independent.co.uk/news/world/europe/spanish-train-crash-driver-facing-80-homicide-charges-but-rail-bosses-cleared-a6686951.html.

Dugast, Jérôme, and Thierry Foucault. 2018. Data Abundance and Asset Price Informativeness. *Journal of Financial Economics* 130 (2): 367–391. https://doi.org/10.1016/j.jfineco.2018.07.004.

Ferguson, Sally A., Nicole Lamond, Katie Kandelaars, Sarah M. Jay, and Drew Dawson. 2008. The Impact of Short, Irregular Sleep Opportunities at Sea on the Alertness of Marine Pilots Working Extended Hours. *Chronobiology International* 25 (2–3): 399–411.

Filzmoser, Peter, Ricardo Maronna, and Mark Werner. 2008. Outlier Identification in High Dimensions. *Computational Statistics & Data Analysis* 52 (3): 1694–1711. https://doi.org/10.1016/j.csda.2007.05.018.

Folkard, Simon, Karen A. Robertson, and Mick B. Spencer. 2007. A Fatigue/Risk Index to Assess Work Schedules. *Somnologie-Schlafforschung Und Schlafmedizin* 11 (3): 177–185.

Grundgeiger, Tobias, Penelope M. Sanderson, and R. Key Dismukes. 2015. Prospective Memory in Complex Sociotechnical Systems. *Zeitschrift Für Psychologie*.

Helmreich, Robert L. 2000. Culture and Error in Space: Implications from Analog Environments. *Aviation, Space, and Environmental Medicine* 71 (9): NaN–NaN.

Herrera-Restrepo, Oscar, Konstantinos Triantis, William L. Seaver, Joseph C. Paradi, and Haiyan Zhu. 2016. Bank Branch Operational Performance: A Robust Multivariate and Clustering Approach. *Expert Systems with Applications* 50: 107–119.

Huber, Peter J. 1985. Projection Pursuit. *The Annals of Statistics* 13 (2): 435–475. http://www.jstor.org/stable/2241175.

Hubert, Mia, Peter J. Rousseeuw, and Karlien Vanden Branden. 2005. ROBPCA: A New Approach to Robust Principal Component Analysis. *Technometrics* 47 (1): 64–79. https://doi.org/10.1198/004017004000000563.

Hulme, Adam, Neville A. Stanton, Guy H. Walker, Patrick Waterson, and Paul M. Salmon. 2019. What Do Applications of Systems Thinking Accident Analysis Methods Tell Us about Accident Causation? A Systematic Review of Applications between 1990 and 2018. *Safety Science* 117 (August): 164–183. https://doi.org/10.1016/j.ssci.2019.04.016.

Jin, Xiaolong, Benjamin W. Wah, Xueqi Cheng, and Yuanzhuo Wang. 2015. Significance and Challenges of Big Data Research. *Big Data Research, Visions on Big Data* 2 (2): 59–64. https://doi.org/10.1016/j.bdr.2015.01.006.

Keeney, Ralph L., and Howard Raiffa. 1993. *Decisions with Multiple Objectives: Preferences and Value Trade-Offs*. Cambridge.

Kitchin, Rob. 2014. Big Data, New Epistemologies and Paradigm Shifts. *Big Data & Society* 1 (1): 2053951714528481.

Kriegel, Hans-Peter, Peer Kröger, Erich Schubert, and Arthur Zimek. 2009. LoOP: Local Outlier Probabilities. In *Proceedings of the 18th ACM Conference on Information and Knowledge Management, 1649–1652*, CIKM'09. New York: ACM.

Kroes, Peter, Maarten Franssen, Ibo van de Poel, and Maarten Ottens. 2006. Treating Socio-Technical Systems as Engineering Systems: Some Conceptual Problems. *Systems Research and Behavioral Science* 23 (6): 803–814.

Leveson, Nancy G. 2011. Applying Systems Thinking to Analyze and Learn from Events. *Safety Science* 49 (1): 55–64. http://www.sciencedirect.com/science/article/pii/S0925753510000068.

Levin, Alan, and Harry Suhartono. 2019. Pilot Who Hitched a Ride Saved Lion Air 737 Day Before Deadly Crash. *Bloomberg, March* 19: 2019. https://www.bloomberg.com/news/articles/2019-03-19/how-an-extra-man-in-cockpit-saved-a-737-max-that-later-crashed.

Lohr, Steve. 2012. The Age of Big Data. *New York Times* 11 (2012).

Maronna, Ricardo A., R. Douglas Martin, Victor J. Yohai, and Matías Salibián-Barrera. 2019. *Robust Statistics: Theory and Methods (with R)*. Wiley.

National Transportation Safety Board. 2016. Amtrak Train Collision with Maintenance-of-Way Equipment, Chester, Pennsylvania, April 3, 2016. In *Accident Report NTSB/RAR-17/02*. Washington DC: NTSB.

Osorio, Carlos A., Dov Dori, and Joseph Sussman. 2011. COIM: An Object-Process Based Method for Analyzing Architectures of Complex, Interconnected, Large-Scale Socio-Technical Systems. *Systems Engineering* 14 (4): 364–382.

O'Sullivan, Arthur, and Steven M. Sheffrin. 2007. *Economics: Principles in Action*. Boston, MA: Pearson/Prentice Hall.

Rasmussen, Jens. 1997. Risk Management in a Dynamic Society: A Modelling Problem. *Safety Science* 27 (2): 183–213.

Reiman, Teemu, and Pia Oedewald. 2007. Assessment of Complex Sociotechnical Systems – Theoretical Issues Concerning the Use of Organizational Culture and Organizational Core Task Concepts. *Safety Science* 45 (7): 745–768.

Roets, Bart, and Johan Christiaens. 2019. Shift Work, Fatigue, and Human Error: An Empirical Analysis of Railway Traffic Control. *Journal of Transportation Safety & Security* 11 (2): 207–224. https://doi.org/10.1080/19439962.2017.1376022.

Roets, Bart, Marijn Verschelde, and Johan Christiaens. 2018. Multi-Output Efficiency and Operational Safety: An Analysis of Railway Traffic Control Centre Performance. *European Journal of Operational Research*. https://doi.org/10.1016/j.ejor.2018.04.045.

Rousseeuw, Peter J. 1984. Least Median of Squares Regression. *Journal of the American Statistical Association* 79 (388): 871–880.

Rousseeuw, Peter J., and Mia Hubert. 2018. Anomaly Detection by Robust Statistics. *WIREs Data Mining and Knowledge Discovery* 8 (2): e1236. https://doi.org/10.1002/widm.1236.

Russell Neuman, W., Guggenheim Lauren, S. Mo Jang, and Soo Young Bae. 2014. The Dynamics of Public Attention: Agenda-Setting Theory Meets Big Data. *Journal of Communication* 64 (2): 193–214. https://doi.org/10.1111/jcom.12088.

Salmon, Paul M., Neville A. Stanton, Guy H. Walker, Daniel Jenkins, Darshna Ladva, Laura Rafferty, and Mark Young. 2009. Measuring Situation Awareness in Complex Systems: Comparison of Measures Study. *International Journal of Industrial Ergonomics* 39 (3): 490–500. https://doi.org/10.1016/j.ergon.2008.10.010.

Salmon, Paul M., Guy H. Walker, and Neville A. Stanton. 2016. Pilot Error versus Sociotechnical Systems Failure: A Distributed Situation Awareness Analysis of Air France 447. *Theoretical Issues in Ergonomics Science* 17 (1): 64–79.

The Aircraft Accident Investigation Bureau of Ethiopia. 2019. Preliminary Accident Investigation Report of B737-8 (MAX), Registered ET-AVJ. Accident Report AI-01/19.

Tjahjono, Soerjanto. 2018. Preliminary Accident Investigation Report of PT. Lion Mentari Airlines Boeing 737-8 (MAX); Registered PK-LQP. Accident Report KNKT.18.10.35.04. Jakarta: Komite Nasional Keselamatan Transportasi (KNKT).

Topcu, Taylan G., and Bryan L. Mesmer. 2018. Incorporating End-User Models and Associated Uncertainties to Investigate Multiple Stakeholder Preferences in System Design. *Research in Engineering Design* 29 (3): 411–431.

Topcu, Taylan G., Konstantinos Triantis, and Bart Roets. 2019. Estimation of the Workload Boundary in Socio-Technical Infrastructure Management Systems: The Case of Belgian Railroads. *European Journal of Operational Research* 278 (1): 314–329.

Wilson, John R. 2000. Fundamentals of Ergonomics in Theory and Practice. *Applied Ergonomics* 31 (6): 557–567.

Wold, Svante, Kim Esbensen, and Paul Geladi. 1987. Principal Component Analysis. *Chemometrics and Intelligent Laboratory Systems* 2 (1–3): 37–52.

Wong, M. Anthony, and Tom Lane. 1983. A Kth Nearest Neighbour Clustering Procedure. *Journal of the Royal Statistical Society. Series B (Methodological)* 45 (3): 362–368. http://www.jstor.org/stable/2345405.

Dynamic Causal Hidden Markov Model Risk Assessment

Michael Sievers and Azad M. Madni

Abstract Understanding system vulnerabilities to risk factors during operation is essential for developing dependable systems. By implication, assessing in-use risk factors requires monitoring system parameters that contribute to making probabilistic inferences. We argue, however, that naïve use of statistical data without regard to causality can yield surprising and often erroneous risk predictions. Making reliable risk predictions is further complicated by lack of full awareness of system states and the existence of unobservable parameters in complex systems. Overly conservative risk assessment leads to increased life-cycle cost and reduced system availability resulting from overly aggressive preventive maintenance or replenishment strategies, while overly optimistic risk assessment can lead to even higher life-cycle cost and potential harm when otherwise preventable failures occur. This paper discusses a causality-aware, dynamic risk assessment model based on hidden Markov model construct. This model employs the concept of hidden system states that account for otherwise unexplainable observations. The model is continuously evaluated during system operation and updated when new observations warrant reevaluation.

Keywords Risk assessment · Markov model · Causality · Causal modeling · Probabilistic inference

1 Introduction

> Telling a risk analyst to 'just specify the likelihood,' is like telling a homeless person to 'just get a house' (Ferson 2005).

M. Sievers (✉) · A. M. Madni
University of Southern California, Los Angeles, CA, USA
e-mail: michael.sievers@usc.edu

A. M. Madni et al. (eds.), *Recent Trends and Advances in Model Based Systems Engineering*, https://doi.org/10.1007/978-3-030-82083-1_13

Nomenclature

a State transition probability
β Covariance of two system events
O Set of observations
π Initial state distribution
PN Probability necessity
S Finite set of states

Risk analyses methods generally predict events that can potentially occur and the impact of those events on system behavior. A commonly used approach by NASA and the US Department of Defense assesses and tracks risk by assigning qualitative values to the likelihood and severity of events (DoD 2014; NASA 2010). An obvious question is whether the relevant data needed for risk assessment is available, especially when the events of interest may have minimal or no prior history of occurrence (Huff 1954). Furthermore, the events selected for analysis are usually based on analyst judgment and reflect analyst's biases, not on hard evidence. When complex systems are involved, uncertainty in the models used may mask system realities, thereby resulting in questionable potentially misleading conclusions.

Several authors have noted that Bayesian modeling methods can potentially help in understanding the true nature of events and event impacts (Ferson 2005; Homayoon 2009; NASA 2010; Baru 2016). When applied appropriately, Bayesian models can accurately converge on the right parameters which influence or are indicators of risk. This is an expected outcome in that Bayesian models account for both parameter and model uncertainties. However, the proper application of the Bayesian approach needs to clearly distinguish between correlation and causation.

Briefly, Bayes' theorem, $P(A \;\&\; B) = P(A|\, B)P(B)$, has been successfully applied in multiple, disparate domains. Of course, if improperly applied, it can lead to surprising, potentially erroneous conclusions. Consider the oft cited example of ice cream sales and drowning. If the data collected considers only number of ice cream sales and number of drownings, then Bayes shows a correlation between increased ice cream sales and people drowning. That is, if A is "drowning" and B is "ice cream sales," then as the priori, $P(B)$, and likelihood, $P(A|\, B)$, increase so does the apparent correlation $P(A \;\&\; B)$. Obviously, this is flawed reasoning because both ice cream sales and people swimming increase in hot weather. Bayes is not at fault here; rather, it is that the "wrong" data set was used in the analysis. While finding correlations is relatively easy, understanding causality is far more difficult (Pearl 2001, Pearl 2009).

At the heart of most forms of risk assessment are so-called weak assertions of the form: if event A occurs, then event B occurs. If B occurs, then there is a higher probability that A also occurs. Weak assertions are expressed by Bayes:

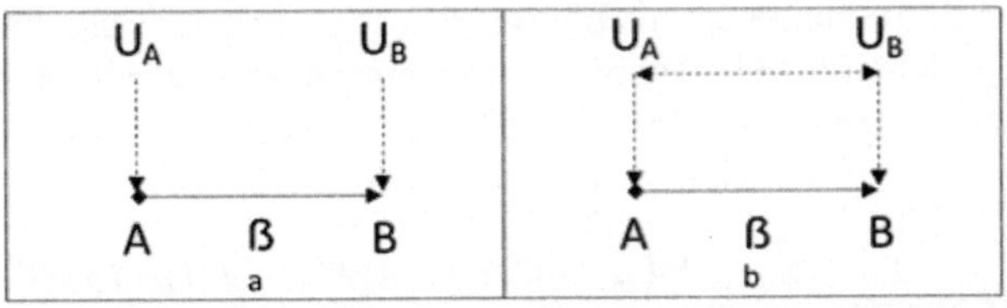

Fig. 1 Two simple models of causality in which exogenous variables U_A and U_B are connected to each other and to endogenous variables A and B with dashed lines. ß represents the direct effect A has on B

$$P(A|B) = P(A)\frac{P\left(B|A\right)}{P\left(B\right)} \tag{1}$$

and are the basis for probabilistic risk assessment (PRA) (NASA 2010).

When considering risk though, simply observing a collection of parameters without understanding causality can lead to false alarms or misinterpretation of potentially hazardous situations. While this seems reasonable, unfortunately, in complex systems, requisite observations and state knowledge for making sound decisions may not be available. Moreover, systems that degrade with time may invalidate priors that frequentists depend on or prevent proper updating of subjectivists' beliefs.

Pearl (2009) describes a causality construct that represents the probability that an event, $B=b$, will occur whenever action, $A=a$, is enforced over the entire population as $P(B = b|\, do(A = a))$. In essence, $do(A)$ implies a controlled experiment with randomized A.

Pearl shows that causality can be associated with directed graphs in which nodes represent observed or unobserved system factors connected by a term that represents the causal effect of one factor on another. The model comprises so-called exogenous variables that are not influenced by other system variables but have an impact on other system variables called endogenous variables. Figure 1 shows two simple examples based on Pearl's paper. In Fig. 1a, $Cov(A, B) = \beta$ and in Fig. 1b, $Cov(A, B) = \beta + Cov(U_A, U_B)$. Note that in some situations $Cov(U_A, U_B) = 0$ in which case the covariance is β as in Fig. 1a.

The evaluation of probabilities needed for causality needs more care than simply collecting data and looking at frequencies of occurrence. For example, Bayesian analyses are strongly influenced by the assumptions made on prior probabilities as shown in Eq. 1. As sample size increases, the sensitivity to those priors is reduced. However, in the case of causality, sensitivity to prior causal assumptions remains strong regardless of sample size. Moreover, hidden and indirect effects confound faithfully representing the relationships between events and actions. Recalling the example of eating ice cream and drowning, a naïve statistical analysis will conclude a strong correlation.

One solution for reducing the likelihood of false correlations uses Pearl's concept of probability necessity, PN. Under the assumption that event, A, is monotonic relative to action, B, then

$$PN = \frac{P(A|B) - P\left(A|B'\right)}{P(A|B)} + \frac{P\left(A|B'\right) - P\left(A|do\left(B'\right)\right)}{P(B, A)} \tag{2}$$

Equation 2 subtracts the likelihood that event, A, occurs even when action B does not. In the case of eating ice cream, the likelihood of drowning will be roughly the same regardless of ice cream consumption which eliminates confounding and incorrect bias.

2 Risk

Loosely, risk assesses the likelihood some event will occur and the impact that event has on a system or on a system's environment. Assessments run the gamut of subjective analyses by subject matter experts (SMEs) to more rigorous and formal mathematical constructs. SME risk assessments are essential during system formulation, design, and test phases because hard data are usually not available. While far from perfect, methods have been created that help mitigate the impact of SME bias and incorrect assumptions that often underlie subjective assessments. Also, while some systems are heavily instrumented for post-deployment data collection, that data may not always be useful for evaluating the cause of a particular event or the probability that an unexpected and dangerous event is likely to occur in the near future.

Dynamic assessment of system state from post-deployment data can provide insights into design weaknesses and aid in scheduling maintenance and replenishment activities. This is not a matter of simply collecting large quantities of data and doing a statistical analysis because dependencies in complex systems can be difficult to untangle. For example, suppose there is a risk of event occurring when exogenic variable $P1 > x$ but never when endogenic variables $P2 < y$ and $P3 = true$. A correct assessment of event risk depends on knowing the correlation of $P1$ to $P2$ and $P3$. Knowing only that an event occurs based on the value of $P1$ is akin to correlating increased ice cream sales to drowning while disregarding the correlation with increased swimming and summer temperatures.

2.1 *Hidden Markov Causality Risk Model*

Traditionally, risk assessments are used by managers and engineers in tradespace and early design evaluations and focus attention on design changes needed for removing or reducing the likelihood of serious, undesirable future events. While using risk assessments in the design process is essential, it is equally important to understand post-deployment system vulnerabilities. That is, systems must be monitored during operation so that risks of serious or dangerous events can be estimated. As previously noted, naïve data collection without consideration of causality will not suffice as a reliable predictor of risk. What is needed is the creation of models in which prior probabilities account for probability necessity as in Eq. 2.

A state machine construct is a natural model for evaluating system risk in which states represent a notion of vulnerability, transitions occur as the result of system events, and outputs result from system state and system events. In an ideal world, the Markov property holds, i.e., the conditional probability distribution of future states depends only on the present state and events and not on the trajectory taken to arrive at that state. When the Markov property holds, we can create an understanding of the state space and the probability distribution for future states based on straightforward statistical analysis of observations made.

In the real world, there might be ambiguity in the knowledge of system state or uncertainty that an event will have the expected result. That is, some aspects of a system may be unobservable or hidden. Hidden Markov models (HMMs) accommodate uncertainty by including hidden states associated with initially unknown observation and transition distributions. HMMs are trained during system use and over time refine distributions by evaluating how well a model predicts system behavior. In essence, continuous model updates are a Bayesian process that improves HMM priors and consequently improves the reliability of the predictions made by the model.

HMMs for real-time vulnerability assessment of network cyberattacks are not new (Årnes et al. 2005; Liu and Liu 2016). Conceptually, these assessments involve the creation of a HMM-based attack graph in which states represent a method of attack and transition probabilities reflect the difficulty or vulnerability of an attack causing an unwanted operational change. Monitors collect an observation sequence that is used for evaluating how well the HMM predicts that sequence but can also be used in determining a state probability distribution (belief state). That is, given an observation sequence, O, it is possible to predict $P(O|\,model)$ using the *forward algorithm* as well as the belief state distribution after observing the sequence, O. Additionally, the state sequence can be determined using the Viterbi algorithm which determines the most probable path the model takes as each observation is made. The state sequence is useful in understanding the events that caused the system to arrive in its current state, that is, it provides the notion of causality. Additionally, if $P(O|\,model)$ is below a threshold, then it is likely that there is a new hidden state at play or the model parameters need adjustment.

In a similar vein, a more general risk model can be created. This model comprises known states that represent system conditions, transitions, and observations associated with system conditions. The model is augmented with hidden system conditions and initially unknown transition and observation probabilities. For example, Fig. 2 shows a HMM comprising two known and two hidden system conditions. The

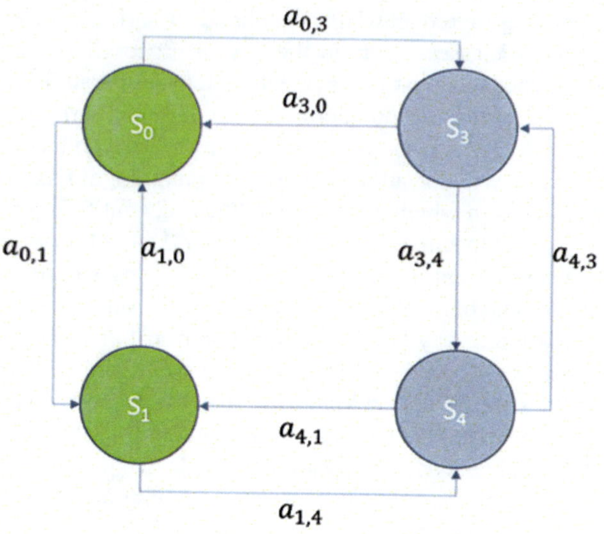

Fig. 2 A four-state Markov model comprising two observable states, S_0 and S_1, and two hidden states, S_3 and S_4

transition and observation probabilities associated with the hidden states must be nonzero but can be arbitrarily small.

A HMM is conventionally defined by:

- A finite set of states, $S = \{s_0, s_1, \ldots s_{n-1}\}$; the state at time, t, is q_t.
- A set of observations, $O = \{O_0, O_1, \ldots O_{T-1}\}$.
- State transition matrix, A, in which element $a_{i,j} = P(q_{i+1} = s_j \mid P(q_t = s_i)$.
- Observation distribution matrix, B, in which $b_j(k) = P(o_k | q_t = s_j)$ where $0 \le j \le n - 1$ and
  $0 \le k \le T$.
- An initial state probability distribution, π, in which $\pi_j = P(q_0 = s_i)$ for $0 \le i \le n - 1$.

2.2 *Assessing Risk*

In evaluating risk, HMM states represent hazard conditions, e.g., the condition that pressure in a tank exceeds a specified threshold. State transitions are determined by substituting the HMM parameters into Eq. 2. Equation 3 computes $a_{i,j}$ by considering whether there is a causal link between s_i and s_j if observation, o, occurs while in s_j:

$$a_{i,j} = \frac{P(s_j|s_i, o) - P(s_j|s_i', o)}{P(s_j|s_i)} + \frac{P(s_j|s_i', o) - P(s_j|do(s_i', o))}{P(s_j, s_i)} \tag{3}$$

The HMM is developed by choosing a set of hazard conditions either randomly or through analyses such as fault tree or branch termination. Hidden states are then added and connected to the initial state set. Hidden state transition and observation probabilities are assigned nonzero, but low values so that they do not exert undue influence on model parameter initialization. However, the consequence of overly high values is that model parameter convergence could take more iterations.

The initial risk algorithm comprises five steps:

1. Determine initial values for A, B, and π; these may be set randomly if initial values are unknown.
2. Collect observations and update the initial model Baum-Welch (Baum and Petrie 1966) using Eq. 3 for evaluating transition updates.
3. Given an observation sequence, O, compute $P(O|\ model)$ using the forward algorithm, i.e., determine whether the observations match a risk scenario predicted by the model.
4. Given the state distribution determined in Step 3, use the model to predict the probability of transitioning to another risk.
5. Go back to Step 2 until $P(O|\ model)$ exceeds a threshold.

The algorithm changes once $P(O|\ model)$ is above a threshold. That is, observations are made, $P(O|\ model)$ is computed, and a risk prediction is made. $P(O|\ model)$ below a threshold is an indication of a novel condition that requires returning to Step 2. Figure 3 shows the risk algorithm flow diagram.

3 Observation Clusters

Making observations to evaluate risk in a real system is more complicated than simply collecting data from monitors. Factors such as noise, faulty monitors, and transient events can potentially create variances that need accommodation without necessarily adding to an already large state space. Moreover, some monitors may have greater influence on the state space than others in certain system operational modes. For example, a fault in an entry-descent-landing (EDL) subsystem during the early cruise phase to Mars is less important than the same fault occurring in the EDL activity.

Dealing with transient effects and certain types of noise is readily managed by requiring persistence on monitor samples. Modal information is collected as a data point in an observation. That is, rather than trust that a commanded mode has been achieved, for the purposes of risk, we depend on correlation of mode-dependent variables that represent the mode the system is actually in. Note too that variations in mode-dependent variables are also likely.

However, normal variations in samples imply the need for n-dimensional clustering in which each snapshot of monitor values is compared with a distance to observation clusters. Snapshots within a cluster are characterized by the cluster centroid. Ambiguous or unassignable snapshots are considered novel and trigger

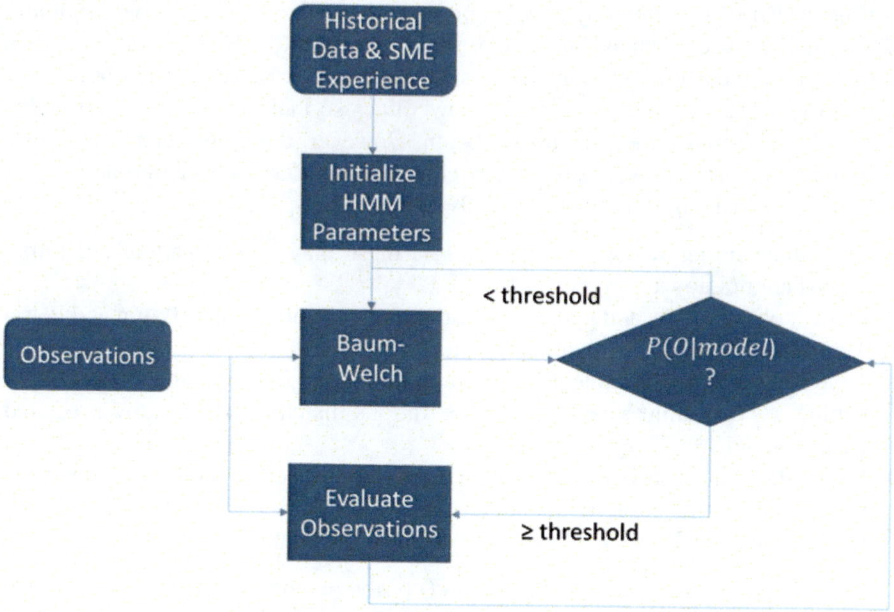

Fig. 3 Basic risk evaluation algorithm including learning iteration and observation-based risk evaluation

both a reevaluation of the cluster space and, as needed, execution of the Baum-Welsh algorithm for updating model parameters.

Not shown in Fig. 3 is the cluster step that occurs during initial observations made for updating the initial HMM parameters. Because clusters may not be known initially, clustering is performed using an expectation maximization-Gaussian mixture model (GMM) (Dempster et al. 1977). Briefly, GMM is initialized by choosing a set of clusters and randomly assigning a mean and distribution to each cluster. After initialization, the probability that each data point belongs to that cluster is computed by evaluating the proximity to the cluster centroid. The results are then used to update the clusters and repeat the probability evaluations until the probability distributions convergence.

During operation, the Mahalanobis distance from the observation to the clusters determines whether an observation belongs to a cluster. An observation is subsequently classified by the mean of the cluster it belongs to. We should note too that clusters likely will change with time, especially as the system encounters new usage, new environments, and changes. For this reason, when practical, offline GMM is periodically performed to update the cluster definitions. In this regard, it may be necessary to include heuristics for assessing the importance of certain monitor values when offline GM is not practical. Using the EDL example from above, it might be necessary to "disable" certain clusters when they no longer apply, e.g., during EDL any cluster related to cruise operation is not applicable, and any observation that would fall into a cruise-mode cluster now falls into an observation associated with an active fault or a fault vulnerability.

4 Conclusions and Future Prospects

The prevalence of autonomous systems in automotive, aircraft, military, space, and commercial sectors is increasing making human-in-the-loop assessment of risk less and less viable. Moreover, with increasing complexity, classical diagnosis methods such as fault dictionaries that match syndromes to cause become less reliable due to false or unaccounted-for correlations. The upshot is that maintaining future systems will either become unacceptably expensive due to false alarms or, worse yet, systems will become vulnerable to serious but unpredicted risks.

In this paper, we have defined a modeling construct and assessment algorithm that, once trained, will provide a causality-aware assessment of risk. Our approach has the advantage of reducing the influence of false correlations, thereby enabling a more accurate understanding of system health. Moreover, there is a built-in learning process that adds new hidden states or adjusts model parameters when needed to explain a novel set of observations.

A distinguishing feature of our approach is that it relies less on individual observations and more on whether a sequence of observations fits a causality pattern. When a pattern is recognized, the model can provide a probability of the pattern as well as the probability of escaping to another pattern. Given both pieces of information, system operators can then decide whether and when repair or replacement is needed. For example, is it necessary to ground an airplane now due to a high probability of a near-term, serious fault condition, or can the airplane complete a mission and receive service later? Additionally, knowing with confidence failure risk simplifies maintenance scheduling and acquisition of spares.

It is well-known, however, that state-based models can be very large and difficult, if not practically impossible, to analyze. General approaches to managing large models comprise breaking them up into smaller models and/or using heuristics that approximate completely rigorous analyses. An issue we have not yet addressed is the impact on causality when decomposition or approximations are used. We understand that practical use of our approach will necessitate a thorough evaluation of state-space explosion.

Our primary work-to-go is to apply this concept to a realistic problem. To that end we have created an unmanned aerial vehicle (UAV) simulation in which we can "fly" multiple UAVs that are tasked with completing a reconnaissance mission. The simulation allows an arbitrary number of monitors and also allows injecting noise, transient upsets, and failures (Madni 2019).

To test-drive these concepts on realistic problems, we have created a minimum viable testbed (Madni 2019). The testbed employs an open-source infrastructure, multiple modeling and simulation methods, a library of components for rapid scenario development, software and hardware building blocks, and an open-source repository. The testbed employs an open, extensible architecture, with the ability to incorporate both virtual models and physical systems. This testbed is different from traditional hardware-in-the-loop testbeds that employ proprietary models and focus on specific system instantiation. We intend to report our findings from testbed experimentation in a follow-on paper.

References

Årnes, A., K. Sallhammar, K. Haslum, T. Brekne, M.E.G. Moe, and S.J. Knapskog. 2005. Real-Time Risk Assessment with Network Sensors and Intrusion Detection Systems. In *Computational Intelligence and Security. CIS 2005*, Lecture Notes in Computer Science, ed. Y. Hao et al., vol. 3802. Berlin/Heidelberg: Springer.

Baru, S. 2016. Bayesian Network Based Dynamic Operational Risk Assessment. *Journal of Loss Prevention in the Process Industries* 41.

Baum, L., and E. Petrie. 1966. Statistical Inference for Probabilistic Functions of Finite State Markov Chains. *The Annals of Mathematical Statistics* 37 (6): 1554–1563.

Dempster, A., N. Laird, and D. Rubin. 1977. Maximum Likelihood from Incomplete Data via the EM Algorithm. *Journal of the Royal Statistical Society, Series B* 39 (1): 1–38.

Department of Defense Risk Management Guide for Defense Acquisition Programs, 7th Edition, 2014

Ferson, S. 2005. Bayesian Methods in Risk Assessment. Unpublished Report Prepared for the Bureau de Recherches Geologiques et Minieres (BRGM), New York.

Homayoon, D., et al. 2009. Bayesian Inference for NASA Probabilistic Risk and Reliability Analysis. NASA Technical Report NASA/SP-2009-569, June 2009.

Huff, D. 1954. *How to Lie with Statistics*. New York: W.W. Norton & Company.

Liu, S., and Y. Liu. 2016. Network Security Risk Assessment Method Based on HMM and aTtack Graph Model. In *Proceedings of the 17th IEEE/ACIS International Conference on Software Engineering, Artificial Intelligence, Networking and Parallel/Distributed Computing*, 517–522. New York: IEEE.

Madni, A.M. 2019. Minimum Viable MBSE Testbed for Exploring Models and Algorithms for System Resilience and Risk Assessment, SAE-TR-01/05/2020.

NASA Risk-Informed Decision-Making Handbook, NASA/SP-2010-576, 2010

Pearl, J. 2001. Causality Models, Reasoning, and Inference, Cambridge University Press, ISBN 0-521-77362-8.

———. 2009. Causal Inference in Statistics: An Overview. *Statistical Surveys* 3: 96–146.

Part III
Use of Ontologies in MBSE

Minimum Viable Model to Demonstrate Value Proposition of Ontologies for Model-Based Systems Engineering

Azad M. Madni

Abstract With increasing connectivity and digitalization, systems continue to become increasingly more complex. To meet this challenge, the model-based systems engineering community has begun exploring the use of ontologies to scope the modeling effort and demonstrate the value of MBSE without resorting to full-blown modeling. To this end, this paper presents a minimum viable model (MVM) approach to system modeling. In the MVM approach, a system model with the requisite structure and just enough semantics is created to resolve semantic inconsistencies in the model, achieve interoperability, and answer a few key questions at the right level of detail posed by stakeholders from the systems acquisition and engineering communities. The larger intent is to have potential customers buy into the viability of an ontology-enabled approach to MBSE.

Keywords Ontology · Metamodel · Semantic model · Model syntax · Digital engineering · MBSE

1 Introduction

Systems engineering (SE) is undergoing a transformation in response to the growing complexity of systems resulting from increasing system scale, interconnectedness, and digitalization of enterprises. Two potential enablers of such transformation are ontologies (Madni et al. 2001; Kaiya and Saeki 2005; Mayk and Madni 2006; Madni et al. 1998, 1999, 2002; Gruninger and Lee 2002; Wand 1996; Orellana and Madni 2014; Sievers 2019) and metamodels (Saeki and Kaiya 2002). An *ontology* is a thesaurus of words (which represent concepts), the relations among them, and the rules that help with model correctness checking. In other words, these model-checking rules help with identifying lack of model elements (i.e., gaps) and

A. M. Madni (✉)
University of Southern California, Los Angeles, CA, USA
e-mail: azad.madni@usc.edu

A. M. Madni et al. (eds.), *Recent Trends and Advances in Model Based Systems Engineering*, https://doi.org/10.1007/978-3-030-82083-1_14

semantic inconsistencies. A *metamodel* defines the abstract syntax (i.e., grammar) of model description languages. For example, the UML metamodel defines the abstract syntax for various UML diagrams. More generally, metamodels express the logical syntactical structures that domain-specific models need to conform to for scalability, reuse, and extensibility. However, metamodels do not specify the semantics of the model. In this regard, the work of Saeki and Kaiya (2002) shows how ontologies can provide the semantics for domain-specific models and metamodels.

2 Minimum Viable Model (MVM)

In the past few years, the SE community has begun exploring the use of ontologies in MBSE to reduce complexity and facilitate interoperability and reuse. However, the few ontology-related SE initiatives that were undertaken in the past few years did not produce the expected outcome. A key lesson learned from these projects is that modeling a complex system with no simplifying assumptions or guiding paradigm can quickly become a never-ending modeling task with no tangible results to show. What is needed is a more pragmatic approach with clear goals and reduced modeling scope to show tangible results. The minimum viable model (MVM) approach is intended to accomplish this objective. A MVM is a model with just enough semantics and structure to demonstrate ability to resolve semantic inconsistencies in the model, demonstrate interoperability, and answer customer questions at the right level of detail posed by stakeholders without resorting to full-blown modeling. One target customer for MVM is the systems acquisition and engineering communities. The larger intent is to have potential customers buy into the viability of an ontology-enabled approach to MBSE.

The MVM approach is concerned with identifying the essential system elements and their relationships that underpin representative use cases (which define and limit the scope of the modeling problem) and help answer questions associated with them. The key idea with MVM is to focus on developing the requisite structure and minimum vocabulary needed to represent the selected use cases. The use cases, for example, can be defined in conjunction with representatives from the DoD systems acquisition and systems engineering communities, as well as from industry. What uniquely differentiates this approach is the emphasis on a minimum vocabulary set and a minimum requisite representation (using the minimum vocabulary set) that can represent the core concepts and their relationships described or implied in the use cases. The value proposition can then be demonstrated in terms of ability to (a) search and query models; (b) perform correctness analysis; (c) add new concepts and relationships; (d) achieve model scalability, interoperability, and reuse; and (e) generate custom documentation on demand.

The MVM approach is essentially concerned with creating a system representation that is a slice of the total system model to demonstrate the value of ontologies in MBSE. It leverages ontologies and metamodels to realize a scalable system model with several desirable properties (e.g., reduced modeling effort, semantically and syntactically correct model, potential for reuse, interoperability).

So, what is a minimal system representation? A minimal system representation is one that has just enough semantics (vocabulary) to represent the system, facilitate search, and answer questions associated with the selected use cases in the domains of interest. It has just enough structure and grammar to organize words and symbols in a logical way. The reason we need a minimal representation is to limit the modeling effort and minimize model complexity when demonstrating the value proposition of an ontology-enabled approach to SE.

3 Ontologies and Metamodels in MVM

Ontologies have been used in a variety of applications such as requirements analysis (Kaiya and Saeki 2005), metamodeling (Wand 1996), enterprise integration (Fox and Gruninger 1994), interoperability (Gronmo and Oldevik 2005), model-driven engineering (Guarino and Welty 2000), ontology merging (Noy and Musen 2000), formal concept analysis (Stumme 2005), architecture development (Terrasse et al. 2002), and systems engineering (Madni et al. 2001; Mayk and Madni 2006; Madni et al. 1998, 1999, 2002; Van Ruijven 2014). The interest in ontologies has surged in the past 3 years as systems have become more increasingly complex and as digitalization is being pursued in large-scale enterprises.

Ontology and metamodel are complementary and synergistic concepts with respect to system modeling (Parreiras et al. 2007). Quite simply, an *ontology* represents concepts (i.e., classes) and their relationships, along with rules for model correctness checking (Sowa 2001). The model-checking rules help with identifying lack of model elements (i.e., gaps) and semantic inconsistencies. *Metamodeling* is concerned with defining the symbols and structure for a predefined class of problems, along with rules that operate on the symbols. These properties allow the instantiation of a model from a metamodel. A metamodel is itself a model that is used to describe a predefined class of problems using a modeling language. In other words, a metamodel defines the general structure, constraints, and symbols that can be used to model a system. A metamodel by itself has no practical value until it is used to create a model. That is, a metamodel does not assign semantics (i.e., meaning) to the symbols and rules. The latter are under the purview of an ontology. Thus, ontologies and metamodels are complementary concepts. The following discussion should further clarify these two concepts.

An ontology can represent concepts and relationships formally using the structure provided by the metamodel. While ontologies are not required to have a metamodel, those that do are more formal and have certain desirable properties (e.g., scalability, reuse, interoperability). Not having a metamodel for an ontology is akin to saying you can write sentences in a domain using an informal language that may work fine for the limited problem at hand but that may not be grammatically correct (i.e., abide by the metamodel).

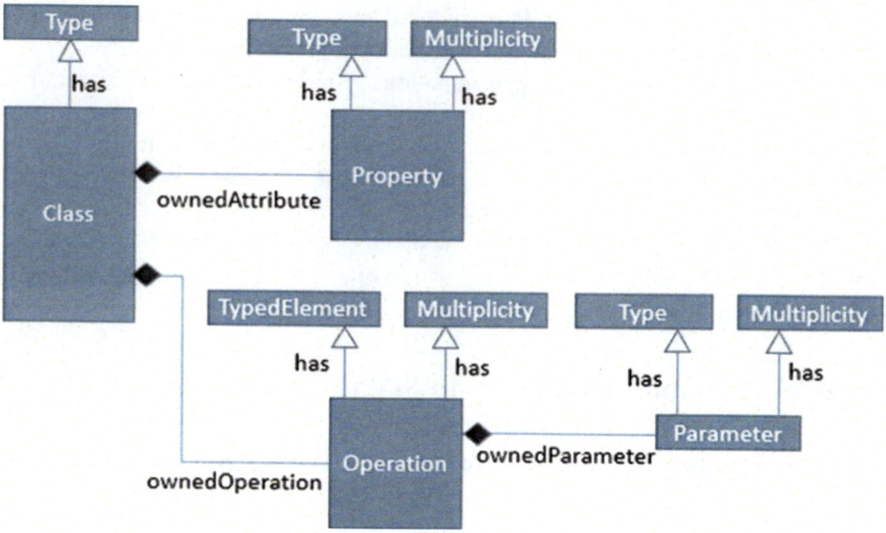

Fig. 1 Metamodel for a generic programming language

The following example clarifies the difference between a metamodel and an ontology. Consider the symbols used to create a map. They typically include lines, arrows, dots, boxes, colors, etc. These symbols and the rules for connecting them are under the purview of the metamodel. However, by themselves, these symbols have no meaning. For example, if we were to draw a map without a legend, we would not understand the meaning or purpose of the diagram. The same lines, colors, boxes, etc. could be used to define a variety of maps such as a voting map, bicycle routes, or population density. When we assign meaning to the symbols, we are creating an ontology for the domain. Moreover, we can use formal languages for creating ontologies. Such languages allow semantic reasoning and analysis.

Figure 1 presents a metamodel for a generic programming language. This metamodel consists of basic programming concepts that can be used in any software program. In this diagram, class has type, property has type and multiplicity, operation has typed element and multiplicity, and parameter has type and multiplicity. Property is an owned attribute of class, and parameter is an owned attribute of operation. And, finally, operation is an owned operation of class.

Now, let us apply this metamodel to create domain-specific ontologies (Fig. 2). In this example, the voltage divider assigns electrical interpretations to the general metamodel construct to create a voltage divider class. The electrical ontology in this case consists of electrical parameters. Similarly, the physics ontology shown in the figure comprises concepts of mass, acceleration, and force and implements Newton's second law. The ontology in this case comprises physics parameters.

As should be evident from the preceding examples, a metamodel defines syntactical constructs. Consequently, it is not applicable to reasoning within a semantic domain. Thus, it is important to understand the applicability and limitations of both

Fig. 2 Examples of metamodel used to create two domain-specific ontologies

Class: VoltageDivider

```
divide (v, r1, r2){
float v;
float r1;
float r2;
float vout =
v*r1/(r1+r2);
return vout;
}
```

Ontology consists of electrical parameters

Class: SecondLaw

```
2ndLaw (m,a){
float m;
float a;
float force = m*a;
return force;
}
```

Ontology consists of physics parameters

metamodels and ontologies. To begin with, it is difficult to create a metamodel or ontology that covers all possible architectures. For example, system models based on the SysML-centric construct suffer from the same limitations as SysML. In particular, the current metamodel does not appear helpful in developing system-of-systems (SoS) architectures in which connectivity, functionality, and purpose are not known until the need arises. Moreover, with a SysML foundation, the current modeling construct is not useful for creating architectures for hard real-time systems, dynamically reconfigurable systems, nondeterministic systems, systems comprising transient functions, complicated state models, and human-system models.

It is important to note that this limitation may not surface if the architecture domain is limited to what SysML can model today. That is, define an ontology for a specific type or class of architecture in which the goal is to define the system's appearance. For example, suppose we want to build a spacecraft. An essential aspect of the architecture is "grounding." Other important characteristics are total incident dose (TID), single-event effect (SEE) tolerance, thermal interfaces and thermal architecture, propulsion architecture, attitude determination, and control architecture (not just the Automated Data Capture System (ADCS) boxes, but the mathematical architecture), for example. Conversely, the metamodel could be used for defining structures but due to its SysML foundation lacks rigorous semantics for time, time synchronization, and models of computation. Given the scope and size of the required metamodel, it is not entirely obvious what value it can provide beyond that of being an academic exercise. Without adequate tools that facilitate architecture creation using the metamodel and then reasoning and analyzing those architectures, there would be little to gain for the effort expended to create a usable metamodel.

4 MVM Modeling Heuristics

In light of the foregoing, the MVM approach is focused on reducing the complexity of the modeling problem by creating a narrowly scoped ontology that is sufficiently rich to demonstrate the value proposition of ontologies for MBSE. In this regard, modeling heuristics can help in circumscribing the modeling scope and in choosing the right problem. The following heuristics can be used to inform and guide problem selection and model scoping.

Heuristic #1: If you do not know the domain and you do not know what you want to know about the domain, you are not ready to start.

If one knows the domain and the concerns associated with performing work in that domain, then one can define the questions that need to be answered; otherwise, one is not ready to start. With domain knowledge, stakeholder use cases can be defined and employed to develop an ontology. Preferably the ontology conforms to a metamodel.

Heuristic #2: Define the simplest possible problem/system that is sufficiently rich to convey the benefits of an ontology-enabled SE approach over the status quo.

This heuristic assures that system model development does not become a never-ending exercise, and the benefits can be demonstrated in a relatively short time.

Heuristic #3: Employ selective fidelity in models to limit the modeling effort.

Not all sub-models require the same level of fidelity in the overall model to provide useful results. What this means is that certain sub-models can be abstracted, while others need more details. For example, in an electromechanical system, if the intent is to study the behavior of the mechanical system, then the electrical system can be abstracted.

Heuristic #4: Choose a well-defined objective and a "small" problem to demonstrate impact in 4 to 6 months.

The problem selected should be of interest to the customer organization, be off the critical path of the organization's core processes (i.e., failure does not adversely affect customer organization's ongoing operations), and have high leverage (i.e., success in one area can be leveraged in other areas because it is a frequently occurring problem).

Heuristic #5: The solution for the status quo should be derivable readily, or by brute force, if necessary, to provide a comparison yardstick.

A comparison yardstick is essential to convey the value proposition. Comparison against the status quo is the most straightforward way to convey the value proposition of the use of ontologies in MBSE.

Heuristic #6: Define the metrics that will be used for comparison ahead of time.

It is much easier to define the metrics for comparison of MVM against the status quo ahead of time than after the fact.

Heuristic #7: Probe model development objectives to ensure that objectives are not being confused with options.

People inadvertently can adopt an option as an objective when developing a system model. As a result, one can end up solving a subproblem. To prevent this from happening, it is important to probe the initial objective for a higher-level objective. If such an objective is found, adopt that as the new objective. Then repeat the process until no higher-level objective is discovered. At that point, it can be concluded that one has the right objective.

Heuristic #8: Explicitly state assumptions and other factors that define problem context before defining use cases.

This requirement will ensure comprehensive formulation of use cases with no hidden implications.

Heuristic #9: Capture stakeholder concerns in viewpoints that create views into the model.

Viewpoints are associated with views into the model. There is typically a hierarchical structure behind the views which allows abstraction or going into depth in the models. Viewpoints represent information that stakeholders need and imposes constraints on what they want to view.

Heuristic #10: Explicitly identify the dimensions in which the model should scale, and make sure that the MVM will scale along those dimensions.

The key dimensions along which the model should scale include the number of nodes, number of links, and number of components/subsystems. Also, there should be loose binding between function and technology to ensure ability to incorporate new technology.

Heuristic #11: Focus on demonstrating value proposition through unique capability, not accuracy of results.

Accuracy of results requires high-fidelity models which requires significant effort. When the focus is on rapidly conveying the value proposition, it pays to have the minimum model to convey value.

Heuristic #12: Make sure the use cases explicate hidden or implied trade-offs.

This capability can be a key differentiator of the MVM approach if these trade-offs can be illuminated without full-blown system model development.

In the MVM approach, the above exemplar heuristics can be used to choose and simplify an appropriate problem in the domain of interest. The basic idea then is to show the solution with the status quo, and then with MVM, and compare the two using appropriate metrics (e.g., model scalability, model reusability, model extensibility, model verification speed, and interoperability with third-party ontologies). The MVM, in essence, is a "thin slice" of the total capability of the envisioned system that can convince potential customers of the value proposition of ontology-enabled SE.

5 A Real-World Interoperability Problem

Today system modeling is becoming an increasingly important activity within companies that build systems and products. In fact, detailed system models are viewed as valuable knowledge assets and a source of competitive advantage. As digitalization (i.e., the conversion of text, pictures, and sound into a digital form) and the need to automate processes and facilitate automated data exchange continue within customer organizations, a tension has emerged between these organizations and application providers/software tool vendors that support them. This tension is rooted in the need for customer organizations to disclose proprietary information models to vendors to interoperate with their tools. On the flip side, vendors are reluctant to share the information models underlying their tools with customers. Each perceives sharing such information as a loss of competitive advantage in their respective markets because of the concern that their respective information models can be reverse engineered. Despite this tension, economic forces are such that systems companies have no choice but to use the software tools provided by the software tool vendors.

The following true story conveys the magnitude of problems that can arise. Organization XYZ had developed a significant simulation that was heavily tied to the semantics of a particular vendor's tool. Documentation of the structures underlying the tool was sparse and, in some cases, turned out to be incorrect. However, after significant back-and-forth between Organization XYZ and the vendor, Organization XYZ was able to build the simulation. This simulation was part of an effort to accommodate V&V continuously with model and design development. About halfway through the project, the vendor released an upgraded tool that promised more efficiency for simulations. After switching to the upgraded tool, Organization XYZ discovered that none of their existing simulations worked because the underlying model semantics had changed. As with the previous version of the tool, the vendor's documentation wasn't clear about the new semantics. Therefore, it became impossible for Organization XYZ to determine whether it was possible to update their simulation. As a result, Organization XYZ was forced to go back to the previous version of the tool. Such horror stories exist in various industries.

To address this problem, customer organizations concluded that they need to create a semantic interface, a separation layer, between the semantics of their knowledge assets/data and the semantic model of the software tool vendors. With such a semantic interface, customer organizations would be able to define their own data views, while the tool vendors would be able to integrate their tools within the

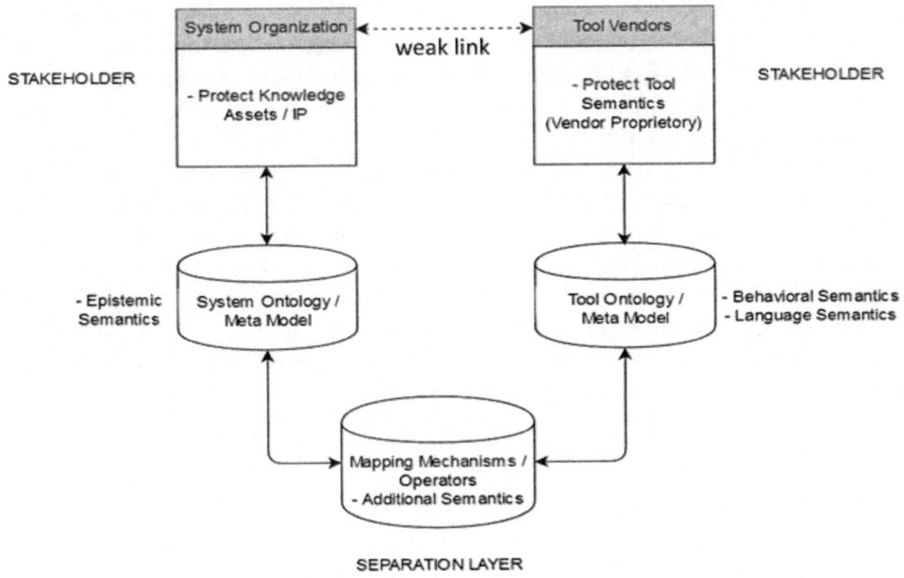

Fig. 3 Semantic mapping between customer knowledge assets and vendor tool semantics

customer's information environment without exposing the behavioral semantics of their tools. To ensure that the two views map to each other requires developing a semantic mapping between customer and vendor ontologies, thereby achieving interoperability. The problem of ontology mapping to achieve alignment between the customer and vendor ontologies is currently performed manually making it a labor-intensive and error-prone activity.

Borrowing on concepts from the Semantic Web and other advances in semantic technologies, several researchers have begun working on developing techniques for semiautomatic creation of semantic mappings between ontologies. The key challenge in semantic mapping is determining the correspondence between the elements of the two ontologies. Semantic correspondence has been defined as the "glue" that holds ontologies together, thereby achieving interoperability. Machine learning is one promising approach to determine the correspondence among ontology elements of two different ontologies (Doan et al. 2004).

Figure 3 shows how a separation layer, in the form of a semantic interface, can be realized in practice and how customer and vendor semantics can be mapped to each other through this semantic interface. Here, we distinguish between *epistemic semantics* and *behavioral and language semantics*. Epistemic semantics is about naming things and what we wish to view. This is where epistemic ontology needs to be created on large projects within customer (systems) organizations. Behavioral and language semantics come into play within vendor tools which employ vendor-proprietary semantics and modeling language. Accomplishing this mapping requires additional semantics that need to be "tied" in some way to the epistemic ontology or mapped to it. Ontologies of both the customer (i.e., systems organization) and vendor have to be populated, and then customer and vendor ontologies are mapped onto the semantic interface. Representing this separation layer formally is essential for customer organizations (i.e., systems companies) to avoid getting inadvertently "locked into specific vendors."

As noted earlier, metamodeling is a complementary construct to ontology. It pertains to the schema for the semantic data and the language needed to express semantic extensions to existing information mechanisms to assure that the tools can interoperate with a broad class of models at run time. The key takeaway from this discussion is that the focus should be on the intentions of the stakeholders, which then leads to the need for representing their respective ontologies and developing a mapping between the two to achieve semantic interoperability without either having to disclose their proprietary information to the other.

6 Concluding Remarks

For the last few years, the introduction of ontologies into SE, and more specifically MBSE, has been recognized as a promising direction in that it has the potential of reducing modeling complexity, assuring semantic consistency, and achieving interoperability among heterogeneous systems. However, ontology initiatives in SE have not yielded the expected results. In hindsight, the SE community believes that these initiatives did not have a clear scope and expectations. This recognition led to the creation of the MVM construct to demonstrate the potential value of ontologies for MBSE. A MVM is defined as a "thin slice" representation of the system with just enough semantics (vocabulary) and structure to answer questions associated with specific stakeholder use cases.

The paper discusses the complementarity and synergism between ontologies and metamodels and provides specific examples of how they complement each other. The paper also presents an important interoperability challenge problem (i.e., protection of proprietary knowledge of customer organization while seeking to interoperate with third-party tools) and discusses how ontologies can be exploited to achieve required interoperability. This situation frequently arises when customer organizations wish to integrate software vendor tools into their development or IT environments. Both customer organizations (e.g., aerospace companies, automotive companies, national labs) and software tool vendors wish to achieve interoperability without having to disclose their proprietary knowledge assets and tool behavioral semantics, respectively. To this end, they can employ ontologies to represent their respective knowledge and tool assets and then create a semantic interface that enables mapping these ontologies to each other through that interface. This approach assures that neither has to disclose proprietary information to the other while accomplishing the necessary tool integration in the customer environment.

It is hoped that the MVM approach, MVM heuristics, and recommendations in this paper will encourage researchers to undertake well-scoped ontology-enabled SE initiatives that help further formalize and extend MBSE capabilities.

Acknowledgments The author acknowledges productive discussions on ontologies and MBSE with Michael Sievers of the University of Southern California and helpful feedback of Carla Madni of Intelligent Systems Technology, Inc., who reviewed the final draft. I also want to acknowledge Shatad Purohit of the University of Southern California who reviewed earlier drafts of this paper.

References

Doan, A., J. Madhavan, P. Domingos, and A. Halevy. 2004. Ontology Mapping: A Machine Learning Approach. In *Handbook on Ontologies. International Handbook on Information Systems*, ed. S. Staab and R. Studer. Berlin/Heidelberg: Springer.

Fox, M., and M. Gruninger. 1994. Ontologies for Enterprise Integration. In *Proceedings of the 2nd Conference on Cooperative Information Systems*, Toronto, May.

Gronmo, R., and J. Oldevik. 2005. An Empirical Study of the UML Model Transformation Tool (UMT). In *Proceeding of the 1st International Conference on Interoperability of Enterprise Software and Applications*, Switzerland.

Guarino, N., and C. Welty. 2000. Towards a Methodology for Ontology-Based MDE. In *Proceedings of the First International Workshop on MDE.*

Kaiya, H., and M. Saeki. 2005. Ontology Based Requirements Analysis: Lightweight Semantic Processing Approach. In *QSIC 2005, Proceedings of the 5th International Conference on Quality Software*, 223–230.

Madni, A.M., C.C. Madni, and W. Lin. 1998. IDEON™/IPPD: An Ontology for Systems Engineering Process Design and Management. In Proceeding of the 1998 IEEE International Conference on Systems, Man, and Cybernetics, San Diego, CA, Oct. 11–14, pp. 2597–2602.

Madni, A.M., W. Lin, and C.C. Madni. 1999. IDEON™: An Ontology for Modeling and Managing Distributed Enterprises. In *Proceedings of the 13th International Conference on Systems Engineering*, Las Vegas, Nevada, Aug 10-12, pp. SE199-SE204.

———. 2001. IDEON™: An Extensible Ontology for Designing, Integrating, and Managing Collaborative Distributed Enterprises in Systems Engineering. *Systems Engineering* 4 (1): 35–48.

———. 2002. Human-agent Collaboration: Ontology and Framework for Designing Adaptive Human-agent Collaboration Architectures, Proc. of 12th INCOSE International Symposium, Las Vegas, NV, July 28 – Aug. 1.

Mayk, I., and A.M. Madni. 2006. The Role of Ontology in System-of-Systems Acquisition. In *Proceedings of the 2006 Command and Control Research and Technology Symposium*, San Diego, CA, June 20–22.

Noy, N., and M.A. Musen. 2000. PROMPT: Algorithm and Tool for Automated Ontology Merging. In Proceedings of the National Conference on Artificial Intelligence (AAAI), USA.

Orellana, D.W., and A.M. Madni. 2014. Human System Integration Ontology: Enhancing Model Based Systems Engineering to Evaluate Human-System Performance, Conference on Systems Engineering Research (CSER 2014) (Eds. Azad M. Madni, et al.) Redondo Beach, CA, March 21–22.

Parreiras, F.S., S. Staab, and A. Winter. 2007. *On Marrying Ontological and Metamodeling Technical Spaces, ESEC/FSE'07, September 3–7*. Croatia: Cavtat near Dubrovnik.

Saeki, M., and H. Kaiya. 2002. On Relationships among Models, Meta Models, and Ontologies, DSM 2006 Gruninger, M. and Lee, J. Ontology: Applications and Design. Communications of ACM, 45(2).

Sievers, M., 2019. Several Personal Discussions.

Sowa, J.F. 2001. Signs, Processes, and Language Games — Foundations for Ontology. In Proceedings of the 9th International Conference on Conceptual Structures, ICCS'01.

Stumme, G. 2005. Ontology Merging with Formal Concept Analysis. In Y. Kalfoglou, M. Schorlemmer, A. Sheth, S. Staab, and M. Uschold, eds., Semantic Interoperability and Integration, no. 04391 in Dagstuhl Seminar Proceedings.

Terrasse, M.N., M. Savonnet, G. Becker, and E. Leclercq. 2002. A UML-Based Meta-Modeling Architecture with Example Frameworks, WISME, Workshop on Software Model Engineering, Dresden, Germany, Available at URL http://www.metamodel.com/wisme-2002/terrasse.pdf.

Van Ruijven, L.C. 2014. Ontology for Systems Engineering as a Base for MBSE., INCOSE IS.

Wand, Y. 1996. Ontology as a Foundation for Meta-Modelling and Method Engineering. *Information and Software Technology* 38 (4): 281–288.

Ontological Modeling of Time and Time-Based Reasoning for Systems of Systems

Surya Vamsi Varma Sagi and Leonard Petnga

Abstract This paper explores the critical issue of temporal modeling and reasoning for successful systems of systems (SoS) architecting and operations. The increasing complexity of missions results into needs to leverage capabilities of constituent systems (CSs) in multiple domains, distributed geographically and temporally (different time zones) too. This introduces issues capable of hindering mission success including clock drifts and synchronization as well as communication delays. To assure correctness of SoS functionality in the face of these challenges, we develop and introduce a new ontological framework for modeling and time-based reasoning in SoS. Knowledge representation of time and temporal semantics in SoS modeling are discussed with a focus on the central role description logics (DL) and interval-based time semantics play in the development of the new framework. The latter consists of a DL-backed theoretical foundation providing formalisms to ontological models encapsulating temporal knowledge on top of which time-based modeling and reasoning applications for SoS can be built. A prototype implementation with a military-directed SoS has been illustrated and is currently under development.

Keywords Ontology · Temporal semantics · Systems of systems · Reasoning · Systems engineering

1 Introduction

This paper introduces an ontological framework for time modeling and time-based reasoning in systems of systems (SoS). These systems are defined as "set or arrangement of systems that results when independent and useful systems are integrated into a larger system that delivers unique capabilities" (Department of

S. V. V. Sagi (✉) · L. Petnga
University of Alabama in Huntsville, Huntsville, AL, USA
e-mail: ss0293@uah.edu

A. M. Madni et al. (eds.), *Recent Trends and Advances in Model Based Systems Engineering*, https://doi.org/10.1007/978-3-030-82083-1_15

Defense (DoD) 2004). Based upon the way that constituent systems (CSs) interact with each other, a SoS can be categorized into four different types: (1) *virtual SoS*, which lacks a central management authority and a centrally agreed-upon purpose; (2) *collaborative SoS*, where each entity knows its predefined role and interacts with the other to fulfill agreed-upon central purposes; (3) *acknowledged SoS*, where each constituent system maintains its independent status, but a designated manager is in charge of ensuring achievement of recognized objectives; and (4) *directed SoS*, where CSs maintain an ability to operate independently, but a central manager is in charge of decision-making to fulfill specific purposes (DeLaurentis 2007). SoS (especially acknowledged, collaborative, and directed ones) generally involve the coordination of multiple, safety-critical CS spatially distributed across multiple time zones, domains, and stakeholders. Thus, establishing and maintaining the temporal order of events and their synchronization as well as processes across a multiplicity of distributed CS while accounting for delays in communication between them are critical to effective decision-making and mission success.

In this work, we investigate the formal representation of domains across disciplines to extend a base ontology of time in order to include core concepts relevant to its semantic modeling in SoS context. A special attention is paid to three aspects critical to the description and handling of time in SoS: time zone, delays, and clock synchronization. Description logics (DL) is found to be the most appropriate knowledge representation formalism for our framework, and extensions as well as mapping to the web ontology language (OWL) are performed to ensure decidability of SoS time reasoning. In Sect. 2 we review existing models and semantics of time in SoS and make the case for a new framework. The latter is introduced and discussed in Sect. 3. Its capabilities are illustrated in a prototype implementation involving time-based reasoning in a military-directed SoS in Sect. 4.

2 Models, Frameworks, and Semantics for Knowledge Representation of Time in Systems of Systems

2.1 Time Modeling in SoS: Need for Semantics and Existing Frameworks

Need for a Framework for Time Modeling and Reasoning in SoS In order for formal verification approaches for SoS to be effective, there is a need for formal models to capture each of the domains involved and their multi-hierarchy interactions at the appropriate levels of *granularity*. This is particularly important for meta-domains such as time and space. They play a central role in safety-critical SoS, i.e., those for which correct and precise answer at all time to the question of "What is going on Where and When?" is critical to the correct planning and execution, thus success of complex missions. Answering such a question in an accurate, precise, and consistent way in the temporal dimension in the context of

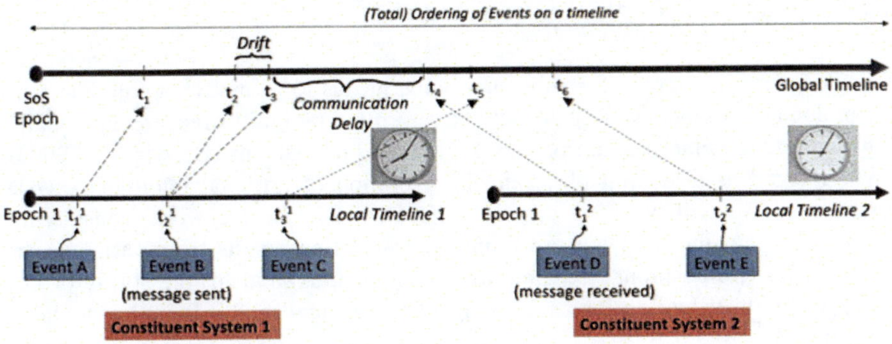

Fig. 1 Total temporal ordering of events on a global timeline

SoS requires (1) mechanisms for mapping and aligning *local times (*from internal clocks of CS as shown in Fig. 1) to a *global time* with account for *time zones*; (2) procedures for identification, processing, and managing *clock drift* as well as *delays* in communication between CSs; and (3) total ordering of temporal entities for better understanding of relationships between *events* across the SoS. These needs add to some well-known challenges in modeling time stemming from its frequent *under-specification* in natural language expressions.

Overview of SoS-Related Time Frameworks Researchers and engineers have recognized the need for mechanisms to handle the complex, distributed nature of time in SoS. For instance, in *acknowledged SoS* such as air traffic control (ATC), the need to ensure safe operation of thousands of aircrafts and users of the airspace across continents has resulted to the adoption of the Universal Coordinated Time (UTC) as a reference for clocks across systems (Fayadh and Hasson 2019). Mechanisms for handling clock synchronization have been found central in the operation of *Collaborative SoS*, as procedures are needed to ensure effective coordination of operations between constituent systems. The multiplicity of algorithms (e.g., Cristian and Berkley algorithms, network time protocol) for clock synchronization across platforms in distributed computer network systems testifies of the researcher's interest (Leela et al. 2018). Other frameworks have pursued a broader perspective of time in SoS. Campbell et al. (Campbell et al. 2005) introduce a time model based on a State Model Object (SMO) to support time simulation and analysis of any number of systems including cross-platform dependencies and across SoS missions. Similarly, among large-scale projects, the Architecture for Multi-criticality Agile Dependable Evolutionary Open System-of-Systems (AMADEOS) effort stands as the only one to explicitly provide a comprehensive model of time for SoS in the form of a taxonomy (AMADEOS Consortium 2016). However, this framework, as the others, fails to provide an approach for formal representation of the time domain and mechanisms for formal reasoning about SoS time.

2.2 Description Logics (DL) Semantics

Knowledge representation formalisms provide means to formally capture and represent domains knowledge (e.g., time) in a rigorous, ambiguity-free, and systematic way. Out of the numerous approaches, logic-based formalisms (Baader et al. 2003) have emerged as a leading player in the evolution of artificial intelligence (AI) formalisms. We will be using description logics (DL) in this work, considering that (i) some results for DL were found by translating results from variations of modal logics (propositional dynamic logics, μ-calculus) and (ii) the ability of DL to support multi-values attributes formalization (a critical need in complex domains description).

In DL, knowledge is represented through the description of domains in terms of concepts (classes in OWL), roles (properties, relationships), and individuals (objects) (Petnga and Austin 2016). Universal (∀), existential (□), intersection (⊓), union (⊔), and negation (¬) operators are used for restriction specifications to make the language decidable with low complexity. In DL, semantics are defined by interpretations. An interpretation *I* is defined as follows.

$I = (\Delta^{\mathbf{I}}, \cdot^{\mathbf{I}})$, where $\Delta^{\mathbf{I}}$ is the domain of interest (non-empty set) and, $\cdot^{\mathbf{I}}$ is an interpretation function that maps:

- Concept name *C:* a subset $C^{\mathbf{I}}$ of $\Delta^{\mathbf{I}}$
- Role name *R*: a binary relation $R^{\mathbf{I}}$ over $\Delta^{\mathbf{I}}$

In DL, interpretations are conceived as potential "realities" or "worlds" and need in no way comply with the actual reality. However, in the context of SoS (i.e., safety-critical systems), this is a nonnegotiable requirement as failure of the reasoner to properly capture systems (temporal) reality can lead to disastrous consequences. Additional information on interpretation and a summary of DL concept constructors can be found in (Rudolph 2011). Atomic concepts (A) in the attribute language (AL) DL can be extended to support arbitrary concepts (C), thereby enabling the description of any domain of interest and leading to ALC (Attributive Language with Complements) or (ALC where we allow transitivity statements) DL on which interpretations *I* are defined. Therefore, the latter are the means through which concepts, roles, and individuals build up to the DL knowledge base K <T, A> of a domain D. Here, T is a set of terminological Tbox axioms and *A* is a set of assertional Abox axioms; x,y are individual names. Further extension of the ALC or S DL with role (R), nominal (O), inverse (I), and qualified cardinality restriction (Q) leads to the SROIQ DL whose mapping to OWL 2 has been shown to ensure the decidability of the latter (Petnga and Austin 2016). We will use OWL 2 as our ontology language in this work.

3 A Time-Based Modeling and Reasoning Framework for SOS

3.1 Concepts and Calculus for SoS Time Ontological Modeling and Reasoning

Core Concepts for SoS Time Ontology and Reasoning The challenges of addressing the complexity of handling time in SoS require successful framework and models to formally define and describe concepts beyond the ones associated with current basic time concepts. The left side of Fig. 2 introduces some of those concepts, the relationships among them and with base time concepts. Standards such as the *ISO/IEC/IEEE 24765E:2017* provide recognizable and accepted definitions for most concepts. However, architectural definitions framework such as AMADEOS can help fill in the gap while also providing comprehensive temporal domain-based definitions. For illustration, in the AMADEOS framework, a *timeline* is a dense line denoting the independent progression of time from the past to the future, while an *event* is viewed as a happening at an *instant*. The latter is a cut of the timeline, while an *interval* is a section of the timeline between two instants. A *timestamp* of an event is the state of a selected clock at the instant the event occurs. A *clock* is an autonomous system consisting of an oscillator and a register. As such, it is subject to *drift* and generally indicates the *local time* in a specific *time zone*. Coordination of distributed clocks across CS of an SoS requires some *synchronization* mechanisms.

Allen's Temporal Interval Calculus (ATIC) Support for Reasoning with Time Several studies have shown that models are built over interval-based temporal logics such as Allen's Temporal Interval Calculus or Moszkowski's Interval Temporal Logic as the most appropriate for formal analysis (including reasoning tasks) involving time-dependent behavior (Allen 1983; Moszkowski et al. 1984). At the core of the ATIC is the relationship between time intervals (as introduced above and represented in Fig. 1) constructed over a given timeline. Thus, given two time intervals I_1 and I_2, a time point t, and a proposition Φ, we might ask a variety of questions over the time domain such as (1) *mereological* or "part-of"

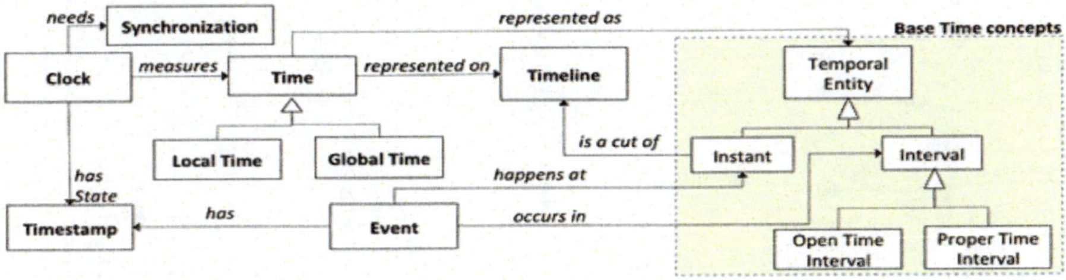

Fig. 2 Base time concepts and relationships with SoS temporal concepts

questions (e.g., *Doesn't occur within I_1?*), (2) *topological* or "connects" questions (e.g., *Does interval I_1 happen before or after interval I_2?*), and (3) *logical* or "rules-based" questions (e.g., *Does proposition Φ hold within the interval I_1?*). The ATIC identifies and specifies 13 relationships between any ordered pair of "convex" time intervals including with the main 7 relationships being "intEquals," "intBefore," "intMeets," "intStarts," "intFinishes," "intDuring," and "intOverlaps." In previous work (Petnga and Austin 2013), we have illustrated these relationships and demonstrated ATIC's powerful ability to support time-based reasoning. Also, for decidability of reasoning, the intervals must be proper time intervals (see Fig. 2).

3.2 *System Architecture and Description*

The system architecture for framework is shown in Fig. 3. It consists of three layers where each layer uses the layer below and builds upon it.

Layer 1: Theories and Standards This layer provides the foundational theories and formalisms on which the SoS time model and the corresponding reasoning mechanism are built. The *DL formalism* supports explicit, formal, and ambiguity-free representation of temporal knowledge and related axioms. Concepts are

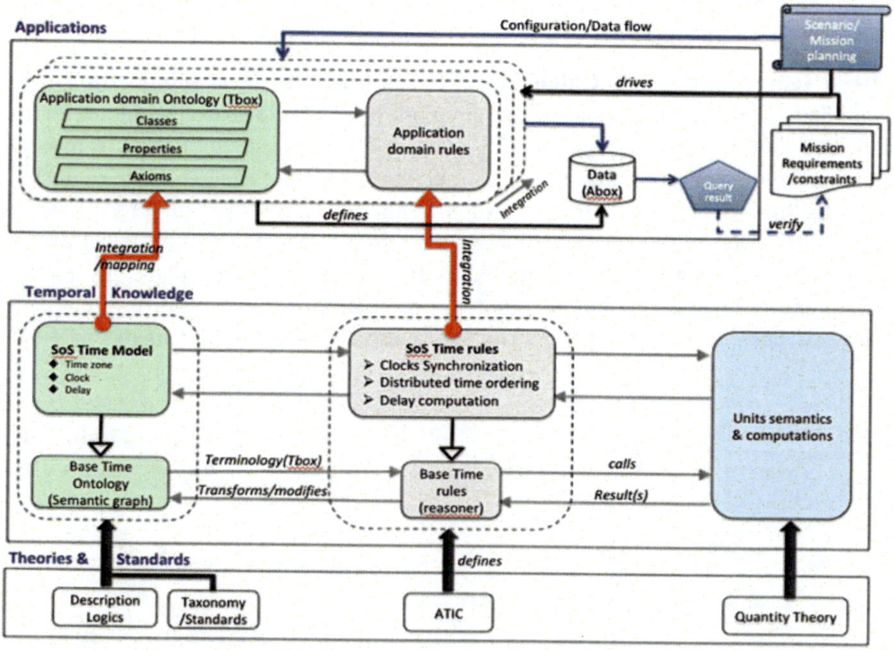

Fig. 3 Time-based reasoning framework for SOS

borrowed or expanded from well-established *standards* or architectural description framework's *taxonomy* providing temporal domain-based definitions. In this work, we build from a subset of the time view of the AMADEOS framework as introduced in Sect. 3.1. Thanks to its ability to manipulate and relate intervals and express temporal properties and their evolution over those intervals, the *Allen's Temporal Interval Calculus (ATIC)* is used as the main algebra for reasoning about time in our framework. However, in the case of SoS, it's critical for intervals to be fully specified using the granularity of a time as represented on a SoS global timeline with the proper mapping to local timelines of CS. This mapping accounts for clock drift, daylight time saving, and communication delays across the SoS. These, coupled with system-wide synchronization of clocks, provide a solid backbone for time-based reasoning in SoS. One key benefit of such approach is the formulation of restricted axioms which, when expressed in an ontology language, will ensure that time reasoning is decidable. Finally, *physical quantity theory* provides the mathematical foundations needed to enable equivalence (via conversion) between temporal entities represented using various units (e.g., hours, minutes, seconds) and date time representations.

Layer 2: Temporal Knowledge This layer leverages the formalisms from Layer 1 to capture and represent SoS temporal knowledge in its complexity and in an integrated manner. Three types of interrelated entities are used to that aim (from left to right): ontologies, rules, and computations modules. In this work, ontologies are expressed in a DL-compliant web ontology language (OWL). OWL-Time, the base time ontology for the semantic web published by World Wide Web Consortium (W3C) (World Wide Web Consortium 2006), is one such ontology, and it is expressed in OWL DL. Thus, we extend this base ontology with SoS time-related concepts as discussed in Sect. 3.1 and summarized in the left side of Fig. 2.

SoS time rules (e.g., clock synchronization, timestamp ordering, delay computations) add to ATIC axioms and rules from basic time domains. The ontology provides the terminological Tbox (as in Sect. 2.2) needed by the rules to transform the SoS time semantic graph using assertions (Abox) constructed from various temporal instances (timestamped events) across the SoS. During the execution of rules, calls to a unit semantics and computation platform allow for the resolution of temporal units differences across temporal entities in the semantic graph. This computation platform is also responsible for the ordering of events' time instants along a timeline of interest. A time synchronization module (not shown) is responsible for the resolution of time zones and clock drifts across the SoS to a single (global) timeline.

Layer 3: Applications This is the layer where we implement the application utilizing the time model and realization of the ontology specific for the problem at hand. Thus, all the application domains involved are captured in ontologies and integrated with the time ontology for full description of system-relevant events with their timestamps. The latter are the main data from the various clocks used to construct time instances (Abox) stored in a data store for query or further processing. Mission requirements and constraints as well as scenarios of interest are used to drive the formulation of application-level rules that are further mapped or integrated with SoS time rules. Data from scenario and mission planning are used to configure and initialize the semantic graph. Thus, the ontology, populated with minimal amount of data, will be transformed through controlled application of set of basic and SoS time rulès as well as domain rules for effective time-based decision-making across the SoS.

4 Prototype Implementation in Military-Directed SoS: Eye in the Sky

4.1 Overview

For the purpose of the illustration of the framework, let us consider a simplified SoS time-based modeling and reasoning scenario adapted from the movie "Eye in the sky" (Hood 2016). The military mission aims at taking down a terrorist cell congregating and operating in Nairobi, Kenya. A directed SoS is assembled and used for a military mission to that aim. The unique capabilities of four main CSs responding to a central command (mission control) are leveraged and coordinated to safely and timely destroy the cell with limited collateral damages. The resulting directed SoS is depicted in the scenario in Fig. 4b. The operation of the SoS spans four different time zones across three different continents (Fig. 4a). This makes issues of communication delays (e.g., between the drone and its operation center), clock drifts (mission control timeline), and synchronization (e.g., via a GPS clock) across multiple time zones critical to the success of the mission. The SoS time modeling and reasoning framework and its temporal knowledge layer capabilities are used to (1) collect CSs selected events relevant for SoS-level decision-making and, simultaneously, (2) collect and instantiate CSs local clock time as timestamps for the events and then synchronize them as per predefined SoS and mission time rules (Fig. 4c). An application graphical user interface (GUI) is used to configure the framework, set up scenarios, perform analyses, and control the time-based reasoning goals of the mission (Fig. 4d).

4.2 Illustration of a Simplified Time-Based Reasoning Scenario

Mission Threads on SoS Global Timeline We consider and track core events on two critical mission threads to be integrated for its success: (A) accurate and timely identification and movement tracking of terrorists and (B) mission "kill chain," i.e., combination of means to actually take them down. From a temporal prospective, the success of the mission is dependent on the proper sequencing of events (past and future) as they unfold along the SoS global timeline as well as the

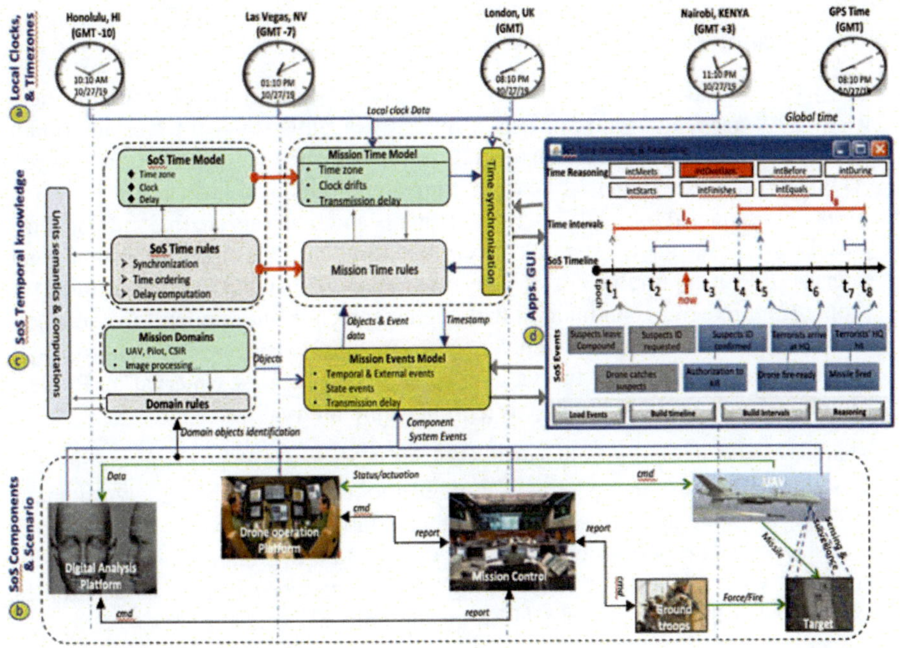

Fig. 4 Prototype application for time-based reasoning in SoS

establishment (and maintenance) of proper relationships between intervals of time across the various mission threads. Figure 4d depicts, from bottom to the top, the representation of past events (in gray) and ones expected (in blue) and their mapping and time synchronization to the SoS time given by the GPS (Fig. 4a and c) on the global timeline (Fig. 4d). In thread (A), the group of unidentified suspects has left a compound (at t_1) and is expected to arrive and stay at their suspected headquarter (at t_5) just after their IDs are all confirmed (at t_3) by Hawaii. In between, in thread (B), mission control will secure (at t_4) the authorization to terminate the (now confirmed) terrorists and will get the drone operation command to fire the missile (at t_7) to eliminate the threat (at t_8). The two main intervals constructed for both threads (i.e., I_A and I_B) result into an *IntOverlaps* relationship between them, one that must be established for the success of the mission.

Time-Based Reasoning for Decision Support in the Military SoS For the sake of simplicity, we assume that (A_1) each event timestamp on the global timeline already accounts for possible communication delays and drifts in CS clocks and (A_2) the mission control manages the SoS clock and constructs the intervals as needed. With these foundations, we can construct the temporal reasoning scheme and use it to reason about mission success based on the "intOverlaps" relationship between temporal intervals across the two threads. First, a minimal ATIC-compliant

SROIQ DL knowledge base (KB) is defined from the taxonomy in Fig. 2 as follows (subset in first-order logic – FOL):

- Rbox (R): before, after, begins, ends (for ordering of temporal entities); intEquals, intBefore, intMeets, intStarts, intFinishes, intDuring, intOverlaps (main ATIC relationships); intOverlaps ≡ intOverlaps- ("The overlaps relation between ATIC intervals is symmetric, i.e., it is equivalent to its own inverse")
- Tbox (T): Instant ⊓ Interval ⊑ ⊥ (i.e., "Instants and intervals are disjoints"); TemporalEntity ≡ Instant ⊔ Interval ("Every temporal entity is either an instant or an interval"); ProperTimeInterval ⊑ Interval ⊓ ∃begins.Instant ⊓ ⊑ends.Instant ("A proper time interval is an interval that begins and ends with nonnull instants")
- Abox (A): Instant(t_i) ("Individual t_i is an instant" with i ε {1,2,3,4,5,6,7,8}); ProperTimeInterval(I_A) ("Individual I_A is a proper time interval"); ProperTimeInterval(I_B) ("Individual I_B is a proper time interval")

Thanks to the mapping between the SROIQ DL and OWL 2, the knowledge base is translated into an SoS time ontology model [[KB]] with its own domain rules (Fig. 4c). One such rule is the one that establishes the ATIC "intOverlaps" relationship between two proper time intervals I and J beginning and ending with instants w,x,y,z. The rule of overlapping relation between I and J is as follows (implemented with Jena ontology API (Apache Jena 2013)):

```
[OverlapsInterv:(?I rdf:type te:ProperTimeInt)(?J rdf:type
te:ProperTimeInt)(?w te:begins ?I)(?x te:ends ?I) (?y te:begins
?J)(?z te:ends ?J)(?w te:before ?y)(?x te:before ?z) =>
(?I overlaps ?J) ]
```

where "`te`" is a prefix abbreviating the namespace (e.g., http://www.petnga.org/time#) for the resources (individual, concept, and role names) in the SoS time ontology. In [[KB]], concepts are captured as classes and relations as object or data properties as defined in the Tbox and Rbox of KB. Unique namespaces are used to differentiate and keep track of the type of various entities for expressiveness (human and machine readability) and effective reasoning. The latter requires the correct interpretation of the DL concepts. As for every DL ontology, we base our ontology on three finite sets of signature symbols to be used to drive the interpretation $I = (\Delta^{I}, ^{.I})$, where Δ^{I} is the SoS time domain: a set N_I of individual names (we use names of events), a set N_C of concept names, and a set N_R of role names. In this case, they are defined as follows: N_I = {Suspects_leave_coumpound, Suspects_ID_requested, Drones_catches_suspects, Suspects_ID_confirmed, Authorization_to_kill, Terrorist_arrive_at_HQ, Missile_fired, Terrorists_HQ_hit, Tracking_thread_A, Tracking_thread_B} (all individual names of entities in the Abox); N_C = {Instant, Interval, ProperTimeInterval} (all individual names in the Tbox); and N_R = {begins, ends, before, after, intEquals, intBefore, intMeets, intStarts, intFinishes, intDuring, intOverlaps} (all individual names in the Rbox). Using these elements, we define the interpretation I over KB and the corresponding function as follows (emphasis on formulations that are relevant to threads A and B):

- Concepts/classes: Instant^I = {t_1,t_3,t_4,t_5,t_7,t_8}; ProperTimeInterval^I = {I_A, I_B}
- Individuals: Suspects_leave_coumpound^I = t_1; Suspects_ID_confirmed^I = t_3; Terrorist_arrive_at_HQ^I = t_5; (thread A) ; Authorization_to_kill^I = t_4; Missile_fired^I = t_7; Terrorists_HQ_hit^I = t_8; (thread B)
- Relationships: before^I = $\{\langle t_1, t_2\rangle, \langle t_2, t_3\rangle\langle t_4, t_5\rangle\langle t_5, t_6\rangle\langle t_6, t_7\rangle\langle t_7, t_8\rangle\}$ (these appear as facts in [[KB]], stemming from the partial ordering of entities in the time domain); begins^I = $\{\langle t_1, I_A\rangle\langle t_4, I_B\rangle\}$; ends^I = $\{\langle t_5, I_A\rangle\langle t_8, I_B\rangle\}$

The latter pair of relationships leads to four facts created by the reasoning system automatically through the tracking of events as they (expect to) unfold and appear with their timestamps on the SoS global timeline (as per assumptions A_1 and A_2). We note that, for operation purpose, the location of the "now" instant on the timeline determines the order in which each pair is created as future time instants can't be part of the interval. This is less relevant for simulation purpose. Thanks to the transitivity of the relation "before," the reasoner infers $\langle t_5, t_8\rangle$ and adds the corresponding fact to the before^I set. With this, we have all the conditions on the left-hand side of the `OverlapsInterv` rule above satisfied. Thus, the rule will fire and an "intOverlaps" relationship created between I_A and I_B with interpretation intOverlaps^I = $\{\langle I_A, I_B\rangle\}$.

5 Conclusion and Future Work

This work has introduced an ontological framework for time modeling and reasoning for systems of systems (SoS) architecting. The framework is shown appropriate for SoS for which mission success is dependent on the correct time-based prediction of the future state of the system. It is equipped with mechanisms to account for spatially and temporarily distributed, drift-prone clocks across multiple time zones. The needs for clock synchronization and account for communication delays are also discussed. These results have been achieved, thanks to Allen's Temporal Interval Calculus (ATIC) as well as description logics (DL) formalisms backing the decidability of our reasoning framework. An illustration of its usage has been described on a directed military SoS example. A Java-based implementation is currently under development.

References

Allen, J.F. 1983. Maintaining Knowledge about Temporal Intervals. *Communications of the ACM* 26 (11): 832–843.

AMADEOS Consortium. 2016. AMADEOS conceptual model, Deliverable D2.3, Revised.

Apache Jena., http://www.jena.apache.org, 2013.

Baader F., D.L. McGuinness, D. Nardi, and P.F. Patel-Schneider. 2003. *The Description Logic Handbook: Theory, Implementation, and Applications*. Cambridge.

Campbell J.E., D.E. Longsine, D. Shirah, and D.J. Anderson. 2005. System of Systems Modeling and Analysis, SAND Report #SAND2005-0020, Sandia National Laboratories, January 2005.
DeLaurentis, D. 2007. *System of Systems Definition and Vocabulary*. School of Aeronautics and Astronautics: Purdue University, West Lafayette, IN.
Department of Defense (DoD). 2004. Defense Acquisition Guidebook Ch. 4 "System of Systems Engineering," Washington, DC: Pentagon.
Fayadh A.M., and S.T. Hasson. 2019. Insights of Models for Air Traffic Management System. *International Journal of Recent Technology and Engineering (IJRTE)* ISSN: 2277-3878, 8(3S3), November 2019.
Hood, G. Eye in the Sky, Drama/Thriller, 1h 43m, Released April 1, 2016 (USA).
Leela, M., Kumar D. Manoj, and G. Bhavana. 2018. Clock Synchronisation in Distributed Systems: A Review. *International Journal of Recent Engineering Research and Development (IJRERD)* 03 (04): 14–17.
Moszkowski, H., et al. 1984. Reasoning in interval Temporal Logic and Tempura. In *Proceeding ofthe AMC/NCF/ONR Workshop on Logics of Programs,* volume 164 of LNCS, pp. 371 – 383. Springer, 1984.
Petnga, L., and M.A. Austin. 2013. Ontologies of Time and Time-based Reasoning for MBSE of Cyber-Physical Systems, 11th Annual Conference on Systems Engineering Research (CSER 2013). Atlanta, GA, March 19–22.
———. January 2016. An Ontological Framework for Knowledge Modeling and Decision Support in Cyber-Physical Systems. *Advanced Engineering Informatics* 30 (1): 77–94.
Rudolph, S. 2011. *Foundations of Description Logics*. Germany: Karlsruhe Institute of Technology.
World Wide Web Consortium (W3C), Time Ontology in OWL, accessible at: http://www.w3.org/TR/owl-time/, 2006.

Ontology-Enabled Hardware-Software Testbed for Engineering Adaptive Systems

Edwin Ordoukhanian and Azad M. Madni

Abstract Adaptive systems are class of systems that change their behavior in response to external or internal disrupting events. These kinds of system are of interest to government, industry, and research community. An instance of an adaptive system is a multi-UAV swarm operating in open, dynamic environment. Such systems are employed to carry out missions such as search and rescue and disaster relief. Developing adaptive systems requires a hardware-software testbed to explore system behavior and make adjustments to achieve desired behaviors. This paper presents an ontology-enabled approach for developing integrated hardware-software testbed for engineering adaptive systems. Multi-UAV operation is used as an illustrative example of an adaptive system that stands to benefit from a hardware-software experimentation testbed.

Keywords Ontology · Hardware-software testbed · Multi-UAV operation · Adaptive systems

1 Introduction

Robot-assisted missions have gained considerable interest in recent years such as search and rescue, reconnaissance, or surveillance missions. In this context, various types of robots such as unmanned aerial vehicles (UAVs) or unmanned ground vehicles (UGVs) are used. For instance, UAVs perform tasks such as surveillance, medicine delivery, or area mapping, while UGVs perform tasks such as inspection. With advances in swarm technology, multiple robots are deployed to assist humans in such missions (e.g., mapping an area using multiple UAVs). Operating multiple vehicles (e.g., UAVs) simultaneously often requires flexible allocation of requirements to multiple vehicles to reduce operational complexity and increase

E. Ordoukhanian (✉) · A. M. Madni
Systems Architecting and Engineering, Viterbi School of Engineering, University of Southern California, Los Angeles, CA, USA
e-mail: ordoukha@usc.edu

A. M. Madni et al. (eds.), *Recent Trends and Advances in Model Based Systems Engineering*, https://doi.org/10.1007/978-3-030-82083-1_16

overall mission coverage. Each UAV collects information from multiple sources (using onboard sensors), share that information with other members, and execute actions in coordinated fashion in multiple locations. This capability brings more time efficiency into mission execution as vehicles perform assigned or negotiated tasks in parallel to fulfil mission objectives. The latter is important for time-critical mission such as medicine delivery.

Operational missions that employ multi-UAV systems are typically carried out in dynamic and open environment subject to disrupting events. The combination of multi-UAV system and human needs to jointly adapt to changing circumstances. Adaptation is the capacity of a system to change its state in response to change in the environment. Adaptive systems change their behavior given some external perturbation to optimize or maintain their condition within that environment. Adaptation occurs in different aspects of a system. For example, system structure, sensing capabilities, or control law are some of the aspects that can be adapted due to changes in the environment.

Designing adaptive multi-UAV systems requires in-depth exploration and experimentation. Such exploration and experimentation requires a platform where system designer can rapidly test-drive various algorithms, model and simulate different system behavior, and draw sensible conclusion from experimentation. Such platforms for experimentation and exploration should not be built in an ad hoc fashion. Instead, there should be a well-thought-out methodology, enabled by an ontology, behind developing such sophisticated platforms.

This paper focuses on developing a hardware-software testbed for engineering adaptive multi-UAV systems. While researchers have explored simulation techniques to demonstrate adaptive behavior of multi-UAV systems, these systems have not been tested in real world. Some researchers have developed application-specific testbeds and performed experimentation. However, these tend to be point solutions. Therefore, there is a need for an integrated and extendable hardware-software testbed that can enable exploration and experimentation of adaptive system behaviors.

This paper reviews the state of the art on hardware and software approaches for multi-UAV systems and discusses current gaps in this area. It then introduces an ontology-enabled approach for developing integrated hardware-software testbed for adaptive systems with discussion of how it impacts multi-UAV system design. This paper is organized as follows. Section 2 presents literature review on the current hardware and software approaches. Section 3 identifies current gaps. Section 4 discusses ontology-enabled approach for developing testbed for adaptive systems. Section 5 discusses the way forward.

2 Review of Current Approaches to Engineering Multi-UAV Systems

Researchers over the years have tried to develop various control techniques for developing adaptive multi-UAV systems. Many of these approaches have adopted different techniques from domain outside engineering to develop adaptive multi-UAV systems. While these techniques perform incredibly well in simulated environment, their applicability in the real world remains questionable. Many of these techniques either neglect key aspects such as dynamics of the environment and the vehicle or get into very specific and specialized aspects of the system that ignore the bigger picture and the fact that UAVs are required to perform a variety of missions.

Table 1 summarizes some of the current techniques that are successfully demonstrated in the simulated environment and are promising approaches for the real world. However, these approaches have not been tested in real-world application.

Several researchers have been working on advancing hardware testbeds for multi-UAVs. However, many of these approaches solve a piece of the problem. Purta et al. (2013) introduce dynamic data-driven application system. In this structure, real data is shared between hardware (actual vehicle) and simulation. Control orders are being sent to the environment as well as getting sensory data from it. It consists of the command module where all the main swarm control algorithms are stored and operated, the GUI module (allows users to switch between environments and checking current status of vehicles), the environment module (simulation or actual

Table 1 Current simulation techniques

Ref.	Techniques	Application
Lidowski (2008)	Behavior rules	Multi-UAV search mission in 2D, primarily focused on communication
Milam (2004)	Behavior rules	No specific target area * utilizing genetic algorithm
Price (2006)	Behavior rules	Formation flight *using self-organizing approach
Pack and Mullins (2003)	Behavior rules	Search missions, applied to robots but can be extended to UAVs
Barnes et al. (2007)	Gradient vector	Formation flight
Liu et al. (2008)	Graph theory	No specific target area, primarily studying leader-follower
Vincent and Rubin (2004)	Mathematical patterns	No specific target area
WangBao and XueBo (2008)	Mathematical patterns	Formation; applied to robots but can be extended to UAVs
Parunak et al. (2002)	Artificial pheromones	Use potential field for drone formation; mostly used for search applications
Gaudiano et al. (2005)	Artificial pheromones	Search and search and destroy mission

hardware), and the middleware module (service-oriented architecture to enable communication between modules). Using MASON multi-agent simulation library as a backbone of many swarming techniques, MASON does most of the heavy lifting while also being able to connect to another simulation environment to acquire data or actual drones for acquisition of real-world data.

O'Neil et al. (2005) developed a simulator based on a game engine to simulate real world. Their approach is primarily applied to wireless communication with specific focus on human interaction. It describes TATUS which is a computing simulator. TATUS is novel in that it removes the need for experimenters to develop games level code, while retaining a large level of flexibility in the scenarios that can be readily developed by researchers.

D'Andrea and Babish (2003) developed hybrid environment specifically designed for RoboFlag game. Some aspects are simulated such as communication link; however, the robots are real. Their testbed has four main components: vehicles, global sensor information, centralized control, and human interfaces and communications network. In their approach communications network subsystem is completely simulated. Bandwidth and latency limitations are all simulated in the corresponding software module. For example, errors in transmissions are captured by the transmission latency. The simulation occurs at a relatively high level.

Montufar et al. (2014) developed a testbed for multi-UAV primarily focusing on moving objects. Their testbed gives experimenter two environments to choose from, simulation and/or real world. Schmittle et al. (2018) developed an open-source, cloud-based simulation testbed for UAVs. Their testbed utilizes commercially available tools and cloud servers to reduce time to setup. It is easy to use and specifically addresses scalability and usability. Palacios et al. (2017) develop a heterogenous testbed based on a specific drone model and demonstrate its application in two different scenarios. It enables users to select four different environments to test their control algorithms. Afanasov et al. (2019) developed a testbed called FlyZone. In their methodology they use tags for vehicle localization. Onboard computer is responsible for giving high-level commands. They have decoupled testbed with the real application. Their approach only considers static obstacles and wind disturbances as disruptions. Their testbed also utilizes vehicle dynamics model to determine the influence of winds on the UAV's behavior.

Michael et al. (2008) identified robustness and reliability as one of the main requirements of the hardware-software testbeds. Scalability and capability of estimating system states are among other key requirements for an integrated hardware-software testbed (Michael et al. 2008). Michael et al. (2008) successfully demonstrate application of formation control algorithm on ground vehicles by utilizing specific type of a ground vehicle platform called Scarab.

3 Gaps in Current Approaches

Many of the current hardware-software testbed approaches for adaptive multi-UAV systems consider only specific type of vehicles and furthermore focus on a specific model such as Parrot AR.Drone 2.0 (Purta et al. 2013; Montufar et al. 2014). While this is valuable, it still doesn't show the applicability of the current approaches to other vehicles with sophisticated sensor suites. After all, Parrot AR.Drone is a specific type of a quadcopter, and findings cannot be generalized unless rigorous testing is performed with different types of quadcopters.

Environment is often a misused terminology. Some researches (Purta et al. 2013) look at environment as either simulation (software) or hardware. But at the same time, the term environment can also be associated with operational environment where system is performing the mission. Some approaches in hardware-software testbed development (Purta et al. 2013) do not focus on a specific mission and only focus on a portion of the overall operation. While this divide and conquer technique is valuable to address the bigger problem, however, often times the link between the smaller portion of the problem and the bigger problem is not identified, which in turn creates misunderstandings.

Some testbeds focus only on the simulation aspect (O'Neill et al. 2005; Montufar et al. 2014) with almost no discussion of physical aspect of the system. In simulations the vehicle is often considered as a point mass which performs some basic actions. Neglecting dynamical aspects of the system may decrease simulation fidelity which can lead to performing insufficient analysis on the outcome.

In some cases (O'Neill et al. 2005) human interactions have been the focus of the testbed with no onboard sensor systems. Some approaches (D'Andrea and Babish 2003) have put vehicle's high-level control on a workstation and not specifically on the robot. While this is a good approach for rapidly testing an algorithm, it lacks enough flexibility for real-world application. Such approaches put extra emphasis on the communication among the workstation and the vehicle to ensure commands are sent and sensory data are received on time.

Current approaches in hardware-software development lack system engineering processes. In that, key steps such as requirement development are explicitly missing. While some researchers have identified a set of the requirement, those requirements are simple and are only applicable to the instance of the testbed they have developed. Furthermore, requirements usually are not validated or verified (Montufar et al. 2014; Schmittle et al. 2018; Palacios et al. 2017).

Current approaches while letting researchers change their control algorithm do not explicitly discuss adaptive behavior of the systems. Particularly, what is missing in the current approaches is how the operational environment changes in the laboratory or the facility where the vehicle is being tested. In many of the current approaches, for instance (Schmittle et al. 2018; Palacios et al. 2017), there is no explicit ontology that defines key concepts and relationships. Many approaches jump directly into an instance to solve their problem without proper consideration of the overall requirements and underlying ontology.

In summary, current methods tend to jump into a specific instance of a testbed without exploring other alternatives, primarily because requirements are not identified or are ill-defined. For instance, many researchers select a specific model of an UAV or vehicle (Michael et al. 2008) without showing any justification for why specific model has been chosen. Currently, there is no explicit discussion of formal ontology for creating hardware-software testbed. Furthermore, current literature in hardware-software testbed fails to address adaptive behavior of multi-UAV systems. Most of the current methods do not consider realistic scenarios and do not consider or overly simplify the operational context. Many applications involving development of hardware-software testbeds have been focused on simple scenarios such as formation fly with no complicating factors. Often times, safety constraints are neglected or implied. In hardware-software testbeds, two types of safety requirements need to be considered, experimenter's safety and vehicle's safety. Both have been lacking in the current literature and partly lack of formal ontology is to be blamed.

4 Ontology-Enabled Approach

The approach introduced and discussed in this paper is focused on developing hardware-software testbed for adaptive systems with application to multi-UAV systems. Most testbeds for multi-UAV systems tend to be ad hoc and point solution. Thus, this area can benefit from such ontological approach. This approach takes a step back and defines key concepts and their relationships. Then, by defining an ontology and reasoning about it, a comprehensive, scalable, and robust testbed can be created.

Ad hoc approaches for developing hardware-software testbed tend to be costly in the longer run since they lack requisite flexibility and scalability. An ontology-enabled approach in developing such testbeds will have significant impact on cost reduction by eliminating unnecessary rework and ensuring requisite flexibility and scalability are considered during architecting process of the testbed.

A testbed which is developed based on an explicit ontology will have requisite formal architecture to support testing of adaptive behavior of systems. In essence, what is needed is a systematic approach to develop an integrated hardware-software testbed by considering the bigger picture and all interacting components (Madni et al. 2019).

Adaptive systems change their behavior given some external perturbation in order to optimize or maintain their condition within operational environment (Madni and Jackson 2009; Ordoukhanian and Madni 2019). In doing that, they employ adaptation logics to deal with disrupting events (Ordoukhanian and Madni 2019). Adaptation logics can affect different aspects such as adapting mission objectives or adapting control laws (Ordoukhanian and Madni 2019). An adaptive system has both hardware and software components that interact with each other. Hardware components are those that directly interact with the environment. Software components are primarily responsible for processing the information and send out control commands.

Any adaptive system is deployed in the operational environment to satisfy set of mission requirements (Ordoukhanian and Madni 2019). These requirements are dictated by the stakeholders and the specific operational environment that the system will operate in. Often times, operational environment is open and dynamic which can be source of many disrupting events (Madni and Jackson 2009; Ordoukhanian and Madni 2019). Disruptions can be categorized into three main groups (Madni and Jackson 2009). External disruptions are associated with environmental obstacles and incidents. They are often random and with unknown severity and duration (Madni and Jackson 2009). Systemic disruptions happen when an internal component's functionality, capability, or capacity causes performance degradation (Madni and Jackson 2009). They are easily detectable in technological systems. Human-triggered disruptions are associated with human operators inside or outside of system boundary impacting system performance. In general, these disruptions can be predictable or random (Madni and Jackson 2009).

Hardware-software testbed supports adaptive system development by enabling an environment where multiple adaptation schemes can be tested. Hardware or software components of the adaptive system can be tested in a physical world (hardware) or virtual world (software) (Madni et al. 2019). A world is a geospatial region which supports experiments (Madni 2019). It may be as simple as an empty room in which a vehicle (e.g., quadcopter) operates, or it may be more complex such as a mountainous terrain (at reduced scale) for a quadcopter search and rescue mission. The world can be virtual (i.e., exist only in simulation), physical (i.e., exist in the laboratory), or a combination of the two. The latter would be the case for an experiment involving both physical and simulated vehicles. The vehicle can be a quadcopter, ground vehicle, or fixed-wing aircraft. Each vehicle can be under manual control (i.e., a human is operating a radio controller) or autonomous. To facilitate autonomous behavior, vehicles can incorporate controllers consisting of single-board computers such as Raspberry Pi.

An experiment is a scientific procedure undertaken to test a hypothesis or demonstrate a capability. An experiment sets a scenario in motion and logs the resulting activity. An experiment can be conducted purely in simulation, entirely in hardware with vehicles moving in the physical laboratory, or in a mixed environment with both simulated and physical elements (Madni 2019). An experiment can be viewed as a specific realization of a scenario. In other words, multiple experiments can be performed based on the same scenario. The results of different experiments will vary in general, for a variety of reasons, which include random differences in initial conditions, equipment differences such as changed battery state, use of random numbers in control algorithms, or algorithm changes due to learning from previous experiments (Madni 2019).

A *scenario* provides the contextual backdrop for experiments. It incorporates elements such as *terrain*, *weather*, and *time*. A scenario also has specific *mission/goals* which *adaptive system* needs to perform. The mission is part of larger set of *requirements* that needs to be fulfilled by the adaptive system. The adaptive system has two parts: human agent (e.g., operator, troops, civilian) and system/devices (e.g., ground system, UAV, handheld device). Mission goals (or sub-goals) can be assigned to one or more systems (or agents) which will be achieved through planned or opportunistic collaboration (Madni 2019). In every scenario, systems (or agents) start in specified initial conditions and operate under the control of various control algorithms. The scenario also includes elements related to various *events*, particularly disrupting events.

An integrated hardware-software testbed is facilitated by a *middleware*. Middleware is responsible for connecting all components together in the testbed. Hardware-software testbed is also comprised of hardware and software components, simulation environments, and user-friendly dashboard. A customizable dashboard program will allow the user to import scenarios. This will result in a dashboard view which provides default capabilities such as a plan view (or "mission view") of the world showing all vehicles, views from one or more vehicle cameras, state indicators for all vehicles of interest, and controls to start and stop an experiment (Madni 2019). It will be possible to customize the dashboard layout either graphically (by moving, scaling, adding, and deleting elements) or by modifying the layout script. A Scenario Creation Tool facilitates the creation of new scenarios and the modification of existing scenarios. The most complex part of scenario creation is the creation of the virtual world (i.e., the world model). For UAVs such as quadcopters, terrain and boundaries are the key elements. Figures 1 and 2 show these main concepts and their relationships.

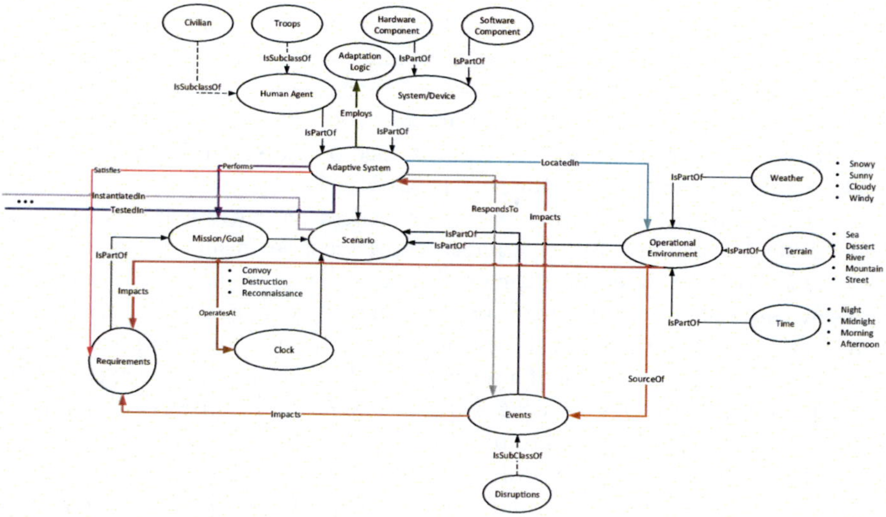

Fig. 1 Overall ontology for hardware-software testbed (part a)

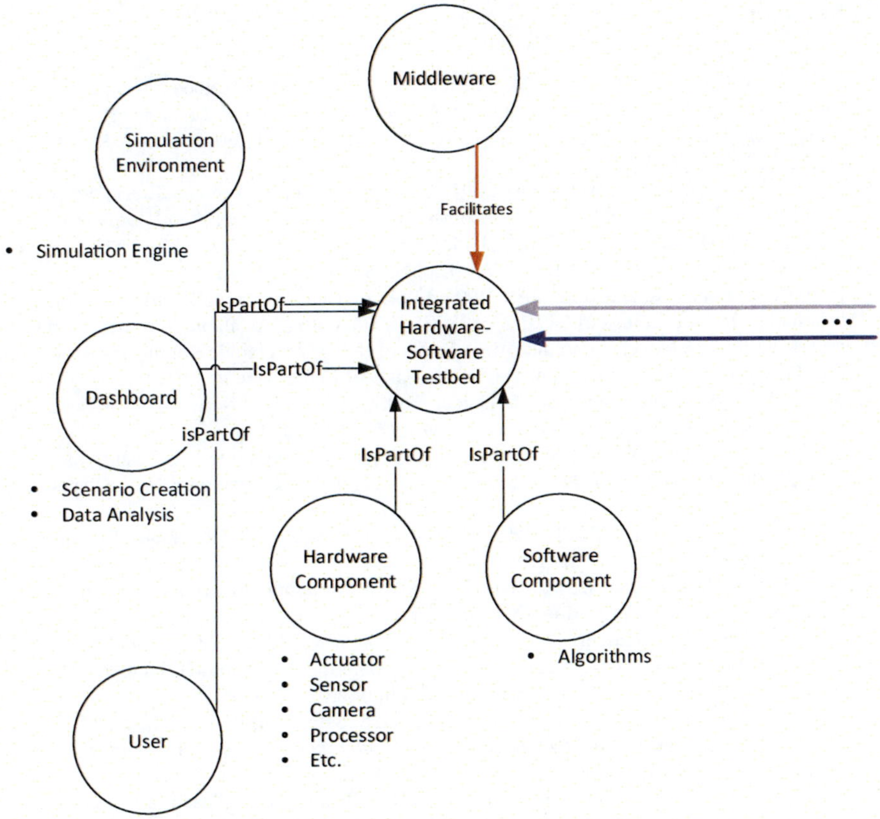

Fig. 2 Overall ontology for hardware-software testbed (part b)

5 Summary and Future Work

In this paper authors reviewed the state-of-the-art approaches for developing hardware-software testbeds for multi-UAV systems. The strength and limitation of current approaches were identified along with the key gaps in them. To fill the current gap, in this paper authors presented an ontology-enabled approach for developing integrated hardware-software testbed for multi-UAV systems. This approach explicitly identifies key components and their relationships. Furthermore, it explicitly takes into account adaptive behavior of such systems. In future work, the ontology developed in this paper will be instantiated within a sufficiently complex scenario for "what-if" experimentation and exploration. Additionally, a process will be created to develop models from actual hardware so accurate information can be used in the simulation.

References

Afanasov, M., A. Djordjevic, F. Lui, and L. Mottola. 2019. FlyZone: A Testbed for Experimenting with Aerial Drone Applications. In T*he 17th Annual International Conference on Mobile Systems, Applications, and Services (MobiSys '19)*, June 17–21, 2019, Seoul, Republic of Korea. ACM, New York, 13 pages.

Barnes, L., M. Fields, and K. Valavanis. 2007. Unmanned Ground Vehicle Swarm Formation Control Using Potential Fields. In *Mediterranean Conference on Control and Automation.*

D'Andrea, R., and M. Babish. 2003. The Roboflag Testbed. In *Proceedings of the 2003 American Control Conference, 2003.* 2003 Jun 4 (Vol. 1, pp. 656-660). *IEEE.*

Gaudiano, P., E. Bonabeau, and B. Shargel. 2005. Evolving Behaviors for a Swarm of Unmanned Air Vehicles, In Proceedings of IEEE 2005, IEEE, 2005, online. Available from IEEEXplore.

Lidowski, R.L. 2008. A Novel Communications Protocol Using Geographic Routing for Swarming UAVs Performing a Search Mission, Master's thesis, Air Force Institute of Technology.

Liu, B., et al. 2008. Controllability of a Leader Follower Dynamic Network With Switching Topology. *IEEE Transactions on Automatic Control* 53: 1009–1013.

Madni, A.M. 2019. Ontology-Enabled Testbed for Developing, Integrating, and Testing of Adaptive Cyber-Physical-Human Systems. Systems Architecting and Engineering, White Paper, July 10, 2019.

Madni, A.M., and S. Jackson. 2009. Towards a Conceptual Framework for Resilience Engineering. *IEEE Systems Journal.* 3 (2): 181–191.

Madni, A.M., C.C. Madni, and S.D. Lucero. 2019 Mar. Leveraging Digital Twin Technology in Model-Based Systems Engineering. *Systems.* 7 (1): 7.

Michael, N., J. Fick, and V. Kumar. 2008. *Experimental Testbed for Large Multirobot Teams.*

Milam, K.M. 2004. Evolution of Control Programs for a Swarm of Autonomous Unmanned Aerial Vehicles, Master's thesis, Air Force Institute of Technology.

Montufar, D. I., F. Munoz, E. S. Espinoza, O. Garcia, and S. Salazar. Multi-UAV Testbed for aerial Manipulation Applications. In *2014 International Conference on Unmanned Aircraft Systems (ICUAS)* 2014 May 27 (pp. 830–835). IEEE.

O'Neill, E., M. Klepal, D. Lewis, T. O'Donnell, D. O'Sullivan, and D. Pesch. 2005. A Testbed for Evaluating Human Interaction with Ubiquitous Computing Environments. In *First International Conference on Testbeds and Research Infrastructures for the Development of Networks and Communities 2005 Feb 23* (pp. 60–69). IEEE.

Ordoukhanian, E., and A.M. Madni. 2019. Model-Based Approach to Engineering Resilience in Multi-UAV Systems. *Systems 7*: 11.

Pack, D.J., and B.E. Mullins. 2003. Toward Finding an Universal Search Algorithm for Swarm Robots, in: Proceedings of the 2003 IEEE/RSJ International Conference on Intelligence Robots and Systems, 2003, pp. 1945–1950.

Palacios, F.M., E.S. Quesada, G. Sanahuja, S. Salazar, O.G. Salazar, and L.R. Carrillo. 2017. Test Bed for Applications of Heterogeneous Unmanned Vehicles. *International Journal of Advanced Robotic Systems.* 14 (1): 1729881416687111.

Parunak, H.V.D., M. Purcell, and R. O'Connell. 2002. Digital Pheromones for Autonomous Coordination of Swarming UAVs, Tech. rep., American Institute of Aeronautics and Astronautics.

Price, I.C. 2006. Evolving Self-Organized Behavior for Homogeneous and Heterogeneous UAV or UCAV Swarms, Master's thesis, Air Force Institute of Technology.

Purta, R., M. Dobski, A. Jaworski, and G. Madey. 2013. A Testbed for Investigating the UAV Swarm Command and Control Problem Using DDDAS. *Procedia Computer Science* 18: 2018–2027.

Schmittle, M., A. Lukina, L. Vacek, J. Das, C.P. Buskirk, S. Rees, J. Sztipanovits, R. Grosu, and V. Kumar. 2018. Openuav: A uav Testbed for the cps and Robotics Community. In *2018 ACM/IEEE 9th International Conference on Cyber-Physical Systems (ICCPS)* 2018 Apr 11 (pp. 130–139). IEEE.

Vincent, P., and I. Rubin. 2004. A Swarm-Assisted Integrated Communication and Sensing Network. In *Battlespace Digitization and Network-Centric Systems IV*, SPIE, SPIE Digital Library, ed. R. Suresh, vol. 5441, 48–60.

WangBao, X., and C. XueBo. 2008. Artificial Moment Method for Swarm Robot Formation Control. *Science in China Online.*. Available from http://www.springerlink.com.

An Ontology for System Reconfiguration: Integrated Modular Avionics *IMA* Case Study

Lara Qasim, Andreas Makoto Hein, Sorin Olaru, Marija Jankovic, and Jean-Luc Garnier

Abstract System reconfiguration (SR) is essential in system management, as it is an enabler for system flexibility and adaptability, attendant *ilities* being reliability, availability, maintainability, testability, safety, and reuse of system entities and technologies. Within current industrial practice, the development of reconfiguration tools is a real challenge. The development of these tools demand clear identification of reconfiguration data. In this paper, key concepts of the reconfiguration process, and relations among them, are represented in the form of the *OSysRec* ontology. These concepts are applied to the integrated modular avionics case study to test the proposed ontology within the aerospace domain.

Keywords System reconfiguration · Ontology · Model-based systems engineering · Operation

1 Introduction

Companies are currently concerned with designing and developing systems, which have considerable lifetimes and diversity of employments. Systems can possibly undergo evolutions in response to changing operational contexts and environments. Configuration management is key to ensuring effective management of an evolving

---

L. Qasim (✉) · A. M. Hein · M. Jankovic
Industrial Engineering Research Department (LGI), Ecole CentraleSuepelc, France

Signal & Systems Research Department (L2S), Ecole CentraleSuepelc, France

Thales technical directorate, Thales Group, Palaiseau, France
e-mail: lara.qasim@centralesupelec.fr

S. Olaru
Signal & Systems Research Department (L2S), Ecole CentraleSuepelc, France

J.-L. Garnier
Thales technical directorate, Thales Group, Palaiseau, France

A. M. Madni et al. (eds.), *Recent Trends and Advances in Model Based Systems Engineering*, https://doi.org/10.1007/978-3-030-82083-1_17

system during its life cycle (ISO/IEC:15288 2015; Walden and Roedler 2015). Thus, system configuration management plays a crucial role in systems engineering.

Systems engineering sustains activities to satisfy internal and external stakeholders requirements (ISO/IEC:15288 2015; Walden and Roedler 2015). In systems engineering, system configuration is defined as a set of elements that compose a system in terms of hardware devices, software, interfaces, human profiles, and processes (Walden and Roedler 2015). Hence, diverse aspects (economic, environmental, legal, operational, behavioral, structural, and social aspects) can impact system configuration. These aspects are critical for capability demonstration. Any change in these aspects can lead to system reconfiguration (SR) to maintain the operational effectiveness of the system. In accordance, Qasim et al. (2019b) define system reconfiguration in systems engineering as subsequent changes of system configurations with the objective of maintaining or adapting (increasing or decreasing) the capabilities provided by the system. SR is of value for stakeholders. It allows for system performance, effectiveness, and affordability improvement to ensure increased reliability, availability, maintainability, testability, safety, reusability, and reuse of system entities and technologies. To support SR in the context of a large international company in aerospace, space, ground transportation, defense, and security domains, the model-based systems engineering (MBSE) approach is proposed. MBSE aims to reduce cost and time via models use and reuse, often including ontology development to support domain knowledge formalization. An initial SR ontology (*OSysRec*) development has been discussed and presented in Qasim et al. (2019a). SR relies on three aspects: The structural aspect concerns resources constituting the system, their functions, and the connections between them. Likewise, SR relies on the dynamic aspect which describes mechanisms of transitions and their causes and effects. The management aspect deals with optimizing the existing resources to meet the requirements of systems context and mission. In practice, this aspect is key for an efficient reconfiguration process. The *OSysRec* ontology synthesizes these aspects (structure, dynamics, and management) in one ontology.

This paper aims at applying the *OSysRec* ontology to a case study from the aerospace domain. For this purpose, the integrated modular avionics case study is used. The remainder of this paper is organized as follows: Section 2 gives the literature review related to ontologies or models used for reconfiguration purposes within different domains. Section 3 presents an overview of the *OSysRec* ontology. Section 4 displays the IMA case study to illustrate and test the proposed ontology. The last section discusses the results and draws conclusions.

2 Literature Review

Runtime reconfiguration (or reconfiguration during system operation) has been addressed and researched by several scientific domains proposing related definitions of this concept. These domains include *control engineering*, *software systems and computing*, *reconfigurable manufacturing systems*, *embedded systems*, and *autonomous systems* (Provan and Chen 1999; Rodriguez et al. 2009; Alsafi and Vyatkin 2010; Saxena et al. 2010; Krichen and Zalila 2011; Regulin et al. 2016). The definitions provided by these domains discuss system reconfiguration either in terms of transitions between modes or changing software or hardware elements. Thus, reconfiguration is referred to as a transition between system modes: functional or failure modes. Qasim et al. (2019b) have attempted to define SR in systems engineering as the subsequent changes of the system configurations with the objective of maintaining or adapting (increasing or decreasing) the capabilities provided by the system. Based on the definition of system configuration in systems engineering, we identified three main aspects that are essential for the management of systems when considering SR (Qasim et al. 2019a). These aspects include structure, dynamics, and management. The structural aspect integrates the resources constituting the system, their functions, and connections between them. The mechanisms of transitions and their causes and effects are referred to as the dynamic aspect. On the other hand, optimizing the existing resources with regard to the considered context and the mission is referred to as the management aspect.

A model-based approach that has risen to become a current trend in systems engineering is model-based systems engineering (MBSE) (Madni and Sievers 2017). MBSE aims at reducing the cost and time via models using and reusing (Wymore 1993). In MBSE, ontology is the commonly used approach toward domain knowledge construction (Medina-Oliva et al. 2014) and formalization. In practice, they provide the vocabulary for a domain and relations among them using formal language (Nadoveza and Kiritsis 2014). The object-oriented model is a way of representing ontologies in which they build on classes, properties of the classes, and instances. Classes represent the domain concepts, and the association relations represent their interactions (Gruber and Özsu 2009). Ontologies have already been used to describe different domains as presented in Obitko and Márík (2002) and Liang et al. (2011). Likewise, several ontologies aiming at supporting SR exist (Walsh et al. 2005, 2006; OMG 2010; Ali et al. 2011; Bermejo-alonso et al. 2011; Krichen and Zalila 2011; Witt et al. 2013; Gogniat et al. 2013; Meyer et al. 2013; Hernández et al. 2015; Bermejo-Alonso et al. 2016). In our research, we aim to propose an ontology for SR synthesizing the three essential aspects which we have previously identified and upon which relies the reconfiguration process from systems engineering perspective: structure, dynamics, and management. Correspondingly, we used these aspects to analyze and classify the existing ontologies. To our knowledge, no model or ontology which integrates these aspects exists (Qasim et al. 2019a). However, from the systems engineering perspective, a comprehensive model integrating all three aspects is needed for system management via reconfiguration. To address this gap, we proposed to integrate these aspects into a synthesizing SR ontology (*OSysRec*) to ensure effective and efficient system management (Qasim et al. 2019a). Section 3 gives an overview of the *OSysRec* ontology.

3 Overview of the OSysRec Ontology for System Reconfiguration

This section gives an overview of the *OSysRec* ontology which integrates the three aspects of SR conjointly. The extensive survey of the literature and the analysis of use case scenarios have allowed us to extract the fundamental concepts of the ontology and their characteristics (Qasim et al. 2019a). We have classified the identified concepts into three main categories representing the main three aspects of the reconfiguration process: structure, management, and dynamics.

Systems generally exist to satisfy a Mission which is the purpose to which all resources should be directed (see the management package of the *OSysRec* ontology, Fig. 1). The *Mission* definition is linked to the applied *Strategy*. A *Mission* can have a single or multiple *Mission Phases*. Each of these phases has its *Objectives* that can be achieved by realizing the mission *Tasks*. System reconfiguration can

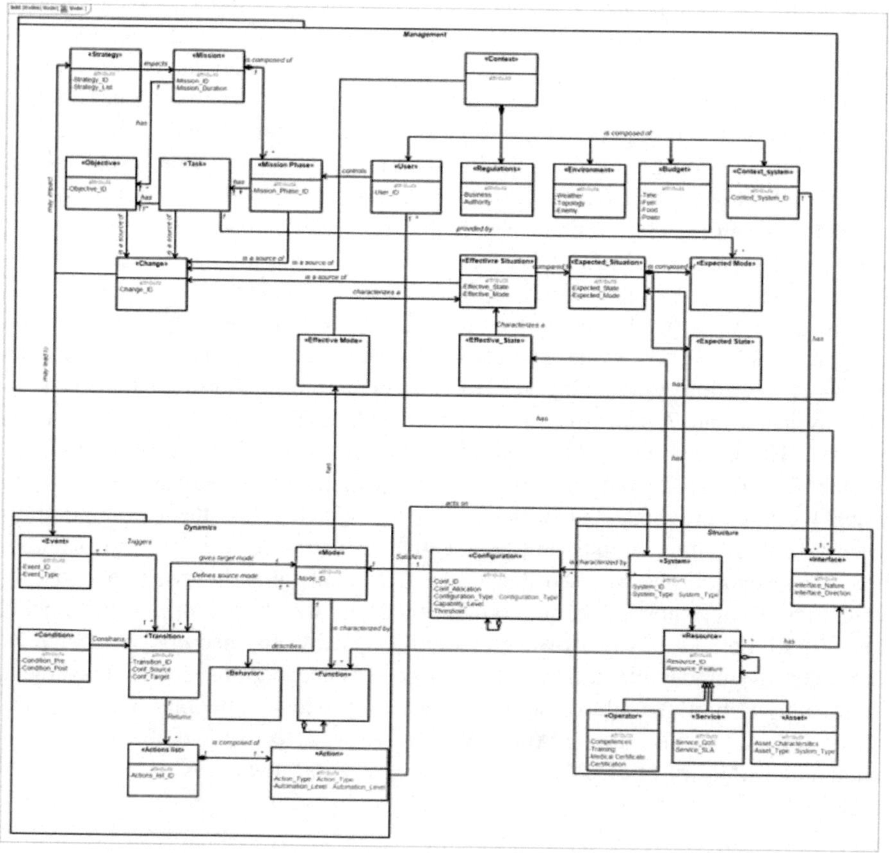

Fig. 1 *OSysRec* ontology

happen to adapt the system to its *Context* changes. Thus, it is crucial to understand what Context elements can impact systems, as systems do not work in isolation from their contexts. In the *OSysRec* management package, the *Context* is composed of *Regulations*, *Environment*, *Budget*, *Context System*, and *User*. If the *Mission* and the *Context* of use are constant, the system can still undergo a process of reconfiguration to adapt itself in case of faults. Therefore, information about the correctness of the expected behavior (*Effective Mode*) and the health state of the system (*Effective State*) should be reported to the management part. The *Effective Mode* and the *Effective State* characterize the *Effective Situation*. The *Effective Situation* is compared to the *Expected Situation* to detect changes. When a *Change* (mission, context, or situation change) is detected, an evaluation should be done, and actions need to be taken. *Actions* can either be at the management level by changing the *Strategy* or at the dynamic and structural levels by generating an *Even*t.

At the dynamic level, the *Event* triggers a *Transition* from one *Mode* to another when *Conditions* are met. Knowing the source and target *Modes*, an *Action List* is generated. Actions ensure the transition between modes and act directly on the system. The *Mode* is satisfied by a *Configuration* and it is characterized by *Functions*. Thus, we refer to *Configuration* as the allocation of *Functions* on *Resources*. Hence, it is obvious that *Configuration* represents the core element that links the dynamic and the structural levels.

The structural level shows that the *System* is characterized by *Configurations*. The *System* is composed of *Resources (*physical *Assets*, *Services*, or human *Operators)* and *Interface*s. Resources can be an aggregation of other resources. Similarly, their *Configurations* can be aggregated from the *Configuration* of the engaged *Resources.*

4 Case Study: Integrated Modular Avionics (IMA)

In this section, we demonstrate the application of the *OSysRec* ontology to integrated modular avionics (IMA) system which is currently deployed in modern aircraft architectures for both civil and military applications such as the Airbus A380 and the Rafale (Personnic 2002). IMA is a real-time computer network airborne system. IMA consists of a number of common function modules (CFM) populated in racks to allow for the replacement of the different modules. Six different CFM exist for IMA systems: (1) data processing module (DPM) for data-dependent processing activities, (2) signal processing module (SPM) for data-independent processing activities, (3) graphics processing module (GPM) for image composition and formatting, (4) mass memory module (MMM) for nonvolatile storing, (5) network support module (NSM) for network and protocol control, and (6) power conversion module (PCM) to allow for two-stage power conversion to 48V. These CFM will be used to run different application processes depending on the type of the application (program). A functional configuration in this context refers to a mapping of application processes to CFM. The functional configuration can have a logical configuration and a physical configuration (ASAAC 2004). The logical configuration defines the application requirements in terms of types and numbers of CFM, number of processing elements, and communication channels. The physical configuration, on the other hand, is one implementation of a logical configuration which is translated (instantiated) for a specific physical configuration.

In order to operate properly, the system will require a number of tested, verified, and certified configurations that should have been made available for operating the aircraft. During operation, the crew or the operator requests a system mode which can be satisfied by one or more logical and physical configurations. Mode selection is done via an application manager, and the reconfiguration is realized by the generic system management element. Within IMA systems, each mode can be satisfied by a nominal configuration, meaning that the required applications run on the main CFM. For each nominal configuration, one or more safety configurations exist. These configurations allow for running the same applications but on different CFM. In this case, the visible functionality does not change. Furthermore, for each nominal configuration, one or more degraded configurations exist. The degraded configurations provide a limited functionality of the IMA system. Changing between these associated logical or physical configurations will not impact the measured functionality level (external visible functionality). Changing from one mode to another may require a reconfiguration through the execution of action predefined in the runtime model. Civil and military avionics have different natures of reconfiguration. Civil IMA undergo static reconfiguration where only changing the hardware element is involved. On the contrary, IMA in the military applications can reconfigure SW elements dynamically.

The types of events that can be encountered within the IMA systems are (1) equipment fault, (2) software (application) fault, (3) change mode, and (4) global relaunch of IMA systems. Depending on the detected event and the current configuration, the system management can propose different actions. These actions may or may not require a reconfiguration. The possible actions include:

- No action: In this case, the processing on the considered modules continues.
- Program re-initialization: the program run on a given module is stopped and relaunched. In this case, no module reconfiguration.
- Program stop: in case of a software error, the program is stopped. When stopping a program, a change of the execution module is necessary.
- Module re-initialization: all the programs run on one module are stopped and relaunched.
- IMA re-initialization: all the programs on all modules are stopped and relaunched.
- Module stop: in this case, the IMA system will detect an absence of this module and hence a reconfiguration is necessary.

To illustrate the ontology described in Sect. 3, we use a simplified example of an IMA system with only four process modules and eight applications. Time intervals refer to events sequencing. Initially, at time t_0, the IMA system is organized to ensure the operation of the take-off (Fig. 2a). For this case, the needed applications for this logical configuration are 1, 2, 3, 4, 6, and 8 (see Fig. 2b). Applications 2

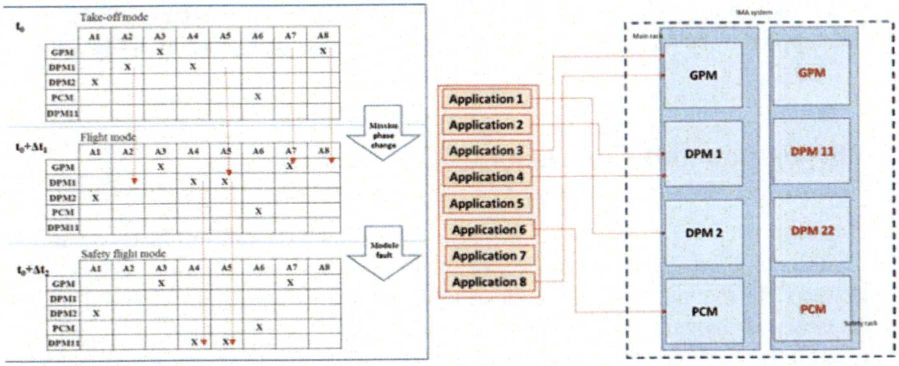

Fig. 2 (**a**) Mapping of applications on available modules of IMA. (**b**) Reconfiguration of IMA based on different events

and 4 are run on DPM1. Application 1 is run on DPM2. Application 6 is run on the PCM. Finally, the GPM is used to run applications 3 and 8. In this context, the applications and the modules are considered the *Resource*s. The mapping of the applications on modules represents the *Configuration*s. The Configuration satisfies a functional *Mode*, in this example the nominal take-off mode.

At time t0+Δt1, when the take-off is completed, the pilot requests the flight mode. The *User* controls the *Mission Phase.* For the flight mode (mission phase), the resources and their configuration are different. Hence, this mission change is considered as a *Change*. After evaluation with regard to the new mission objectives in terms of required resources, the result would be to generate an *Event* at the dynamic level in order to implement the Flight *Mode*. Technically, being in the take-off mode and receiving the *Event* to pass to the flight mode will trigger a *Transition* to implement the configuration corresponding to the Flight Mode (application 1 on DPM2, applications 4 and 5 run on DPM1, application 6 on the PCM, and applications 3 and 7 on the GPM). A *Condition* on this transition would be the availability of all the required applications and modules (*Resources*). The resulting *Actions list* can include the following *Actions*: (1) stop application 2 on DPM1, (2) run application 5 on DPM1, (3) stop application 8 on the GPM, and finally (4) run application 7 on the GPM. This procedure is illustrated in Fig. 2b.

At time t0+Δt2, a detection of a fault in the module DPM1 (*Resource*), while in the flight nominal mode (Effective_mode), will be reported to the management part via the *Effective_State*. Thus, comparing the IMA effective state for the current flight mode (*Effective_Situation*) to the expected one will lead to a change at the management level. This change when evaluated a signal of failure (SYSFAIL) is considered an event demanding a reconfiguration of the IMA at the dynamic level. This event triggers a transition toward the safety flight mode where DPM11 replaces DPM1. In this case, the condition on this transition is the availability of DPM11. In order to achieve the transition (reconfiguration), the actions needed are (1) stop application 5 on module DPM1, (2) stop module DPM1, (3) relaunch DPM11, and (4) run applications 4 and 5 on DPM11. This procedure is illustrated in Fig. 2b.

5 Discussion and Conclusions

The work presented in this paper is part of ongoing research initiated to support SR within industry. In a previous work, we proposed an ontology for SR. The proposed *OSysRec* ontology integrates the three aspects which are essential for the management of systems via reconfiguration: structure, dynamics, and management. This ontology is considered as a comprehensive conceptual framework within which it is possible to study and support dynamic evolutions of systems.

This paper seeks to go beyond the proposal of the *OSysRec* ontology and its validation within the defense domain. Thus, we propose to extend the validation process by using an additional case study from the aerospace industry. The scenario-based test and validation is used to demonstrate that the proposed ontology is sufficiently generic and can be applied to a variety of cases. For this purpose, we use a case study from our industrial context. In this case study, we discuss the integrated modular avionics (IMA) systems and we use it to test the *OSysRec* ontology. To do so, we used a simplified example of IMA systems. The demonstration shows that the *OSysRec* ontology is able to represent the chosen reconfiguration scenario of IMA systems.

The overarching research objective is to explore and propose a tool for SR (reconfiguration engine) based on the *OSysRec* ontology. Ongoing research activities are investigating the possibility of using multi-agent systems to develop such a tool. The main feature we seek to obtain from this tool is the evaluation of a system's ability to accomplish missions based on the system's health and context. Moreover, we expect this feature to support system users by informing them on actions that are necessary to ensure effectiveness.

Many issues and challenges can be encountered when developing reconfiguration features. In order to reconfigure systems based on models, it is necessary to use models which reflect reality exhaustively and precisely. Thus, future works should address model fidelity by developing formalisms and formal languages to allow for better representation of the real world. Systems which have the reconfiguration feature pass through transition states and modes. To master reconfiguration, we should be able to capture the system states and modes at any moment. For this reason, we believe that future work should address the states and modes determinism. Reconfiguration implies implementing configurations which have not been previously tested and validated via the IVVQ process. Thus, certifying is a real issue that needs to be addressed.

References

Ali, A.B.H., et al. 2011. Safe Reconfigurations of Agents-Based Embedded Control Systems. In *IECON Proceedings (Industrial Electronics Conference)*. IEEE, pp. 4344–4350. https://doi.org/10.1109/IECON.2011.6120023.

Alsafi, Y., and V. Vyatkin. 2010. Ontology-Based Reconfiguration Agent for Intelligent Mechatronic Systems in Flexible Manufacturing. *Robotics and Computer-Integrated Manufacturing* 26 (4): 381–391. https://doi.org/10.1016/j.rcim.2009.12.001.

ASAAC. 2004. *ASAAC Final Draft of Proposed Guidelines for System Issues Document reference: ASAAC2-GUI-32450-001-CPG.*

Bermejo-alonso, J., et al. 2011. Engineering An Ontology for Autonomous Systems – The OASys Ontology, pp. 47–58. https://doi.org/10.5220/0003634600470058.

Bermejo-Alonso, J., C. Hernandez, and R. Sanz. 2016. Model-Based Engineering of Autonomous Systems Using Ontologies and Metamodels. In *ISSE 2016 – 2016 International Symposium on Systems Engineering – Proceedings Papers*. https://doi.org/10.1109/SysEng.2016.7753185.

Gogniat, G., et al. 2013. Dynamic Applications on Reconfigurable Systems: From UML Model Design to FPGAs Implementation. In *2011 Design, Automation & Test in Europe*. IEEE, pp. 1–4. https://doi.org/10.1109/date.2011.5763315.

Gruber, T., and M. Özsu. 2009. *Encyclopedia of Database Systems. Ontology.*

Hernández, C., et al. 2015. Model-Based Metacontrol for Self-adaptation. In *International Conference on Intelligent Robotics and Applications*. Springer, pp. 643–654. 10.1007/978-3-319-22879-2_58.

ISO/IEC:15288. 2015. *ISO/IEC/IEEE/15288: Systems and Software Engineering – System Life Cycle Processes.*

Krichen, F., and B. Zalila. 2011. Towards a Model-Based Approach for Reconfigurable DRE Systems Towards a Model-Based Approach for Reconfigurable DRE Systems. In *European Conference on Software Architecture*. Springer, pp. 295–302. https://doi.org/10.1007/978-3-642-23798-0.

Liang, Q., et al. 2011. Ontology-Based Business Process Customization for Composite Web Services. *IEEE Transactions on Systems, Man, and Cybernetics Part A: Systems and Humans* 41 (4): 717–729. https://doi.org/10.1109/TSMCA.2011.2132710. IEEE.

Madni, A.M., and M. Sievers. 2017. Model-Based Systems Engineering: Motivation, Current Status, and Needed Advances. In *Disciplinary Convergence in Systems Engineering Research*, pp. 311–325. https://doi.org/10.1007/978-3-319-62217-0_22.

Medina-Oliva, G., et al. 2014. Predictive Diagnosis Based on a Fleet-Wide Ontology Approach. In *Knowledge-Based Systems*. Elsevier B.V., 68, pp. 40–57. https://doi.org/10.1016/j.knosys.2013.12.020.

Meyer, F., et al. 2013. An Approach for Knowledge-Based IT Management of Air Traffic Control Systems. In *2013 9th International Conference on Network and Service Management, CNSM 2013 and Its Three Collocated Workshops – ICQT 2013, SVM 2013 and SETM 2013*, pp. 345–349. https://doi.org/10.1109/CNSM.2013.6727856.

Nadoveza, D., and D. Kiritsis. 2014. Ontology-Based Approach for Context Modeling in Enterprise Applications. *Computers in Industry* 65 (9): 1218–1231. https://doi.org/10.1016/j.compind.2014.07.007.

Obitko, M., and V. Mářík. 2002. Ontologies for Multi-agent Systems in Manufacturing Domain. In *Proceedings – International Workshop on Database and Expert Systems Applications, DEXA*. IEEE, 2002–Janua, pp. 597–602. https://doi.org/10.1109/DEXA.2002.1045963.

OMG (Object Management Group). 2010. UML profile for MARTE Object Management Group.

Personnic, G. 2002. Asaac: The Way to Flying Military Open Systems. *3rd European Systems Engineering Conference Systems Engineering: A focus of European Expertise Pierre Baudis Congress Centre, Toulouse, 21st–24th May 2002.*

Provan, G., and Y.-L. Chen. 1999. Model-Based Diagnosis and Control Reconfiguration for Discrete Event Systems: An Integrated Approach. In *Proceedings of the 38th IEEE Conference on Decision and Control*, pp. 1762–1768. https://doi.org/10.1109/CDC.1999.830888.

Qasim, L., A.M. Hein, S. Olaru, et al. 2019a. An Overall Ontology for System Reconfiguration Using Model-Based System Engineering. *Submitted to IEEE Transactions on Systems, Man, and Cybernetics: Systems*.

Qasim, L., A.M. Hein, M. Jankovic, et al. 2019b. Towards a Reconfiguration Framework for Systems Engineering Integrating Use Phase Data. In *Proceedings of ICED 2019, the 22nd International Conference on Engineering Design: Responsible Design for Our Future, Delft, Netherlands, 05.-08.08. 2019*.

Regulin, D., et al. 2016. Model Based Design of Knowledge Bases in Multi Agent Systems for Enabling Automatic Reconfiguration Capabilities Of Material Flow Modules. In *IEEE International Conference on Automation Science and Engineering*, pp. 133–140. https://doi.org/10.1109/COASE.2016.7743371.

Rodriguez, I.B., et al. 2009. A Model-Based Multi-level Architectural Reconfiguration Applied to Adaptability Management in Context-Aware Cooperative Communication Support Systems. In *2009 Joint Working IEEE/IFIP Conference on Software Architecture and European Conference on Software Architecture. WICSA/ECSA 2009*: 353–356. https://doi.org/10.1109/WICSA.2009.5290829.

Saxena, T., et al. 2010. Enabling Self-Management by Using Model-Based Design Space Exploration. In *Seventh IEEE International Conference and Workshops on Engineering of Autonomic and Autonomous Systems*, pp. 137–144. https://doi.org/10.1109/EASe.2010.22.

Walden, D., and G. Roedler. 2015. *INCOSE Systems Engineering Handbook: A Guide for System Life Cycle Processes and Activities*, 4th ed. Available at: http://eu.wiley.com/WileyCDA/WileyTitle/productCd-1118999401.html.

Walsh, D., F. Bordeleau, and B. Selic. 2005. A Domain Model for Dynamic System Reconfiguration. pp. 553–567. https://doi.org/10.1007/11557432_42.

———. 2006. Change Types of Dynamic System Reconfiguration. in *Proceedings of the International Symposium and Workshop on Engineering of Computer Based Systems*, pp. 3–12. https://doi.org/10.1109/ECBS.2006.28.

Witt, R., et al. 2013. Implementation of Fault Management Capabilities for the Flying Laptop Small Satellite Project through a Failure-Aware System Model. pp. 1–16. https://doi.org/10.2514/6.2013-4661.

Wymore, A.W. 1993. *Model-BASED Systems Engineering*. CRC Press.

Reducing Design Rework Using Set-Based Design in a Model-Centric Environment

Shawn Dullen, Dinesh Verma, and Mark Blackburn

Abstract Digital engineering (DE) provides an opportunity to reduce engineering design rework. However, this potential depends significantly on the approach used in exploring the design tradespace. A classical approach, using a traditional point-based design (PBD), has the likelihood of creating engineering rework even within the context of digital engineering. To address limitations of PBD, several researchers have proposed the use of set-based design (SBD). However, there is no formal definition of SBD and there is limited guidance on how to effectively implement it in a DE environment. To address such concerns, a literature review was conducted around the following questions: (1) What is the current state of SBD application – approaches and models? (2) What are the strengths and limitations to these approaches and models? (3) How is knowledge developed, captured, and reused to cause convergence of sets? (4) What DE tools were recommended/used to enable SBD?

Keywords Rework · Set-based design · Lean product and process development · Product development · Systems engineering · MBSE · Model-centric engineering

1 Introduction

Engineering design rework has been a persistent problem in product/system development for several decades. Most design efforts have significant rework, accounting for 30–67% of the total design hours on a project. It has a detrimental impact on a program's cost and schedule as indicated by several Government Account-

S. Dullen (✉)
Combat Capabilities Development Command (CCDC) Armaments Center, Picatinny Arsenal, NY, USA
e-mail: shawn.m.dullen.civ@mail.mil

D. Verma · M. Blackburn
Systems Engineering Research Center (SERC), Stevens Institute of Technology, Hoboken, NJ, USA

A. M. Madni et al. (eds.), *Recent Trends and Advances in Model Based Systems Engineering*, https://doi.org/10.1007/978-3-030-82083-1_18

ability Office (GAO) reports (GAO 2011, 2014). Based on the literature review conducted by Dullen et al. (2019), most of the research addressing rework has been on understanding the impact of information exchange and organizational structure related to task dependencies, process execution, project complexity, design complexity, information evolution, and information completeness. Several modeling approaches have been used to understand the nature of rework. These include design structure matrix models (Cho and Eppinger 2001), system dynamic models (Taylor and Ford 2006), dynamic network models (Braha and Bar-Yam 2007), discrete-event simulation models (Yang et al. 2014), graphical evaluation and review technique network models (Nelson et al. 2016), and agent-based models (Lévárdy and Browning 2009). The primary reasons for engineering rework have been identified to be downstream sensitivity to upstream preliminary information, upstream sensitivity to downstream feedback, information uncertainty (e.g., design mistakes, requirement changes, model fidelity, knowledge level, feedback delays, etc.), information ambiguity (e.g., novel design), and organizational misalignment. The varying modeling approaches used by the researchers help with understanding and analyzing variables influencing rework, but they do not provide means to improve these variables (task and design dependencies, uncertainty, sensitivity, and ambiguity). Potential improvements can include desensitizing activities to information changes, improving knowledge evolution, and effectively reusing knowledge.

Model-centric engineering (Blackburn et al. 2018), more formally known as digital engineering (DE) (Bone et al. 2018), can address some of these design rework concerns. DE has the potential to allow better communications (Bone et al. 2018) and accelerate learning across a team (West and Blackburn 2018), improve knowledge discovery (Bandaru et al. 2017), and improve data reuse (Li et al. 2018; Noy 2004). However, a classical approach to converging too early to a point design (point-based design – PBD) can get in the way of these potential benefits (Kennedy et al. 2014). PBD requires decisions to be made with uncertain information (Canbaz et al. 2014), and when new information becomes available, the design needs to be reworked. The sequential approach to PBD also suffers from the potential of constraints being imposed by one design team (e.g., thermal analysis) to cause design rework (Kennedy et al. 2014) by another team (e.g., structures). Singer et al. (2009) propose the use of set-based design (SBD) as an alternative to traditional PBD. In SBD, engineers develop a set of design alternatives in parallel; as design progresses, they gradually narrow their prospective set of alternatives based on additional information until they converge to a final solution (Sobek II et al. 1999). SBD has been identified as a methodology that is robust to changes in information (Kennedy et al. 2014). Limited experimental results from Thomas A McKenney et al. (2011a) indicate SBD is robust to requirement changes. B. M. Kennedy et al. (2014) identified the positive impact of SBD on engineering and design rework. This is because the engineering team is able to (1) accelerate learning using rapid prototyping and limit curves to represent knowledge; (2) use sets (instead of point values) for stakeholder needs, system specification, and sub-

subsystem specifications, where final specifications emerge from the convergent learning process; and (3) reuse knowledge developed and captured for narrowing the tradespace.

Conceptually, implementing DE using SBD methodology can significantly reduce rework. The engineering and design community can benefit from a methodology that encapsulates the precepts of SBD to allow its promulgation. This seems to be a limitation. There is little evidence of such a methodology within the literature (Al-Ashaab et al. 2013; Ghosh and Seering 2014). Case studies reflecting the implementation of SBD are sparse in the open literature. This further limits an understanding of how the conceptual underpinnings of SBD can be translated into a pragmatic set of methods (Ammar et al. 2017; Dag S Raudberget 2011). In addition, there seems to be a misconception of what SBD is resulting in confused discussions regarding its similarities and differences vis-à-vis multi-attribute utility theory and multi-objective optimization (Thomas Abbott McKenney 2013). Irrespective, there has been a significant research interest on SBD over the past last 20 years given its potential benefits. The remainder of this paper is as follows: (2) Background, (3) Literature Review, and (4) Conclusion and Future Work.

2 Background

SBD's first literature appearance is in Ward (1989) dissertation on "A Theory of Quantitative Inference for Artifact Sets, Applied to a Mechanical Design Compiler." Allen Corlies Ward (1989) made inferences on sets of artifacts to eliminate the sets that would not perform, rather than to identify optimal solutions. Allen C Ward et al. (1990) demonstrated constraint propagation of intervals can be used to reason about a set of physical objects. What distinguishes SBD from the more widely practiced PBD is the emphasis on reasoning about sets of design options (Braha et al. 2013). Allen C Ward and Seering (1993) used an SBD approach to apply their theory of quantitative inference on a mechanical design problem. SBD is also known as set-based concurrent engineering (SBCE). A. Ward et al. (1995) define SBCE as an approach where designers focus on design alternatives at both the conceptual and parametric levels. As elimination of inferior alternatives occurs, the sets gradually narrow until there is convergence upon a final solution.

In the early 1990s, several studies conducted by Allen Ward, Jeffery Liker, John Cristiano, and Durward Sobek generated more literature development breakthroughs for SBD. They surveyed 92 Japanese and 119 US automotive companies (Liker et al. 1996) and performed empirical testing to show that high-performing companies use a set-based approach, as opposed to their lower-performing counterparts. The results of the studies revealed Toyota demonstrated a more advanced set-based approach than any of the other Japanese and US automotive companies (Ward et al. 1995). There were four key observations made about Toyota's SBCE approach (Ward et al. 1995): (1) system and subsystem solutions were defined as sets; (2) exploration of subsystems was done in parallel to systems solutions; (3)

sets of solutions slowly converged to a point solution using analysis, design rules, and experiments; (4) teams stayed committed to a converged point solution and did not change unless it was absolutely necessary.

Toyota's principles of SBCE were later discovered and documented by Sobek II et al. (1999) during their subsequent research of Toyota product development process. The first principle is to map the design space which includes defining feasible regions, exploring trade-offs by designing multiple alternatives, and communicating the sets of possibilities. The second principle is to integrate by intersections such as looking for intersections of feasible sets, imposing minimum constraints, and seeking conceptual robustness. The third principle is to establish feasibility before commitment by narrowing sets gradually while increasing the details for the design, staying within sets once committed, and maintaining control by managing uncertainty at process gates. Ward (2007) further expanded upon the previous work on SBCE in his manuscript on Lean Product and Process Development (LPPD) that was published in 2007. This work focused on emergent learning. M. Kennedy (2008) later goes on to develop the learning first product development model that integrates the knowledge value stream with the product value stream.

3 Literature Review

Limited guidance on how to apply SBD is available in the literature. To narrow this gap, this literature review focused on the state of the art for SBD procedural models. Procedural models are models that convey best practices intended to guide real-world situations (Wynn and Clarkson 2018). This section provides a brief overview of the literature published in the past 10 years that intends to provide guidance in the form of procedural models for SBD, SBCE, and LPPD. A comparison between the models is in Table 1 (at the end of Sect. 3.1). The following questions are addressed: (1) What are the strengths and limitations to these approaches and models? (2) How is knowledge developed, captured, and reused to cause convergence of sets? (3) What DE tools were recommended/used to enable SBD?

3.1 Procedural Models

In support of a collaborative European research project titled "Lean Product and Process Development," Al-Ashaab et al. (2010) developed the Conceptual Lean Product and Process Development (cLPPD) model, expanding upon the previous research on Toyota's principles of SBCE (Sobek II et al. 1999). M. Khan et al. (2013) further refined the cLPPD model to include five core enablers: (1) SBCE process, (2) chief engineer technical leadership, (3) value-focused planning and development, (4) knowledge-based environment, and (5) continuous improvement culture. To allow for more practical application of the cLPPD, M. Khan et al. (2011)

Table 1 SBD process model comparison

Source	What are the strengths and limitations to these approaches and models?	How is knowledge developed, captured, and reused to cause convergence of sets?	What DE tools were recommended/used to enable SBD?
Ström et al. (2016)	The process model provides general guidance on methods that can be used in a day to generate and evaluate alternatives. The process is aligned to two of three SBCE principles (mapping the design space and integrating by intersection). There are limited ties to SE technical processes. The tools recommended for model implementation are limited to innovation techniques and Pugh matrix. No mention of rapid prototyping	How knowledge will be developed, captured, and reused was not defined	No evidence of DE
Dag S Raudberget (2011)	The process model provides a rich set of activities for developing alternatives. The model does not have a clear tie to the SE processes. The model lacks detail for stakeholder needs, requirements definition, architecture definition, interface definition, decision analysis, rapid prototyping, and how sets are established and narrowed. Very limited guidance on tools and methods to implement model. No use of rapid prototyping	Knowledge will be developed by testing. How knowledge will be captured and reused was not defined	No evidence of DE

(continued)

Table 1 (continued)

Source	What are the strengths and limitations to these approaches and models?	How is knowledge developed, captured, and reused to cause convergence of sets?	What DE tools were recommended/used to enable SBD?
Wade (2018)	The process model provides a rich understanding of the design parameters that drive the overall value of the system through the use of data visualization and statistical tests. The process has limited ties to the SE technical processes. No guidance was provided on how to narrow sets while improving the level of abstraction of the system. No use of rapid prototyping. There was limited guidance on tools and methods to enable process	Knowledge is developed using performance and cost models with Monte Carlo simulations. The knowledge was captured in the form of trade-off curves. How knowledge will be reused was not defined	Trade study performed using Engineered Resilient Systems Tradespace Toolkit
B. M. Kennedy et al. (2014)	The process model provides ties to the SE technical processes. There is a strong emphasis on rapid prototyping. Limited guidance was provided on how to map the design space, develop sets, communicate sets, and narrow sets. There was limited guidance on the tools and methods to enable process	Knowledge is developed using rapid prototyping and captured in the form of limit curves. Knowledge reuse was not clearly defined	No evidence of DE

(continued)

developed a SBCE process model with five phases. Each phase is decomposed into lower-level steps (e.g., 1.1). M. S. Khan (2012) dissertation proposes the finalized LPPD model with more refined lower-level sub-steps (e.g., 1.1.1) for each of the phases. Maulana et al. (2016) provide a detailed (step-by-step) application case study instantiating M. Khan et al. (2011) SBCE process model on a surface jet pump and highlight the use of specific methods and tools.

Table 1 (continued)

Source	What are the strengths and limitations to these approaches and models?	How is knowledge developed, captured, and reused to cause convergence of sets?	What DE tools were recommended/used to enable SBD?
Frye (2010)	The process model includes developing organization structure, highlights the difference in narrowing sets at different abstraction levels (e.g., system, subsystem, and component), and distinguishes measure types (e.g., discrete vs. continuous). There is no clear tie to the SE technical processes (e.g., stakeholder needs, requirements definition, architecture definition, and verification). No use of rapid prototyping or methods for developing alternatives	Knowledge is developed from DOEs. How the information will be captured and reused was not clearly defined	Performance data was developed using computer code (software/tool not specified). Statistical analysis was performed using JMP
Mebane et al. (2011)	The process model directly ties to the TMRR phase of the DOD life cycle. The model provides clear flow of detailed activities to include their interrelationships. The tools and methods to enable the process are not well defined. The process did not include activities for functional and logical architecture. No mention or use of rapid prototyping. There was limited guidance on the tools and methods that should be used to enable the process. Model lacks detail for how to narrow sets while improving the level of abstraction	Knowledge is developed from DOEs. How the information will be captured and reused was not clearly defined	The only tool explicitly defined was DOORS. The use of analysis, data management, and statistical tools were mentioned, but no specific tool was defined

(continued)

Table 1 (continued)

Source	What are the strengths and limitations to these approaches and models?	How is knowledge developed, captured, and reused to cause convergence of sets?	What DE tools were recommended/used to enable SBD?
Garner et al. (2015)	The process model incorporates the following SE technical processes and technical management processes: mission analysis, stakeholder needs, and decision management. The process lacks details for alternative generation, functional and logical architecture, and how to narrow sets while improving the level of abstraction. No use of rapid prototyping. It was not clear what tools and methods should be used to enable process	Knowledge is developed from DOEs. Knowledge captured as regression models and trade-off curves. How knowledge will be reused was not defined	Performance data was developed using Advanced Evaluation Tool and Rapid Ship Design Environment. Statistical analysis performed using JMP. Trade study performed using Engineered Resilient Systems Tradespace Toolkit
Ammar et al. (2017)	The process model ties nicely with the SE technical processes during the early phases of product development. There is limited guidance on how to execute process. The tools and methods are not clearly defined. There is no guidance on how to narrow sets while improving the level of abstraction. No use of rapid prototyping. Verification and validation were listed as activities; however the techniques were not (modeling and simulation, testing, or the combination of both)	Requirements are captured in tables, and architecture information is captured in SysML. There is no evidence of design data being developed, captured, or reused	SysML was utilized for architecture

(continued)

Table 1 (continued)

Source	What are the strengths and limitations to these approaches and models?	How is knowledge developed, captured, and reused to cause convergence of sets?	What DE tools were recommended/used to enable SBD?
Gumina (2019)	The process model provides clear guidance on how to create sets, communicate sets, integrate sets, and narrow sets. Sets were created using design of experiments. The process lacks details for alternative generation, architecture definition, and how to narrow sets while improving the level of abstraction of the system. There are limited ties to SE technical processes. No use of rapid prototyping	Knowledge is developed using low-fidelity modeling of the measures of effectiveness and measures of performance using Latin hypercube sampling. Knowledge is captured in graphical format (e.g., trade-off curves) and communicated using action reports. How knowledge will be reused was not defined	No evidence of DE
Ammar et al. (2018)	The process model provides general guidance on mapping the design space, integrating by intersection, and establishing feasibility before commitment. There are limited ties to SE technical processes: stakeholder needs, requirements definition, architecture definition, design definition, integration, verification, and validation. No use of rapid prototyping. The tools and methods to enable the process were not clearly defined	Knowledge is developed from DOEs. The results are captured in the form of trade-off curves. Requirements are captured in tables, and architecture information is captured in SysML. How knowledge will be reused was not defined	DOE results came from CAE models that were developed and integrated using Modelica. Optimization was performed using ModelCenter. Architecture was developed in SysML. This approach was effective at facilitating communication between functional team

(continued)

Table 1 (continued)

Source	What are the strengths and limitations to these approaches and models?	How is knowledge developed, captured, and reused to cause convergence of sets?	What DE tools were recommended/used to enable SBD?
M. S. Khan (2012)	The process model is very detailed and incorporates some aspects of systems engineering activities with SBD such as stakeholder needs, requirements definition, design definition, and integration. For each of the sub-steps, there is a recommended list of tools to enable activities and guidance for innovation, conceptual robustness, and producibility. It is not clear how the model is integrated into a concept development process. The model lacks detail for architecture definition, interface definition, decision analysis, rapid prototyping, and how to narrow sets while improving the level of abstraction	Knowledge will be developed for life-cycle cost, quality, and performance using simulations, prototyping, and testing for each design alternative. The knowledge will then be captured in graphical formats (e.g., limit curves and trade-off curves). This knowledge will be communicated using an A3 report. A3 reports will be reused for later decisions and future projects	Limit curves, trade-off curves, and A3 reports were stored in knowledge database (product lifecycle management (PLM)). Computer-aided design (CAD) and computer-aided engineering (CAE) were mentioned but no specific tool defined. There was no evidence of the use of the Systems Modeling Language (SysML) or integration of CAE models
Al-Ashaab et al. (2013)	The process model has a clear tie to implementing SE using SBD during TMRR phase of DOD life cycle. The model includes applicable SE technical processes and technical management processes. Tools and methods were listed for each step of the process without clear guidance on how to implement them. Model lacks detail for how to narrow sets while improving the level of abstraction	Knowledge is developed from design of experiments (DOEs) and functional means analysis. The information is documented in the form of trade curves and decision matrices. The process identifies the reuse of information from previous projects but not what type of information or how it would be facilitated	No evidence of DE

(continued)

Table 1 (continued)

Source	What are the strengths and limitations to these approaches and models?	How is knowledge developed, captured, and reused to cause convergence of sets?	What DE tools were recommended/used to enable SBD?
Dobrovolskyte (2015)	The process model has the same strengths as Al-Ashaab et al. (2013) with the additional benefit of using terms more familiar for design engineers. No clear guidance on how to execute model or the tools to implement model. Model lacks detail for how to narrow sets while improving the level of abstraction	Knowledge is developed, captured, and reused as Al-Ashaab et al. (2013)	Knowledge is developed, captured, stored, and reused using SBD Navigator. SBD Navigator includes CAD, CAE, CAE model integration, A3 reports, requirements, architecture, etc. The tool is utilized at the system and subsystem level

In 2013, Al-Ashaab et al. (2013) developed a new RR-LeanPD model that integrates M. S. Khan (2012) LPPD model with Rolls-Royce System Design and Integration (SD&I) model. The RR-LeanPD model was developed for the first two SD&I system design reviews (SDR). The first SDR is focused on defining customer value and understanding system requirements. The second SDR is focused on developing system-level specification and assuring system concepts are capable of meeting system-level specification. The RR-LeanPD model uses an integrated systems engineering approach (Walden et al. 2015) with SBCE, intended for early product development equivalent to the Technology Maturation and Risk Reduction (TMRR) phase of the DOD acquisition life cycle (Defense 2015). As a part of the Configuration Optimization of Next Generation Aircraft (CONGA) project initiative (Al-Ashaab et al. 2014), the RR-LeanPD model was further refined to align with Rolls-Royce vision of SBD (Dobrovolskyte 2015). The new procedure model is the Rolls-Royce Set-Based Design (RR-SBD) model. The modifications allowed for improved SBD collaboration within Rolls-Royce (Dobrovolskyte 2015). The CONGA project also defined a phased-gate model that uses a model-centric tool, Knowledge Shelf, that captures, compares, and reuses key project design information to support designers with knowledge they need to implement SBD (Al-Ashaab et al. 2014). In 2016, Araci et al. (2016) developed a process to generate trade-off curves, further improving the SBCE process developed by M. S. Khan (2012).

In 2010, as part of a 3-year joint-venture study between Swedish industry, the School of Engineering in Jönköping University, and Swerea IVF Research Institute, D. Raudberget (2010) investigated if SBCE (Sobek II et al. 1999) could improve the efficiency and the effectiveness of a company's development process. His research identified barriers for implementing SBCE, the need for a methodology or model to help implement SBCE, and recommendations for implementing/introducing SBCE. Next, Dag S Raudberget (2011) developed a ten-step framework to implement SBCE for system-level design and detailed-level design. In 2014, D. S. Raudberget et al. (2014) integrated functional means modeling for platform design with SBCE. Later in 2016, Ström et al. (2016) developed an instant set-based design (ISBD) framework considering just two principles of SBCE (Sobek II et al. 1999). They emphasized utilizing generic methods to enable teaching and implementation of SBD in 1 day. In 2017, Ammar et al. (2017) developed a proposed methodology called DMIV (Development, Mapping, Integration, and Verification) that has 2 phases and 21 steps. The next year, Ammar et al. (2018) developed a new SBD process model to include verification and validation. The process was applied to an electronic throttle body using the Systems Modeling Language (SysML).

The US Navy and their affiliated universities, the University of Michigan, Naval Postgraduate School, and Massachusetts Institute of Technology, have performed a significant amount of research on SBD. One of the most cited papers on SBD for the Navy that has sparked a lot of interest in the research on SBD was written in 2009 by Singer et al. (2009). This paper documented general implementation guidance on SBD, described early research conducted at the University of Michigan (1999–2003) on SBD, and described how SBD can help the Navy design ships during the early phases of product development, during the Material Solution Analysis (MSA) phase of the DOD acquisition life cycle (Defense 2015).

More guidance on implementing SBD can be found in Frye (2010) publication on a generic SBD framework model, which is based on lessons learned from the Ship-to-Shore Connector (SSC) project. The process includes six high-level steps. This research indicates that a definition of the organizational structure must occur prior to executing SBD. The following year Mebane et al. (2011) generated a more detailed SBD process model integrating SBD into the TMRR phase of the DOD life cycle for the SSC project. Thomas A McKenney et al. (2011b) developed a limited-scope SBD process model to study the impact of requirement changes to the SBD process. A set reduction process was later introduced by T. McKenney et al. (2012) for SCC that incorporated an SBD rigor standard to evaluate design activity against five SBD elements: (1) characterization, (2) flexibility, (3) convergence, (4) communication, and (5) facilitation. In 2013, Thomas Abbott McKenney (2013) published dissertation proposes a decision support framework that introduces the use of design space mapping, longest path problem (LPP) as a Markov decision process, and preference change simulations.

In response to the Secretary of Defense direction to the Department of the Navy, the Small Surface Combatant Task Force (SSCTF) developed a SBD process to submit alternate proposals for procurement of a capable and lethal small surface combatant (Garner et al. 2015). The authors describe the process of developing a model-centric environment to implement aspects of their SBD process using tools such as Advanced Surface Ship and Submarine Evaluation Tool (ASSET), Rapid Ship Design Environment (RSDE), and Engineered Resilient Systems (ERS) Tradespace Toolkit. Using an example use-case illustration, Parker et al. (2017)

provide guidance on set reduction during the MSA phase of the DOD life cycle. Toshon et al. (2017) expanded upon T. McKenney et al. (2012) set reduction process using Taguchi robust design methods for Naval shipboard power systems. In 2019, in her dissertation, Gumina (2019) proposes an "improved SBD process" due to a speculation that the Navy process models are missing intermediate steps.

B. M. Kennedy et al. (2014) propose a process model for the front end of the SE Vee model (Walden et al. 2015). Yannou et al. (2013) developed a SBD process model that integrates physics-based models with usage models. The usage scenarios consider the physical surroundings, social surroundings, temporal surroundings, task definition, and antecedent states. In his thesis, Wade (2018) proposed a convergent SBD process. The process is iterative, where the fidelity and sample size increase with increased narrowing of the design space. Specking et al. (2018) proposed a SBD process for early design tradespace exploration that is integrated with model-centric engineering.

The procedural models with the most details were selected to address the following questions: (1) What are the strengths and limitations to these approaches and models? (2) How is knowledge developed, captured, and reused to cause convergence of sets? (3) What DE tools were recommended/used to enable SBD? The results in Table 1 are ordered by progressive level of detail from least to most descriptive.

4 Conclusion and Future Work

Based on this literature review, the Rolls-Royce System Design and Integration model (Al-Ashaab et al. 2013) provided the most detail on how to implement SBD during the equivalent TMRR phase of the DOD life cycle. This model is clearly tied to the SE technical processes and technical management processes. The tools, methods, and templates to execute the process were clearly defined and readily available for those within the Rolls-Royce internal organization. A more complete picture would have included how the tools and methods are used to define sets and reason about them, as well as descriptions on how knowledge is developed and reused. There exists an opportunity to solidify guidance on how to implement SBD, how to narrow sets while improving the level of abstraction of the design, how to define and reason about sets, and how to reuse knowledge. There also exist opportunities to improve the connection of SBD to SE technical processes.

The most detailed illustration of DE was the SBD Navigator (Dobrovolskyte 2015). The SBD Navigator was developed to enable the activities defined in the RR-SBD model (Dobrovolskyte 2015). This approach was missing an industrial application, a way to define and reason about sets, and how to narrow sets while improving the level of abstraction and analysis. Therefore, there exists an opportunity to develop a DE approach to execute SBD for knowledge development, capturing, and reuse. The approach should include use of multi-fidelity models, multi-physics models, and their integration. An observation made based on literature review was there was no distinction of developing knowledge for ambiguous problems versus uncertain problems. Another observation was there was no indication that the models used for developing knowledge were verified or validated; in order for this to be a viable option for deductive reasoning, the models need to have some level of model validation.

Based on this literature review, the following research questions are proposed:

- What is the most effective way to develop, capture, and reuse knowledge during the TMRR phase of the DOD life cycle?
- What level of model fidelity is required for SBD?
- When do critical long-lead items become PBD decisions?
- Can ambiguous problems be handled the same way as uncertain problems for SBD?
- How can models be shared in a noncooperative environment?
- Does a noncooperative environment impede SBD?
- What challenges exist that prevent SBD from being applied to industrial applications?

References

Al-Ashaab, A., E. Shehab, R. Alam, A. Sopelana, M. Sorli, M. Flores, et al. 2010. The Conceptual LeanPPD Model. In *New World Situation: New Directions in Concurrent Engineering*, 339–346. Springer.

Al-Ashaab, A., M. Golob, U.M. Attia, M. Khan, J. Parsons, A. Andino, et al. 2013. The Transformation of Product Development Process into Lean Environment Using Set-Based Concurrent Engineering: A Case Study from an Aerospace Industry. *Concurrent Engineering 21* (4): 268–285.

Al-Ashaab, A., M. Golob, J. Oyekan, Z.C. Araci, M. Khan, D. Deli, and E. Al-Ali. 2014. Flying Into Aerospace's Next Generation. *Industrial Engineering 46* (10): 38–43.

Ammar, R., M. Hammadi, J.-Y. Choley, M. Barkallah, J. Louat, and M. Haddar. 2017. *Architectural Design of Complex Systems Using Set-Based Concurrent Engineering.* Paper presented at the Systems Engineering Symposium (ISSE), 2017 IEEE International.

Ammar, R., M. Hammadi, J.-Y. Choley, M. Barkallah, and J. Louati. 2018. *Mechatronic System Design with Manufacturing Constraints Using Set-Based Concurrent Engineering.* Paper presented at the 2018 Annual IEEE International Systems Conference (SysCon).

Araci, Z.C., A. Al-Ashaab, and M. Maksimovic. 2016. *Knowledge Creation and Visualisation by Using Trade-off Curves to Enable Set-Based Concurrent Engineering.* Cranfield University.

Bandaru, S., A.H. Ng, and K. Deb. 2017. Data Mining Methods for Knowledge Discovery in Multi-objective Optimization: Part A-Survey. *Expert Systems with Applications 70*: 139–159.

Blackburn, M., Verma, D., Dillon-Merrill, R., Blake, R., Bone, M., Chell, B, and Giffin, R. (2018). *Transforming systems engineering through model centric engineering.* Stevens Institute of Technology Hoboken, United States.

Bone, M.A., M.R. Blackburn, D.H. Rhodes, D.N. Cohen, and J.A. Guerrero. 2018. Transforming Systems Engineering Through Digital Engineering. *The Journal of Defense Modeling and Simulation* 1548512917751873.

Braha, D., and Y. Bar-Yam. 2007. The Statistical Mechanics of Complex Product Development: Empirical and Analytical Results. *Management science 53* (7): 1127–1145.

Braha, D., D.C. Brown, A. Chakrabarti, A. Dong, G. Fadel, J.R. Maier, et al. 2013. *DTM at 25: Essays on Themes and Future Directions.* Paper presented at the ASME 2013 international design engineering technical conferences and computers and information in engineering conference.

Canbaz, B., B. Yannou, and P.-A. Yvars. 2014. Preventing Design Conflicts in Distributed Design Systems Composed of Heterogeneous Agents. *Engineering Applications of Artificial Intelligence 28*: 142–154.

Cho, S.-H., and S.D. Eppinger. 2001. *Product Development Process Modeling Using Advanced Simulation.* Paper presented at the ASME 2001 Design Engineering Technical Conferences and Computers and Information in Engineering Conference.

Defense, D.o. (2015). DoD Instruction 5000.02: Operation of the Defense Acquisition System. In: Author Washington, DC.

Dobrovolskyte, N. 2015. *Preliminary Design Enhancement by Incorporating Set-Based Design Principles and a Navigator.* Cranfield University.

Dullen, S., D. Verma, and M. Blackburn. (2019). *Review of Research into the Nature of Engineering and Development Rework: Need for a Systems Engineering Framework for Enabling Rapid Prototyping and Rapid Fielding.* Paper presented at the 17th Annual Conference on Systems Engineering Research (CSER), Washington, DC.

Frye, M.C. 2010. *Applying Set Based Methodology in Submarine Concept Design.* Massachusetts Inst of Tech Cambridge.

GAO. 2011. *Trends in Nunn-McCurdy Breaches and Tools to Manage Weapon Systems Acquisition Costs.* Retrieved from (GAO Publication No. GAO-10-106). Washington, DC: U.S. Government Printing Office.

Government Accountability Office. (2013). Where Should Reform Aim Next. (GAO Publication No. GAO-14-145T). Washington, DC: U.S. Government Printing Office.

Garner, M., N. Doerry, A. MacKenna, F. Pearce, C. Bassler, S. Hannapel, and P. McCauley. 2015. *Concept exploration methods for the Small Surface Combatant.* Paper presented at the World Maritime Technology Conference.

Ghosh, S., and W. Seering. 2014. *Set-Based Thinking in the Engineering Design Community and Beyond.* Paper presented at the Proceedings of the ASME 2014 International Design Engineering Technical Conferences & Computers and Information in Engineering Conference.

Gumina, J.M. 2019. *A Set-Based Approach to Systems Design* Ph. D. thesis, Naval Postgraduate School.

Kennedy, M. 2008. *Ready, Set, Dominate: Implement Toyota's Set-Based Learning for Developing Products and Nobody Can Catch You.* Oaklea Press.

Kennedy, B.M., D.K. Sobek II, and M.N. Kennedy. 2014. Reducing Rework by Applying Set-Based Practices Early in the Systems Engineering Process. *Systems Engineering 17* (3): 278–296.

Khan, M.S. 2012. *The Construction of a Model for Lean Product Development.* Cranfield University.

Khan, M., A. Al-Ashaab, A. Doultsinou, E. Shehab, P. Ewers, and R. Sulowski. 2011. Set-Based Concurrent Engineering Process Within the LeanPPD Environment. In *Improving Complex Systems Today*, 433–440. Springer.

Khan, M., A. Al-Ashaab, E. Shehab, B. Haque, P. Ewers, M. Sorli, and A. Sopelana. 2013. Towards Lean Product and Process Development. *International Journal of Computer Integrated Manufacturing 26* (12): 1105–1116.

Lévárdy, V., and T.R. Browning. 2009. An Adaptive Process Model to Support Product Development Project Management. *IEEE Transactions on Engineering Management 56* (4): 600–620.

Li, Z., X. Zhou, W. Wang, G. Huang, Z. Tian, and S. Huang. 2018. An Ontology-Based Product Design Framework for Manufacturability Verification and Knowledge Reuse. *The International Journal of Advanced Manufacturing Technology* 99: 1–15.

Liker, J.K., D.K. Sobek, A.C. Ward, and J.J. Cristiano. 1996. Involving Suppliers in Product Development in the United States and Japan: Evidence for Set-Based Concurrent Engineering. *IEEE Transactions on Engineering Management 43* (2): 165–178. https://doi.org/10.1109/17.509982.

Maulana, M., J.W. Flisiak, A. Al-Ashaab, Z.C. Araci, P.W. Lasisz, N. Beg, and A. Rehman. 2016. The Application of Set-Based Concurrent Engineering to Enhance the Design Performance of Surface Jet Pump. *WSEAS Transactions on Business and Economics 13*: 634–643.

McKenney, T.A. 2013. *An Early-Stage Set-Based Design Reduction Decision Support Framework Utilizing Design Space Mapping and a Graph Theoretic Markov Decision Process Formulation*. Ph. D. thesis, University of Michigan.

McKenney, T.A., L.F. Kemink, and D.J. Singer. 2011a. Adapting to Changes in Design Requirements Using Set-Based Design. *Naval Engineers Journal 68*: 3.

———. 2011b. Adapting to Changes in Design Requirements Using Set-Based Design. *Naval Engineers Journal 123* (3): 67–77.

McKenney, T., M. Buckley, and D. Singer. 2012. *The Practical Case for Set-Based Design in Naval Architecture*. Paper presented at the International Marine Design Conference.

Mebane, W.L., C.M. Carlson, C. Dowd, D.J. Singer, and M.E. Buckley. 2011. Set-Based Design and the Ship to Shore Connector. *Naval Engineers Journal 123* (3): 69–82. https://doi.org/10.1111/j.1559-3584.2011.00332.x.

Nelson, R.G., A. Azaron, and S. Aref. 2016. The Use of a GERT Based Method to Model Concurrent Product Development Processes. *European Journal of Operational Research 250* (2): 566–578.

Noy, N.F. 2004. Semantic Integration: A Survey of Ontology-Based Approaches. *ACM Sigmod Record 33* (4): 65–70.

Parker, M., Garner, M., Arcano, J., and Doerry, N. (2017). Set-based requirements, technology, and design development for SSN (X). Warship 2017. Submarines & UUVs, 14-15. June 2017.

Raudberget, D. 2010. Practical Applications of Set-Based Concurrent Engineering in Industry. *Strojniski Vestnik 56* (11): 685–695.

Raudberget, D.S. 2011. *Enabling Set-Based Concurrent Engineering in Traditional Product Development*. Paper presented at the DS 68-1: Proceedings of the 18th International Conference on Engineering Design (ICED 11), Impacting Society through Engineering Design, Vol. 1: Design Processes, Lyngby/Copenhagen, Denmark, 15.-19.08. 2011.

Raudberget, D. S., M.T. Michaelis, and H.L. Johannesson. 2014, December 9–12. *Combining Set-Based Concurrent Engineering and Function—Means Modelling to Manage Platform-Based Product Family Design*. Paper presented at the 2014 IEEE International Conference on Industrial Engineering and Engineering Management.

Singer, D.J., N. Doerry, and M.E. Buckley. 2009. What is Set-Based Design? *Naval Engineers Journal 121* (4): 31–43.

Sobek, D.K., II, A.C. Ward, and J.K. Liker. 1999. Toyota's Principles of Set-Based Concurrent Engineering. *MIT Sloan Management Review 40* (2): 67.

Specking, E., G. Parnell, E. Pohl, and R. Buchanan. 2018. Early Design Space Exploration with Model-Based System Engineering and Set-Based Design. *Systems 6* (4): 45.

Ström, M., D. Raudberget, and G. Gustafsson. 2016. Instant Set-Based Design, an Easy Path to Set-Based Design. *Procedia CIRP 50*: 234–239.

Taylor, T., and D.N. Ford. 2006. Tipping Point Failure and Robustness in Single Development Projects. *System Dynamics Review: The Journal of the System Dynamics Society 22* (1): 51–71.

Toshon, T., R. Soman, C. Wiegand, M. Israel, M. Faruque, and M. Steurer. 2017. *Set-Based Design for Naval Shipboard Power Systems Using Pertinent Metrics from Product Development Tools*. Paper presented at the 2017 IEEE Electric Ship Technologies Symposium (ESTS).

Wade, Z. 2018. *Convergent Set-Based Design in Integrated Analysis of Alternatives: Designing Engineered Resilient Systems*. University of Arkansas.

Walden, D.D., G.J. Roedler, K.J. Forsberg, D. Hamelin, and T.M. Shortell. 2015. *INCOSE Systems Engineering Handbook: A Guide for System Life Cycle Processes and Activities*. 4th ed. Wiley.

Ward, A.C. 1989. *A Theory of Quantitative Inference for Artifact Sets, Applied to a Mechanical Design Compiler*. Massachusetts Institute of Technology.

Ward, A. C. 2007. Lean Product and Process Development Cambridge, MA: The Lean Enterprise Institute Incorporation.

Ward, A.C., and W.P. Seering. 1993. Quantitative Inference in a Mechanical Design 'Compiler'. *Journal of Mechanical Design 115* (1): 29–35.

Ward, A.C., T. Lozano-Perez, and W.P. Seering. 1990. Extending the Constraint Propagation of Intervals. *AI EDAM 4* (1): 47–54.

Ward, A., J.K. Liker, J.J. Cristiano, and D.K. Sobek. 1995. The Second Toyota Paradox: How Delaying Decisions can make Better Cars Faster. *Sloan management review 36*: 43–43.

West, T.D., and M. Blackburn. 2018. Demonstrated Benefits of a Nascent Digital Twin. *INSIGHT 21* (1): 43–47.

Wynn, D.C., and P.J. Clarkson. 2018. Process Models in Design and Development. *Research in Engineering Design 29* (2): 161–202.

Yang, Q., T. Lu, T. Yao, and B. Zhang. 2014. The Impact of Uncertainty and Ambiguity Related to Iteration and Overlapping on Schedule of Product Development Projects. *International Journal of Project Management 32* (5): 827–837.

Yannou, B., P.-A. Yvars, C. Hoyle, and W. Chen. 2013. Set-Based Design by Simulation of Usage Scenario Coverage. *Journal of Engineering Design 24* (8): 575–603.

Knowledge Representation and Reasoning in the Context of Systems Engineering

Hanumanthrao Kannan

Abstract Large-scale systems engineering projects may be construed essentially as multi-agent problems wherein decisions are made by several people (stakeholders, managers, designers, etc.) across the organizational hierarchy. All agents in an enterprise possess knowledge in one form or the other, be it the knowledge gained from requirements gleaned from stakeholders, domain-specific knowledge, knowledge of rules and regulations, knowledge gained from experience on other projects, etc. Lack of a formal means of representing the knowledge shared among these agents often results in miscommunication which in turn results in poor decision-making and, thereby, schedule delays and cost overruns. Such issues can hinder the competitive advantage in a mission-critical environment. It is equally important to capture the knowledge possessed by systems engineers, who have a great deal of experience having worked on multiple long-term and large-scale complex projects, who are leaving the workforce. This paper focuses on formally capturing knowledge that exists in various phases of systems engineering lifecycle by leveraging epistemic modal logic. The approach in this paper aims to address some of the issues with the traditional document-centric approaches.

Keywords Knowledge representation · Reasoning · Systems engineering · Formal logic · Epistemic modal logic

1 Motivation and Introduction

Current systems engineering practices use document-centric approaches to represent and communicate information within the organization (examples include requirements documents, interface control documents, etc.) (NASA 2007). With the rising complexity of large-scale complex engineered systems, document-centric

H. Kannan (✉)
Virginia Polytechnic Institute and State University, Blacksburg, VA, USA
e-mail: hkannan@vt.edu

A. M. Madni et al. (eds.), *Recent Trends and Advances in Model Based Systems Engineering*, https://doi.org/10.1007/978-3-030-82083-1_19

approaches bring with them a host of challenges. A major limitation is the difficulty of such approaches in ensuring traceability to the origin of the information. Version control is another big challenge in cases like having to update a diagram or a table that recurs in a document or across multiple documents. In short, document-centric approaches are cumbersome, open to misinterpretation, difficult to update, and difficult to be reused and have extremely limited reasoning capabilities. With the advent of digital engineering, it is all the more critical to transition from document-centric approaches to model-centric approaches.

Model-based systems engineering (MBSE) is a new approach to system development that treats a system "model" as the sole source of truth (Madni and Sievers 2018). The model is essentially a central repository that stores all the system-related information, and it continually evolves as the system development progresses. This approach has the capability to create documents representing the current state of the system tailored to different stakeholders/agents. It provides a common visualization of the evolving system, thereby improving communication and reasoning based on the shared knowledge. It aims to improve quality while reducing costs and time to market. However, MBSE is still in its infancy and has yet to provide a significant demonstrable improvement in terms of elimination of rework, cost and time reduction, etc. It requires methodological advances and development of supporting tools and processes (Madni and Sievers 2018). The biggest roadblock to widespread adoption of MBSE in the systems engineering community, however, is the deeply ingrained culture of using document-centric approaches.

Some researchers have adopted model-centric approaches, to represent knowledge, specifically in the context of requirements (Fraga et al. 2015, 2019), evaluate inconsistencies in MBSE (Herzig 2015), etc. These approaches, although promising, have limited reasoning capabilities. Some of the capabilities that the current approaches lack include assessing what an agent knows and does not and what an agent knows about the knowledge of other agents, thereby identifying knowledge gaps and potential knowledge gathering activities, making inferences with existing knowledge, etc. Formal knowledge representation is not a new research area. Researchers in computer science have used multiple approaches to represent knowledge – rule-based systems, semantic nets, frames, and epistemic modal logic (widely used in distributed systems).

This paper focuses on expanding epistemic modal logic (Ditmarsch et al. 2015) to enable representation and reasoning of knowledge in systems engineering. The approach provided in this paper will facilitate the following:

- Assessing the state of knowledge of agents
- Evaluation of inconsistencies in knowledge in a multi-agent organization
- Communication of knowledge to other entities in the organization
- Inferences from the knowledge base

The paper is organized as follows. Section 2 provides a discussion of knowledge in systems engineering, and the necessary formalism for knowledge representation is provided. In Sect. 3, descriptive examples are used to demonstrate the capabilities of the knowledge representation formalism, followed by the conclusion and future work in Sect. 4.

2 Knowledge Representation in Systems Engineering

Any organization that acquires, develops, and operates large-scale complex engineered systems is comprised of a number of entities (subsystem teams, disciplines, contractors, etc.), wherein decisions are made by several people (stakeholders, team managers, engineers, etc.). Knowledge is essential to solve complex problems in systems engineering. Knowledge about the system exists and evolves at all phases of the system lifecycle, including but not limited to functional and structural knowledge, properties, conditions, assumptions, constraints, knowledge about interfaces and the environment (physical and organizational), etc. Any knowledge pertinent to the system is only as good as its reusability. In order to be reusable, knowledge needs to be stored in a repository that facilitates easy access, understanding, and reasoning, across the organizational hierarchy.

The knowledge of agents in a multi-agent organization developing large-scale complex engineered systems can be broadly categorized into *descriptive* and *procedural* knowledge. *Descriptive* knowledge involves basic facts, whereas *procedural* knowledge deals with "if... then... " rules. For example, the proposition "satellite antenna gain is directly proportional to signal-to-noise ratio" is *descriptive* knowledge, whereas "if antenna material is aluminum, then use aluminum fixtures" is *procedural*.

A major motivation behind representing knowledge in a formal manner is reasoning. Reasoning includes assessing the state of knowledge of individuals in a multi-agent organization, making inferences with existing knowledge, and evaluating inconsistencies in knowledge. Specifically assessing what an agent knows and does not know, what an agent knows about the knowledge of other agents, and vice versa is helpful in identifying the knowledge gaps that can be fulfilled through knowledge gathering activities. Making inferences with existing knowledge is applicable in a number of scenarios including feasibility analysis, making decisions consistent with the knowledge, etc. It is also critical to identify and address inconsistencies in the knowledge base of a single agent and multiple agents, as a failure to do so may lead to expensive rework in the later stages of the system lifecycle. One can argue that a formal framework that can enable representation and reasoning of knowledge in a systems engineering context must have all the reasoning capabilities discussed above. It is for these reasons that epistemic modal logic has been adopted in this paper. The following subsection provides the necessary background for epistemic modal logic.

2.1 Epistemic Modal Logic

Syntax for Epistemic Modal Logic

As with any formal language, the syntax for epistemic modal logic consists of a non-empty set (Φ) of atomic propositions that represent basic facts about the situation under consideration and are usually denoted by p, q, r, etc. "It is raining" and "Mars is a planet" are examples of atomic propositions. Compound sentences or formulas, typically represented using Greek symbols φ, ψ , etc., can be formed by closing off under conjunction and negation. Additionally, we have an additional modal operator (K_i) that represents if agent (e.g., stakeholder) i *knows* something or not. The epistemic or knowledge operator (K) can be applied to atomic propositions as well as to formulas. For example, if φ is a formula, then $K_1(\varphi)$ means "agent 1 knows φ." Another example of a compounded formula using K can be seen in Eq. 1:

$$K_1\left(K_2\left(\varphi\right)\right) \wedge K_2\left(\neg K_1\left(\psi\right)\right) \tag{1}$$

Equation 1 means "Agent 1 *knows* that agent 2 *knows* φ and agent 2 *knows* that agent 1 does not *know* ψ." The set of primitive propositions, all formulas that can be formed using the atomic propositions, connectives and the modal operators K,and the agent number (denoted by subscript), together define language L. Such a formal language L that enables representation and reasoning of knowledge can be defined formally by the following *Backus-Naur/Backus normal form* (BNF)(McCracken and Reilly 2003):

$$\varphi := p \mid \neg\varphi \mid (\varphi \wedge \psi) \quad \mid K_i\left(\varphi\right) \tag{2}$$

Semantics for Epistemic Modal Logic

In epistemic modal logic, knowledge is evaluated based on the classical *possible-worlds* approach. In the case of the knowledge operator, an individual *knows* a statement if and only if the statement is true in all the states he considers possible. Such a concept behind knowledge can be represented and evaluated using Kripke models (Ditmarsch et al. 2015). A Kripke model for a single-agent epistemic logic (M) is a structure given by the tuple shown in Eq. 3, where i corresponds to the agent:

$$M = (\Omega, P_i, \pi) \tag{3}$$

In Eq. 3, Ω is the set of all states, also sometimes called the domain of M. For example, in the case of an agent rolling a die, this can be considered as the sample space, $\Omega = \{1,2,3,4,5,6\}$. In the context of large-scale systems, Ω can be considered as the set of possible alternatives or values a decision-maker considers possible for a particular attribute or a vector of attributes. In Eq. 3, P_i is a binary relation on

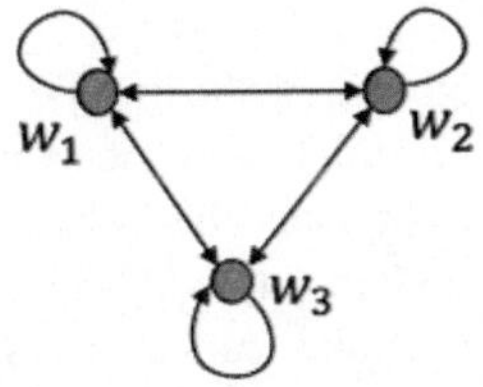

Fig. 1 Possible worlds

Table 1 Properties of accessibility relation

Properties	Description	Class of Kripke models
Reflexive	If $(w, w) \in P_i$ for all $w \in \Omega$	kT
Symmetric	If $(u, v) \in P_i$ implies that $(v, u) \in P_i$ for all $u, v \in \Omega$	kB
Transitive	If $(u, v), (v, w) \in P_i$ implies $(u, w) \in P_i$, for all $u, v, w \in \Omega$	$k4$
Equivalence	P_i is reflexive, transitive, and symmetrical	$S5$

Ω, called the possibility relation or accessibility relation. P_i can be considered as a function representing the states that an agent considers possible when in a particular state. This can be visualized using a directed graph as shown in Fig. 1. The arrows in the figure indicate states that an agent considers possible in a particular state. For instance, at w_1, the agent considers w_1, w_2, and w_3 to be the possible worlds. Formally P_i can be defined as follows:

$$\begin{aligned} &P_i(w) = \{w' : (w, w') \in P_i\} \\ &P_i \subset \Omega \times \Omega \end{aligned} \tag{4}$$

The class of all Kripke models is denoted by k. A number of classes of Kripke models can be defined in terms of the properties of the possibility or accessibility relations P_i. These properties (Fagin et al. 2004, Ditmarsch et al. 2015) define how an agent perceives these cognitive attitudes. Some of the properties that can be defined on P_i to capture the agent's perceptions are given in Table 1.

Considering the possibility relation to be *reflexive* is the primary property that distinguishes knowledge from belief. *Reflexive* means that the agent always considers the actual state (w) to be in the set of possible states (P_i). In Eq. 3, π is a valuation function, as defined in Eq. 5, that assigns truth values to each of the atomic propositions in Φ (set of all atomic propositions) at each state, i.e., $\pi(w, p) = \text{TRUE}$ means that the proposition p is true at state w. The state w is emphasized here as the truth assignment changes when the state changes:

$$\pi(w) : \Phi \rightarrow \{TRUE, FALSE\} \text{ for each state } w \in \Omega \tag{5}$$

With the elements of the Kripke structure defined, the semantic relation can be recursively defined as $(M, w) \vDash \varphi$ which can be read equivalently as "φ is true in structure M at state w" or "structure M satisfies φ at state w." Equation 6 states that atomic proposition p is TRUE at state w in structure M, if and only if π assigns *TRUE*. This is the same as in propositional logic. The atomic propositions represent basic facts about the domain we are interested in reasoning about. For example, "The system passed the test" may be considered as an atomic proposition. Compound formulas can be constructed using the atomic propositions, conjunction ($\wedge$), and negation ($\neg$) terms, as seen in Eq. 7, which represents the satisfaction of compound formulas involving conjunction and negation:

$$(M, w) \models p \; iff \; \pi\,(w, p) = \text{TRUE} \tag{6}$$

Also, we have

$$\begin{aligned} &(M, w) \models \varphi \wedge \psi \; iff \; (M, w) \models \varphi \text{ and } (M, w) \models \psi \\ &(M, w) \models \neg\varphi \; iff \; (M, w) \not\models \varphi \end{aligned} \tag{7}$$

Similarly, "(M, w) satisfies the proposition that *agent i knows* φ" is defined as follows:

$$(M, w) \models K\,(\varphi) \iff \left(M, w'\right) \models \varphi \text{ for all } w' \in P_i(w) \tag{8}$$

This means that the agent knows φ if and only if φ is true in all the states the agent considers possible at state w in structure M. Here w' represents all the worlds that are considered possible by the agent, as dictated by the possibility or accessibility relation P_i.

The properties of the notion of knowledge are further characterized through some of the following axioms and rules of inference, which represent some form of idealizations (Fagin et al. 2004, Ditmarsch et al. 2015) (Table 2).

Axiomatic logic systems can be formed by carefully selecting the axioms and rules of inference to reflect the problem of interest. The validity of these axioms depends on how we represent the possibility relation. For instance, when we consider P_i to be an equivalence relation (reflexive, symmetric, and transitive), then we can see that axiom T follows from P_i being reflexive, axiom 4 from transitive, and axiom 5 from symmetric and transitive. Various axiom systems can be constructed to capture how we want the possibility relation to look like. For instance, system **S5**, which includes axioms K, T, 4, and 5, is typically used to capture properties of knowledge.

The following section provides simple descriptive examples that demonstrate the reasoning capabilities of epistemic modal logic in a systems engineering context.

Table 2 Axioms and rules of inference

Axioms/rules of inference	Mathematical representation	Description
All instances of propositional tautologies		All axioms and rules of inference associated with propositional logic
Distribution axiom (K)	$K_i\varphi \wedge K_i(\varphi \Rightarrow \psi)) \Rightarrow K_i\psi$	Agents are powerful reasoners who know all the logical consequences of their knowledge. For example, if an agent *knows* φ and *knows* that φ *implies* ψ, then he *knows* ψ
Knowledge generalization rule (N)	From φ *infer* $K_i\varphi$	This means that if a statement φ is true in all the states the agent considers possible, then the agent *knows* φ. This does not mean that the agent knows all statements that are true
Knowledge or truth axiom (T)	$K_i\varphi \Rightarrow \varphi$	Agent only *knows* things that are true
Positive introspection axiom (4)	$K_i\varphi \Rightarrow K_iK_i\varphi$	Axioms 4 and 5 mean that the agent has introspection of his own knowledge base, i.e., the agent knows what he knows and does not know
Negative introspection axiom (5)	$\neg K_i\varphi \Rightarrow K_i \neg K_i\varphi$	
Modus ponens	From φ and $\varphi \Rightarrow \psi$ *infer* ψ	

3 Descriptive Examples

3.1 State of Knowledge of Single Agent

Let us consider two atomic propositions – *p: Electrical system is verified* and *q: Mechanical system is verified.* Assuming that the agent has no information available to her at the moment, she considers multiple situations (i.e., worlds) to be possible that are consistent with the information (i.e., no information). The possible worlds considered by the agent are represented in Fig. 2a, where in w_1, both p and q are true; in w_2, p is true and q is false; and so on. The arrows indicate the accessibility/possibility relation. Suppose the agent receives information from the mechanical systems engineer that all the verification activities in the mechanical system have passed. This information makes it possible for the agent to know q and thereby reduces the set of possible worlds to just w_1 and w_2 as shown in Fig. 2b.

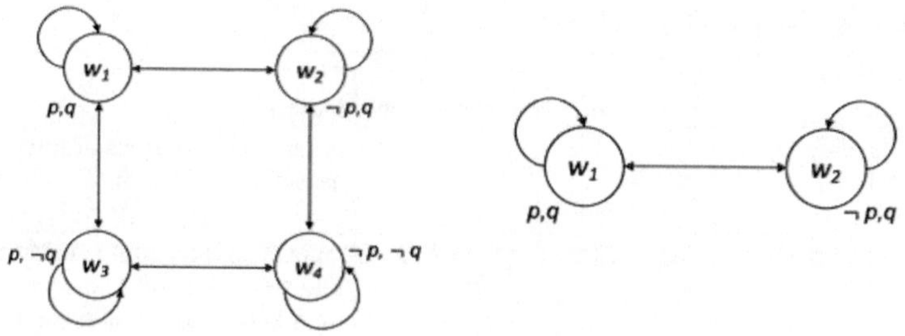

Fig. 2 (**a**) State of knowledge – single agent with no information. (**b**) With information on q

The following represents the state of knowledge of the agent at w_1:

- At w_1, the agent knows q, since q is true in all worlds that are accessible from w_1 – represented by the following equation:

$$(M, w_1) \models K(q) \tag{9}$$

- Also, at w_1, the agent does not know p, as represented by the following equation:

$$(M, w_1) \models \neg K(p) \tag{10}$$

As mentioned earlier, one of the major advantages of using epistemic logic in representing knowledge is to reason about what an agent knows and does not know at a particular moment. Knowing this "state of knowledge" of an agent is useful in identifying steps that aid in achieving a specific knowledge state from an initial state.

3.2 State of Knowledge of Multiple Agents

Let us consider two atomic propositions – q*: Design activity x done on time* and p*: Design activity y done on time* – and two agents A (designer A) and B (designer B) responsible for design activities x and y, respectively. Designers A and B belong to different teams and are working on separate components the need to be integrated. Assuming that p and q are the only propositions in consideration, the state of knowledge of both the designers can be visualized using the graph in Fig. 3.

The following can be said about the state of knowledge of designers A and B.

Designer A:

- Designer A knows q, $(M, w_1) \models K_A(q)$.
- Designer A does not know p, $(M, w_1) \models \neg K_A(p)$.

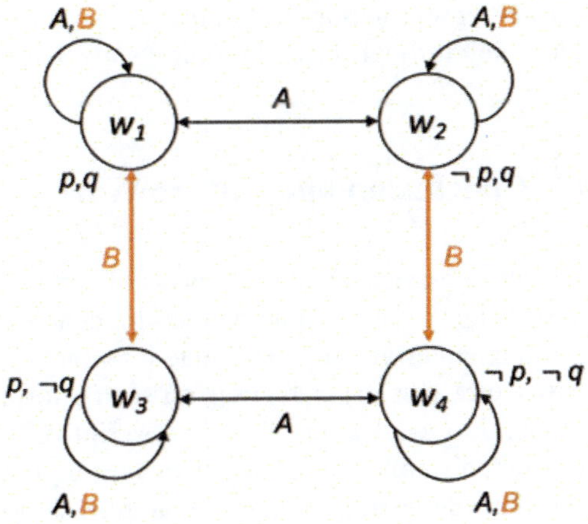

Fig. 3 State of knowledge – multiple agents

- Designer A knows that B knows whether p is true or not, $(M, w_1) \vDash K_A(K_B(p) \vee K_B(\neg p))$.

Designer B:

- Designer B knows p, $(M, w_1) \vDash K_B(p)$.
- Designer B does not know q, $(M, w_1) \vDash \neg K_B(q)$.
- Designer B knows that A knows whether q is true or not, $(M, w_1) \vDash K_B(K_A(q) \vee K_B(\neg q))$.

3.3 Evaluation of Inconsistencies

Let us consider a scenario wherein the stakeholder (S) and the system architect (A) are in the process of finalizing the requirements. The stakeholder has three requirements represented as propositions, *p: The satellite shall image Earth in IR band*; *q: The satellite shall image Earth in X band*; and *r: The cost of the satellite shall be under $100M*. The state of knowledge of the architect can be represented as follows:

- $K_A(K_S(p \wedge q \wedge r))$.
- However, the architect knows that it is impossible to have cost under $100M if both p and q are to be satisfied. This is represented as $K_A((p \wedge q) \rightarrow \neg r)$.

From these two knowledge statements, one can see that the knowledge of the two agents are contradictory, i.e., there is an inconsistency. The architect can then bring this to the attention of the stakeholder and request for a revision of requirements. This simple example demonstrates the cognitive process of the architect. For a case with only a few requirements, it may be straightforward to reason without a formal representation. But in the case of an actual system development, cognitively reasoning through several requirements may be not adequate.

4 Conclusion and Future Work

Knowledge exists in all the aspects of system development. It is crucial to represent knowledge in a formal manner to enable reuse and effective communication, avoid misinterpretation, facilitate inferences, evaluate inconsistencies in knowledge bases, etc. This paper focuses on such a formal representation of knowledge using epistemic modal logic. Simple descriptive examples are used to demonstrate the use of epistemic modal logic in representing and reasoning of knowledge in a systems engineering context. Examples show that such a representation will aid in reasoning about the state of knowledge of agents, inferences related to knowledge, and evaluation of inconsistencies in knowledge bases. As seen in the examples, an epistemic modal logic-based representation and reasoning of knowledge is applicable to any phase in the system lifecycle, i.e., architecture, design, verification, etc.

The process of translating the existing knowledge in the form of models, diagrams, texts, etc. to logical statements is challenging. Future work will focus on developing a framework to facilitate such translation. Another research direction will be to explore representation and reasoning of uncertainties within the context of epistemic modal logic. Information and evaluation are two main aspects in any decision-making activity. This paper is a step toward representing information as knowledge. In order to make decisions (or evaluate), one needs an additional layer. Future work will focus on having requirements and/or preferences as an additional layer for evaluation. Future work will also focus on integrating the proposed approach in SysML.

References

Ditmarsch, H., J.Y. Halpern, W. van der Hoek, and B.P. Kooi. 2015. *Handbook of Epistemic Logic*. College Publications.

Fagin, R., J.Y. Halpern, Y. Moses, and M. Vardi. 2004. *Reasoning About Knowledge*. MIT Press.

Fraga, A., J. Llorens, L. Alonso, and J.M. Fuentes. 2015. *Ontology-Assisted Systems Engineering Process with Focus in the Requirements Engineering Process. Complex Systems Design & Management*, 149–161. Springer.

Fraga, A., J. Llorens, and G. Génova. 2019. Towards a Methodology for Knowledge Reuse Based on Semantic Repositories. *Information Systems Frontiers* **21** (1): 5–25.

Herzig, S.J. 2015. *A Bayesian Learning Approach to Inconsistency Identification in Model-Based Systems Engineering*. Georgia Institute of Technology.

Madni, A.M., and M. Sievers. 2018. Model-Based Systems Engineering: Motivation, Current Status, and Research Opportunities. *Systems Engineering* 21 (3): 172–190.

McCracken, D.D., and E.D. Reilly. 2003. Backus-naur form (bnf).

NASA. 2007. *NASA Systems Engineering Handbook*. National Aeronautics and Space Administration: Washington, DC.

Ontology-Driven Knowledge Modeling and Reasoning for Multi-domain System Architecting and Configuration

Leonard Petnga

Abstract Our work is concerned with the development of model-based systems engineering (MBSE) procedures for multi-domain system architecting, configuration, and reasoning. This class of problems is characterized by the presence of multiple domains (which could be cyber, physical, or hybrid), each with their rules and constraints that need to be integrated in a correct-by-design systems and processes in order to satisfy stringent constraints on performance, safety, and effective management of system functionality. To that aim, there is a strong need for formal methods of analysis that can enable effective assembly of components into correct-by-design and provable systems capable of delivering desired capabilities. Thus, this paper discusses semantics and their central role in the development of a new ontology-driven knowledge modeling and reasoning approach for multi-domain system architecting and configuration. Three interdependent modules supporting each other are integrated to make up the system. A foundation module (1) with description logics (DL) as core knowledge representation formalism provides the necessary mathematical foundations to a semantic platform module (2) constituted of semantic blocks (i.e., ontology, rules, and specialized computational capabilities). A configurator module (3) later assembles systems from instantiated semantic blocks as per architectural configurations of interest to generate valid system design alternatives. An implementation of the system on rule-based generation of architectural configurations for satellite robotic arms demonstrates the capability of our approach.

Keywords Ontology · Architectural configuration · Multi-domain · Semantics Web · Model-based systems engineering (MBSE)

L. Petnga (✉)
University of Alabama in Huntsville, Huntsville, AL, USA
e-mail: leonard.petnga@uah.edu

A. M. Madni et al. (eds.), *Recent Trends and Advances in Model Based Systems Engineering*, https://doi.org/10.1007/978-3-030-82083-1_20

1 Introduction

This paper examines knowledge representation and modeling and their use for effective multi-domain system architecting and configuration. The central role ontologies play in formally capturing and representing the domains of interest across disciplines is discussed. A particular attention is paid to augmenting the ontologies with associated domain and design rules as well as needed computational capabilities in order to enable the systematic creation of system semantic models consistent to selected/desired configurations, from libraries of configurations and components. We review and use description logic (DL) as the knowledge representation formalism for our framework. DL extensions as well as mapping to the web ontology language (OWL) are investigated to ensure decidability of reasoning within and across domains. We develop and introduce a three-part modular architecture comprising DL formalisms, semantic platform, and a system configurator. The mathematical foundations provided by the DL formalisms to a reconfigurable semantic platform enable it to empower a system configurator in creating and assembling a full, multi-configuration design space. We illustrate the capability of our approach on a problem that involves the successful rule-based generation of a two-configuration design space for a satellite robotic arm using a small library of components (e.g., arms and tool).

2 System Ontology and Description Logic Semantics Support for Knowledge Representation and Reasoning

2.1 Engineering Systems: Overview and Conceptual Representation

Our view of a system is as a collection of interconnected components having system-level functionality (as a whole) that is beyond the execution capabilities of individual components. In other words, the system is more than the sum of its parts. In this work, we focus on man-made (e.g., a satellite) engineering systems that either are specifically designed to provide a set of predefined functions (i.e., single stand-alone systems) or are assembled on demand from independent systems working collaboratively (i.e., system of systems) to realize a given mission. Either way, describing a system would most likely involve common concepts/characteristics: *Boundary* (i.e., the perimeter separating it from the environment), *Inputs and Outputs* (i.e., the entities that enter and exit the system), *Components and (Sub)systems* (i.e., elements or group of elements performing a set of functionality), *Interface* (i.e., the point where two components, subsystems, or systems meet to interact by exchanging inputs and/or outputs), *Connectivity* (i.e., contact or noncontact connections between components/subsystems), and *Environment* (i.e., elements external to the system but that might influence its functioning and/or

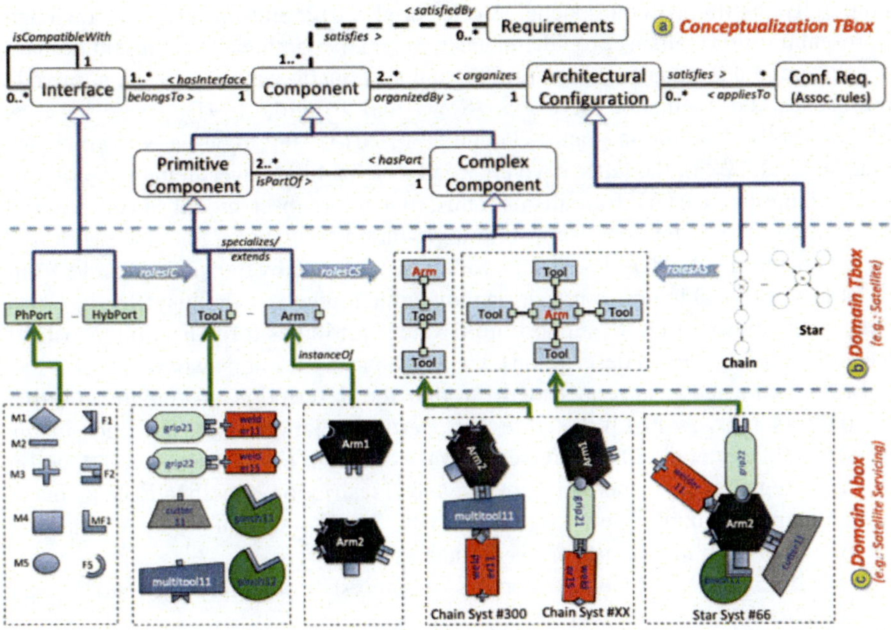

Fig. 1 A simple ontological view of the system domain: (**a**) as a conceptual Tbox, (**b**) an example domain Tbox extension, and (**c**) domain Abox (instance)

performance). Additional concepts would have to be defined when considering the process through which the system is created, i.e., the system development lifecycle (SDLC) as in model-based systems engineering (MBSE). Thus, one might need to specify the *Requirements* (i.e., the needs and wants) to be satisfied by the system, its *Configurations* (i.e., the forms of the arrangement of elements), as well as corresponding *Rules*. The latter encapsulate the set of domain-related constraints (e.g., environment, design) applicable to the system of interest and/or the domain of the design. The necessary scope and depth of the description of the domain are dependent on the usage needs. For the purpose of this work, we will use the subset of concepts and relationships represented in Fig. 1a.

2.2 *Knowledge Representation Formalisms and Description Logic (DL) Semantics*

Multi-domain systems are heterogeneous by nature. Thus, reconciliation of differences between the domains and handling of meta-domains such as time in complex systems (e.g., cyber-physical systems) modeling requires sound semantics. The latter are also needed for effective modeling of distributed behaviors in spatially and temporally distributed components (Derler et al. 2012; Eidson 2010). Knowledge representation formalisms provide means to formally capture and represent domain knowledge in a rigorous, ambiguity-free, and systematic way. Out of the numerous approaches, i.e., semantic networks (Sowa and Borgida 1991), frame systems (Hayes 1980), description graphs (Pavlic et al. 2013), and logic-based formalisms (Baader et al. 2003), the latter have emerged as a leading player in the evolution of artificial intelligence (AI) formalisms. Thus, two of its subset, modal and description logics, appear to be the most appealing logic-based formalisms for framework like ours. Considering that (i) some results for DL were found by translating results from variations of modal logics (propositional dynamic logics, μ-calculus) (Schild 1994) and (ii) the ability of DL to support multi-values attributes formalization (a critical need in complex domain description), we will be using it in this work.

In DL, knowledge is represented through the description of domain in terms of concepts (classes in OWL), roles (properties, relationships), and individuals (objects) (Baader et al. 2003). Universal (∀), existential (∃), intersection (⊓), union (⊔), and negation (¬) operators are used for restriction specifications to make the language decidable with low complexity. In DL, semantics are defined by interpretations. An interpretation *I* is defined as follows:

$\mathrm{I} = (\Delta^{\mathbf{I}}, \cdot^{\mathbf{I}})$, where $\Delta^{\mathbf{I}}$ is the domain of interest (non-empty set) and, $\cdot^{\mathbf{I}}$ is an interpretation function that maps:

Concept name *C:* a subset $C^{\mathbf{I}}$ of $\Delta^{\mathbf{I}}$
Role name *R*: a binary relation $R^{\mathbf{I}}$ over $\Delta^{\mathbf{I}}$

A summary of DL concepts constructors can be found in Fig. A.11 of (Petnga and Austin 2016). Atomic concepts (A) in the attribute language (AL) DL can be extended to support arbitrary concepts (C), thereby enabling the description of any domain of interest and leading to ALC DL on which interpretations *I* are defined. Therefore, the latter are the means through which concepts, roles, and individuals build up to the DL knowledge base *K* <*T, A*> of a domain D. Here, *T* is a set of terminological Tbox axioms and *A* is a set of assertional Abox axioms; x,y are individual names.

2.3 DL Extensions for the Web Ontology Language (OWL)

Building knowledge models capable of addressing the challenges of multi-domain modeling stated earlier in Sect. 2.2 requires the backing of ontologies that have well-defined semantics and support for formal reasoning. Fortunately, DLs can provide these formal foundations to the web ontology language (OWL) (Baader et al. 2005). However, the ALC DL is missing important pieces to that aim (Baader et al. 2005). Notable required extensions are role hierarchy (H), nominals (O), inverse and transitive roles (I), cardinality/number restriction (N), qualified number restrictions (Q), role restrictions (R), and concrete domains. In (Petnga and Austin 2017), we show how these extensions can be organized and mapped to semantics for the OWL sublanguages. As a result, *OWL1-DL* is found to be a first-order logic (FOL) restriction based on SHOIN DL (S is a simplified, yet equivalent, notation for ALC). This DL is decidable, thanks partially to well-defined semantics and proven reasoning algorithms. However, it presents some weaknesses that reflect on the expressiveness power of *OWL1-DL.* To that extent, OWL2 (standardized in 2009) overcomes a number of those weaknesses (e.g., relational expressivity, syntax deficiencies, species definitions) through its mapping with the SHROIQ DL. Leveraging the full power of these formalisms for effective modeling and reasoning for system architecting and configuration requires sound decidable reasoning capabilities as enablers. These capabilities need to be provided by reasoners that can derive, through inferencing, additional facts about the concepts of the domain(s) of interest. Among the key reasoning services needed are *satisfiability* (i.e., for a given concept C, $\exists I$, $I \models T$ such that $C^I \neq \varnothing$), *subsumption* (i.e., concept C is subsumed by D, i.e., $C \sqsubseteq_T D$ with C, D $\in \mathbf{C}$ if for all interpretations I, if $I \models T$, then $C^I \subseteq D^I$), *equivalence* (i.e., two concepts C and D $\in \mathbf{C}$ are equivalent with respect to T if for all interpretations I, if $I \models T$, then $C^I = D^I$), and *disjointness* (i.e., two concepts C and D $\in \mathbf{C}$ are disjoint with respect to T if for all interpretations I, if $I \models T$, then $C^I \cap D^I$). Also, the decidability of the SHROIQ DL is formally established in Appendix C of (Petnga and Austin 2016). Hence, given its mapping with OWL2-DL, the latter will be the language of development of our ontological framework that we introduce in the next section.

3 An Ontological Framework for Multi-domain System Architecting and Configuration

The system architecture for our ontological framework is shown on Fig. 2. It's comprised of three integrated modules: domain and knowledge management foundations, semantic platforms, and system configuration platform.

3.1 Module 1: Mathematical Foundations

The primary role of this module (see Fig. 2a) is to provide the mathematical foundations needed to support the formal description of domains and meta-domain knowledge. Thus, we need knowledge representation formalisms to that aim. Description logic semantics introduced and described in Sect. 2.2 have been shown appropriate to support the formal description of the domains involved. Selected theories are used to inform the formulation of the description as per the modeling needs of the system/domains of interest. Specifically, meta-theories (for known cross-cutting domains such as time, space, communication, etc.) are distinguished from plain ones, which could be well-accepted domain standards

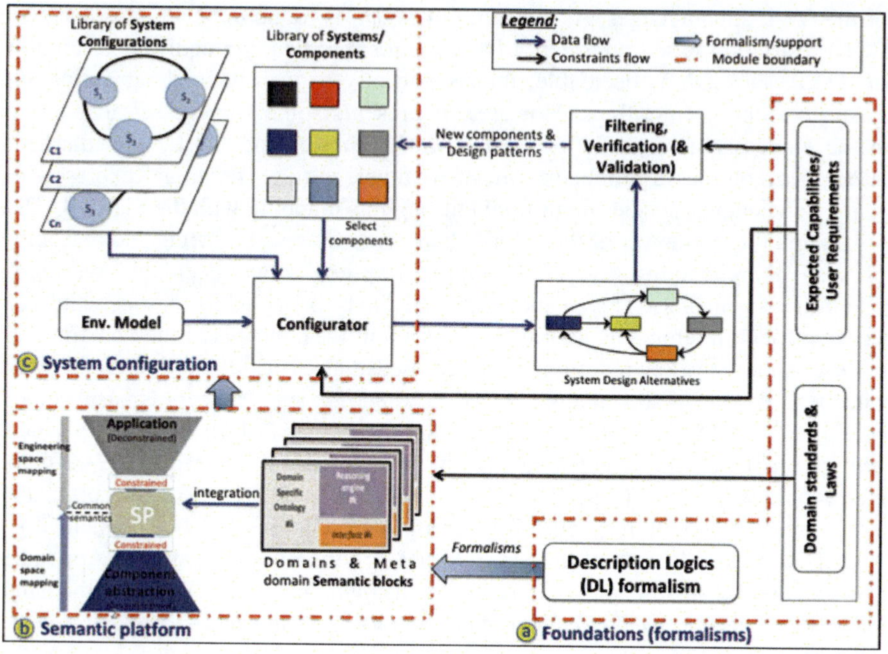

Fig. 2 High-level architecture of the ontology-driven framework for multi-domain system architecting and configuration and its core modules: (**a**) mathematical foundations, (**b**) DL-powered semantic platform, and (**c**) system configuration framework

(e.g., construction, electrical, etc.). Unlike the former, the latter might not always be available for the problem at hand. In previous work, we have shown that both the Region Connectedness Calculus (RCC-8) and Allen Temporal Interval Calculus (ATIC) were effective theories for the formal description of spatial and temporal domains (Petnga and Austin 2017). This module is completed with users' expected capabilities and requirements (in a separate compartment) that are needed to inform the proper configuration of the platform and the validation of selected design alternatives at the application level.

3.2 *Module 2: Semantic Platform*

The basic premise of our approach is that *any complex system (CS) application can be developed using limited domain-specific (DS) components provided they are consistent with the relevant CS of interest semantic platform*. The latter encapsulates and integrates design and domain knowledge (not just raw data) in models (i.e., common semantics) that are determinate, provable (ambiguity-free), executable, and with semantic parameters and rule engines that can work and reason with physical quantities. As such, DL formalisms provide means through which domains and meta-domains semantics can be captured in formal and reusable models. The building blocks of platforms are *semantic blocks* which encode knowledge in the form of a three-part "knowledge block" made of (1) an ontology, (2) set of rules, and (3) interfaces that enable cross-block communication and linking with computation platforms via customized (built-in) functions (Petnga and Austin 2017). The flexibility and generality of this representation scheme allows for the capture of system (geometric) configurations along with its corresponding constraints as well as organizational or operational policies, standard operating procedure/process (SOP), or directives that are applicable to the system at hand in part or as a whole. The OWL2-DL is shown to be effective in the implementation and integration of the semantic blocks, thanks to its DL extensions as introduced in Sect. 2.3.

3.3 Module 3: System Configurations

We build from the actionable semantic foundations provided by the SP module to design (via composition) systems models of various levels of complexity that can be formally verified against (user) requirements. The ontology, logic, proof, and trust stemming from the DL foundations introduce vocabularies, logical reasoning, establishment of consistency and correctness, and evidence of trustworthiness into the framework. Thus, we employ semantic descriptions via semantic graphs (i.e., network of semantic resulting from the rule-based integration of semantic blocks) of application domains and configurations and use ontologies and rule-based reasoning to enable a semantic configurator to generate candidate system design alternatives from the set of available components/systems and library of configurations for the environment model of interest. The design alternatives are then validated against domain and meta-domain laws and (user) requirements as pictured in the top part of Fig. 2. Semantic graphs are used to model architectural configurations, design requirements and their relationships, and the attributes of the CS components as illustrated in Fig. 2a. Synthesized graph models are employed for libraries of design requirements and component systems, as well as the mapping of requirements to design solutions.

4 Prototype Implementation: Knowledge-Driven Design Space Generation for Satellite Robotic Arms

4.1 Overview

To illustrate the use and effectiveness of the ontological framework for multi-domain system architecting and configuration, we consider the problem of rule-based generation of architectural configurations for satellite robotic arms. Recent research at NASA and universities across the USA has shown promise in robotic satellite servicing by articulating the huge economical and environmental benefits of sending a satellite into space to robotically service a large number of older, but mostly functional satellites (Knizhnik et al. 2017). In previous work, researchers have investigated graph transformations for downselecting the number of design options (Nassar and Austin 2013) and applications to trade-space downselection for servicing spacecraft (Knizhnik et al. 2017). This prototype illustrates how our semantic framework takes the next step forward, with the demonstration of how the semantic graph could support the generation of candidate design alternatives across multiple possible system architectural configurations. For the sake of simplicity, we consider hypothetical robotic arms – as illustrated in the right-hand side of Fig. 1b and c – to be mounted on a servicing satellite.

4.2 Multi-configuration and Multi-domain Generation of Satellite Robotic Arm Design Alternatives: Prototype

Robotic Arm Architectural Configurations The set of possible architectural (structure viewpoint) configurations of the robotic arms to be included in the library of system configuration are considered. We identify and group configurations in three types (as ontologically defined in Fig. 3): centralized, decentralized, and hybrid (any combination of both). Decentralized configurations are characterized by a leaderless organization of elements into a regular geometry (e.g., circle). In centralized architectural configurations, there is one leader, whose position is central to the distinction between configurations. Two such configurations – Chain and Star – are illustrated in the far right-hand side of Fig. 1b. An "X" is used to indicate the leader (i.e., the main arm in our case), while other elements are simple nodes in the graph representation. Each architectural configuration represents a "blueprint" for the assembly of components. Architectural rules specify additional constraints that provide a context to the configuration as well as necessary clarifications. Instances of resulting elements and systems are shown in Fig. 1b.

Ontology Models for the Semantic Platform The system components (e.g., arms, effectors, tools) and the architectural configurations (e.g., Chain, Star) are all modeled as "semantic blocks" (see example in Fig. 7 of (Petnga and Austin 2016)).

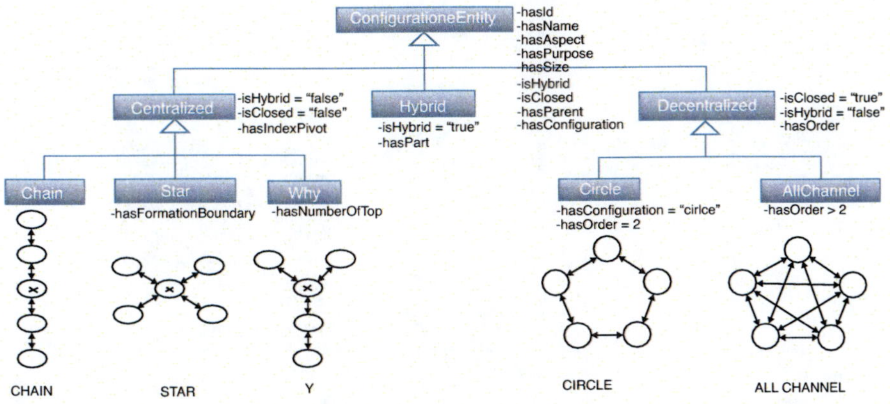

Fig. 3 Ontological perspective of sample architectural configurations

These are compact knowledge representations that leverage the DL formalisms in the foundation layer (Fig. 2a) to encapsulate meta-domain and generic domain knowledge. We create a semantic block for the configuration domain and one for the satellite domain where concepts stem from extensions of a general component ontology as pictured in Fig. 1. Note that interfaces are specialized to ensure correct assembly of components via rule-driven checking of compatibility between interfaces, beyond the predefined association constraints imposed by architectural configurations. For instance, in a chain-based robotic arm system, the arm (leader) connects to the end effector, which connects to a tool, and the tool interacts with the client satellite. Ontology models in the semantic blocks are created with OWL2, while rules are implemented in Jena by a (decidable) reasoner, and the assembly and integration of heterogeneous domain data/knowledge are performed using Java.

System Configuration and Reasoning for Design Alternatives Generation For the purpose of this example, acceptable design alternatives are ones that satisfy domain, meta-domain, and configuration rules. First, we instantiate both the satellite and the architectural configurations Tbox with corresponding sample data borrowed from (Knizhnik et al. 2017). This is done via creation of scenarios as illustrated in Fig. 4a and b. Thus, a library of semantically enabled elements is created as instance of the Tbox, resulting into an Abox of components as illustrated in the left side of Fig. 1c. For the sake of simplicity, the interface types are treated as attributes of components (class). For instance, `grip21` is a `Tool` with two distinct interfaces (i.e., `M1` and `F2`). Similarly, there are two instances of the entity `Arm` in the library: `Arm1` and `Arm2`. The latter presents two `F1`, one `M5`, one `F2`, and one `F5` interfaces. With the individual primitive component ready, the configurator can assemble them according to predefined architectural configurations as created in Fig. 4b. We note here that the number of nodes in the configuration graph can be adjusted to accommodate the designer need. This results into a semantically annotated library of components and architectural configurations. Next, a Java-

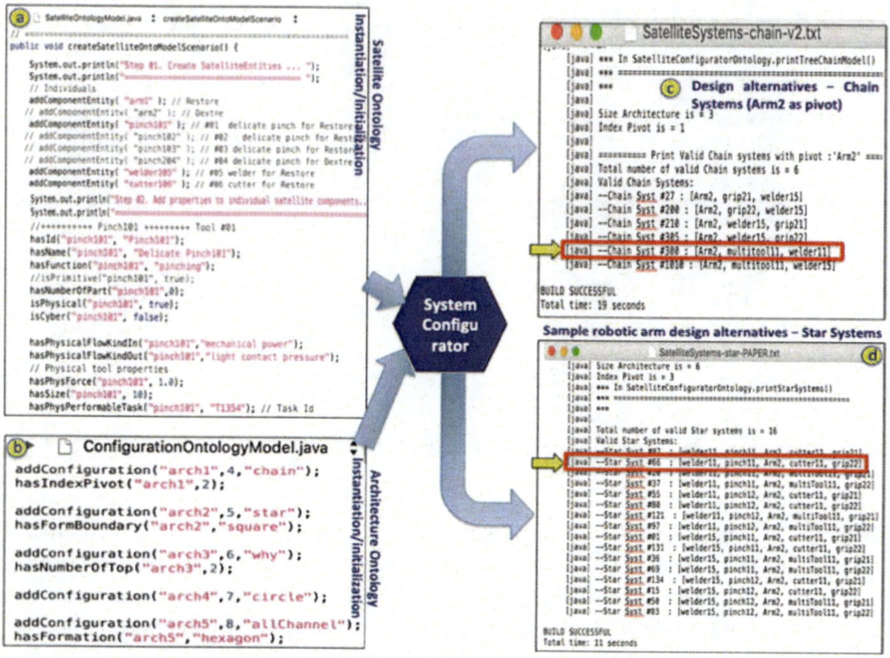

Fig. 4 Illustration of the generation of valid Chain and Circle design alternatives from input libraries of configurations and components using the configurator for the satellite robotic arm case study

implemented configurator generates configuration-compliant design alternatives through careful orchestration of inference rules first at the domain levels and then for each of the candidate configurations. It starts by generating an unconstrained design space as baseline. With the execution of each rule, the baseline is progressively constrained, and designs that do not satisfy expressed constraints are removed, leaving only qualified alternatives. Figure 4c and d illustrates resulting acceptable designs, respectively, for a three-component chain-based arm system with `Arm2` as leader (pivot) at index 1 of the list and a five-component star-based system (for the sake of illustration purpose only) encompassing an arm. Two of the systems illustrated in Fig. 1c, i.e., `Chain Syst♯300` and `Star Syst♯66`, are pointed out in Fig. 4c and d, respectively.

5 Conclusion and Future Work

In this work, we have introduced, described, and prototyped a new ontology-driven knowledge modeling and reasoning approach for multi-domain system architecting and configuration. Our framework supports ontological and multi-domain rule-based generation of complex system design alternatives as per predefined configurations. Its capability includes mechanisms to integrate multiple domains, each captured as semantic blocks made of (meta)-domain ontologies, rules, and specialized computations modules. These results have been achieved, thanks to description logic (DL) formalisms supporting the decidability of our reasoning framework and enabling a configurator to correctly assemble components from a library according to a given architectural configuration. An illustration of the implementation of the system on rule-based generation of architectural configurations for satellite robotic arms has shown promising results. Future work would include extensions to enable trade studies based on system-level measures of effectiveness, semiautomatic procedures for loading data for scenarios generation, as well as support for more complex architectural configurations, especially hybrid ones.

References

Derler, P., E.A. Lee, and A. Sangiovanni. 2012. Modeling Cyber-Physical Systems. *Proceeding of the IEEE* 100 (1): 13–28.

Eidson. 2010. *A time-centric Model for Cyber-Physical Applications*. MoDELS 2010 ACES-MB Workshop Proceedings, Oslo October 4, 2010.

Sowa, J.F., and A. Borgida. 1991. *Principles of Semantic Networks: Explorations in the Representation of Knowledge*, ed. John F. Sowa.

Hayes, P.J. 1980. The logic of frames. In *Frame Conceptions and Text Understanding*, ed. D. Metzing, 46–61. Berlin: de Gruyter.

Pavlic, M., A. Mestrovic, and A. Jakupovic. 2013. *Graph-Based Formalisms for Knowledge Representation.* 17th World Multi-Conference on Systemics, Cybernetics and Informatics, July 9–12, Orlando, Florida, USA.

Baader, F., D.L. McGuinness, D. Nardi, and P.F. Patel-Schneider. 2003. *The Description Logic Handbook: Theory, implementation, and applications*. Cambridge: Cambridge University Press.

Schild, K. 1994. Terminological Cycles and the Propositional With Mu- Calculus. In *4th Int. Conference on the Principle of Knowledge Representation and Reasoning(KR-94),* eds. J. Doyle, E. Sandewall, and P. Torasso, 509–520.

Baader, F., I. Horrocks, and U. Sattler. 2005. Description Logics as Ontology Languages for the Semantic Web. In Mechanizing Mathematical Reasoning: Essays in Honor of Jrg Siekmann on the Occasion of His 60th Birthday, number 2605 in *Lecture Notes in Artificial Intelligence*, eds. Dieter Hutter and Werner Stephan, 228–248. Springer.

Petnga, L., and M.A. Austin. 2016. An Ontological Framework for Knowledge Modeling and Decision Support in Cyber-Physical Systems. *Advanced Engineering Informatics* 30 (1): 77–94.

———. 2017. *Semantically-enabled Model-based Systems Engineering of Safety-critical Network of Systems.* 27th Annual INCOSE International Symposium (IS 2017), Adelaide, Australia, July 15–20, 2017.

Knizhnik, J., M.A. Austin, and C. Carignan. 2017. *Robotic Satellite Servicing Trade Space Down Selection.* 27th Annual INCOSE International Symposium (INCOSE 2017), Adelaide, Australia, July 15–20, 2017.

Nassar, N., and M. Austin. 2013. *Model-Based Systems Engineering Design and Trade-off Analysis with RDF Graphs*. Conference on Systems Engineering Research. GA, US: Atlanta.

Part IV
MBSE Processes and Languages

A Literature Review of the Integration of Test Activities into the Product Development Process

Aksel Elkjaer, Geir Ringen, and Cecilia Haskins

Abstract The purpose of this paper is to investigate product development test processes. A literature review examines research on test activities in product design, product development and systems engineering research fields. The publications reviewed have been categorized based on the stage in development and placed into a proposed test process framework. The proposed framework sets an agenda of functions and characteristics important for the integration of test processes into model-based systems engineering. The findings presented are of interest to researchers by structuring test activities from product development, systems engineering and prototyping research into a context for the design process. The findings also allow practitioners to identify research at the level of planning and development stage relevant to their test processes.

Keywords Product development · Test activity integration · Literature review

1 Introduction

Testing is an essential component of the development process; however, it's integration into the design process has received limited attention (Engel 2010; Tahera et al. 2018). A test of a design provides the possibility either to confirm a rationale or to learn from the discovery of unknown (and unexpected) outcomes, offering vital information to the design process. From this viewpoint, the purpose of a test in product development can be considered as a method to reduce uncertainty (Bjorkman et al. 2013). The reduction of uncertainty supports decision-making at any stage in development, but the value of information and new knowledge is inversely proportional with time (Kennedy 2008). Reduced uncertainty offers two outcomes. If the performance is as expected, the result is evidence which confirms

A. Elkjaer (✉) · G. Ringen · C. Haskins
Norwegian University of Science & Technology, Trondheim, Norway
e-mail: aksel.elkjaer@ntnu.no

A. M. Madni et al. (eds.), *Recent Trends and Advances in Model Based Systems Engineering*, https://doi.org/10.1007/978-3-030-82083-1_21

quality in a context. If the performance is not as expected (or expectation unknown), the result provides a pathway to new understanding.

Experience tells us that test activities incur a significant financial burden, typically accounting for a substantial proportion of total development costs (Tahera et al. 2017). In contrast, even greater costs from a late-stage failure are often attributed to the deficiency of untested decisions (Kukulies and Schmitt 2018). The planning of test activities during development requires carefully balancing the potential benefits of new knowledge relevant to risk mitigation against the applicable programmatic constraints.

Model-based systems engineering (MBSE) promotes the utilization of models throughout the development process, from needs analysis and requirement definition to the end of a product life cycle, to enhance development (INCOSE 2007). Such models offer the potential for the explicit connection of test processes to design, supporting traceability to both knowledge gains and risk mitigations. Model-based testing has arisen from the need to test formalized models generated from analysis and simulations. The research on how MBSE and model-based testing should be integrated, especially in a context outside of software development, is lacking. Raz et al. (2018) have used Design of Experiments methodology to link system architecting and the system design space through formal models showing the potential for the further integration of test processes. In this study perspectives from research on test activities of products and systems are presented to support the integration of testing perspectives into MBSE.

The motivation of this paper is to orient the reader to research into test activities and to provide a synthesis of research into test processes related to early development stages. A literature review has been conducted to classify research on test activities based on their aims, perspectives to planning and stage in development. The objective of the review is to develop a framework of test activity that supports integration into the development process. Two research questions were proposed as a foundation for establishing the framework.

1. In which stages of product development have the integration of test activities into the development process been researched?
2. What perspectives are taken in the planning of test activities?

The paper is structured as follows: Section 2 Methodology presents the literature search strategy; the findings are given in Sect. 3 Results, analysed in Sect. 4 Discussion and summarized in Sect. 5 Conclusion.

2 Methodology

2.1 Search Strategy

The review was performed with guidance from procedural literature by Machi and McEvoy (2016). The SCOPUS electronic database was searched for any combination of product development keyword AND testing keyword shown in Table 1. The search was conducted within the title, abstract and keywords. The search was limited to the journal sources listed in Table 2 with no restriction on year of publication. The sources were selected due to their focus on technical engineering design and their high reputation within the field, but are not considered exhaustive. Sources focused on engineering management were excluded to concentrate on the technical implementation of testing rather than related business and management topics.

2.2 Inclusion and Exclusion Criteria

The discovered articles were processed in three subsequent steps to remove articles not relevant to the intention of the study. First articles were removed based on title, then abstract and finally after reading the paper. The sample population for each stage is shown in Fig. 1 with a final population of 34 articles.

Table 1 Keyword search terms

Topic	Keywords
Product development	"product development" OR "product design" OR "system development" OR "design and development" OR "system design" OR "design method*" OR "design theory" OR "system engineering" OR "v model" OR "development process" OR "design process" OR "design for" OR "robust design" OR "knowledge based engineering" OR "knowledge management" OR "organi*ational learn" OR "model-based" OR "set*based"
Testing	"test and evaluation" OR "test plan" OR "test definition" OR "test specification" OR "test verification" OR "test validation" OR "test management" OR "verification activities" OR "physical test" OR "virtual test" OR "test activities" OR "testing" OR "set*based test" OR "prototyp*"

Table 2 Journal sources

Journals	Research in Engineering Design, Journal of Engineering Design, Systems Engineering, Concurrent Engineering, CIRP Journal of Manufacturing Science and Technology, Journal of Mechanical Design

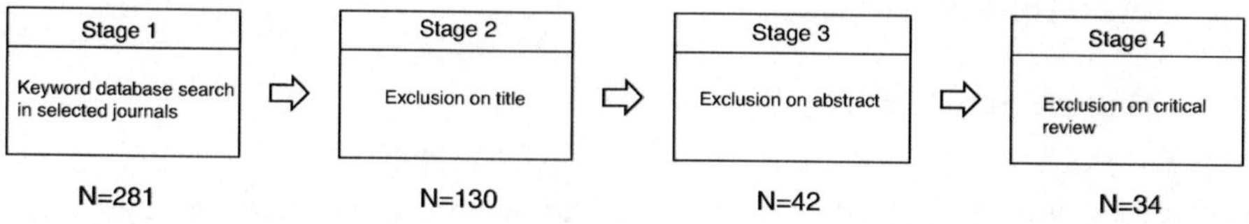

Fig. 1 Search processing stages

The article was included if the research addressed:

- The engineering design process, new product development practice, system engineering/design, development activities or design approach.
- Test activities or processes were integral to the paper's research question/thesis.
- Test activities were discussed in relation to the design process (e.g. use of test results contributing to design process/decisions).
- Tests were analysed as a method of verification/validation or source of discovery/learning/reusable knowledge.

The article was excluded if:

- "Test" used in reference to testing paper's hypothesis and not product development test activity. This includes the "testing" of a new design method – if test activities are not relevant to that design method.
- Studies addressing solely software products or construction projects.
- Studies focusing solely on the design improvement of a specific product for the benefit of that specific product – as opposed to development process/design methodology in general.
- "Design methodology" referred to as the methodological design of the study itself (i.e. not product development methodology).
- Virtual prototype, virtual testing, simulation or analysis research not discussed in a context of impact on the design/development process (i.e. research into improving specific modelling technique for a design problem was not included).

3 Results

Results from the literature review are summarised in the following two subchapters. First, the number of articles addressing each stage in development is presented, followed by their categorization into a model of perspectives on product development testing. Appendix A provides a complete list of the reviewed literature showing the classification of each article according to the presented frameworks.

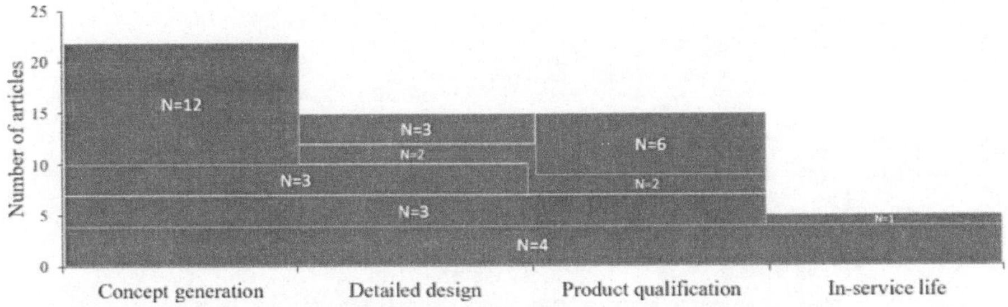

Fig. 2 Stage of product development in which testing was being investigated

3.1 Stage of Product Development

The research on test activities identified in this paper has been categorized with respect to the stage of the development process being investigated. Four stages, unique in respect to test activities, have been defined by the authors covering the fundamental stages from textbook literature (Ulrich and Eppinger 2012). The stages are concept generation, detailed design, product qualification and in-service life. The number of articles addressing each stage in development is shown in Fig. 2.

A test process is not necessarily unique to only one stage in development. It is possible for research to be conducted solely on testing during concept generation (one stage) or with a broader perspective of the development process covering multiple stages. Figure 2 therefore includes boxes spread across the applicable stages with the number of articles for each box specified.

3.2 Test Process Perspective

The perspective of testing that was studied was categorized into three areas of activity suggested by the literature. In test design, the research addressed the best way to perform a specific test. A second area was defined as test objectives and focused on determining what to test. Finally, test strategy is where a test campaign of defined test objectives was studied for optimization. The number of articles discovered for each perspective is shown in Fig. 3.

4 Discussion

Splitting the development process into stages allows the objective of testing in each phase to be discretized. In essence, testing with respect to stage in development can be considered as progressive investigations to discover: *Will the idea work?*

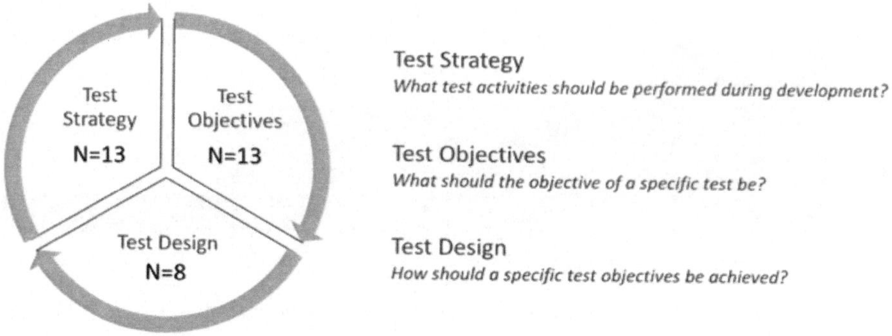

Fig. 3 Test process perspectives

Will the solution work? Does the product/system work? And finally *how well did the product/system work?* The greatest number of research articles addressed the earliest phase, concept development, which reflects the importance of starting with the right idea by frontloading activities and generating knowledge when it is most valuable. Whilst the understanding and maturity of the product is often limited in the concept phase, expanding the potential to answer the subsequent questions as early as possible achieves compounding benefits. Furthermore, it is specifically these compounding benefits that are the focus of integrating test activities into the development process.

Three test process perspectives are proposed as relevant for the integration of test activities in the development process. The following discussion highlights the key insights from the identified research from each perspective. The purpose is to distil the important considerations for the integration of testing into the development process.

4.1 *Test Strategy*

The category "test strategy" considers a holistic perspective of test activities during development. It concerns overall aspects of development, such as the duration, cost, quality or risk management. This requires analysing the test activities collectively and establishing the appropriate approach.

The systems engineering processes of verification and validation align with this perspective. Testing is technically a method of verification and validation. However due to testing's critical role in the process, it is considered a test strategy in the context of the proposed framework. Several articles were discovered modelling the set of activities in a verification and validation plan to compare and query development approaches.

Engel and Barad (2003) and Engel and Shachar (2006) developed a quantitative methodology of modelling the cost and duration of activities in a verification plan. Subsequently, Hoppe et al. (2007) performed a multi-case study analysis of the quantitative methodology, along with qualitative guidelines, and showed how frontloading of test verification activities was critical for the development life cycle. This advocates for the close integration of test strategy in the design phase. Similar models, which estimate the characteristics of a set of test activities for the mathematical assessment of optimal solutions, have been developed by Salado (2015) and Shabi et al. (2017). These approaches achieve optimized test strategies for given designs which, although useful for comparison of different designs, do not influence directly the design process. They view an optimized test strategy as output for a given design problem. This is similar to Tahan and Ben-Asher (2005) who investigated specifically the ideal number of incremental stages for verification for a design.

Two studies which did explicitly integrate test strategy into the design process were (Tahera et al. 2018) and (Shin et al. 2017). Tahera et al. (2018) addressed the importance of incremental stages of testing with direct dependences to design iterations, and Shin et al. (2017) modelled the sequence of design tasks and test activities to establish the process with the shortest duration. These studies emphasized the efficiency and effectiveness to overall development which testing can provide when considered early and throughout development.

A different area of research concerning test strategy was research on prototyping. The following three studies examined the role of prototypes on a strategic level: Barkan and Iansiti (1993), Camburn et al. (2015) and Lauff et al. (2018). They all addressed aspects of using prototypes throughout development, such as timing, scope and prototype characteristics, to understand how they influenced the process.

4.2 Test Objectives

Articles in this category investigated the objective of specific test activities. Three general approaches were discovered in the identified research. Test objectives could be defined from either (i) evaluation of design uncertainty, (ii) supporting virtual/simulation activities or (iii) leveraging benefits of physical tests.

Goh et al. (2007) proposed a method of modelling uncertainty to inform design decisions. The method addressed the trend of increased analysis during development by structuring the uncertainty of design decisions based on untested assumptions. In a similar manner, Bjorkman et al. (2013) and Kukulies and Schmitt (2018) have both established test objectives by evaluating the performance uncertainty of functional product characteristics. Whilst those studies prioritized attention to design parameters with greatest uncertainty, Sanders et al. (2013) examined unforeseen and low-probability aspects with high consequences. These potential high-consequence characteristics required discovering them in very early testing before necessary changes were unfeasible.

Another key area of focus within the test objectives category was physical versus virtual testing. Research in this category was addressing directly the role of simulation to reduce the need for physical testing or prescribing physical testing for model correlation (to achieve even greater use of virtual models). Sutcliffe and Gault (2004) exposed potential benefits that can be achieved by integrating virtual tests, from CAD models to augmented reality, highlighting possible objectives during development from such test methods. Mejía-Gutiérrez and Carvajal-Arango (2017) reported on the latest integration of development software in a case study directly linking virtual prototypes into systems engineering modelling software.

The final area of study defining objectives for test activities focused on leveraging the benefits of physical testing. Viswanathan has published a number of studies (Viswanathan et al. 2014; Viswanathan and Linsey 2012, 2013) investigating the effect of design fixation and sunk costs on physical models. Design fixation considers an unnoticed integration of test activities, which prevents improved solutions being discovered. Campbell et al. (2007) on the other hand showed that physical models provide the best understanding to the customer and therefore allow the best feedback to be gathered.

The research into test objectives considered a greater level of integration into the design process than the other framework perspectives due to the strong link between key design parameters and definition of tests.

4.3 Test Design

Research in the final category, "test design," investigated methodology for a defined objective. This category would be broad and extensive if it was to consider methodology for specific applications, e.g. what test methodology is best for measuring the health of batteries or the performance of passive dampers. However, the nature of such application-specific methodology has been excluded, according to the criteria defined in Sect. 2.2, as it is independent to the development process (or at most only applicable to development of a specific product type).

The eight articles identified in this category discuss the methodology in relation to the development process which means the methodology is generalizable to different development contexts. Two general areas of test design were discovered. The first was a Design of Experiments or Robust Engineering approach where the test process is structured statistically around the identification of parameters that will have greatest impact on performance. The second topic was related to understanding customer needs or, in a wider context, stakeholder analysis.

The impact of Design of Experiments theory on the design process was presented in case studies by Herrmann (2007, 2009). The case studies show how implementation of Taguchi methodology identifies the parameters with greatest effect on performance. This allows for informed concept tradeoffs to be performed based on maximizing intended performance and minimizing undesired effects.

The second area of test design methodology researched addressed the less quantitative field of stakeholder analysis. Research by Deininger et al. (2019), Starkey et al. (2019) and Wall et al. (1992) all investigated methodology relating to the influence of product representations on conclusions gathered. In contrast, Tovares et al. (2014), Artacho et al. (2010) and Engelbrektsson and Söderman (2004) emphasized the importance of capturing stakeholders' perceptions and preferences.

5 Conclusion

A broad search was conducted on a limited set of prominent sources within product development research. The resulting literature was not exhaustive for the discovered topics yet achieved a wide overview covering all stages in the product development process. The purpose of the review was to uncover the literature addressing the definition and utilization of testing throughout the development life cycle.

The literature review identified methods from the start of the development process to qualification testing and in-service life. This answers the first research question by showing that integration of test activities is important and has been considered in all stages of development. Research focusing on the integration of testing was most prominent in concept development where the greatest impact is achievable.

Answering the second research question, *What perspectives are taken in the planning of tests?*, prompted the first author to devise a framework of test planning perspectives that established three areas of importance. The research was categorized based on the contribution in defining: what key objectives can be realized by testing, how such tests should be designed and how to establish the overall strategy of test activities in development. The proposed framework with classification into three areas provides insight into different levels of detail needed during planning of tests in the development process. The dependencies and overlap between these groups highlighted both their sequential and iterative nature as represented in the framework in Fig. 3.

This study provides an overview of relevant research structured in a framework to assist the future analysis and development of test processes for integration into MBSE approaches.

Acknowledgements This study has been conducted as part of the KPN project VALUE, supported by the Norwegian Research Council and the industry partners Hydro and Alcoa.

Appendix A. Literature Review Study Sources, Evidence Categories and Aims

Study	Evidence category[a]	Stage of development[a]	Aim
Salado and Kannan (2019)	TS	2	Formalize the application of Bayesian networks to verification problems to facilitate instruction and communication among verification engineers and with researchers from other domains
Tahera et al. (2018)	TS	1–4	Establish importance of testing in product development to inform the development of pragmatic support methods
Shabi et al. (2017)	TS	2,3	Propose a method for determining the optimal verification activities with respect to product quality/risk
Shin et al. (2017)	TS	3	Demonstrate that model-based integration of T&E process and system safety process reduces development time
Salado (2015)	TS	3	Demonstrate the benefit of trade space exploration in the optimization of test strategy
Hoppe et al. (2007)	TS	1–4	Develop a generic verification, validation and testing methodology guideline and an economic VVT process model in order to realize improved product quality
Engel and Shachar (2006)	TS	2,3	Measure systems quality cost/times in a typical development project as well as suggest ways to optimize it in order to meet business objectives
Tahan and Ben-Asher (2005)	TS	3	Demonstrate that incremental integration offers both time and cost benefits vs single stage integration
Engel and Barad (2003)	TS	3	Propose a novel approach for modelling VVT strategies as decision problems
Lauff et al. (2018)	TS	1–3	Define the roles of prototypes in industry
Camburn et al. (2015)	TS	1,2	Provide a method to repeatedly enhance the outcome of prototyping efforts
Barkan and Iansiti (1993)	TS	1–4	Examine roles which prototyping plays in product development

(continued)

Study	Evidence category[a]	Stage of development[a]	Aim
Isaksson et al. (2000)	TS	1–3	Evaluate alternative design strategies and methods with respect to their impact on the development process time
Kukulies and Schmitt (2018)	TO	3	Investigate the use of uncertainty modelling to support design verification
Bjorkman et al. (2013)	TO	3	Present a methodology that uses an MBSE framework and Monte Carlo simulation to define uncertainty reduction goals for test planners to use in developing test strategies and detailed test designs for evaluating technical performance parameters
Sanders et al. (2013)	TO	1	Propose model for discovery of low probability events in the formative stages of the requirements definition and risk management planning activities in order establish the safety requirements and responsive conceptual designs for mitigation
Goh et al. (2007)	TO	4	Create framework for organizing uncertainty in product development simulation results therefore improving understanding between simulations and tested results for the purpose of assisting design decisions
Takala (2005)	TO	1	Propose a concept that bridges the gap between physical and virtual domains prototyping
Sutcliffe and Gault (2004)	TO	1	Propose guidelines for configuring virtual engineering technology and design of requirements analysis sessions
Mejía-Gutiérrez and Carvajal-Arango (2017)	TO	1,2	Investigate the usefulness of integrated virtual design verification simulation
Kiefer et al. (2004)	TO	1,2	Present the design development of a new product that explored many of the different prototyping technologies
Wang and Chen (2011)	TO	1	Introduce users' participation in the conceptual design stage of product development to avoid interpreting biased from marketers' information
Viswanathan et al. (2014)	TO	1	Study how physical models can assist novices in mitigating design fixation on undesirable features

(continued)

Study	Evidence category[a]	Stage of development[a]	Aim
Viswanathan and Linsey (2013)	TO	1	Investigate physical modelling role in idea generation and design fixation
Viswanathan and Linsey (2012)	TO	1	Investigate if physical models supplement designer's mental models and if physical models induce design fixation
Campbell et al. (2007)	TO	1–3	Demonstrate that physical models are the single presentation format that is readily understood by most customers
Wall et al. (1992)	TD	1–4	Develop a systematic method of evaluating prototyping processes in order to determine the best process for a given situation
Engelbrektsson and Söderman (2004)	TD	1	Investigate the use and perceptions of methods and product representations in Swedish companies and its possible impact on problems associated with late-discovered customer requirements
Starkey et al. (2019)	TD	1	Investigate the impact of prototype fidelity, concept creativity and risk aversion on perceived riskiness and concept selection
Deininger et al. (2019)	TD	1	Provide insights into how prototype type, group membership (stakeholder characteristics) and question type can influence stakeholders' perceptions of a design concept and the resulting feedback they provide
Tovares et al. (2014)	TD	1	Develop a method to elicit, capture and model consumer preference through experiential preference judgements
Artacho et al. (2010)	TD	1	Analyse how slight changes might affect users' perception as well as influence their intention to purchase a product
Herrmann (2009)	TD	2	Demonstrate that the successful use of Taguchi test methodology provides efficient and reliable design knowledge
Herrmann (2007)	TD	2	Demonstrate the successful use of Taguchi test methodology to support system design

[a]*TS* test strategy, *TO* test objectives, *TD* test design, *1* concept generation, *2* detailed design, *3* product qualification, *4* in-service life

References

Artacho, M.A., A. Ballester, and E. Alcántara. 2010. Analysis of the Impact of Slight Changes in Product Formal Attributes on User's Emotions and Configuration of an Emotional Space for Successful Design. *Journal of Engineering Design* 21: 693–705.

Barkan, P., and M. Iansiti. 1993. Prototyping: A Tool for Rapid Learning in Product Development. *Concurrent Engineering* 1: 125–134.

Bjorkman, E.A., S. Sarkani, and T.A. Mazzuchi. 2013. Using Model-Based Systems Engineering as a Framework for Improving test and Evaluation Activities. *Systems Engineering* 16: 346–362.

Camburn, B., B. Dunlap, T. Gurjar, C. Hamon, M. Green, D. Jensen, R. Crawford, K. Otto, and K. Wood. 2015. A Systematic Method for Design Prototyping. *Journal of Mechanical Design* 137: 081102.

Campbell, R.I., D.J. De Beer, L.J. Barnard, G.J. Booysen, M. Truscott, R. Cain, M.J. Burton, D.E. Gyi, and R. Hague. 2007. Design Evolution Through Customer Interaction with Functional Prototypes. *Journal of Engineering Design* 18: 617–635.

Deininger, M., S.R. Daly, J.C. Lee, C.M. Seifert, and K.H. Sienko. 2019. Prototyping for Context: Exploring Stakeholder Feedback Based on Prototype Type, Stakeholder Group and Question Type. *Research in Engineering Design* 1–19.

Engel, A. 2010. *Verification, Validation, and Testing of Engineered Systems*. Hoboken, United States: Wiley.

Engel, A., and M. Barad. 2003. A Methodology for Modeling VVT Risks and Costs. *Systems Engineering* 6: 135–151.

Engel, A., and S. Shachar. 2006. Measuring and Optimizing Systems' Quality Costs and Project Duration. *Systems Engineering* 9: 259–280.

Engelbrektsson, P., and M. Söderman. 2004. The Use and Perception of Methods and Product Representations in Product Development: A Survey of Swedish Industry. *Journal of Engineering Design* 15: 141–154.

Goh, Y.M., C.A. McMahon, and J.D. Booker. 2007. Development and Characterisation of Error Functions in Design. *Research in Engineering Design* 18: 129–148.

Herrmann, D.K. 2007. Taguchi Optimization in the Design of a Printer Registration System. *Journal of Engineering Design* 18: 1–11.

———. 2009. Application of Multiparameter Optimization for Robust Product Design. *Journal of Mechanical Design* 131.

Hoppe, M., A. Engel, and S. Shachar. 2007. SysTest: Improving the Verification, Validation, and Testing Process—Assessing Six Industrial Pilot Projects. *Systems Engineering* 10: 323–347.

INCOSE. 2007. Systems Engineering Vision 2020. INCOSE-TP-2004-004-02.

Isaksson, O., S. Keski-Seppälä, and S.D. Eppinger. 2000. Evaluation of Design Process Alternatives Using Signal Flow Graphs. *Journal of Engineering Design* 11: 211–224.

Kennedy, M. 2008. *Ready, Set, Dominate: Implement Toyota's Set-Based Learning for Developing Products and Nobody Can Catch You*. Oaklea Press.

Kiefer, S., L. Silverberg, and M. Gonzalez. 2004. A Case Study of Prototyping Methods and Design for Manufacture: Electrostatic Window Blinds. *Journal of Engineering Design* 15: 91–106.

Kukulies, J., and R. Schmitt. 2018. Stabilizing Production Ramp-Up by Modeling Uncertainty for Product Design Verification Using Dempster–Shafer Theory. *CIRP Journal of Manufacturing Science and Technology* 23: 187–196. https://doi.org/10.1016/j.cirpj.2017.09.008.

Lauff, C.A., D. Kotys-Schwartz, and M.E. Rentschler. 2018. What is a Prototype? What Are the Roles of Prototypes in Companies? *Journal of Mechanical Design* 140: 061102.

Machi, L.A., and B.T. McEvoy. 2016. *The Literature Review: Six Steps to Success*. Corwin Press.

Mejía-Gutiérrez, R., and R. Carvajal-Arango. 2017. Design Verification Through Virtual Prototyping Techniques Based on Systems Engineering. *Research in Engineering Design* 28: 477–494.

Raz, A.K., C.R. Kenley, and D.A. DeLaurentis. 2018. System Architecting and Design Space Characterization. *Systems Engineering* 21: 227–242. https://doi.org/10.1002/sys.21439.

Salado, A. 2015. Defining Better Test Strategies with Tradespace Exploration Techniques and Pareto Fronts: Application in an Industrial Project. *Systems Engineering* 18: 639–658.

Salado, A., and H. Kannan. 2019. Elemental Patterns of Verification Strategies. *Systems Engineering* 22: 370–388.

Sanders, G.A., S. Sarkani, and T. Mazzuchi. 2013. High Consequence Systems Phenomenological Characterization: A Tutorial. *Systems Engineering* 16: 464–472.

Shabi, J., Y. Reich, and R. Diamant. 2017. Planning the Verification, Validation, and Testing Process: A Case Study Demonstrating a Decision Support Model. *Journal of Engineering Design* 28: 171–204.

Shin, Y.-D., S.-H. Sim, and J.-C. Lee. 2017. Model-Based Integration of Test and Evaluation Process and System Safety Process for Development of Safety-Critical Weapon Systems. *Systems Engineering* 20: 257–279.

Starkey, E.M., J. Menold, and S.R. Miller. 2019. When Are Designers Willing to Take Risks? How Concept Creativity and Prototype Fidelity Influence Perceived Risk. *Journal of Mechanical Design* 141: 031104.

Sutcliffe, A., and B. Gault. 2004. The ISRE Method for Analyzing System Requirements with Virtual Prototypes. *Systems Engineering* 7: 123–143.

Tahan, M., and J.Z. Ben-Asher. 2005. Modeling and Analysis of Integration Processes for Engineering Systems. *Systems Engineering* 8: 62–77.

Tahera, K., C. Earl, and C. Eckert. 2017. A Method for Improving Overlapping of Testing and Design. *IEEE Transactions on Engineering Management* 64: 179–192.

Tahera, K., D.C. Wynn, C. Earl, and C.M. Eckert. 2018. Testing in the Incremental Design and Development of Complex Products. *Research in Engineering Design*. https://doi.org/10.1007/s00163-018-0295-6.

Takala, R. 2005. Product Demonstrator: A System for Up-Front Testing of User-Related Product Features. *Journal of Engineering Design* 16: 329–336.

Tovares, N., P. Boatwright, and J. Cagan. 2014. Experiential Conjoint Analysis: An Experience-Based Method for Eliciting, Capturing, and Modeling Consumer Preference. *Journal of Mechanical Design* 136: 101404.

Ulrich, K.T., and S.D. Eppinger. 2012. *Product Design and Development*. 5th ed. New York: McGraw-Hill.

Viswanathan, V.K., and J.S. Linsey. 2012. Physical Models and Design Thinking: A Study of Functionality, Novelty and Variety of Ideas. *Journal of Mechanical Design* 134: 091004.

———. 2013. Role of Sunk Cost in Engineering Idea Generation: An Experimental Investigation. *Journal of Mechanical Design* 135: 121002.

Viswanathan, V., O. Atilola, N. Esposito, and J. Linsey. 2014. A Study on the Role of Physical Models in the Mitigation of Design Fixation. *Journal of Engineering Design* 25: 25–43.

Wall, M.B., K.T. Ulrich, and W.C. Flowers. 1992. Evaluating Prototyping Technologies for Product Design. *Research in Engineering Design* 3: 163–177.

Wang, C.-H., and R.C.-C. Chen. 2011. A MPCDM-Enabled Product Concept Design Via User Involvement Approach. *Concurrent Engineering* 19: 19–34.

Implementing a MOSA Decision Support Tool in a Model-Based Environment

Michael Dai, Cesare Guariniello, and Daniel DeLaurentis

Abstract The Modular Open Systems Approach (MOSA) is a DoD initiative that requires major defense acquisition programs to employ modular architectures using widely accepted standards. In order to realize the benefits of modular and open architectures, program stakeholders must successfully navigate various technical and programmatic decisions throughout the acquisition life cycle. Our observation is that many programs do not have sufficient methods and tools to perform analysis, assess trades, and produce evidence for decisions that produce good program outcomes in general and in specific respect to modularity. This paper presents a model-based approach to rigorously collect and present acquisition context data and data from analysis tools in a Decision Support Framework (DSF). Through an example multi-domain mission engineering problem, we demonstrate how the DSF enables comparison of modular/non-modular mission architectures in terms of cost and performance. In addition, an MBSE enterprise architecture model is used to implement the DSF and is shown to (1) provide detailed visualizations of alternative architecture solutions for better comparison; (2) allow traceability between features of the architecture and organizational requirements to better document adherence to MOSA principles; and (3) lay the groundwork for continued model-based engineering development downstream of the Analysis of Alternatives activity to the rest of the acquisition life cycle.

Keywords System of systems · Model-based systems engineering · Mission engineering

M. Dai (✉) · C. Guariniello · D. DeLaurentis
Purdue University, West Lafayette, IN, USA

A. M. Madni et al. (eds.), *Recent Trends and Advances in Model Based Systems Engineering*, https://doi.org/10.1007/978-3-030-82083-1_22

1 Introduction

1.1 The Modular Open Systems Approach (MOSA)

MOSA is a strategy to implement "highly cohesive, loosely coupled, and severable modules" into system designs that can be "competed separately and acquired from independent vendors" (DDR&E(AC) n.d.). This is to be achieved using widely supported and consensus-based (i.e., "open") standards, as they are available and suitable (ODASD 2017). The Office of the Under Secretary of Defense for Research and Engineering states that the approach intends to realize the following benefits (DDR&E(AC) n.d.):

1. *Enhance competition* – open architectures with severable modules allow components to be openly competed.
2. *Facilitate technology refresh* – new capabilities or replacement technology can be delivered without changing all components in the entire system.
3. *Incorporate innovation* – operational flexibility to configure and reconfigure available assets to meet rapidly changing operational requirements.
4. *Enable cost savings/cost avoidance* – reuse of technology, modules, and/or components from any supplier across the acquisition life cycle.
5. *Improve interoperability* – severable software and hardware modules to be changed independently.

MOSA compliance has become a mandate by law for all major defense acquisition programs (10 USC §2446a). However, effective tools for DoD programs to conduct analysis and produce evidence of MOSA implementation are still lacking. This, combined with inconsistent understanding on how to balance MOSA with other trade variables, has produced a situation where many programs struggle to make effective choices and document the rationale for these decisions.

1.2 Barriers to Achieve MOSA Benefits

While MOSA promises great benefits, challenges remain to successfully realize them. Through workshops and interviews, we have interacted with expert practitioners from industry, military, and DoD to better define these challenges (see DeLaurentis et al. 2017, 2018). Practitioners identified among the challenges the need to understand how modular and open architecture alternatives modify technical trades on system life cycle cost, development schedule, performance, and flexibility toward changing mission requirements. There are also many programmatic difficulties associated with the adoption of MOSA. For example, data rights and intellectual property often incentivize vendors from sharing detailed design information that may impact other modules. Selection of working groups, compartmentalization of information, and other organizational structure features can also have a strong impact on how modular systems are successfully realized.

Thus, guidance is needed to navigate the technical trade space regarding modular architectures and the organizational requirements that would best enable modular system designs. A Decision Support Framework (DSF) is being developed to address these challenges.

2 A Decision Support Framework to Guide MOSA Implementation

Figure 1 shows the basic concept of the Decision Support Framework. The idea is to create an executable software that can provide key information to program managers and other stakeholders to guide MOSA-related decisions throughout the acquisition life cycle.

2.1 *DSF Inputs – Mission Engineering and Early-Stage Acquisition Contexts*

The inputs to the software are the parameters of the mission engineering problem that would surround an acquisition: a mission Concept of Operations, a description of capability gaps to be fulfilled, and a library of candidate systems to be selected

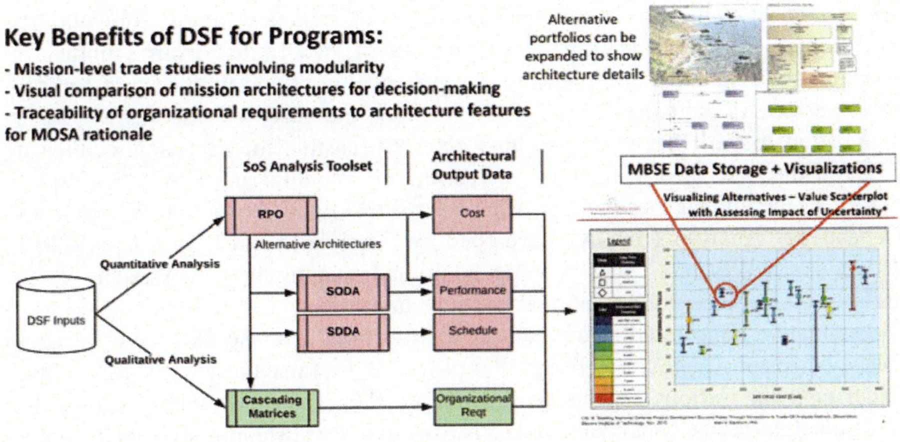

Fig. 1 Workflow of the Decision Support Framework concept. The decision-making scatterplot is a figure taken from SERC research efforts led by Blackburn, described in Bone et al. (2018). RPO, SODA, and SDDA are tools from the Purdue Analytic Workbench described in DeLaurentis et al. (2016)

and integrated to achieve mission objectives. This library represents the capabilities of current and to-be-acquired systems, each involving varying degrees of modularity. The alternative systems in the library will allow us to explore how the selection of modular and open systems will impact the mission architecture as a whole.

The user will also input the mission-level requirements that all solution SoS architectures must satisfy. These requirements will be based on mission capability metrics relevant to the mission category. For example, an amphibious assault mission may require SoS capabilities like naval superiority, air superiority, tactical bombardment, and land seizure. Rigorously defining metrics for this level of capability is challenging and is outside the scope of this paper.

2.2 *DSF Analysis – Quantitative and Qualitative Analysis Threads*

The DSF analysis is divided into two threads: a quantitative analysis addressing the technical trade-offs associated with modularity and a qualitative analysis addressing programmatic considerations for MOSA.

The quantitative analysis could eventually use a set of tools that are most familiar and trusted by a particular program, as long as they are configurable to represent choices related to modularity and openness. Our prototype DSF employs tools from Purdue's SoS Analytic Work Bench, described in DeLaurentis et al. 2016. Robust Portfolio Optimization (RPO), detailed in Davendralingam and DeLaurentis 2013, generates alternative architectures and analysis of cost and performance. In RPO, hierarchies of systems are modeled as nodes on a network that work cohesively to fulfill overarching capability objectives. Capabilities (outputs) from existing nodes connect to fulfill requirements (inputs) of other nodes, amidst compatibility constraints. The end goal is to generate a set of "portfolios" from a library of constituent systems (or components) that are pareto-optimal with respect to SoS-level performance goals and constraints, under measures of uncertainty. In its application to the DSF, the portfolios represent feasible mission architectures in terms of their constituent systems, including both modular and monolithic assets. Systems Operational Dependency Analysis (SODA) and Systems Developmental Dependency Analysis (SDDA), developed by Guariniello and DeLaurentis (2013, 2017), are AWB tools that provide additional quantitative assessment of the architectures in terms of operational and schedule risks.

Qualitative considerations on MOSA architectures are also analyzed. In many cases, our prior research has found that programs first need a way to explore and understand the various aspects of modularity, their interplay with key program cost-schedule-performance outcomes, and long-term sustainment considerations (DeLaurentis et al. 2017, 2018). This context, along with the initial sparsity of data in early life cycle phases, makes the qualitative thread important in the benefits of the DSF. The qualitative portion uses quality function deployment (QFD) techniques

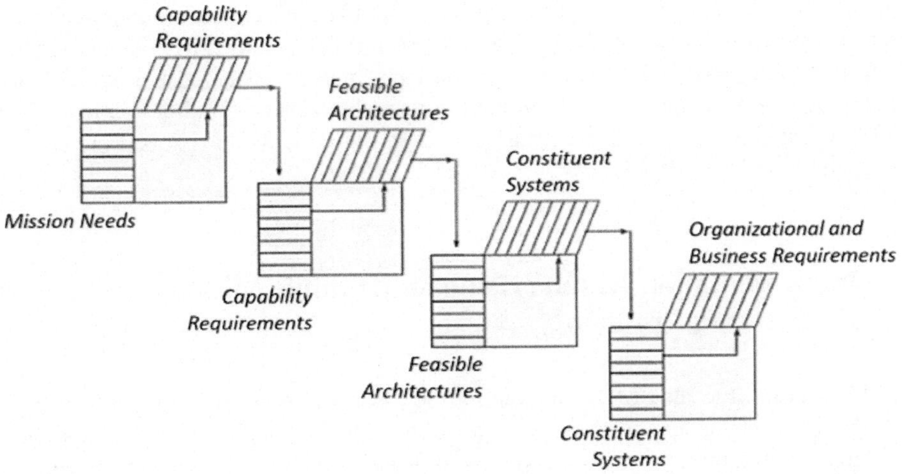

Fig. 2 Cascading matrices in the DSF are used to trace mission needs through alternative architectures to organizational and business requirements. Waterfall representation is adapted from a figure by the American Society for Quality in Revelle (2004)

and cascading matrices to trace features of alternative architectures to the ideal organizational requirements associated with them (Fig. 2). The idea is to use a series of matrices to map mission-level capability needs to certain mission-level requirements. These requirements are then mapped onto the alternative architectures (identified by RPO) which satisfy them. Finally, alternative architectures are mapped to the organizational and MOSA-related resources needed to achieve them.

2.3 DSF Outputs – Integrated Decision-Making Views in a Model-Based Environment

Finally, projected outputs of the DSF software will display the implications on cost, schedule, and risk of modular architectures and relationships between features of system solutions and the organizational structures that would best support them. We are presently assembling a comprehensive synthetic problem to exercise and demonstrate both the qualitative and quantitative tracks of the DSF. A simplified multi-domain battle scenario problem is used in this work to demonstrate the use of MBSE to support the application of RPO in the DSF.

We apply the concept of an MBSE system model with visualizations to the implementation of the DSF. Using a model-based environment allows DSF inputs and outputs to be collected and linked together in a central database. This will enable an integrated means to visualize pertinent data and facilitate DSF decision-making. Our application of MBSE to this problem domain is inspired by SERC research efforts led by Blackburn and leverages principles and best practices in model-centric engineering from his work. Bone et al. (2018) and Blackburn (2019) demonstrate how integrating data from various engineering analysis tools in a model-based workflow can be achieved and presented in a single decision-making window. His decision-making window is shown as part of in Fig. 1 as the envisioned means of displaying the DSF outputs.

3 Implementation Results Using an Example Mission Engineering Problem

In this section, the idea of implementing the Decision Support Framework using MBSE concepts is expounded upon and illustrated through a simple example problem. In it, a multi-domain mission is to be performed using five concept roles: a surveillance system, a communications system, an air superiority system, a power supply system, and a maritime superiority system. They interact as a network to achieve generic SoS capabilities.

3.1 MBSE Views Establishing Mission Context for the DSF

The general premise of the DSF is to allow the analysis and comparison of modular/non-modular architecture solutions in a given mission context. Thus, before the analysis is performed, the mission engineering problem must be clearly stated. The enterprise architecture model is therefore instantiated with high-level operational concept information expressed in OV-1 and OV-2 DoDAF models (Fig. 3). The OV-1 shows the general concept roles that will later be filled by alternative system solutions, in addition to notional dependency relationships. The OV-2 specifies the intended flow of information, energy, and material entities between the general system categories (Fig. 3b). Creating these views is helpful for validating to-be architectures against the original mission needs and assessing their ability to adapt to changes in mission configuration.

3.2 Identification of Feasible Architectures with RPO

To identify feasible architectures, a library of component systems is collected along with their individual performance, requirements, compatibility constraints, and associated uncertainties, shown in Appendix A. Among the candidate component systems are modular and non-modular options. The tool is used to generate the set

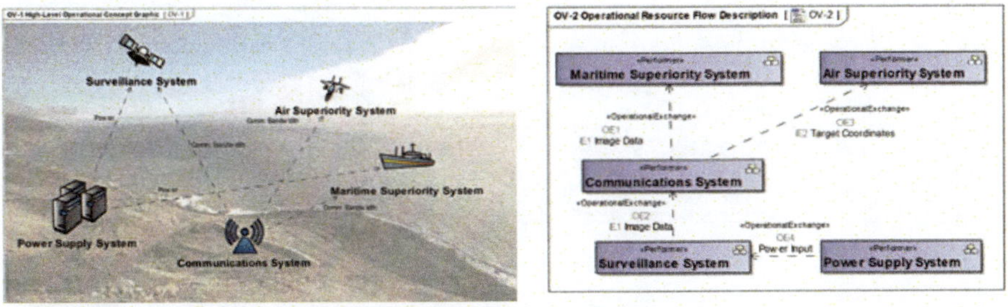

Fig. 3 (**a**) OV-1 high-level operational concept view of the mission and capability roles to be fulfilled. (**b**) OV-2 operational resource flow description detailing how each performer concept will exchange information, energy, and material flows (see DoD CIO 2010)

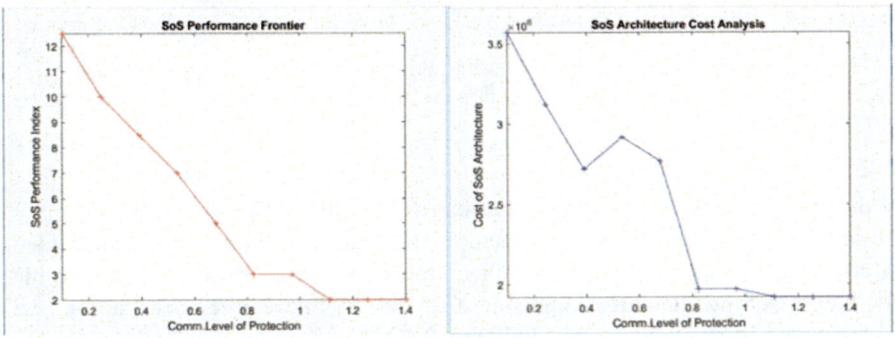

Fig. 4 Pareto frontier of SoS portfolios identified by RPO in terms of SoS performance, cost, and risk aversion toward constraint violations on the communications requirements

of architectures that are optimal with respect to SoS-level capabilities, cost, and risk protection against constraint violations.

Running the optimizer results in the pareto-optimal portfolios shown in Fig. 4. Here the SoS capabilities have been consolidated into a single "SoS Performance Index" metric. Each portfolio has a different performance index, cost, and level of protection with respect to communications constraints. This level of protection is essentially an inverted measure of how likely node-level communication bandwidth requirements are to be violated due to uncertainties in system communication capabilities.

3.3 MBSE Representations of Output Data

After feasible architectures have been identified, they are filtered through three QFD cascade matrices. The first matrix maps user-selected mission needs to mission performance requirements. The second matrix connects these SoS-level requirements

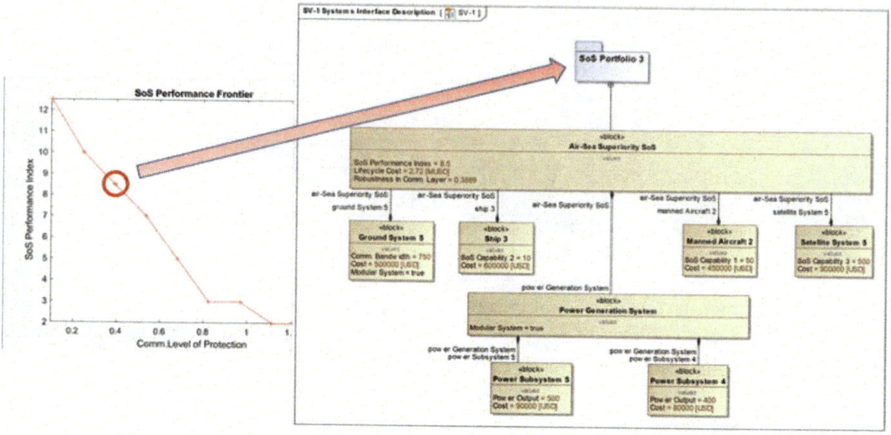

Fig. 5 The selected SoS portfolio is expanded to show its composition details in a SysML block diagram. This level of granularity reveals that the architecture incorporates a modular power generation system

to the RPO feasible architectures that can fulfill them. The third cascade maps the features of the alternative architectures to organizational requirements needed to effectively realize them. The result of the QFD cascades is a set of feasible alternative SoS architectures. Imposing a single mission requirement that the SoS Performance Index ≥ 5, the results from RPO identify four architectures on the Pareto frontier as the final set of alternatives.

The result of the analysis portion of the DSF is a listing of alternatives that can be compared in terms of architecture cost, acquisition timeframe, and performance (but here, only cost and performance). The Pareto frontier plot allows these architectures to be visually compared at a high level. More detailed architecture information can be obtained by linking each pareto-optimal point to a MBSE representation of that alternative. In Fig. 5, a block definition diagram is used to model a selected portfolio, containing information on the systems comprising it and how each collectively contributes to the SoS-performance objectives. In this example, the model and visualization were made manually for the selected portfolio. While steps toward automated diagram generation and linkage are described in the closing section, this is reserved for future work.

In addition to showing portfolio composition, MBSE can also represent the results of the QFD cascades by showing the traceability from system alternatives to organizational requirements using a SysML requirements diagram. For example, additional organizational requirements may be specifically tied to modular features of a system. Representing this traceability enables programs to provide evidence of MOSA principles in their design decisions. This is illustrated in Fig. 6. Specifying these organizational requirements will be based on MOSA case study data – however, for now, they remain notional.

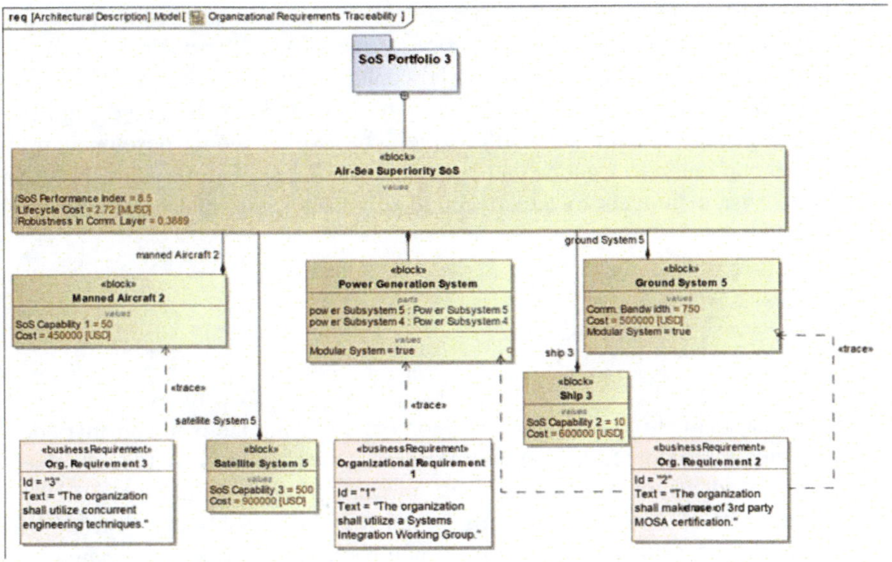

Fig. 6 Requirements diagram showing the traceability of system features to organizational requirements

4 Summary

4.1 Key Takeaways

This paper examined how the Decision Support Framework can be used as a tool toward better achieving the benefits of modularity, as motivated by the MOSA initiative. The DSF addresses key challenges concerning MOSA implementation, primarily those related to evaluating technical trades involving modularity, and tracing modular architectural features to organizational requirements needed to enable them. The simplified example problem demonstrated the use of Robust Portfolio Optimization in the DSF to enable comparison between architectures with modular/non-modular system alternatives, in terms of cost and performance. In addition, the example portrays how the DSF utilizes an MBSE enterprise architecture model to store and visualize mission context, architecture structural features, and requirement traceability.

MBSE adds value to the DSF in three ways:

1. Linked visualizations can provide more detailed architecture data upon user inquiry. This allows decision-makers to quickly jump between levels of granularity when comparing alternative architectures.
2. MBSE system models allow clear traceability between mission requirements, selection of modular/non-modular alternatives, and MOSA-relevant organiza-

tional requirements. This digital thread records the rationale behind programmatic decisions and can provide evidence for use of MOSA principles.
3. Creating a high-level architecture model capturing results from the Analysis of Alternatives (AoA) activity paves the way for continued model-based engineering throughout the acquisition life cycle. The MBSE model developed from the DSF can come to establish an authoritative source of truth, allowing more detailed system models to directly build off the mission and enterprise level models.

4.2 Future Work

There are many directions for future work on this project. One is to consider additional metrics to compare mission architectures. Other analysis tools in Purdue's AWB will be added to the DSF to assess flexibility and schedule metrics. With more dimensions of comparison, having an integrated decision-making window as a DSF output, such as that shown in Fig. 1, will become even more important.

A second area of ongoing work is to understand what organizational requirements are necessary for different kinds of modular mission architectures. This work is being performed through case study analysis of successful MOSA programs and through collaboration with ongoing partner programs.

A third area of future work is in creating a digital linkage directly from RPO Pareto fronts (or other decision windows) to the MBSE portfolio visualizations. Practically, this would entail being able to click on a certain portfolio in the decision-making windows in Figs. 1 or 5 and having its visualizations directly be generated.

Acknowledgments This material is based upon work supported, in whole or in part, by the U.S. Department of Defense through the Systems Engineering Research Center (SERC) under Contract HQ0034-19-D-0003 WRT-1002. SERC is a federally funded University Affiliated Research Center managed by Stevens Institute of Technology. We further acknowledge the contributions of research collaborators Gary Witus, Charles Domercant, and Thomas McDermott.

A.1 Appendix A: RPO Input Data for Example Mission Engineering Problem

In this example, candidate systems are labeled generically (e.g., Satellite System 1–5) and use notional data. The data in Fig. A.1 can be read as follows: Satellite System 1 contributes 100 to "SoS Capability 3" and requires 75 [units] in communication bandwidth and 95 [units] in power input. Likewise, Power System 3 offers no SoS or communication capabilities, but is capable to supply 300 [units] of power to other systems. Each capability is subject to an uncertainty that may result in violating node input requirements. This information is reflected in a risk aversion metric shown

		SoS Capabilities (Outputs)			Capabilities (Outputs)		Requirements (Inputs)		Cost	Uncertain. Capabilities (+/- delta)					Num
No.	System Name	SoS CAP 1	SoS CAP 2	SoS CAP 3	Comm.	Power.	Req 1	Req 2	($)	CAP 1	CAP 2	CAP 3	CAP 4	CAP 5	Links
1	GroundSystem 1	0	0	0	150	0	0	0	10000	0	0	0	30	0	1
2	GroundSystem 2	0	0	0	300	0	0	0	20000	0	0	0	60	0	1
3	GroundSystem 3	0	0	0	450	0	0	0	300000	0	0	0	150	0	1
4	GroundSystem 4	0	0	0	600	0	0	0	400000	0	0	0	300	0	1
5	GroundSystem 5	0	0	0	750	0	0	0	500000	0	0	0	500	0	1
6	Satellite System 1	0	0	100	0	0	75	95	500000	0	0	0	32	0	1
7	Satellite System 2	0	0	200	0	0	125	150	650000	0	0	0	12	0	1
8	Satellite System 3	0	0	300	0	0	150	250	750000	0	0	0	14.6	0	1
9	Satellite System 4	0	0	400	0	0	175	350	850000	0	0	0	8.4	0	1
10	Satellite System 5	0	0	500	0	0	185	450	900000	0	0	0	49	0	1
11	UAV 1	20	0	0	0	0	100	0	200000	0	0	0	30	0	1
12	UAV 2	30	0	0	0	0	200	0	300000	0	0	0	10.5	0	1
13	Manned Aircraft 1	40	0	0	0	0	300	0	400000	0	0	0	10.9	0	1
14	Manned Aircraft 2	50	0	0	0	0	120	0	450000	0	0	0	10.3	0	1
15	Manned Aircraft 3	60	0	0	0	0	300	0	500000	0	0	0	13.4	0	1
16	Ship 1	0	5	0	0	0	50	0	500000	0	0	0	25	0	1
17	Ship 2	0	10	0	0	0	150	0	600000	0	0	0	35	0	1
18	Ship 3	0	20	0	0	0	200	0	700000	0	0	0	50	0	1
19	Power System 1	0	0	0	0	100	0	0	50000	0	0	0	0	10	1
20	Power System 2	0	0	0	0	200	0	0	60000	0	0	0	0	20	1
21	Power System 3	0	0	0	0	300	0	0	70000	0	0	0	0	40	1
22	Power System 4	0	0	0	0	400	0	0	80000	0	0	0	0	80	1
23	Power System 5	0	0	0	0	500	0	0	90000	0	0	0	0	150	1

Can Have Options		Select Limit
1	5	1
6	10	2
11	15	1

Must Select Options		Select Limit
16	18	1
19	23	2

Fig. A.1 Example of RPO input data for the example problem

on the horizontal axis of Fig. 4. Finally, compatibility and selection constraints are set in the input spreadsheet as well. Here, the optimizer can select one option from systems 1–5 (Ground Systems) and 11–15 (Aerial Systems) and up to two options from systems 6–10 (Satellite Surveillance Systems). Likewise, the optimizer is constrained to select one Naval System (16–18) and two Power Systems.

References

Blackburn, M. 2019. *Transforming Systems Engineering through Model Centric Engineering*. Systems Engineering Research Center Workshop.

Bone, M., M. Blackburn, B. Kruse, J. Dzielski, T. Hagedorn, and I. Grosse. 2018. Toward an Interoperability and Integration Framework to Enable Digital Thread. *Systems* 6 (4): 46.

Davendralingam, N., and D. DeLaurentis. 2013. A Robust Optimization Framework to Architecting System of Systems. *Procedia Computer Science* 16: 255–264.

DeLaurentis, D., N. Davendralingam, K. Marais, C. Guariniello, Z. Fang, and P. Uday. 2016. An SoS Analytical Workbench Approach to Architectural Analysis and Evolution. *OR Insight* 19 (3): 70–74.

DeLaurentis, D., N. Davendralingam, M. Kerman, and L. Xiao. 2017. *Investigating Approaches to Achieve Modularity Benefits in the Acquisition Ecosystem*. (Report No. SERC-2017-TR-109). Retrieved from the Systems Engineering Research Center.

DeLaurentis, D., N. Davendraligam, C. Guariniello, J.C. Domercant, A. Dukes, and J. Feldstein. 2018. *Approaches to Achieve Benefits of Modularity in Defense Acquisition*. (Report No. SERC-2018-TR-113). Retrieved from the Systems Engineering Research Center.

Directorate of Defense Research and Engineering for Advanced Capabilties (DDR&E(AC)). (n.d.). *Modular Open Systems Approach*. Retrieved from Office of the Under Secretary of Defense for Research and Engineering: https://ac.cto.mil/mosa/

DoD Deputy Chief Information Officer. 2010. *The DoDAF Architecture Framework Version 2.02*. Retrieved from Chief Information Officer- US Department of Defense: https://dodcio.defense.gov/library/dod-architecture-framework/

Guariniello, C., and D. DeLaurentis. 2013. Dependency Analysis of System-of-Systems Operational and Development Networks. *Procedia Computer Science* 16: 265–274.

———. 2017. Supporting Design via the System Operational Dependency Analysis Methodology. *Research in Engineering Design* 28 (1): 53–69. https://doi.org/10.1007/s00163-016-0229-0.

Office of the Deputy Assistant Secretary of Defense. 2017. MOSA Definition. FY17 National Defense Authorization Act (NDDA), 10 USC §2446a.

Revelle, J.B. 2004. *Quality Essentials: A Reference Guide from A to Z*. ASQ Quality Press. Retrieved from American Society for Quality: https://asq.org/quality-resources/qfd-quality-function-deployment.

Change Management Processes in MBSE

Isabeta Rountree, Victor Lopez, and L. Dale Thomas

Abstract Changes are intrinsic to system development. These changes regularly cause cost and schedule overruns due to design rework. Existing techniques seem unable to provide an adequate change management process. Model-based systems engineering (MBSE) allows for the creation of new change management processes that improve on the traditional methods. A change management process for projects developed in a MBSE environment is outlined. This process depicts the advantages of using MBSE to reduce time spent on a project when faced with high-level system changes. This process was then applied in the development of a CubeSat project and reduced the time spent on changes compared to the traditional process.

Keywords SysML · MBSE · Change

1 Introduction

Project changes are recognized as key concerns (Stare 2011) relating to schedule delays and cost overruns while being considered intrinsic to system development (Stare 2011). Traditional project management methods seem unable to properly address changes in these complex systems (Saynisch 2010). Dynamic environments are a contributing factor in the difficulty of managing complex systems. Environments are constantly changing, making it hard to predict changes during early planning phases in projects (Stare 2011). Traditional systems engineering methods seem inadequate to account for this inability to predict factors of change. While changes have negative programmatic consequences, such as schedule excess or incurred cost, the benefit to system development from a change can sometimes outweigh the repercussions. Thus, identifying which changes will ultimately result in a better system becomes critical.

I. Rountree · V. Lopez · L. D. Thomas (✉)
The University of Alabama in Huntsville, Huntsville, AL, USA
e-mail: ldt0001@uah.edu

A. M. Madni et al. (eds.), *Recent Trends and Advances in Model Based Systems Engineering*, https://doi.org/10.1007/978-3-030-82083-1_23

2 Origin of Changes

Introducing change to a system creates a new design space, with alternatives that need to be compared and evaluated. This comparison is usually performed by looking at the impact of the change in the performance and capabilities of the system as well as the cost and schedule impacts. In traditional, document-based, systems engineering practice, when a change is proposed, subject-matter experts are contacted to evaluate the impact of the change by rerunning analysis and simulations. Oftentimes, physical and behavioral system models must be reorganized based on the change incurred, and an iterative approach must be used to find the new feasible solution. This iterative process typically includes multiple non-integrated physical simulations and is resource-intensive.

During the iterative analysis and design process, unforeseen issues may arise. The introduction of a new solution may interfere with an existing design solution or system interface. Existing interfaces and solutions must then be evaluated and corrected to ensure system success. One change can lead to many smaller changes that may have less apparent or perceived impact and therefore are easier to overlook. These changes have the potential to cause issues during later system development.

Changes can manifest due to controllable and uncontrollable factors. McMahon identifies five modes of design change, two of which involve types of change that may arise during system development due to uncontrollable factors (Mcmahon 1994). It is the uncontrollable factors that are most difficult to anticipate. The first of these two modes introduces changes due to an improved understanding of the development of analytical or experimental techniques. The second introduces changes due to external factors resulting in a change to the product design specification such as changes in the market conditions and stakeholder interests (Mcmahon 1994). The other modes outline change necessitated by the ability to create a more optimized design through parameter space exploration, modification of the feasible design space, and the selection of alternative designs. Changes can incur significant schedule cost, but this loss may be outweighed by system performance and capability enhancements.

3 Current Change Management Methods

Changes can be dealt with in a couple of ways. When change is not managed, changes may be made with little to no regulation. This approach leads to a great number of changes that make system integration and development difficult due to constant rework of interfaces and designs. Change is usually managed with a configuration change management process in traditional systems engineering.

Change management processes vary depending on the organization and project. As an example, NASA's configuration change process is shown in Figure 1 (National Aeronautics and Space Administration 2016). NASA defines configura-

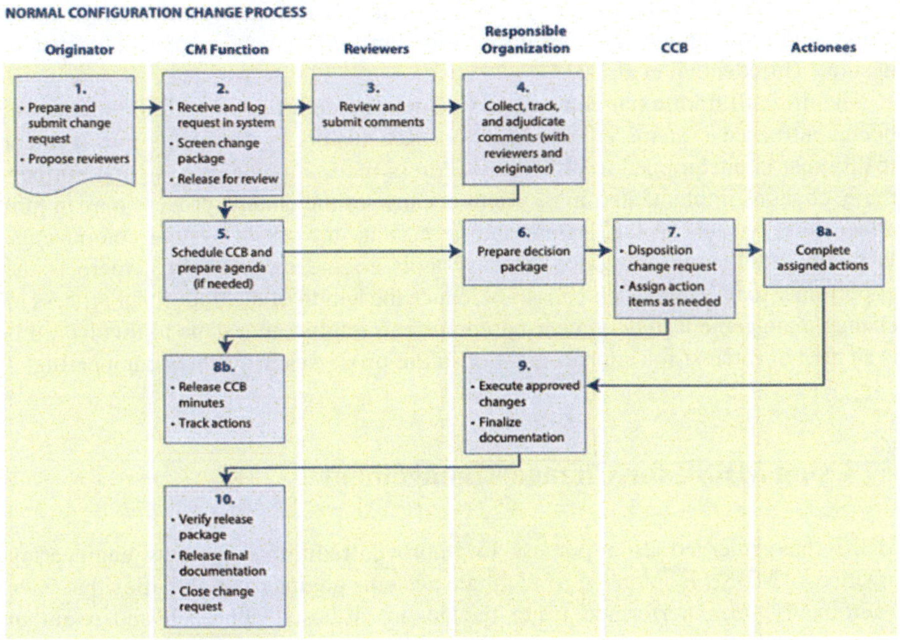

Fig. 1 Illustration of the normal NASA configuration change process (National Aeronautics and Space Administration 2016)

tion change management as the "systematic proposal, justification, and evaluation of proposed changes followed by incorporation of approved changes and verification of implementation" (National Aeronautics and Space Administration 2016).

The NASA change management process outlined in the figure involves many steps designed to moderate the number of changes made and ensure there is more benefit than detriment to the system.

This process depicts the participation of six parties, four of which involve multiple people and actions. The coordination of this amount of people, the compilation of a decision package, and iteratively evaluating design alternatives are all very involved and time-consuming processes. The traditional change control process is lengthy and can be difficult to implement cleanly and effectively. Additionally, even with the effective implementation of change management processes like the afore outlined one, schedule detriment due to incurred system changes is still a significant issue. This is due to the large number of steps and involved parties in the development and delivery of a change and its respective repercussions.

Other limitations with document-based approaches include limitations in "The completeness, consistency, and relationships between requirements, design, engineering analysis, and test information are difficult to assess since this information is spread across several documents" (Friedenthal et al. 2015) which translate into "to poor synchronization between system-level requirements and design and lower-level hardware and software design. It also makes it difficult to maintain or reuse the system requirements and design information for an evolving or variant system design." (Friedenthal et al. 2015).

The difficult traditional change management process paired with the traditional document-based systems engineering approach results in a systems management reality that is unequipped to efficiently address modern complex systems. Furthermore, changes implemented in a system because of the change process may in turn affect other aspects of the system and necessitate rework of system components. With the emergence of model-based systems engineering (MBSE), there is the opportunity to re-evaluate and possibly reduce the length of the traditional process of change management in document-based systems engineering. Due to this fact, this is an area of interest for improvement of future processes in systems engineering.

4 Using MBSE for Change Management

MBSE has emerged as a process to improve traditional systems engineering practices; "MBSE is intended to facilitate systems engineering activities that have traditionally been performed using the document-based approach and result in enhanced specification and design quality, reuse of system specification and design artifacts, and communications among the development team" (Friedenthal et al. 2015). This paper will focus on how MBSE can be used to reduce schedule impacts associated with change management in large complex systems.

When implementing MBSE, "The system model can also be integrated with engineering analysis and simulation models to perform computation and dynamic execution" (Friedenthal et al. 2015). This capability allows management of all compatible engineering analysis and simulation models in a singular location that can be executed simultaneously when faced with changes to a system. The advantages of MBSE could potentially include "Increased productivity; Faster impact analysis of requirements and design changes, more effective exploration of trade-space, reuse of existing models to support design evolution, reduced errors and time during integration and testing, and automated document generation" (Friedenthal et al. 2015).

The Systems Modeling Language (SysML) is a widely used language for MBSE. This language "supports the practice of model-based systems engineering (MBSE) that is used to develop system solutions in response to complex and often technologically challenging problems" (Friedenthal et al. 2015). There is extensive documentation on this language in texts such as *A Practical Guide to SysML: The Systems Modeling Language* by Friedenthal, Moore, and Steiner (Friedenthal et al. 2015), and it is well known as an extension of the Unified Modeling Language (UML). For this reason, the implementation of MBSE using SysML is explored towards improving the change management process to reduce time spent managing change when faced with new requirements in complex systems.

It should be noted, however, that while MBSE does have the potential to realize the stated benefits, there are some obstacles. First the implementation of MBSE usually requires the creation of a system model which requires proficiency in both the model's language, usually SysML, and the software tool used, No Magic's Cameo or IBM's Rhapsody for example. Needing to be proficient would mean having to train current systems engineers and might steepen the learning curve for new hires. This necessitates a significant time investment prior to being able to effectively develop a system model.

Additionally, for large complex projects, an integrated system model would likely be developed by multiple system modelers. Different approaches to model development could lead to consistency or compatibility issues. However, the software development industry has encountered similar issues in software modeling resolved through methodologies and tools such as Agile software development: "...agile methods aim to answer a need to develop software quickly, in an environment of rapidly changing requirements" (Greer & Hamon 2011). Similar approaches are and will need to be developed for systems modeling (Schindel & Dove 2016).

In order to take full advantage of the MBSE approach, a system model should include integrated engineering analysis and simulation models. This integration is not trivial, especially when dealing with very specialized analyses, legacy tools, or geographical and organizational barriers. Although some work and tools are aimed at improving this situation, there is still work to be done until this can become standard practice.

Finally, different common processes in system development such as contracting, acquisition, trade studies, scheduling, and communication rely on traditionally produced documentation. The change from document-based to model-based systems engineering would require an adaptation of all these processes.

5 Methodology

Change management practices with respect to MBSE have been introduced as a topic of interest. However, little research has been done on the process itself. Authors Friedenthal, Moore, and Steiner introduce the idea that MBSE needs processes to manage change in their text *A Practical Guide to SysML: The Systems Modeling Language*: "Disciplined processes and associated tools must be put in place to manage changes to models" (Friedenthal et al. 2015). Further in the text, configuration management approaches with regard to MBSE models are discussed. The general configuration management responsibilities include: "Manage the set of artifacts (often called configuration items) being developed, including managing access to the current working set of artifacts (often called a configuration) and archiving significant versions of that working set (called baselines)," and "Ensure that products built from a project baseline are complete and consistent, including the identification of different variants of system components and the compatibility between them" (Friedenthal et al. 2015). Regarding change management, the configuration management responsibilities implied personnel were to "Manage changes to that working set, including enforcing a consistent change control process, for example based on change requests, and analyzing the impact of changes to configuration items. Tools that fulfill this function are typically called change management tools, and often incorporate configuration management functions" (Friedenthal et al. 2015). This text also includes a reference to the Object Management Group (OMG) Meta Object Facility (MOF) versioning standard (https://www.omg.org/spec/MOFVD/2.0/PDF) as a possible framework for configuration management of models.

The OMG MOF versioning specification was designed to "manage the co-existence of multiple versions of such metadata in a Meta Object Facility and their inclusion in different configurations (for example, a specific version of a Platform Independent Model (PIM), the corresponding version of the derived Platform Specific Model (PSM), and the corresponding version of the generated system)" (https://www.omg.org/spec/MOFVD/2.0/PDF). This specification focuses on the management of specific configurations (model variations and baselines) rather than the process of change management itself. There is otherwise little documentation on a change management process for systems using MBSE. A change management process will be proposed that supports a MBSE-centered environment and highlights the time saving advantage of using a model when faced with system changes.

In a MBSE environment, changes occur just as in a document-based system engineering environment. However, there are some noticeable differences in the management of system artifacts. The following is an outline of a MBSE change process. Likewise, to traditional processes, assume a change request is submitted by someone on the project. This change order would go to a configuration management representative who would then have it reviewed. Then, in preparation for the decision package, the change request would be disseminated to a model owner. The model owner would then execute the change in a branch model and record highlighted changes based on modelled metrics and traceability. This information would then be sent to the change control board (CCB). Following the traditional method, then the CCB would approve or deny the change. Since a model allows the capability of uniformly updating model artifacts through all model documentation and accompanying simulations by a singular model editor, the "actionees" step of the traditional method is skipped. The change control board would send the change decision to the configuration management representative, who would then notify the model owner of approved changes. The model owner would then be able to cleanly and uniformly implement the changes in the model. An illustration showing this potential shortened methodology is shown in Figure 2. This method was developed to be tested using the accompanying case study example.

The traditional NASA change management process depicted in Figure 1 involves many steps. These steps include preparation and submittal of a change request, reception and distribution of the change request by configuration management, a review of the change request by a panel, compilation by the organization, scheduling

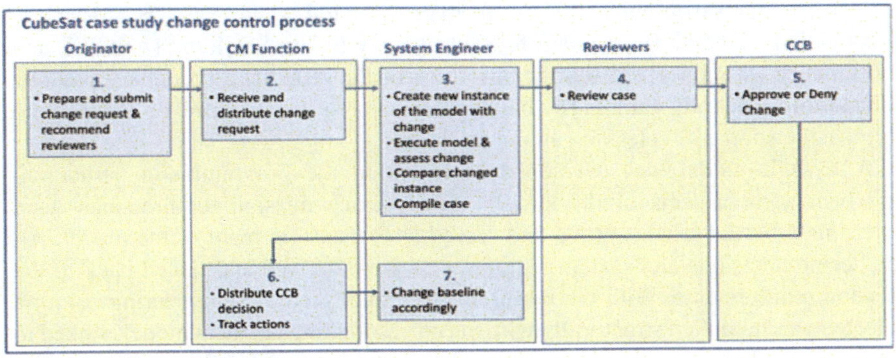

Fig. 2 Potential change control process for a MBSE project

of a change control board meeting, preparation of a decision package, an analysis of tasks associated with the change, performance of designated tasks, tracking of task completion, execution of approved changes, preparation of final change documentation, and official change documentation release.

The change management methodology shown in Figure 2 varies from the methodology in Figure 1 in a few key aspects. Firstly, it removes the creation and handling of documents between different actors. Information normally depicted in documents is contained within the model. Secondly, it reduces the actions generated by the change. The model being linked to most physics models, the subsequent actions needed from the change are automatically generated before the change goes to the CCB. Finally, it allows the CCB to make a more informed decision. Since the model automatically assesses the impact of the change to other parts or metrics, the CCB can use that information to make their decision instead of the information being generated as an action after the change has been approved.

A model allows changes to be implemented quickly and uniformly through all model documentation. This saves time in updating model artifacts and increases document consistency as it is uniform and automatic. Additionally, engineering analysis models will automatically update when tied to model parameters, and affected subsystems can be easily identified. These capabilities significantly reduce the time taken in the traditional method due to the centrally located definition of model artifacts and simulations. A singular model owner has the capability to evaluate a change, see what the change affects, and make the change which will update related model artifacts and simulations.

6 Case Study

The planned NASA Artemis 2 mission solicited proposals in the fall of 2019 for secondary payload CubeSat missions aimed at addressing NASA's strategic knowledge gaps. The Alabama Space Grant Consortium has organized a cooperative project between seven universities in Alabama to build a CubeSat. The mission is being called Alabama Experiment for Galactic-ray In-situ Shielding (AEGIS), and it originally was conceived – prior to release of the Artemis 2 secondary payload solicitation – to study gamma ray bursts in cislunar space to improve the ability to localize the source of celestial gamma ray bursts.

A SysML model was developed representing the programmatic, structural, and behavioral aspects of the system. The science mission requirements were determined by the science group and served as the starting point of the model. All the design decisions and system requirements were derived and traced back to the mission requirements. With the requirements established the engineering solution selected was modelled structurally with the corresponding physical models linked to it. Finally, the behavior of the system was determined and captured in a set of SysML diagrams. Using the behavioral and structural representation, different performance parameters of the system were calculated. The MBSE approach taken allowed for every part of the system to be traced back to the mission requirements.

The design was started before NASA released the specific criteria by which the missions would be selected for secondary payloads. After the official solicitation was released, it was determined that the planned scientific mission was not well aligned with the solicitation criteria and as such was inadequate for the project. The science mission was then modified to study the shielding properties of lunar regolith in a cislunar environment to be more responsive to the Artemis 2 secondary payload solicitation criteria. This change in mission scope would then change the mission requirements and cause multiple changes across different aspects of the design.

In a traditional system one would have to either go into a long change process for every single change or redo the entire design process from the start. The first option would then result in a long process where different changes have individual change processes. These individual changes would have been extremely resource expensive. Additionally, these changes would have been accompanied by the risk that an approved change might have to be reworked after another change is implemented. If the design were to be restarted, the whole documentation would then need to be remade, and all the work previously done would have been useless. In either case, when the change resulted in a parameter change used in one of the physical analyses, the new parameter would need to be sent to the analyst. This analyst would do his analysis and send back the result. Since this project has very strongly coupled analysis, this would result in an iterative loop until a feasible solution is found.

Instead, since the system was under development in a MBSE environment, the requirements that changed were able to be changed in the model; from there it was easy to see what design decisions and what requirements were derived since there was a traceable connection. This was then followed until the complete impact of the design change was evaluated. The changes were translated into changes in the behavioral and structural diagrams. The main difference is that the parametrized physics models relating to model elements did not require rework. When all the changes were made, all the models were able to be run by the systems engineer without the need of the individual analyst, thereby greatly reducing the time to evaluate the impact of the change, although some specialized analyses that were not included in the model had to go through the traditional change process.

Currently, additional physics models are being parametrized and included in the model. The generation of the flight software will also be incorporated into the model through the implementation of F Prime. The introduction of these aspects into the model will allow a quantitative way to measure the benefit of having physics models included in the MBSE approach. An analysis will be done comparing how long some changes took at the beginning of the project against at the end of it. This difference in time could then be used to determine if the benefit of the MBSE approach outweighed the time and resources needed to implement the approach.

7 Conclusion

The AEGIS project encountered change during system development. The MBSE change management approach outlined was applied to the project to assist in managing these changes. The result was a CM process with fewer numbers of steps and individuals involved in the process, thereby reducing the time taken to address changes. This implementation allowed for faster change response and continuous system development.

Change is inevitable in large systems; the ability to properly manage them can be the difference between successful and failed systems. Existing traditional change management processes seem unable to properly address the changes of new systems resulting in cost increases and schedule overruns. These effects originate in part from a lengthy process involving multiple parties, people, and actions. MBSE opens the door for new change management processes that could improve on the traditional processes. MBSE allows for the systems engineer to quickly create a different version of the model with the modified physics results. This can then be used to evaluate figures of merit to decide if the change should happen or not. The ability to reduce the number of people involved in the change process coupled with lower analysis time allows reduction in the overall schedule and cost impact incurred due to a change. These new processes could be faster and more reliable and thus could contribute to designing better systems, faster and at a lower cost. Looking forward, a proper change management process should be studied to learn exactly the best way to handle change in a MBSE environment.

Further study is needed to assess the applicability of a MBSE change management process like the one outlined to larger complex projects. Larger projects are likely to include several model-based systems engineers all operating in an integrated model space. With the introduction of more modelers, more steps may be necessary in model management procedures. Further development and analysis of a MBSE change management process applying to variously sized projects needs to be done to broaden the applicability of a general MBSE change management process and validate its effectiveness. The current CM methodologies are the result of years of evolution through many system developments; it can be expected that as new projects are developed in a model-based environment the CM will also evolve to meet a system or organization's needs.

References

Friedenthal, Sanford, Alan Moore, and Rick Steiner. 2015. *A Practical Guide to Sysml: The Systems Modeling Language*. Waltham: Morgan Kaufmann.

Greer, D., and Y. Hamon. 2011. *Agile Software Development 41* (9): 943–944. https://doi.org/10.1002/spe.1100.

Mcmahon, Christopher A. 1994. Observations on Modes of Incremental Change in Design. *Journal of Engineering Design* 5.3: 195–209. Web.

National Aeronautics and Space Administration. 2016. *NASA Systems Engineering Handbook. Rev 2*. Washington, DC: National Aeronautics and Space Administration. Print.

Object Management Group. "Meta Object Facility (MOF) Versioning and Development Lifecycle Specification, v2.0.". https://www.omg.org/spec/MOFVD/2.0/PDF.

Saynisch, M. 2010. Mastering Complexity and Changes in Projects, Economy, and Society via Project Management Second Order (PM-2). *Project Management Journal* 41 (5): 4–20. https://doi.org/10.1002/pmj.20167.

Schindel, Bill, and Rick Dove. 2016. *Introduction to the Agile Systems Engineering Life Cycle MBSE Pattern*. 26th Annual INCOSE International Symposium (IS 2016) Edinburg, Scotland, UK, July 18–21, 2016, www.parshift.com/s/160718IS16-IntroToTheAgileSystemsEngineeringLifeCycleMBSEPattern.pdf.

Stare, Aljaz. 2011. Reducing Negative Impact of Project Changes with Risk and Change Management. *Zagreb International Review of Economics & Business* 14.2: 71–85. Web. https://search-proquest-com.elib.uah.edu/docview/875892187/9D0A393C3B564AF3PQ/1?accountid=14476.

The Need for Semantic Extension of SysML to Model the Problem Space

Paul Wach and Alejandro Salado

Abstract Requirements in natural language, like shall statements, while commonly used, present inherent limitations in terms of accuracy and precision. Modeling requirements within a model-based systems engineering MBSE) framework shows promise to cope with these issues. Common approaches include either the definition of textual requirements as model objects or the flagging of system models as requirements. The first approach inherits the weaknesses of natural language. We show in this paper that the second approach necessarily leads to a poor set of requirements. We therefore argue that modeling languages, in particular SysML, need to be semantically extended to adequately model the problem space. We demonstrate with a specific example that simply flagging model elements as requirements is not effective to model the problem space. In fact, we show that such an approach produces a deficient definition of the problem space, since it inherently discards solutions that could otherwise be potentially acceptable to solve the problem that is being addressed. In addition, we leverage this example to discuss potential semantic extensions of SysML that could enable adequate modeling of the problem space that fulfills the formal conditions of good requirements.

Keywords Model-based requirements · SysML · Model-based systems engineering · MBSE · Modeling semantics

P. Wach
Virginia Tech, Blacksburg, VA, USA

A. Salado (✉)
The University of Arizona, Tucson, AZ, USA
e-mail: alejandrosalado@arizona.edu

A. M. Madni et al. (eds.), *Recent Trends and Advances in Model Based Systems Engineering*, https://doi.org/10.1007/978-3-030-82083-1_24

1 Introduction and Background

Defining the problem space, usually in the form of requirements, is a critical activity in system development (Buede 2009). Failure to correctly frame the problem to be solved has led to the development of systems that are not fit for purpose once deployed (Bahill and Henderson 2005). The problem space is arguably most commonly defined in systems practice through the use of requirements captured in natural language (e.g., shall statements) (INCOSE 2012; Buede and Miller 2016). However, natural language poses inherent limitations in terms of accuracy and precision (Pennock and Wade 2015). This inherent ambiguity of *shall* statements contributes to major problems in acquisition programs (GAO 2016; Gilmore 2011), even if good practices of writing requirements are followed (such as those defined in (INCOSE 2012; Tjong et al. 2006; Salado and Nilchiani 2014a)). In order to attenuate this problem, some may opt to increase the amount of words employed in a requirement statement to reduce ambiguity when formulating requirements (Salado and Wach 2019). However, this effort is usually fruitless because leveraging this flexibility provided by natural language also leads to deviating from good requirement writing practices (INCOSE 2012; Salado and Nilchiani 2014a; Salado and Nilchiani 2017) and results in poor requirement sets overall (Salado and Nilchiani 2017; Salado and Nilchiani 2014b).

A potential solution to the ambiguity problem of natural language consists of substituting traditional *shall* statements by models of requirements within a model-based systems engineering (MBSE) framework (Schneidera et al. 2012; Helming et al. 2010; Mordecai and Dori 2017; Fockel and Holtmann 2014; Soares and Vrancken 2008; Adedjouma et al. 2011). In systems modeling in general, and in the Systems Modeling Language (SysML) in particular, modeling requirements has primarily taken two paths. In the first path, the modeling environment directly incorporates requirement elements. For example, SysML incorporates object types called *requirement element* and *requirements diagram* (Friedenthal et al. 2015). These are intended to model the requirements the system is expected to fulfill. This path fails to fulfill its objectives however, "because the resulting requirement elements are no more than an encapsulation of a textual requirement in natural language as a model element" (Salado and Wach 2019).

In the second path, behavioral and structural models of the system of interest are used (or flagged) as requirements. That is, the specific models captured in the diagram themselves become the requirements, without the need for a statement in natural language that contains a *shall* clause (Soares and Vrancken 2008; Adedjouma et al. 2011; Pandian et al. 2017). While this approach is promising because "it directly leverages behavioral and structural models of the system without requiring statements in natural language" (Salado and Wach 2019), we show in this paper that using system models directly as requirements necessarily leads to a poor set of requirements because of an unnecessary constraining of the problem space.

In this paper, we argue that modeling languages, in particular SysML, need to be semantically extended to adequately model the problem space. We suggest that this extension is necessary if requirements in natural language are to be truly substituted by requirements in model form. We demonstrate with a specific example that simply flagging model elements as requirements is not effective to model the problem space. In fact, we show that such an approach produces a deficient definition of the problem space, since it inherently discards solutions that could otherwise be potentially acceptable to solve the problem that is being addressed. In addition, we leverage this example to discuss potential semantic extensions of SysML that could enable adequate modeling of the problem space that fulfills the formal conditions of good requirements (Salado et al. 2017).

2 Theoretical Justification for Semantic Extension

Consider first a formal definition of system model. Building on systems theory, a system can be generally described (modeled) as a transformation of input trajectories into output trajectories (von Bertalanffy 1969), since closed systems do not exist in nature, at least in the context of engineered systems. This paradigm is applicable regardless of the type of system model that is defined, be it a continuous system model (Sterman 2000) or a discrete system model (Wymore 1993).

Definition 1 A system model is an explicit definition of a transformation of input trajectories into output trajectories.

The purpose of working in the problem space is to derive or define the conditions by which some systems will be considered acceptable as solutions to the problem that is being addressed and some solutions will be considered inacceptable. In this paper, we call the set of acceptable solutions the solution space, and requirements refer to the conditions that enable evaluating a system solution as acceptable or not acceptable. Note that, for the purpose of this paper, and for simplicity and without loss of generality, we consider indifference between all acceptable system solutions.

Definition 2 A problem space is a set of conditions.

Definition 3 A solution space is a set of systems that fulfill a predefined set of conditions.

By the definitions, it follows that:

Proposition A problem space yields a solution space.

Hence, a problem space yields a set of systems, specifically all those that fulfill the conditions that define the problem space. By extension, it follows that a critical aspect of a model of the problem space is that it must yield a set of systems (or system models). We note that a solution space can be empty. This specific case does not change the theoretical implications of the discussion in this section.

Theoretical work in how requirements affect the solution space indicate that a good model of the problem space would include all those solutions that belong to the problem space it models (Salado et al. 2017). This insight is supported by practice, which indicates that requirements should be solution-independent (INCOSE 2012).

Failing to do so unnecessarily shrinks the solution space (Salado et al. 2017) and could lead to automatically rejecting system solutions that could be preferred over those that remained in the solution space (Salado and Nilchiani 2015).

Consider now the approach of flagging a system model as a requirement, as predominantly proposed in the literature for adopting model-based requirements (Soares and Vrancken 2008; Adedjouma et al. 2011; Pandian et al. 2017). If such a system model is used as the model of the problem space, then the solution space will consist of all systems that are homomorphic to such system model. In plain words that means that the solution space will only include systems that refine (are like) the system model that is used as the problem space. In this case, since a model of a specific system solution is used as the starting point to define the problem space, the resulting problem space is necessarily solution dependent. As discussed previously, however, a solution-dependent problem space is considered a deficient model of the problem space, both from the perspective of systems engineering practice and of systems theory.

A good model of the problem space should yield a solution space that allows for including heterogeneous models that are not necessarily homomorphic images of each other. In fact, we suggest that a homomorphic relationship between any solution in the solution space and the problem space should be infeasible because the former captures a transformation of inputs into outputs and the latter captures conditions that discriminate acceptance of transformations of inputs into outputs.

Therefore, we propose that flagging system models as requirement should be avoided. Instead, modeling the problem space necessitates its own distinct semantics. Capturing conditions that yield sets of heterogeneous system models instead of system models themselves enables modeling solution-independent requirements. Formal modeling could yield higher precision and accuracy than natural language (Salado and Wach 2019).

3 Application Example: Operational Scenarios vs System States

3.1 Problem Statement

Consider a need for a system to detect fires in certain area. Two different operational conditions may be present: *Clear Sky* and *Rain*. Regardless of the operational condition, the system is required to continuously capture images of the surveilled area and send them to a station. In addition, the system is required to detect fires and send alarms to a station, at least when the operational condition is *Clear Sky*. Therefore, the problem space is defined by four conditions:

1. Imaging of surveilled area in all weather conditions.
2. Detection of fires in surveilled area in Clear Sky condition.
3. Provision of gathered image data in all weather conditions.

4. Provision of a fire alert for detected fires in Clear Sky condition.

In a natural language requirements paradigm, those conditions would be typically reworded as *shall* statements. We show in the next section how the conditions may be captured in the form of a system model flagged as a requirement and show later the weaknesses of such approach.

It should be noted that the formulation of the conditions has been simplified for presentation purposes. However, this limitation does not affect the generality of the results presented in this paper.

3.2 System Model Flagged as a Requirement

Consider the system model given in Fig. 1, using the form of a state machine diagram. The system has two states, *S1* and *S2*, which are triggered as a function of the operational conditions *clearSky* and *rain*. Given the operational condition of *clearSky*, the system will *imageArea&detectFire* (which is defined as a behavior in which the system will provide gathered images, as well as an alert to the station). Given the operational condition of *rain*, the system will *imageArea* (which is defined as a behavior in which the system will only provide gathered images to the station). Since the model captured in Fig. 1 seems consistent with the conditions given in Sect. 3.1, we assume that the system model is directly flagged as a requirement. No other requirement in natural language (i.e., *shall* statement) is given.

To be compliant with this model-based requirement, a system should exhibit two different states. When in one of them, the system will need to provide image data and fire alerts; and when in the other one, the system will need to provide just image data. In addition, the system will have to be in the first state only when *Clear Sky* conditions are present and in the second state only when *Rain* conditions are present.

Consider now the system model given in Fig. 2. We assess if such a system is compliant to the requirements dictated by the system model in Fig. 1. The system

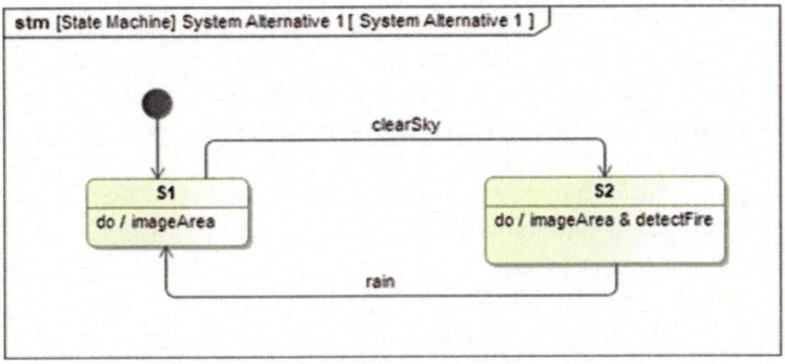

Fig. 1 System model used as a model of the problem space

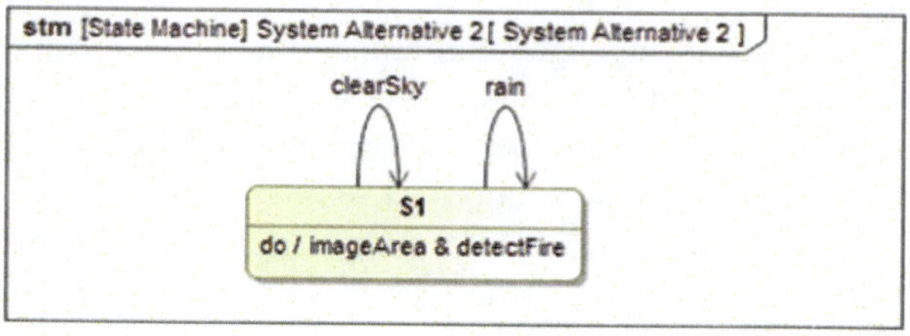

Fig. 2 System model compliant with the required conditions in Sect. 3.1, but not with the problem space model yielded by Fig. 1

exhibits only a single state, in which, regardless of the operational condition, it will provide image data and fire alerts. Formally, it can be shown that the system model in Fig. 1 is not a homomorphic image of the system model in Fig. 2. Clearly, the behavior of such a system is different than the behavior of the system model in Fig. 1: under same input conditions, both systems provide different outputs. Therefore, the solution in Fig. 2 is not compliant to the requirements captured by the system model in Fig. 1, and, as a result, the system in Fig. 2 cannot be considered an acceptable solution.

However, let's examine more deeply if the system in Fig. 2 fulfills the conditions originally specified in Sect. 3.1. If it does, then it follows that using the system model in Fig. 1 unnecessarily constraints the solution space and, therefore, it would be a poor requirement.

Given the conditions in Sect. 3.1, we need the following test vectors:

1. Provide reference spectral features representative of surveilled area with no figure when in *Rain* condition and confirm the system provides images of such input.
2. Provide reference spectral features representative of surveilled area with no figure when in *Clear Sky* condition and confirm the system provides images of such input.
3. Provide spectral features of a fire when in *Clear Sky* condition and confirm the system provides a fire alert.

Table 1 summarizes the required conditions described in Sect. 3.1, the test conditions mentioned above, and the results obtained with the system models in Figs. 1 and 2 when applying such test conditions. As can be seen, both system models successfully pass the various test vectors, and, therefore, both system models fulfill the conditions defined in Sect. 3.1. However, as previously discussed, the system model in Fig. 2 yields a different behavior than that displayed by the system model in Fig. 1. That means that, when the model in Fig. 1 is used as the system requirement, the system solution modeled in Fig. 2 would be assessed as being not compliant with the requirement, as previously indicated. Consequently, using the model in Fig. 1 unnecessarily constraints the solution space by discarding the system model in Fig. 2. Since this is a characteristic of a poor requirement (INCOSE 2012; Salado et al. 2017), we suggest this system model in particular, and any system model in general, should not be used as a model of the problem space.

Table 1 Verification of system models in Figs. 1 and 2

Required condition	Test vector	System model Fig. 1	System model Fig. 2
1. Imaging of surveilled area in all weather conditions	1 2	S1. Provide image S2. Provide image	S1. Provide image S1. Provide image
2. Detection of fires in surveilled area in Clear Sky	3	S2. Provide alert	S1. Provide alert
3. Provision of gathered image data in all weather conditions	1 2	S1. Provide image S2. Provide image	S1. Provide image S1. Provide image
4. Provision of a fire alert for detected fires in Clear Sky	3	S2. Provide alert	S1. Provide alert

3.3 Potential Extension of SysML to Model Different Required Operational Conditions in the Problem Space

Based on the discussion in the previous section, a model of the problem space should be able to capture, among others, the requirement for a system to fulfill different sets of requirements (conditions in general) depending on the presence of different operational conditions. In requirements jargon, one could think of this as requirements that are applicable only in certain conditions, but not always. We show a small example of how SysML could be semantically extended to model effectively this aspect of the problem space while overcoming the two main limitations of existing approaches, as previously identified: first, without falling into the trap of enforcing specific solutions; second, without leveraging natural language as the main vehicle to convey information and meaning.

We leverage prior work (Salado and Wach 2019) and use *mode requirements*, which extend SysML to capture sets of requirements that do not have to be fulfilled simultaneously necessarily. An example is provided in Fig. 3. Note that the diagram should not be interpreted as the standard *state machine diagram* in SysML; this is why we have purposefully omitted the formal identification of the diagram in the figure. We are representing a semantic extension of SysML. Hence, while we use some in this paper existing SysML objects, the model represented in Fig. 3 has a different meaning than that of a state machine. Similarly, other elements in Fig. 3 have been modeled according to the larger true model-based requirements (TMBR) SysML extension, of which *mode requirements* are part. The same idea regarding model interpretation applies therefore for other elements in Fig. 3. Details about how to fully *read* the model-based requirements depicted in Fig. 3 are provided in (Salado and Wach 2019).

Coming back to *mode requirements*, the model in Fig. 3 captures the idea that the system must fulfill two different sets of requirements (*Clear Sky requirements* and *Nominal requirements*) depending on the presence of different operational conditions, although not necessarily simultaneously. The two yellow boxes capture this aspect: each box indicates the existence of a unique set of requirements that need to be fulfilled; they do not represent system states, as a state machine diagram would do in SysML. Sequence diagrams are used to model the specific requirements that

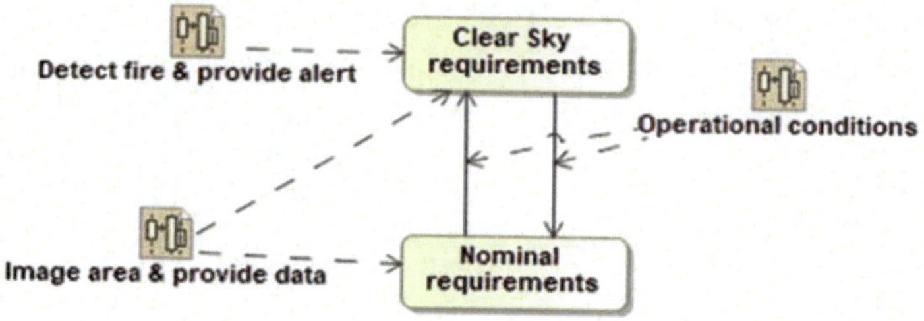

Fig. 3 Potential semantic extension of SysML to capture different operational conditions

Table 2 Explanation of how the system model in Fig. 3 captures the problem presented in Sect. 3.1

Required condition	Model element
Imaging of surveilled area	Elaborated in sequence diagram *Image area & provide data*
Detection of fires in surveilled area	Elaborated in sequence diagram *Detect fire & provide alert*
Provision of gathered image data	Elaborated in sequence diagram *Image area & provide data*
Provision of a fire alert	Elaborated in sequence diagram *Detect fire & provide alert*
Clear Sky conditions	Described/elaborated in sequence diagram *Operational conditions*
In Clear Sky conditions; **In** all Weather	Two different sets of requirements are informed, captured by yellow boxes *Clear Sky requirements* and *Nominal requirements*, respectively. The linkage between the sequence diagram *Operational conditions* and the transitions between the yellow boxes capture when each set of requirements is applicable
In Clear Sky conditions, do (2) and (4)	Sequence diagram *Detect fire & provide alert* is linked to yellow box *Clear Sky requirements*
In All weather, do (1) and (3)	Sequence diagram *Image area & provide data* is linked to both yellow boxes

apply under the different operational conditions. The transitions between the yellow boxes, mapped to another sequence diagram, capture the conditions that make each set of requirements applicable. Table 2 describes how the model in Fig. 3 captures all the problem conditions (requirements) listed in Sect. 3.1.

The model does not imply that the system solution will have two states, each one with a specific behavior and each one being activated depending on certain conditions. In fact, the model has no implication as to the number or type of states the system will have. This is because Fig. 3 models the problem space, not a particular system. The two yellow rounded rectangles simply act as collectors of the subsets of requirements that need to be satisfied simultaneously. Hence, the diagram reads that the requirements captured in *Nominal requirements* must not

Table 3 Consistency between system alternatives 1 and 2 and the requirement model in Fig. 3

Required conditions in requirement model in Fig. 3	Can Syst. Alt. 1?	Can Syst. Alt. 2?
In not Clear Sky, image area and provide data	Yes, through S1	Yes, through S1
In Clear Sky, image area and provide data	Yes, through S2	Yes, through S1
In Clear Sky, detect fire and provide alert	Yes, through S2	Yes, through S1

necessarily be fulfilled while also fulfilling those captured in *Clear sky requirements*. No inference about potential system solutions is made. As an example, the two solutions shown in Figs. 1 and 2 are compliant with the requirement model presented in Fig. 3 (ref. Table 3).

It should be noted that this diagram is not intended to substitute the state machine diagram that is used to model system behavior. Instead, the diagram in Fig. 3 should be interpreted as a visualization of an additional model, visually similar, but semantically different to the SysML state machine diagram. The model in Fig. 3 should however substitute all requirement elements and diagrams in standard SysML, since it offers an alternative way to capture requirements. Specifically, it enables capturing requirements in a true model-based form instead of as textual requirements encapsulated as a model element (Salado and Wach 2019), potentially improving precision and accuracy when modeling the problem space.

4 Conclusion

We have demonstrated in this paper that using system models directly as requirements should be considered as bad practice in problem formulation. We have supported this claim with a theoretical argumentation and a specific application example. Since a problem space yields a solution space, using a specific system model as a model of the problem space could unnecessarily constrain the solution space. This effect is well known in requirements engineering practice as formulating design-dependent requirements: by using a system model as the requirement, such a solution is enforced in the form of requirements.

This major weakness of using system models as requirements (more generally, as models of the problem space) informs the need for a semantic extension of SysML (and likely other modeling languages). New model elements and model diagrams and/or additional semantics that can extend the interpretation of existing model elements and diagrams are necessary to formally, and correctly, model the problem space. Given that problem formulation is a key activity in systems engineering, we suggest that the semantic extension we propose is paramount for a successful implementation of MBSE.

In this paper, we have shown the effects of using a system model as a model of the problem space (i.e., as a requirement) by specifically addressing how to capture sets of requirements that are applicable (i.e., that need to be fulfilled simultaneously) under different conditions. In the example, we have leveraged the *state machine diagram* in SysML for such purpose. We have also described a potential semantic extension, based on our prior work, which allows for capturing such type of requirements without enforcing any specific design solution on the states that the system must exhibit.

Acknowledgments This material is based on work sponsored by the Department of the Navy, Naval Engineering Education Consortium, award number N00174-19-1-0012. Any opinions, findings, and conclusions or recommendations expressed in this material are those of the authors and do not necessarily reflect the views of the Naval Engineering Education Consortium.

References

Adedjouma, M., H. Dubois, and F. Terrier. 2011. *Requirements exchange: From specification documents to models*, in *16th IEEE international conference on engineering of complex computer systems*, 350–354. Las Vegas.

Bahill, A., and S. Henderson. 2005. Requirements development, verification, and validation exhibited in famous failures. *Systems Engineering* 8 (1): 1–14.

Buede, D. 2009. *The engineering design of systems: Models and methods*. Hoboken: Wiley.

Buede, D., and W. Miller. 2016. *The engineering design of systems: Models and method*. Hoboken: Wiley.

Fockel, M., and J. Holtmann. 2014. *A requirements engineering methodology combining models and controlled natural language*, in *2014 IEEE 4th International Model-Driven Requirements Engineering Workshop (MoDRE)*, 67–76. Karlskrona.

Friedenthal, S., A. Moore, and R. Steiner, eds. 2015. *A practical guide to SysML – The systems modeling language*. 3rd ed. Waltham: Morgan Kaufman.

GAO. 2016. *Defense acquisitions – Assessments of selected weapon programs*. General Acconting Office.

Gilmore, J.M. 2011. Key issues causing program delays in defense acquisition. *ITEA Journal* 32: 389–391.

Helming, J., et al. 2010. *Towards a unified requirements modeling language*. in *2010 Fifth International Workshop on Requirements Engineering Visualization*. Sydney.

INCOSE. 2012. *Guide for writing requirements*. The International Council of Systems Engineering.

Mordecai, Y., and D. Dori. 2017. *Model-based requirements engineering: Architecting for system requirements with stakeholders in mind*, in *2017 IEEE International Systems Engineering Symposium (ISSE)*, 1–8 Vienna, Austria.

Pandian, M.K.S., et al. 2017. *Towards industry 4.0: Gap analysis between current automotive MES and industry standards using model-based requirement engineering*, in *2017 IEEE International Conference on Software Architecture Workshops (ICSAW)*, 29–35. Gothenburg.

Pennock, M.J., and J.P. Wade. 2015. The top 10 illusions of systems engineering: A research agenda. *Procedia Computer Science* 44: 147–154.

Salado, A., and R. Nilchiani. 2014a. A categorization model of requirements based on max-neef's model of human needs. *Systems Engineering* 17: 348–360.

———. 2014b. *Categorizing requirements to increase the size of the solution space: Moving away from NASA and ESA's requirements categorization models*. in *6th international systems & concurrent engineering for space applications conference*. Stuttgart.

———. 2015. On the evolution of solution spaces triggered by emerging technologies. *Procedia Computer Science* 44: 155–163.

———. 2017. Reducing excess requirements through orthogonal categorizations during problem formulation: Results of a factorial experiment. *IEEE Transactions on Systems, Man, and Cybernetics: Systems* 47 (3): 405–415.

Salado, A., and P. Wach. 2019. Constructing true model-based requirements in SysML. *Systems* 7 (2): 19.

Salado, A., R. Nilchiani, and D. Verma. 2017. A contribution to the scientific foundations of systems engineering: Solution spaces and requirements. *Journal of Systems Science and Systems Engineering* 26 (5): 549–589.

Schneidera, F., H. Naughtona, and B. Berenbach, 2012. *New challenges in systems engineering and architecting*, in *Conference on Systems Engineering Research (CSER) 2012*. St. Louis: Elsevier B.V.

Soares, M.D.S., and J. Vrancken. 2008. Model-driven user requirements specification using SysML. *J. Softw* 3 (6): 57–68.

Sterman, J. 2000. *Business dynamics: Systems thinking and modeling for a complex world*. New York: Irwin MGraw-Hill.

Tjong, S.F., N. Hallam, and M. Hartley. 2006. *Improving the quality of natural language requirements specifications through natural language requirements patterns*. in *Proceedings of the Sixth IEEE International Conference on Computer and Information Technology*. Seoul.

von Bertalanffy, L. 1969. *General systems theory – Foundations, development, applications*. New York: George Braziller, Inc.

Wymore, A.W. 1993. *Model-based systems engineering*. Boca Raton: CRC Press.

Variant Modeling for Multi-perspective, Multi-fidelity Systems Simulation

Ryan Colletti, Ahsan Qamar, Sandro Nuesch, William Bailey, and Christiaan Paredis

Abstract Current methods of Model-Based Product Line Engineering (MBPLE) do not seamlessly extend to the management of variants of system simulation models. Both architectural exploration activities and product line verification and validation analyses could be streamlined by applying MBPLE to simulations. However, this new application requires careful consideration to mitigate new challenges associated with simulation variant management, such as consistency between architecture and simulation models, the addition of simulation context and model fidelity as new areas of variation, and management of both plant and controller models. This paper presents an in-depth literature review of previous work on variant management of simulation models, a discussion of the complexity of the problem, and a preliminary proposal for an approach of simulation model variant management in which SysML and variant modeling capabilities are used to define 150% black box representations of simulation models in order to automatically create full simulation models.

Keywords Integration of MBSE with simulation technology · Model-Based Systems Engineering · Model-Based Product Line Engineering · Variant modeling · Model-centric engineering · Model-based verification and testing

1 Introduction

As system development tools and methods have progressed and systems become more complex, so the customer's expectations have heightened. Customers not only demand robust systems that are safe, reliable, and full of features, but they also

---

R. Colletti (✉) · C. Paredis
Clemson University International Center for Automotive Research, Greenville, SC, USA
e-mail: racolle@clemson.edu

A. Qamar · S. Nuesch · W. Bailey
Ford Motor Company, Dearborn, MI, USA

A. M. Madni et al. (eds.), *Recent Trends and Advances in Model Based Systems Engineering*, https://doi.org/10.1007/978-3-030-82083-1_25

value a customized system that fits their specific needs. This is no truer anywhere than it is in the automotive industry. Standard practice is to allow the customer to select from a plethora of options, including engine size, transmission type, interior and exterior colors, and upholstery material, along with technology-packed features such as adaptive cruise control and automatic parking. All of this must be available while also meeting ever-increasing standards for safety and environmental impact. This current design landscape calls for Product Line Engineering (PLE) methods that facilitate the development of a family of products instead of only individual products. Modeling variants and capitalizing on reuse of elements, subsystems, and behaviors are essential for product lines as this greatly reduces the amount of time and effort necessary to develop the system. Current system modeling tools allow for the implementation of reusable assets, but most of these tools are focused in the architecture definition space. The lack of variant management tools and practices in the simulation space has led to a bottleneck in the development process at the system analysis step, which costs companies valuable development time.

Variant management methods have proven beneficial in Model-Based Systems Engineering (MBSE) by making opportunities for reuse more obvious and hastening verification and validation activities of architectural variants within a product line. It is possible that similar benefits could be seen in the system analysis phase by applying variant management techniques to simulation models. Other potential benefits include clearer guidelines for the creation of new simulation models, faster parameterization of existing simulation models, and increased consistency between specification and simulation models. This research aims to investigate this possibility by exploring the relevant problem space.

Before a method for applying variant management techniques to simulation models can be found, a greater understanding of the challenges surrounding variant management of simulation models is required. Therefore, the motivating question for this paper is "what are the current challenges of variant management of simulation models?" Once the problem space is better understood, an approach can be formed. The corresponding research question is "what is a good way to use Model-Based Product Line Engineering to manage variants of simulation models?" It is important to note that this research is ongoing, so this paper only presents a preliminary idea for an approach. Further research intends to refine this idea and test it with a case study.

The rest of the paper is organized as follows: In Section II, we give a brief history of variant modeling and show the gaps this paper aims to fill. Next, we detail the challenges of variant management that are specific to the simulation space as compared to the architecture design space in Section III. In Section IV, a potential technique of 150% modeling and feature modeling as part of Model-Based Product Line Engineering (MBPLE) to perform variant management of simulation models is proposed. Finally, a brief summary of the paper and future work that could stem from this research is presented in Section V.

2 Literature Review

2.1 *MBSE, SysML, and Early PLE*

Several product line engineering methods have emerged from the practice of Model-Based Systems Engineering. MBSE is a method of formalizing the application of modeling to support all steps of the systems engineering development life cycle, including requirements, design, analysis, verification, and validation (INCOSE 2007). Traditional MBSE uses a system modeling language to create a cohesive model of the system, which allows for components to be designed using domain-specific tools and brought back together for validation (Friedenthal et al. 2014). The most common language used to create these models is SysML, a graphical system modeling language that facilitates MBSE by enabling the creation of a cohesive and consistent model of the system (ISO 2017).

The inaugural variant modeling practice is often referred to as the "clone-and-own" approach, where the first product in a family is developed, then the design documents are copied and modified for subsequent products (Schulze et al. 2013). While this works for a simple system of two or three variants, it has proven to be inefficient with significant amounts of duplicated work and difficulty of traceability (Schulze et al. 2013). The first formalized variant modeling method was the Carnegie Mellon University's Software Engineering Institute's (SEI) Product Line Practice (PLP), which consists of domain engineering, application engineering, and the management of both (Clements and Northrop 2002). Domain engineering includes the development of a common platform for a product line to be designed and implemented, as well as a set of reusable artifacts that support the various features of the product line (Krueger and Clements 2013). Application engineering is the derivation of a product variant from the set of artifacts and product line platform developed in the domain engineering phase (Deelstra et al. 2005). The goal of PLP is for the benefits of rapid product variant development to outweigh the additional effort of creating reusable assets.

SEI also developed Feature-Oriented Domain Analysis (Kang et al. 1990). FODA allows for variants in a product line to be described by their features, or the various functions a product performs, regardless of the specific hardware or software that realizes those functions. A feature model organizes features into hierarchies and uses symbols to denote cardinality and relationships, such as mandatory or optional features (Kang et al. 1990). Features from the feature model are connected to elements in the domain model; this denotes which of the system's elements implements each feature (Czarnecki and Kim 2005). In later work, SEI extended FODA into the Feature-Oriented Reuse Method (FORM) (Kang et al. 2002). This method not only supports architecture design and object-oriented component development, but it also incorporates a marketing perspective, facilitating analysis and design problems from a marketing standpoint (Kang et al. 2002). This is important as marketing is the engineer's connection to the customer, and the customer ultimately decides which features a system should have, thus dictating product line variants. The majority of SEI's work in PLP, FODA, and FORM deals with the management of variations in an architecture model. This research aims to apply similar variant modeling methods to a simulation model.

2.2 *MBPLE*

A challenge with using PLP, FODA, and FORM is the modeling of variability between core assets of the domain model and their applications in the variant model. Other research describes four approaches to modeling variability – parameterization, information hiding, inheritance, and variation points – and it introduces a method called the Variation Point Model that allows application engineers to extend components at variation points specified by domain engineers (Webber and Gomaa 2004). Currently, one of the most common methods for variant modeling and management is a combination of PLP and FODA using variation points called Model-Based Product Line Engineering (MBPLE) (Pohl et al. 2005). In this method, system engineers first build a reference architecture model which acts as a central starting point for the development of a product line. The reference architecture formally defines the key subsystems of the product line and maps the interactions between them (Branscomb et al. 2013). The system architect also works with stakeholders to decide which features the product line will provide, and he maps these features onto a feature model (Thiel and Hein 2002). Concurrently, domain engineers further define the reference architecture into a 150% model which encompasses the entire variability spectrum within the product line. In the 150% model, a superset of all subsystems and components are defined such that any variant of the product could be built from a subset of the 150% model (Grönniger et al. 2008). Variation points are used in this model to distinguish which of the model's elements are needed to provide each feature. These variation points supply the connection between the feature model and the 150% architecture model, and they include a logical expression that determines the behavior of the associated element in the superset model based on the state of one or more features (Böckle et al. 2005). Variants of the product line are defined by selecting a state for each feature of the product line, i.e., the inclusion, omission, or kind of each feature for a given variant (Krueger 2008). Since a variant is defined by its features, and the variation points in the 150% model distinguish which components are required for each feature, the 150% model can automatically be pared down to a variant model using a model transformation tool (Krueger 2008). Finally, specification documents are generated from the variant model that guide the programmers in the development of code for the system, or, in some instances, code is automatically generated (Schaefer and Hähnle 2011).

MBPLE has been used extensively to manage variants of architecture description models. The majority of research has focused on the management of variations in the structural aspect of the system model, but recent work has been looking into variant management of the behavioral side of these models. Proper organization and abstraction of the system's behaviors have been identified as keys to recognizing opportunities for reuse of behavioral elements of the model (Karban et al. 2016a). In particular, the use of modeling patterns helps improve consistency and reduce the amount of effort needed to create and maintain models. Two complimentary variant modeling patterns are the 150% and encapsulation patterns (Colletti et al. 2020). In the 150% pattern, all actions and pins for all variants of an activity are represented together in a diagram, and variation points are used to indicate which actions and pins are present for a given variant of the activity. In the encapsulation pattern, each variant of an activity is modeled as its own activity, and then all variants of that activity are gathered in a wrapper activity which contains a superset of the input and output pins and uses variation points to indicate the appropriate activity for each variant. Selection of the best pattern for each modeling context facilitates the modeling process by reducing the total number of modeling elements used to define the 150% model and separating parts of the model that are maintained by different modelers or teams, ultimately reducing the total cost of modeling a product line by improving manageability of the model, increasing consistency across the model, and promoting reuse of assets throughout the product line (Colletti et al. 2020). These patterns could possibly be extended to simulation models to realize similar benefits, and this research intends to investigate this proposition. The Executable System Engineering Method (ESEM) takes advantages of modeling patterns that involve structural, behavioral, and parametric diagrams to integrate requirements and executable behavior and performance models for certain system level analyses (Karban et al. 2016b). However, ESEM patterns and processes do not account for variants within a product line as the 150% and encapsulation patterns do.

2.3 *SysML and Simulation Modeling*

SysML is primarily intended for creating architecture description models, with structural decompositions, behavior models, requirements tracing, and parametric relationships (Friedenthal et al. 2014). SysML can also be used to support basic simulation activities, such as observing behavioral models by executing activity diagrams and modular analyses based on constraints modeled in parametric diagrams (Peak et al. 2007). However, SysML does not support modeling continuous system dynamics using differential algebraic equations. Modelica, an object-oriented modeling language, is well suited for simulating continuous system dynamics based on energy transfer among system components because it is an equation-based language that does not require the user to assign causality (Modelica Association 2014). Simulink, on the other hand, is a causal, signal-flow modeling environment that is often used to model control algorithms (The MathWorks, Inc 2019). It is not uncommon to see a Simulink model of a system outsource the plant models to a language like Modelica and represent the control models natively to execute a full system simulation; however, SysML is rarely brought into this mix for either architecture management or variant modeling.

It is possible to use SysML to support modeling system dynamics with an extension called SysML4Modelica. The custom profile allows a modeler to use a representation of the most common Modelica language constructs in SysML; then, a Triple Graph Grammar is used to link the SysML and Modelica meta-models (Paredis et al. 2010). Doing this not only encourages consistency between the SysML and Modelica models, but it also allows for an integration of simulation experiments with other SysML constructs to support MBSE activities (Johnson et al. 2012). Another method for using SysML to analyze dynamic system behavior is with a SysML extension developed by NIST called SysPhS (Object Management Group 2018). More accurately, SysPhS and its accompanying translator are intended to enable both physical interaction and signal flow simulation contexts in SysML. The creators of SysPhS identified the core simulation constructs common to most simulation tools, and then they designed SysPhS to extend four SysML constructs to include the information necessary for simulation modeling using only SysML (Matei and Bock 2012). To increase SysPhS's capabilities, the same authors developed a translator to transform a SysML model extended with SysPhS into simulation files for Modelica or Simulink, and the translator can go the other direction as well (Barbau et al. 2019). Neither SysPhS nor SysML4Modelica tackles the challenge of variant modeling; however, since both use SysML as the root of their approach, MBPLE methods could likely be used in conjunction with these approaches to accomplish variant management. This research intends to investigate this conjecture as a possibly viable method.

While Modelica is well-suited for physics-based simulations, Simulink is currently the standard for signal flow simulations. Several approaches for handling variability within Simulink models exist, each with their own strengths and weaknesses. To facilitate the 150% modeling pattern, Simulink includes conditional elements such as *If* blocks, *Switch* blocks, and logical operators. To support the encapsulation modeling pattern, Simulink supplies *Model Variant* blocks and *Variant Subsystem* blocks (Weiland and Manhart 2014). However, these constructs are rather unwieldy to use for variant modeling, and they provide no connection to the system architecture model in SysML. As an alternative, a Simulink extension called Delta-Simulink has been developed to enable 150% modeling in Simulink based on the concept of delta modeling (Haber et al. 2013). Delta-Simulink is a transformational approach which applies color-coded deltas to a base model to indicate changes that are made to the model for different variants, similar to variation points (Kolassa et al. 2015). However, Delta-Simulink does not include a rigorous method for relating the deltas in the Simulink model to a feature model as is done in MBPLE with SysML. One way of making this connection is with an ontology-assisted approach which uses the variant modeling capabilities within Simulink and the System Entity Structure ontology to synchronize an external variant model with the dynamic system models in Simulink (Pawletta et al. 2014). Similarly, this research plans to use a feature model developed in SysML using MBPLE to manage the dynamic system models.

Some research has been done in the area of using MBPLE to manage variants in domain-specific simulations. One approach uses SysML as a central repository to generate a source model where a trade study is defined and requirements are represented. Then, through either automatic code generation, model transformation, or model parameterization, the trade study is executed in a domain-specific simulation tool, and the results are sent back to the SysML model (Ryan et al. 2014). However, since this approach focuses on architecture trade studies, it only supports variations in selected components within the architecture, not variations in the architecture itself. Alternatively, other work has developed a framework for using a reference architecture model in SysML to manage the block interfaces and signal flow in the Simulink model (Bailey et al. 2018). While this framework effectively used SysML to manage variants – even architectural variants – in Simulink, much of the modeling effort must still be accomplished manually. It did not employ MBPLE methods such as feature models and variation points in SysML to automate much of the process.

3 Additional Challenges to Variant Management of Simulation Models

MBPLE has tackled many of the challenges of modeling variants of a product line in an architecture model, including identifying opportunities for reuse, relating variable features to the components that realize them, and managing variations in components, interfaces, and architectural structures across the product line. Consequently, applying MBPLE to simulation models will require the identification and mitigation of additional challenges in order for an approach to be effective. One challenge will be maintaining consistency between the architecture and simulation models. As seen in Sect. 2.3, consistency management has been achieved through several approaches. The most popular approaches are a direct mapping between model elements of two languages and an abstraction or extension of one language to another.

Another challenge will be the management of the multitude of analysis models. In the architecture description model, there is already the challenge of managing variants of the product line, the features they provide, and the hardware and software that implement each feature. MBPLE successfully manages these by employing 150% models, feature models, and variation points. However, the simulation space must manage not only variants of a product line but also the simulation context (environmental conditions), the simulation test (the analysis to be run on the system), and the various levels of fidelity for each plant and controller model used in the simulation (Sinha et al. 2001). As an example in the automotive domain, a particular variant of a vehicle product line could be simulated doing an acceleration test, a braking test, a drive cycle, or any number of maneuvers, and each of these tests could occur in hot or cold temperatures, sunny or rainy conditions, and on various road surfaces. In addition, there is not only one vehicle simulation model that is used for each simulation test but likely a different vehicle model for each test, where each plant and controller model may be at a different level of fidelity to maximize the execution speed of the simulation without sacrificing accuracy (Sinha et al. 2001). Varying fidelity could also vary a simulation model's structure, so an effective variant management approach must take this into account. Finally, synchronizing the interfaces of interacting plant and controller models will be required at each fidelity level. Managing all these additional variables could quickly become cumbersome and must be considered in an approach for variant management of simulations.

A third challenge will be the management of both plant and controller simulation models. Current MBPLE practices already manage variants in the system structure, and this will likely translate well to the management of the system's plant models, but this becomes much more difficult for the controller models. As stated before, MBPLE has mostly been used to manage variations of the structural aspect of architecture models, and only minimal research has been completed in managing variations of the behavioral models. This is because behaviors are often much more tightly coupled together than structural components. Learning how to properly abstract controller models such that they can be built and combined in a modular fashion at various levels of fidelity will be the key to this aspect of variant management (Paredis et al. 2001).

4 Overview of Proposed Approach

This section introduces an approach to variant management of simulation models that the researchers are currently in the process of investigating. Unfortunately, no quantifiable results are ready at time of writing; however, based on the presented literature review and the authors' experience in the field of variant management, the approach seems promising. The proposed variant management approach is in three steps: 150% and feature modeling in SysML, composable simulation modules creation in Simulink, and derivation of the full simulation in Simulink using the SysML 150% and feature models.

First, SysML is used to create a 150% architecture description model of the entire product line of interest as well as a feature model to link the components of the description model to variable features offered by the product line. A variant can be instantiated by selecting a set of features, and the variation points automatically pare down the 150% model to a variant model. All of these steps are already a part of traditional MBPLE practices. To include management of simulation models, the SysML architecture model will require two more parts – additional features and a reference architecture for the simulation model. In order to use the SysML model to manage variants of the simulation model, features corresponding to variation among simulation models of the product line of interest must be added to the existing feature model in SysML. These features will represent the objective of the simulation model, such as predicting fuel economy or testing autonomous functionality, which would be analyzed through one or more simulation tests. Each simulation objective feature would also have several levels of fidelity associated with it that correspond to the milestones within the development process. Describing the simulation models in this way allows for traditional MBPLE methods to be used to manage both variants and levels of fidelity of the simulation models. Additionally, a reference architecture for the simulation model serves as a baseline for how the simulation modules should be put together to form a full simulation model. Each plant or controller module is represented by a black box in SysML, showing only the inputs and outputs of each module and abstracting away the internals. In this way, all of the connections among the simulation modules can be modeled in SysML, taking advantage of existing variant management techniques. A 150% model of the simulation model can be built and managed in SysML such that no 150% modeling is necessary in Simulink.

Concurrently, simulation modules corresponding to the black boxes in the SysML simulation reference architecture are developed in Simulink. As stated in Sect. 3, these modules will need to be abstracted and organized such that they can be combined in the same way they are in the SysML black box representation. This approach allows for the system dynamics to be modeled in Simulink instead of in SysML, minimizing both duplication of modeling effort and the need for any modeler to work in both SysML and Simulink. Additionally, the only synchronization between the SysML and Simulink models will be the inputs and outputs of each simulation module; the simulation module's internals can be updated without affecting the SysML reference architecture.

Finally, once the SysML 150% architecture description model, SysML 150% simulation reference architecture, SysML architecture and simulation feature model, and Simulink plant and controller model library are built, they can be combined to form a full simulation model. A variant is instantiated in the SysML model, and the 150% model is pared down to a variant model. That variant model is then used to combine the simulation modules to match the simulation reference architecture. Ideally, this last step would require as little human effort as possible since all of the information necessary to build the simulation already exists.

5 Summary

The objective of this research is to determine a good way to apply and extend MBPLE practices to manage variants of simulation models. The purpose of this paper is to, through an in-depth literature review, identify the unique challenges of this endeavor and propose a potential method for simulation variant management. This research is ongoing, and the researchers are currently in the process of building a test case using existing simulation models at Ford Motor Company. As this space is further explored, additional challenges may become apparent that must be dealt with, and the proposed approach will be duly modified to best support both architecture and simulation modelers in their respective tasks.

Acknowledgments The research for this paper was funded in part by Ford Motor Company.

References

Bailey, W.C., J. Che, P. Tsou, and M. Jennings. 2018. A framework for automated model interface coordination using SysML. *Journal of Computing and Information Science in Engineering* 18 (3).

Barbau, R., C. Bock, and M. Dedfarnia. 2019. Translator from extended SysML to physical interaction and signal flow simulation platforms. *Journal of Research of the National Institute of Standards and Technology* 124: 1–3.

Böckle, G., K. Pohl, and F.J. van der Linden. 2005. A framework for software product line engineering. In *Software product line engineering*, 19–38. Berlin, Heidelberg: Springer.

Branscomb, J.M., C.J.J. Paredis, J. Che, and M.J. Jennings. 2013. Supporting multidisciplinary vehicle analysis using a vehicle reference architecture model in SysML. *Procedia Computer Science* 16 (1): 79–88.

Clements, P., and L. Northrop. 2002. *Software product lines: Practices and patterns*. Vol. 3. Reading: Addison-Wesley Reading.

Colletti, R., A. Qamar, S. Nuesch, and C.J.J. Paredis. 2020. Best practice patterns for variant modeling of activities in model-based systems engineering. *IEEE Systems Journal*.

Czarnecki, K., and C.H.P. Kim. 2005. Cardinality-based feature modeling and constraints: A progress report. In *Proceeding of the international workshop software factories*, 16–20. San Diego.

Deelstra, S., M. Sinnema, and J. Bosch. 2005. Product derivation in software product families: A case study. *Journal of Systems and Software* 74 (2): 173–194.

Friedenthal, S., A. Moore, and R. Steiner. 2014. *A practical guide to SysML: The systems modeling language*. San Mateo: Morgan Kaufmann.

Grönniger, H., H. Krahn, C. Pinkernell, and B. Rumpe. 2008. Modeling variants of automotive systems using views. In *Proc. of Workshop Modellbasierte Entwicklung von eingebetteten Fahrzeugfunktionen (MBEFF)*, 76–89. Berlin.

Haber, A., C. Kolassa, P. Manhart, P.M.S. Nazari, B. Rumpe, and I. Schaefer. 2013. First-class variability modeling in Matlab/ Simulink. In *Proceedings of the 7th international workshop on variability modelling of software-intensive systems*, 11–18. New York.

INCOSE. 2007. Systems engineering vision 2020. In *International Council on Systems Engineering, ver. 2.03*. Seattle.

ISO, ed. 2017. *Information technology: Object management group systems modeling language (OMG SysML)*. 1.4.1 ed ISO/IEC 19514:2017.

Johnson, T., C.J.J. Paredis, and R. Burkhart. 2012. Integrating models and simulations of continuous dynamics into SysML. *Journal of Computing and Information Science in Engineering* 12 (1): 135–145.

Kang, K.C., S.G. Cohen, J.A. Hess, W.E. Novak, and A.S. Peterson. 1990. *Feature-oriented domain analysis (FODA) feasibility study*. Pittsburgh., CMU/SEI-90-TR-021: Software Engineering Institute.

Kang, K.C., J. Lee, and P. Donohoe. Jul-Aug 2002. Feature-oriented product line engineering. *IEEE Software* 19 (4): 58–65.

Karban, R., N. Jankevičius, and M. Elaasar. 2016a. ESEM: Automated systems analysis using executable SysML modeling patterns. In *26th Annual INCOSE International Symposium*. Edinburgh.

Karban, R., F.G. Dekens, S. Herzig, M. Elaasar, and N. Jankevičius. 2016b. Creating systems engineering products with executable models in a model-based engineering environment. In *Proc. Modeling, Systems Engineering, and Project Management for Astronomy VI*, 99110B1–99110B16. Edinburgh.

Kolassa, C., H. Rendel, and B. Rumpe. 2015. Evaluation of variability concepts for Simulink in the automotive. In *48th Hawaii International Conference on System Sciences*, 5373–5382. Kauai.

Krueger, C. 2008. The BigLever software gears unified software product line engineering framework. In *12th international software product line conference*, 353. Limerick.

Krueger, C., and P. Clements. 2013. Systems and software product line engineering. *Encyclopedia of Software Engineering* 2: 1–14.

Matei, I., and C. Bock. 2012. SysML extension for dynamical system simulation tools. *National Institute of Standards and Technology, NISTIR* 7888.

Modelica Association. 2014. *Modelica Language Specification*. [Online]. Available: https://www.modelica.org/documents/ModelicaSpec33Revision1.pdf

Object Management Group. 2018. *SysML extension for physical interaction and signal flow simulation*. [Online]. Available: https://www.omg.org/spec/SysPhS/1.0/PDF

Paredis, C.J.J., A. Diaz-Calderon, R. Sinha, and P.K. Khosla. 2001. Composable models for simulation-based design. *Engineering with Computers* 17 (2): 112–128.

Paredis, C.J.J., Y. Bernard, R.M. Burkhart, H. de Koning, S. Friedenthal, P. Fritzson, N.F. Rouquette, and W. Schamai. 2010. An Overview of the SysML-Modelica Transformation Specification. *INCOSE International Symposium* 20 (1): 709–722.

Pawletta, T., D. Pascheka, A. Schmidt, and S. Pawletta. 2014. Ontology-assisted system modeling and simulation within MATLAB/Simulink. *SNE Simulation Notes Europe* 2: 59–68.

Peak, R.S., R.M. Burkhart, S.A. Friedenthal, M.W. Wilson, M. Bajaj, and I. Kim. 2007. Simulation-based design using SysML part 1: A parametrics primer. *INCOSE International Symposium* 17 (1): 1516–1535.

Pohl, K., G. Böckle, and F.J. van der Linden. 2005. *Software product line engineering: Foundations, principles and techniques*. Springer.

Ryan, J., S. Sarkani, and T. Mazzuchi. 2014. Leveraging variability modeling techniques for architecture trade studies analysis. *Systems Engineering* 17 (1): 10–25.

Schaefer, I., and R. Hähnle. Feb. 2011. Formal methods in software product line engineering. *Computer* 44 (2): 82–85.

Schulze, M., J. Mauersberger, and D. Beuche. 2013. Functional safety and variability: Can it be brought together? In *Proceedings of the 17th International Software Product Line Conference*, 236–243. Tokyo.

Sinha, R., C.J.J. Paredis, V. Liang, and P.K. Khosla. 2001. Modeling and simulation methods for design of engineering systems. *Journal of Computing and Information Science in Engineering* 1 (1): 84–91.

The MathWorks, Inc. 2019. *Simulink Release Notes*. [Online]. Available: https://www.mathworks.com/help/pdf_doc/simulink/rn.pdf

Thiel, S., and A. Hein. Jul-Aug 2002. Modelling and using product line variability in automotive systems. *IEEE Software* 19 (4): 66–72.

Webber, D.L., and H. Gomaa. 2004. Modeling variability in software product lines with the variation point model. *Science of Computer Programming* 53 (3): 305–331.

Weiland, J., and P. Manhart. 2014. A classification of modeling variability in Simulink. In *Proceedings of the 8th international workshop on variability modelling of software-intensive systems*, 7–14. Sophia Antipolis.

An Executable Systems Modeling Language (ESysML): Motivations and Overview of Language Structure

Matthew Amissah and Holly Handley

Abstract In this paper we offer an overview of concepts and implementation of the Executable Systems Modeling Language (ESysML). ESysML is a domain-specific language (DSL) developed in response to challenges regarding executability and support for time-based simulation models, which are often necessary for analysis in a Model-Based Systems Engineering (MBSE) effort. ESysML is loosely based on the Systems Modeling Language (SysML). It downsizes the language schema of SysML in favor of model execution. Consequently, only language concepts such as Block, Ports, Activity, Events, etc. which are essential for defining system structure and behavior are retained in ESysML. While this presents a tradeoff of language expressivity for execution, the objective here is to offer precise semantics for a kernel of SysML constructs which can be extended to support fit-for-purpose applications in other domains.

Keywords SysML · Model-Based Systems Engineering · Simulation

1 Introduction

The Systems Modeling Language (SysML) (OMG 2017a, b) is an adaptation of the Unified Modeling Language (UML) (OMG 2015) aimed at offering a UML profile tailored to modeling engineered systems in general. A SysML model supports specification of requirements for a designed system. Additionally, it offers a representation of system components, their interconnections, interfaces, constraints, and the resulting behavior they exhibit. SysML provides an underlying schema that

M. Amissah (✉)
Worcester, MA, USA
e-mail: mamissah@wpi.edu

H. Handley
Norfolk, VA, USA

A. M. Madni et al. (eds.), *Recent Trends and Advances in Model Based Systems Engineering*, https://doi.org/10.1007/978-3-030-82083-1_26

supports an improved data management strategy particularly regarding exchange and reuse of data in systems architecture and design models.

The primary challenge with SysML model execution is one of language formality and standardization. While there are myriad approaches and tools offered both by the research community and tool vendors for model execution, there remains the challenge of a uniform implementation of the language. This is in part due to the complexity of the language infrastructure and its arcane specification, most of which stems from similar challenges with the UML (Cook 2012).

SysML like the UML consists of several independently derived modeling formalisms (i.e., use cases, state charts, activity models, etc.) which have been co-opted into a single modeling framework. This, together with the lack of an overarching meta-model that specifies the relationships and rules governing the use of various modeling constructs beyond their respective diagrams, has resulted in at best a semiformal language specification, with adverse implications for uniformity of language implementation and execution across modeling tools.

In response to these challenges, we propose the Executable Systems Modeling Language (ESysML), an executable domain-specific language (DSL) aimed at refining SysML. Analogous to the foundational UML (fUML) (OMG 2018) and its textual action language, i.e., Alf (OMG 2017a, b), ESysML essentially reimagines SysML as a simulation language. It prescribes an approach for time advance, execution of time-based actions, and logging of the resulting change in model attributes. ESysML is an internal DSL implemented in Python. Internal DSLs are languages embedded in a host programming language. This affords the capability to reuse syntax and existing runtime, as well as access to third-party libraries which reduces the language development effort (Ghosh 2010). A relatively well-known example of an internal DSL is the SystemC (Arnout 2000).

Given current trend towards increasingly interconnected systems such as Internet of Things and Artificial Intelligence, there is a need for model-driven engineering languages and methods that support formal architecture description and analysis for such highly interconnected real-time systems. This work leverages the relatively popular and accessible graphical syntax of SysML to support a formal model-driven engineering process. The aim here is to support a consistent systematic approach for realizing conceptual models and corresponding executable models useful for architecture analysis and decision-making.

Subsequent sections of the paper are organized as follows: Section 2 discusses motivation and design goals for ESysML. Section 3 discusses an overview of the major language constructs, textual syntax, and implementation in structural and behavioral modeling perspectives and concludes with a discussion on the approach for simulation execution and model logging. We conclude in Sect. 4 with proposed future work and the implications of this work and similar efforts on MBSE practice.

2 Motivation and Language Design Goals

As formerly mentioned, the primary challenges with SysML are a lack of precise semantics, complexity of language structure and support for modeling and execution of time-driven behavior. Notwithstanding, SysML offers useful constructs for conceptual modeling in systems engineering (SE). Additionally, it retains essential features from UML for extending language breadth and depth (i.e., through profiles and opaque expressions respectively).

ESysML was designed with the aim of leveraging SysML's relatively popular and accessible modeling constructs and features for language extension to support development of executable models within the MBSE framework. Given executability as the primary goal of this effort, the initial emphasis here is on simplicity in favor of expressivity or breadth. As such a number of SysML constructs have been removed from ESysML while preserving the language's essentials of structural, behavioral, and parametric modeling perspectives (Friedenthal et al. 2014).

The fourth perspective pertinent to Requirements is not supported here since ESysML is primarily aimed for modeling and simulation of architecture and design concepts; this is essentially descriptive contrary to the prescriptive nature of Requirements engineering. We recognize that the outcomes of Requirements engineering are necessary inputs for model validation; in ESysML this may be implemented using constructs for constraint modeling to support model-based validation of requirements. A more complete approach for requirement modeling leveraging language features for extension is proposed as future work. The following points summarize the design goals for this effort:

- Ontological foundation: Primary language concepts shall be based on SysML with the aim of supporting structural, behavioral, and parametric modeling perspectives.
- Execute ability: An implementation that serves as a standard demonstration of language semantics shall be offered.
- Support for time: The language shall offer constructs to reference time, as well as event-based simulation time advance and logging of dynamic attributes in a model.
- Extensibility: The language shall offer constructs to enable extension through new meta-constructs and specializing of existing model elements, i.e., language profiles and model libraries respectively.
- Model governance: The language shall underlie a framework of tools and methods for efficient model development in MBSE. This entails offering a development environment with capabilities for model checking, code completion, model libraries, etc.
- Graphical syntax: The textual syntax shall be compatible with SysML graphical syntax where possible to support automated/semiautomated generation of graphical models.

- Simplicity: The language shall offer a concise/simplified schema relative to SysML. This should be measurable using metrics such as total number of language constructs and traceability of derived constructs from primitive ones.

3 ESysML: Taxonomy of Primary Concepts

An ESysML comprises model elements and properties. A model element here is parallel to the concept of fundamental concept of Thing traceable to the Bunge-Wand-Weber (BWW) ontology (Wand and Weber 1990). A model element may own zero or more properties, which specify its relation to other model elements. Properties are implemented here as unidirectional relations with a single source and zero or more target model elements. Figure 1 illustrates the concept of model element and property using UML notation.

Model elements are further categorized under the two main types of instance and type. Instances reference real or notional things present in the world. Types are definitional elements, used to specify a template for creating instances and defining action executions. A distinction is made between real instances with spatiotemporal extent, which are typed by block and notional/conceptual ones typed by data type. Data instances serve as attributes of block instances. Attributes here are observer imputed properties useful for exposing the nature of real things. Block (i.e., physical) and data (i.e., conceptual) instances are differentiated from each other solely by the time attribute. Time is defined here as a Value (i.e., has a unit and value property with a singular instance).

Activity is the primary construct for modeling behavior. It specifies the rules by which model elements are created, destroyed, or transformed. Constraints, events, and opaque actions are special kinds of activities in ESysML. Constraints specify restrictions on the values the properties of model elements may assume. Events refer to time- or condition-based change triggers defined for activation of actions. Opaque actions represent an abstraction for black box behavior. To enable execution, the implementation details for opaque actions may be specified using a prescribed executable formalism. Figure 2 illustrates a hierarchy of the main model element categories.

Properties specify relations between model elements. These are broadly categorized into dependency and characterization relationships. Characterization is a property between types. Characterization properties are further specialized into

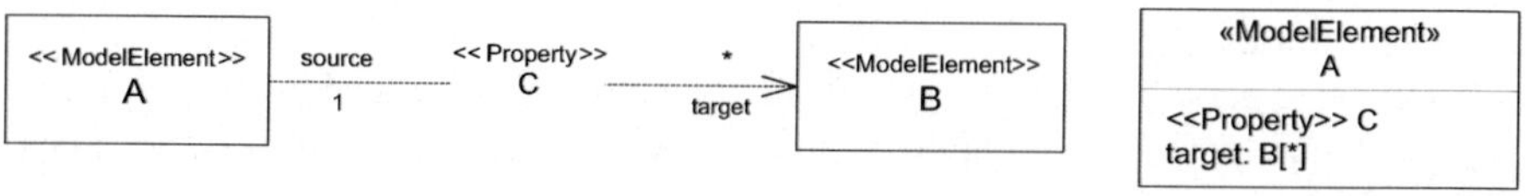

Fig. 1 UML notation for primary constructs

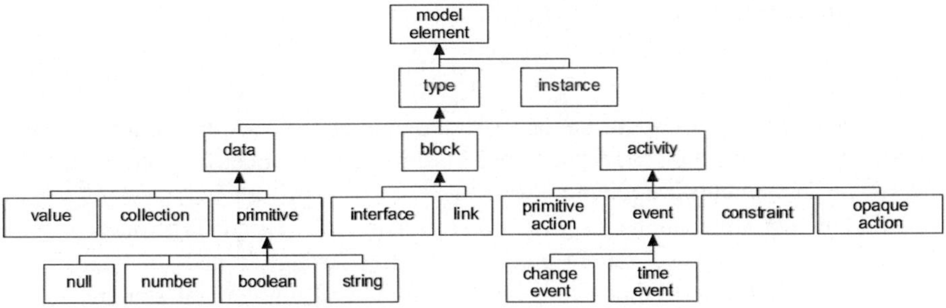

Fig. 2 Hierarchy of model elements

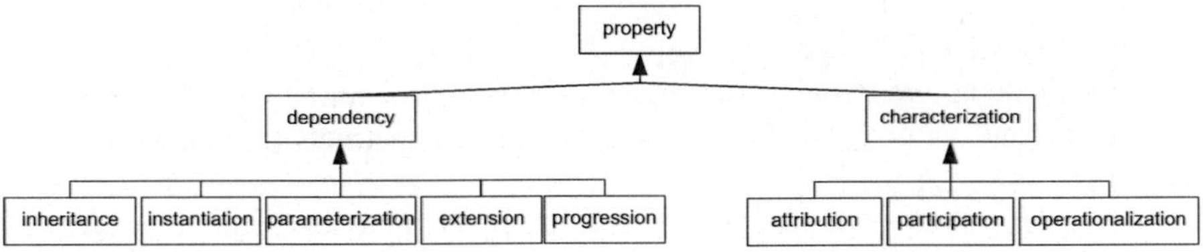

Fig. 3 Hierarchy of property classes

attribution, operationalization, and participation. In an attribution relation, the element at the target end of the relation serves as a descriptor to the source element. Operationalization is a relation between a type and one or more actions, which prescribe how the properties of the type change/evolve in a model. Participation is a relation between block instances. It specifies a whole-part relationship between block instances. Parts on the boundary of a block are termed Ports. These may be typed by the specialized interface block which offers inbuilt functionality for exchange across blocks via links.

The Dependency property is used broadly to specify logical dependence relations between model elements. This is specialized into inheritance, instantiation, parameterization, extension, and progression properties. Instantiation is a relation between an instance and its type; this enables implementation of the instance based on rules specified in the type. Inheritance is a relation between types that implies the element at the target end may exhibit all of the properties of the element at the source.

A progression property denotes an ordering relation between actions which specifies precedence or parity of action execution sequence. The parameterization property specifies relations between an action and model elements required for its execution (i.e., inputs) or model elements created or transformed as a result of its execution (i.e., output). Figure 3 illustrates the hierarchy of property types in ESysML.

3.1 Overview of Textual Syntax

ESysML adopts a C-style syntax in order to support some level of continuity with fUML and Alf. As such, language statements and blocks are delimited with semicolons and curly braces respectively. Additionally C syntax for comments "if" and "while" statements are retained.

Unlike most C-like languages, ESysML is dynamically typed. Type declarations can be optionally included and enforced similar to a strongly typed language. This feature is retained from Python as it generally engenders simpler and more readable code without necessarily trading off simplicity for a more robust specification. Additionally Python syntax for creation and access of collections is retained. Specifically in ESysML, Bag, Sequence, and Set are equivalent to and use the syntax of Dictionary, List, and Set respectively.

The main departures from this general style are in the syntax for Actions, Events, Collections, and Value types. Further detail on these constructs is provided in later sections.

Finally the textual syntax maps to SysML graphical notation with minor modifications. Based on this ESysML model elements may be visualized using block and activity diagrams. Depending on the scope and type of language elements featured, block and activity diagrams may be used to visualize abstract domain concepts and actual physical connections and relationships as well as behavior. Ultimately the goal is to support automated generation of fit-for-purpose visualizations of model data.

3.2 Structural Modeling

With regard to structural modeling, ESysML is aimed at supporting three primary levels of specification, namely, meta-modeling/language specification, ontological/-domain reference specification, and instance/physical concept specification. Meta-modeling use cases develop model libraries that support new language constructs in line with the language profiling mechanism of UML. Ideally meta-models should offer novel language concepts that transcend more than one application context as well as provide reusable functionality for specifying novel concepts. The primary language construct that supports meta-modeling is the extension property.

Ontological/reference specification entails models of categories, concepts, relationships, assumptions, etc. relevant to characterizing the model referent or problem domain. The constructs of block and data and their associated properties are the language constructs typically applied at this level. Instance models/physical concept specifications are models of real or notional entities with interconnections, constraints, and behaviors which are bound by the laws of a real or notional universe. Block diagrams may be used to illustrate both the high-level domain concepts

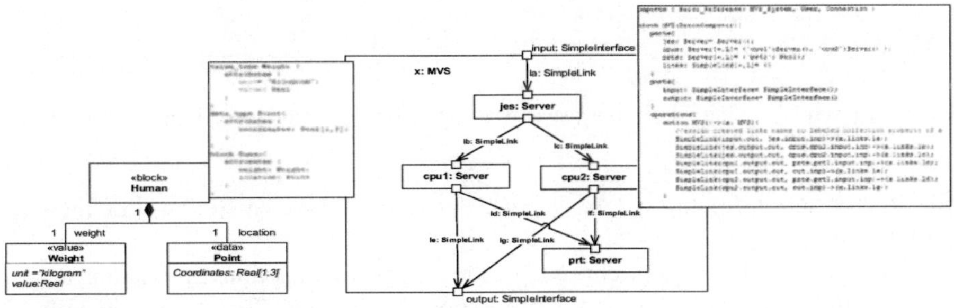

Fig. 4 Modeling domain and physical constructs with blocks

and instantiation of structural constructs which are typically implemented using activities. Figure 4 illustrates textual and graphical models specifying domain level and instance level constructs.

3.3 *Behavioral Modeling*

Behavioral modeling entails specifying rules that govern change in model structure. Change here can be characterized under creation, transition, or destruction of model elements. The goal in behavioral modeling is to offer underlying constructs that support a coherent construction of models under various perspectives of how change occurs in a system. In line with SysML, (1) state/event, (2) process, and (3) parametrics are the three main views supported in behavioral modeling. Unlike SysML, in ESysML constraint modeling and parametrics are considered behavioral, as this aligns with the given definition of behavior as rules governing change in model structure.

The primary behavioral modeling construct in ESysML is the activity. Specializations of activity include primitive actions, opaque expressions, events, and constraints. Similar to primitive types, primitive actions are predefined model elements with user-specified slots. Examples of primitive actions are start, final, instance creation, assignment, conditionals, and loops.

The order of action execution is determined by either progression relations between action calls, which may be specified using the control and object flow notation of SysML activity diagrams. Actions do not have the typical return statement such as in C or Python. Action outputs specify zero or more elements that are assigned in the action method (i.e., activity) and available to the caller of the action once the output is assigned in the method.

As an example the *gen_request* action definition in Fig. 5a specifies a name *request* that must be assigned a value of type *FlowItem*. The first statement in the action's body assigns the request name to the returned instance generated by the

```
action gen_request()->(request: FlowItem){
  FlowItem()->(request);
  increment(numRequests)->(numRequests); }
```

```
opaque pyrandint(x: integer, y:integer)->(z: integer){
  "import random \n"
  "z = random.randint(int(x).(int(y))"
}
```

```
action send_request()->(){
  output.send();
  time_event(genRateExpo)-> gen_request;}
```

Fig. 5 (**a**) Activity definition; (**b**) opaque activity definition; (**c**) action invocation via time_event

constructor action for FlowItem. Opaque expressions may be similarly defined as actions; this is shown in Fig. 5b.

Time-events and *change-events* enable the invocation of actions after a time delay or a specified change in a model's properties respectively. Time-events specify a trigger which is a number or an expression that evaluates to a number, while change-events must specify a Boolean-valued trigger. Figure 5c illustrates the syntax for action invocation with events.

3.4 Simulation Execution and Data Logging

To support event scheduling and time advance, an implementation of a simulator based on Tocher's three-phase simulation worldview is proposed (2018). In the three-phase simulation approach, changes to a system's state can only occur under two conditions for change, i.e., changes that are driven solely by the passage of time and those that are triggered by conditions other than time. These two alternatives for state change are termed as bound events (B's) or conditional events (C's) respectively. In ESysML these are represented by the change and time event constructs.

Simulation software can then be organized under Simulation Executive/Simulator and Model. The Simulation Executive entails underlying infrastructure that handles event scheduling. The executive maintains a global time variable and updates it to the timestamp of the next imminent event(s). For each iteration the simulator executes the following: (1) Advance simulation clock to time of next imminent B event(s). (2) Execute B's due at this time. (3) Scan C's and execute entities with conditions evaluating to true. This is illustrated in Fig. 6a.

Blocks in the model maintain a due_list, c_list, and local time variable. These are abstract attributes which are inherited by all blocks; additionally these are automatically updated during simulation run based on user-defined time event, change_events, and constraints. Similar to change events, constraints are evaluated and updated with every time step.

Finally, to support logging of simulation data, the construct of a model Observer is introduced. Model observation can be considered here as a research and experimentation viewpoint. As such model questions/queries can be developed along with appropriate simulation run and data collection parameters. Blocks in the model can access a global *observe* action, which enables modelers to specify properties to be logged during simulation, as well as log_events which essentially prompts an update of a log file during simulation run. The model execution architecture is illustrated in Fig. 6b.

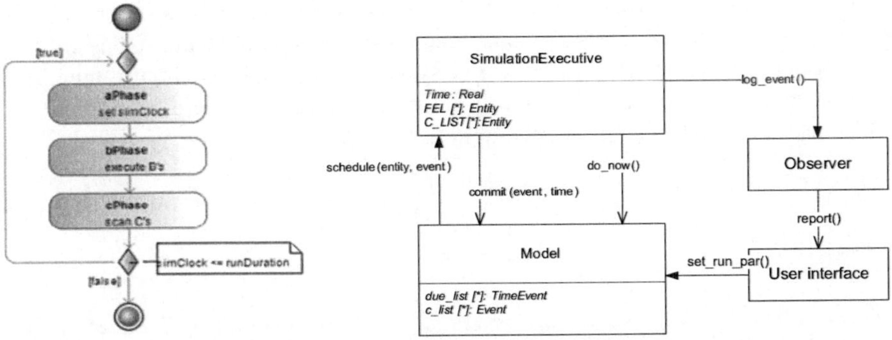

Fig. 6 (**a**) Execution process of a three-phase simulation (Pidd 2004) (**b**) ESysML model execution architecture

4 Conclusion and Future Directions

This work introduced ESysML, a modeling language that simplifies and refines UML/SysML modeling constructs. The primary goal here is to leverage a widely accepted standard in SyML, to support executable model specification in Model-Based Systems Engineering (MBSE) practice. This work lays an essential foundation for a model development framework that enables a uniform and efficient transition from high-level conceptual models to detailed computational models which are often necessary for architecture analysis and verification.

Future research is aimed at offering a framework of tools and methods along with the underlying language to enable model continuity and reuse across the various levels of model specification. This requires research effort that extends ESysML constructs to support implementation of novel and domain-specific concepts. Particularly, an extension of the framework to support formal specification of system requirements and their relation to current modeling constructs such as constraints is proposed. Such an extension of the language will provide a more rigorous approach to modeling the user/stakeholder domain, requirements, and architecture definition.

The time advance and model execution strategy offered here is essentially sequential. An extension of this approach optimized for parallel and distributed model execution is proposed to enhance scalability of the approach. Additionally, further research is required to extend ESysML's support for integration of other programming languages in order to leverage opaque expressions to implement libraries available in new languages. Besides programming languages, standards that support graph visualization such as graph description language (Gansner and North 2000) and data import/export libraries (XML, JSON, etc.) are required. Finally a web-based repository of sample models of typical architecture patterns and SE idioms is proposed; this is aimed at contributing to model reuse and collaboration in the systems modeling community.

Regarding the immediate future of MBSE practice, the next generation of tools will be able to leverage a growing resource of structured architecture, design, and operational data to facilitate self-design systems. This will essentially automate much of the design function particularly activities such as architecture specification, trade space analysis, design specification, testing, etc.

References

Arnout, G. 2000. SystemC standard. In *Proceedings 2000. Design Automation Conference*, IEEE Cat. No. 00CH37106, 573–577. IEEE.

Cook, S. 2012. Looking back at UML. *Software & Systems Modeling* 11 (4): 471–480.

Friedenthal, S., A. Moore, and R. Steiner. 2014. *A practical guide to SysML: The systems modeling language*. Morgan Kaufmann.

Gansner, E.R., and S.C. North. 2000. An open graph visualization system and its applications to software engineering. *Software: Practice and Experience* 30 (11): 1203–1233.

Ghosh, D. 2010. *DSLs in action*. Manning Publications Co.

Object Management Group. 2015. *OMG Unified Modeling Language*. Version 2.5. Retrieved from http://www.omg.org/spec/UML/2.5

———. 2017a. *OMG Systems Modeling Language*. Version 1.5. Retrieved from http://www.omg.org/spec/SysML/20161101

———. 2017b. *Action Language for Foundational UML (Alf): Concrete Syntax for a UML Action Language*. Retrieved from http://www.omg.org/legal/tm_list.htm.

———. 2018. *Semantics of a Foundational Subset for Executable UML Models (fUML):* Version 1.4. Retrieved from https://www.omg.org/spec/FUML/20180501/fUML_Syntax.xmi

Pidd, M. 2004, December. Simulation worldviews-so what? In *Proceedings of the 2004 winter simulation conference, 2004*, vol. 1. IEEE.

Wand, Y., and R. Weber. 1990. Mario Bunge's Ontology as a formal foundation for information systems concepts. *Studies on Mario Bunge's Treatise* 18: 123.

Quantitative System Reliability and Availability Analysis Using SysML

Jaron Chen, Michael Hailwood, and Myron Hecht

Abstract A SysML library and method for calculating system reliability and availability is described. The method can be used to model and predict reliability and availability early in the design and continue through detailed design and system integration. Values for failure rates, restoration rates, recovery probabilities, and other parameters are stored in specialized reliability blocks that are defined from a single parent reliability block, and equations for reliability block configurations (series, parallel, active/standby, and k out of n) have been implemented and included in the library. The paper includes an example demonstrating the application of the library model elements and how the library can be extended to include additional system redundancy configurations.

Keywords System reliability · System availability · MBSE · SysML

1 Introduction

To create systems that have operationally acceptable Reliability, Availability, and Maintainability (RAM) attributes, developers must consider design issues associated with these attributes from the beginning of the design process. System engineers should define system-level RAM requirements based on user needs, allocate these requirements to the functional architecture, and further allocate them into the physical architecture and then to the lower-level subsystems, assemblies, and components (both hardware and software) (Society of Automotive Engineers (SAE) 2008). These requirements should drive design decisions as to what level of rigor should be required of software development, what types of components should be acquired, how much redundancy, monitoring and diagnostics capabilities, prognostic indicators, the physical configuration of components to allow for repair

J. Chen · M. Hailwood · M. Hecht (✉)
The Aerospace Corporation, El Segundo, CA, USA
e-mail: myron.j.hecht@aero.org

A. M. Madni et al. (eds.), *Recent Trends and Advances in Model Based Systems Engineering*, https://doi.org/10.1007/978-3-030-82083-1_27

and replacement accessibility, the nature of external support (test equipment, spares, storage, training, personnel) required.

Although no system developer or product vendor would want to admit that RAM design attributes are ignored during development until the late phases of the design, it is common practice. The quantitative and qualitative analyses needed to support design decisions early in the design are not performed, often because analytical capabilities are not available or simply because the design leads are not aware of problems. Model-Based Systems Engineering (MBSE) is intended to support system requirements, design, analysis, and verification and validation activities beginning from the conceptual design throughout the development and subsequent phases (International Council on Systems Engineering 2007). SysML is a widely used design language which supports MBSE (Object Management Group 2017). As such, it can be used through all phases of development and even into operation and sustainment.

This paper describes how SysML and an accompanying library of modeling elements can be used to perform quantitative system reliability analysis. With this library, RAM engineering can be integrated into the MBSE process from the time of user needs analysis and be carried forward through all the later phases of the development effort. With this capability, more informed decisions and precise tradeoff analyses can be performed. The net result will be systems that are operationally suitable for their intended use.

Previous work on the use of SysML for quantitative reliability analysis includes a method for deriving series/parallel relationships using SysML Internal Block Diagrams (Liu et al. 2013). However, this work did not describe how system reliability results could be produced from this analysis from within the SysML tool. Earlier work by David et al. (David et al. 2009) described how probabilities stored in an external database could be used in conjunction with SysML to determine priority risks in Failure Modes, Effects, and Criticality analysis. The Object Management Group is developing a SysML reliability and availability standard for a profile and library (Biggs et al. 2019). However, that work does not include reliability block diagrams. Work describing how SysML constraint blocks could be used to calculate reliability has also been described, but only for a series system (Myron Hecht 2014). This paper extends that earlier work by adding model elements for additional system configurations (passive parallel, passive k out of n, and active/standby) and describes how this work can be further extended for more types of reliability blocks.

2 Methodology

Our method uses SysML parametric diagrams and an interpreter built into the SysML tool we used to perform the calculations. Parametric diagrams allow for modeling of low-level system constraints/requirements with values. While this is beneficial, the SysML language is still limited by being a descriptive language and needs a modeling tool with simulation capabilities for the realization of constraint testing. Stemming from our use of Cameo Systems Modeler (CSM) (Cameo Systems Modeler 2019) for our MBSE environment, we decided to use Dassault Systemes' commercial "plugin" Cameo Simulation Toolkit (Cameo Simulation Toolkit 2019). This plugin specifically adds multiple layers of functionality to CSM, but the ability to calculate parametric diagrams and pass objects across the SysML Binding Connector Element is the most important for our reliability analysis. The definition of any applicable SysML elements with practicality to our work is shown in Table 1.

As noted above, parametric diagrams are used to model components in a system and the configuration (series, parallel, k out of n, or active/standby). In our implementation, part properties are used to hold the necessary reliability values, and these values are updated when their corresponding part properties are passed through constraint blocks via constraint parameters. The values on the lowest level components are provided beforehand by the user. Every constraint block has two constraint parameters, one that stores the input values and one that stores the calculated output. Each constraint parameter has a corresponding multiplicity, which determines how many corresponding part properties a constraint parameter can have. Depending on the constraint block, an input constraint parameter can accept a differing amount of part properties, but all the output constraint parameters will allow for only one connected part property.

Each constraint block has a constraint embedded in them that updates the availability and failure rate of its corresponding block. Constraints in Cameo usually involve a set of mathematical expressions that are written up in a scripting language. Jython, which is an implementation of Python designed to run on Java platforms like CSM, was used as the scripting language for the constraints, and they evaluate the system reliability equations shown in Table 2. To evaluate Jython or any other scripting language, Cameo Simulation Toolkit uses Java scripting API, which allows for embedding scripting language code into Java applications.

The Jython scripts perform the reliability calculations using basic arithmetic for the series, parallel, and k out of n blocks. The other constraint blocks, namely, the redundancy ones, were more complex, involving matrix manipulation to solve Markov models. In general, constraint blocks will take in as input at least one reliability block and extract the values it needs, such as failure rate and availability, to calculate overall availability and failure rate. The constraint block will then output the calculated availability and failure rate into another reliability block, which will represent the owning system of the inputs. This can allow for reliability analysis to be captured across the system hierarchy. In order to allow reliability blocks to be passed as inputs and outputs, the values within the reliability blocks need to be updated within the script. Fortunately, Cameo Simulation Toolkit offers a way to get and set structural feature values like the ones found on part properties and the reliability block through the Action Language Helper (ALH) class, which is used within the scripts. A summary of the available constraint blocks can be found in Table 2.

Table 1 Summary of SysML elements

Model element	Description
SysML block	This is the foundational element of SysML that will represent our system components in an MBSE environment. A general SysML block includes 7 different compartments. This library and profile use two of them: Value and part properties
Abstract reliability block	The reliability block is a SysML block that has been given a stereotype <<Reliability Block>>. General SysML blocks to inherit value properties associated with reliability from the Abstract Reliability Block required for reliability and availability prediction. The relationship is structurally created with a generalization (inheritance) relationship of system component to the reliability block. Consequently, any instance of a system component is of type Reliability Block, which is an important precondition for constraint blocks
Value type	A value type defines a kind of value (numerical, character, or more complex). Value types used in the reliability profile and library are: Repair rate (restorations per hour, applicable to both automated and manual restoration actions) Availability (proportion of uptime, a decimal number with a range between 0 and 1) Failure rate (failures per hour) k – Number of system components necessary to operate in a k out of n system, integer n – Number of replicated system components available, integer Switchover probability (active standby)
Value properties	In SysML, a value property is a quantity in a block. For reliability analysis, value properties are defined by the value types identified above
Part properties	In SysML, a part property is a lower-level component of a higher-level block, i.e., the lower-level blocks (parts) are owned by the higher-level block. In the reliability profile and library, the value properties in the component blocks are combined to yield a system-level reliability
Instances:	In SysML, instances are data structures contained in a model and defined by blocks whose value properties have been assigned values. For the reliability profile and library, instances contain the values used in system reliability calculations
Constraint blocks	In SysML, constraint blocks contain mathematical or logical formulas. Inputs and outputs are passed into a constraint block using a port-like structure called a parameter. Parameters are connected to value properties using binding connectors (see below). In the reliability profile and model, constraint blocks contain equations for series, parallel, k out of n, and active/standby redundancy
Binding connectors:	This element connects two SysML parts and sets each end of a connection between two elements equal. This element goes hand in hand with our constraint blocks because the binding connectors are connected to constraint parameters to make each end equal

Table 2 Summary of reliability constraint blocks

Name	Type	Availability expression
Series	Series of n components	$A_s = \prod_{i=1}^{n} A_i$
Parallel	n parallel components	$A_S = 1 - \prod_{i=1}^{n} (1 - A_i)$
k out of n	Homogenous k out of n	$A_S = \sum_{k}^{n} \binom{n}{k} A^k (1 - A)^{n-k}$
Active/standby	Active/standby parallel system, imperfect switching	$A_S = P_0 + P_1$ $P = S^{-1}U$ $S = \begin{pmatrix} 1 & 1 & 1 \\ (1+c)\lambda & -(\mu+\lambda) & 0 \\ (1-c)\lambda & \lambda & -\mu \end{pmatrix}$ P_0 = Probability of no component failed P_1 = Probability of 1 component failed λ = Component failure rate μ = Component restoration rate c = Recovery success probability

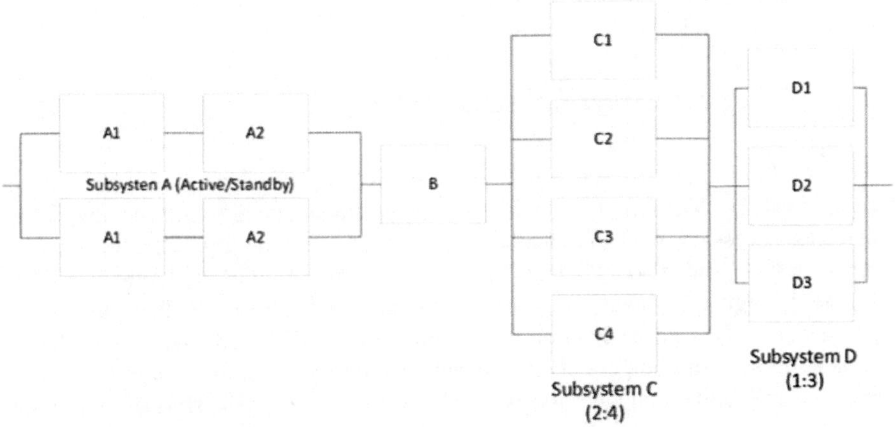

Fig. 1 Reliability block diagram of example system

3 Example

This example will demonstrate application of the reliability profile to an example system. Figure 1 shows the example system in the form of a reliability block diagram (RBD). There are four subsystems, A, B, C, and D, all in series. Subsystem A is an active/standby redundant system with two channels, each having two components in series (A1 and A2). Subsystem B is a non-redundant subsystem. Subsystem C has four channels, with two required in a passive parallel configuration, and Subsystem D has three channels, with only one required in a passive parallel configuration.

The SysML block definition diagram (BDD) in Fig. 2 represents (models) the hierarchical composition of the system. Although this hierarchical composition does not show the interconnection among the lower-level components, it is not

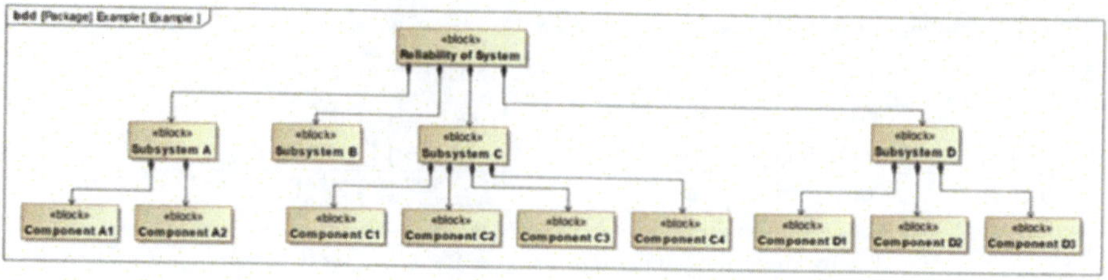

Fig. 2 System composition

necessary. For the RAM analysis methodology shown here, the use of parametric diagrams with constraint properties is sufficient and averts a limitation in SysML and the Cameo Systems Toolkit used for the quantitative analysis.[1] However, it is required that the analyst understand the series, parallel, or other configurations of the component blocks in order to correctly perform the analysis.

3.1 Using the Abstract Reliability Block

As noted above, the abstract reliability block is used as a parent to provide the general SysML blocks that are part of the system analysis to inherit the value properties necessary for our calculations such as "Availability," "Failure Rate," "Repair Rate," and also "k" and "n" recovery probabilities and others shown in Table 1. Figure 3 shows the general SysML blocks are specialized to inherit these features. This generalization to the reliability block is created for all blocks representing system components, so all blocks gain the value types and properties needed for reliability analysis. However, this step only creates the slots that will hold the values. Only when we instantiate the blocks in the process of creating parametric diagrams described below will we be able to add values.

[1]The parametric diagram's functionality inherits limitations from both SysML and the CST, and as such we must make an assumption. Part properties on each composition association's composite have a multiplicity of 1. For any valid configuration of a part requiring a higher multiplicity, the reliability MBSE profile will satisfy this requirement by using multiple connections or changing slot values equating to value properties instantiations on the parts. While this may not affect small systems, larger systems with higher levels of complexity will need to take this into consideration. Our example configuration in Figure 1 is relatively simple and will not be affect by the previously mentioned limitations.

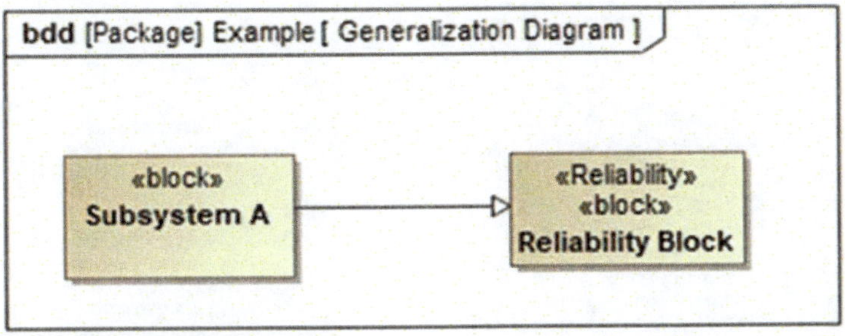

Fig. 3 Specialization relationship

Table 3 Application of constraint blocks

Reliability/availability analysis operation	Applicable constraint block
Rollup of Component A1 and Component A2 in series in each channel of Subsystem A	Series
Rollup of active and standby channel in Subsystem A with 0.1 probability of switchover failure	Active/standby
Rollup of the Subsystem D components (Component D1, Component D2, component D3) in parallel	Parallel
Rollup of Subsystem C which has four channels in parallel and only two are required	K out of n
Rollup of system-level reliability (Subsystems A, B, C, and D)	Series

3.2 *Reliability/Availability Modeling*

The next step in the process is the application of the constraint blocks. Reliability and availability analyses are performed "bottom-up," i.e., component values are calculated, followed by the channels and subsystems. The final calculation is taken the subsystem reliability and availability probabilities and combining (usually by multiplication because independence is assumed) to find the system level success probability (reliability or availability, depending on the application of the system and the nature of the requirements). Table 3 shows the allocation of the constraint blocks listed in Table 2 to the system described in Fig. 1.

The constraint blocks are brought into the model by declaring them using the block definition diagram shown in Fig. 4.

3.3 *Calculation of Results*

The next step is instantiating the blocks and constraints and blocks defined in Figs. 2, 3, and 4 on a parametric diagram and binding the parameters (i.e., ports) on the constraint blocks to value properties in the system blocks as shown in Fig. 5.

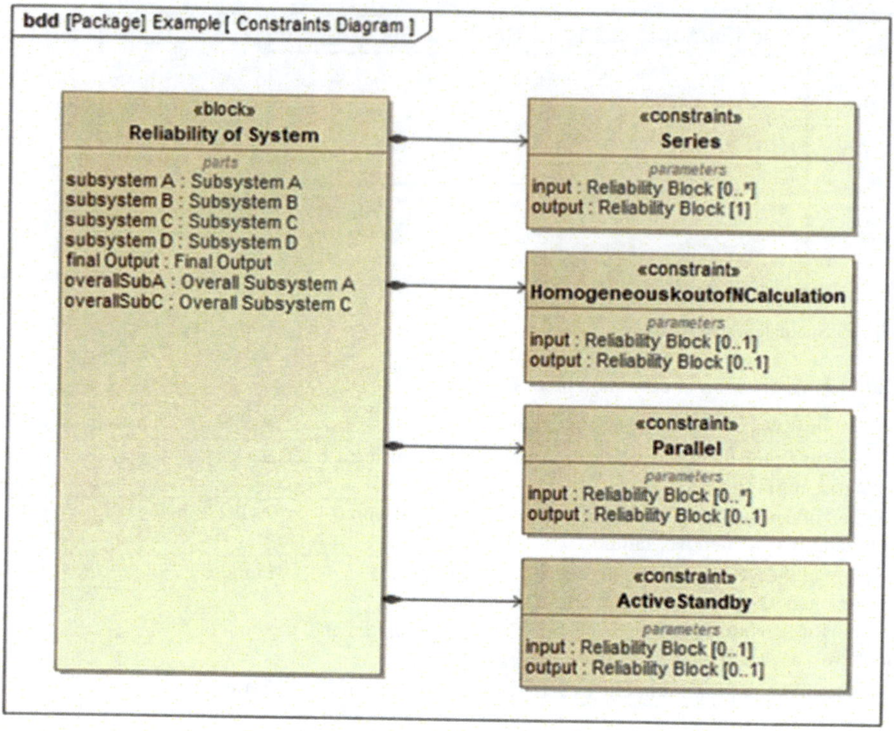

Fig. 4 Constraint blocks definition

After an instance is created, the failure rate and repair rate are manually added for the lowest level elements' slot values represented by value properties. For the purposes of the demonstration, all component failure rates (Components A1, A2, B, C1, C2, C3, C4, D1, D2, D3) were set to 0.01 (MTBF of 100 h), and all restoration times were set to 1 h (MTTR of 1 h).

When the slots are filled with these values for those instances, the Cameo Simulation Toolkit (CST), the interpreter within a module of the CSM tool, calculates all the other instance slots cascading up to the highest level (channel, subsystem, and system availability). Table 4 shows the results.

The system availability has been calculated to be 0.987 as shown on the top line of Table 4.

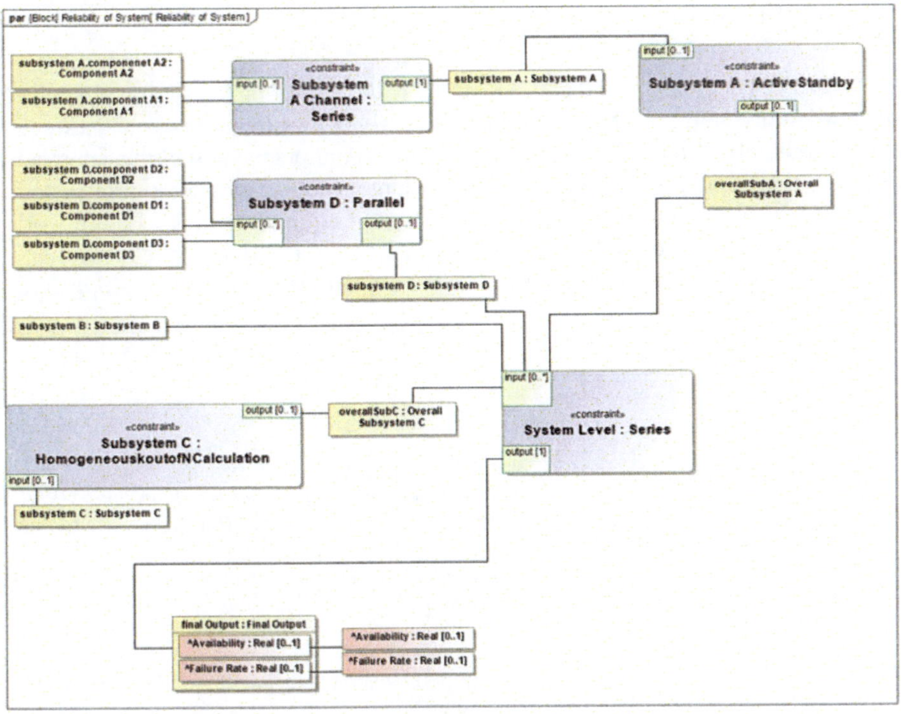

Fig. 5 System reliability/availability parametric diagram

Table 4 Results of model calculations

Name	Availability	Failure rate	Repair rate	k	n
System	**0.9874968901**	**0.01262861**	**1.00**	**1**	**1**
System.overallSubA	0.9973766705005475	0.00263022	1.00	1	1
System.overallSubC	0.9999961464688174	2.4E-5	1.00	1	1
System.subsystem A	0.9802960494049603	1.99E-2	1.00	1	2
System.subsystem A.component A2	0.9900990099	0.01	1.00	1	1
System.subsystem A.component A1	0.9900990099	0.01	1.00	1	1
System.subsystem B	0.9900990099	0.01	1.00	1	1
System.subsystem C	0.9900990099	0.01	1.00	2	4
System.subsystem D	0.9999990294098517	1.00E-6	1.00	1	1
System.subsystem D.component D1	0.9900990099	0.01	1.00	1	1
System.subsystem D.component D2	0.9900990099	0.01	1.00	1	1
System.subsystem D.component D3	0.9900990099	0.01	1.00	1	1

4 Conclusions

This paper has demonstrated that it is possible to develop complex system reliability and availability within SysML using a small library of constraint blocks and a single abstract reliability block. The method is scalable. Authors have used this approach to model the reliability and availability of a large satellite ground control system consisting of more than 40 virtual machines and 4 equipment racks. Future work includes (1) adding more constraint blocks to model more types of reliability and availability configurations and (2) relaxing the assumption of a single multiplicity.

References

Biggs, Geoffrey, Andrius Armonas, Tomas Juknevicius, Kyle Post, Nataliya Yakymets, and Axel Berres. July, 2019. *OMG standard for integrating safety and reliability into MBSE: Core Concepts and Applications*. International Council on System Engineering (INCOSE) International Symposium, Orlando, FL.

Cameo Simulation Toolkit. 2019. [Online]. Available: https://www.nomagic.com/product-addons/magicdraw-addons/cameo-simulation-toolkit#intro.

Cameo Systems Modeler. 2019. [Online]. Available: https://www.nomagic.com/products/cameo-systems-modeler.

David, Pierre, Vicent Idasiak, and Frederic Kratz. 2009. *Improving reliability studies with SysML*. IEEE Reliability and Maintainability Symposium.

International Council on Systems Engineering. 2007. *Systems Engineering Vision 2020 (INCOSE-TP-2004-004-02)*, September.

Liu, Xaio, Yi Ren, Zili Wang, and Linlin Yiu. 2013. Modeling method of SysML-based reliability block diagram. In *International Conference on Mechatronic Sciences, Electrical Engineering, and Computing (MEC)*. IEEE: Shenyang.

Myron Hecht. 2014. *SysML reliability modeling of ground based systems with virtualized architectures*. Los Angeles: Ground Systems Architecture Workshop. Available at https://gsaw.org/wp-content/uploads/2014/03/2014s05hecht.pdf.

Object Management Group. 2017. *OMG SysML Version 1.5 Specification*. Available online at http://www.omg.org/spec/SysML/1.5/.

Society of Automotive Engineers (SAE). 2008. *SAE-GEIA-STD-009, Reliability Program Standard for System Design, Development, and Manufacturing*, August.

Part V
Advances in MBSE

Towards Making the Business Case for MBSE

Shatad Purohit and Azad M. Madni

Abstract In the face of ever-increasing system and program complexity, several aerospace, automotive, and defense organizations have already begun transitioning to model-based systems engineering (MBSE). A key challenge that organizations face is determining whether it makes business sense and whether it is technically feasible to transition to MBSE given legacy and budgetary constraints. This paper presents a methodological framework for analyzing whether an organization is likely to benefit from MBSE implementation on large-scale system programs. In this approach, MBSE implementation is characterized in terms of: system complexity, environment complexity including regulatory constraints, and system lifespan. Twelve major industry sectors are evaluated to determine whether MBSE can be of benefit to that sector. Then cost-benefit analysis is used to justify the decision to invest in MBSE. The approach is generic and can be applied to different industry sectors.

Keywords Model-based systems engineering · Return on investment · Economic analysis · MBSE methodology · MBSE tools · Document-centric systems engineering

1 Introduction

Industries which can potentially benefit from model-based systems engineering (MBSE) can be grouped into 12 major sectors: transportation and mobility, aerospace and defense, industrial equipment, energy and utilities, architecture and construction, life sciences, high-tech, marine and offshore, financial and business services, consumer goods and retail, natural resources, and consumer packaged goods and retail. The different industry sectors need the ability to quantify the

S. Purohit (✉) · A. M. Madni
University of Southern California, Los Angeles, CA, USA
e-mail: shatadkp@usc.edu

A. M. Madni et al. (eds.), *Recent Trends and Advances in Model Based Systems Engineering*, https://doi.org/10.1007/978-3-030-82083-1_28

benefits of MBSE investment to justify making an investment in MBSE. Despite the dearth of such quantitative data, some industry sectors (e.g., aerospace, energy, and automotive) with systems engineering (SE) practices in place have begun to transition to MBSE. Others (e.g., consumer electronics, healthcare, and construction) who have not adapted SE thus far are turning to SE to manage both development and management complexity in their respective markets. Yet others have begun exploring possible "on-ramps" to accelerate MBSE implementation that ensure that their ongoing operations are not disrupted. However, most industries, with traditional SE practices in place, are looking for ways to justify MBSE investment to upper management.

Systems in the 12 industry sectors identified above differ along 3 dimensions: system complexity, characteristics of the operational environment in which they operate, and system lifespan. For example, an aircraft has high complexity and long lifespan and operates in a strict regulatory environment. On the other hand, consumer electronics has relatively low complexity, short lifespans, and a relatively loose regulatory environment. Since industry sectors differ from each other along these dimensions and the amount of investment and potential gains can be expected to vary across industry sectors, the MBSE strategy needs to be industry specific.

2 Approach

2.1 Primary Methodology

The primary methodological flow begins with an analysis of industry sectors. We investigated whether MBSE has been successful in the specific industry sector. If the MBSE has not been successful in the given industry, then we examined the reasons. If the causes are systemic, then MBSE adoption is not advised; on the other hand, if the problem is related to implementation, then limitations and constraints related to implementation need to be analyzed and incorporated in the constraint analysis along with other restrictions related to legacy data, technology, cost, and schedule. In the next step, we propose to investigate the systems engineering (SE) processes associated with a specific company, which gives insights on whether the company is ready to adopt the MBSE or SE processes need to be adjusted. Figure 1 presents the methodological flow.

2.2 Detailed Approach

Our methodology is based on a multi-criteria evaluation framework for analyzing the costs and benefits of MBSE adoption. The inputs to the evaluation framework include (a) specifications and reports on MBSE tools; (b) lessons learned from

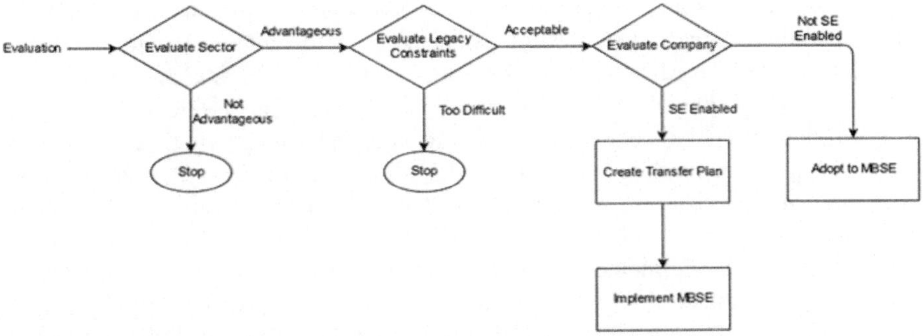

Fig. 1 Methodological flow

industry experts who have personal experience in MBSE implementation; and (c) literature on the value proposition and economic justification of MBSE. The literature on MBSE tools is needed to understand the amount and type of training needed in general engineering, systems engineering, modeling languages, and methodology. MBSE tool specifications and reports provide pertinent information on the effort, cost, and talent required to implement MBSE. In addition, keynote talks at various conferences and workshops from government and industry experts provide a source of lessons learned along with anecdotal information pertaining to investments and returns associated with traditional SE and MBSE.

The overall methodology takes into account interest rates, inflation rates, and risk premiums when evaluating viability of MBSE adaptation and implementation. The evaluation framework comprises: (a) SE life cycle processes; (b) guidelines for tailoring SE life cycle processes for different industries; (c) key problem characteristics such as system complexity, environment complexity, and system lifespan that help assess MBSE value proposition for the different industries; (d) parametric cost curves for traditional systems engineering and MBSE; (e) popular MBSE tools and their coverage of the SE life cycle; (f) Process Structure Matrix (PSM) approaches for process analysis, process sequencing, and database of lessons learned on major government programs. The analysis of system life cycle processes is essential to ensure that the framework is sufficiently general and applicable to a wide range of industry sectors. Since each industry sector has somewhat different processes, our approach calls for analyzing processes from the perspectives of flow, dependencies, and parallelism and evaluating their temporal characteristics. These factors are used to calculate the net present worth of investments and returns. For example, there could be several real-world situations in the transition to MBSE. In such cases, the framework can be applied to ensure maximum return on already made investments.

In our approach, we parameterize the MBSE implementation problem using the system's internal, external, and temporal (i.e., lifespan) characteristics. These characteristics were selected because they enable comparison of different industries in terms of the MBSE value proposition for each industry sector. This parameterization revealed that some industry sectors may not benefit significantly from MBSE implementation in their current circumstance, but with rapidly changing internal and external characteristics of systems in those sectors, MBSE implementation may well turn out to be beneficial at some point in the future. The aforementioned three problem characteristics also provide insights into the amount of time needed in relative terms to accomplish MBSE implementation in the different industry sectors. With this parameterization, we are able to cluster industry sectors based on their similarities. These clusters can then help identify which sectors stand to potentially benefit from MBSE implementation. Upon applying this MBSE implementation problem characterization to each sector, it became apparent that not all sectors derived the same benefit from MBSE implementation. This was quite expected because based on the systems in each sector, the MBSE implementation problems are different and vary in difficulty. The clustering of industry sectors also allowed us to address MBSE implementation for each cluster as a whole. Thus, a MBSE approach successfully employed in one industry sector within a cluster can also be successfully applied to other industry sectors in the same cluster.

Evaluation metrics are typically associated with those variables whose values we want to estimate using the methodology. The metrics encompass both generic measures such as effort, cost, and cycle times and domain-specific measures. The results of the evaluation are presented in the form of graphs and charts to facilitate comprehension. Often, these results can contribute to actionable insights into when, where, and what degree MBSE should be adopted to economically justify MBSE investments and returns for a particular system development project or a family of projects.

The key steps in the overall methodology begin with characterizing the MBSE implementation problem in terms of system complexity, environment complexity, and system lifespan. These attributes are used to characterize different industry sectors in terms of their unique characteristics. The evaluation results along with budgetary constraints are used to perform cost-benefit tradeoff analysis. Investments, payback, and their time frame are used in evaluating both benefits and costs. If the benefits justify the cost of MBSE investment, then the next steps in the methodology are to plan, schedule, and cost the effort to transition to MBSE. After ensuring that the costs are within budget constraints, the plan is executed. If on the other hand the benefits do not justify the cost, the organization stays with the status quo (i.e., traditional SE practices).

2.3 Economic Analysis of Traditional SE Approach

Traditional document-centric systems engineering approaches tend to have hidden costs. To illuminate the economics of document-centric SE approaches, we present an important curve (Fig. 4) (Madni and Purohit 2019). The percentage cost for each life cycle stage for systems engineering projects was evaluated based on a statistical analysis conducted on the US Department of Defense, as reported by the Defense

Acquisition University (Forsberg et al. 2015). From the evaluation, we deduced that investments in the early phases of system life cycle are comparatively lower than they are in the later phases. Furthermore, concept, design, and development phase cumulatively account for 20% of the total life cycle cost. The remaining 80% of the cost occurs in the production, testing, operations, support, maintenance, and disposal phases.

From an analysis of cost over the system life cycle for traditional SE, we can create an approximate cost profile for SE initiatives as shown in Fig. 4. The red curve indicates the traditional SE project cost across the system life cycle. In this figure, the ordinate (i.e., the vertical axis) represents normalized system life cycle cost, and the abscissa (i.e., the horizontal axis) represents time. As shown in this figure, SE investment is relatively low in the early stages of the system life cycle. Specifically, for conceptual design and preliminary design, investments are relatively low, but for detailed design, manufacturing, production, and operation, the investment increases sharply.

2.4 *Economic Analysis of MBSE Approach*

While cost curves for SE projects are described, it is difficult to find similar cost curves for MBSE projects. This is because MBSE is a relatively more recent development, and therefore, economic analysis data for MBSE initiatives is not yet available. Therefore, we took a different approach in our study to approximate cost curves for MBSE projects. Specifically, we combined historical data (evidence) from a space mission with current MBSE coverage of the system life cycle processes using the ISO 15288 process standard. One source of the former was a keynote address at the 2016 No Magic World Symposium. The keynote speaker spoke specifically about the scale of the investment to implement MBSE to "save" NASA's Mars mission with an overall project cost of $327 M (Technology & Enterprise Architecture 2016). According to this speaker, an initial MBSE investment of $5–$10 M from the project budget could have immeasurably helped that project. We gathered additional evidence from a project manager from the industrial equipment sector.

He indicated that for a $40 M equipment development project, basic MBSE implementation took less than 1% additional investment to cover requirement management, functional allocation, traceability, document generation, verification, and validation. However, in this instance his team did not employ MBSE to cover system analysis, simulation, and detailed integration with the design and development processes of the system life cycle. From these sources, we were able to make an informed inference about the relative cost of MBSE investment for a system development project. Even so, it is not feasible to identify how such costs should be distributed over the system life cycle. Therefore, we need to analyze MBSE investment over the system life cycle based on the coverage of the system life cycle by MBSE tools. To this end, we examined MBSE tool coverage of system life cycle processes. According to "System Life Cycle ISO/IEC/IEEE 15288 2015(E): Processes in System Life Cycle," there are four major processes that characterize the system life cycle: technical processes, technical management processes, organizational project-enabling processes, and agreement processes (IEEE 2012). These four major processes consist of 30 sub-processes. Each sub-process consists of activities, and each activity consists of tasks. We evaluated the coverage of MBSE tools for system life cycle processes for each task using four popular MBSE tools (e.g., No Magic Cameo Systems Modeler, IBM Rational Rhapsody, Eclipse Papyrus, and Eclipse Capella) that are representative of the life cycle coverage provided by MBSE tools. The data associated with SE life cycle process coverage by MBSE tools provides important insights into the temporal spend profile for MBSE investment over the system life cycle.

2.5 *System Life Cycle Processes*

In our study we evaluated the major MBSE tools against ISO 15288 processes (which define the system life cycle and coverage requirements). Once again, we decomposed the processes into activities and activities into tasks (Table 1). There are 14 technical processes, 54 activities, and 240 tasks that resulted from the decomposition (IEEE 2012). It was rather straightforward to determine whether or not the exemplar MBSE tools addressed each task. Based on the total number of tasks in a given process and the number of tasks covered by the MBSE tool, we were able to calculate the percentage coverage by each tool of the corresponding technical process. Since technical processes are the primary focus of MBSE tools, our study focused mostly on technical processes. The remaining processes (i.e., technical management, organizational project-enabling processes, and agreement processes) were analyzed using system life cycle process dependency structure matrix (DSM) shown in Fig. 2.

Our evaluation is based on research papers from the SE literature, our hands-on experience with MBSE tools, available documentation on MBSE tools, and models that we specially developed to complete the evaluation. To this end, our study took into account plugins which allow these MBSE tools to integrate with third-party tools related to requirement engineering, PLM, CAD integration, analysis, simulation, design optimization, real-time and embedded development, publishing, reviewing, and collaboration (No Magic Documentation 2018; Eclipse Capella Documentation 2018; Eclipse Papyrus Documentation 2018; IBM Rational Rhapsody Documentation Library 2018). Figure 3 presents the percentage coverage of technical processes by MBSE tools.

Table 1 System life cycle ISO/IEC/IEEE 15288 2015(E): Processes in system life cycle (IEEE 2012)

Technical Processes	Technical Management Processes	Organizational Project-Enabling Processes	Agreement Processes
1. Business or Mission Analysis Process 2. Stakeholder Needs & Requirement Definition Process 3. System Requirements Definition Process 4, Architecture Definition Process 5. Design Definition Process 6. System Analysis Process 7. Implementation Process 8. Integration Process 9. Verification Process 10. Transition Process 11. Validation Process 12. Operation Process 13. Maintenance Process 14. Disposal Process	1. Project Planning Process 2. Project Assessment and Control Process 3. Decision Management Process 4. Risk Management Process 5. Configuration Management Process 6. Information Management Process 7. Measurement Process 8. Quality Assurance Process	1. Life Cycle Model Management Process 2. Infrastructure Management Processes 3. Portfolio Management Process 4. Human Resource Management Process 5. Quality Management Process 6. Knowledge Management Process	1. Acquisition Process 2. Supply Process

2.6 *Sequencing System Life Cycle Processes*

After quantifying the coverage provided by MBSE tools of system life cycle processes, we identified where these tools are employed in the system life cycle. To this end, we created a Simulink® model of all 30 processes and their interfaces using the processes identified in the INCOSE SE Handbook (Madni and Purohit 2019). We then converted this model in to a Process Structure Matrix (PSM) shown in Fig. 2. From the PSM, it became possible to evaluate the parallel, sequential, and coupled processes. We sequenced processes in accord with their coupling, dependency, and parallelism. After sequencing the PSM, it became evident that MBSE tools today are focused primarily on the front end of the system life cycle. Furthermore, for SE projects that use an MBSE approach, projects can be expected

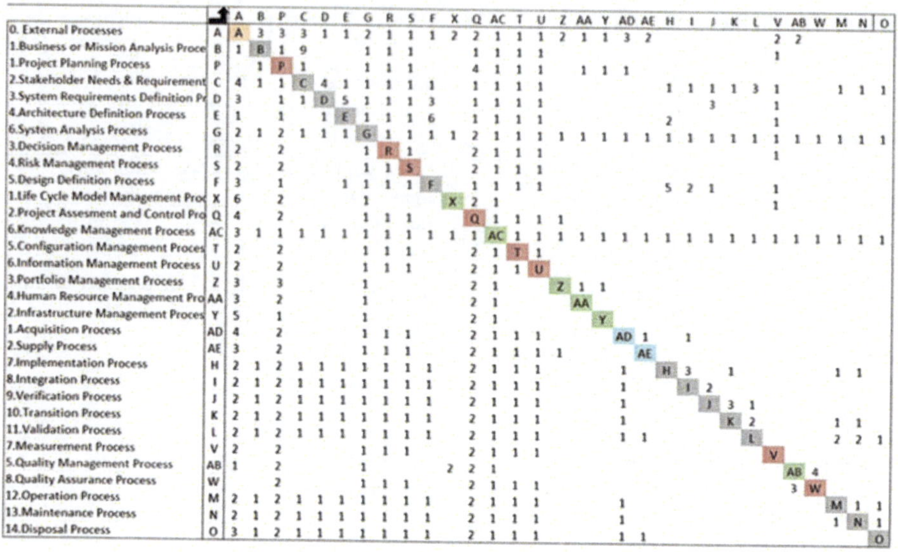

Fig. 2 Sequenced process dependency structure matrix processes. (Madni and Purohit 2019)

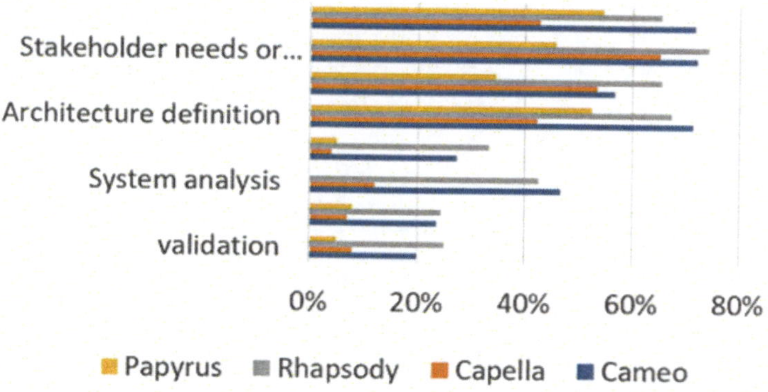

Fig. 3 Percentage coverage of system life cycle processes by the four MBSE tools that we selected. (Madni and Purohit 2019)

to make higher investment in conceptual design and preliminary design stages. In these upfront engineering phases, considerable investment goes into the preliminary design stage. Figure 3 indicates percentage coverage of MBSE tools for system life cycle processes. In addition, with MBSE, substantial cost savings can be expected during the latter stages of design. Since the MBSE approach is likely to reduce the number of defects in latter stages of design, it follows that the cost in these later stages will be significantly lower in comparison with the traditional SE approach. The blue curve in Fig. 4 presents an approximate cost curve associated with the MBSE spend profile based on the life cycle analysis and interaction with

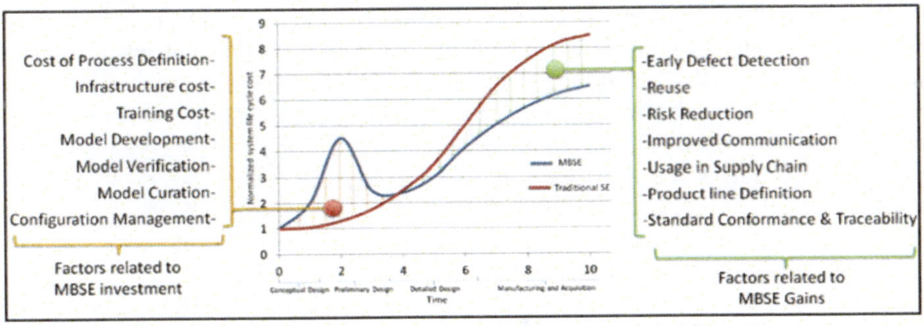

Fig. 4 MBSE factors related to investments and gains. (Madni and Purohit 2019)

industry practitioners. As seen in this figure, MBSE cost is greater in conceptual and preliminary design phases. In fact, the majority of the cost associated with MBSE implementation can be expected to be in the preliminary design phase.

2.7 *Cost Curves for SE Programs with MBSE Approach*

The comparison of investments in MBSE and in traditional (i.e., document-centric) SE is depicted in Fig. 4. At the conceptual and preliminary design stages, MBSE projects require greater investment than traditional SE projects. However, in the latter stages of design MBSE projects require substantially lower investment than traditional SE projects. For economic justification, the net present worth (NPW) of investment should be strictly less than the NPW of gains resulting from the MBSE approach in the system life cycle.

As noted earlier, with MBSE, organizations invariably spend more money upfront and early in the life cycle. Due to the time value of money, additional expenditures early in the system life cycle should be considerably less than the gains from the latter stages in the system life cycle. Discounting and converting gains into present worth provide an accurate measure of return on investment (ROI). Clearly, gains should be significantly more than investments to make MBSE implementation successful. From approximate calculations on sample projects, the NPW of traditional SE projects and MBSE projects is found to be comparable.

However, there exists a tipping point in MBSE implementation, where the NPW of investment exceeds the gains from MBSE. This point could happen because of excess investment early in the life cycle or lower than expected gains later in the life cycle. The ROI also depends on the timing of investments and gains. The investment NPW for traditional SE and MBSE is calculated from the following equations.

Investment NPW approximation for traditional SE P $= \sum_{k=0}^{n} F_k(1+\mathrm{i})^{-k} \approx$ 36.6597C. Investment NPW approximation for MBSE P $= \sum_{k=0}^{n} F_k(1+\mathrm{i})^{-k} \approx$ 34.7564C, where C is assumed to be a scaling constant, n is the number of time interval for investments considered to be 10 from the cost curves, i is overall interest rate for time interval considered to be 3.5%, and F_k is investment made after k time intervals depicted in the cost curves as normalized system life cycle cost. Because approximate NPW investment for traditional SE and MBSE is comparable in our study, the results are inconclusive about which is the preferred option. This recognition suggests the need for a deeper analysis of factors related to MBSE investment and gains.

2.8 *Economic Analysis of MBSE Implementation*

As discussed in the previous section, SE initiatives that employ an MBSE approach require greater upfront investment than is needed with traditional SE. However, such initiatives can be expected to produce greater gains in the latter stages of the system life cycle (Architecture and Systems Engineering: Models and Methods to Manage Complex Systems 2017). Therefore, an analysis of factors related to early investment in MBSE and factors related to gains in the latter stages from MBSE can be expected to provide useful insights for economic justification of MBSE implementation.

2.9 *Factors Related to MBSE Investment*

MBSE investment covers costs associated with MBSE process definition, infrastructure, training, and model-related expenses. Each cost is briefly described next.

Process definition cost is a key consideration in calculating the cost of adopting an MBSE methodology for organization-wide implementation. MBSE process definition depends on the MBSE methodology selected. There are several MBSE methodologies, such as INCOSE Object-Oriented Systems Engineering Method (OOSEM), Object-Process Methodology (OPM), and JPL State Analysis (SA) (Estefan 2007). Implementing a particular methodology and generating models using that methodology have different costs associated with them. Infrastructure cost is another cost category. Infrastructure cost consists of licenses, equipment, environments, processing, and collaboration. Infrastructure cost for this study is assumed to be specific to projects. In general, infrastructure cost is incurred at the enterprise level, but that can be shared among projects to quantify costs specific to each project. Training cost is a distinct category. It involves training on tools, training on modeling languages, and training employees in systems engineering and MBSE. Training cost involves the cost for learning curves and organization-level resistance to change. Model-related costs are a major source of costs. Model development efforts include identifying the goal, purpose, and scope of the model, improving model capabilities, defining the intent of use, and configuring the model for the intended number of users. Building federated, centralized, or hybrid models have different costs (Architecture and Systems Engineering: Models and Methods to Manage Complex Systems 2017). The scale of the model is also a cost modifier. Unifying the model format has its own cost (Madni and Sievers 2018). Maintaining model consistency is a source of costs. Building models are not enough; model verification needs to be performed at each stage of model development to ensure model correctness and credibility (Architecture and Systems Engineering: Models and Methods to Manage Complex Systems 2017). Identifying criteria of model verification and determining the right level of model abstraction have their own costs (Carroll and Malins 2016). Identifying opportunities of model trading and executing it in the right context also require effort. In the case of MBSE, an increase in the number of stakeholders working on central digital models makes the models more prone to errors and requires model curation efforts. Model curation consists of gauging model characteristics, selecting models for implementation, and devising model policies (Architecture and Systems Engineering: Models and Methods to Manage Complex Systems 2017). Configuration management efforts include maintaining ownership, managing data rights and distribution of rights, reuse, or copyrights.

2.10 Factors Associated with MBSE Gains

The factors that relate to MBSE gains include early defect detection, reuse, product line definition, risk reduction, improved communication, usage in supply chain, and standards conformance (Architecture and Systems Engineering: Models and Methods to Manage Complex Systems 2017; Estefan 2007; Madni and Sievers 2018; Carroll and Malins 2016). It is well known that the later in the system life cycle that defects are detected, the greater the cost of correcting the defects. The ability of MBSE to identify defects early in the system life cycle can contribute to significant cost savings and thereby provide significant gains (Architecture and Systems Engineering: Models and Methods to Manage Complex Systems 2017). Another key factor is the ability to use legacy models and data during the latter stages of system life cycle to eliminate extraneous activities and reduce rework. The ability of MBSE to exploit legacy models and data also contributes to economic justification. Since individual projects may not have the luxury of time to work on data reuse, additional measures are needed at the enterprise management level to incentivize projects to achieve this reuse. In particular, commercial aircraft and automobiles which have product lines with variants can benefit from model and data reuse. MBSE can be expected to provide significant gain to define product line characteristics because of single, centralized configurable system models (Estefan 2007). The adoption of MBSE can be expected to increase confidence while reducing risks related to both processes and products. Also, with increasing system complexity, it is becoming increasingly difficult to account for emergence. Therefore, risk reduction in complex systems is a significant gain (Carroll and

Malins 2016). With centralized digital repositories and data analysis capability, MBSE promises to increase communication efficiency and reduce feedback loops. These characteristics can potentially provide important gains (e.g., speed up the process) during the system life cycle. MBSE also provides opportunities to share data and knowledge in timely fashion, which translates into increased supply chain efficiency (Madni and Sievers 2018). Additionally, MBSE provides cost savings through quick traceability mechanisms that are built into today's modeling environments. However, the ever-increasing complexity of systems in multiple domains is making it increasingly difficult to conform to standards.

From the analysis of factors related to both investments and returns/gains, it became apparent that MBSE needs to derive value from a variety of systems engineering activities across the system life cycle to produce convincing economic justification. In other words, using model-based approach solely for requirements engineering is not enough. The use of integrated digital models for data analysis, trade studies, decision support, and communication is necessary throughout the detailed design, implementation, production, maintenance, retrofit, operation, and retirement stages of the system life cycle.

2.11 Assessing MBSE Implementation Benefit

The factors associated with MBSE investment and expected gains depend largely on the system's intrinsic characteristics such as complexity, operational environment characteristics, and system lifespan. The system's complexity can be considered to be a function of the number of unique components in the system, the interactions between them, the amount of knowledge required for developing the system, and the amount of information needed to describe the system (De Weck et al. 2011; Kolmogorov 1983). A system's environment can be characterized in terms of the number of stakeholders and the number of external entities the system needs to interact with which are outside the system's boundary and the regulatory environment. In other words, the system's environment consists of stakeholders, external systems which impose regulatory and interface constraint, and applicable standards that the system needs to conform to. This factor includes system interaction with humans or entities outside the system boundary. A system's lifespan can be characterized by the system's useful life. The system life cycle spans the time required for concept formulation, development, production, operation/utilization, and retirement stages. Before transitioning to operation, a system must conform to the various applicable standards.

As discussed earlier, we parameterized MBSE implementation problem using systems complexity, environment complexity, and system lifespan as distinguishing characteristics. We then identified those industry sectors in which maximum gains are likely to be achieved with the adoption of MBSE. Figure 5 presents MBSE implementation problem in terms of system complexity, environment complexity,

	System Complexity	Environment Complexity	Lifespan
Transportation and Mobility	High	High	High
Aerospace and Defense	High	High	High
Industrial Equipment	High	Medium	High
Energy, Processes & Utilities	High	High	High
Architecture, Engineering & C	Meduim	Medium	High
Life Sciences	High	Medium	High
High-Tech	Medium	Low	Low
Marine & Offshore	High	High	High
Financial & Business Services	Medium	Low	High
Consumer Goods & Retail	Low	Low	Low
Natural Resources	High	High	High
Consumer Packaged Goods &	Low	Low	Low

	System Complexity	Environment Complexity	Lifespan
Transportation and Mobility	9.5	9.8	9.8
Aerospace and Defense	9.8	9.9	9.8
Industrial Equipment	8.2	3.5	9
Energy, Processes & Utilities	9.5	9.9	9.8
Architecture, Engineering & C	5.5	3.8	9.8
Life Sciences	7	4.5	8.5
High-Tech	6.5	2.8	2
Marine & Offshore	7.5	7.5	9.8
Financial & Business Services	6.5	1.5	7
Consumer Goods & Retail	1.5	1	1
Natural Resources	9.5	9.5	9.8
Consumer Packaged Goods &	2	1	1

Fig. 5 Industry sectors on approximate scale of system complexity, environment complexity, and lifespan. (Madni and Purohit 2019)

and system lifespan for 12 industry sectors. The constructed scale of system complexity, environmental complexity, and lifespan are divided into three segments: high, medium, and low. For example, aerospace systems such as airliners have more than ~100,000 parts (high), which is more complex than high-tech industry sector systems like copying machines which have ~2000 parts (medium). The systems used by household consumers on a daily basis have up to ~300 parts (low) (Miller 1956). Here, complexity of systems is determined on the basis of the number of unique components, interactions between components, amount of knowledge required to develop the system, and amount of information needed to describe the system. High, medium, and low scales of environmental complexity are evaluated based on the number of stakeholders, external systems which impose regulatory and interface constraint, and applicable standards that the system needs to conform to related to a particular industry sector. Systems with a relatively short useful lifespan of 0–1 year are attributed to the low score in parameterization compared to systems with a lifespan of more than 30 years.

3 Conclusion

This paper has presented a methodological framework to justify the business case for MBSE implementation. MBSE is compared to traditional SE approaches in terms of upfront investment and expected gains. Compared to traditional SE, MBSE is shown to require greater upfront investment with gains showing up in the latter stages of the system life cycle. The different sectors are characterized in terms of implementation problem characteristics (i.e., system complexity, environment complexity, and system lifespan). These factors are the key discriminators for evaluating MBSE implementation in the different sectors. Specifically, these factors are used to evaluate the different industry sectors in terms of MBSE costs and MBSE returns (Fig. 6).

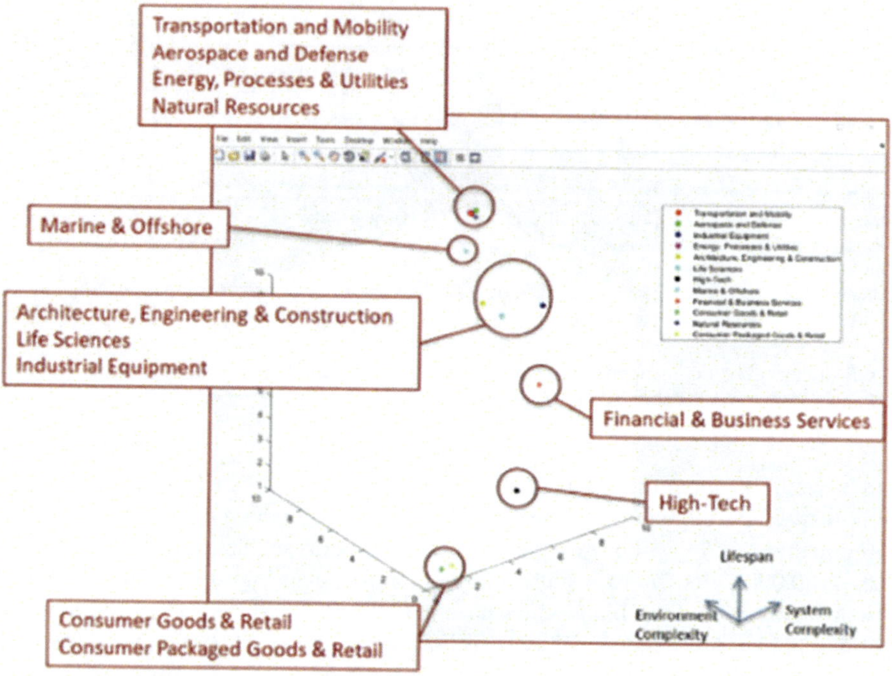

Fig. 6 3D visual representation of MBSE implementation problem in industry sectors. (Madni and Purohit 2019)

By parameterizing the MBSE implementation problem in terms of system complexity, regulatory and operational environment complexity, and system lifespan, we were able to identify those industries that stand to derive the most value from the adoption of the MBSE approach. As shown in Fig. 6, MBSE can be expected to greatly benefit the transportation and mobility, aerospace and defense, energy, processes and utilities, and natural resources industries. With increasing complexity of systems in the marine and offshore sector, architecture, engineering and construction, life sciences, and industrial equipment sectors, MBSE can potentially play an important role throughout the system life cycle. Industry sectors such as financial and business services and high-tech are not likely to benefit from MBSE as much as the other industries. From our study, systems related to home, lifestyle and fashion, consumer packaged goods, and retail are likely to benefit less than the other sectors.

In sum, the MBSE value proposition for a particular industry and the affordable transition from the current practice to a model-based approach are the two key challenges that need to be addressed in quantifiable terms before committing to an MBSE implementation. By identifying when and where in the system life cycle processes MBSE is likely to have a major impact and then quantifying the factors that contribute to MBSE investments and gains, organizations can make informed decisions. By parameterizing individual SE projects in this way, we can identify the leverage points where maximum gains are likely to be achieved with the adoption of MBSE. Future research can employ this framework with additional factors to provide more comprehensive economic justification for MBSE in the different industries. Finally, MBSE is itself evolving with a great emphasis on experiential storytelling where economic analysis can be used to engage broader stakeholder communities (Madni et al. n.d.).

References

Architecture and Systems Engineering: Models and Methods to Manage Complex Systems. 2017. *Course 3 Model-Based Systems Engineering: Documentation and Analysis*. Massachusetts Institute of Technology Online Professional Education Program. Available online: https://sysengonline.mit.edu/course-3/. Accessed on 7 Aug 2017.

Carroll, E., R. Malins. 2016. *Systematic Literature Review: How Is Model-Based Systems Engineering Justified?* Sandia National Laboratories. Available online: https://www.incose.org/docs/default-source/enchantment/161109-carrolled-howismodel-basedsystemsengineeringjustified-researchreport.pdf?sfvrsn=2&sfvrsn=2. Accessed on 4 Dec 2018.

De Weck, O., D. Roos, C. Magee, and C. Vest. 2011. *Engineering Systems Meeting Human Needs in a Complex Technological World*. Cambridge, MA: MIT Press.

Eclipse Capella Documentation. Available online: https://www.polarsys.org/capella/. Accessed on 4 Oct 2018.

Eclipse Papyrus Documentation. Available online: https://www.eclipse.org/papyrus/documentation.html. Accessed 4 Oct 2018).

Estefan, J.A. 2007. Survey of model-based systems engineering (MBSE) methodologies. *INCOSE MBSE Focus Group* 25: 1–12.

Forsberg, K., R.D. Hamelin, G.J. Roedler, T.M. Shortell, D.D. Walden, et al. 2015. *Systems Engineering Handbook: A Guide for System Life Cycle Processes and Activities*. Hoboken: John Wiley & Sons Inc.

IBM Rational Rhapsody Documentation Library. Available online: https://www-01.ibm.com/support/docview.wss?uid=swg27041286. Accessed on 4 Oct 2018.

IEEE Guide—Adoption of ISO/IEC TR 24748-2:2011 Systems and Software Engineering—Life Cycle Management—Part 2: Guide to the Application of ISO/IEC 15288 (System Life Cycle Processes) (24748-2-2012). USA: IEEE, 2012. Web.

Kolmogorov, A.N. 1983. Combinatorial Foundation of Information Theory and the Calculus of Probability. *Russ. Math. Surv.* 38: 29–40.

Madni, A.M., and S. Purohit. 2019. Economic Analysis of Model-Based Systems Engineering. *Systems* 7 (1): 12.

Madni, A.M., and M. Sievers. 2018. Model Based Systems Engineering: Motivation, Current Status, and Research Opportunities. *Syst. Eng.* 21: 172–190.

Madni, A.M., M. Nance, M. Richey, W. Hubbard, and L. Hanneman. Towards an Experiential Design Language: Augmenting MBSE with Technical Storytelling in Virtual Worlds. *Procedia Computer Science* 28: 848–856.

Miller, G.A. 1956. The Magical Number Seven, Plus or Minus Two: Some Limits on Our Capacity for Processing Information. *Psychol. Rev.* 63: 81–97.

No Magic Documentation. Available online: https://www.nomagic.com/support/documentation. Accessed 12 Dec 2018.

Technology & Enterprise Architecture: Economical Modeling: Minimizing Effort, Maximizing Modeling Return on Investment (ROI), Speaker: Michael Vinarcik—Booz Allen Hamilton, May 24th, 2016 Keynote from the No Magic World Symposium. 2016. Available online https://vimeopro.com/nomagic/no-magic-world-symposium-2016-presentations/video/170588353. Accessed on 4 Dec 2018.

COSYSMO 3.0's Improvements in Estimating and Methodology

James P. Alstad

Abstract The COSYSMO 3.0 systems engineering cost estimating model was introduced at CSER 2019 (Alstad 2019). That introduction summarized the sources for the model, notably COSYSMO 1.0 (Valerdi 2005), and explained its updated and new features; however, the actual impact of the new features was not analyzed. Specifically, these impacts were not discussed: the differential impact on larger programs of having Process Capability as a scale factor, rather than a cost driver; the quantitative results of the model's solution to the impact-of-a-step-is-too-large problem; the possible numerical impact on size estimates due to COSYSMO 3.0's greater emphasis on (exponential) scale factors versus (multiplicative) cost drivers; the comparative impact of the four size drivers; and an overall comparison of model features in COSYSMO 3.0 versus previous models.

Keywords Systems engineering economics · Cost modeling · Systems modeling · Cost estimation

1 Introduction

1.1 COSYSMO 3.0's Improvements

The COSYSMO 3.0 systems engineering cost estimating model was introduced at CSER 2019 (Alstad 2019). That introduction summarized the sources for the model, notably COSYSMO 1.0 (Valerdi 2005), and explained its updated and new features; however, the actual impact of the new features was not analyzed. Specifically, these impacts were not discussed: the differential impact on larger programs of having Process Capability as a scale factor, rather than a cost driver; the quantitative results of the model's solution to the impact-of-a-step-is-too-large problem; the possible numerical impact on size estimates due to COSYSMO 3.0's greater emphasis on

J. P. Alstad (✉)
Center for Systems and Software Engineering, University of Southern California, Los Angeles, CA, USA
e-mail: jalstad@usc.edu

A. M. Madni et al. (eds.), *Recent Trends and Advances in Model Based Systems Engineering*, https://doi.org/10.1007/978-3-030-82083-1_29

(exponential) scale factors versus (multiplicative) cost drivers; the comparative impact of the four size drivers; and an overall comparison of model features in COSYSMO 3.0 versus previous models.

In addition, some innovative methods in model development were used in creating COSYSMO 3.0; these also have not been published in conjunction with COSYSMO. These are the use of covariance and ad hoc prior information in Bayesian computations of model parameters; a brief discussion of some difficulties in model fitting; and the impact of symmetrical cost driver values.

1.2 Review of the Key Elements of COSYSMO 3.0

COSYSMO 3.0 is a parametric cost estimating model. That means that the model requires a user to rate aspects of an anticipated systems engineering project; for example, an aspect such as CONOPS and Requirements Understanding may be rated on a Very Low/Low/Nominal/High/Very High (Likert) scale, or it may be an anticipated quantity of some element of the system, such as the number of system requirements. Given these ratings, the model provides an estimated effort (in person-hours) for the project's systems engineering labor.

Figure 1 gives COSYSMO 3.0's top-level estimating equation. The aspects that need to be rated are:

- The adjusted size, which includes anticipated numbers of each of the following, with each rated both on an Easy/Nominal/Difficult scale provided by the model and on a scale for degree of reuse from a previous project:
  - System requirements
  - System interfaces
  - Algorithms
  - Operational scenarios

The adjusted size is expressed as an equivalent number of Nominal system requirements (eReqs).

- The exponent E is computed from ratings for the project aspects Risk/Opportunity Resolution, Process Capability, and Requirements Volatility; these exponential aspects are called "scale factors." The value of E is the sum of a base value plus model-defined numerical values for the scale factor ratings.

$$PH = A \cdot (AdjSize)^{E} \cdot \prod_{j=1}^{15} EM_j$$

Fig. 1 The top-level estimating equation of COSYMO 3.0 In this equation, PH is the estimated person-hours, A is a calibration parameter, AdjSize is the adjusted size of the project, E is the exponent (a sum of terms), and each EM is an effort multiplier

- Each of the *EM*s (Effort Multipliers) is computed from the Likert rating of a Cost Driver aspect; for each Cost Driver, the model provides a definition and a specific rating scale. Example Cost Drivers are Tool Support, Technology Risk, and Level of Service Requirements. Together the 13 Cost Drivers cover these aspects of the project: Understanding, Complexity, Operational aspects, Personnel, and (project) Environment.

  COSYSMO 3.0 is based on these earlier models and model elements:

- COSYSMO 1.0 (Valerdi, 2005), which established the basic form of COSYSMO estimating models
- Model elements addressing reuse, notably work by Wang (Wang et al. 2008; Fortune 2009; Wang 2016)
- A model element addressing requirements volatility (Pena 2012)
- A submodel covering estimating in a system-of-systems context (Lane 2009) (this part of COSYSMO 3.0 is not further discussed in this paper)

2 Impact of New Features of COSYSMO 3.0

2.1 Impact of Process Capability as a Scale Factor

Process Capability was a cost driver in COSYSMO 1.0, but it was changed to be a scale factor in COSYSMO 3.0. The motivation was to increase the impact of Process Capability ratings on larger projects, as that effect was believed to be more realistic. The result is shown in Fig. 2.

2.2 Ensuring the Impact of a Step Is Not Too Large

An influential paper in the COSYSMO community was (Wang 2008), wherein the authors argued that taking a single step – changing the rating of one cost driver from, say, Nominal to High – caused too large an impact (23% reduction in estimated effort, in their example). The goal of COSYSMO 3.0 was to generally reduce the impacts of such single steps in cost driver ratings. One way of assessing overall step size impact is to consider a model's Total EMR measure. The EMR (effort multiplier ratio) of a single cost driver is the ratio of its numerically maximum effort multiplier value to its numerically minimum effort multiplier value; this is a good measure of the impact of that cost driver. The Total EMR of a model is given by multiplying together the EMRs of all cost drivers of the model; it is therefore a measure of the total impact of a model. The Total EMR of the COSYSMO 1.0 model is 44,095.

The COSYSMO 3.0 model reduced its Total EMR in two ways. One was to reduce the number of cost drivers from 14 to 13 (the Documentation cost driver

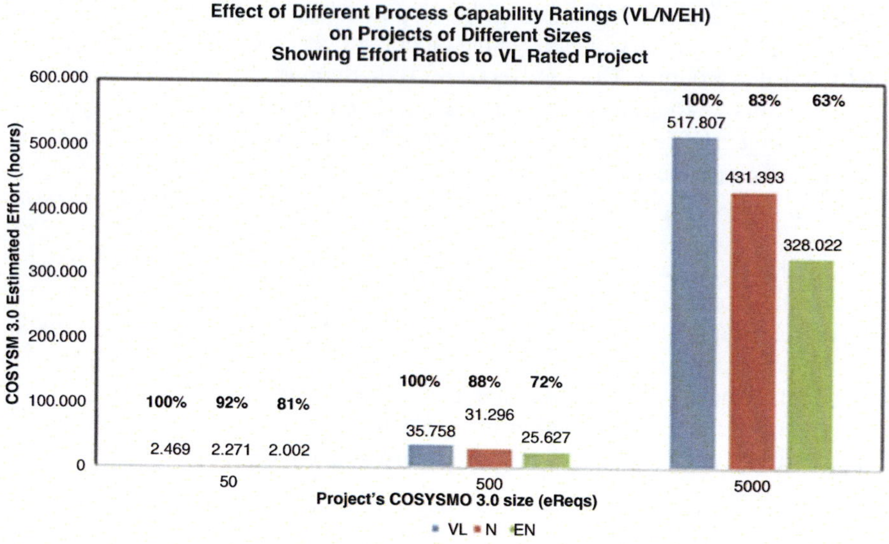

Fig. 2 The increasing impact of Process Capability rating
As the project size increases from 50 to 500 to 5,000 eReqs, the effort ratio due to better process capability (comparing Very Low to Extra High ratings) goes from 81% to 72% to 63%. (All other parameters are Nominal in this example)

was eliminated; the Process Capability cost driver was changed to be a scale factor; however, the Developed for Reuse cost driver was introduced). The second way was to generally reduce the EMRs of all cost drivers via the technique described in Section III.A. The result is that the Total EMR of COSYSMO 3.0 is 23,001. (Compare the Total EMR row of Table 1.)

2.3 *Impact of Shifting Emphasis from Cost Drivers Toward Scale Factors*

The Total EMR reduction just discussed tends to reduce the impact of Cost Drivers in a COSYSMO 3.0 estimate. However, there are two other changes in emphasis that also affect estimate sizes.

First, the size of A, the calibration parameter (see Fig. 1), has been reduced from 38.55 to 26.33. Under some circumstances, this will mean that the COSYSMO 3.0 estimate is smaller than that for COSYSMO 1.0. For example, for a project of size 1500 eReqs with all Cost Drivers Nominal, the COSYSMO 1.0 estimated effort is 89,677 hours. With all scale factors having numerically minimum values, the COSYSMO 3.0 estimate is 50,355 hours; even with all scale factors having Nominal ratings, the COSYSMO 3.0 estimate is only 6% above COSYSMO 1.0, at

95,313 hours. (However, with all scale factors at numerically maximum values, the COSYSMO 3.0 estimate is 152,731 hours.)

Second, under COSYSMO 3.0, scale factor ratings can now have a strong effect on estimated effort (in COSYSMO 1.0, the exponent is constant at 1.06, so there is no effect of scale factors). (Scale factors and the resulting exponent E are explained in Section I.B.) On the 1500 eReq example just given, changing the scale factor ratings can have an effect of a factor of 3.03. As project size increases, the effect of scale factor ratings increases. Consider now a project of 25,000 eReqs. If all scale factor ratings are Nominal, the value of the COSYSMO 3.0 exponent is 1.12, which has the effect of multiplying the estimate by a factor of 3.39 (the corresponding factor for COSYSMO 1.0 is 1.84). Furthermore, scale factor ratings have a range in the exponent of 0.151; for the 25,000 eReq project, this corresponds to a factor of 4.63. (Another aspect of scale factor impact was covered in Section II.A.)

2.4 Comparative Impact of the Four Size Drivers

Table 1 This shows the relative impacts of the four elements of size (eReqs) for a particular data set

Calculated Sizes in eReqs, by Size Driver					
Statistic	SysReq	SysIntf	Alg	OpScen	Total
Mean	299.14	45.01	56.33	57.51	457.99
Std. Dev.	476.51	109.12	185.65	75.57	655.56
Median	136.02	16.38	12.59	31.26	221.73
Mean %	65%	10%	12%	13%	100%

This data set consists of 68 projects. For this data set, almost 2/3 of the size comes from system requirements. Other data sets will have different percentages

2.5 Overall Comparison of COSYSMO 3.0 Features Versus Previous Models

Table 2 compares some model features across models in the COSYSMO family.

3 New Techniques Used in Developing the Model

3.1 Use of Covariance and Ad Hoc Prior Information in Bayesian Computation

COSYSMO 3.0, like most validated cost estimating models, needs to be calibrated against a data set of actual project data. Such a data set will include several actual, completed projects; data for each project will consist of the project's actual effort

Table 2 Comparison of features across the COSYSMO family

Parameter	COSYSMO 1.0	Previous proposals (together)	Expert-based C3	Final (v4) COSYSMO 3.0
# Cost drivers	14	-	14	13
CD DOCU?	Yes	Yes	No	No
CD PROC?	Yes	Yes	No	No
CD interoper-ability?	No	Yes	Yes	No
CD DFR?	No	No	Yes	Yes
# Scale factors	0	2	3	3
SF RVOL?	No	Yes	Yes	Yes
SF PROC?	No	No	Yes	Yes
SF ROPM?	No	Yes	Yes	Yes
Total EMR	44,095	-	145,745	23,001
A (productivity factor)	38.55	-	38.55	26.33
Nominal exponent	1.060	1.060	1.116	1.120
Multiplier, on 25,000 eReq project	1.84	1.84	3.24	3.39
Max-min exponent	0.000	0.038	0.153	0.151
Multiplier, on 25,000 eReq project	1.00	1.47	4.73	4.63
Size drivers	Yes	Same	Same	Same
Reuse size factors	No	Yes	Yes	Yes

This table summarizes the key features of the COSYSMO models that were considered and specified during the development of COSYSMO 3.0. The Previous Proposals column encompasses the papers and the intermediate complete models of COSYSMO. The Expert-Based model was developed entirely from my Delphi sessions gathering expert opinion on parameter values; the Final model is COSYSMO 3.0. One use of this table is as a (summary) historical record of COSYSMO model development

Here are the meanings of the abbreviations: *CD* cost driver; *SF* scale factor; *EMR* effort multiplier ratio (see section II.B); *eReq* equivalent Nominal system requirements (see section I.B); *DOCU* document; *PROC* Process Capability; *DFR* development for reuse; *RVOL* requirements volatility; *ROPM* risk/opportunity management

and its actual ratings on the variables (parameters) of the model. Standard statistical techniques, such as linear regression, are used on the data set to yield the values of coefficients in the model. Some model calibrations use a more sophisticated "Bayesian" analysis to allow expert opinion to influence the coefficient values.

This subsection describes how calibration proceeded in COSYSMO 3.0, emphasizing the use of newer techniques. The statistics used were based on Chapter 14 of (Gelman 2013).

Previously, some Bayesian computations in models (COCOMO II, in Section 4.5.2.3 of (Boehm 2000), and COSYSMO 1.0 (Valerdi 2005)) used the "single-variable" approach to computing posterior means and variances from data and expert opinion. Each variable was considered independently of the other variables; a mean and variance of the variable were determined from the data, a mean and variance were determined from the expert opinion, and they were combined to produce a mean (and variance) for the posterior value of that variable's coefficient. Then the resulting means were taken to be the posterior values of the variables' coefficients.

That approach does not account for the correlations between variables that may be present in both the data and the expert opinions. The rest of this subsection lays some groundwork and then explains how to account for such correlations in the resulting variable values.

Our approach uses matrix arithmetic to compute its results; we used the free matrix manipulation package R (Unknown 2003) to do our computations. If M is a matrix, then M^T denotes its transpose, and M^{-1} denotes its inverse. We use the terminology in Fig. 3 (some of which will be explained later).

The usual computation for determining the calibrated coefficients is given in the left side of Fig. 4; this gives the same result as a linear regression.

Suppose one had prior information about one of the variables, say β_j, specifically its mean and covariance. A key insight (Gelman 2013) is that this information can be treated as an additional data point – an additional element of y and an additional row of X. The added element of y would be the prior mean of β_j; the added row of X would include the prior covariance of β_j, plus "ratings" having a 1 under β_j and zeroes elsewhere. The intuition is that this new row is an additional "vote" on

n = The number of projects in the calibration dataset

k = The number of variables in the model

X = The matrix of project ratings (n × k)

y = The vector of actual efforts (n × 1)

$\hat{\beta}$ = The estimate of coefficients (k × 1)

V_β = The variance-covariance matrix of coefficients (k × k)

I_m = The m × m identity matrix (1 on diagonol, otherwise 0)

Σ_y = The variance-covariance matrix of the ratings, when available (n × n); can use I_n when unavailable

Fig. 3 Terminology for matrix approach to calibration

When Σ_y is not available:	When Σ_y is available:
$\hat{\beta}=\left(X^T X\right)^{-1} X^T y$	$\hat{\beta}=\left(X^T \Sigma_y^{-1} X\right)^{-1} X^T \Sigma_y^{-1} y$
$V_\beta=\left(X^T X\right)^{-1}$	$V_\beta=\left(X^T \Sigma_y^{-1} X\right)^{-1}$

Fig. 4 Formulas for calculating the calibrated coefficients and the corresponding covariance matrix The formulas on the left are used when covariance input is not available; it assumes that all coefficients have the same variance and no covariance. The formulas on the right are used when covariance input (Σ_y) is available

When prior information (means β_0, covariance Σ_β) is available:

Use these enhanced matrices:

$$y_* = \begin{pmatrix} y \\ \beta_0 \end{pmatrix}$$

$$X_* = \begin{pmatrix} X \\ I_k \end{pmatrix}$$

$$\Sigma_* = \begin{pmatrix} \Sigma_y & 0 \\ 0 & \Sigma_\beta \end{pmatrix}$$

Then apply the "when Σ_y is available" formulas.

Fig. 5 How to add additional rows to the input matrices to include prior information in a calibration computation

the value and covariance of β_j; the ordinary rows can also be regarded as votes – complex ones – on the means and covariance of the results.

For COSYSMO 3.0, the expert opinion data from Delphi sessions was provided in the form of ballots, with each expert providing one ballot with her opinions about all the coefficients. Some experts tended to give higher values for all coefficients; others tended to give lower values; therefore, the coefficient votes covary. So the COSYSMO 3.0 calibration data included a covariance matrix for the (prior) Delphi data.

The upshot is that the project data matrices can be enhanced per Fig. 5, and those enhanced matrices can be fed directly into the formulas on the right side of Fig. 4 to yield the posterior coefficients in a single step. (Compare the additional step for the Bayesian computation in previous model calibrations.)

As mentioned in section I.B, it was desired to reduce the impact of a step size. This was easily accomplished by adding prior rows (via the technique just discussed) that tended to lower the impact of step sizes.

3.2 Comparison of Least-Squares Versus Absolute Deviation Model Fitting

In the course of fitting the COSYSMO 3.0 model coefficients, I discovered that there were some opposing tendencies at work. One was between data-only fits and Bayesian fits; with a data-only fit, the data tended to be close to the fitted model, but some of the coefficients had non-credible[1] values; with a Bayesian fit, the data was somewhat dispersed from the fitted model, but all the coefficients were credible. Another opposing tendency was between least-squares fits (e.g., via linear regression or the F statistic) and absolute-deviation fits (e.g., via MMRE or PRED). With a least-squares fit, data far from the middle have more influence on the fit, and the logarithmic form of the estimating equation had to be used to match the assumption of a Gaussian distribution; with an absolute-deviation fit, each datum has an equal influence on the fit, and the estimating equation can be used directly.

I was able to resolve these tendencies by searching for appropriate coefficient values using a hill-climbing procedure (specifically, by using R's *nlm* built-in function). This resulted in a fully credible fit that was reasonably close to the data (specifically, PRED (.30) was greater than 50%).

3.3 Use of Symmetrical Cost Driver Ratings

COSYSMO 1.0's cost driver ratings were set, in part, by expert input. Those experts felt that it was more accurate to allow a cost driver to have different step sizes below and above Nominal (though all a cost driver's step sizes below Nominal were the same and likewise above Nominal; in fact, such below/above step sizes didn't differ very much). The negative feature of this approach was that each cost driver really has two coefficients that need to be fit: the step size below Nominal and the step size above Nominal. This doubled the number of degrees of freedom for cost drivers in doing the fit, thereby reducing the significance of the fit, or otherwise introduced noise into the fitting process.

[1]"Non-credible" here means that the coefficient had a value that was obviously wrong. As an example of non-credibility, consider the cost driver CONOPS and Requirements Understanding. As the rating here goes up, the estimated cost should obviously go down. However, with a data-only fit, an increased rating caused the estimated cost to go up.

To avoid such difficulties, I made each cost driver in COSYSMO 3.0 symmetrical: its step size was the same above and below Nominal.

References

Alstad, J.P. 2019. Development of COSYSMO 3.0: An Extended, Unified Cost Estimating Model for Systems Engineering. *Procedia Computer Science 153*: 55–62.

Boehm, B.W. 2000. *Software Cost Estimation with COCOMO II*. Upper Saddle River: Prentice-Hall, Inc.

Fortune, J. 2009. *Estimating Systems Engineering Reuse with the Constructive Systems Engineering Cost Model (COSYSMO 2.0)*. Los Angeles: USC.

Gelman, A.e. 2013. *Bayesian Data Analysis*. 3rd ed. Boca Raton: Chapman & Hall/CRC Texts in Statistical Science.

Lane, J.A. 2009. Cost Model Extensions to Support Systems Engineering Cost Estimation for Complex Systems and Systems of Systems. In *7th Annual Conference on Systems Engineering Research*. CSER.

Pena, M. 2012. *Quantifying the Impact of Requirements Volatility on Systems Engineering Effort*. Los Angeles: USC.

Unknown. 2003, November 23. *R (Programming Language)*. (wikipedia, Editor) Retrieved November 1, 2019, from wikipedia: https://en.wikipedia.org/wiki/R_(programming_language)

Valerdi, R. 2005. *The Constructive Systems Engineering Cost Model*. Los Angeles: University of Southern California.

Wang, G.V. 2008. Proposed Modification to COSYSMO Estimating Relationship. *INCOSE International Symposium 18* (1): 249–262.

Wang, G. 2016. The Generalized Reuse Framework—Strategies and the Decision Process for Planned Reuse. *INCOSE 2016*. INCOSE.

Wang, G., A. Ankrum, R. Valerdi, and C. Millar. 2008. COSYSMO Reuse Extension. In *18th INCOSE Symposium*. Utrecht: INCOSE.

Assurance Case Property Checking with MMINT-A and OCL

Nick L. S. Fung, Sahar Kokaly, Alessio Di Sandro, and Marsha Chechik

Abstract Assurance cases are a means to argue about the safety, security, etc. of software systems in critical domains. In previous work, we presented a tool called MMINT-A to automate change impact assessment of assurance cases given system design changes. In this paper, we argue that applying model-driven techniques to assurance case development allows safety engineers and assessors to ask questions about these artifacts and answer them using automated tool support – something not achievable with traditional document-based approaches. To support this argument, we present a library of well-formedness constraints on assurance cases structured in the Goal Structuring Notation (GSN). The constraints are formalized using OCL and implemented in MMINT-A. We also discuss other types of constraint checks that are useful in the automotive domain given the ISO 26262 standard and internal company processes.

Keywords Assurance cases · Model-based engineering · Property checking · Tool support

1 Introduction

Assurance cases are structured, evidence-based arguments that a system satisfies a particular property for a given application in a given context (SCSC Assurance Case Working Group 2018). For example, the use of assurance cases is generally accepted as best practice in various industries for demonstrating the safety of safety-critical systems, such as in healthcare (Health Foundation (UK) 2012). However, managing assurance cases is also generally nontrivial due to their potential complexity; an exemplar assurance case for the safety of infusion pumps contains around

N. L. S. Fung (✉) · S. Kokaly · A. Di Sandro · M. Chechik
Software Engineering Group, Department of Computer Science, University of Toronto, Toronto, ON, Canada
e-mail: nlsfung@cs.toronto.edu

A. M. Madni et al. (eds.), *Recent Trends and Advances in Model Based Systems Engineering*, https://doi.org/10.1007/978-3-030-82083-1_30

100 claims (Larson 2014). Furthermore, an assurance case may require frequent maintenance to reflect any updates to the system design.

To manage this complexity, we advocate the use of model-based tools and techniques to create and maintain assurance cases along with the corresponding system designs. Di Sandro et al. (2015) presented the generic tool MMINT to support the management of collections of related models, and Fung et al. (2018) presented a set of extensions to MMINT (collectively called MMINT-A) to automate the change impact assessment of assurance cases. In this paper, we focus on formalizing the properties of well-formed assurance cases, such as the absence of circular arguments, allowing them to be automatically enforced in MMINT-A and other such tools. More specifically, we present a library of well-formedness constraints that were derived from the relevant assurance case and safety standards (i.e., GSN and ISO 26262). They are formalized using OCL, which facilitates portability to any assurance case tool based on UML (or MOF). The constraints have been incorporated into MMINT-A and validated using a case study in the automotive domain.

The rest of this paper is structured as follows. We present relevant background in Sect. 2 and the well-formedness constraints in Sect. 3, and we conclude in Sect. 4.

2 Background

2.1 GSN: Goal Structuring Notation

A commonly used graphical standard for modelling assurance cases is the Goal Structuring Notation (GSN) (SCSC Assurance Case Working Group 2018), which is specified in natural language English and is the main source for our well-formedness constraints. In general, arguments in GSN are organized into a tree of six types of elements (Fig. 1): goals, strategy, solution, context, justification, and assumption. The root is the overall goal to be satisfied by the system, and it is gradually decomposed into other goals or solutions, possibly via some strategies. Goals and strategies may also be associated with some contextual information (including justifications and assumptions), which, along with solutions, form the leaves of the tree.

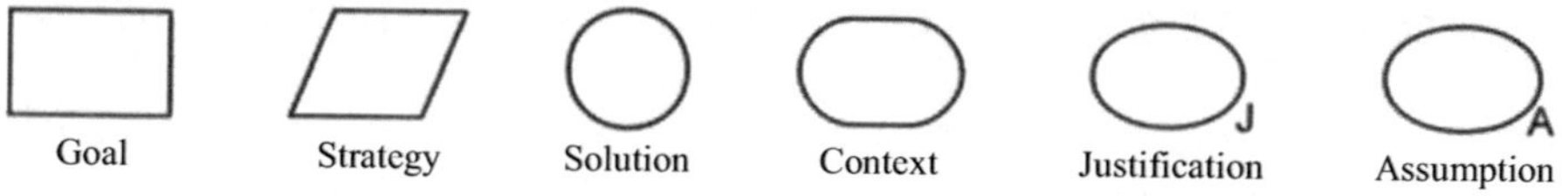

Fig. 1 The six core elements in GSN. (SCSC Assurance Case Working Group 2018)

2.2 ISO 26262 Standard

We apply our research to the automotive domain, which presents additional requirements on assurance cases. In particular, according to the ISO 26262 standard (ISO 2011), safety hazards and, therefore, safety goals must be assigned an *automotive safety integrity level* (ASIL) to indicate its risk and required degree of assurance. ASIL levels are QM (indicating the least stringent measures), A, B, C, or D (most stringent).

2.3 MMINT: Model Management INTeractive

MMINT (https://github.com/adisandro/MMINT) is built on top of the Eclipse platform and designed for managing collections of related models (Di Sandro et al. 2015). In particular, MMINT allows the user to work at both the "instance" level, in which models and their relations are instantiated and manipulated, as well as the "type" level, in which the necessary metamodels, relation types, and model operators are defined.

For example, to manage assurance cases, Fung et al. (2018) presented a collection of extensions at the "type" level to form MMINT-A. Specifically, MMINT-A includes a metamodel and an editor for creating and visualizing assurance cases (Fig. 2) as well as some slicers and a workflow to perform change impact assessment on them. Furthermore, Di Sandro et al. (2019) showed how the Eclipse VIATRA framework can be incorporated into MMINT to perform queries on automotive system models and assurance cases.

2.4 Assurance Case Metamodel

OCL constraints must be expressed in the context of a specific metamodel; for this paper, we adopt the same assurance case metamodel as the one used in MMINT-A, which is built using the Eclipse variant of OMG's meta-object facility (MOF) standard and is based on the GSN standard. Thus, as shown at the top of Fig. 3, an assurance case is modelled to contain goals, strategies, solutions, contexts, justifications, and assumptions, all of which are connected to each other by either "supported-by" or "in-context-of" relations. However, it is extended for the automotive domain by including ASILs (bottom left of Fig. 3) and by distinguishing between two types of goals (basic and independence goals) as well as two types of strategies (basic and ASIL decomposition strategies). To support impact assessment, assurance case elements can also be associated with an impact annotation (bottom of Fig. 3) indicating whether they can be reused or must be revised or rechecked for validity (Fung et al. 2018).

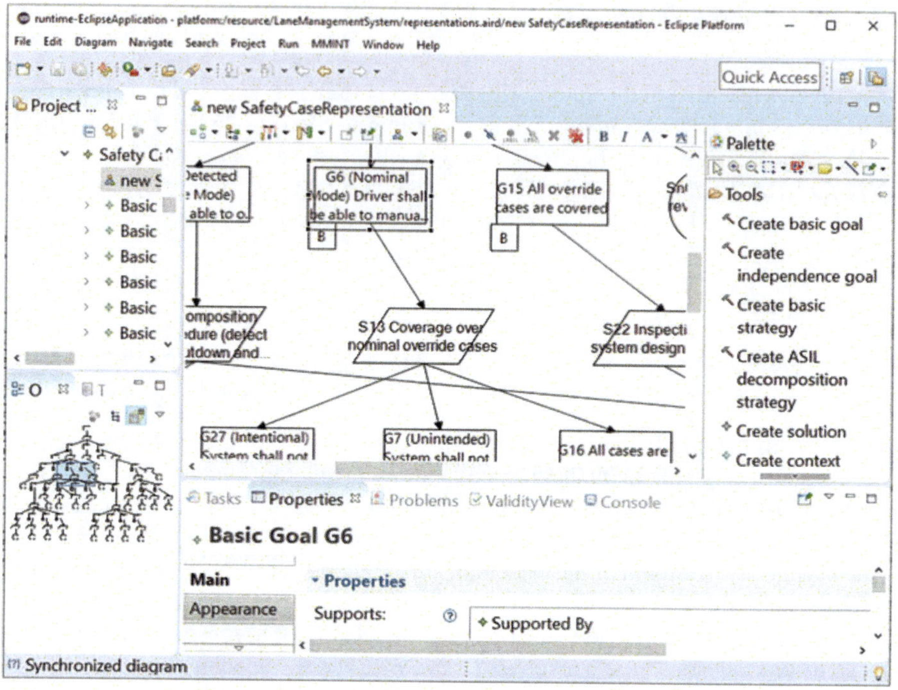

Fig. 2 Screenshot of MMINT-A showing a portion of an assurance case

2.5 OCL: Object Constraint Language

UML (including MOF) is generally insufficient to express all relevant aspects of a model (Object Management Group 2014). Therefore, OCL was developed as a formal language for specifying expressions on models. These expressions do not have side effects and are generally used for querying models or specifying constraints over them, a typical example of which looks as follows:

```
context Goal inv:
 self.supportedBy -> forAll(s | s.premise.oclIsKindOf(Goal) or
 s.premise.oclIsKindOf(Strategy)or s.premise.oclIsKindOf
(Solution));
```

`context Goal inv:` specifies that the accompanying OCL expression is an invariant that applies to all instances of `Goal` in the metamodel; thus, `self` in the subsequent line refers to the `Goal` model element being checked. `self.supportedBy` traverses the `supportedBy` relation in goals, which results in a set of `SupportedBy` model elements (see Fig. 3). The collection operator `->` and the `forAll` iterator cause an iteration to be performed on the `SupportedBy` model elements, returning true if they all satisfy the condition

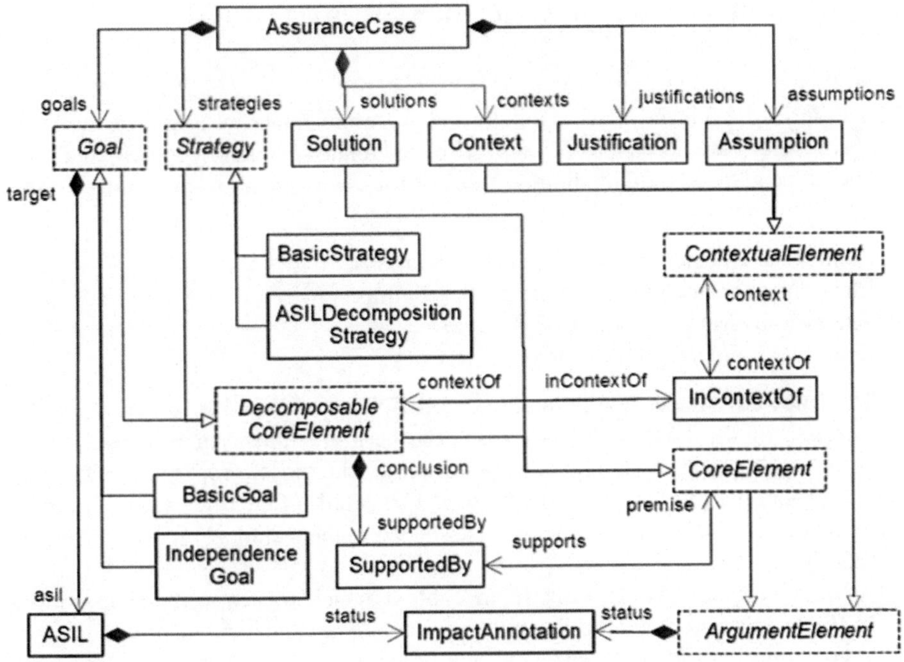

Fig. 3 Simplified class diagram showing the main parts of the adopted assurance case metamodel. Rectangles with italic text and dashed borders represent abstract classes; the rest are concrete

specified inside the `forAll` iterator. In this case, the condition is that the premise of the `SupportedBy` element conforms to type `Goal`, `Strategy`, or `Solution`.

2.6 LMS: Lane Management System

As part of our evaluation process, we used MMINT-A to verify the well-formedness of an assurance case for an automotive lane management systems (LMS). The LMS (Blazy et al. 2014) is a safety-critical system for preventing a vehicle from straying from its lane. It comprises multiple sub-systems such as the Lane Departure Warning System (LDWS), which warns the driver of any possible unintentional lane changes, as well as the Lane Keeping System (LKS), which can take control of the vehicle to keep it within its lane. To assure its safety, the LMS assurance case consists of 26 goals, 28 strategies, 20 solutions, and 2 contexts, the full details of which can be found in Fig. 10.4 of Kokaly 2019. In this paper, only the relevant excerpts will be shown to illustrate various OCL constraints.

3 OCL Constraints for Assurance Cases

Overall, we identified and formalized 16 constraints. As summarized in Table 1, 12 are derived from the GSN standard and therefore relate to assurance cases in general, while 4 are derived from ISO 26262 and relate specifically to ASILs. These constraints can also be divided into groups of four, with each group relating to one of the following:

- Elements connected by "supported-by" relations
- Elements connected by "in-context-of" relations
- The overall structure of an assurance case
- Inheritance and decomposition of ASILs

The rules have been incorporated into MMINT-A and used to verify the well-formedness of the LMS assurance case. Furthermore, intentional changes were made to create incorrect assurance cases in order to expose errors and ensure that the corresponding rules are correctly flagged. For details of the rules, including their implementation and test cases, the user is referred to the technical report by Fung et al. (2019).

As an example, consider Rule 15 in Table 1 which relates to the propagation of ASILs from a parent goal to its children goals. In general, evidence supporting a child goal must be sufficiently strong to support its parent goal; thus, the child should inherit the same ASIL value as its parent. Figure 4a shows an excerpt of the LMS assurance case that satisfies this rule, with goal G6 (ASIL B) being decomposed into goals G27 (B), G7 (B), and G16 (B), while Fig. 4b shows an example that violates it, with goal G1 (ASIL B) being decomposed into goals G2 (B) and G3 (A). The red cross in Fig. 4b indicates that the rule is violated by G3.

This requirement can be formalized into the OCL constraint shown in Fig. 5, which can be thought of as comprising the following steps:

- Retrieve all parent goals directly supported by the selected goal.
- Retrieve all parent goals supported by the selected goal via a basic strategy.
- Check that the ASIL of each parent is inherited properly:
  - If the parent has no ASIL, then it is trivially true.
  - If the child has no ASIL (but the parent does), then it is trivially false.
  - Otherwise, the child ASIL must be higher than or equal to the ASIL of every parent.

Note that Rule 15 only concerns goals that support other goals either directly or indirectly via a basic strategy. When a goal is supported by an ASIL decomposition strategy, it means that redundancy is added into the system, which, in accordance with ISO 26262, allows the redundant elements (and therefore children goals) to be supported by weaker evidence.

However, such decomposition is only valid if the safety of the redundant elements is supported by completely independent goals and solutions (Rule 16). For example,

Table 1 Overview of 16 well-formedness constraints on assurance cases.

Category	Rule no.	Description	Source
"Supported-by" relations	1	Goals can only be supported by goals, strategies, and solutions	GSN
	2	Strategies can only be supported by goals and solutions	
	3	Solutions cannot be supported by any assurance case element	
	4	Contextual elements cannot be supported by any assurance case element	
"In-context-of" relations	5	Goals can be in the context of contexts, assumptions, and/or justifications	
	6	Strategies can be in the context of contexts, assumptions, and/or justifications	
	7	Solutions cannot be in the context of any assurance case element	
	8	Contextual elements cannot be in the context of any assurance case element	
Overall structure	9	There cannot be any "supported-by" cycles	
	10	There can only be one root in an assurance case	
	11	The root of an assurance case must be a basic goal	
	12	The leaves of an assurance case must be solutions, not goals nor strategies	
ASIL decomposition	13	An ASIL decomposition strategy must be supported by one independence goal	ISO 26262
	14	An ASIL decomposition strategy must be supported by two basic goals	
	15	Any child goal that supports a parent goal directly or via a basic strategy must have the same ASIL or stronger as the parent goal (if any)	
	16	The two basic goals supporting an ASIL decomposition strategy cannot share any common descendant goal or solution	

Fig. 6a shows a goal (G5) relating to the LMS alarm system being decomposed with an ASIL decomposition strategy into three goals (G19, G20, and G18) relating to two independent alarms: a visual and an audible alarm. Although not shown, these goals are supported by completely nonoverlapping descendants; thus, G19 is permitted to have a less stringent ASIL (viz., QM) than its parent (A). On the other

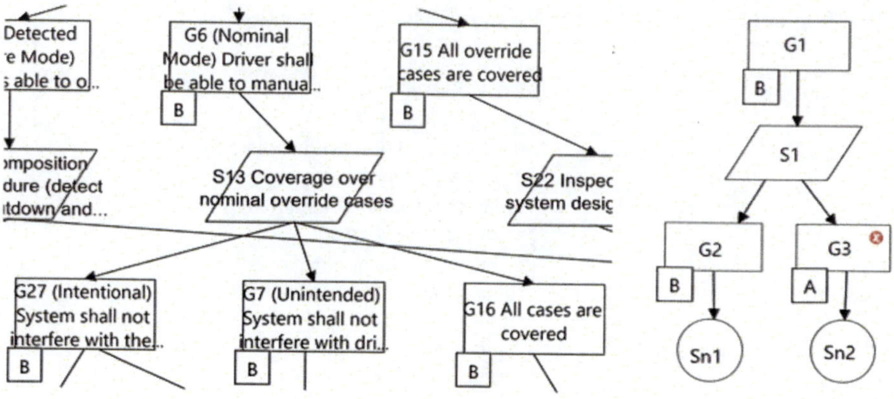

(a) (b)

Fig. 4 An assurance case that (**a**) passes and (**b**) fails Rule 15. The pass case is extracted from the LMS assurance case

```
context Goal inv:
let directParents : Set(Goal) = self.supports.conclusion ->
      select(d |d.oclIsKindOf(Goal)).oclAsType(Goal) -> asSet(),
    indirectParents : Set(Goal) = self.supports.conclusion ->
      select(d|d.oclIsTypeOf(BasicStrategy)).supports.conclusion ->
      select(d |d.oclIsKindOf(Goal)).oclAsType(Goal) -> asSet()
in indirectParents -> union(directParents) -> forAll(g |
  if g.asil = null then true
  else if self.asil = null then false
  else g.asil.value = ASILLevel::QM or
     (g.asil.value.toString() <= self.asil.value.toString() and
             self.asil.value <> ASILLevel::QM) endif endif);
```

Fig. 5 OCL constraint for Rule 15

hand, Fig. 6b shows two children goals (G2 and G3) sharing the same solution (i.e., evidence); thus, the ASIL decomposition rule is violated.

As shown in Fig. 7, such requirement can also be formalized in a similar manner as for Rule 15. In this case, OCL's closure operator is used to extract all descendants of the relevant goal, not just its immediate children. After extracting the descendants of both children goals, OCL's intersection operator is used to ensure that they are completely independent.

4 Conclusions and Future Work

In this paper, we described constraint checks on assurance cases with the purpose of checking well-formedness. We checked all of these constraints on our LMS assurance case. While we did not uncover any problems because the assurance

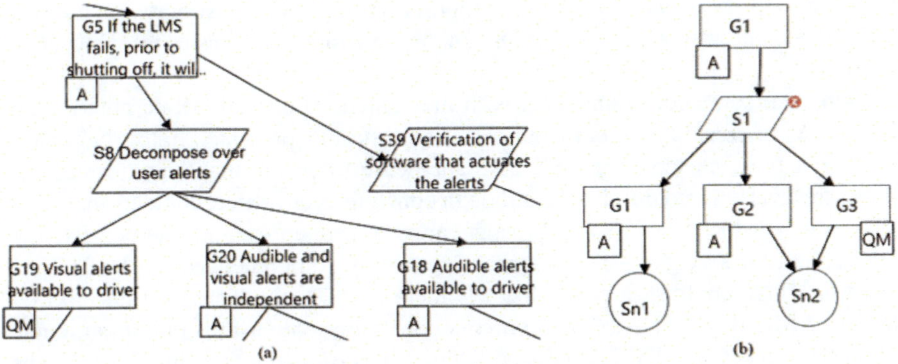

Fig. 6 An assurance case that (**a**) passes and (**b**) fails Rule 16. The pass case is extracted from the LMS assurance case and shows a user alert being decomposed into two independent alerts, visual and audible

```
context ASILDecompositionStrategy inv:
let goalSeq: Sequence(CoreElement) = self.supportedBy.premise
      -> select(p | p.oclIsTypeOf(BasicGoal)),
    g1Descendants : Set(CoreElement) = goalSeq -> at(1) ->
      closure(c | if c.oclIsKindOf(DecomposableCoreElement) then
           c.oclAsType(DecomposableCoreElement).supportedBy.premise
        else null endif),
    g2Descendants : Set(CoreElement) = goalSeq -> at(2) ->
      closure(c | if c.oclIsKindOf(DecomposableCoreElement) then
           c.oclAsType(DecomposableCoreElement).supportedBy.premise
        else null endif)
in g1Descendants -> intersection(g2Descendants) = Set{};
```

Fig. 7 OCL constraint for Rule 16

case has been extensively validated, we believe that the library of constraints we present provides a starting point to construct more complex and interesting properties that are domain-specific, standard-specific, or company-specific. As the library grows, it will enable safety engineers and reviewers of assurance cases to automatically answer complex questions regarding the quality of the assurance cases (e.g., completeness, correctness, compliance to standards, etc.) they are working with.

As part of our ongoing research, we intend to define a constraint check taxonomy to help practitioners work with constraints in a systematic manner. For example, all the constraint checks presented in this paper are *intra-model* checks as they are checked on a single model (the assurance case). Examples of *inter-model* constraints are those which rely on the existence of traceability mappings between models. For example, a check such as "Is there a fault tree analysis (FTA) conducted for all hazards that are ASIL B and above?" is a multi-level inter-model constraint because each "hazard" element in the Hazard Analysis document is mapped to an entire FTA document. Our MMINT tool already supports writing such constraints (Di Sandro et al. 2019), but we are currently investigating how to make it easier for the user to do so.

Moreover, some constraints deal with the *content* of artifacts while others with the *process* to create them. For example, the constraints presented in this paper are *well-formedness* checks of a GSN assurance case to ensure that it is meaningful. These are clearly *syntactic* checks, although constraints on artifact contents may also be *semantic*, which can check a broader range of properties, e.g., the consistency between artifacts (e.g., "Are all system functions comprehended in the Hazard Analysis?"). In contrast, a check like "Is there an FTA conducted for all hazards that are ASIL B and above?" is a *process* constraint that ensures that a particular artifact (in this case, FTA) is produced. Other kinds of distinctions that are relevant to a taxonomy include constraints to check *correctness* vs. *completeness*, constraints that are *necessary* vs. *sufficient* conditions, *existential* vs. *universal* constraints, etc. The criteria for including a category in the final taxonomy can be based on how it is relevant to supporting the safety process.

Acknowledgments The work reported in this paper has been funded by General Motors (GM) and NSERC Canada. The authors thank their collaborators at GM and the McMaster Centre for Software Certification for many useful discussions.

References

Blazy, B., A. DeLine, B. Frey, and M. Miller. 2014. *Software Requirements Specification (SRS): Lane Management System*. Michigan State University.

Di Sandro, A., R. Salay, M. Famelis, S. Kokaly, M. Chechik. 2015. *MMINT: A Graphical Tool for Interactive Model Management*. ACM/IEEE 18th International Conference on Model Driven Engineering Languages and Systems (MoDELS), demo.

Di Sandro, A., S. Kokaly, R. Salay, M. Chechik. 2019. *Querying Automotive System Models and Safety Artifacts with MMINT and Viatra*. 2nd Workshop on Modeling in Automotive Software Engineering (MASE).

Evidence: Using safety cases in industry and healthcare, Health Foundation (UK), 2012.

Fung, N.L.S., S. Kokaly, A. Di Sandro, R. Salay, M. Chechik. 2018. *MMINT-A: A Tool for Automated Change Impact Assessment on Assurance Cases*. 6th International Workshop on Assurance Cases for Software-intensive Systems (ASSURE).

Fung, N.L.S., S. Kokaly, A. Di Sandro, and M. Chechik. 2019. *Technical Report: Syntactic Checks for Safety Cases*. University of Toronto. Available at: https://www.cantab.net/users/nick.ls.fung/mminta/OclConstraints.pdf. Last accessed: Sep. 20, 2019.

ISO 26262: Road Vehicles – Functional Safety, 1st version, International Organization for Standardization, 2011.

Kokaly, S., 2019. *Managing Assurance Cases in Model Based Software Systems*. PhD dissertation, McMaster University.

Larson, B. 2014. *Open PCA Pump Assurance Case*. Kansas State University.

Object Constraint Language, Version 2.4, Object Management Group, 2014.

SCSC Assurance Case Working Group. 2018. *Goal Structuring Notation Community Standard Version 2*.

Interpretation Discrepancies of SysML State Machine: An Initial Investigation

Ben Cratsley, Siwani Regmi, Paul Wach, and Alejandro Salado

Abstract Model-Based Systems Engineering (MBSE) is expected to improve communication and consistency in system development over document-based approaches. As systems become more complex, modeling languages increase the information content of their semantics to simplify modeling construction and visualization. We hypothesize in this paper that such increase in the complexity of the semantics may be detrimental to the MBSE objectives of facilitating communication and consistency. We present the results of an initial survey in which we asked systems engineers individually to interpret the behavior captured by several models in the form of Systems Modeling Language (SysML) state machines. Significant discrepancies in their answers were found. In addition, we present a qualitative assessment of the potential implications of such interpretation discrepancies for system development, which include need for rework, modeling gaps, inefficient solutions, and solutions that are not fit for purpose.

Keywords Model interpretation · SysML · Model-based systems engineering · MBSE · Modeling semantics

1 Introduction

Literature indicates that the construction and visualization of system models can be simplified by embedding (or hiding) part of the complexity within the semantics of the modeling language with which the model is created (Wach and Salado 2019). We suggest that such increase of the semantic complexity may lead to discrepancies in how a model is interpreted by different engineers. If this were the case, promised benefits of Model-Based Systems Engineering (MBSE) such as improved communication (Piaszczyk 2011; Andersson et al. 2010) or consistency in

B. Cratsley · S. Regmi · P. Wach · A. Salado (✉)
Virginia Tech, Blacksburg, VA, USA
e-mail: asalado@vt.edu

A. M. Madni et al. (eds.), *Recent Trends and Advances in Model Based Systems Engineering*, https://doi.org/10.1007/978-3-030-82083-1_31

system development (Gregory et al. 2019; Vipavetz et al. n.d.) could be challenged. We present the results of an initial investigation that addresses this hypothesis. First, a theoretical assessment of intricate aspects of the semantics of SysML was performed. Second, several models with potential semantic misinterpretation were created and a survey with systems engineering practitioners conducted. SysML was chosen due to its widespread within academia and the community of practice.

2 Interpretation Discrepancies in SysML: Theoretical Insights

2.1 Semantic Vulnerabilities

Semantically rich modeling techniques have been instrumental to make modeling of complex systems possible. For example, state machines had to evolve from the mathematically rigorous (yet semantically simple) Moore (Bisht and Dhami 2015; Moore 1956) and Mealy (Bisht and Dhami 2015; Mealy 1955) state machines to Harel's statecharts (Harel 1987), which were semantically richer but lacked a mathematical description, in order to effectively handle state explosion. By increasing the semantic expression of the model, Harel statecharts could, relative to Mealy and Moore state machines, reduce the size of the model elements that the modeler needed to define while capturing the same behavior. As with Harel statecharts, UML (and by extension SyML) state machines (Friedenthal et al. 2015) leverage semantic complexity to simplify diagrammatic complexity. In fact, UML and Harel employ the same diagrammatic elements and understanding (actually, a one-to-one mapping of model elements), even though their underlying semantics are different. This means that, while a Harel and a UML state machines may be identical diagrammatically (i.e., visually), their execution behavior may be different (Crane 2005). Failing to recognize (know) such semantic specifications could therefore lead to misinterpreting the behavior captured by the model.

We define *semantic vulnerability* of a model as an unknown knowledge gap that results from the precision with which the model captures aspects of an actual system. For example, given a model of the behavior of a system, do two modelers interpret the same behavior? If this were not the case, then there is a potential for an identified conflict of understanding, leading to unfounded assumptions about expected system behavior. Potential consequences to system development could include omission of necessary functionalities/features, overlapping/inefficient implementation of necessary capabilities, and interoperability problems, among others. When a large-scale system is considered, where several engineers collaboratively build larger models, the risk of semantic vulnerability explodes.

2.2 Example of Potential Semantic Misinterpretations

We illustrate the problem of semantic vulnerability due to interpretation discrepancy with two examples. Consider the SysML state machine depicted in Fig. 1a. This diagram is consistent with the SysML specification and is correct. It captures three states of a vehicle, with transitions between them triggered by inputs and/or events. At the start of operation, the vehicle goes to the Idle state. Consider the following question. Assume the vehicle is in *Braking* state. How does the system behave when *releaseBrake* and *speed* = *0* are activated or occur simultaneously? Clearly, there is no transition defined for such a situation in the diagram. Furthermore, if one would start with one of the conditions randomly, for example, by first considering *releaseBrake*, the system would transition to *Accelerating/Cruising*, but once there, there is no transition defined for the event *speed* = *0*. Similarly, if the vehicle would transition to *Idle* first, there would be no transition defined on that state for the event *releaseBrake*. However, the model has a specific behavior for that case, regardless of whether we can tell from the diagram or not. Such is the effect of semantics. For example, the actual behavior could be:

1. The vehicle does not execute any transition if a specific transition for the combined event is defined. That means, when *releaseBrake* and *speed* = *0* are activated or occur simultaneously, the vehicle will remain in *Braking* state.
2. The state of the vehicle is not controlled for undefined events. That means, we do not know the state the vehicle will be in when *releaseBrake* and *speed* = *0* are activated or occur simultaneously. It could even be a hidden state.
3. Conditions rule over events. Hence, when *releaseBrake* and *speed* = *0* are activated or occur simultaneously, the vehicle will transition to *Idle* state.
4. Events rule over transitions. Hence, when *releaseBrake* and *speed* = *0* are activated or occur simultaneously, the vehicle will transition to *Accelerating/Crusing* state.
5. The model is incomplete/incorrect as all possible conditions for transitions have not been defined.
6. The vehicle cannot experience both the condition and the event at the same time.

A similar situation occurs if we consider the system to be, for example, in the *Braking* state and ask how the system behaves if *accelerate* is activated. Will the system stay in *Braking* state, or will it cause an unknown transition? Consider now the state machine captured in Fig. 1b. Light goes on when turned on and goes off when turned off. The model is correct and consistent with the SysML specification. However, it is unclear, from the diagram, if there is any time associated to the transitions. For example, if this would be a Moore machine, transitions are defined to occur with no lapse of time; they are instantaneous. But, how should this SysML state machine be interpreted? A modeler may generate different interpretations:

1. Transitions are instantaneous.
2. Time is undefined explicitly; hence, transitions are considered instantaneous.
3. Time is undefined explicitly; hence, the model is incomplete/incorrect.
4. Time is undefined explicitly; hence, transition times are uncontrolled.

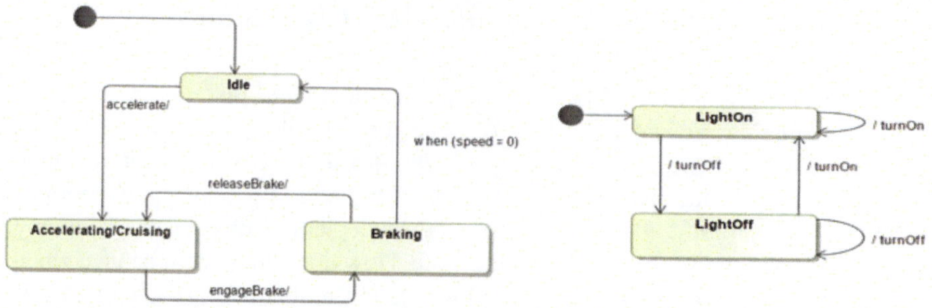

Fig. 1 **(a)** SysML state machine capturing the behavior of a vehicle; **(b)** SysML state machine capturing the behavior of a light switch

These two examples show how certain behavioral aspects are embedded semantically in the models to simplify visualization and construction. However, they play a significant part in describing the behavior of the systems the models capture. Such role is though not easily apparent. A deep understanding of the language specification is necessary to correctly interpret the diagram.

3 Interpretation Discrepancies in SysML: Empirical Evidence of State Machines

3.1 Design

Three instruments were developed and administered sequentially through an online platform: (1) a consent form, (2) a demographic survey, and (3) a modeling survey. The demographic survey was used to gather demographic information of the participant, including experience and competence in systems engineering and system modeling. The modeling survey was used to assess the participants' interpretations of various system models using multiple-choice questions. Given a model, the participant was asked to choose the behavior captured by the model out of a set of predefined options. The specific questions are presented in Table 1, and they refer to the models presented on Table 2. Models were consistent with SysML specification (hence, they were not wrong), but only visual representations of the models were presented, with no other information. In order to account for the case in which a participant felt the model was incomplete and more information was necessary to describe the behavior of the model, all questions had the choices *Other* and *I do not know*.

Table 1 Survey modeling questions

Question
Q1. The state machine depicts the nominal behavior of a car when it is in operation. Consider the system is in the Braking state. How does the system behave when accelerate is activated? (a) The system will not receive accelerate activation while in Braking state. (b) Such situation is out of the scope of SysML. (c) The system automatically defaults to an available transition. (d) The system remains in Braking state. (e) This is a non-nominal behavior and needs to be defined in a separate state machine.
Q2. The state machine depicts the nominal behavior of a car when it is in operation. Consider the system is in the Braking state. How does the system behave when *releaseBrake* and *speed = 0* are activated/occur simultaneously? (a) The system will not receive *releaseBrake* and *speed = 0* simultaneously while in Braking state. (b) Such situation is outside the scope of SysML. (c) The system automatically defaults to one of both transitions. (d) This is a non-nominal behavior and needs to be defined as a separate state machine.
Q3. Assume the system enters Example State C. Choose the statement that best describes the behavior by the state machine diagram. (a) The system will execute *Initialize* behavior. Once the behavior is finalized, the system will execute *gatherData* behavior. Once the behavior is finalized, the system will execute *Finalize* behavior. (b) The system cannot execute *Finalize* behavior because there is no transition exiting the state. (c) Behaviors *Initialize*, *gatherData*, and *Finalize* occur in parallel. (d) Only one of the three behaviors (*Initialize*, *gatherData*, or *Finalize*) will be executed, which will depend on the state from which Example State C is entered.
Q4. During the transition between the light switching on and off, how much time or delay should be assumed? (a) An amount of time determined by each user when using the system. (b) A time that is determined by the system, depending on the system design. (c) The time is undefined. (d) Defining the time that it takes a transition between states is outside of the scope of SysML. (e) Zero time.
Q5 and Q6. Choose the statement that best describes the behavior captured by the state machine diagram. (a) *setTemp* is a condition of the system, measured internally by the system. (b) *setTemp* is an internal transition of the system. (c) *setTemp* is a condition external to the system measured by the system. (d) *setTemp* is an external input to the system that causes an internal transition.

3.2 *Participants*

Complete responses to the survey were obtained from a total of 11 participants who completed the study, although 49 started the survey. Participants were recruited using the LinkedIn platform, specially addressing professional groups in systems engineering and system modeling, and directly through the sponsor. Participation was voluntary and had no explicit benefits. Participants conducted the study at their own location, using an online survey. Only practicing systems engineers and

Table 2 Summary of survey results

Model	Interpretation
Q1 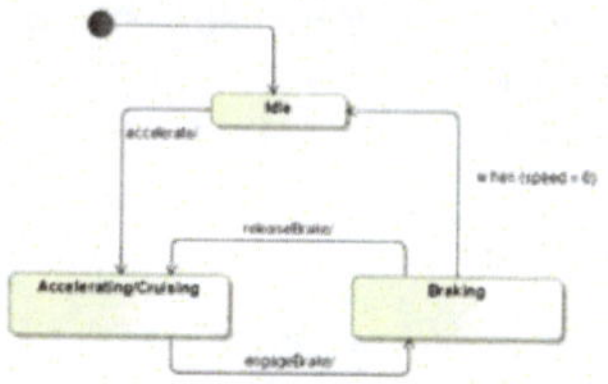	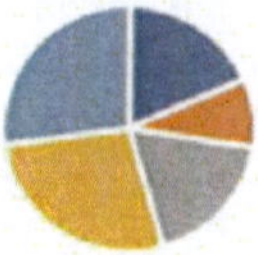 **Dark blue.** The system will not receive releaseBrake and speed = 0 simultaneously while in Braking state **Orange.** The system automatically defaults to one of both transitions **Grey.** This is a non-nominal behavior and needs to be defined in a separate state machine **Yellow.** Other **Light blue.** I do not know
Q2 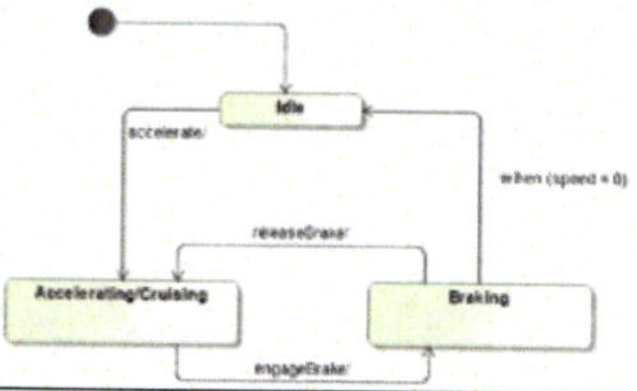	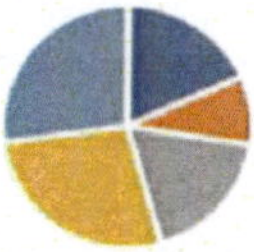 **Dark blue.** The system will not receive releaseBrake and speed = 0 simultaneously while in Braking state **Orange.** The system automatically defaults to one of both transitions **Grey.** This is a non-nominal behavior and needs to be defined in a separate state machine **Yellow.** Other **Light blue.** I do not know
Q3 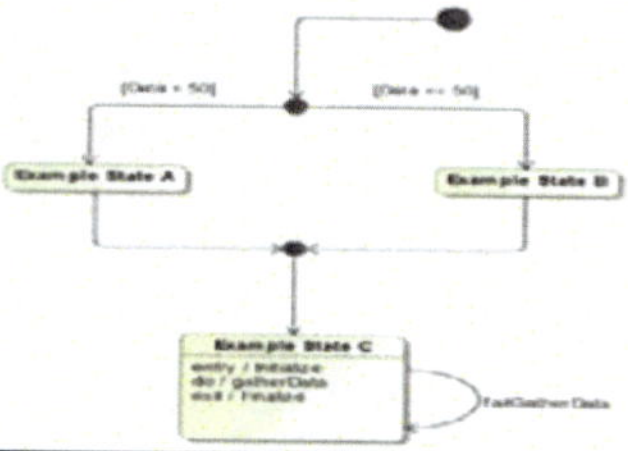	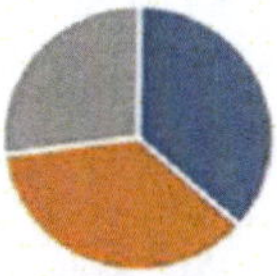 **Dark blue.** The system will execute Initialize behavior. Once the behavior is finalized, the system will execute gatherData behavior. Once the behavior is finalized, the system will execute Finalized behaviour **Orange.** The system cannot execute Finalize behavior because there is no transition exiting the state **Grey.** Other
Q4 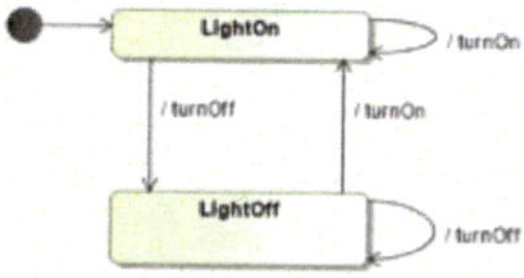	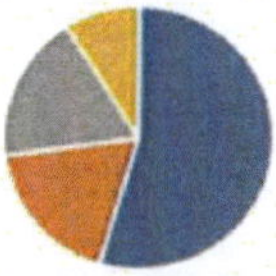 **Dark blue.** An amount of time determined by each user when using the system **Orange.** A time that is determined by the system, depending on the system design **Grey.** The time is undefined. **Yellow.** Defining the time that it takes a transition between states is outside of the scope of SysML
Q5 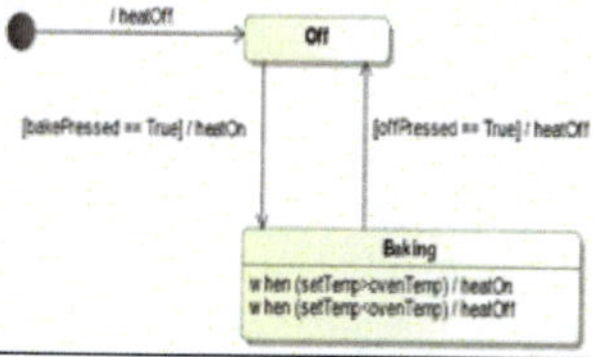	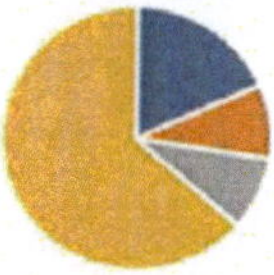 **Dark blue.** setTemp is a condition of the system, measures internally by the system **Orange.** setTemp is an internal transition of the system **Grey.** setTemp is a condition external to the system measured by the system **Yellow.** setTemp is an external input to the system that causes an internal transition
Q6 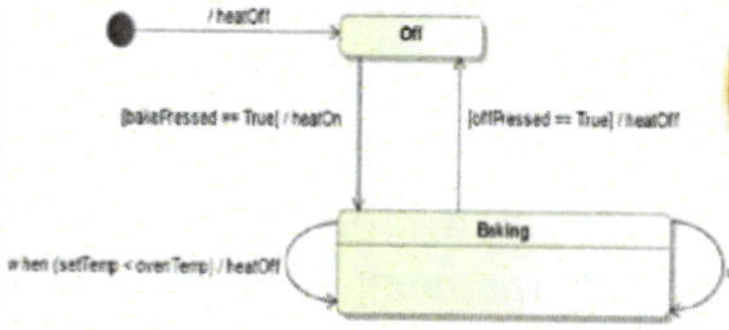	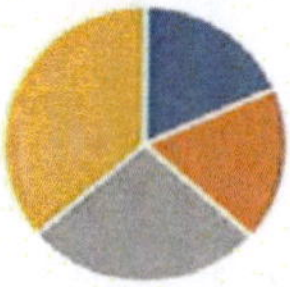 **Dark blue.** setTemp is a condition of the system, measures internally by the system **Orange.** setTemp is an internal transition of the system **Grey.** setTemp is a condition external to the system measured by the system **Yellow.** setTemp is an external input to the system that causes an internal transition

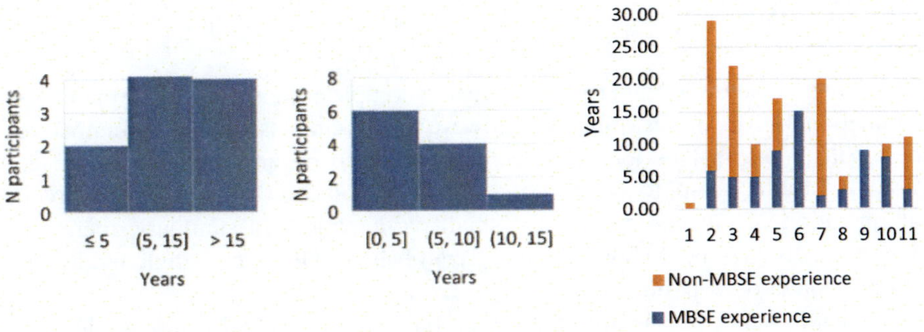

Fig. 2 (**a**) Participants with years of experience in systems engineering; (**b**) participants with years of experience using MBSE; (**c**) balance of systems engineering experience and MBSE experience of each participant

graduate students in systems engineering were invited to participate in the study. A practicing systems engineer was defined as an individual who executes some or all the activities described by the International Council of Systems Engineering (INCOSE) as part of his/her professional work. Determination of compliance of participants to selection criteria was performed by the researchers by evaluating participants' responses to demographic questions. Therefore, we cannot guarantee accuracy of the participants' answers to demographic questions. Goodwill of the participants in responding to the survey has been assumed.

Demographic data is shown in Fig 2. Figure 2a shows the years of experience that participants had in the field of systems engineering. Although participation was low, participants uniformly represented a breadth of experience, spanning from 1–5 years of experience to more than 26 years of experience. Figure 2b shows the years of experience that participants had in using MBSE. The pattern of experience in this case changes, with a higher number of respondents on the low end of the spectrum and just one respondent with over 10 years of experience in applying MBSE. We suggest that this is a good representation of the MBSE population, given that, although initial efforts in MBSE started several years ago, widespread adoption has begun more recently. This can also be visualized by showing the specific MBSE experience of each respondent, with respect to their systems engineering experience (ref. Fig. 2c). It is shown that around half of the participants have significantly engaged in MBSE, while the other half used it less intensively. In terms of professionalism in MBSE, four participants indicated that they had used SysML and/or MBSE as part of their education, and ten of them indicated that they had used MBSE and/or SysML in professional practice. In addition, one participant had an OCSMP Model Builder Advanced certification.

3.3 Results

Participants took an average of 32 minutes to complete the survey, with the fastest completion in 16 minutes and the slowest in 78 minutes. Table 2 summarizes the results (note that options that were not selected by any participants are not shown). Results seem to support our initial hypothesis about the diversity of interpretation of models, despite the small participation. In summary, there was significant disagreement about the interpreted behavior for every single model that was presented to the participants.

Evaluating the specific different interpretations selected by the participants provides additional insights beyond the quantitative comparison. In particular, they show the potential risks to system development that can stem from these various interpretations. These are discussed individually for each model and question below.

Question 1 The model in this question referred to the behavior of a vehicle. About half of the participants assumed that the vehicle would never receive a command to accelerate while it was in *Braking* state, while around the other half of the participants assumed that the vehicle would receive the command but ignore it. This different interpretation is not subtle, but completely changes both the context and the expectations for the system. First, both groups interpreted opposing assumptions about the problem space (receiving vs not receiving an acceleration command when in *Braking* state). Second, one group assumed an embedded behavior, whereas the other group considered it inapplicable. As a result, whereas one group of participants would further a design by implementing a given behavior in the *Braking* state, such a design feature would be purposefully omitted by the other group.

Question 2 The model in this question remains the vehicle. In this case, the diversity of opinions was even higher than for Question 1. Given a condition of the inputs to the system and the state of the system, participants interpreted different behaviors of the system. As in Question 1, such interpretations would lead to vehicles exhibiting different functionality, depending on the engineer that read the model.

Question 3 The model in this question does not capture any specific type of system; it captures the behavior a notional system. Results show that participants uniformly disagreed about the execution by the system of certain actions or behaviors that occur once the system reaches a given state. Consider now two such engineers, one creating the model under the assumption that all behaviors would execute sequentially (as given by the blue answer) and one using it to produce the detail design of the system, interpreting that the system cannot execute one of the actions (as given by the orange answer). In the best case, the second engineer would raise the concern to the first, and both would reach a consistent understanding. In the worst case, the second engineer would design the system without the possibility to execute the last action, which would be deficient with respect to how the system was initially conceptualized in the model by the first engineer. Regardless, both cases lead to rework, which negatively impacts the success of a system's development.

Question 4 The model in this question captures a light switch. While around half of the participants interpreted that the time to transition from one state to the other was determined by each used, other participants declared that such time was determined by the system, undefined, or that it had to be defined outside of the scope of the model. This is a common situation in system development that leads to a modeling gap, where everyone assumes that the variable of interest (time in this case) will be determined by someone else, and hence this reciprocal assumption leads to the variable remaining undefined. These gaps eventually surface as the system development progresses, usually leading to rework as well.

Questions 5 and 6 The model in these questions captures an oven-like system, which is used to bake something. Results show that participants disagreed on whether the condition that generates the state transition (in this case, the variable *setTemp*) was internal of the system or external to the system. Clearly, these two interpretations have considerable implications for system development. If the condition is internal to the system, there is no need for an external interface to receive the temperature or the information about the temperature. However, there will be a need to design the capability to measure or determine the condition internally. On the contrary, if the condition is external to the system, then the system will need to be able to receive an external input (either the condition directly or information about the condition). Falling into this misaligned interpretation can easily lead, as discussed, to systems that are either inconsistent with the system context (i.e., that are unable to interact with external systems as required) or unable to provide the desired capabilities (i.e., that are not capable to identify the internal condition). Regardless, the system solution would be inefficient, since either the system would implement internal functionality that was not required, or it would implement external interfaces that would remain unused.

4 Conclusion

The results of this study seem to support the hypothesis put forth at the beginning of the paper: A lack of knowledge of SysML semantics leads to model interpretation discrepancies. We suggest that such discrepancies in model interpretation could generate problems during system development, including need for rework, modeling gaps, inefficient solutions, and solutions that are not fit for purpose; we recognize though that this needs to be empirically demonstrated in actual applications of MBSE. Hence, while semantic sophistication is useful for simplifying model construction and visualization, semantically rich modeling languages may actually increase the risk of miscommunication and inconsistent system development. This is contrary to current thinking in the field of MBSE, where communication and consistency have been highly cited as key strengths of the approach. However, because the results presented in this paper are based on a small sample, we suggest that a continuation of the study with a higher sample is necessary to confirm the hypothesis.

Acknowledgments This material is based on work sponsored by the Northrop Grumman Undergraduate Research Experience in Industrial and Systems Engineering at Virginia Tech project. Any opinions, findings, and conclusions or recommendations expressed in this material are those of the authors and do not necessarily reflect the views of the Northrop Grumman Corporation.

References

Andersson, H., et al. 2010. Experience from introducing Unified Modeling Language/Systems Modeling Language at Saab Aerosystems. *Systems Engineering* 13 (4): 369–380.

Bisht, R.K.D., and H.S. Dhami. 2015. *Discrete Mathematics*. Oxford: Oxford University Press.

Crane, M.L. 2005. *On the Syntax and Semantics of State Machines*. Kingston: Queen's University.

Friedenthal, S., A. Moore, and R. Steiner, eds. 2015. *A Practical Guide to SysML – The Systems Modeling Language*. 3rd ed. Waltham: Morgan Kaufman.

Gregory, J., et al. 2019. Early Validation of the Data Handling Unit of a Spacecraft Using MBSE. In *2019 IEEE Aerospace Conference*.

Harel, D. 1987. Statecharts: A visual formalism for complex systems. *Science of Computer Programming* 8 (3): 231–274.

Mealy, G.H. 1955. A method for synthesizing sequential circuits. *The Bell System Technical Journal* 34 (5): 1045–1079.

Moore, E.F. 1956. *Gedanken-experiments on Sequential Machines*, Automata Studies, Annals of Mathematical Studies. Vol. 34. Princeton: Princeton University Press.

Piaszczyk, C. 2011. Model Based Systems Engineering with Department of Defense Architectural Framework. *Systems Engineering* 14 (3): 305–326.

Vipavetz, K., D. Murphy, and S. Infeld. Model-Based Systems Engineering Pilot Program at NASA Langley. In *AIAA SPACE 2012 Conference & Exposition*.

Wach, P., and A. Salado. 2019. Can Wymore's Mathematical Framework Underspin SysML? An Initial Investigation of State Machines. In *Conference on Systems Engineering Research (CSER)*. Washington, DC.

Fuzzy Multicriteria Optimization for System Engineer's Design of Myoelectric Prostheses

Kenneth W. Garner and Kamran Iqbal

Abstract Fuzzy logic provides a means to address uncertainty in selecting optimal elements in the systems design process. Using mathematical algorithms coupled with existing system engineering tools has the potential to increase stakeholder satisfaction and reduce lifecycle cost of the system. In this study we use Fuzzy TOPSIS algorithm to demonstrate realization of top-level design choices as well as enhancement of stakeholder satisfaction in the case of designing a prosthetic arm. Although prosthetic industry is taken as an example, this research is valuable in any situation where the exact intent of the stakeholder is uncertain; nevertheless, the systems engineer seeks an optimal product design that would meet stakeholder needs.

Keywords Prosthetic arm · Multicriteria optimization · Fuzzy TOPSIS algorithm · Top-level design

1 Introduction and Background

In this study, the initial phases of the systems engineering design process in the case of a prosthetic arm are presented. These include stakeholder requirement analysis, system requirements, and top-level attribute and design selection. This information is used to generate the systems architecture that sets a baseline for specialization and procurement of system elements. The trade space for a prosthetic arm defines a group of options with various performance ratings and costs. An optimal selection of subsystem elements will greatly reduce the lifecycle costs of the system. In the case of prosthetic arm, this aim is achieved through multicriteria decision-making (MCDM) by using Fuzzy TOPSIS algorithm. This research aims to

---

K. W. Garner · K. Iqbal (✉)
Department of Systems Engineering, University of Arkansas Little Rock, Little Rock, AR, USA
e-mail: kxiqbal@ualr.edu

A. M. Madni et al. (eds.), *Recent Trends and Advances in Model Based Systems Engineering*, https://doi.org/10.1007/978-3-030-82083-1_32

advance the study of systems engineering processes within the domain of advanced mathematical optimization algorithms.

Throughout history, man has needed prosthetics to overcome limb loss. In the United States, prosthetics first appeared after the civil war. With a large number of amputees from World War II, a market for prosthetics began to grow. Focusing on user satisfaction paved the way for plastics, resins, polycarbonates, and laminates that were lightweight and easier to clean. Today, high-performance prosthetic designs are changing the way prosthetics are viewed. Prosthetic arms are of three main types: passive, body-powered, and electrical. Passive arm is mostly cosmetic. Body-powered prosthetic uses a cable system, body harness, and existing muscles to move the elbow and hand functionally. Myoelectric prosthetics are considered the most psychologically natural yet remain the most expensive. A myoelectric upper limb prosthetic arm uses motors with an external power source controlled by the EMG signals from the existing limb via a microprocessor.

A myoelectric prosthetic arm is a system that can be broken down into three main subsystems: mechanical, electrical, and communications. Each subsystem has an extensive set of elements that in turn have attributes. Attributes of the device are highly dependent on user preference, lifestyle, financial standing, and insurability. Choosing the proper configuration of attributes and elements is paramount for a quality prosthetic arm to thrive in its intended environment and increase user satisfaction. The complexity of the device is directly correlated to cost; whether it be the cost of consultation, manufacturing, learning, or maintenance complexity is the problem. The systems engineering process is tailored to manage complexity and achieve value for all stakeholders. Therefore, adroitness optimization techniques in selecting attributes and elements could minimize cost and maintenance while maximizing user performance and satisfaction.

Fuzzy logic, as opposed to Boolean logic, includes imprecise information in the decision-making process. In Fuzzy logic, the truth value of a proposition may be any real number between 0 and 1. Fuzzy logic emerged in the context of fuzzy set theory introduced by Zadeh (1965). A fuzzy set contains elements with varying degrees of membership. Fuzzy relations in MCDM problems are a common model for analyzing alternatives with Pareto-optimal solutions. In the prosthetics industry, stakeholders have conflicting preference structures and receive value from the product in different ways, whether they are the doctor, insurance company, third party, or prosthetics engineers. An optimal system should satisfy all stakeholders. From the amputation type to daily use to maintenance costs of a prosthetic arm, many system characteristics can be optimized to achieve more value.

From the insurance companies' perspective, realizing patient's needs assumes paramount importance especially when it comes to expensive prosthetic devices. Functional needs that are likely to be met include gripping, releasing, holding, and coordinating movement during primary activities of daily living. Prosthetics should meet the requirement of patients' need for control, durability (maintenance), function (speed, work capability), and usability. As technology increases, options for amputees also increase, but not all options are covered under an insurance plan. Many times, insurance companies will not pay for a prescribed arm due to the lack of medical necessity. For example, at BlueCross BlueShield a certain number of conditions must be met for the prosthetic device to be covered; these are:

- *Remaining muscle meets minimum microvolt threshold to allow operation of device.*
- *The patient is free from any neuromuscular disease.*
- *Amputation is at the wrist or above.*
- *Body-powered prosthetics are insufficient in meeting functional needs.*
- *A functional evaluation indicating that the individual's functional needs will be met.*

The main objective of this study is to use systems engineering tools in sequence with fuzzy multicriteria decision optimization algorithms to develop stakeholder needs, systems requirements, systems architectures, element trade space, and final design definition. Classic systems engineering process models do not address uncertainty in stakeholder preference structure which can lead to an inaccurate selection of sub-elements in the design. Our approach is to use fuzzy logic in addition to existing mathematical tools to address stakeholder ambiguity in their needs. A problem from the prosthetic industry will be used to show the benefits of using fuzzy logic within the decision analysis framework.

2 Multicriteria Optimization

Oftentimes a system has many attributes that must be optimized; this is formulated as a multi-attribute objective function. For example, an engineering firm tasked to build a race car would want more speed and more range. An objective function that can handle both attributes is given as:

$$f(x) = \sum_i w_i v_i (a_i)$$

where $v_i(a_i)$ is a value function of the attribute a_i and w_i is a weight. This way the decision-maker (DM) can have his/her value of individual attributes addressed. Yet, there are times that the designer might want to maximize speed and range while minimizing noise and cost. The extra function can be used as a constraint. Thus, a multicriteria optimization problem is defined as:

$$\max\ f(x) = \sum_i w_i v_i \left(a_i, a_j\right) \text{ such that } c_i \left(a_i, a_j\right) \geq r_i$$

where r_i is the satisficing level for c_i (Chen 2000 & Chu and Lin 2003). In multi-objective optimization problems, dominance is used to determine the best solution. Therefore, if x_1 is no worse than x_2 in all objectives and x_1 is at least better than x_2 in one objective, then x_1 is considered as the nondominated solution, i.e., it is Pareto-optimal.

2.1 Fuzzy TOPSIS Algorithm

The Technique for Order of Preference by Similarity to Ideal Solution (TOPSIS) introduced by Hwang and Yoon (1981) is a multicriteria decision analysis technique for establishing order preference in a multi-DM environment. It is based on the concept that the selected alternative should be nearest to a certain positive ideal solution (PIS) and farthest from a negative ideal solution (NIS). The TOPSIS algorithm was extended to include linguistic variables and fuzzy sets by Chen (2000). In fuzzy TOPSIS method (Chu and Lin 2003), linguistics terms are used to represent DMs' appraisal of alternatives. Thus, a fuzzy-PIS represents the best performance descriptors of an alternative, whereas a fuzzy-NIS represents the worst performance descriptors.

A fuzzy number represents a convex and normalized fuzzy set (Lee 2005). Triangular fuzzy numbers are commonly used to represent linguistic preferences. Using fuzzy numbers in the TOPSIS method for criteria analysis makes it natural to apply in group decision-making. A triangular fuzzy number, represented as a triplet: $\overline{a} = (a_1, a_2, a_3)$, defines a membership function $\mu_{\overline{a}}(x)$ that maps a range of values over the universe of discourse to a real number in the interval $[0, 1]$. Further, let $\overline{a} = (a_1, a_2, a_3)$ and $\overline{b} = (b_1, b_2, b_3)$ represent two triangular fuzzy numbers; then, their geometrical distance is defined as:

$$d\left(\overline{a}, \overline{b}\right) = \sqrt{\left[(a_1 - b_1)^2 + (a_2 - b_2)^2 + (a_3 - b_3)^2\right]/3}$$

As an example, the fuzzy numbers corresponding to commonly used linguistic evaluation may be defined as (Table 1):

To illustrate the aggregation of fuzzy numbers, suppose vendors of coffee cups are evaluated for purchase by a two-member committee. The decision-makers (DM_1, DM_2) assess two alternatives (A_1, A_2) on criteria: $C_1 - C_3$ (e.g., size, material, and cost) in linguistic terms. They also assign weights and risk to each evaluation criteria, again in linguistic terms. Suppose fuzzy numbers $\overline{a} = (a_1, a_2, a_3)$ and $\overline{b} = (b_1, b_2, b_3)$ describe the DM's assessment of alternative A_1 on criteria C_1; then, an aggregated assessment of A_1 on C_1 is obtained as: $\overline{c} = (c_1, c_2, c_3)$, where $c_1 = \min(a_1, b_1)$, $c_2 = \text{mean}(a_2, b_2)$, $c_3 = \max(a_3, b_3)$.

In multi-attribute decision-making (MADM), all available alternatives are assessed on each criteria by the DMs and the results compiled in an evaluation matrix (of size $n \times m$, where n is the number of alternatives and m is the number of criteria). In fuzzy MADM, their evaluations are translated into an aggregated

Table 1 Example of fuzzy numbers assigned to criteria evaluation in linguistic terms

Linguistic term	Very poor (VP) Very low (VL)	Poor (P) Low (L)	Fair (F) Medium (M)	Good (G) High (H)	Very Good (VG) Very High (VH)
Membership function	$<1, 1, 3>$	$<1, 3, 5>$	$<3, 5, 7>$	$<5, 7, 9>$	$<7, 9, 9>$

Table 2 Example of fuzzy numbers assigned to criteria weights in linguistic terms

Linguistic terms	LOW	MEDIUM	HIGH
Membership function	<0.01, 0.20, 0.40 >	<0.11, 0.30, 0.60 >	<0.25, 0.50, 0.80 >

fuzzy decision matrix $\left[\overline{D}_{ij}\right]$ (of size $n \times m$) using fuzzy numbers. Next, the entries in the fuzzy decision matrix are normalized for comparison. Normalization using maximum benefit and minimum cost compares each element in the decision matrix to its best criteria value. For the max benefit case, we divide each alternative matrix by the max benefit for its specified criteria. For the min cost, we multiply the min cost by the inverse of each grouped evaluation matrix element. The fuzzy numbers in the normalized fuzzy decision matrix are restricted to [0, 1].

The DMs also assess their preference toward the assessment criteria on each alternative. The linguistic appraisal of criteria is similarly aggregated as fuzzy numbers and assembled as $W = [w_1, w_2, \ldots, w_m]$. As an example, the criteria weights may be represented as (Table 2):

The DMs may additionally attach a risk assessment to the criteria weights. The risk is denoted by $\lambda \in [0, 1]$, where $\lambda = 0$ represents "risk-averse," $\lambda = 0.5$ is "risk-neutral," and $\lambda = 1$ may represent "risk-prone" criteria. The assigned weights can then be modified by the risk involved.

The normalized fuzzy decision matrix is column-wise multiplied by the aggregated weights assigned by the DMs to define the weighted normalized fuzzy decision matrix $[V_{ij}]$. The FPIS and FNIS are defined as row minimums and row maximums on the normalized matrix.

The resulting ideal solution space is represented by two vectors: the maximum benefit at minimum cost for each alternative is represented as $A^+ = \left\{v_1^+, v_2^+, \ldots, v_m^+\right\}$; the minimum benefit at maximum cost is represented as: $A^- = \left\{v_1^-, v_2^-, \ldots, v_m^-\right\}$. These vectors establish a Euclidean space so the alternatives may be compared geometrically between the upper and lower bounds of the solution space. Using the distance formula, the distance of each alternative from the positive and negative ideal solutions is given as:

$$d_i^+ = \sqrt{\sum_j \left(v_{ij} - v_j^+\right)^2}, \quad d_i^- = \sqrt{\sum_j \left(v_{ij} - v_j^-\right)^2}, i = 1, \ldots, n$$

A closeness coefficient index allows the ability to rank each alternative based on the distance from the positive and negative ideal solutions. The alternative with the highest closeness coefficient is considered the best choice for the DMs preference structure. The closeness coefficient is computed as:

$$cc_i = \frac{d_i^+}{d_i^+ + d_i^-}, i = 1, \ldots, n$$

3 Application to Prostheses Design

3.1 System Requirements

The system requirements were obtained from stakeholder surveys (customers and insurance providers) covering device abatement and un-insurability constraints. The following is a list of system requirements:

- *Prosthetics system must use human cognition to control.*
- *Human control shall be observable.*
- *Prosthetics system user shall not fatigue the user.*
- *Prosthetics system shall be usable for a normal days' time.*
- *Prosthetics system shall look and feel like a natural appendage.*
- *Prosthetics system down time shall be reduced.*
- *Prosthetic system shall be able to interact in its environment.*
- *Prosthetic system shall have a minimal learning curve.*
- *Prosthetics system shall maintain a comfortable temperature.*
- *Prosthetic system shall be able to withstand normal wear.*
- *Prosthetic system maintenance shall be minimized.*
- *Prosthetic system shall attach to the user easily.*
- *Prosthetic system shall have natural movement.*
- *Prosthetic system shall provide a useful grip.*

3.2 Requirement Analysis and Criteria Selection

System requirements were refined using Quality Function Deployment (QFD) diagram and N2 diagram. The relevant data taken from insurance surveys, prosthetic device forecasts, and historical device abatement information was used in a QFD to extrapolate the system requirements from the stakeholder requirements. The N2 diagram revealed how each element interfaced within the system. In addition to the N2 diagram, the interface matrix shows the exact nature of interface between elements.

Criteria selection was divided into four categories: functional ability, lightweight, cost, and maintenance. Evaluation showed a high degree of importance for the following criteria from the second-tier QFD: degrees of freedom, reaction latency, total system weight, and operating temp. Likewise the N2 diagram showed the following functional high-risk couplings: classifying data, processing information, lifting, carrying, containing all elements, and attaching with comfort. Additionally, N2 physical elements high-risk couplings were microprocessor, motor actuator, rigid structure, natural appearance, and housing assembly. Criteria weights were selected per each subsystem attribute.

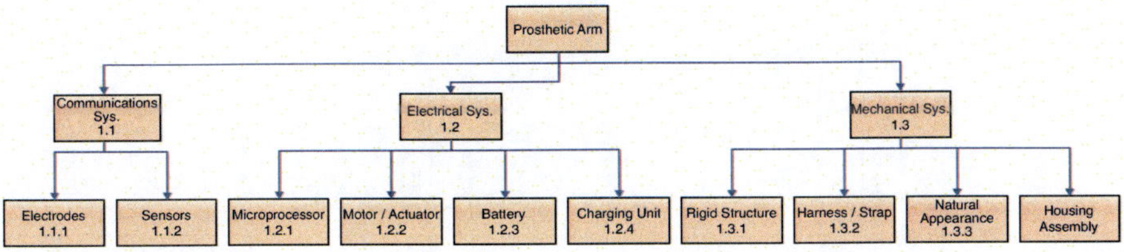

Fig. 1 A generic top-down hierarchy for prostheses design

The system architecture for the preliminary design of prostheses is shown in Fig. 1. The figure shows top-level breakdown into the communications, electrical system, and mechanical system units. The sub-assemblies in each configuration unit are shown underneath.

3.3 Fuzzy TOPSIS Analysis

The fuzzy evaluation matrix for the prosthetic is shown in Table 3. Each subsystem element has three alternatives that are measured against cost-and-benefit criteria. Fuzzy linguistic terms are used for evaluation, and their numerical definition can be seen next to the linguistic term entry. Additionally, the DM's criteria weights for each subsystem element are seen within the table.

Using the Fuzzy TOPSIS algorithm with lambda cuts for risk analysis, a closeness coefficient index matrix was formed (Table 4), which shows how each alternative ranks within the context of closeness to the positive ideal solution. Furthermore, the lambda cuts for risk assessment allows the DM to view their alternative selection from a risk-prone, risk-neutral, and risk-averse perspectives.

Once each subsystem alternative has been evaluated and ranked, the attributes are ready to be selected for the next phases of the systems development process. Table 5 shows the effect lambda cuts have on the decision-making process. Dependent on how lambda is shifted from risk-prone to risk-averse, optimal alternatives may be switched in favor of the DM's preference.

3.4 System Architecture

The final top-down hierarchy is created for each risk preference class within our mathematical evaluation framework. The risk-neutral hierarchy (Fig. 2a) establishes a baseline for a safe prosthetic architecture that should add value to any patient. The risk-prone hierarchy (Fig. 2b) adds a metal rigid structure to the prosthesis. The risk-averse hierarchy (Fig. 2c) deviates from risk-neutral in two areas: addition of

Table 3 Fuzzy evaluation matrix for determination of prosthetic architecture

EVALUATION MATRIX	Evaluation Criteria		BENEFIT CRITERIA								COST CRITERIA							
			Functional Ability				Light Weight				Cost				Maintenance			
			C1				C2				C3				C4			
ELECTRODE CRITERIA WEIGHTING			LL	0.01	0.2	0.4	MM	0.11	0.3	0.6	HH	0.26	0.5	0.8	HH	0.25	0.5	0.8
Communications electrodes	A1_1	SURFACE	F	3	5	7	P	1	3	5	L	1	3	4	H	5	7	9
	A1_2	FINE WIRE	G	5	7	9	G	6	7	9	M	3	5	7	L	1	3	4
	A1_3	WIRELESS	VP	1	1	3	P	1	3	4	H	5	7	9	M	3	6	7
SENSORS CRITERIA WEIGHTING			LL	0.01	0.2	0.4	MM	0.11	0.3	0.6	HH	0.25	0.5	0.8	HH	0.25	0.5	0.8
Communications sensors	A2_1	ANGULAR POSITION	F	3	5	7	G	5	7	9	M	3	5	7	M	3	5	7
	A2_2	GYRO	VG	7	9	9	F	3	5	7	VH	7	9	9	L	1	3	4
	A2_3	ACCELETOMETER	P	1	3	4	G	5	7	9	M	3	5	7	H	5	7	9
MICROPROCESSOR CRITERIA WEIGHTING			HH	0.25	0.5	0.5	MM	0.11	0.3	0.6	HH	0.25	0.5	0.8	HH	0.25	0.5	0.8
Electrical microprocessor	A3_1	HIGH PERFORMANCE	VG	7	9	9	F	3	6	7	VH	7	9	9	M	3	5	7
	A3_2	MID PERFORMANCE	G	5	7	9	G	5	7	9	H	5	7	9	M	3	5	7
	A3_3	LOW PERFORMANCE	F	3	5	7	VG	7	9	9	M	3	5	7	M	3	5	7

MOTOR/ACTUATOR CRITERIA WEIGHTING			*HH*	*0.25*	*0.5*	*0.8*	*LL*	*0.01*	*0.2*	*0.4*	*LL*	*0.01*	*0.2*	*0.4*	*LL*	*0.01*	*0.2*	*0.4*
Electrical-MOTOR/ACTUA-TOR	*A4_1*	*HIGH PERFORMANCE*	*VG*	*7*	*9*	*9*	*G*	*5*	*7*	*9*	*VH*	*7*	*9*	*9*	*L*	*1*	*3*	*4*
	A4_2	*MID PERFORMANCE*	*G*	*5*	*7*	*9*	*F*	*3*	*6*	*7*	*H*	*5*	*7*	*9*	*L*	*1*	*3*	*4*
	A4_3	*LOW PERFORMANCE*	*F*	*3*	*5*	*7*	*F*	*3*	*5*	*7*	*M*	*3*	*5*	*7*	*M*	*3*	*5*	*7*
CHARGING UNIT CRITERIA WEIGHTING			*HH*	*0.25*	*0.5*	*0.8*	*LL*	*0.01*	*0.2*	*0.4*	*MM*	*0.11*	*0.3*	*0.6*	*LL*	*0.01*	*0.2*	*0.4*
Electrical-CHARGING UNIT	*A5_1*	*THERMAL*	*F*	*3*	*5*	*7*	*F*	*3*	*5*	*7*	*H*	*5*	*7*	*9*	*H*	*5*	*7*	*9*
	A5_2	*AC TO DC*	*VG*	*7*	*9*	*9*	*F*	*3*	*5*	*7*	*L*	*1*	*3*	*4*	*VL*	*1*	*1*	*3*
	A5_3	*MAGNETIC INDUCTION*	*G*	*5*	*7*	*9*	*G*	*5*	*7*	*9*	*VH*	*7*	*9*	*9*	*M*	*3*	*5*	*7*
BATTERY TYPE CRITERIA WEIGHTING			*MM*	*0.11*	*0.3*	*0.6*	*HH*	*0.25*	*0.5*	*0.5*	*LL*	*0.01*	*0.2*	*0.4*	*MM*	*0.11*	*0.3*	*0.6*
Electrical BATTERY TYPE	*A6_1*	*LITHIUM ION*	*VG*	*7*	*9*	*9*	*G*	*5*	*7*	*9*	*H*	*5*	*7*	*9*	*L*	*1*	*3*	*4*
	A6_2	*NICKEL CADIUM*	*G*	*5*	*7*	*9*	*F*	*3*	*5*	*7*	*H*	*5*	*7*	*9*	*L*	*1*	*3*	*4*
	A6_3	*LEAD ACID*	*P*	*1*	*3*	*4*	*P*	*1*	*3*	*4*	*VL*	*1*	*1*	*3*	*L*	*1*	*3*	*4*

(continued)

Table 3 (continued)

EVALUATION MATRIX	*Evaluation Criteria*		*BENEFIT CRITERIA*								*COST CRITERIA*							
			Functional Ability				*Light Weight*				*Cost*				*Maintenance*			
			C1				*C2*				*C3*				*C4*			
RIDGED STRUCTURE CRITERIA WEIGHTING			*HH*	*0.25*	*0.5*	*0.8*	*LL*	*0.01*	*0.2*	*0.4*	*LL*	*0.01*	*0.2*	*0.4*	*MM*	*0.11*	*0.5*	*0.6*
Mechanical RIGID STRUCTURE	*A7_1*	*3D PRINT*	*G*	*5*	*7*	*9*	*G*	*5*	*7*	*9*	*L*	*1*	*3*	*4*	*M*	*3*	*5*	*7*
	A7_2	*METAL*	*VG*	*7*	*9*	*9*	*P*	*1*	*3*	*4*	*VL*	*1*	*1*	*3*	*L*	*1*	*3*	*4*
	A7_3	*HARD POLYMER*	*G*	*5*	*7*	*9*	*F*	*3*	*5*	*7*	*M*	*3*	*5*	*7*	*L*	*1*	*3*	*4*
HARNESS STRAP CRITERIA WEIGHTING			*MM*	*0.11*	*0.5*	*0.6*	*LL*	*0.01*	*0.2*	*0.4*	*HH*	*0.25*	*0.5*	*0.5*	*HH*	*0.25*	*0.5*	*0.5*
Mechanical-HARNESS STRAP	*A8_1*	*CLOTH*	*F*	*3*	*5*	*7*	*G*	*5*	*7*	*9*	*VL*	*1*	*1*	*3*	*M*	*3*	*5*	*7*
	A8_2	*NYLON*	*G*	*5*	*7*	*9*	*G*	*5*	*7*	*9*	*L*	*1*	*3*	*4*	*L*	*1*	*3*	*4*
	A8_3	*PLASTIC*	*P*	*1*	*3*	*4*	*F*	*3*	*5*	*7*	*L*	*1*	*3*	*4*	*M*	*3*	*5*	*7*
EXTERIOR COVER CRITERIA WEIGHTING			*HH*	*0.25*	*0.5*	*0.8*	*LL*	*0.01*	*0.2*	*0.4*	*LL*	*0.01*	*0.2*	*0.4*	*MM*	*0.11*	*0.3*	*0.6*
Mechanical-EXTERIOR COYER	*A9_l*	*SILICON*	*VG*	*7*	*9*	*9*	*G*	*5*	*7*	*9*	*H*	*5*	*7*	*9*	*H*	*5*	*7*	*9*
	A9_2	*ALUMINUM*	*P*	*1*	*3*	*5*	*F*	*3*	*5*	*7*	*M*	*3*	*5*	*7*	*M*	*3*	*5*	*7*
	A9_3	*HARD POLYMER*	*F*	*3*	*5*	*7*	*F*	*3*	*5*	*7*	*M*	*3*	*5*	*7*	*M*	*3*	*5*	*7*

Table 4 The closeness coefficients for risk-prone, risk-neutral, and risk-averse design of prostheses

			RISK-PRONE	*RISK-NEUTRAL*	*RISK-AVERSE*
ELECTRODE ALTERNATIVE SCORE					
Communication electrode	*A1_1*	*SURFACE*	*0.3701*	*0.4706*	*0.3473*
	A1_2	*FINE WIRE*	*0.4927*	*0.6291*	*0.4987*
	A1_3	*WIRELESS*	*0.169*	*0.1919*	*0.2602*
SENSORS ALTERNATIVE SCORE					
Communications sensors	*A2_1*	*ANGULAR POSITION*	*0.5958*	*0.5038*	*0.4778*
	A2_2	*GYRO*	*0.4609*	*0.6159*	*0.4357*
	A2_3	*ACCELETOMETER*	*0.2958*	*0.5633*	*0.4606*
MICROPROCESSOR ALTERNATIVE SCORE					
Electrical microprocessor	*A3_1*	*HIGH PERFORMANCE*	*0.6087*	*0.5794*	*0.4821*
	A3_2	*MID PERFORMANCE*	*0.5918*	*0.5726*	*0.49*
	A3_3	*LOW PERFORMANCE*	*0.5*	*04785*	*0.4549*
MOTOR/ACTUATOR ALTERNATIVE SCORE					
Electrical MOTOR/ACTUATOR	*A4_1*	*HIGH PERFORMANCE*	*0.7037*	*0.6257*	*0.5299*
	A4_2	*MID PERFORMANCE*	*0.6632*	*0.5639*	*0.4775*
	A4_3	*LOW PERFORMANCE*	*0.5631*	*0.454*	*0.597*
CHARGING UNIT ALTERNATIVE SCORE					
Electrical- CHARGING UNIT	*A5_1*	*THERMAL*	*0.5583*	*0.4708*	*0.3549*
	A5_2	*AC TO DC*	*0.7059*	*0.6178*	*0.525*
	A5_3	*MAGNETIC INDUCTION*	*0.663*	*0.6072*	*0.4885*

(continued)

Table 4 (continued)

			RISK-PRONE	RISK-NEUTRAL	RISK-AVERSE
BATTERY TYPE ALTERNATIVE SCORE					
Electrical- BATTERY TYPE	*A5_1*	*LITHIUM ION*	*0.5618*	*0.6278*	*0.5306*
	A6_2	*NICKEL CADIUM*	*0.5163*	*0.5673*	*0.4439*
	A6_3	*LEAD ACID*	*0.2779*	*0.3396*	*0.3047*
RIDGED STRUCTURE ALTERNATIVE SCORE					
Mechanical- RIGID STRUCTURE	*A7_l*	*3D PRINT*	*0.6624*	*0.604*	*0.4741*
	A7_2	*METAL*	*0.6919*	*0.5717*	*0.4853*
	A7_3	*HARD POLYMER*	*0.6599*	*0.5662*	*0.4565*
HARNESS STRAP ALTERNATIVE SCORE					
Mechanical- HARNESS STRAP	*A8_l*	*CLOTH*	*0.437*	*0.4744*	*0.434*
	A8_2	*NYLON*	*0.5825*	*0.5836*	*0.4928*
	A8_3	*PLASTIC*	*0.3802*	*0.3391*	*0.3293*
EXTERIOR COVEX ALTERNATIVE SCORE					
Mechanical EXTERIOR COVER	*A9_1*	*SILICON*	*0.6872*	*0.6241*	*0.5427*
	A9_2	*ALUMINUM*	*0.418*	*0.32*	*0.3388*
	A9_3	*HARD POLYMER*	*0.5564*	*0.4515*	*0.4252*

Table 5 Subsystem attribute alternative ranking for prostheses design

Attribute alternative ranking			RISK-PRONE	RISK-NEUTRAL	RISK-AVERSE
ELECTRODE ALTERNATIVE SCORE					
Communications electrodes	A1_2	FINE WIRE	0.4927	0.6291	0.4987
SENSORS ALTERNATIVE SCORE					
Communications sensors	A2_1	ANGULAR POSITION			0.4778
	A2_2	GYRO	0.4609	0.6159	
MICROPROCESSOR ALTERNATIVE SCORE					
Electrical-microprocessor	A3_1	HIGH PERFORMANCE	0.6087	0.5794	0.4821
MOTOR/ACTUATOR ALTERNATIVE SCORE					
Electrical-MOTOR/ACTUATOR	A4_1	HIGH PERFORMANCE	0.7037	0.6257	0.5299
CHARGING UNIT ALTERNATIVE SCORE					
Electrical-CHARGING UNIT	A5_2	AC TO DC	0.7059	0.6178	0.526
BATTERY TYPE ALTERNATIVE SCORE					
Electrical-BATTERY TYPE	A6_1	LITHIUM ION	0.5618	0.6278	0.5306
RIDGED STRUCTURE ALTERNATIVE SCORE					
Mechanical-RIGID STRUCTURE	A7_1	3D PRINT		0.604	
	A7_2	METAL	0.6919		0.4853
HARNESS STRAP ALTERNATIVE SCORE					
Mechanical-HARNESSSTRAP	A8_2	NYLON	0.5825	0.5836	0.4928
EXTERIOR COVER ALTERNATIVE SCORE					
Mechanical-EXTERIORCOVER	A9_1	SILICON	0.6872	0.6241	0.5427

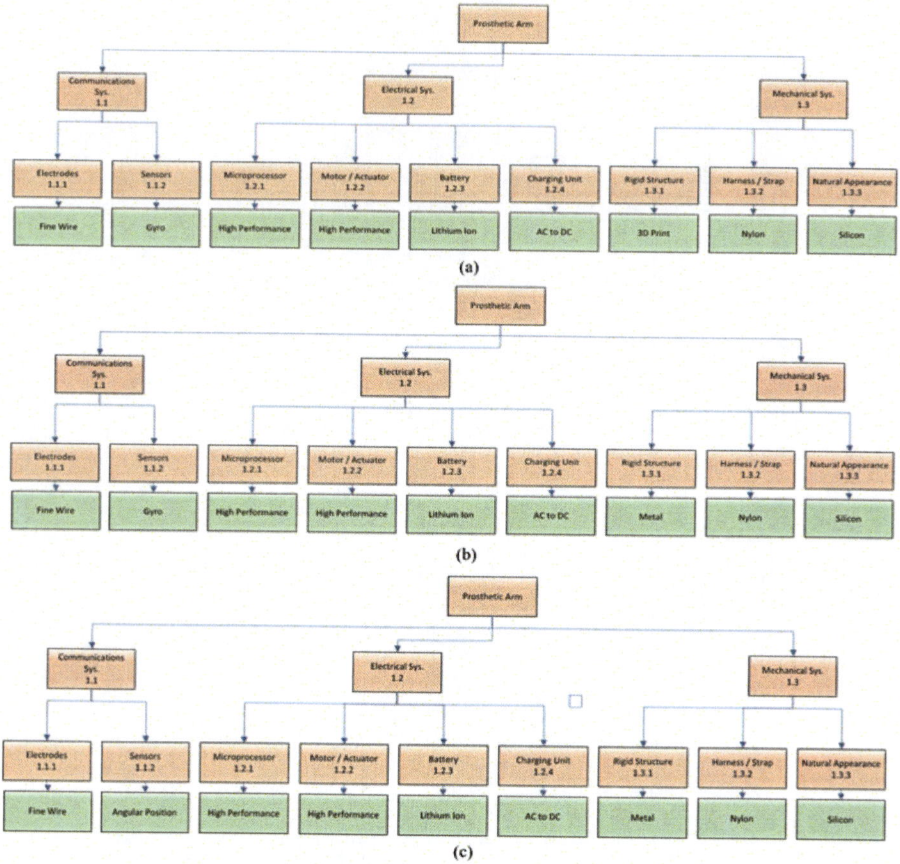

Fig. 2 The top-level system architecture for the following categories: (**a**) risk-neutral; (**b**) risk-prone; and (**c**) risk-averse

angular position sensor and metal rigid structure. What is interesting is both risk-prone and risk-averse found the metal rigid structure optimum in the design.

In conclusion, the study's main focus was to establish novel mathematical techniques to fill the gaps in uncertain information within the stakeholder needs and expectations of the actual trade space of the design process. The specific Fuzzy TOPSIS method utilizes the availability of using several decision-makers. That gives all the potential decision-makers an opportunity for input in the evaluation matrix and criteria weights. The Fuzzy TOPSIS model can be applied to MCDM in other technological fields.

Acknowledgments The authors would like to acknowledge the support from the Systems Engineering Department and the Donaghey College of Science, Technology, Engineering, and Mathematics at UA Little Rock.

References

Chen, C.-T. 2000. Extension of the TOPSIS for group decision-making under fuzzy environment. *Fuzzy Sets and Systems* 114: 1–9.

Chu, T.-C., and Y.-C. Lin. 2003. A fuzzy TOPSIS method for robot selection. *International Journal of Advanced Manufacturing Technology* 21: 284–290.

Hwang, C.L., and K. Yoon. 1981. *Multiple attributes decision making methods and applications*. Berlin/Heidelberg, Germany: Springer.

Lee, Kwang H. 2005. *First course on fuzzy theory and applications*. Berlin/Heidelberg: Springer.

Zadeh, Lotfi A. 1965. Fuzzy sets. *Information and Control* 8 (3): 338–353.

Chen 2000 & Chu and Lin 2003

Functional Decomposition: Evaluating Systems Engineering Techniques

Cal M. Cluff and Dinesh Verma

Abstract The functional decomposition allows description of complex system functionality with a hierarchy of simpler sub-functions. Determining the best set of sub-functions for any given function should be an orderly and methodical process. It is important that the method used captures the system without overlooking any necessary functionality. Of the many techniques, this paper will examine six (6) decomposition methods (operating modes, inputs and outputs, Hatley-Pirbhai template, processing rates, organizational structure, and matching the physical architecture). The LN-39, a standard inertial navigation system (INS), first deployed in the A-10 and the F-16, is used as a sufficiently complex exemplar for applying each method. Finally, the resulting LN-39 functional architectures are compared from a number of perspectives in the product development lifecycle to illustrate strengths and weaknesses of each decomposition technique. Each of the decomposition methods revealed aspects of the system function that were important to include in the functional architecture, and no single method was clearly the best. Finally, it is recommended that multiple methods be used to reveal a comprehensive set of elementary functions. Thereafter, it is recommended that these elementary functions are clustered and composed to develop the functional architecture.

Keywords Functional decomposition composition complex systems

C. M. Cluff (✉)
Nothrop Grumman, Woodland Hills, CA, USA
e-mail: cal.cluff@ngc.com

D. Verma
Stevens Institute of Technology, Hoboken, NJ, USA

A. M. Madni et al. (eds.), *Recent Trends and Advances in Model Based Systems Engineering*, https://doi.org/10.1007/978-3-030-82083-1_33

1 Introduction/Background

Developing a functional decomposition of a system is an important systems engineering activity. It is crucial for describing what the system does for the customer and stakeholders and for use throughout the system lifecycle. Larson et al. (2009) identify 11 different functional decomposition techniques for determining subfunctions. Each technique has strengths and weaknesses, which draw out different aspects of a system's functional capabilities and how they are organized (Larson 151). According to (Summers et al. 2013), "Rather than develop a single, unified definition of function, each approach has its own strengths and weaknesses; each approach is useful and particularly well suited for different reasoning applications and domains yet the transference across these being difficult at best. Therefore, a set of comparative benchmarks that can be [applied] with the different modeling approaches ... to discover which elements of the representations and vocabularies are most conducive for different elements of functional thinking" (Summers 2).

For this paper, the LN-39 Inertial Navigation System (INS) by Litton Company, the USAF standard navigator (ENAC 77–1), is used as an exemplar to decompose according to six of the techniques. The six techniques were chosen (operating modes, inputs and outputs, Hatley-Pirbhai template, processing rates, organizational structure, and matching physical architecture) because they each draw out an important aspect of the LN-39 functional design that is not necessarily evident when using the other methods. The INS has a single well-defined function – navigate. That means to keep track of its position and orientation. The design to accomplish this is sufficiently complex, and this complexity will help to better understand the strengths and weaknesses to use as a basis for comparing the different functional decomposition techniques (Komoto et al. 2011).

2 System Description

The LN-39 is an avionics instrument designed for high-performance aircraft. It is equipped internally with accelerometers to sense linear acceleration and gyroscopes to sense torque. An internal processor integrates acceleration for velocity and position. The angular rotation is integrated for orientation. Input from the aircraft barometric altitude device is integrated into the vertical solution to dampen instability due to the accumulated effect of changes in gravity over the course of the aircraft flight path. The internal 6-DOF (six degrees of freedom) solution is transformed into the various navigation parameters for pilot display.

A description of the LN-39 from the Litton sales brochure is as follows. "The LN-39 inertial navigation system employs mature inertial navigation system technology in a modern, high reliability design compliant with the Air Force Standard INS requirements. The LN-39 provides high accuracy and reliability consistent with low initial and life cycle costs for multiple applications. Performance options ranging

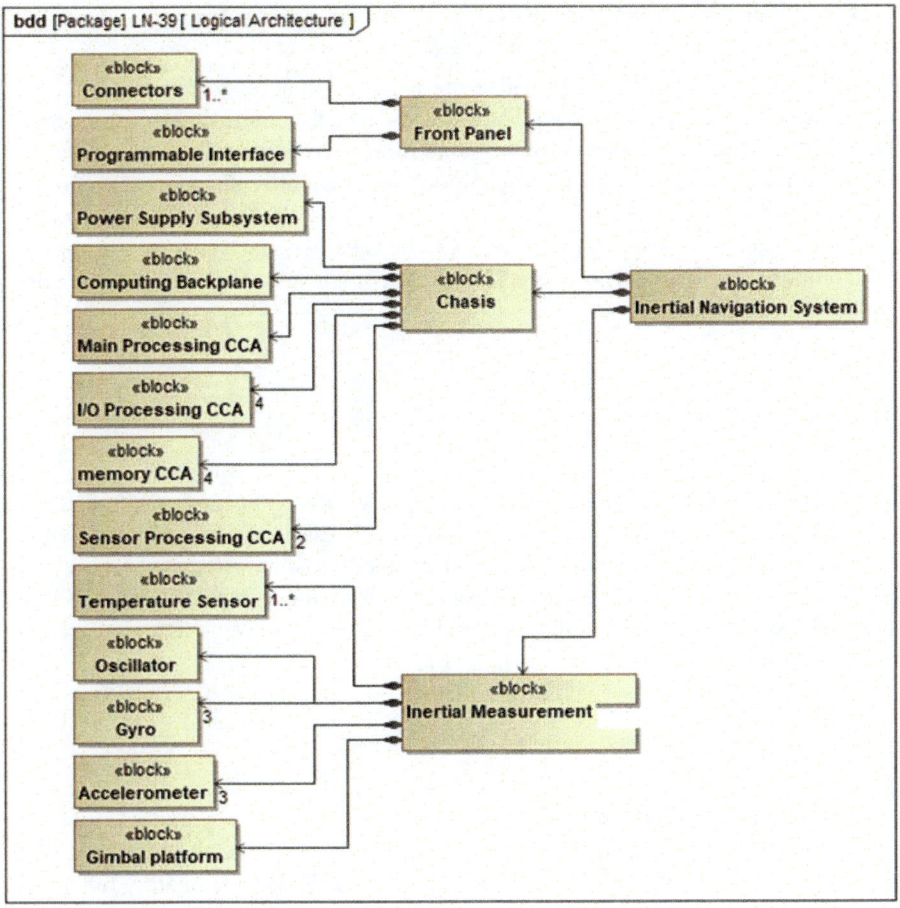

Fig. 1 LN-39 (genericized) Object Decomposition

from medium to high are available with the rapid-reaction necessary for military environments."

2.1 *Object Decomposition*

Figure 1 LN-39 (genericized) object decomposition shows a genericized object decomposition of the LN-39. The object decomposition depicts the component pieces; we need the accompanying functional description to write the system requirements. The object and functional architecture hierarchy diagrams were created using Cameo Enterprise Architecture 2019.

3 Functional Decomposition

The functional decomposition provides the top-level function "navigate" and the sub-functions necessary to accomplish the top-level function. What each decomposition is describing is essentially the same thing. The terminology, organization, and detail for a given level of decomposition are dictated by the approach. Strictly speaking, the decomposition is recursive from abstract to concrete with as many levels as necessary to describe the system. For this analysis, each approach is applied to decompose "navigate" (zeroth level) to the second level representing elementary functions.

3.1 Evaluation Method

A functional decomposition is produced using each decomposition method. Then observations, pros and/or cons, using an approach that considers three different dimensions: representation, cognition, and enabled reasoning (Kruse et al. 2014, 37–38). The three dimensions are considered from the perspective of different phases of the product lifecycle: designer architect, product development, integration and test, and sustainment.

Evaluation Dimensions

- Representation Evaluation Dimension.

  This dimension is used to evaluate how well the decomposition captures the "elementary functions" of the system. According to Kruse, elementary functions are the terminus (last level of decomposition) of the decomposition and are reusable in other contexts (37). In practical terms, it is the point at which we would transition from "what" the system does to "how" the system is to accomplish it. Decomposition of sub-functions to the second level is evaluated based on how well they represent the system.
- Cognition Evaluation Dimension.

  This dimension is used to evaluate the ability to decompose across abstraction levels and commitment or decisions when information is available. In a hierarchical approach, functions are decomposed into more detailed and concrete sub-functions relative to the level above it. Kruse introduces the concept of "the abstraction gradient, meaning the minimum and maximum levels of abstraction that can be represented" and "how intuitive the resulting model is" (38). The decomposition of sub-functions will be evaluated based on how abstract is the resulting second level. The decomposition method is evaluated based on how understandable the sub-functions are. The concept of level of information available for "commitment or decisions" is difficult to apply in this analysis since the system used for the example is well known.

- Enabled Reasoning Evaluation Dimension.

  This dimension is used to evaluate consistency and precision across engineering activity, domains, and expertise. Kruse uses an example related to this evaluation dimension regarding how well a model is received by the other engineering disciplines applied throughout the lifecycle of the product (38). For this evaluation, the various decompositions are looked at from the perspective of the engineering activity and their dependency on the original design team to accomplish their tasks.

Product Lifecycle Perspectives

The functional architecture is one of the artifacts developed by the architect of a system. Four product lifecycle activities (engineering roles) are used to provide a perspective on the functional decompositions.

- Designer Architect

  The designer architect develops the functional decomposition. The decomposition approach evaluation is based on the ease with which the decomposition is developed.
- Product Development

  The product development activity creates the system that implements the design. The decomposition approach evaluation is based on how well it conveys information necessary for system development.
- Integration and Test

  The integration and test (I&T) activity verifies that the system does what the designer architect specified. System verification is done by verifying requirements, but the functional requirements are created in the context of the functional architecture. The decomposition approach is evaluated on how well the decomposition lends to the test approach.
- Sustainment.

  The sustainment activity keeps the system operational when fielded. The designer architect may have the product development and I&T team in mind as the consumers of the architecture products, but it takes more disciple to consider the sustainment activities when designing a system. The decomposition approach evaluation is based on how well sustainment concepts are supported with the design.

3.2 Functional Decomposition by Operating Modes

The LN-39 has several operating modes that it must transition through before it is operational, and the pilot can begin his/her mission. The performance of accelerometers for sensing acceleration and gyros for sensing torque is temperature-sensitive. The degree of sensitivity varies from part to part, even in the same production lot. Each LN-39 is calibrated over the full operational temperature range. The calibration coefficients are calculated with off-line software and uploaded to the system for use when deployed. When the LN-39 is first powered up, it goes through an initialization cycle and then waits for the operator to align the IMU. The alignment defines level, heading, and local gravity (accelerometers measure the vector acceleration, including acceleration due to gravity, which is subtracted before integrating velocity and position). After alignment, the INS integrates the IMU data to compute the position, velocity, and orientation of the vehicle based on changes from the aligned position and orientation.

Figure 2 shows the second-level functional decomposition to sub-functions accomplished in each operating mode. Table 1 lists the pros and cons from each perspective and along each evaluation dimension.

3.3 Functional Decomposition by Inputs and Outputs

The LN-39 has well-defined inputs and outputs. Inputs include data from other aircraft sensors for altitude. Barometric pressure altitude is identified here. The physical motion of the aircraft is considered input as three-dimensional linear acceleration, roll, pitch, and yaw of the aircraft. The outputs are the navigation products that drive the pilot's airspeed, altitude, heading, and artificial horizon displays. The LN-39 is too complex for this approach. The designer architect and the product development team will find the resulting functional architecture to be insufficient. Figure 3 shows the sub-functions for processing the INS inputs and outputs. Table 2 lists the pros and cons from each perspective and evaluation dimension.

3.4 Functional Decomposition by Hatley-Pirbhai Template

The Hatley-Pirbhai template approach for functional decomposition is the most comprehensive of the decomposition methods evaluated in this paper. It incorporates the inputs and outputs as well as the operating modes for controlling the operations of the system. It appears to contain all of the sub-functions of input and outputs as well as sets the stage for the operating modes, albeit at the next level of

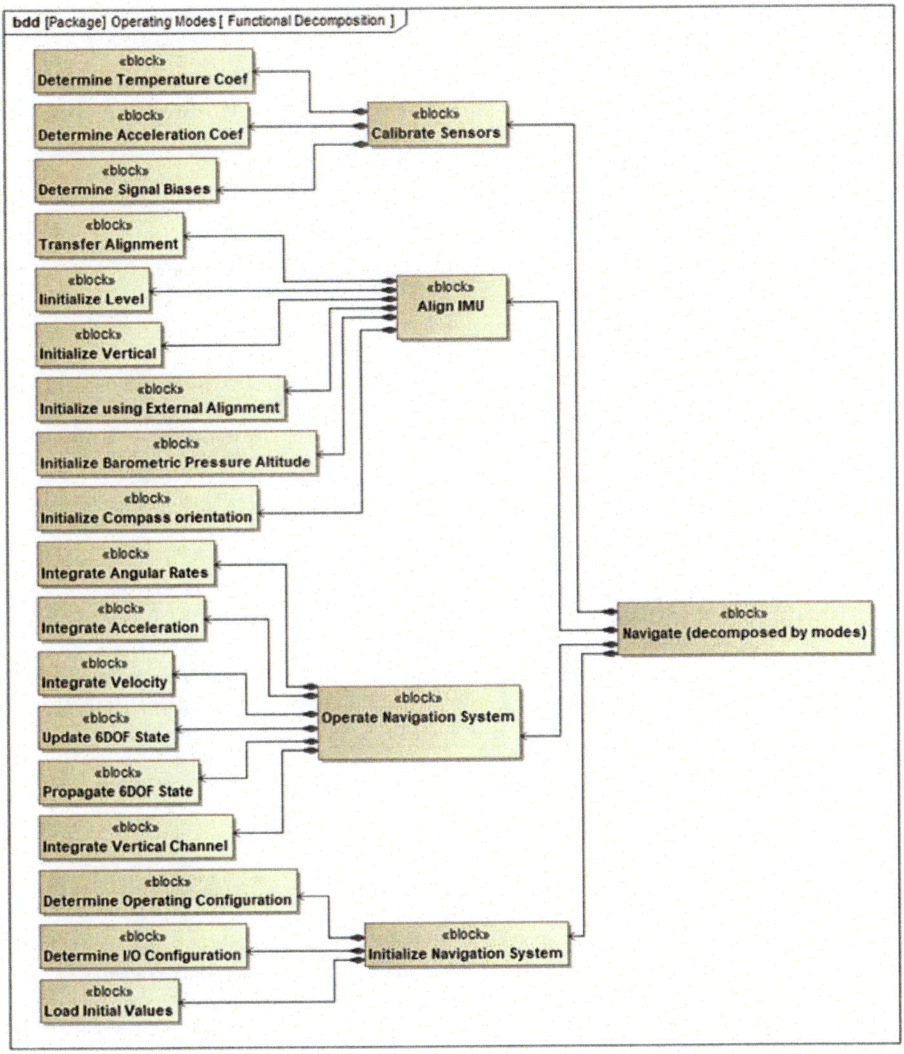

Fig. 2 Functional decomposition by operating modes

decomposition. In addition, the Hatley-Pirbhai template explicitly incorporates functions to support sustainment.

Figure 4 shows the sub-functions associated with each category of the Hatley-Pirbhai template. Table 3 lists the pros and cons from each perspective and evaluation dimension.

Table 1 Functional decomposition by operating mode observations

Engineering role	Representation	Cognition	Enabled reasoning
Designer architect	Pros: Was able define elementary functions quickly because the operating modes are functional in nature Cons: Potential for redundancy for sub-functions that operate during multiple modes	Pros: It is easy to think about what the system does during each mode. Concrete sub-functions were established at a high level making the overall decomposition flat	Pros: Communicates INS expertise effectively Cons: Missing the input/output processing need for integrating INS with the larger system
Product development	Pros: Sub-functions are well compartmentalized suitable for software development teams Cons: Hardware functionality is hidden. The same HW development team would have responsibilities with every operating mode	Pros: Easy for software team to understand and transition to "how" Cons: The decision regarding the allocation of functionality between HW and SW is postponed, unnecessarily to lower levels of the decomposition	Cons: While the process flow is critical, much it is implied. Organization of blocks indicates sequence, but that can be misleading
Integration & test	Pros: Easy mapping between sub-functions and test cases Cons: Missing input/outputs will make it difficult to develop test procedures	Pros: Easy to see what to test as sub-functions are concrete	Pros: The test engineer is not dependent on the original designer to create test cases/procedures
Sustainment	Pros: Calibration activities are clearly represented Cons: Sub-functions do not represent replaceable components	Pros: The sub-functions lend themselves to diagnostics	Cons: Failed components would disable many sub-functions and require designer architect expertise to decipher

3.5 *Functional Decomposition by Processing Rates*

The LN-39 software is real time embedded. The real-time operating system manages software processing with a scheduler. There are three periodic schedulers (process is automatically scheduled at the next periodic time) and intermittent scheduler (processes are scheduled for "as soon as possible" for a single execution). Processes in one scheduler can trigger processing in the other schedulers. During instantiation, the initialize functions are loaded into the intermittent scheduler to "bootstrap" the processing. By decomposing the "navigate" function by processing rate, there is a one-to-one mapping of the functionality to the processing schedulers.

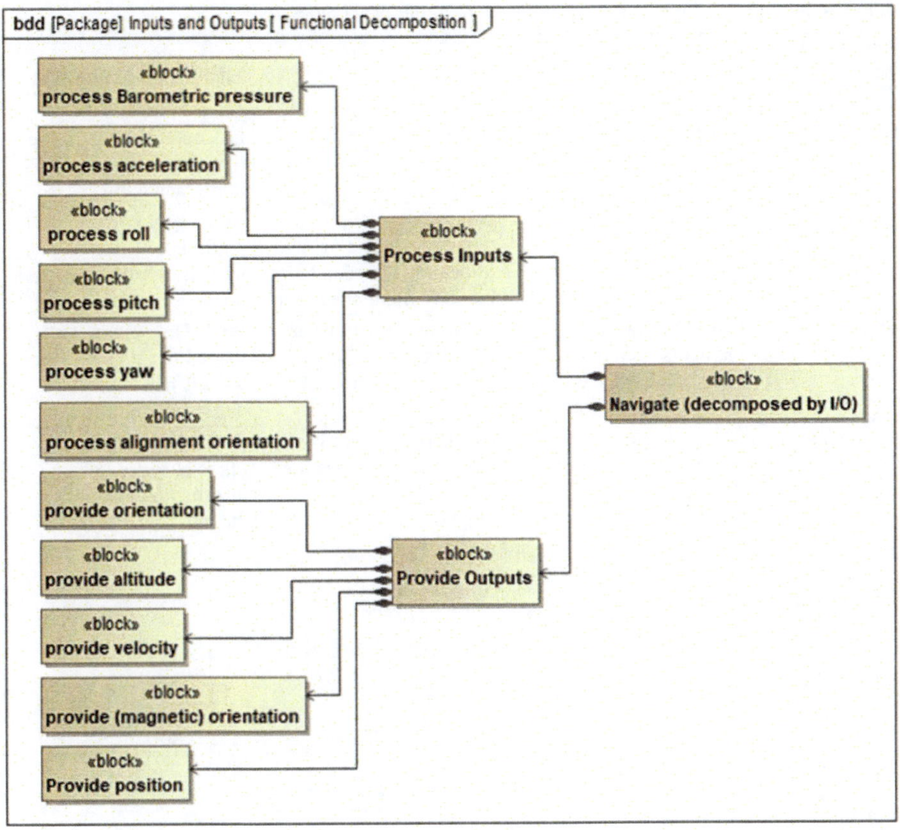

Fig. 3 Functional decomposition by inputs and outputs

Larson states that hardware functions and software functions are separated by processing rates (164); while hardware is necessary for high-rate processing, the LN-39 temperature and barometric pressure sensors are low-rate processing inputs. This method did not facilitate hardware and software functional allocation during the decomposition by processing rate. Figure 5 shows the sub-functions associated with the internal processing rates of the real-time system. Table 4 lists the pros and cons from each perspective and evaluation dimension.

3.6 Functional Decomposition by Organizational Structure

The Litton Systems, Inc., and its successor corporation are organizationally a matrix. The engineering organization is partitioned by function and engineers are assigned to programs as needed. Typically, each program will work with engineering to create

Table 2 Functional decomposition by inputs and output observations

Engineering role	Representation	Cognition	Enabled reasoning
Designer architect	Cons: Does not represent processing elements	Cons: Difficult to relate inputs to outputs	Cons: Assumes the processing elements are trivial or left up to the implementation
Product development	Cons: Does not provide enough details to implement	Cons: Does not provide a basis for decision, i.e., allocation between HW and SW implementation	Cons: Highly dependent on designers to transition from what to how
Integration & test	Pros: Provides the elements need to construct test cases	Pros: Sub-functions are described in terms of known data elements	Pros: The system is a black box and provides what is needed to understand interfaces
Sustainment	Cons: Sub-functions focus on data elements; do not represent replaceable components.	Cons: The system is a black box, provides no information for fault diagnostics	Cons: The system is a black box, provides no information for fault diagnostics

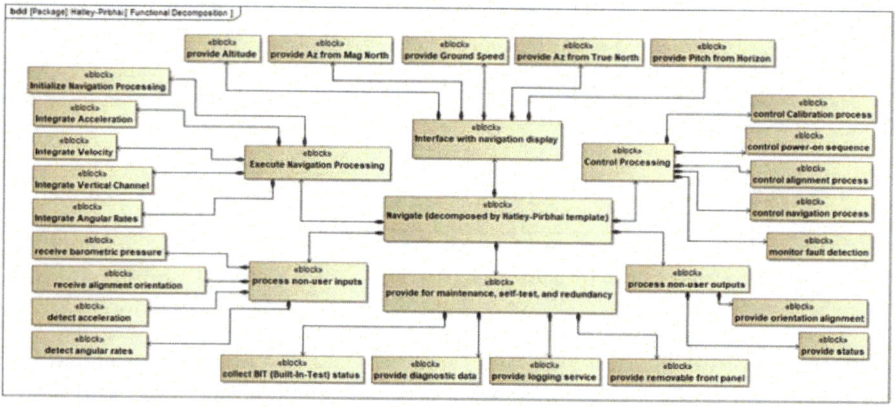

Fig. 4 Functional decomposition by Hatley-Pirbhai template

a staffing plan to identify what type of engineer and when they are needed during the execution of the program. The engineering organization manages the assignments of each engineer moving from one program to another in order to fulfill the various staffing plans at any given time. The LN-39 program needs engineers from five of the broad engineering disciplines, hardware, software, firmware, mechanical, and sustainment. In addition, a sixth discipline, systems engineers are needed, but they are not included in the decomposition.

Figure 6 shows the sub-functions associated with the internal organizational matrix structure of Litton Systems, Inc. Table 5 lists the pros and cons from each perspective and evaluation dimension.

Table 3 Functional decomposition by Hatley-Pirbhai template observations

Engineering role	Representation	Cognition	Enabled reasoning
Designer architect	Pros: Provides comprehensive perspective on elements of a system Cons: Implies recursive decomposition to for concrete sub-functions	Cons: Control and processing model elements are too abstract; an extra level of decomposition is needed. Compartmentalizes sustainment early in the decomposition	Pros: Addresses I&T and sustainment in the framework
Product development	Pros: Provides context for sub-functions with an implied processing structure	Pros: Organized in modular categories for modular implementation	Cons: The processing and control models are too abstract to implement
Integration & test	Pros: Provides input and output elements need to construct test cases	Pros: Sub-functions are described in terms of known input/out data elements	Pros: The system is a black box and provides what is needed to understand interfaces
Sustainment	Pros: Explicit elements for sustainment Cons: Sub-functions do not represent replaceable components	Pros: Concrete elements for sustainment Cons: Hidden relationship of sustainment sub-functions to other system sub-functions	Pros: Treats sustainment as stakeholder, allows for greater role in system specification

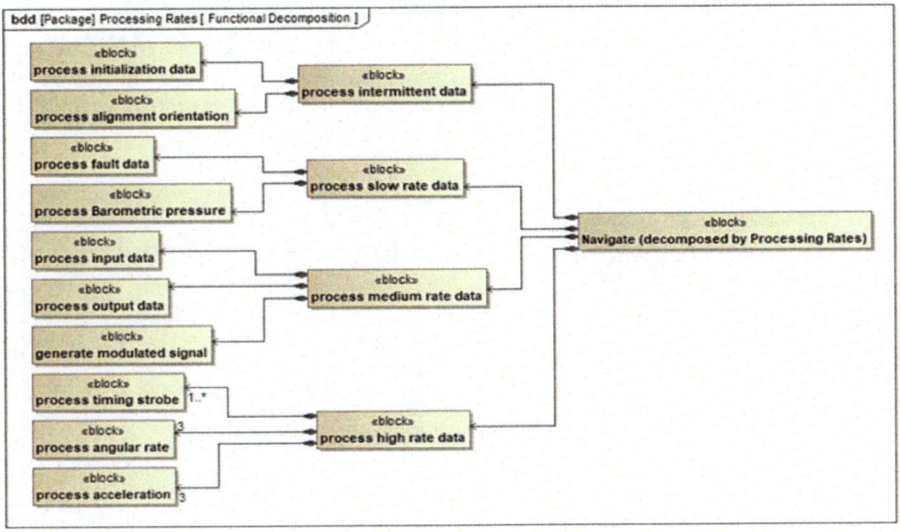

Fig. 5 Functional decomposition by processing rates

Table 4 Functional decomposition by processing rate observations

Engineering role	Representation	Cognition	Enabled reasoning
Designer architect	Cons: Too much focus on data, not enough on function	Cons: Sub-functions are too abstract at this level. Processing rates add artificial decomposition level	Pros: Both development and test teams would benefit from the processing rate allocations
Product development	Pros: Provides a clear representation of the real-time behavior in non-behavior diagram	Pros: Provides initial real-time process scheduling architecture Cons: Too abstract, needs another level of decomposition	Cons: Still relies on designer architects for function gaps
Integration & test	Pros: Provides description of data elements needed to construct test cases	Pros: Sub-functions are described in terms of data processing	Cons: Still relies on designer architects for functions related to output data
Sustainment	Cons: Sub-functions do not represent replaceable components	Cons: Sub-functions do not relate to sustainment activities	Cons: This functional decomposition is not beneficial for sustainment

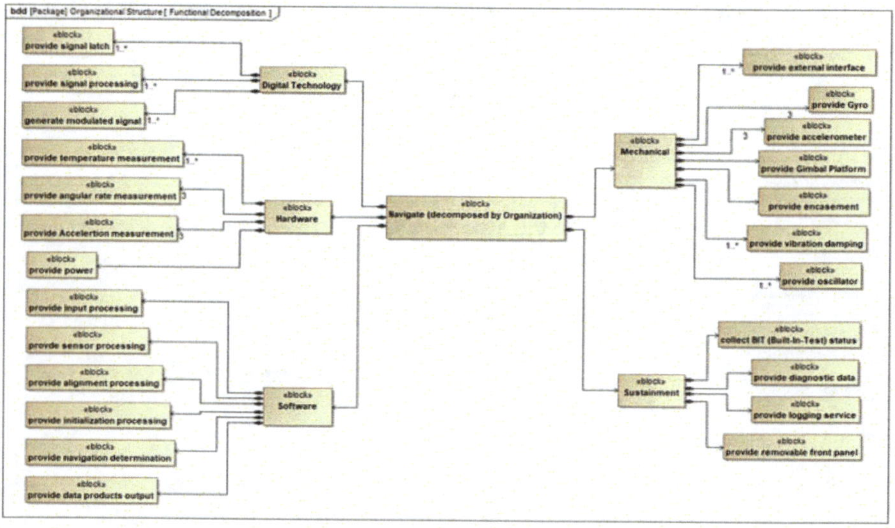

Fig. 6 Functional decomposition by organizational structure

Table 5 Functional decomposition by organizational structure observations

Engineering role	Representation	Cognition	Enabled reasoning
Designer architect	Pros: Explicitly includes function aspects of hardware and mechanical Cons: Not clear (too abstract) for software function	Cons: Sub-functions are too abstract at this level. Possibly because it includes broader scope	Pros: Includes sub-functions for all engineering disciplines involved
Product development	Pros: Explicitly allocates between hardware, software, and firmware	Cons: Too abstract, needs another level of decomposition	Pros: Good hand off between engineering disciplines
Integration & test	Cons: I&T is not separate organization in matrix, does provide data needed to construct test cases	Pros: Sub-functions are concrete enough to establish test categories, if not test cases	Cons: Organizational structure does not provide system context, need additional designer architect for how sub-functions relate to one another
Sustainment	Pros: Explicit elements for sustainment Cons: Sub-functions do not represent replaceable components	Pros: Concrete elements for sustainment Cons: Hidden relationship of sustainment sub-functions to other system sub-functions	Pros: Treats sustainment as stakeholder, allows for greater role in system specification

3.7 Functional Decomposition by Matching Physical Architecture

The physical architecture of the LN-39 shown in Fig. 2 1 LN-39 (genericized) object decomposition provides the basis for the functional decomposition by matching physical architecture. The inertial measurement unit (IMU) and the front panel are components that were on multiple systems in the LN 3x product line. The IMU was also a component on products not included in the INS product lines. The decomposition by matching physical architecture promotes reuse of architecture products on different products and different product lines by treating them as subsystems within the functional decomposition. This approach works well as a reverse engineering activity.

Figure 7 Functional Decomposition by Matching Physical Architecture shows the sub-functions for associated with the internal organizational matrix structure of Litton Systems, Inc. Table 6 Functional Decomposition by Matching Physical Architecture Observations lists the pros and cons from each perspective and evaluation dimension

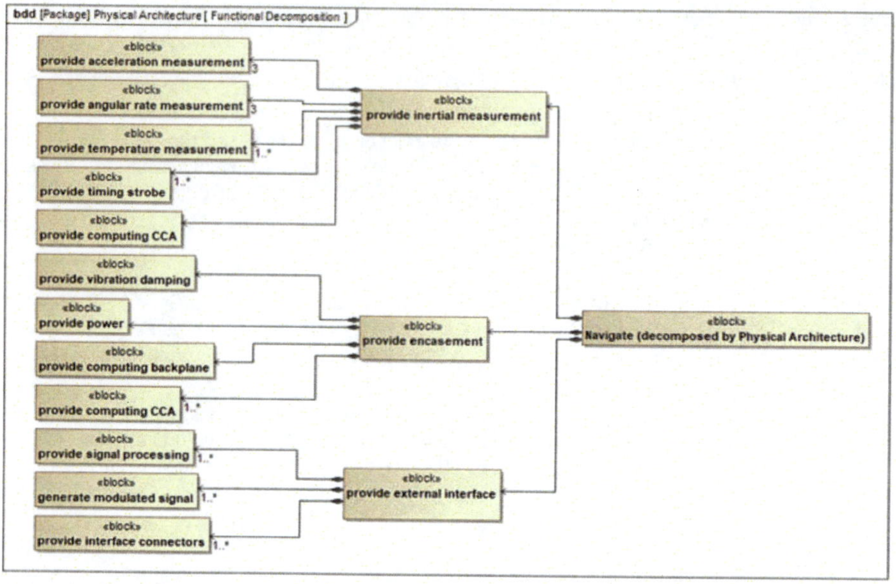

Fig. 7 Functional decomposition by matching physical architecture

Table 6 Functional decomposition by matching physical architecture observations

Engineering role	Representation	Cognition	Enabled reasoning
Designer architect	Pros: Provides functional purpose for each physical component. Addresses hardware and mechanical	Cons: Sub-functions are too abstract; an extra level of decomposition is needed	Pros: Includes sub-functions for all components. Possible to use branch as subsystem
Product development	Cons: Hardware oriented. Not clear for software functions	Cons: Sub-functions are too abstract; not clear where software function decomposition will take place	Cons: Software development dependent on designer architects
Integration & test	Pros: Provides sub-functions for interfacing. Cons: Does not represent the data on the interfaces	Pros: Each physical piece has a sub-function allocated. Test case coverage ensured	Pros: Information provided for the top-level test cases
Sustainment	Pros: Sub-functions represent replaceable components	Cons: Not clear where the sustainment function will reside	Cons: Sustainment activities are dependent on additional information from architecture

4 Recommendations and Conclusions

Each of the decomposition methods had strengths and weaknesses. Some of the weaknesses are addressed easily by including additional diagram/artifacts with the architectural description. For instance, the processing rates, which are not apparent unless explicitly decomposed by processing rate, can be described with concentric processing cycles depicted with an activity diagram. Other weaknesses associated with the functional architecture as a description of "what" the system does may be remedied with additional levels of decomposition. However, not every approach results in covering the entire system, even when developed with further levels of decomposition.

4.1 Observations by Engineering Role Perspectives

The determination of a strength or weakness is given the perspective of the consumer of the functional architecture. Table 7 Functional Decomposition Method Preference by Engineering Perspective shows the different engineering perspectives used for evaluating the functional decomposition methods and highlighting the preferred method.

Each method has a different perspective for looking the functionality of the system. While some were better than others were, each of the decomposition methods revealed aspects of the system function that were important to include in the functional architecture. There was no clearly preferred method.

4.2 Recommendation

Each decomposition method is a framework/template for thinking about the functionality of the system. As the system functionality is decomposed, the framework guides the architect's thoughts for describing what is necessary to accomplish the abstract/parent function. "Systems engineers and architects have traditionally decomposed their systems; recently, they've also started using composition" (Larson 154). By employing multiple decomposition methods and given the elementary functions that result, engage in a functional composition to create the functional architecture.

4.3 Functional Composition

The functional decomposition of the LN-39 by 6 different methods resulted in a collection of 100+ sub-functions on the second level. Following the rule-of-thumb of six sub-functions (Larson 154), sometimes seven sub-functions if the

Table 7 Functional decomposition method preference by engineering perspective

Engineering Role	Preference
Designer architect	The designer architect of the LN-39 would prefer describing the functionality using the decomposition by operating mode approach. The approach flowed naturally from a stream of consciousness description of what the LN-39 does to "navigate." each operating mode sets the stage for the next, included the calibrate mode which must be addressed in the system design and accomplished by the sustainment organization as the last step in production before deployment
Product development	The product development team is the first consumer to take the architecture provided by the designers. The software developers create their top-level design partitioning the functionality by processing rates. Then they identify the interface between function with dissimilar rates to address specific implementation for each interface. The software development team would prefer if the functional architecture is provided in the form decomposed by processing rates so that they did not need to "refactor" the architecture while creating their top-level design. The hardware and firmware developers would prefer the decomposition by organization as it clearly identifies the functionality they are responsible. The mechanical developers would prefer the decomposition by matching physical architecture as it clearly maps the physical components to the allocated functionality
Integration & test	The I&T team is less concerned with the internal functionality of the INS except for the broad category of test configurations that are usually provided in the environmental requirements. The I&T team would prefer in the functional architecture was provided decomposed by inputs and outputs to keep a clear black box representation of the system under test. The Hatley-Pirbhai template method also contains decomposition categories for inputs and outputs
Sustainment	The sustainment organization would prefer a functional decomposition using the Hatley-Pirbhai template method. This method specifically identifies the functions supporting maintenance, self-test, and redundancy, although it is still lacking in associating functionality to replaceable components. The sustainment engineers participate in the peer-review activities of the architecture during design phase. By explicitly including sustainment in the decomposition, the review process would be more efficient and beneficial to the design architecture team. This provides an opportunity for the sustainment engineers to contribute to the initial architecture

decomposition is not too complex (Larson 155), the sub-functions developed in Sect. 3 were grouped into seven groups forming the first (middle) tier of the function architecture of the system. See Fig. 9. Functional Composition of Navigation Function for the groups, now the first tier of the functional architecture developed by composition.

Of the 100+ sub-functions, some were duplicates; for instance, the decomposition method by inputs and outputs yields similar sub-functions as the method by Hatley-Pirbhai template, which also include inputs and outputs. Duplicate sub-functions were consolidated, and similar sub-functions were combined for a composition that represented each function identified, resulting in a functional architecture as follows in Figs. 8 and 9:

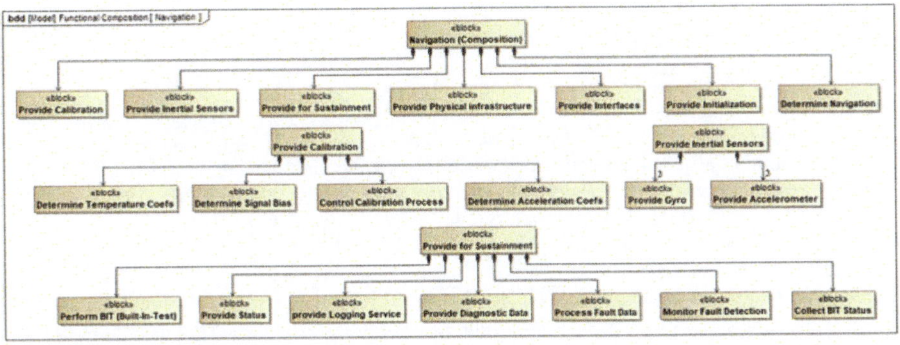

Fig. 8 Functional composition of navigation function

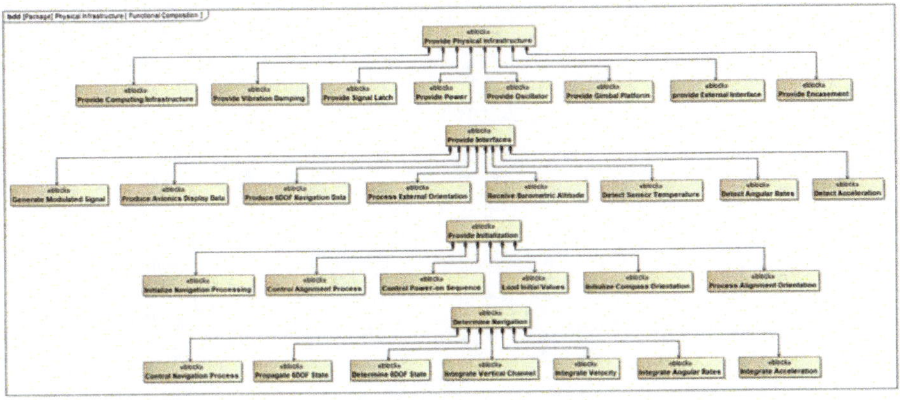

Fig. 9 Functional composition of navigate function (continued)

4.4 Conclusions

Trying to find a single decomposition approach best suited for a specific system becomes more difficult the more complex the system of interest is. Trying many different methods was instructive as it revealed elementary functions from a number of different perspectives. The composition approach allowed us to construct the functional architecture from the collection of decomposed elementary sub-functions. The approach of first decomposing using multiple methods and then composing the functional architecture from the decomposed elementary sub-functions allowed us to select the best of each approach, resulting in the best overall functional architecture.

References

Cameo Enterprise Architecture. 2019. No Magic, Inc.

Komoto, Hitoshi, and Tetsuo Tomiyama. August 2011. A theory of decomposition in system architecting. In *International conference on engineering design, ICED11, proceedings*.

Kruse, Benjamin, et al. 2014. Systematic comparison of functional models in SysML for design library evaluation. In *24th CIRP design conference proceedings*, 34–39. CIRP.

Larson, Wiley J., et al. 2009. *Applied space systems engineering*. McGraw-Hill.

Sales Brochure, Litton Systems Inc., Guidance & Control Systems, 5500 Canoga Avenue, Woodland Hills, CA., 1984.

Summers, Joshua D., C. Eckert, and A. Goel. August 2013. Function in engineering: Benchmarking representations and models. In *International conference of engineering design, ICED13, proceedings*.

Part VI
MBSE Applications

Model-Driven Safety of Autonomous Vehicles

N. Annable, A. Bayzat, Z. Diskin, M. Lawford, R. Paige, and A. Wassyng

Abstract We make the case that since model-based development of complex software-intensive systems has proven to be so effective, a model-based paradigm that encompasses assurance of the system makes excellent sense and will result in more rigorous, less ad hoc approaches to the development and maintenance of assurance cases. This will become especially clear in the manufacturing of autonomous motor vehicles. Adequate demonstration of the safety of autonomous vehicles is a huge challenge. Doing it once for a single vehicle is difficult. Doing it for multiple vehicles in a product family and coping with incremental changes in design from one model version to the next without redoing the complete safety analysis is even more difficult. We show that a comprehensive, rigorous model-driven approach to development and assurance holds the promise of more efficient and more effective assurance in general and also provides a mechanism for incremental assurance. We also briefly compare that with one of the current staples for documenting assurance cases – Goal Structuring Notation.

Keywords Model-based development · Model-based assurance · Assurance cases

1 Introduction

Model-based development (MBD) of complex automotive systems is well established and has proven to be the preferred approach. Analysis of models together with model management and correct-by-construction software generation are convincing reasons to use a model-based approach. The approach to the associated safety assurance has lagged somewhat, but it makes good sense to utilize model-based approaches for those same reasons. In addition, safety assurance needs to be planned

N. Annable (✉) · A. Bayzat · Z. Diskin · M. Lawford · R. Paige · A. Wassyng
McMaster Centre for Software Certification, Department of Computing and Software, McMaster University, Hamilton, ON, Canada
e-mail: annablnm@mcmaster.ca

A. M. Madni et al. (eds.), *Recent Trends and Advances in Model Based Systems Engineering*, https://doi.org/10.1007/978-3-030-82083-1_34

ahead of development and tightly integrated with development as it proceeds. One of the huge assurance hurdles we have to overcome is the inability to perform incremental safety assurance. Incremental design is now the norm with manufacturers routinely modifying existing vehicle (or vehicle product-line) designs for their next model year. In doing this, manufacturers must be able to do the same with the associated safety assurance. In other words, they need to be able to produce safety assurance based on that of the previous model and changes in design as well as changes in regulations, operating conditions, etc., without redoing the assurance from scratch.

This is challenging right now with the current proliferation of Advanced Driving Systems and introductory autonomous features. It is going to be even more difficult with more autonomous features leading to full Level 5 autonomy (i.e., features that completely replace the driver) (SAE J3016 2018), exacerbated by the fact that the required levels of safety increase as the level of autonomy increases. A common mitigating factor for early autonomous features is that there is a human driver who can take over control of the vehicle. This will not be true for Level 5 autonomy, drastically increasing the level of safety required for these future vehicles.

We believe that existing notations and tools for safety assurance fall far short of what we need for achieving the necessary safety levels for (current and) future vehicles. Together with an automotive partner, we have developed an approach to safety assurance that better integrates the assurance processes and development processes; is much more rigorous than existing techniques; is less ad hoc; includes extremely comprehensive traceability; is compatible with current model management techniques; and facilitates incremental assurance.

The remainder of this paper briefly introduces one of the most popular assurance case notations used today, introduces our new methodology, and very briefly compares them. Readers can find excellent publications on various aspects of assurance/safety cases, for example, in (Rushby et al. 2015 and Rhinehart et al. 2015).

2 Goal Structuring Notation (GSN)

GSN is one of the most popular graphical notations for the presentation of assurance cases (Kelly 1998). It presents its "argument" by stating *Goals* (representing claims) and *Strategies* (reasons for decomposing goals into sub-goals) and supports terminal goals using *Solutions* (representing evidence). It has a rich supporting notation including *Assumptions*, *Contexts*, and *Justifications*.

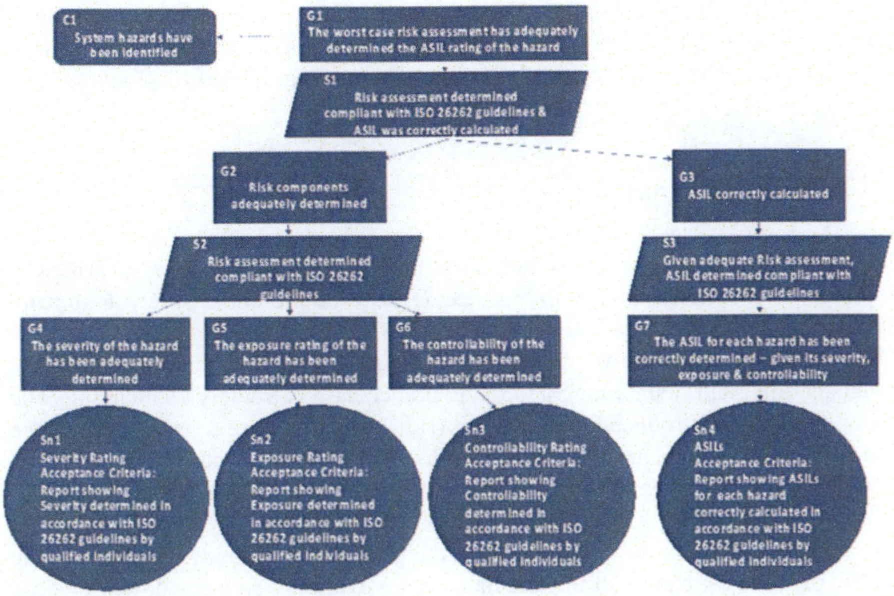

Fig. 1 GSN ASIL example

2.1 A GSN Example

Figure 1 provides an example of a GSN fragment representing the calculation of ASILs (Automotive Safety Integrity Levels) for all identified system-level hazards as described in ISO 26262 [ISO 26262, 2018]. Briefly, an ASIL captures the risk associated with a system-level hazard and guides the development of its mitigation. The context C1 is included to emphasize that this fragment would follow demonstration that system hazards were adequately determined. There is no space here to include other support nodes, such as assumptions.

The top-level claim G1 is decomposed into G2 and G3 as described in S1. G2 is further decomposed and supported by evidence that the severity, exposure, and controllability ratings have all been correctly determined in accordance with ISO 26262 guidelines. G3 is supported by a calculated ASIL rating that utilizes the correctly assigned severity, exposure, and controllability ratings.

2.2 GSN Benefits

GSN is intuitive. People understand it readily, and it seems to enable people to ask critical questions related to the (somewhat) implicit argument presented by GSN. It has motivated many developers and certifiers to consider seriously what must be demonstrated to ensure safety. There is significant work on automating aspects of GSN, including safety case construction and formal approaches to dealing with evidence. An excellent source for this is https://ti.arc.nasa.gov/profile/edenney/.

2.3 GSN Challenges

GSN promotes an ad hoc approach to structuring an assurance case. There is nothing in GSN itself that helps us decide how to present a safety argument. Patterns and experience are the basis of good GSN assurance cases. The tree-like structure, while intuitive, results in cross-cutting concerns that make creating, understanding, and maintaining a GSN assurance case extremely challenging. The major challenges introduced by GSN are: (i) it leads to a false sense of confidence because the reasoning in GSN is about why/how claims are decomposed, not as to why premises (grounded in evidence) support parent claims, and (ii) rigorous safety impact analysis is extremely difficult, bordering on impossible. The inherent traceability in GSN is through arcs connecting nodes, and this will not detect the impact of changes in parts of the tree that are not explicitly connected.

3 Workflow$^+$

Workflow$^+$ (WF+) is a modelling framework which aims to provide a way for all information necessary for safety assurance to be captured in a single model. This model shows relationships between development processes, assurance processes, development outputs, assurance outputs, and the environment of the system(s) of interest. WF+ grew out of our work with an automotive partner; an overview of WF+ modelling and its core mechanisms can be found in (Diskin et al. 2019).

WF+ uses metamodels that define workflows to be followed during the development of domain-specific safety-critical systems, complete with all process definitions, data definitions, control flow, data-to-data and data-to-process traceability, and constraints over processes and data. These core mechanisms allow all necessary validation, verification, checks, and reviews to be modelled and included. When the process defined in the metamodel is executed, an instance of this metamodel documents the development of the real-world system produced. This includes details of the system data, reports generated by tasks within the process, etc.

A metamodel can be based on prevailing standards such as ISO 26262 (ISO 26262, 2018), best practices, internal company procedures, etc. and can also be used to check compliance with the different types of guidelines mentioned. A metamodel can be checked to see that it is well-formed based on rules suggested by the mathematical foundations of these models. These checks on well-formedness result in assurance steps, and these assurance steps can be viewed in different ways – one

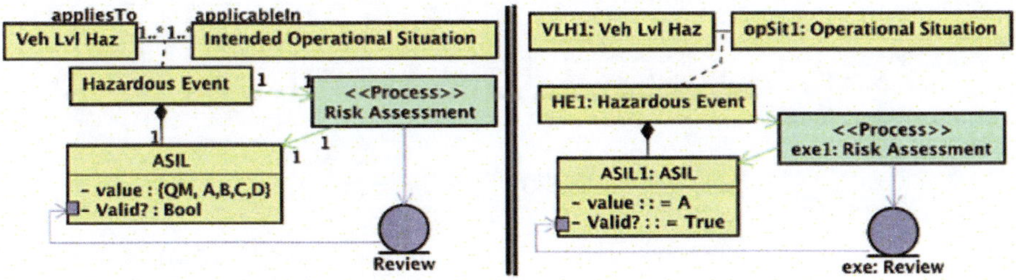

Fig. 2 A simple WF+ example (metamodel on the left, instance on the right)

important view is a GSN-like structure. Metamodels can be thought of as templates for development and/or assurance.

The WF+ metamodels presented in this paper are built using a profile of UML class diagrams with the following features: (i) two types of classes, process and data classes; (ii) two types of associations – dataflow associations (green) from data to process classes and back and static data associations (black) from data to data classes; (iii) a special type of process class to model reviewing; and (iv) several constraints on the interactions of the features mentioned above, the most important of which is that processes and their dataflow form a hierarchy, i.e., a directed acyclic graph.

Figure 2 shows a simple example of a WF+ metamodel of the risk assessment process described in ISO 26262 and an instantiation. In this example, the metamodel (left) specifies that when executed, the Risk Assessment process takes in a hazardous event, which is a pair of an operational situation and vehicle-level hazard, and outputs an ASIL classification for that hazardous event. The output data are connected to the input data, shown as a composition (black diamond) association from Hazardous Event (HE) to ASIL. The multiplicities dictate that this process is to be executed once for each HE and that each execution produces one ASIL. There is also a review process (purple), which evaluates the execution of Risk Assessment and the validity of its data. The instantiation (on the right) shows the documentation of an execution of Risk Assessment for a particular hazardous event HE1, which was determined to be ASIL A, and the output of a review process that validates this ASIL classification for HE1.

3.1 A WF+ Example

Figure 3 is a refined version of the example metamodel in Fig. 2. (The added argument elements will be explained later.) In this example, the input data definition has been refined to capture the types of Intended Operational Situations that can be part of a HE (operational situations are from (SAE J2980, 2015)) and the Consequence(s) of HEs. The definition of the Risk Assessment process itself has

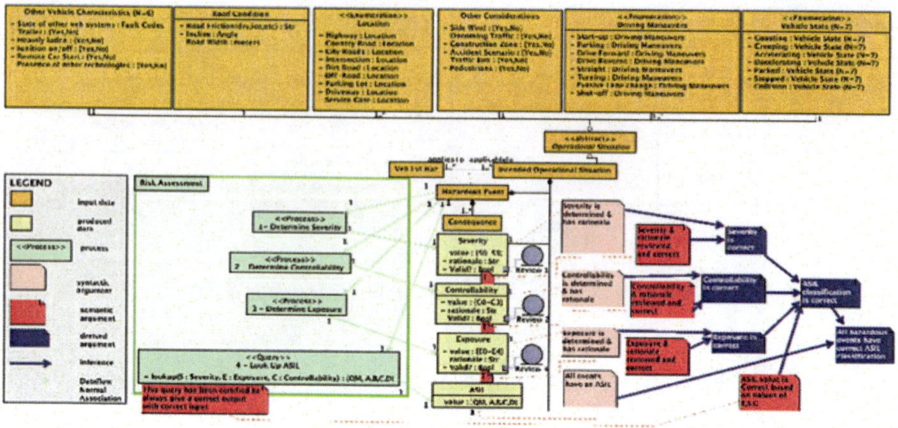

Fig. 3 Refined version of the WF+ example

been refined by decomposing it into four steps. When executed, steps 1 and 2 take in one HE, including Consequences, and produce values for the Severity and Controllability of that HE.

When executed, Step 3 takes in one HE and outputs Exposure as an attribute of the Intended Operational Situation of that HE. This HE, which will have its Severity, Controllability, and Exposure assigned, is then input to Step 4. Step 4 is a query (i.e., automatic process) that, when executed, assigns an ASIL classification to that HE. As steps 1, 2, and 3 define processes that are to be executed by humans, they have accompanying review processes to assess the validity of their output. As Step 4 is a Query, its output does not require validation (more on this in 4.6 Automation).

3.2 *Building Arguments Over a WF+ Example*

When a WF+ metamodel is built, its creators include constraints designed to ensure that its instances will be (i) syntactically correct (i.e., properly structured) and (ii) semantically correct (i.e., valid). Assurance-related arguments can be created based on these constraints almost mechanically. Light and dark pink elements in Fig. 3 are natural-language arguments based on syntactic and semantic constraints on the data definitions in the metamodel. Syntactic constraints (light pink), such as the multiplicity constraint that each HE must have exactly 1 Severity attribute, are put in place in the process definition to ensure that instances documenting execution of the metamodel are always properly structured. Violations of these syntactic constraints in the instantiation of the metamodel could be detected automatically by tooling. We can also set the tool to make such checks in the process of building the instance so that the instance is syntactically correct by construction.

Semantic constraints (dark pink), such as the constraint T on the attribute Valid? of Severity, are put in place to ensure that all (critical) data in an instance are validated by a review. In an instance documenting the execution of a process, if the value of a Valid? attribute is False or missing, that constraint is violated.

Since ASIL is produced by an automated Query, we want to certify that the Query itself will produce the correct result given correct inputs, rather than manually review the Query's output every time it is executed. This is represented by the pink node attached to Step 4 (see 4.6 Automation).

By composing syntactic and semantic constraints, we can derive higher-level constraints and add higher-level assurance-related arguments for those derived constraints (blue). For example, when combining the syntactic and semantic constraints on Severity, we add the argument "Severity is correct." That is, we claim that if a severity value is present *and* it has been reviewed, then it is correct, as shown by the blue arrows in Fig. 3. Derived constraints can be formed by combining syntactic, semantic, and/or derived constraints. In a GSN setting, a *strategy* must be included to explain the logic as to why sub-claims support their parent claim(s). In a WF+ setting, derived constraints (and their corresponding arguments) are based solely on the logical composition of constraints and thus do not require a strategy. The logical soundness of this approach enables us to ensure that any valid execution of the process definition will satisfy all constraints and thus all arguments in the process definition will hold for any valid instance (see (Diskin et al. 2019) for details). This is useful as it allows process definitions to be used as templates for the development of assurance cases (see 4.7 Templates). It is worth mentioning that while WF+ itself does not ensure that considerations for safety are included in some particular workflow, it does, however, provide a setting in which workflows can be planned and evaluated by experts to ensure that we are adequately confident they will result in safe systems. In safety-critical embedded systems, including AVs, this usually amounts to ensuring that we must (i) demonstrate that the requirements specification will result in a safe system; (ii) demonstrate that the system satisfies its requirements; and (iii) demonstrate the system does not implement any behavior not in the requirements specification.

4 Advantages of WF+ for Model-Based Assurance

4.1 Making Assurance Less Ad Hoc

Assurance of AVs using GSN can be ad hoc and reliant upon how individual safety experts interpret the safety requirements, the certification standards, and the GSN syntax itself. More specifically, there is no precisely defined methodology for assurance when using GSN; it relies on individual expertise. There are benefits to this, as engineers can optimize the process of constructing an assurance case based on their experience with AVs, but this comes at a price in terms of repeatability and learnability. An important benefit of using WF+ is the clear methodology that is to be followed when building assurance arguments; this will enable engineers to more readily learn the techniques and for the steps to be repeatable.

By comparison with GSN, the semantics for each element of WF+, as well as the role each element plays in the assurance process, are precisely defined. This should lead to fewer opportunities for misinterpretation. Also, WF + 's structure makes managing large assurance cases more systematic, repeatable, and inexpensive.

4.2 Improved Traceability

Detailed traceability is essential for assurance cases of complex systems such as AVs, because it improves understandability and facilitates following all pertinent links to an argument. In particular, change impact analysis is completely dependent on accurate and complete traceability. Current approaches lack mechanisms to include direct traceability and often rely on implicit (i.e., assumed) traceability between arguments. For example, in Fig. 1, the traceability between hazards and their respective severity, controllability, and exposure is left implicit through the wording of the diagrammatic elements. This implicit traceability is easy to identify and follow in this simple example, but in large-scale industrial safety cases it is often much more difficult to identify and understand, especially when cross-cutting concerns branch over multiple argument legs. This places an undue burden on independent reviewers to discover this implicit structure on their own and, when compounded with the ad hoc structure typical of GSN-style safety cases, can lead to significant misunderstanding of the intended argument. This prevents independent reviewers from being able to review an assurance case with sufficient confidence and may result in potentially dangerous flaws in arguments.

WF+ was developed from the ground-up with this in mind and enables detailed traceability. All traceability necessary between data and processes underlying arguments is maintained explicitly and allows for cross-cutting concerns to be accurately and explicitly represented. WF+ facilitates improved understandability and independent reviewing and provides an excellent basis for change impact analysis.

4.3 Change Impact Analysis

When dealing with highly complex embedded systems such as AVs, it can be difficult to determine the impact of incremental design changes on the system's assurance case. As AV systems continue to increase in complexity, even experienced engineers have trouble keeping up with the thousands of connections between design and their respective elements of an assurance case. The model-based nature of WF+ provides the necessary foundations for change impact analysis to be automated as much as is possible. The detailed granularity and traceability possible in WF+ metamodels allow for tools to be built that can automatically follow traceability links to all related data and their associated arguments, directing engineers to areas of assurance that are affected by changes in design. On top of this, the well-defined semantics of WF+ metamodels and assurance cases allows for an explicit ontology of change propagation that enables a well-defined approach to assurance of incremental changes to systems.

4.4 Integrating Assurance with Development

The model-based approach of WF+ opens up the opportunity for WF+ models to be directly integrated with model-based development or V&V tools. This allows assurance to be built directly over data from development, rather than having an assurance case as a separate document with references to development documentation. With direct access to artifacts from development, some aspects of assurance cases can be generated automatically and validated based on the content of those design artifacts (see 4.6 Automation). While it is possible to integrate GSN approaches with development (Hawkins et al. 2015), integrating WF+ with development will allow for more scalable solutions that are better suited to change impact analysis. Also, its traceability into the environment facilitates dealing with feature interactions that stretch into the environment.

4.5 Automation

As AV systems continue to increase in complexity, it is desirable to automate as much of the assurance case development as possible to reduce development costs. Building WF+ on well-established MBD principles allows tool developers to leverage a wide range of pre-existing techniques for managing assurance cases, including automated querying to search assurance cases, and transformations for applying templates.

The model-based approach of WF+ allows for static syntactic correctness to be checked automatically. As more granularity is added to the WF+ metamodel, some semantically significant properties can be encoded in the structure of the metamodel through the use of constraints. For example, if there are certain structural properties of design-related elements desirable for safety, then the corresponding constraints can be placed on the metamodel to allow these properties to be checked automatically. Table 1 in ISO 26262-6 (ISO 26262, 2018) outlines properties of software architectural design that are desirable for avoiding systematic faults. Many of these properties such as restricted size of interfaces, restricted size of complexity of software components, and loose coupling between software components are all good candidates for automatic checking through constraints over detailed models.

An MBD approach also allows for some processing to be automated, such as Look Up ASIL in Fig. 3. It is possible for these automated tasks to be certified, i.e., have trustworthy outputs given correct inputs. Time is saved by automating the process, and time is saved by not requiring their outputs to be reviewed.

4.6 Templates

As with any safety-critical system, it is necessary to plan for the safety of AVs ahead of development. Assurance case templates aim to specify a nearly-complete assurance case for a particular type of system before development begins (see Wassyng et al. 2015). A template includes sufficiently prescriptive limitations on systems (as determined collectively by experts in the field) but still allows enough flexibility so as to not unduly interfere with the creative design of a system. Assurance case templates specify higher-level argumentation and the overall structure of an assurance case, as well as acceptance criteria for required evidence.

WF+ is well suited for implementing assurance case templates using high-level WF+ metamodels that must be conformed to. For a particular system, this WF+ template can be refined to fit the needs of the system of interest and can then be executed. The modular nature of WF+ allows for assurance case templates to be created hierarchically to produce different versions of the templates to fit different use cases. Benefits of this include repeatability, ease of audit, and potentially increased productivity as tools can be used to carry out refinements and instantiation. Importantly, it also facilitates incremental assurance when changes that eventually occur were already taken into account as options in the metamodel.

5 Conclusion

The modelling effort required by WF+ is substantial. However, the vast majority of the required modelling is of the form "model once – use many times." What do we get from this effort? One of the most important attributes is that it facilitates effective incremental assurance! Traceability links in WF+ are comprehensive and cover planning, development, processes, work products, and the environment. The fact that we start with metamodels that act as templates means we reduce confirmation bias in the assurance process. We also conform to the dictum – design safety into the vehicle, do not add it afterwards. Compared with existing notations/methods, WF+ is extremely rigorous, less ad hoc, facilitates automation of many aspects of safety assurance, and reduces the difficulties associated with cross-cutting concerns.?

Acknowledgments The authors want to thank our industry collaborators for the in-depth discussion and suggestions over the years we have been working on this topic. Input from Joseph D'Ambrosio, Lucian Patcas, Galen Ressler, Ramesh S, and Sigrid Wagner has been invaluable to our research.

References

Diskin, Z., N. Annable, A. Wassyng, and M. Lawford. 2019. *Assurance via Workflow+ Modelling and Conformance (an extended version)*. https://www.mcscert.ca/wp-content/uploads/2019/11/McSCert-Technical-Report-32.pdf.

Hawkins, R., I. Habli, D. Kolovos, R. Paige, and T. Kelly. 2015. Weaving an Assurance Case from Design: A Model-Based Approach. In *IEEE 16th International Symposium on High Assurance Systems Engineering*, vol. 2015, 110–117. Daytona Beach Shores, FL.

International Organization for Standardization. 2018. *ISO 26262: Road Vehicles – Functional Safety*. 2nd ed.

Kelly, T. 1998. *Arguing safety – A systematic approach to managing safety cases*. Ph.D. dissertation, Univ. York, York UK.

Rinehart, D., J.C. Knight, and J. Rowanhill. 2015. *Current practices in constructing and evaluating assurance cases with applications to aviation*. NASA Report NASA/CR–2015-218678.

Rushby, J., X. Xu, M. Rangarajan, and T.L. Weaver. 2015. *Understanding and evaluating assurance cases*. NASA/CR–2015-218802.

SAE International. 2015. SAE J2980. Considerations for ISO 26262 ASIL Hazard Classification.

———. 2018. SAE J3016. Taxonomy and Definitions for Terms Related to Driving Automation Systems for On-Road Motor Vehicles.

Wassyng, A., N. Singh, M. Geven, N. Proscia, H. Wang, M. Lawford, and T. Maibaum. 2015. Can Product-Specific Assurance Case Templates Be Used as Medical Device Standards? *IEEE Design & Test* 32 (5): 45–55.

A Model-Based Engineering Approach for Development of ADAS Features

Arun Adiththan, Joseph D'Ambrosio, Prakash Peranandam, S. Ramesh, and Grant Soremekun

Abstract Advanced Driver Assistance Systems and higher-level automated features are rapidly being deployed in the automotive industry. A common development approach taken for ensuring safe operation of these vehicles is to focus on driving real vehicles in the planned operating environment. This approach has benefits, including helping to identify challenging driving situations a vehicle may encounter and providing evidence of safe operation. However, driving millions of miles during vehicle development does not scale as more features are deployed. A model-based engineering approach can be used to augment real-world driving to provide a more efficient method for developing safe features. This paper describes key elements of such a model-based approach that includes a mission plan containing key parameterized use cases that trace down to simulation scenarios used to identify scenario edge cases to support Safety of the Intended Functionality.

Keywords ADAS · SOTIF · Model-based engineering · Simulation · Model fidelity

1 Introduction

The automotive industry is experiencing a revolution with the introduction of new technologies associated with safe, connected electric vehicles. Specifically, in the context of safety, a wide range of new features associated with Advanced Driver Assistance Systems (ADAS) and higher levels of automation (SAE J3016 2018) are currently being deployed, for example, GM's Super Cruise and Tesla's AutoPilot. These new features must operate in complex environments and must

A. Adiththan · J. D'Ambrosio (✉) · P. Peranandam · S. Ramesh
General Motors Research Laboratories, Warren, MI, USA
e-mail: joseph.dambrosio@gm.com

G. Soremekun
General Motors Research Laboratories, Sunnyvale, CA, USA

A. M. Madni et al. (eds.), *Recent Trends and Advances in Model Based Systems Engineering*, https://doi.org/10.1007/978-3-030-82083-1_35

ensure safe operation of the vehicle. Although traditional approaches to developing safety critical systems, such as ISO 26262 (ISO 26262 2011), do apply, they are not sufficient to ensure that these features operate correctly in such complex environments.

ISO 26262 focuses on functional safety and covers the measures and means to identify and contain risks arising out of failures in electronic subsystems. It does not address issues associated with making design tradeoffs, such as the need to balance false activations vs. lack of activation due to sensing/perception limitations of safety-critical automated features. In addition, a method is needed to assure that the risk of yet to be identified scenario edge-case conditions has been sufficiently minimized. Finally, a rigorous safety case with sufficient evidence must be provided to confirm that the above issues have been addressed.

ISO PAS 21448 Safety of the Intended Functionality (SOTIF) (ISO PAS 21448 2018) specifies a methodology to identify hazards associated with sensor technology limitations and development process activities to help ensure the target feature achieves an acceptable level of risk associated with these hazards. A key goal is to evaluate SOTIF in the context of both potentially hazardous known and unknown scenarios to provide an argument that these categories of scenarios are sufficiently small and residual risk is acceptable.

A model-based engineering (MBE) process, in combination with test results from operating development vehicles in the field, can produce evidence needed to support the argument that acceptable minimal risk has been achieved. Testing of development vehicles in the targeted operating domain, initially for environment data collection/analysis purposes, and then transitioning into testing the evolving feature being developed also support the feature safety case by helping to minimize the number of unknown hazardous scenarios. In addition to driver takeover events that are logged, development vehicle operator can tag challenging events they experience for later analysis, such as unique pedestrian behavior or specific road locations like difficult intersections or sharp turns. Without exposing a development vehicle to the complex real-world operating environment, it is difficult to argue that all possible hazardous operation situations have been identified.

Although operating development vehicles are essential, it is not sufficient for building the SOTIF safety case. Creating repeatable tests is very difficult to achieve with the use of development vehicles given the problems of orchestrating the operating environment to the precise conditions needed for the test. Another challenge is that even though a difficult scenario is identified by development vehicle operations, there still may be even worse associated edge-case conditions than were directly experienced by the test vehicle.

In this paper, we propose a model-based development approach to complement development vehicle testing. The approach makes use of a mission plan that evolves through the development process. Use cases, requirements, design models (both descriptive and analytical), and automated test scenarios are used to support the overall development of the intended functionality as well as the necessary evidence for the SOTIF argument. One key aspect is capturing parameter ranges associated with use case operating conditions. These use cases and associated parameter ranges evolve throughout the development process as new operating conditions are identified, and they are fundamental to the analysis that is performed during each stage of the development process. The approach also makes use of a series of increasingly higher-fidelity analytic models as the development process proceeds. These models support analysis and simulation activities used to derive initial key requirements associated with the feature being developed and are also used to ensure robust operation, confirm scenario edge cases identified in the field, and provide the ability for executing repeatable tests.

This paper is structured as follows. Section 1 provides an introduction, including motivation for a MBE approach. Section 2 describes key elements of our proposed model-based process and how it supports SOTIF and verification/validation. Section 3 describes the simulation approach implemented and requirements on simulation models. Finally, Section 4 provides a summary and discussion.

2 Model-Based Development Approach

2.1 Initial Requirements Development

At the start of development of a complex vehicle feature, an initial set of requirements for the system(s) that will implement the feature can be developed using different strategies, such as utilizing existing company knowledge/know-how from past development efforts. We suggest a data-driven, model-based engineering approach, especially if the feature is unique enough from anything developed previously.

This approach utilizes a Mission Plan (MP) document for the feature. The MP provides a high-level overview of the feature and describes all the use cases the feature needs to execute in a safe, reliable, and robust/resilient manner. Ideally, much of the information gold source for the MP is contained in models. These models are comprised of properties and diagrams (e.g., context, allocation, architectural, decomposition, use case, etc.) that describe various aspects of the feature, the systems that implement the feature, and all the use cases (languages like SysML (Hause 2006) can be used to develop the feature and system models).

The MP document is then generated/updated on demand by extracting relevant information about the feature and use case descriptions from the corresponding models. This methodology helps establish traceability and data consistency across engineering teams as the gold-sourced feature and system models mature during the development process.

The system models can also be linked to various types of analytical models that describe mathematical and physics-based relationships about the system parameters [SEBok]. In the Concept/Requirements development phase, the analytical models are often low fidelity (models with reasonable accuracy that execute quickly). Value ranges for environmental parameters (e.g., road surface type) and vehicle operation

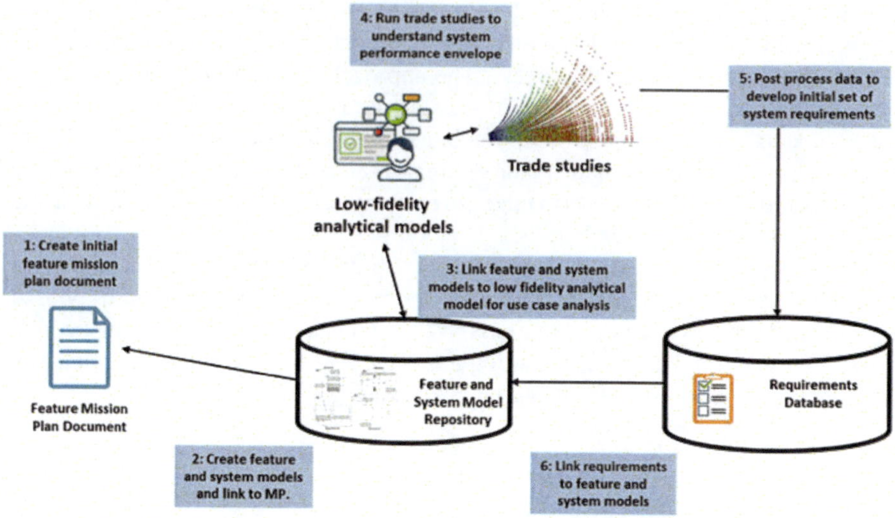

Fig. 1 Model-based approach for requirements and design development

parameters (e.g., min/max acceleration rates, velocities, etc.) defined in the feature and system models can be mapped directly to parameters defined in the analytical models that compute various performance metrics (e.g., vehicle stopping distance). Each combination of parameter values represents a specific use case instance.

At this stage, a data-driven approach can be employed where tens of thousands of use case instances are evaluated using the analytical models. The trade study data can be post-processed to determine the operational envelope of the system(s) so the feature can execute all use cases and an initial set of system requirements. To maintain traceability and data consistency, the requirement specifications are gold-sourced in a database that is linked to the feature and system model repository. The overall approach is shown in Fig. 1.

Consider a Crash Imminent Braking (CIB) feature. Development of CIB would involve creating a mission plan that provides a high-level overview of the feature and its intended functionality. A feature model would be comprised of decomposition diagrams describing different aspects of CIB functionality, and contextual diagrams describing interfaces between CIB and the systems that implement the feature, and CIB and its external environment. System models would describe all the systems needed to implement the feature, including the braking system and sensor system that needs to determine when braking needs to be activated.

Example use cases (described in the system model) are depicted in Fig. 2. The first use case (a) describes the situation where the host (blue) vehicle, which is equipped with the CIB feature, encounters a slower vehicle directly in front of it. The intent of the CIB feature is to automatically apply the vehicles brakes if the host vehicle and stopped vehicle are closing too quickly. The second use case describes the situation where the host vehicle equipped with CIB passes a vehicle

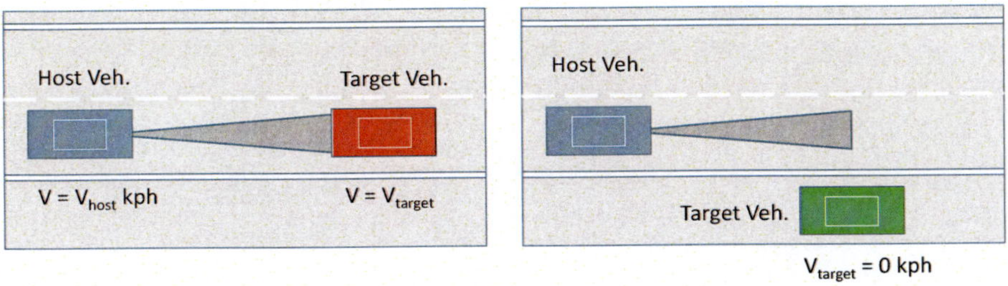

Fig. 2 Two CIB use cases: (**a**) encountering a slower vehicle; (**b**) passing a stopped vehicle

that is stopped along the side of the road. The intent in this case is to not apply the brakes given that the stopped vehicle is not in the path of the host vehicle.

Fundamental to the approach described in this paper is identifying parameters and their associated ranges that characterize each of the above use cases. Low-fidelity analytical models utilize these parameters to compute performance metrics such as braking distance. The analytical models are iterated over a range of the parameters to understand the performance envelope required of sensing and braking systems so that initial requirements can be specified. A primary goal of using this approach is to specify an initial set of requirements as accurately as possible so that the number of design changes that occur during later development phases is minimized.

2.2 *SOTIF Scenarios and Triggering Conditions*

As the development process progresses, SOTIF analysis is performed to ensure that the intended functionality of the feature has acceptable safety risk. SOTIF conceptually divides the relevant driving scenarios for a feature into four categories, as shown in Fig. 3. The categories are (1) Known Safe scenarios, (2) Known Hazardous scenarios, (3) Unknown Hazardous scenarios, and (4) Unknown Safe scenarios (see Fig. 3). A fundamental aspect of SOTIF is to reduce the risk associated with the Known and Unknown Hazardous scenarios by improving the feature or by providing the supporting evidence such that the scenarios can be finally classified as Known Safe scenarios.

To reduce both the Known and Unknown categories, SOTIF requires a systematic analysis of potential triggering conditions that could lead to a hazard. A triggering condition is a specific event or factor of a driving scenario that serves as an initiator for subsequent system reaction. For example, if we consider the CIB situation for SOTIF analysis, then a cutout (late reveal) of a vehicle as shown in Fig. 4(a) and tailgating in Fig. 4(b) are such triggering events. All identified triggering conditions are evaluated against the established acceptance criteria and must be shown to either be acceptable. Otherwise functional modifications to the feature are required to drive

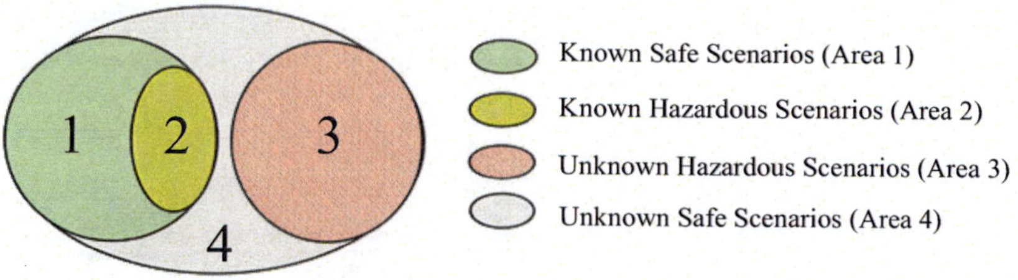

Fig. 3 Categories of SOTIF scenarios

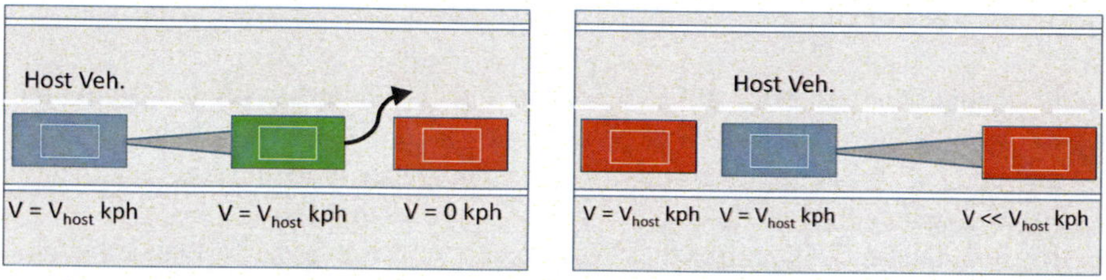

Fig. 4 Challenging use cases: (**a**) cutout; (**b**) tailgating

the assessment of triggering conditions to an acceptable level. For the examples shown in Fig. 4, there may need to be a tradeoff made between stopping fast enough to avoid a collision in the cutout use case versus being rear ended due to stopping too quickly in the tailgating use case.

An MBE approach can support the analysis of triggering conditions through simulation of the scenarios associated with the identified use cases. SOTIF specifies various methods to identify triggering conditions, and these conditions can be evaluated in the context of relevant scenarios to assess the ability of the feature to handle these situations in an acceptable manner. The identified use cases from the MP specify the functional range and desired behavior for a driving situation. These use cases can be refined into scenarios that describe the sequence of actions and events that take place during a certain time interval of interest. Comparing MP use cases versus SOTIF scenarios, a use case is a more general description that may be comprised of multiple relevant scenarios. Note that the parameters available to be controlled in a simulation may not directly match the parameters specified in the use case. In this case, there may be a need to map use case parameters to the related but nonidentical set of simulation scenario parameters.

2.3 *Identifying SOTIF Scenario Edge Cases*

Critical to the assessment of the scenarios is identifying the edge-case conditions associated with scenarios and triggering conditions to help ensure worst-case conditions are handled. In this paper, an edge case is defined as a specific set of scenario parameter values that lead to a worst-case condition. Consider the cutout scenario shown in Fig. 4. To help ensure the worse-case parameter value combinations are identified, multiple simulations are performed with different combinations of parameter values. The performance of the feature is evaluated across these simulations, and the worst case observed in simulation represents the edge case. Note that in some cases, especially for complex scenarios, there may be more than one edge case for the scenario.

Exhaustively exploring the scenario input parameter space is a time- and resource-intensive process and hence not desirable. Non-exhaustive parameter sampling techniques such as random, structural, and statistical are inadequate to effectively identify faulty input combinations in nonlinear and complex systems such as automotive software (D'Ambrosio et al. 2019). Therefore, we employ Quasi-random Technique (QRT) (Chen and Merkel 2007) for generating input combinations. The QRT sequences have low-dispersion and low-discrepancy properties, and hence the sampled points are spread more evenly throughout the hypercube. In contrast, random techniques (such as uniform distribution) might lead to regions where there are clusters of points as well as regions with significantly less points.

2.4 MBE Support for Verification and Validation

Verification and Validation (V&V) of automotive systems involve testing and analysis of the behavior of system features under various scenarios to ensure compliance to specified functional and nonfunctional requirements. In the automotive domain, V&V activity for ADAS features typically involves extensive exposure (or on-road) testing and/or closed course testing. While closed course testing is more controllable and reproducible compared to on-road testing, it is still both time- and resource-intensive. In addition, it is difficult and/or dangerous to create certain conditions in the real world. Simulation of driving scenarios, on the other hand, enables generation of near real-world conditions in simulated environments without safety and cost concerns of real-world testing.

An MBE approach provides the opportunity to perform V&V tasks using simulation to augment exposure or development vehicle testing. Scenario tests should be reviewed and classified with respect to the appropriate testing technique (e.g., test track or simulation). In some cases, requirements testing of the sensors and perception system may be performed in vehicle, while testing of planning and decision-making is done through simulation. This approach helps address current challenges related to sensor models (e.g., radar) that may run significantly slower than real time, thus limiting the amount testing that can be done. For those to be simulated, the simulations can build upon the same scenario models used to identify edge-case conditions for SOTIF. The software targeted for final release is utilized in the simulation environment for the feature behavior. Use of high-performance/cloud computing environments may reduce the overall test time required and provide the opportunity to comprehensively evaluate all known scenarios. The amount of testing required is a function of the number of known scenarios, the possible variability associated with these scenarios, and demonstrating continued robust operation in real-world conditions (lack of identification of new unknown scenarios).

3 Simulation Framework and Model Capabilities

Simulation-based validation of ADAS features requires models of the environment, the feature, the base vehicle, and drivers (both host vehicle and traffic). The models can range from low to high fidelity depending on how early or late in the development process tasks are being performed. The general trend is to improve model fidelity as the design matures. The fidelity of the model may also depend on the scope of simulation such that high-fidelity models may be used at the component level and relatively lower-fidelity models used at the feature/vehicle level. The model fidelity determines the level of effort required to develop the model and the accuracy with which it captures the characteristics of the actual system under test.

Our simulation framework is shown in Fig. 5. Using the QRT sampling method described in Sec. 2.3, we generate hundreds of thousands of variations of a base scenario. The generated combinations are then executed in the simulation platform (e.g., VIRES VTD, CARLA) with controller model in-the-loop. The log data for each scenario variation is collected using the simulation platform APIs. The collected log data is then analyzed using a sensitivity analysis (Saltelli 2002) technique. The sensitivity analysis allows us to identify and rank scenario input parameters according to the effect on the output being tracked. This sensitivity index is then used to identify whether the most sensitive parameters are considered in the existing controller model. If they are not factored with sufficient detail, the model can be refined in order to improve overall system performance. The resulting

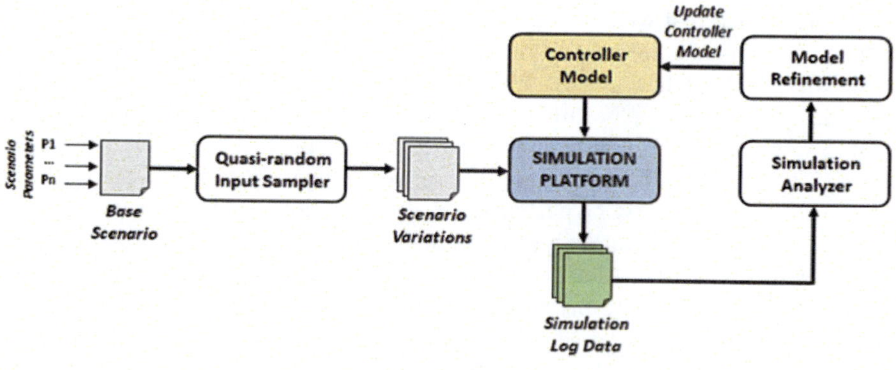

Fig. 5 Simulation-based validation-driven model evolution

higher-fidelity model obtained after each iteration of the simulation-based validation process can further be refined according to the desired level of model complexity and expected performance.

4 Summary and Discussion

The MBE approach described in this paper can be a key element in supporting the overall development of ADAS and higher-level automation features. A mission plan capturing use cases serves as a focal point for coordinating the development activities. The traceability provided by the model-based environment helps ensure that all identified requirements are properly implemented and tested. Also, by making best use of physical assets (e.g., real vehicles) and model-based assets, the development process can transition from being focused on driving real vehicles millions of miles to one more focused on scenario-based testing. Through the combination of model-based analyses to support SOTIF and real-world testing focused on challenging situations (vs. just driving many miles), efficient development of safe ADAS features is possible.

References

Chen, T.Y., and R. Merkel. 2007. Quasi-random testing. *IEEE Transactions on Reliability* 56 (3): 562–568.

D'Ambrosio, J., A. Adiththan, E. Ordoukhanian, P. Peranandam, S. Ramesh, A. Madni, and P. Sundaram. 2019. An MBSE Approach for Development of Resilient Automated Automotive Systems. *Systems* 7 (1).

Hause, M. 2006. *The SysML Modeling Language, Fifteenth European Systems Engineering Conference, September.*

ISO 26262 Road Vehicles- Functional Safety. 2011. *International Organization for Standardization*, Geneva, Switzerland.

ISO PAS 21448 Safety of the Intended Functionality. 2018. *International Organization for Standardization*. Geneva, Switzerland.

SAE J3016. 2018. *Taxonomy and Definitions for Terms Related to Driving Automation Systems for On-Road Motor Vehicles*. Society of Automotive Engineers.

Saltelli, A. 2002. Sensitivity analysis for importance assessment. *Risk analysis* 22 (3): 579–590.

Systems Engineering Body of Knowledge (SEBoK). 2019. https://www.sebokwiki.org/wiki/Types_of_Models

Optimal Management and Configuration Methods for Automobile Cruise Control Systems

Arun Adiththan, Kaliappa Ravindran, and S. Ramesh

Abstract Autonomous systems incorporate varying degrees of adaptation behavior to sustain their operations with acceptable quality of service (QoS). The QoS capability of such highly complex dynamic adaptive systems depends on how well they respond to hostile external events. The paper formulates *model-based assessment* techniques to benchmark the QoS capability of a networked system of cars *S*. We elaborate on this approach with a MATLAB-SIMULINK-based case study of adaptive cruise control (ACC) system in automobiles: first, for in-vehicle CC and then for multi-vehicle coordinated ACC. We employ model-predictive intelligent control methods to dynamically adapt the ACC system configurations to attain optimal behavior.

Keywords QoS of adaptive systems · Model-based assessment · Automobile cruise control systems

1 Introduction

A quantitative assessment of the QoS (quality of service) capability of an embedded software system *S* enables a sustained QoS behavior and/or reduced cost of system operations, in the face of uncontrolled external environment conditions incident on *S* (Gjorven and et al. 2006). In this paper, we study the *autonomic management* of adaptive cruise control (ACC) processes in automobiles (Wiki 2016) based on a model-based system assessment logic.

The managed system is a native in-vehicle CC operating under hostile road conditions E^*: e.g., road slipperiness and elevation, wind forces, air density, etc. E^* depicts the events that are hard to measure but nevertheless significantly impact

A. Adiththan (✉) · S. Ramesh
General Motors R&D, Warren, MI, USA

K. Ravindran
The City University of New York, New York, NY, USA

A. M. Madni et al. (eds.), *Recent Trends and Advances in Model Based Systems Engineering*, https://doi.org/10.1007/978-3-030-82083-1_36

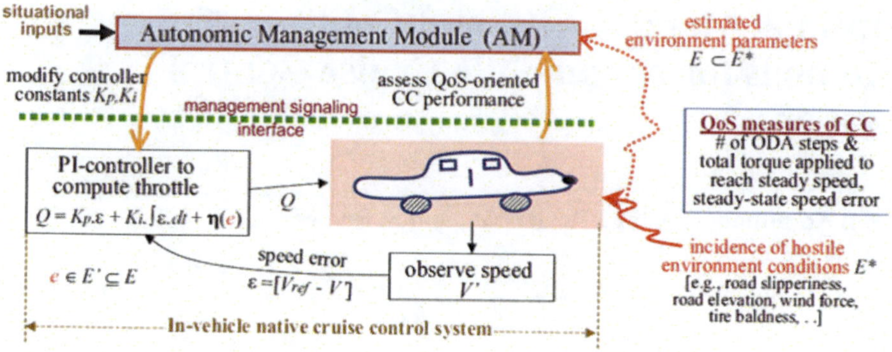

Fig. 1 Functional blocks composed in our ACC system (=CC + AM)

the CC operations. The autonomic manager (AM) is a situation-based control module that optimizes the in-vehicle CC behavior in a context of the sensed external environment conditions $E \subset E^*$. The AM is realized as a hierarchical control module interacting with the in-vehicle CC system via a signaling interface – as advocated in (Ravindran and Ramesh 2014). The paper extends our earlier preliminary study on the infusion of intelligent control behavior by external management modules (Adiththan et al. 2017).

The AM maps the user-level QoS-oriented preferences onto a CC logic. Here, QoS depicts multiple attributes of the speed ramp-up behavior of a vehicle when a reference speed is set, such as fuel consumption for speed ramp-up, latency in reaching a reference speed, and speed jitter in steady-state. The CC logic determines the torque generation from engine components to speed up/down the vehicle in pursuit towards a desired speed while meeting a desired QoS specs. It is often based on a proportional-integral (PI) controller with gain constants: (K_p,K_i) to map the observed speed error onto a torque. We use a CC simulation module developed at CalTech, written in MATLAB-SIMULINK, in our study of AM.

See Fig. 1. The AM consists of two functional modules: (i) *monitor* to assess the QoS performance of CC under the prevailing E^* and (ii) *configuration generator* to determine the modified in-vehicle CC settings (K_p,K_i) to effectively deal with E^*. (i) quantitatively analyzes how the QoS attributes are affected by E^*, whereas (ii) dynamically changes the control algorithm and/or its parameters to achieve a calibrated improvement under E^*. For instance, the torque generated by the in-vehicle CC is increased in the face of higher road elevations and/or wind forces (the latency and steady-state speed are also suitably controlled). The AM dynamically plugs in suitable in-vehicle CC parameters (K_p,K_i), as determined for the sensed environment $E \subset E^*$ under a certain policy. A policy may also factor in other situational inputs: say, road detours and barriers and multi-vehicle coordination needs.

We also consider a multi-vehicle cruise control (MVC), realizable by coordinating the per-vehicle CC modules. The multi-vehicle coordination involves generating vehicle configurations that enforce the inter-vehicle safety spacing constraints while maximizing the collective QoS experienced by the vehicle ensemble.

2 QoS-Oriented View of ACC System

The true model of a vehicle's raw physical process (RPP), i.e., how the engine, tires, and axles perform under various road conditions, is often not known in an exact form. So, the CC logic uses an approximate model of the RPP, as absorbed in a PI-control law (PI: proportional-integral) (CalTech 2014). The difference between the model-expected and observed speeds during a control step is also factored in a decision for the next step about the torque to be applied. An iterative sequence of such control steps leads the CC to a steady state.

The number of observe-decide-act (ODA) steps needed to reach the steady-state (K_{oda}) is a measure of the latency of CC for speed-up, with $\Sigma_{j=1}^{Koda} Q(j)$ being the total torque expended for the speed ramp-up – and hence the fuel consumed – and $[V_{ref} - V'(K_{oda})]$ being the steady-state error (SSE) (Kienke and Nielson 2005). The latency, torque, and SSE are the QoS attributes for the in-vehicle CC. A parameter setting of small values for the PI constants (K_p,K_i), for instance, yields a lower SSE, *albeit* at the expense of a larger latency and/or more fuel consumption.

The controller generates a trajectory of torque values $[Q_j]j = 1,2,\cdots,K_{oda}$ that finally leads to a sustainable speed V', where $V' \approx V$. An optimal QoS q'_{opt} depicts a scenario where the QoS utility of CC as perceived by the drivers is high, i.e., $U(q, q'_{opt}) \approx 1.0$. A control trajectory, as computed oblivious of the road conditions by a static PI-control law, often leads to a less optimal QoS (when compared with its intelligent counterpart). The suboptimality of QoS is more pronounced when the road conditions are light to moderate, because the static controller is constrained to a smaller region of the search space of [speed, torque] tuples for control trajectory generation.

In our model-based approach, the CC employs user-supplied utility functions to map a degradation in each of these QoS attributes onto user-level displeasure – and hence a contribution to the overall cost. Here, a throttle action X contemplated for speed ramp-up factors in the expected changes in [K_{oda}, Q, SSE] – and hence the cost incurred by X. Among various candidate actions $\{X\}$, the controller chooses an X that is expected to incur the lowest cost and then unfolds X to be exercised on the vehicle. Figure 2 illustrates a mapping of the system-level QoS of CC to an user-perceivable utility index: $U(q, q') \in [0,1]$.

Our QoS-centric optimization approach for CC can be contrasted from the control-theoretic optimization approach advocated in (Bauer and Gauterin 2016). The latter too provides for model-predictive computation and horizon-based planning to deal with: (i) unexpected disturbances encountered on a road over short timescales and (ii) control inaccuracies arising from model uncertainty itself. In our approach, however, (i) and (ii) appear as a difference between the model-computed and observed costs, thereby triggering a throttle action to lower the cost

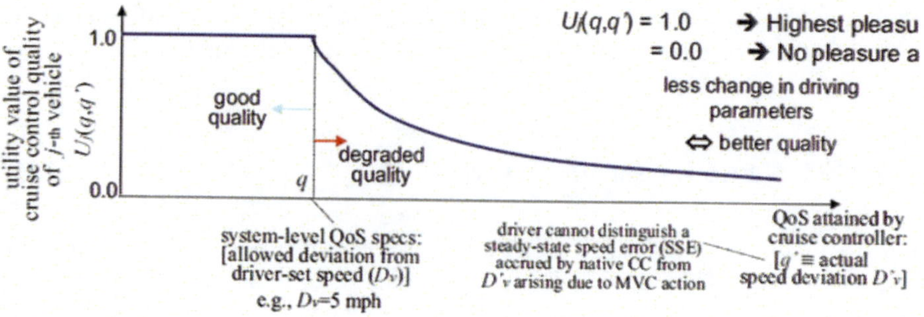

Fig. 2 Usefulness measure of a cruise control system

in a next control step. Because of its reliance on predefined cost relations and utility functions, our approach is more amenable for incorporation as generalizable software-level solutions (say, offloading the expensive model-predictive computations to a cloud).

3 Autonomic Control of In-Vehicle CC

We adjust the K_p and K_i parameters dynamically based on the sensed and/or estimated environment conditions E. For example, if the wind speed increases, K_p is jacked up to generate higher amounts of torque in the throttle actions that would counter the wind forces more effectively (than what a statically set K_p would otherwise do). Likewise, the K_i parameter that captures the accumulated modeling error since the start of a CC action is also jacked up to accelerate the learning. Our dynamic setting of (K_p, K_i) based on the estimating the environment parameters such as road elevation (θ), air density (ρ), and road friction coefficient ($C_r \in [0,1]$) depicts an infusion of controller intelligence with the process level modeling of core components of in-vehicle CC system.

3.1 *Optimal Setting of In-Vehicle CC Parameters*

Our study on how the QoS changes vis-a-vis some representative (K_p, K_i) parameter values of the PI controller reveals two points. First, a different controller configuration can yield a better QoS under the current ρ-C_r conditions. Second, an optimal setting of the controller parameters can change with ρ-C_r. We thus reason that [K_p, K_i] should be adjustable according to the current ρ-C_r conditions.

We comprehensively analyze the impact of [K_p, K_i] on the overall optimal behavior of the CC sub-system of a vehicle. It involves determining how the optimal

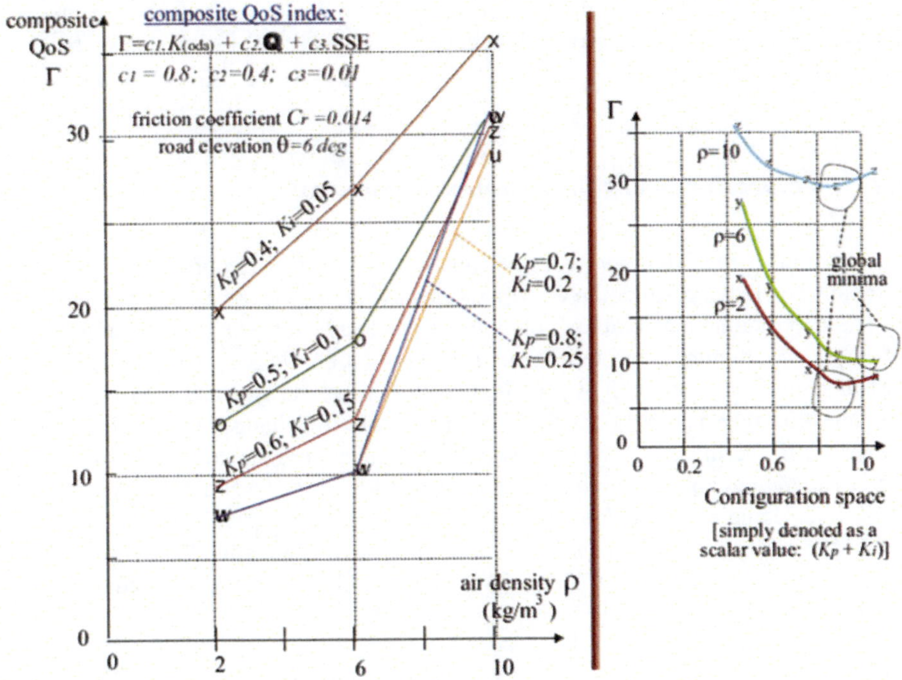

Fig. 3 Simulation results on optimal per-vehicle CC design

controller setting, denoted as $[K_{po}, K_{io}]$, changes under different ρ-C_r conditions. We formulate a composite QoS index γ for the CC sub-system in terms of individual QoS attributes $[K_{oda}, Q, SSE]$ attained with a controller setting$[K_p, K_i]$, as:

$$\gamma = c1.K_{oda} + c2.Q + c3.SSE, \tag{1}$$

where $c1$, $c2$, and $c3$ are positive constants (we use the notations q and γ interchangeably in the paper). The γ depicts a mapping of the QoS attributes onto a scalar space, attaining an optimal value at: $[K_p = K_{po}, K_i = K_{io}]$. The domain-specific interpretation of $[K_{oda}, Q, SSE]$ attributes, namely, lower values mean a better QoS, casts the optimization goal as finding the lowest γ value.

We conducted a gradient search of the $[K_p, K_i]$-space to determine the optimal setting. The composite QoS index γ was computed from the $[K_{oda}, Q, SSE]$ data output by the SIMULINK module for various $[K_p, K_i]$ settings under a given ρ-C_r. Figure 3 shows the γ values for different $[K_p, K_i]$ and ρ-C_r values. It is found that the setting $[K_{p0} = 0.7, K_{i0} = 0.2]$ gives the optimal point, namely, the lowest γ value. The system has a global minimum point, which is detected by the gradient search algorithm in a short number of epochs.

An optimal parameter setting $[K_{p0}, K_{i0}]$ is tied to the external environment conditions: $[\rho, C_r]$. So, a reasonably accurate (and low cost) estimation of the ρ-C_r values is needed.

3.2 Declarative Specs for Autonomic Control

An infusion of the intelligent control capabilities however requires capturing the domain-knowledge pertinent to CC operations by the AM, as described below.

The AM module may employ policy functions to adjust the (K_p, K_i) parameters – c.f. Eq. 1. We assume that the information about $[\rho, C_r, \theta, \ldots]$ is available to the AM in some form (say, with estimators running in the CC system's control plane). The question then is: how does the AM orchestrate the changes to (K_p, K_i)? This is because triggering these changes requires AM to capture the domain-knowledge: such as an increase in θ or ρ requires generating a higher Q. Over a certain operating region, the domain-knowledge can be codified as axioms and rules maintained by the AM to relate a change in $[\rho, C_r, \theta, \ldots]$ to a desired change in (K_p, K_i).

A declarative specs of the AM-level basic rule relates the current value of ρ to the newly sensed value of ρ as:

$$\begin{aligned} &> [\rho_{sen}, \rho_{cur}] \rightarrow \\ incr\_k_p &\left[diff\left(\rho_{sen}, \rho_{cur}\right), \alpha_{(p,1)}\right] \wedge \\ incr\_k_i &\left[diff\left(\rho_{sen}, \rho_{cur}, \delta_\rho\right), \alpha_{(i,1)}\right]; \\ &incr\_k_p\,[x, y] \rightarrow \end{aligned}$$

The applicative functions $incr\_K_p[.]$ and assign $K_p(.)$ are exported by the CC system for use by the AM: first, to realize an increase of K_p subject to the condition $\rho_{sen} > (\rho_{cur} + \delta_p)$ and second, to signal a plug-in of the modified K_p to the PI controller. Similarly, the update rules for K_p based on the changes in C_r and θ can be specified. Likewise, the update rules for K_i can be specified.

A declarative specs of the PI-controller update rules can make the AM oblivious of the domain-knowledge, enabling its reuse (at a meta-level). The AM is basically a symbolic processor that evaluates the truth or otherwise of various relations and then plugs in a new (Kp,Ki) as needed.

4 Multi-vehicle Cruise Control (MVC)

We outline the key tenets of a multi-vehicle cruise control system (MVC) for automobiles. Figure 4 shows the functional modules. The QoS depicts how safely a set of vehicles maintain their set speeds within acceptable deviations while avoiding

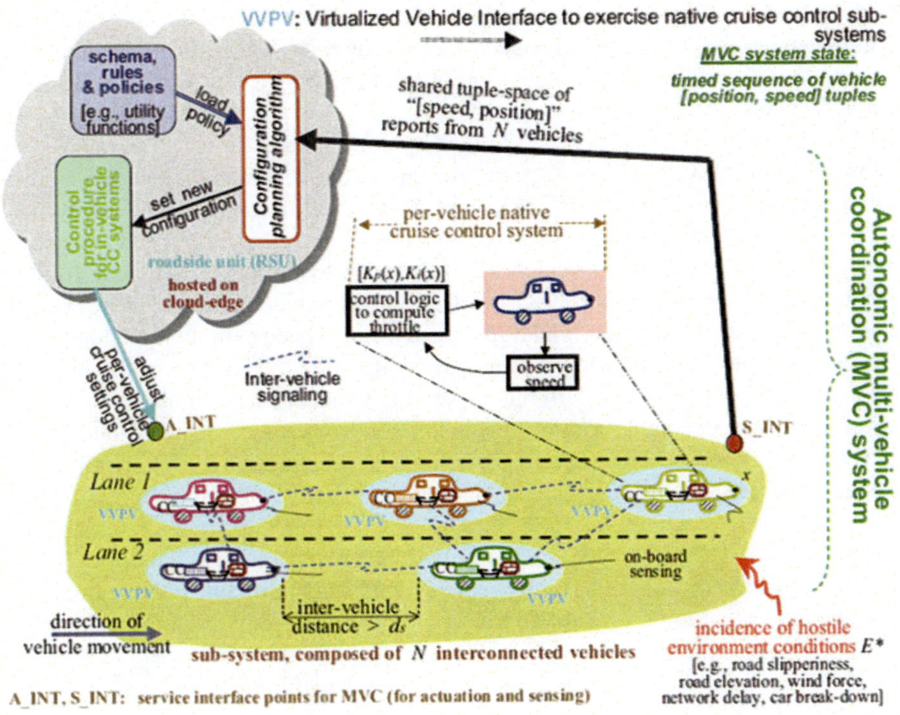

Fig. 4 Functional modules of a MVC system

drastic changes. We provide an empirical view of how an optimal QoS behavior can be determined for a multi-vehicle ensemble.

4.1 Modeling Aspects of MVC

Suppose a vehicle X takes an action to speed up or down, as part of its local CC action. X cannot hit any of the vehicles $\{X'\}$ moving in its vicinity. To satisfy this critical safety requirement, a coordination of the vehicles is needed wherein an intended move of X is evaluated in conjunction with its effects on one or more of the nearby vehicles $\{X'\}$. For instance, a slowdown of X can force a vehicle X' behind also to lower its speed (but may not affect the vehicles ahead). The extent of slowdown of X' is determined by many parameters, such as its distance to X, environment conditions (e.g., road wetness, vehicle traction), occupant characteristics (e.g., comfort level, children and old-aged passengers), etc. We envisage that the multi-vehicle coordination controller C gets hosted on a roadside assist compute node (possibly located at the edge of a compute cloud), with adequate physical and logical sensors and computational resources to enforce the QoS needs of MVC. *C* interacts with the various vehicles under its control realm over wireless signaling channels. Given that each vehicle itself implements an onboard CC module, the coordination of multiple vehicles by *C* is deemed as a hierarchical control.

The QoS in MVC scenario is determined by how far each car is forced to deviate below from the driver expected speed. We define the speed deviation, D_v, as the difference between a driver-set speed and a speed adjustment computed by the MVC controller *C*. One scenario where C may choose to adjust the speed is when a vehicle *V* interacts with other vehicles $\{X'\}$ that have lower locally set speed limits. The D_v is thus different from the in-vehicle *SSE* parameters, but the driver often cannot distinguish between D_v and *SSE*. The adjusted speed computation done at *C* may consider the safe following distance ds set by individual drivers while turning ON the cruise control mode. Different drivers may independently choose d_s values according to the acceptable comfort level while the vehicle is on cruise mode.

The MVC works by potentially overriding the speed limits of individual car drivers, to achieve a collective safe driving of the *N*-vehicle ensemble while striving to keep the D_v of each vehicle low for a given d_s setting (i.e., maximize the per-vehicle QoS of CC). The QoS of MVC is thus gauged by how good the MVC algorithm strives to keep all the *N* vehicles safe with minimal displeasure. An assessment of this MVC capability requires a model of the multi-vehicle ensemble.

4.2 Optimal Configuration of MVC System

A globally optimal configuration for N vehicles is one that minimizes the cost:

$$\Gamma_{\mathrm{m}} = \sum_{k=1}^{N} 1\text{-}U_k \left[D_v(k) - D_v{}'(k) \right]; \tag{2}$$

subject to a constraint that the inter-vehicle distances are higher than the safe-limit d_s. Here, V(k) is the MVC-enforced speed limit of kth vehicle and $D_0(k) = [V_{ref}(k) - V(k)]$ and $U_k(.)$ is the per-vehicle utility function. Here, $D'_v(k) = [V_{ref}(k) - V(k)]$, where $V'(k)$ is the actual speed sustained by the local CC of kth vehicle, with $\varepsilon(k) = [V(k) - V'(k)]$ being the steady-state error (SSE) induced by the native PI-control law. $D'_v(k)$ is not known to the kth driver, unless the MVC signals about the speed override. Here, the computation of Γ_{m} for a single configuration would amount to running *N* instances of SIMULINK module (each instance with potentially a different parameter set) and summing the per-driver displeasure accrued in that configuration.

Determining the globally minimal Γ_{m} is however an NP-complete problem, as the MVC should conduct an exhaustive search of all feasible configurations. If *L* is the number of distinct speed settings possible, the computational complexity of an exhaustive search would then be $O(L^N)$. Therefore, one should employ a greedy (and/or genetic) search algorithm to quickly come up with a reasonable suboptimal configuration. Such an MVC-level QoS definition would require knowing what the global minimum would be – in order to determine how suboptimal a greedy-computed solution is.

In the absence of accurate knowledge about the best configuration G^*_m, we have the MVC specify a cost threshold to compute an acceptably suboptimal configuration G^a. The use of situational knowledge by a management entity would allow specifying $G^a{}_m$ (say, through a GUI). We then use G^a as a *reference benchmark* for analysis purposes. Now, the MVC would run its own heuristics-based search of some representative configurations to quickly come up with its own view of what the (sub-)optimal configuration would be – which we denote as $G^a{}_g$. If $\Gamma(G^a{}_g) > \Gamma(G^a)$, the search gets preemptively stopped; otherwise, the search would end by a designer-set termination condition: say, the cost difference $[\Gamma(G^a) - \Gamma(G^a{}_g)]$ falls below a threshold. Thereupon, the MVC system is deemed to have reached a steady state.

4.3 Configuration Search Algorithms

As outlined in the earlier section, we employed greedy search and genetic search methods to decide a next state of the MVC system that would generate a close-to-optimal trajectory. To determine how suboptimal the trajectory is, we also carry out an exhaustive search of the state space to find the best solution Γ^*_{m} (which is basically a brute-force computation).

For greedy search, the MVC controller treats a subset of the vehicles $\{X\}$ as ignoring its commands, when computing a new configuration (even though $\{X\}$ may be compliant) – where $1 < |X| \ll N$. The non-consideration of vehicles $\{X\}$ in a decision-making by the MVC manifests as allowing them to stay course, as determined by their native CC sub-systems – while the remaining vehicles $(N - |X|)$ are treated as controllable. This leads to a much small-dimensional search space: $O(L^N - |X|)$. A deterministic search therein yields a candidate that incurs the lowest cost among the small set of configurations considered.

A genetic search, on the other hand, occurs over the full-dimensional space of N vehicles but with a random selection of candidate vehicle configurations. A candidate configuration is produced after multiple crossover checks on the model-computed vehicle states. The random mutation of multiple solutions (two or more) occurs on those known to be good for at least a selected subset of the vehicles. The rationale is that the sub-elements of parent configurations known to be good for some vehicles will likely yield a better offspring when combined (i.e., a new configuration good for additional vehicles as well). The evolutionary process leads to a better offspring, albeit probabilistically, as a result of the crossovers of sub-elements from good parents.

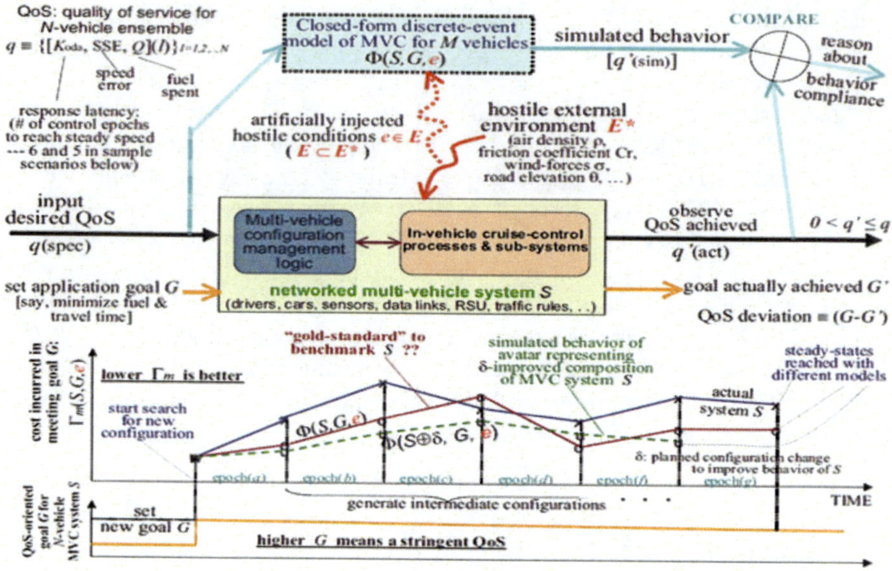

Fig. 5 Avatar-based approach for MVC system benchmarking

4.4 Avatar-Based Benchmarking of MVC System

We employ *avatars* to generate multiple benchmarks for an MVC system S in a simulated world. An avatar is basically a mathematical model of S executed in a virtualized world, which is subjected to simulated environment conditions that are close approximations to the actual conditions incident on S. A model $\Phi(S,E,G)$ representing the current system S is used to benchmark the actual behavior of S. This objectively verifies the goodness of S in meeting its QoS goal G in the face of environment conditions $e \in E$. See Fig. 5.

Different models of S that represent alternate compositions of S from the core functional components, namely, the per-vehicle CC processes and the configuration parameters, may yield different QoS behaviors. For instance, an increase in speed and/or lane change of one or more cars can be parameterized in the sub-system models, thereby computing its effects on the MVC-wide QoS – and the associated configuration cost (S,G,E). Such purported changes are first analyzed by a model-driven simulation of the modified MVC processes $(S + \delta,\ e,G)$ in a virtualized world, before committing a configuration change δ in the actual MVC system. The analysis involves two things: (1) comparing the estimated behavior of model $\Phi(S + \delta,E,G)$ with the currently observed run-time behavior of actual system S and (2) infusion of system-level changes δ, if needed, in the operational processes of S using compositional methods (e.g., changing the K_p/K_i parameters of in-vehicle CC). In the above light, the behavior computed by the model (S, e,G) serves as a "gold-standard": first, to determine how good is the actual behavior of S relative to the expectations and second, to provide the basis for a calibrated change in S. In general, the different benchmarks obtained from alternate compositions of S can be analyzed to meet the desired optimality conditions. The model-driven simulations of $\Phi(S,E,G)$ and $\Phi(S + \delta,E,G)$ can be seen as different avatars of S for benchmarking and reconfigurations of S.

5 Conclusions

Our work adds a *software cybernetics* dimension to the autonomic management of an in-vehicle CC system. The dynamically settable CC logic allows a vehicle to be more responsive to the stringent demands in a multi-vehicle coordination scenario, relative to a statically set CC. Our compositional design of ACC system promotes incremental evolution and reconfiguration of ACC functionality by software and/or model reuse. The paper also described a multi-vehicle coordination functionality composed from the per-vehicle CC models (static or dynamic) – realizable on a roadside unit (say, hosted on a cloud edge). Our adaptive control of vehicles meets a critical functionality requirement for Intelligent Transport Systems, as enunciated in (Koopman and Wagner 2014).

References

Adiththan, A., K. Ravindran, and S. Ramesh. July 2017. Management of QoS-oriented Adaptation in Automobile Cruise Control Systems. In *Proc. Intl. Conf. on Autonomic Computing, (ICAC)*, 79–80. Columbus.

Bauer, K. and F. Gauterin. 2016. *A Two-Layer Approach for Predictive Optimal Cruise Control*. SAE Technical Paper: 2016-01-0634. https://doi.org/10.4271/2016-01-0634.

Caltech Research Group. Cruise Control. 2014. http://www.cds.caltech.edu/murray/amwiki/index.php/Cruise control.

Gjorven, E., F. Eliassen, and J.O. Aagedel. 2006. Quality of Adaptation. In *Proc. SELF: Self Adaptability and Self-Management of Context-Aware Systems, Silicon Valley*.

Kienke, U., and L. Nielsen. 2005. Road and Driver Models. Chap. 11. In *Automotive Control Systems: Engine, Driveline, and Vehicle*. Springer.

Koopman, P., and M. Wagner. Jan 2014. Transportation CPS Safety Challenges. In *NSF Workshop on Transportation Cyber Physical System*. Arlington.

Ravindran, K., and S. Ramesh. 2014. *Model-Based Design of Cyber-Physical Software Systems for Smart Worlds: A Software Engineering Perspective. In workshop on Modern Software Eng*. Hyderabad: Methods for Industrial Automation (MoSEMInA), ACM-ICSE.

Wikipedia. 2016. *Autonomous cruise control system*. https://en.wikipedia.org/wiki/Autonomous cruise control system.

A Systems Modeling Illustration of the Military Academy Doolie Cadet System

Nathan Hasuk Oh and Martin "Trae" Span

Abstract Systems engineering process models provide the framework for visualizing and organizing a system life cycle guiding its design and development. The systems engineering Vee process model is ubiquitous for describing systems. It is the most widely used and often tailored for specific system types. This paper models the doolie cadet as a system of interest using the Vee model. The term doolie refers to a freshman at the United States Air Force Academy. The doolie year is a rigorous year with difficult requirements to complete the transformation of high school graduates into disciplined military academy cadets. This work describes the challenges associated with the doolie year and models the doolie cadet as a system of interest using the systems engineering Vee model. Beginning with decomposition along the left side of the Vee, requirements, subsystems, and components that need to be assessed for an evaluation of performance are identified, progressing up the right side of the Vee; this work details verification events such as exams, athletic assessments, and military tests and training. These events validate the system, demonstrating that the doolie cadet is proficient and ready for acceptance into the cadet wing. This work contributes a new perspective on modeling human performance as a system of interest and new insights into uses of the systems engineering Vee model.

Keywords Systems engineering · Systems engineering process model · Life cycle · Military cadet · Education · first year student · Freshmen

N. H. Oh · M. "Trae" Span (✉)
United States Air Force Academy, Colorado Springs, CO, USA
e-mail: martin.span@usafa.edu

A. M. Madni et al. (eds.), *Recent Trends and Advances in Model Based Systems Engineering*, https://doi.org/10.1007/978-3-030-82083-1_37

1 Introduction

Military academies provide a unique rigorous environment to facilitate and accelerate character growth to develop high school graduates into leaders of hundreds of young servicemen upon graduation. More specifically, the United States Air Force Academy (USAFA) is a challenging institution whose goal is developing leaders to commission as officers within four years. Each USAFA cadet year has a unique set of challenges, rewards, and expectations that culminate into advancement to an officer. There are three commissioning paths into the Air Force as an officer. The first way is to attend the United States Air Force Academy; graduation confers both a Bachelor of Science degree and a commission into the Air Force. The second way is to attend a Reserve Officer Training Corps (ROTC) program at a civilian undergraduate institution. There is no guarantee, however, that one will graduate as an officer, but there is still a military training component to develop graduates into people of character. The final way is to complete Officer Training School. It is a rigorous nine and a half week program that only accepts persons with at least a bachelor's degree to attend (Officer Training School 2018). Upon completion of one of the three programs, a candidate will be commissioned as a second lieutenant in the United States Air Force. This work models the first year of officer development at USAFA using the systems engineering Vee process model. Prior works have been published in applications of systems engineering life cycles to human systems. The health information systems domain has demonstrated benefit from the use of human-centered design through the classical systems engineering method (George and Samaras 2005). SE models provide value to ergonomics by enabling a systematic human design approach. In a related application, this paper demonstrates a systems life cycle model as applied to human systems, a first year undergraduate student at USAFA. Section II introduces the reader with the requirements of doolie year at USAFA in an effort to provide context and background understanding. Section III describes the systems engineering Vee process model as the methodology used to model the doolie cadet system. Section IV presents the doolie cadet system as modeled by the Vee model, and Section V describes the impact of this unique human systems modeling effort. Section VI provides a summary and possible extensions for future work.

2 Background

The first year at the United States Air Force Academy (USAFA), infamously known as "doolie year," begins after the completion of summer basic training and builds on what cadets learned including to become a follower, the first step in leadership development. The separation of classes between a doolie cadet and the upper three classes is extensive. The term "doolie" actually originates from the classical Greek word *doulos* meaning "slave" or "bondservant." While freshmen cadets are not indentured servants, they have significantly limited privileges and extensive military training requirements to balance in addition to their first year of academics. There are strict rules against fraternization or unprofessional relationships between anyone of the upper three classes and a doolie, including social media relationships. There are various ROEs (rules of engagement) that all doolies must follow, for example, running on the long established strips (limited pathways between the dorms and academic buildings), greeting all upperclassmen they pass in close proximity, and other tasks focused on improving their discipline and bearing. Doolie cadets are constantly being observed and corrected by upper-class cadets, enlisted members, and officers. Unlike a civilian university, the USAFA doolie experience is split into three different categories of assessment, grade point average (GPA), military performance average (MPA), and physical education average (PEA). The combined result of a weighted average is called the overall performance average (OPA) which consists of 50% weighted GPA, 40% weighted MPA, and 10% weighted PEA. Each category can be further decomposed into components that comprise each element which are all factored into the validation of the doolie year (United States Air Force Academy 2018).

The mission of the United States Air Force Academy is not only to train but also to educate future leaders of the United States. This includes receiving a Bachelor of Science degree with a focused major and optional minor. The 8-semester curriculum averages 6 classes per semester, and each class is made up of 40 lessons with 53 min per lesson. The student-to-faculty ratio is one of the lowest in the nation at 8:1, meaning there are always instructors willing to help regardless of the time of day or situation. The entirety of classes taken during the doolie year are core requirement classes such as calculus I and II, chemistry, English, and physics and two classes in a foreign language. It is also a time to consider which major to select since one must declare no later than October of sophomore year. There are 27 majors offered with an opportunity to minor in either a foreign language, philosophy, or religious studies. At USAFA, courses are more structured than many other universities with an extensive amount of graded events including participation points, pre- and post-class quizzes, homework, graded reviews (tests), and final exams.

Most upper-class cadets will agree that while doolie year is the easiest academically since most classes are at the introductory level, it is the most challenging as military and physical events occupy almost every non-class hour of the day. The military aspects of doolie year are also very mentally challenging. Military training within the squadron is focused on developing doolies into disciplined followers and military personnel. A primary mechanism for military training in the squadron is training sessions which occur two to three times a week at 1530. What they entail are largely squadron dependent but are primarily focused around physical and mental training. One of the most difficult aspects of doolie year militarily is greeting all upper-class cadets and staff across the wing. A wing is comprised of the four thousand cadets that attend USAFA. Greeting within squadrons is difficult because every upperclassmen's (upwards of 70–80 people) name, job, and rank need to be memorized and recited out loud every time a doolie walks past any cadet of higher rank than them. Another military aspect of doolie year is taking weekly knowledge tests. The questions mostly consist of military quotes, specific information about the Air Force, and military knowledge on aircraft. Passing or failing these will not only affect doolies' stratifications (the individual's numerical rank based on their performance contrasted with their classmates' performance and their leadership's expectations), but they will also affect their upper-class MPA to some degree. Another form of training is called "Minutes." Every squadron does this multiple times a week where the doolies wake up earlier than the rest of the upperclassmen at around 0620 and yell out a pre-established script about the uniform of the day and every meal the mess hall is serving that day. During minutes, every doolie recites military knowledge and must have a current event ready to share upon the request of an upperclassman. As a doolie, freshman cadets are not permitted to have or wear civilian clothes any time the entire year and must wear the uniform of the day (UOD) until TAPS (2300 Sun-Thur, 0130 Fri-Sat) unless going to the gym. Doors in dorms must be open every night until 1950 with rooms in precise SAMI standard at all times. This SAMI standard requires an extreme amount of time to arrange the bedding and clothing to detailed and precise specifications with measurement tolerances required within ¼ inch of preciseness.

As a cadet, everyone must participate in a sport, either an intercollegiate team or a club sport competing against local schools or participating in intramural sports which go against other squadrons within the academy. The actual PEA comprises components such as physical education classes, the physical fitness test (PFT), and the aerobic fitness test (AFT). The PFT consists of pull-ups, long jump, sit-ups, push-ups, and a 600 yard run, whereas the AFT is a 1.5 mile run. There are various PE classes that are mandatory graduation requirements, such as boxing, water survival, and combatives, and also a variety of elective courses such as golf, warrior enhanced yoga, and rock climbing.

Recognition is the most culminating and rewarding event of doolie year, occurring in March. It is a highly anticipated three-day event for all doolies with the purpose of earning the recognition of their upperclassmen as being ready to train the next class of doolies and becoming upperclassmen themselves. It is a rigorous capstone event of intense military and physical training that ends with pinning prop and wings which symbolize the completion of the training aspect of doolie year. After completion, now-recognized doolie cadets earn back their civilian clothes, the ability to walk where they please, not having to greet, and several other privileges taken away since basic training. The background on the doolie year experience provided here provides context to understand the model elements presented in Section IV.

3 Methodology

Systems engineering's most widely accepted definition is from INCOSE:

> Systems Engineering is an interdisciplinary approach and means to enable the realization of successful systems. It focuses on defining customer needs and required functionality early in the development cycle, documenting requirements, then proceeding with design synthesis and system validation while considering the complete problem. (Incose 2018)

Much of early systems engineering was focused on large-scale systems (Ryen 2008) but has progressed into a broad discipline of versatile applications. Systems engineering is primarily focused on the system life cycle and ensuring the resulting product meets the initial requirements and customer needs.

There are numerous process models of the system life cycle such as the waterfall and Vee. The waterfall process model is a linear depiction of the system that progresses in a logical fashion (Powell-Morse 2016). In the 1980s, however, systems engineers decided they needed to modify the waterfall model, bringing the latter half of the components up into the right side of the diagram which paved the way for the new Vee model. This new model was able to verify early developmental stages to the later tested stages and make progressive alignments with the right and left sides as shown in Fig. 1 (Donald 2013). The verification and validation performed in the right side of the model was imperative to the success of new engineering efforts (Blanchard and Fabrycky 2011).

Although the Vee process model has numerous advantages, it does carry faults such as being too simplistic and not iterative, as the figure alludes to a strict step-by-step process. There are some levels where the activities can occur simultaneously or repetitively regardless of the system block. While iteration is not explicitly shown

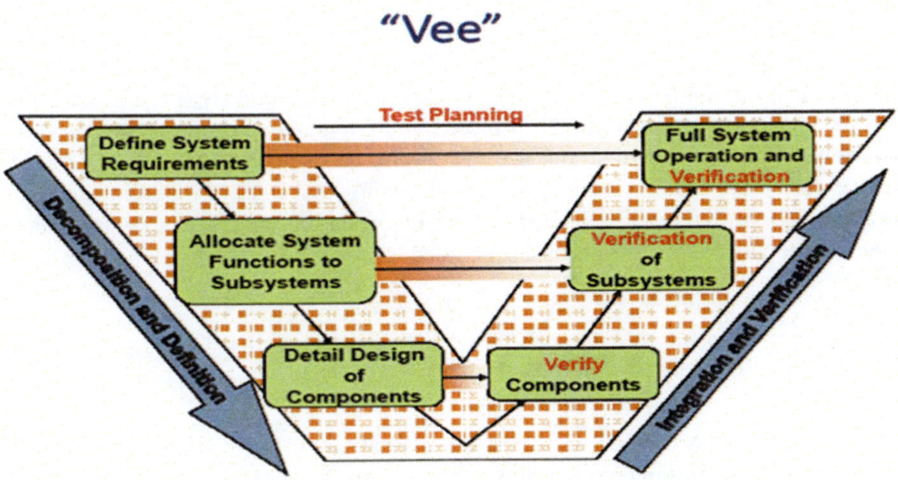

Fig. 1 Systems Engineering Vee Process Model. (Blanchard and Fabrycky 2011)

in the diagram, its value is in the clear representation of the system life cycle incorporating testable activities with their aligned developmental activities (Donald 2013).

This next section will depict the doolie cadet system using the Vee model. The USAFA doolie year has many components and can be modeled as a life cycle following a systems engineering approach with detailed requirements, functions, and components along with their associated verification and validation events.

4 Results

This section splits the Vee model into components and populates the doolie year into respective sections of the system life cycle. There are numerous variants of the Vee model tailored for specific scenarios and focuses. Figure 3 adapts the generic Vee model from Fig. 1 to the doolie year. The details presented within each block of the Vee model will be discussed with the understanding that the reader has been familiarized with USAFA specific requirements and activities as described in the background section of this paper.

4.1 Define System Requirements

The initial step in the Vee model is identifying overall system requirements. These general requirements are typically the "what" in the system life cycle in identifying the key necessities for everything that follows. In the example of the doolie life cycle, the initial requirement for the Vee model is meeting the United States Air Force Academy minimum standards. A cadet cannot fail any class or have more than one academic grade of "D" per semester. An A is a 4.0, an A- is a 3.7, a B+ is a 3.3, a B is a 3.0, a B- is a 2.7, a C+ is a 2.3, a C is a 2.0, a C- is a 1.7, etc. The average GPA must be above a 2.0. If these requirements are not met, then a doolie will be put on academic probation. Their MPA also needs to be above 2.0, or a doolie will be put on aptitude or conduct probation. Finally, the PEA of a doolie needs to be above 2.0, or they will be placed on athletic probation which requires the cadet to attend reconditioning, a rigorous, scheduled, instructor-led workout regimen.

4.2 Allocate System Functions to Subsystems

The next step down the Vee model is the allocation of the system performance goals. Using the requirement definitions, the objectives for performance are created by decomposing the requirements previously identified in the process model. Decomposing system-level functions to subsystems is necessary to further specify

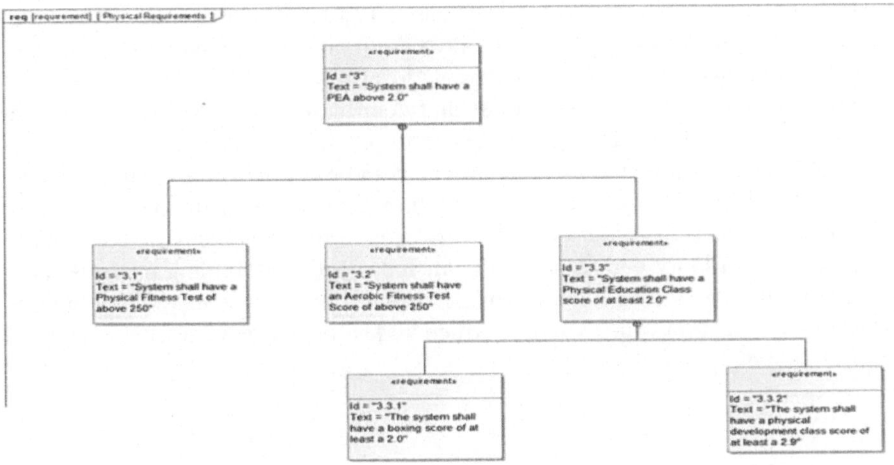

Fig. 2 A SysML physical requirements chart example

how the requirements will be met. In relation to the doolie year, the performance goals at the subsystem level are divided into the three major components as shown in Fig. 2. For the academic aspect, the goal is course completion. Completion of every class is a major aspect as to whether the GPA meets the 2.0 requirement. This course completion goal decomposes the GPA requirement into a lower-level definition of success. The performance goal for the military aspect can be broken down into military competency. Military competency is mostly assessed by upperclassmen who inspect the doolie's aptitude and ability to perform as a member of the US Armed Forces throughout the year. For athletics, the performance goal identified is athletic achievement. There are many components that go into the athletic composite score, but an overall athletic achievement aspect allows for a foundational breakdown of the requirement for PEA. Figure 2 illustrates how the Vee model can be extended with SysML models specifically in this case a SysML requirements diagram. High-level requirements should be decomposed to specific obtainable goals in accordance with the decomposition effort on the left side of the SE Vee. This work expands on this decomposition in the following sections.

4.3 Detail Design of Components

The final decomposition on the left side of the Vee model is the detailed design block. This section breaks down the subsystems presented in the previous segment into more comprehensive and detailed allocation to components. This step details the lowest-level components that will be verified as we work up the right side of the Vee for the doolie cadet system. The GPA requirement and course completion functions are allocated to class-specific knowledge at the component level. Understanding the content for each class and preparing for validations are the lowest-level components of a GPA. The PEA requirement and athletic achievement goals are allocated at the component level to athletic ability. Athletic ability is composed of individual exercise fitness and contributions to a team sport. These lowest-level components aggregate together to result in athletic achievement and the overall PEA. Finally, the MPA requirement and military competency functions are allocated to military knowledge/performance at the component level. Similar to academics, having military knowledge is essential in fulfilling the required military competencies. Military performance also includes actively building character and honor in order to do the right thing. This section is the lowest level and final step in system decomposition which will be verified and validated as we work up the right side of the Vee.

4.4 Verify Components

Verifying the components is the first step up the right side of the Vee diagram. This step ensures that the lowest-level components identified on the left side are being validated and integrated into the system. The specific events verifying the class-specific knowledge component are graded reviews, quizzes, and homework. These specific aspects, along with every academic event that affects the class specific GPA, fulfill the detailed design components. For the physical fitness, specific scores in graded physical tests verify the lowest component of athletic ability. For example, for the PFT the component verification would be the push-up event or the sit-up event. If it was a gym class, it would be a certain graded event in that class such as swimming the timed mile in water survival. Military verifications of components are the knowledge tests, performance in training sessions, minutes, and greeting – all the activities upperclassmen can verify that will be rolled up into the total military performance assessment.

A systems engineering tool that can be implemented within the Vee model is a synthesis, analysis, and evaluation (SAE) loop. This loop can be used to improve each specific component that needs development (Hall 1997). An example of an SAE loop is applied to push-ups. If a physical training session demonstrates that the max push-ups for a certain doolie was below the average, then push-ups may be improved through a SAE loop. The first step is to synthesize the problem and brainstorm potential causes and solutions to remedy the doolie's poor push-up performance. Synthesis would consider if strength, endurance, or another factor is the issue. The next step, analysis, would consider actions that can be taken to fix, improve, or eradicate the problem. Potential analysis for improving push-ups could include daily repetition such as doing 100 push-ups split up into three or four sets every day for a month. Alternatively one may analyze a weight-based (lifting) workout plan tailored to increasing upper body strength. The final portion of the SAE loop, evaluation, is testing the activities implemented as a result of the analysis to evaluate their impact on the problem. In this example a timed assessment of max repetition push-ups would be a fitting evaluation of the performance improvements gained since the initial start of the loop. At this point another SAE loop could begin to continue performance improvements. While applied to a physical scenario in this example, SAE loops could be applied to each aspect of the doolie cadet system.

4.5 Verification of Subsystems

Moving up the right side of the Vee model, verification of subsystems is a higher-level activity combining the results of individual component verification efforts. In this block the system performance goals are verified through a more inclusive assessment of doolie performance across academics, athletics, and military. In relation to the overall life cycle, the academic portions are the midterm and semester grades. These grades culminate the individual academic assessments in *verify components* along with validating the course completion component for academics. The physical aspect aggregates the individual components in athletic assessments and is concerned with the overall scores from the PFT, AFT, and grades from gym classes. These physical component scores reflect the athletic achievement goal identified on the left side of the Vee model. The military aspect has aggregated the knowledge tests, upperclassmen assessments, training sessions, SAMIs, and other military duties into military stratifications. Military stratifications are a major factor into the doolie MPA. The MPA validates the left side performance goals of military competency and allows for a tangible affirmation to the components. A similar synthesis, analysis, and evaluation loop can be executed at the verification of subsystems level. Expanding the original push-up SAE loop, the focus can now be on the entire PFT in general. If the PFT score from the first semester is not as good as the doolie intended, then a loop can be used to try and improve the score. The doolie can synthesize causes of the low score and potential remedies to improve the PFT score, analysis compares potential remediation plans, and the evaluation will be the next PFT the following semester (Fig. 3).

4.6 Full System Operation and Verification

The final step in the Vee model is the full system operation and verification. In this step the entire system is verified at the highest level. The accumulation for academics is the actual GPA at the end of the year. All the individual class GPAs are averaged according to the weight of each class (credits) and are brought together to give the final GPA and class ranking in respect to academics. This final GPA validates meeting the USAFA minimums of a 2.0 GPA and determines whether the doolie is put on academic probation. The accumulation of grades and scores for athletics is the actual PEA at the end of the year along with the class ranking in athletics. It is the accumulation of the PFT, AFT, and gym classes that affect the final

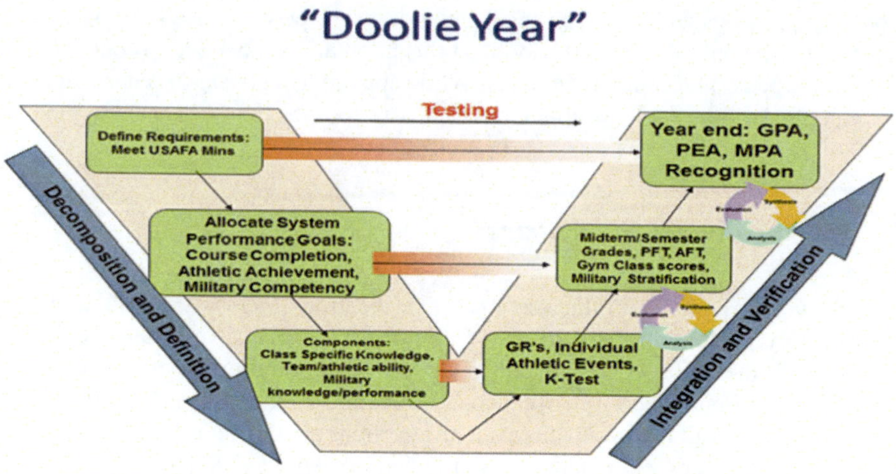

Fig. 3 The doolie year represented using the systems engineering Vee model

PEA for the semester/year. The PEA can determine whether the minimum of a 2.0 PEA was met and if the cadet will be put on athletic probation and reconditioning. The accumulation for military is a combination of MPA and Recognition. If the MPA is over a 2.0, it validates the minimum USAFA requirements, and a doolie must be seen fit by upperclassmen to be recognized during Recognition. This final step completes the Vee diagram with a full validation of the doolie cadet resulting in a promotion to the status of an upper-class cadet as they enter their sophomore year at USAFA.

5 Impact

Understanding this Vee model as applied to a non-technical system allows for multiple applications that can be effective for the development and planning of any system. Using the Vee model to capture a military cadet development system lends us to creating goals and specific objectives at the component levels that would not be otherwise obvious without using the Vee model. Allocating systems to subsystems and further decomposition into performance components enable a specific actionable roadmap of goals to achieve successful doolie cadet performance in athletics military and academics.

By applying the Vee model as we would for a complex technical system, the effort of decomposing requirements, and allocating them to components, lets us create lower-level objectives and goals to better meet the top-level requirements of our human performance objectives. The conventional way of creating requirements for human objectives is to think broadly and make high-level objectives. The limitation is these requirements such as a 2.0GPA do not have actionable events at a lower level to ensure the goal is on track. Applying the Vee model to the doolie cadet example of human performance goals allows for the successful decomposition of the high-level doolie cadet requirements into specific actionable goals and objectives spanning from detailed course test grade benchmarks to class grade objectives to overall GPA. This work application to the doolie cadet system demonstrates new utility of applying the Vee model to human performance objectives. It details the value of using the decomposition of the requirements to form testable objectives for verification efforts up the right side of the Vee. Using the SE Vee model for human performance objectives provides unambiguous actionable goals to ensure the human can attain the high-level overarching objectives.

6 Conclusion

This paper reflects on the military cadet experience describing the doolie cadet year modeled using the systems engineering Vee model. This paper presents a nontraditional example of a system modeled using the systems engineering Vee model. It provides a unique perspective and insight as to how systems engineering can be applied to many different areas beyond traditional software and mechanical systems. This doolie year integration with systems engineering demonstrates the widespread applicability and relevance of the field. More specifically, this paper demonstrates how a human experience and development effort can be modeled using a systems engineering process model. The authors intend to extend this work to include more Model-Based Systems Engineering with many additional SysML illustrations of the content covered in this work. The insights from the doolie system apply not just to USAFA cadets but also freshmen undergraduates on the whole and also high schoolers seeking to prepare for the rigors of a freshman experience especially at a military academy.

Acknowledgments This work was inspired by the Introduction to Systems Engineering class at the United States Air Force Academy. Completion of doolie year inspired the creation of the relationship between systems engineering and cadet life.

References

Blanchard, B., and W. Fabrycky. 2011. *Systems Engineering and Analysis*. Upper Saddle River: Prentice Hall.

Donald, F. "insights," Carnegie Mellon University, 11 November 2013. [Online]. Available: https://insights.sei.cmu.edu/sei_blog/2013/11/using-v-models-for-testing.html. Accessed 09 Nov 2018.

George, R.L.H., and M. Samaras. 2005. A systems engineering perspective on the human-centered design of health information systems. *Journal of Biomedical Informatics* 38 (1): 61–74.

Hall, M. "Mankind," Models for concurrent Engineering design, 15 July 1997. [Online]. Available: https://mankindsoftware.github.io/dissertation/model_27.htm. Accessed 09 Nov 2018.

Incose. *About Systems Engineering.* [Online]. Available: https://www.incose.org/about-systems-engineering. Accessed 10 Nov 2018.

Officer Training School. United States Air Force. [Online]. Available:https://www.airforce.com/education/military-training/ots. Accessed 10 Nov 2018.

Powell-Morse, A. "Airbrake," 16 December 2016. [Online]. Available: https://airbrake.io/blog/sdlc/waterfall-model. Accessed 09 Nov 2018.

Ryen, E. 2008. *Overview of the System Engineering Process*. NDDOT.

United States Air Force Academy. 2018. [Online]. Available: https://www.usafa.edu/. Accessed 10 Nov 2018.

Project Managers and Systems Engineers, "Can two walk together, unless they agree?": Recent Research Findings on Development Projects

Sigal Kordova, Eyal Kats, and Moti Frank

Abstract There are two significant "players" in development projects: the project manager and the systems engineer. They work together with the aim of meeting the technical (execution/performance, quality) and managerial (schedule, costs, and customer satisfaction) goals of the project.

The purposes of the current study (Kordova S, Katz E, Frank M, Syst Eng 22(3). https://doi.org/10.1002/sys.21474) are to identify the management processes shared by project managers and systems engineers in the defense industry; to understand which factors influence the ways in which joint project management is accomplished and how it impacts meeting project goals; and to provide recommendations for joint project management that will lead to project success.

The research method was qualitative, based on 16 semi-structured interviews with project managers and systems engineers in defense companies that deal with the development of technological systems.

The main recommendations for joint project management are: Set a clear distribution of responsibility and delegation of authority between the both parties before starting the project; choose a project manager who was once a systems engineer or who possesses knowledge of engineering; insist on ongoing dialogue between the two professionals; solve/prevent conflicts through discussion and persuasion; and expand the common ground between the project manager and systems engineer's areas of responsibility.

Keywords Systems engineering · Development projects · Project management

---

S. Kordova (✉)
Department of Industrial Engineering & Management, Ariel University, Ariel, Israel

E. Kats
HIT-Holon Institute of Technology, Holon, Israel

M. Frank
IAC-Israel Academic College, Ramat Gan, Israel

A. M. Madni et al. (eds.), *Recent Trends and Advances in Model Based Systems Engineering*, https://doi.org/10.1007/978-3-030-82083-1_38

1 Research Goals

The current study discusses project management methods and processes from the perspectives of both project managers (based on the PMBOK) and systems engineers (based on the SE Handbook). Our objective was to identify the management processes shared by both project managers and systems engineers in the defense industry; to understand which factors influence the ways in which joint management is executed; to define the consequences of joint management for achieving project targets; and to recommend joint project management methods that lead to project success.

2 Literature Review

Systems engineering management is a practice that grows and develops together with systems engineering. Therefore, its standards address the close relationships linking systems engineering management and project management, emphasizing their critical importance for improving project management. While project-related methods have traditionally focused on issues such as scheduling, budget, and scope, systems engineering management centers on managing the project-product and issues related to developing project technology (Sharon 2010; Sharon et al. 2011).

Systems engineering management and project management flow together, a fact addressed in at least three prominent project management and systems engineering books:

1. *The SE Handbook* (INCOSE 2011) includes the technical processes, management processes, and extra-organizational processes relevant to systems engineers. The chapter dealing with project management processes focuses on processes relevant to technical coordination of a project and presents management processes referenced in the PMBOK (Project Management Institute 2013).
2. *NASA Systems Engineering Handbook* (NASA 2007a). Figure 1 is taken from this book and depicts the overlap between systems engineering and project management. According to NASA (2007a, b), systems engineering provides information on the technical aspects in these areas, while project management is concerned with management, costs, and scheduling aspects.
3. *Defense Acquisition Guidebook* (DoD 2010), an Internet source provided by the US Department of Defense, outlining basic project management principles and processes.

In October 2012, PMI and INCOSE conducted a joint survey in an attempt to better understand the roles of project managers and systems engineers and determine the level of integration between the two (Confronto et al. 2013). The survey included a total of 680 systems engineers and project managers. The survey findings showed that:

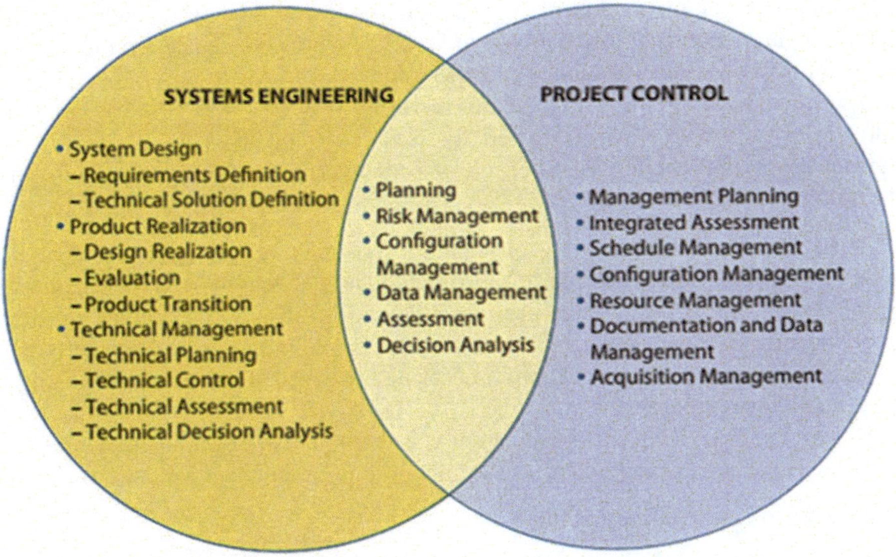

Fig. 1 NASA: Overlaps between systems engineering and project management (NASA 2007a, b)

- A total of 30% of respondents believe there is a certain level of undesirable tension between project managers and systems engineers.
- The three factors leading to this tension are:
  - Lack of an integrated plan for carrying out project management/systems engineering activities
  - Failure to identify the responsibilities related to each area
  - Conflicting project management and systems engineering practices

According to Lang and Stratton (1997) and Langley et al. (2011), the key to successful project management in the twenty-first century is understanding the complex relationships between the disciplines (e.g., project management and systems engineering) and their shared processes and differences and then finding the most efficient way to integrate the various components, thus maximizing project management. Project management, systems engineering, and content supervision are the three essential disciplines required for effective project management (MIT-PMI-INCOSE 2012; Scott et al. 2015).

3 Methodologies

The study design was qualitative and based predominantly on semi-structured interviews with project managers and systems engineers employed in defense companies dealing with the development, integration, and upgrading of technological systems.

The interview sample included 16 experienced project managers and systems engineers, all recognized as experts in their fields, who were selected on the basis on their position, education, and wide experience.

The main study findings, particularly recurring patterns, were first categorized into groups and were then examined for frequency and intensity of repetition. Several procedures were used to ensure the trustworthiness (internal validity) of the qualitative study findings:

- Triangulation: A finding was considered trustworthy/valid if it appeared in at least three interviews. Because this is a qualitative study based on semi-structured interviews, the statistical analyses usually used in quantitative studies are not applicable. In their stead, three interviewees mentioning the same topic is an accepted triangulation procedure for ensuring trustworthiness or internal validity of qualitative research.
- During the interviews, cross-validation was employed to examine the degree of respondents' agreement with the given definitions (respondent validation).

In order to ensure confirmability:

- The objectivity level was examined throughout the study.
- The interview questions were worded clearly and in an unbiased manner.
- Efforts were made to avoid errors due to false first impressions and premature assumptions, so the research findings could be applied to the general population (external validity – fittingness).
- The research sample was not biased towards either of the two domains (project management/systems engineering); as far as was possible, an equal number of project managers and systems engineers were interviewed.
- Known experts from the project management and systems engineering communities were chosen to represent their fields.

4 Research Results

The key objectives of the qualitative analysis were to identify conceptual similarities and discover types, classes, processes, patterns, and key points. Therefore, after transcribing and summarizing all of the interviews, their content was analyzed in an attempt to find the patterns and common themes emerging from the interviews. These patterns helped to illuminate the study questions. The content analysis process included:

1. Mapping the management processes shared by project managers and systems engineers (such as risk management, purchasing processes management, budget management, customer relationship management, etc.)
2. Mapping the factors that influence joint management methods (such as personality, interpersonal relations, interpersonal communication, the project's organizational structure, professionals' background and experience, etc.)
3. Consolidation of recommendations regarding the project's joint management.

The main findings of the study are as follows:

1. The overlapping areas between project management and systems engineering often cause conflicts between the two professionals. These areas include risk management; procurement management; systems engineering-architecture-related decisions; concept definition and integration processes; validation/verification tests (impact on schedule and resources); and work program management issues.
2. In most projects, cooperation between the two professionals is achieved through a peer approach to determine project goals and how to best achieve them. Conflicts relate mainly to professional orientation (meeting requirements and performance demands/compliance with cost and schedule constraints). All organizations strive towards processes that clearly state the responsibilities of each professional involved in the process, with project managers and systems engineers at the top of the management pyramid. If there is good synergy between them, they can help one another and decide together on technical and management-oriented subjects. Lack of effective cooperation between the two may inevitably hamper project success (all interviewees unanimously agreed on this subject).
3. Factors influencing the joint project management method included the personality of the involved professionals; their organizational culture; project managers' familiarity with systems engineering processes; level of mutual trust and professional appreciation; the project's organizational structure (in particular, regarding projects in which systems engineers are under the supervision of project managers); and background (specifically, education and professional experience).
4. The main management problem resulting from overlapping responsibilities is schedule delays, but there were additional consequences related to performance.
5. Recommendations for joint project management:

   5.1. Coordinate responsibilities and authority between the two professionals before the project's onset.
   5.2. Document the responsibility and authority in the PMP and SEMP documents.
   5.3. Define a mechanism for settling disputes.
   5.4. Train project managers in engineering development processes (the project manager is the systems engineer's boss, so he/she is also responsible for technical supervision).
   5.5. Use shared tools to manage the project (e.g., CORE systems engineering software).
   5.6. Ensure that the systems engineer takes cost and schedule considerations into account when making systems engineering decisions.
   5.7. Come to a clear agreement about project goals and how to achieve them.
   5.8. Clarify interrelations between the project manager and the systems engineer during professional training.
   5.9. Appoint a project manager who has previous systems engineering experience.

5 Discussion

The current findings from the interviews indicate that the management processes shared by both project managers and systems engineers are:

- Risk management
- Systems engineering processes
- Schedule management
- Procurement management
- Scope management
- Cost management
- HR management

Conversely, the literature review found that the overlapping management processes appearing in both the PMBOK and the SE Handbook are:

- Management of lifecycle processes
- Requirements management
- Project planning
- Project monitoring and control
- Risk management
- Configuration management
- Information management
- Procurement processes management
- Portfolio management
- HR management

There is indeed some congruence between the findings of current study and those of the literature review, including risk management, requirements management (a top priority for systems engineering processes), procurement processes, HR management, and schedule management (project planning). There is also partial congruence with the work of Scott et al. (2015). Specifically, the shared management areas in both studies are risk management, subcontractor management (procurement processes management), quality management, and product lifecycle planning. The overlapping management processes found in this study but not mentioned in the literature reviewed above include scope management and cost management.

5.1 Overlapping Management Processes Found in the Current Study

Risk Management Project risks include management-related risks (such as cost or organizational expenses), as well as technical/engineering risks (regarding requirements, performance demands, and premature technology). During the project, there is joint discussion about risk in which the relevant risks are ranked, and a risk reduction plan is created. Most of the interviewees mentioned that project managers usually integrate all risks, while systems engineers are commonly responsible for identifying and managing only technical risks. However, technical risks often have managerial consequences, because risk reduction plans generally require allocation of resources (scheduled time, budget), a task usually performed by the project manager.

Systems engineering processes, managed by the systems engineer, include major project decisions such as choosing the best technology for the project (including level of maturity); determining the system architecture; defining the concept behind the integrations and tests; choosing the optimal design; establishing the number of development cycles (versions); and deciding how many prototypes will be built, etc.

These decisions all have significant implications for project success and the ability to meet project goals. Both professionals need be involved, because each of these decisions requires resources and is therefore a subject that must be discussed in depth. All technical requirements are the systems engineer’s responsibility; therefore, attention must be given to each and every requirement and all relevant aspects. If a requirement cannot be met, systems engineers may try to convince their project managers that this is the case. If the project manager is convinced, it will then be his/her responsibility to carry out the necessary negotiations with the customer.

Project schedules aim to provide a framework for time constraints, as dictated by the customer. Schedules for development entail identifying the scope of work (SOW), dividing it into appropriate activities, and understanding the constraints and order of activities involved in the different development processes. Project milestones are usually determined according to project goals and dictated roughly in a top-down manner by the project manager, while a bottom-up schedule is drafted by the systems engineer. Striking a balance between vision and reality is seldom achieved in the first planning cycle. This iterative process requires multiple discussions between the project manager and the systems engineer.

Procurement management requires a detailed listing of all requirements and SOW as part of the subcontractor’s agreement. Systems engineers usually define the requirements, while the statement of work is written jointly with the project manager. The project manager is interested in minimizing scheduling and cost factors; however, this is not always in line with the specifications as defined by the systems engineer. Therefore, there must be dialogue between them about resource limitations and compliance with essential requirements.

Project scope relates to deciding which work packages are incorporated into the project, including the tasks necessary to handle each project issue. *Scope* is derived from project requirements in a process initiated by the systems engineer, while the document listing the project’s detailed work contents is usually initiated by the project manager. Scope is also derived from the system’s architectural design, which clearly defines what the project entails. Thus, scope is deeply related to each system’s engineering process, including the number of tests to be conducted, which development and testing tools will be developed, etc. In technology development projects, the systems engineer leads the systems development process from a technological perspective and plays a large part in defining the SOW and building the work breakdown structure (WBS). The project's scope has an impact on both project resources and project success. Therefore, the project's SOW is also a subject for discussion between the two professionals.

Project cost is managed solely by the project manager. Nevertheless, preparing a rational, accurate cost estimate must be carried out together with the systems engineer, who is usually able to evaluate how much time/scope must be invested in engineering development processes.

HR management involves both professionals in identifying the skills required for the project (defining the professionals needed, etc.) and ongoing manpower management. The level of management varies between the "small" circle (project management team) and the "large" circle (the rest of the matrix). In large development projects, the systems engineer is usually in charge of all aspects of the work done by several additional systems engineers, under his/her direct management.

5.2 *Joint Processes that Failed to Meet the Validity Criteria*

Customer Relations Management According to most of the interviewees, project managers usually play the major role in customer relations. Nevertheless, there are many dialogues around technological issues (design reviews, customer acceptance tests, etc.), in which systems engineers are at the forefront. Both professionals must agree on the messages they want to convey to customers and decide who oversees discussions of each specific subject.

Product Lifecycle Management Project managers are responsible for planning and managing the product's lifecycle. Systems engineers must consider lifecycle cost requirements throughout the development period. Additional project team members are usually involved in managing the project production, system deployment, and maintenance processes.

6 Conclusions and Recommendations

The current study focuses on project-related management methods and processes from perspectives of the project managers and systems engineers. The main factors influencing joint project management methods are personality clashes between the professionals, their background and experience, natural inclinations (the systems engineer towards performance and engineering and the project manager towards budget/schedule considerations), and the project's organizational structure. Factors that have less of an impact are mutual trust and appreciation, the project's organizational culture, power struggles related to project resources (managed by project managers), the extent to which each professional interferes in the other’s area of responsibility, and the ability to see eye-to-eye regarding project goals and how to achieve them (“goal unification”).

In most projects, the two professionals cooperate out of a shared motivation to meet project goals. However, when conflicts arise between the two, it has a negative effect on meeting the project’s goals, primarily in the form of failing to meet schedule/budget targets.

Our main recommendations proposed for successful joint project management are coordinating responsibilities and having a clear delimitation of authority between the two professionals before beginning work on the project; choosing a project manager who has previous systems engineering/engineering experience (if not, the project manager should receive training in engineering development processes); strict adherence to ongoing dialogue between the two professionals; resolving/avoiding conflicts through discussion and persuasion; and expanding the overlap between project managers and systems engineers’ areas of responsibility.

The secondary recommendations are to motivate systems engineers to think about management considerations; train project managers and systems engineers to discuss relevant management interfaces; provide both professionals with mentoring; and appoint suitable project managers and systems engineers.

This study was conducted solely in security development firms and does not include any additional data about the projects (budget, scope, systems complexity, development time, etc.). Therefore, additional studies should be conducted in civilian development companies. These future studies should attempt to integrate quantitative tools to examine the relationship between project data and level of joint project management among systems engineers and project managers.

Acknowledgments The authors would like to thank the Gordon Center for Systems Engineering at the Technion, Israel Institute of Technology, for their support of this study.

References

Confronto, E., M. Rossi, E. Rebentisch, J. Oehmen, and M. Pacenza. 2013. *Survey Report: Improving Integration of Program Management and Systems Engineering*. Boston: INCOSE & PMI.

Department of Defense. 2010. *Defense Acquisition Guidebook*. Defense Acquisition University. https://www.dau.edu/tools/dag.

INCOSE, SE Handbook Working Group. 2011. *Systems Engineering Handbook*. San Diego: International Council on Systems Engineering.

Koral Kordova, S., E. Katz, and M. Frank. 2018. Managing Development Projects – The Partnership between Project Managers and Systems Engineers. *Systems Engineering* 22 (3). https://doi.org/10.1002/sys.21474.

Lang, C., and M. Stratton. 1997. *Project Management, Configuration Management and Systems Engineering: What’s Needed Most for the Next Century?* PMI.

Langley, M., S. Robitaille, and J. Thomas. 2011. Toward a New Mindset: Bridging the Gap Between Program Management and Systems Engineering. In *PMI Global Congress Proceedings*. Dallas.

MIT-PMI-INCOSE. 2012. *The Guide to Lean Enablers for Managing Engineering Programs*. Joint MIT-PMI-INCOSE community of practice on Lean in Program Management.

NASA. 2007a. *NASA Systems Engineering Handbook*. Washington, DC: NASA.

———. 2007b. *NPR 7120.5, NASA Space Flight Program and Project Management Handbook*. Washington, DC: NASA.

Project Management Institute. 2013. *A Guide to the Project Management Body of Knowledge*. 5th ed. Newton, PA.

Scott, E., S. Townsend, and E. Carlos Confronto. 2015. Collaboration Across Linked Disciplines: Skills and Roles for Integrating Systems Engineering and Program Management. In *122nd ASEE Annual Conference*.

Sharon, A. 2010. *A Unified Product and Project Lifecycle Model for Systems Engineering*. HAIFA: Technion.

Sharon, A., D. Weck, L. Oliver, and D. Dori. 2011. Project Management vs. Systems Engineering Management: A Practitioners' View on Integrating the Project and Product Domains. *Systems Engineering* 14 (4): 427–440.

A Plan for Model Curation at the US Army Armaments Center

Christina Jauregui and Mary A. Bone

Abstract Model curation is an explicit requirement under the first goal of the Digital Engineering Strategy. Model curation is the active archival of quality models. Model curation traces to digital and data curation; however, research in this area is within its infancy. A literature search identified the three major sources that are pioneering the field. This paper provides a highlighted overview of the three pioneering sources for model curation. The US Army Combat Capabilities Development Command CCDC) Armaments Center (AC) is prioritizing the transformation to digital engineering through Model-Based Systems Engineering and the Integrated Model-Based Engineering Environment and has begun to investigate how to implement a model curation capability.

Keywords Model curation · Digital engineering · MBSE · Model · Centric engineering

1 Introduction

The office of the Deputy Assistant Secretary of Defense (ODASD) for Systems Engineering issued the Digital Engineering Strategy in June of 2018. Model curation is an explicit requirement under the first goal, which is to implement the use of models to enable enterprise and program decision-making. The Department of Defense (DoD) encourages implementation of digital engineering as a means of modernizing the current engineering practice to enhance the way technologies evolve during the conceptual phase and minimize costly expenditures during testing (Department of defense digital engineering strategy 2018). The US Army

C. Jauregui (✉)
US Army Combat Capabilities Development Command – Armaments Center, Picatinny Arsenal, NJ, USA
e-mail: christina.jauregui.civ@mail.mil

M. A. Bone
Stevens Institute of Technolgoy, Hoboken, NJ, USA

A. M. Madni et al. (eds.), *Recent Trends and Advances in Model Based Systems Engineering*, https://doi.org/10.1007/978-3-030-82083-1_39

Combat Capabilities Development Command (CCDC) Armaments Center (AC) is prioritizing the transformation to digital engineering and has begun to investigate how to implement a model curation capability. Dullen et al. have emphasized a need to develop capabilities that enable reuse of knowledge in engineering analyses (Dullen et al. 2019) – model curation can address this need.

Tom McDermott et al. cite the importance of future research in model curation. They indicate the need to develop a process that controls and manages quality models to enable future reuse. Further, they indicate a need to define standards for model metadata to enable model curation (McDermott et al. 2019). Stimpson discusses the importance of using the right data, or else experiments can go wrong (Stimpson 2019). This is a key principle to consider for model curation; there is no sense in curating a "bad" model or "bad" data, as this will cause experiments to be misrepresented, leading to the wrong answer and wasted time. There must be emphasis placed on developing good criteria for selection, review, and validation of the curated models. Along with this emphasis, a better understanding of what a model is and what are the types of models must be developed and understood. The term model as defined in (DoD 1996) is abstract, and model curation will require explicit definitions of models and types of models (Dod modeling and simulation (m&s) verification, validation, and accreditation (vv&a) number 500.61 1996).

The first step in this endeavor was to develop and document a literature search. The second step was to develop a high-level procedure for AC model curation. Section two provides a background on Digital Engineering at AC. Section three presents some highlights from the literature review. Section four discusses high-level plans to begin implementing model curation. The conclusion and future research are in Section five.

2 Digital Engineering at Armaments Center

Digital transformation calls for product development to occur in a model-centric environment. There is an opportunity to improve research and development at Armaments Center (AC) through digital engineering. Two digital engineering efforts applied at the AC are Model-Based Systems Engineering (MBSE) and the Integrated Model-Based Engineering (iMBE) project. The Interoperability and Integration Framework (IoIF) is a research effort focused on implementing semantic web technologies.

2.1 MBSE at AC

The Systems Engineering Directorate has embarked on the path to implement Model-Based Systems Engineering (MBSE). MBSE at AC is "a Systems Engineering approach that emphasizes the application of integrated structured processes and data, visual modeling principles, and best practices to systems engineering activities throughout the Systems Development Life Cycle. The Systems Engineering Model is the product of MBSE, which includes:

- Requirements Models
- Behavioral Models
- Structural Models
- Logical, Physical
- Parametric Models
- Performance Models
- Item, System, Aggregate
- Technical Baseline

MBSE intends to replace or augment the document-centric approach practiced by systems engineers in the past and to influence the future practice of systems engineering" (Mbe/mbse/mbent/mbde definition and end states 2018).

The MBSE Subject Matter Expert (SME) committee is a cross-competency team that discusses and implements improvements in AC modeling techniques and current trends in the field of MBSE. The MBSE SME committee shares information to the rest of the engineers through an internal online community forum. They have developed products to enable MBSE at AC; these products include a SysML starter model. The starter model elements are the AC template, schema domain, AC profile, sample mortar model, and an SE process model.

2.2 iMBE at AC

Developing armaments systems require modeling for both traditional and nontraditional engineering disciplines. At AC, although modeling is extremely advanced, most information developed from models is shared through PowerPoint reports. Transitioning from this document-centered approach to an integrated model-based sharing approach will require a paradigm shift. The Integrated Model-Based Engineering (iMBE) project is developing the software capability to enable integration of models and the IPTs.

The below is an excerpt from the IMBE Implementation Plan (Integrated model based engineering environment implementation plan 2019):

> The integrated Model Based Engineering (iMBE) initiative will accelerate development of increasingly complex and advanced armament technologies and products by deploying a collaborative engineering environment that integrates a variety of commercial and in-house developed software tools across the technology development lifecycle.
>
> The iMBE Environment will be deployed using a phased approach to enable end-to-end digital continuity through all phases of design and prototyping of AC technologies. The iMBE approach to delivering digital continuity is through the integration of software tools, data, and models across engineering domains at various levels of model fidelity, and across multiple levels of abstraction. Such an environment will facilitate Cross-Functional

Teams (CFTs) with participatory design and optimization activities and advanced analytics. Additionally, teams will be able to search, retrieve and reuse previous design and analysis data. The iMBE initiative is in alignment with the Army Future Command's (AFC) stated goal of accelerating development and delivery of more innovative solutions to meet the future needs of the Warfighter, as well as the DoD Digital Engineering Strategy.

The iMBE project is exploring how the transformation to digital engineering impacts research and development at AC. There are multiple efforts deployed under the iMBE umbrella – one is model curation. Another effort that falls under the iMBE purview is the development of a digital thread schema. The intent of the digital thread schema is to identify a definition for the enterprise digital thread at AC and to provide a means for projects to establish a digital thread, similar to establishing a baseline for each phase of research and development. Model curation would enable the authority and accuracy of a digital thread.

2.3 *InterOperability and Integration Framework (IOIF)*

The InterOperability and Integration Framework (IoIF) is being researched as a semantic web technology framework to enable digital thread capabilities (Bone et al. 2018) at AC. The IoIF is envisioned to take element-level data from multiple digital artifacts to share, reason, and capture the data. Due to usage of open-source tools , IoIF has usage limitations on a secure government network. However, AC recognizes the importance of exploring semantic web technologies and continues to fund the development of the IoIF.

One key question that had been thought to be understood but was identified to have diverging meanings from the IoIF research is, "What is a model?" In Table 2 a list of metadata that needs to be captured with each model is presented. This metadata is relative to the model itself, so the boundaries of the model must be well defined; thus, the list of metadata should be exhaustive in order to enable reuse.

Although an exhaustive metadata list is a requirement for model curation and it is a good start, there is likely a better model-based approach than collecting lists of data. The IoIF research has taken the perspective that capturing a digital thread as the model may be more advantageous than just a single file in a specific tool format. For example, in Cameo Systems Modeler SysML models can import and utilize other SysML model data. Those imported models can also import model data from other models and so on. The data in one model may not make sense or be helpful without data from a model many imports away. Therefore, the data from one model becomes metadata for another model. Creating these boundaries or threads of data has been identified as important research for both curation and digital engineering effort overall. The IoIF would have the capability to gather all the data elements, from multiple tool models, to create a meaningful snapshot of the digital thread model.

There are many open questions, but the IoIF research has exposed that the current view of a model, which is usually defined by one file of some tool format, may be inadequate to achieve the goals of digital engineering. Therefore, more research needs to focus on identifying what a model is and metadata elements that support that concept of a model. Semantic web technology-enabled frameworks like the IoIF can capture data elements across multiple tools, which may be more useful to curate and maintain than a plethora of individual files from tools.

3 Background: Model Curation Literature Review

The intent of the literature review was to identify a definition for model curation and the processes used to apply model curation and to compare model curation to model governance.

3.1 Curation

Using data curation methods for scientific and research purposes is not a novel idea; however, in the transformation to a digital way of practice, curation methods are being extended to accommodate digital data (Ball 2010). The Systems Engineering Advancement Research Institute (SEARI) produced a literature review on model curation and likened it to digital curation and museum curation (Rhodes Lucie Reymondet & Ross 2016). The biomedical engineering discipline uses model curation to promote shareability and reuse by the research community (Le Novere et al. 2006; Walters 2009; Waltemath & Wolkenhauer 2016; Nickerson & et al. 2006). Palmer et al. used a data curation lifecycle approach to enable a means of collecting and curating data from geobiology researchers at Yellowstone National Park (Palmer & et al. 2013).

3.2 Definition of Model Curation

A literature review on the term "model curation" identifies very few sources that distinguish a precise definition on the term. Model curation is a newer field of research and is similar to both data curation and digital curation. Table 1 provides definitions for model curation, digital curation, and data curation. There are words bolded in each definition; the bolded terms are common across the definitions.

The definitions in Table 1 were synthesized to a definition used herein for model curation, which is *active archival of quality models*. Active indicates continuous enhancement of the models to keep them current and avoid obsolescence. Archival indicates that the models have been prepared for preservation and are accessible to

Table 1 Literature on definitions relevant for model curation

Definitions	Sources
"Model curation is the **lifecycle management, control, preservation** and **active enhancement** of models and associated information to **ensure value for current and future use**, as well as **repurposing** beyond initial purpose and context."	Rhodes, D. H. (2018, November 19, 2018). "Model Curation." OMG MBSE Wiki. Retrieved August 27, 2019., 2019. (Rhodes 2018a)
"Digital curation involves **maintaining, preserving** and **adding value** to digital research data throughout its lifecycle."	Centre, D. C. (2004-2019). "DCC Model Curation Lifecycle Model." Retrieved 06/27/2019, 2019. (Data Curation Centre 2004)
Data Curation is "The activity of, **managing** and promoting the use of data from its point of creation, to **ensure it is fit for contemporary purpose** and available for discovery and **re-use.** For dynamic datasets, this may mean **continuous enrichment or updating** to keep it fit for purpose. Higher levels of curation will also involve maintaining links with annotation and with other published materials."	Macdonald, P. L. a. A. (2003). Data curation for e-Science in the UK: an audit to establish requirements for future curation and provision. e-Science Curation Report. T. D. A. C. Limited. 2 Wayside Court, Arlington Road, Twickenham, TW1 2BQ - UK, The JISC Committee for the Support of Research (JCSR): 85. (Lord & Macdonald 2003)

a wide audience. The term "quality models" indicates that the curated models have gone through an elaborate selection and review process to identify their validity and context for future use.

3.3 *Model Curation Pioneers*

Donna Rhodes and her team at SEARI have been pioneers in the field of model curation. During a workshop discussing model-centric engineering, the participants indicated a major gap in modeling practice. The gap was the ability to develop, manage, and maintain key models and their associated information to allow for reuse. Thus, the research topic of model curation kicked off (Ross & Rhodes 2019). SEARI has developed a lexicon for model curation which includes key terms, such as Accession, Acquisition, Cataloging, and Composability (Rhodes 2018a). Through their research, they have identified a need for a "chief model curator" who governs the curated model collection and has deep knowledge of it (Rhodes 2018b; Rhodes 2019). They have also identified challenges to curating models such as inclusion of legacy models and how they will be handled, potential duplicated model efforts, trust and legitimacy of models, and model competency distributed across individuals and organizations (Blackburn et al. 2018).

The Data Curation Centre (DCC) developed the data curation lifecycle, a lifecycle management approach for data curation. There is a starting point (models are conceptualized) and an ending point (models are disposed of). At the center of the model, there is an indication that description information is associated to the data, and community watch and participation are key inputs to the curation and preservation process. The main steps in the process are create; appraise and select; ingest; preservation action; store; access, use, and reuse; and transform. The DCC curation lifecycle model is foundational for data curation and model curation (Data Curation Centre 2004).

The Consultative Committee for Space Data Systems Secretariat published a reference model for an open archival information system (OAIS) as a framework for developing and maintaining preservation archival systems that store referenced materials. The OAIS is developed for the National Aeronautics and Space Administration (NASA 2012), but it can be extended to fit other enterprises. The OAIS is published as a standard by the International Organization for Standardization as ISO 14721:2003 and is considered to be the standard for creating and preserving digital repositories. While the OAIS maintains standards for archival, it does not prescribe how to validate models as part of the review and selection process.

Science Applications International Corporation (SAIC) is a well-known contracting company, which has a major research focus on digital engineering and the transformation that is required to achieve it. They developed a model curation process based on a searchable repository to identify candidates for reuse when developing architecture for new products. Users search the repository to see if they can find a "model" or anything else that can address the tasker at hand, and then there is a decision node for Yes or No. This process is similar to the DCC lifecycle model which contains all the keywords for model curation like appraise, access, use, reuse, transform, create, ingest, execute, etc. (Michael & Vinarcik 2019).

3.4 Comparison of Model Governance to Model Curation

Proponents of digital engineering liken model curation to model governance. There are key similarities between the two such as control, validation, and oversight (Federal Deposit Insurance Corporation 2005). Khatri and Brown say that data governance considers "data as an asset," which is also the case in model curation where models are considered assets with business value. Governance is a practice that allows for a derivation of business value from data assets (Khatri & Brown 2010).

The major difference between governance and curation is their intent. IT governance emphasizes accountability of IT and data and decision-making activities in five domains: IT Principles, IT Architecture, IT Infrastructure Strategies, Business Application Needs, and IT Investment and Prioritization. These IT characteristics translate to data governance (Weill & Ross 2005). Khatri and Brown extend the IT domains to data, with an emphasis on decisions made in the context of the organizational purpose for the data (Khatri & Brown 2010). Whereas curation is a process that injects quality, archival and active enhancement to data and models. Governance is also missing the composability factor, which is that the models are accessible and categorized in a way that enables a more meaningful understanding of a subject than the model alone.

4 Model Curation at Armaments Center

The Systems Engineering Directorate is spearheading the initiation of model curation at AC and the development of an internal digital library specifically housing a curated model catalog. A purpose for curation is to reuse quality models to prevent rework. These quality models have gone through an extensive validation, cleaning, and preservation process to be elevated as enterprise assets in the digital collection. As more digital information becomes available, there is a growing need to identify the works that should be curated and create a knowledge repository. A curated digital library will not only offer new employees a starting point, but it will also serve as a knowledge base so work is not lost as projects terminate. A digital library is under development to enable model curation at AC.

Initial meetings have generated a high-level procedure for AC model curation. This effort is in its initial stage and will continue to be refined with more experience and knowledge gained. The initial stage will attempt to understand the intricacies of model curation for MBSE. The procedure will be peer reviewed, piloted, and iterated several times prior to formal implementation. Upon formal implementation of model curation for MBSE models, the scope will broaden to the unique armaments engineering disciplines within the AC.

4.1 Model Curation Procedure

When developing the first iteration of the process, the SMEs were presented with terms from the model curation lexicon developed by SEARI (Walters 2009). The SMEs confuted the word acquisition immediately, as acquisition professionals for the US Army; this word has a slightly different intention than the definition provided in the lexicon. Since most of the work to be curated is developed under government contracts or by government employees on official duty, the models do not have to be acquired because they already belong to the government. As discussions ensued, it became apparent that the procedure should leverage common terms; they are depicted in Figure 1.

Figure 1 shows the high-level procedure map with the steps that one would follow to perform model curation at AC. This is a draft procedure and will continue to be enhanced as the competency for model curation is developed. The entry criteria indicate that in order for the process to begin, the model identification and planning for receipt of the model are completed. This is in contrast to the idea that

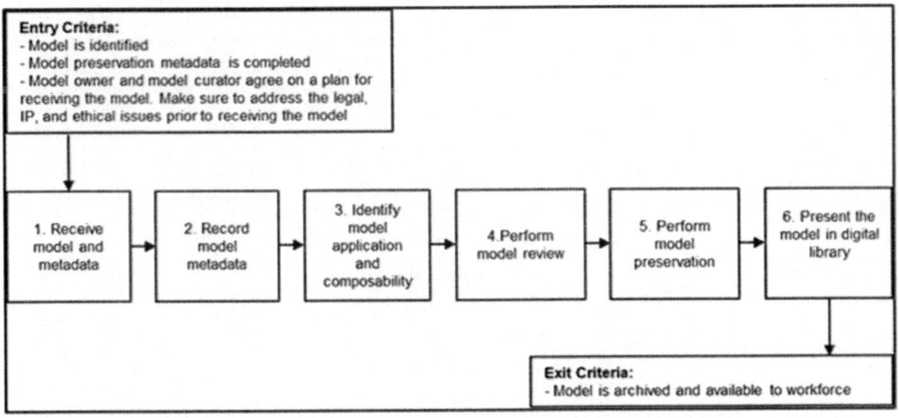

Fig. 1 Model curation procedure map

accessioning occurs in one step; the plan to receive the model and addressing the IP and ownership must be developed prior to receiving the model. After finalizing the preservation activities, the model is archived in the digital library, and the procedure is exited.

Minimum required metadata to begin with for model curation at AC is listed in Table 2.

The next step is to peer review and pilot the procedure. The Systems Engineering Directorate will soon receive a reference architecture for an artillery system as a result from a Lean Six Sigma Green Belt project. The reference architecture handoff will be the first instance of piloting the model curation procedure.

The engineers within the Systems Engineering Directorate are amenable to standing up this capability because access to curated models will improve the systems engineering competencies. Engineers have remarked that the most important benefit of model curation is access to quality models that can be reused. Other identified benefits are:

- Knowledge management
- Tools for teaching new employees
- Enable quicker turnaround of artifacts since the curated models serve as starting points
- Accessibility of good examples
- Archive work that was developed for projects that are terminated
- Archive work that was developed for projects that are transitioned
- A means of storing work that was completed by employees before they leave the organization
- Validated models to represent the digital thread for a system

Table 2 Metadata for model curation

Model curation metadata
POC for model
Model owning IPT, directorate, division and branch
Model creator
Category of model
Type of model
Project/customer/model consumer
Project S&T years
Date published
Repository location
Software used
Software version
How to run the model
External files required and provided
Inputs to the model
Validated by and for what conditions: list validation info
Validation results
Accreditation results
Assumptions
Known limitations
Conditions or bounds of the model
Test data location

5 Conclusion and Future Work

The field of model curation is new, but the meaning of curation is extended from other disciplines. Digital and data curation are two disciplines that technically align with model curation. The top three sources of literature on data and model curation are the Data Curation Centre (DCC), the Systems Engineering Advancement Research Institute (SEARI), and the National Aeronautics and Space Administration (NASA).

At Armaments Center (AC), the model curation procedure begins with the process of identifying a selected model and its metadata; the model is assigned to a collection; a review is performed to confirm its validity; and upon a favorable review, it is preserved and presented in a digital library. Curated models undergo a transformation to enable reuse by those who access the digital library and pull the curated models. The curated models are enhanced as needed, to ensure they remain current and usable.

As more models are identified for curation, continued iteration of the procedure and development of the digital library are expected. The AC will continue to mature the model curation competency by participating in peer-reviewed conferences and publishing findings in peer-reviewed journals. The SEARI research is the cornerstone of model curation, and this research will continue to be leveraged into application at AC.

Future research will identify the role that reference architectures play. The intent is to gain an understanding of what is required and how a reference architecture minimizes rework and increases competency. As the model curation scope broadens outside of MBSE, understanding of what a reference architecture looks like for each type of model must be developed.

Perhaps a more important future research topic is defining "what" a model is and defining its associated information. Being able to articulate and directly state what constitutes a model and specify information associated with that model is a key concept for future research. There is an assumption that model and therefore the data in the model is a known defined "thing," but this is not typically the case. There are research questions that straddle the topic of model curation and MBSE in general that will affect how models are curated:

- What constitutes a model in digital engineering, and what are the different types of models that are of interest to digital engineering?
- What data and metadata is needed to describe/recreate each model?
- Is model data unique to each tool or data format?
- How does a tool get captured with the model so that the model can be useful?
- Can a set of guidelines be developed that allow a model to be expressed independent of a tool?
- How to version control a model that could be made up of multiple models?

References

Alex Ball 2010. *Review of the state of the art of the digital curation of research data*. in "(unpublished).

Blackburn, Mark R., Donna H. Rhodes, Mary A. Bone, David N. Cohen, and Jaime A. Guerrero. 2018. Transforming systems engineering through digital engineering. *Journal of Defense Modeling and Simulation: Applications, Methodoology Technology*: 1–17.

Bone, Mark Blackburn Mary, Benjamin Kruse, John Dzielski, Thomas Hagedorn, and Ian Grosse. 2018. Toward an interoperability and integration framework to enable digital thread. *Systems* 6 (46): 12.

Data Curation Centre. 2004-2019, 06/27/2019. *Dcc model curation lifecycle model*.

(2018). *Department of defense digital engineering strategy*.

(1996). *Dod modeling and simulation (m&s) verification, validation, and accreditation (vv&a) number 500.61*.

Shawn Dullen; Dinesh Verma; Mark Blackburn 2019. *Review of research into the nature of engineering and development rework: Need for a systems engineering framework for enabling rapid prototyping and rapid fielding*. presented at the 17th Annual Conference on Systems Engineering Research (CSER).

Federal Deposit Insurance Corporation. 2005, 9/19/2019. *Supervisory insights - compliance examinations - model goveernance*.

(2019). *Integrated model based engineering environment implementation plan*.

Khatri, Vijay, and Carol V. Brown. 2010. Designing data governance. *Communications of the ACM* 53 (1): 148–152.

Le Novere, Nicolas, et al. 2006. Biomodels database: A free, centralized database of curated, published, quantitative kinetic models of biochemical and cellular systems. *Nucleic Acids Research, Oxford Academic- Oxford Journals* 34: D689–D691.

Philip Lord and Alison Macdonald 2003. Data curation for e-science in the uk: An audit to establish requirements for future curation and provision. In *"e-Science Curation Report," The JISC Committee for the Support of Research (JCSR)*, 2 Wayside Court, Arlington Road, Twickenham, TW1 2BQ - UK2003.

(2018). *Mbe/mbse/mbent/mbde definition and end states.*

Paul Collopy Tom McDermott, Molly Nadolski, Christiaan Paredis 2019. The future exchange of digital engineering data and models: An enterprise systems analysis. Presented at the *17th Annual Conference on Systems Engineering Research.*

P.E. Michael J. Vinarcik 2019. FESD, Hypermodeling discussion.

NASA. 2012. *Recommendation for space data system practices_reference model for an open archival information system (oais)*

David Nickerson et. al. 2006, Toward a curated cellml model repository. In *28th IEEE EMBS Annual International Conference*, New York City, USA.

Carole L. Palmer et. al. 2013. Building a framework for site-based data curation. In *American Society for Information Science and Technology*, Montreal, Quebec, Canada.

Donna H. Rhodes. 2018a, August 27, 2019. *Model curation* [omgwiki.org].

——— 2018b. *Model curation.*

——— 2019. Model curation requisite leadership and practice in digital engineering enterprises. *Conference on Systems Engineering.*

Rhodes Lucie Reymondet, Donna H., and Adam M. Ross. 2016. Considerations for model curation in model-centric systems engineering. *IEEE* 16.

Adam Ross and Jack Reid Donna H. Rhodes 2019. Interactive model-centric systems engineering (imcse) phase 6 tech report.

Daniel E. Stimpson. (2019) So much data, so little time. *Army ALT Magazine, Commentary*. Available: https://asc.army.mil/web/news-alt-jas19-so-much-data-so-little-time/

Waltemath, Dagmar, and Olaf Wolkenhauer. 2016. How modeling standards, software, and initiatives support reproducibility in systems biology and systems medicine. *IEEE Transactions on Biomedical Engineering* 63 (10): 1999–2006.

Walters, Tyler O. 2009. Data curation program development in u.S. Universities: The georgia institute of technology example. *The International Journal of Digital Curation* 4 (3): 83–92.

Weill, Peter, and Jeanne Ross. 2005. A matrixed approach to designing it governance. *MIT Sloan Management Review.*

Executable Modeling of a CubeSat-Based Space Situational Awareness System

Mostafa Lutfi and Ricardo Valerdi

Abstract As systems grow in complexity, systems engineers have embraced Model-Based Systems Engineering (MBSE) to tackle this complexity. The Systems Modeling Language (SysML) is the most commonly used language by the systems engineers to implement MBSE. SysML is not highly capable of expressing conceptual but not executable models. In order to perform requirements/behavior verifications, systems engineers/designers mostly use separate simulation tools. Hence, the efficiency of the systems engineering process is often reduced due to the isolated and consecutive use of both SysML modeling tool and other simulation tools, for example, defining simulation inputs to each simulation tool separately. Hence, executable SysML is the next logical step towards achieving true MBSE support for all systems engineering activities in the life cycle phases – system requirements, analysis, design, implementation, integration, verification, transition, validation, acceptance testing, training, and maintenance. Therefore, various research efforts are being conducted to develop executable SysML modeling approaches. This research develops a SysML Executable Modeling Methodology (SEMM), which is demonstrated by modeling a CubeSat-based Space Situational Awareness (SSA) system in SysML. The SysML SSA-CubeSat system model is made executable by integrating with Commercial-Off-The-Shelf (COTS) simulation software, namely, Systems Tool Kit (STK) and MATLAB, following the approaches defined in the SEMM.

Keywords MBSE · SysML · Executable modeling · Space Situational Awareness

M. Lutfi (✉) · R. Valerdi
Systems and Industrial Engineering, The University of Arizona, Tucson, AZ, USA
e-mail: mostafalutfi@email.arizona.edu

A. M. Madni et al. (eds.), *Recent Trends and Advances in Model Based Systems Engineering*, https://doi.org/10.1007/978-3-030-82083-1_40

1 Introduction

Model-Based Systems Engineering (MBSE) focuses on formalized application of modeling to support systems engineering artifacts development from the conceptual design phase throughout the end of the system of interest (SOI) life cycle (Hart 2015). SysML has emerged as the de facto standard system modeling language for MBSE (Delligatti 2014). MBSE is advancing in a way that intends to combine modeling with its natural next step simulation, to support the definition of system requirements, system design, system analysis, and system verification and validation. Therefore, SysML needs to be an executable language in order to directly support system's life cycle activities (Nikolaidou et al. 2016). This research paper presents a practical approach for enabling execution of models described in SysML. Specifically, the authors modeled a SSA system using CubeSats in SysML and integrated with MATLAB and Systems Tool Kit (A. G. Inc 2013). The approach described in the methodology can be expanded to other simulation tools too.

2 Literature Review

2.1 Recent Research on SysML Executable Modeling

Tsadimas et al. presented the transformation procedure of Enterprise Information System (EIS) SysML models to executable simulation code (Tsadimas et al. 2014). The author used QVT as the transformation tool. The paper also demonstrated how simulation results can be incorporated into the source SysML model using Model-Driven Architecture (MDA). Robinson et al. introduced a new analysis framework to develop executable and object-oriented SysML models through Python programming interface (Balestrini-Robinson et al. 2015). The analysis framework demonstrated a new procedure to enable rapid prototyping through the integration of SysML model and existing diagramming languages. In order to lower the financial implications, the framework used the following open-source software (OSS) – Python, MongoDB, Django, MongoEngine, OpenMDAO, RDFLib, BerkelyDB, and Django Rest Framework.

Chabibi et al. presented a taxonomy of links between SysML and various simulation environments (Chabibi et al. 2015). The paper studied an integration approach of several simulation environments into a common platform and enable two-way transformation between those environments and SysML. Chabibi et al. in another paper proposed an integration of SysML and Simulink utilizing modern techniques of Model-Driven Engineering (MDE) (Chabibi et al. 2016). The proposed integration approach consists of following steps – SysML modeling of a system using SysML4Simulink profile, executable MATLAB code generation from SysML source diagrams, and conducting simulation in order to verify system behavior through Simulink.

Kotronis et al. developed a framework that supports continuous performance assessment of Railway Transportation System (RTS) SysML model (Kotronis et al. 2016). The authors used QVT for generation of executable simulation models from the RTS SysML model. The executable simulation models are simulated in Discrete Event System Specification (DEVS) simulators, and the simulation results are incorporated into the RTS SysML model. Cawasji and Baras studied different methods to perform integration of SysML and other simulation tools (Cawasji and Baras 2018). Then, they proposed a new method by constructing a SysML executable model of a two-room house. They used the Functional Mock-up Interface (FMI) standard to integrate the SysML model with a Modelica model. The authors exported the Modelica model as Functional Mock-up Unit (FMU). Then, they used Simulink as an interface between the FMU and the SysML model. Finally, a tradeoff analysis was run through SysML, in MATLAB, to demonstrate the decision-making capability of the proposed approach for SysML executable modeling.

In order to reduce the gap between high-level modeling and evaluation of system performance through simulation, Gauthier et al. proposed a Model-Driven Engineering (MDE) tooled approach for automatic system requirements validation (Gauthier et al. 2015). The OMG SysML-Modelica working group has officially adopted this integration approach as the "SysML4Modelica" profile.

Karban et al. proposed a new Executable Systems Engineering Method (ESEM) for automatic requirements verification (Karban et al. 2016). ESEM is based on executable SysML modeling patterns and consists of structural, behavioral, and parametric diagrams. The authors used MagicDraw as the SysML modeling tool and Cameo Simulation Toolkit (CST) as the simulation engine. The authors developed an eight-step method to follow for executable SysML modeling – formalize requirements, specify design, characterize components, specify analysis context, specify operational scenarios, specify configurations, run analysis, and evaluate requirement satisfaction.

2.2 *Space Situational Awareness*

Space Situational Awareness is the ability to observe, understand, and predict the physical location and behavior of natural and manmade objects in orbit around the earth (Space Situational Awareness n.d.). Space traffic (both physical and informational) is increasing at an exponential rate. Since the start of space exploration in 1957, about 5500 rockets have been launched into earth orbit, resulting in approximately 5000 satellites and 23,000 debris still in space (esa n.d.). SSA could deliver knowledge of potential threats posed to both space assets and Earth by adversaries and environments, including space weather, space debris, uncontrolled spacecrafts, and space weapons (Gasparini and Miranda 2010; Kennewell and Vo 2013).

3 SysML Executable Modeling Methodology

SysML Models can be made executable by enabling integration of COTS simulation software through scripting languages supported by the MBSE tool. For example, Cameo Systems Modeler by Nomagic supports the following scripting languages – JavaScript, Groovy, Ruby, MATLAB, Python, and BeanShell (N. M. Inc. 2020). So, these scripting languages can take input from the SysML model and return the output/result from the integrated simulation tool into the model (Fig. 1).

In this research paper, CSM with Cameo Simulation Toolkit (CST) plug-in from Nomagic was used to create the SysML model. Moreover, MATLAB and Systems Tool Kit (STK) were two other COTS simulation tools being integrated with the system model to make it executable (A. G. Inc 2013). CST is a plug-in attached to the CSM in order to provide extendable model execution framework based on OMG fUML standards. fUML stands for Foundational Subset for Executable UML Models (Seidewitz and Tatibouet 2015). STK is a physics-based 3D modeling, simulation, and visualization tool used by engineers, space mission analysts, space operators, and decision-makers in order to model and simulate complex land, sea, air, or space systems (A. G. Inc 2013). The following steps describe the procedure to implement the SEMM for any system of interest (SOI). For this research paper, CubeSat-SSA system has been chosen as the SOI.

Model Organization The model is organized by the package names according to the four pillars of SysML (Requirements, Structure, Behavior, and Parametric). For this research paper, Parametrics package was excluded. SysML Requirements diagram falls under Requirements package. SysML Block Definition Diagram and SysML Internal Block Diagram reside inside Structure package. All the behavior diagrams (SysML Use Case, SysML Activity Diagram, and SysML State Machine Diagram) used in the model are being placed inside Behavior package. Further package decomposition was used for organization of the model elements inside these three major packages (Requirements, Structure, and Behavior) (Fig. 2).

Defining System Requirements/Use Cases/Concept of Operations (ConOps) The research paper assumed preliminary stakeholder analysis and customer requirements identification already being conducted to produce the following operational requirements for the CubeSat-SSA system (Fig. 3). The OpReq-03 has been tested for automatic verification later in the paper. Use case diagram is drawn with the aid

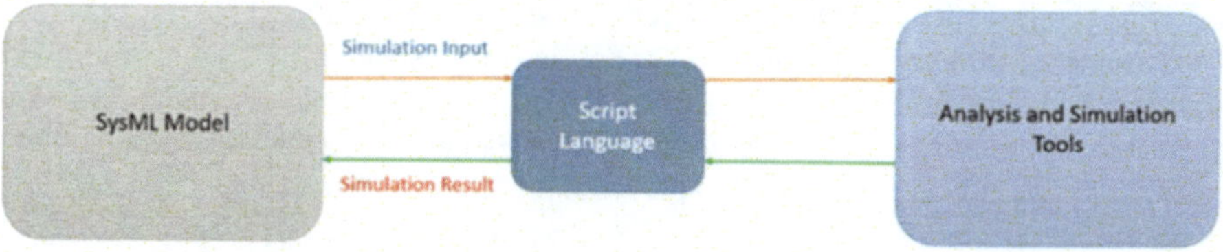

Fig. 1 SysML executable modeling framework

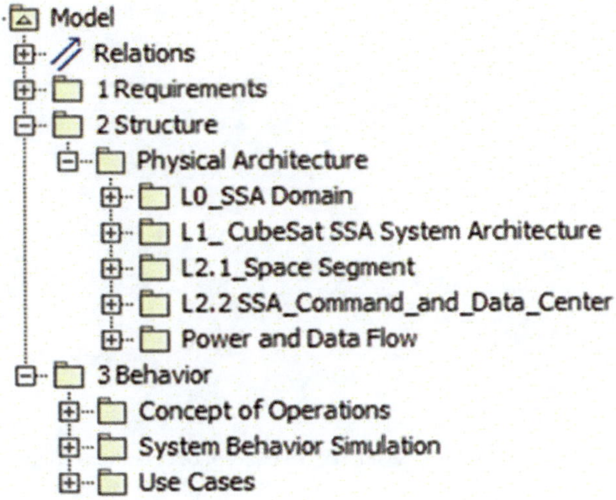

Fig. 2 Model containment tree showing package

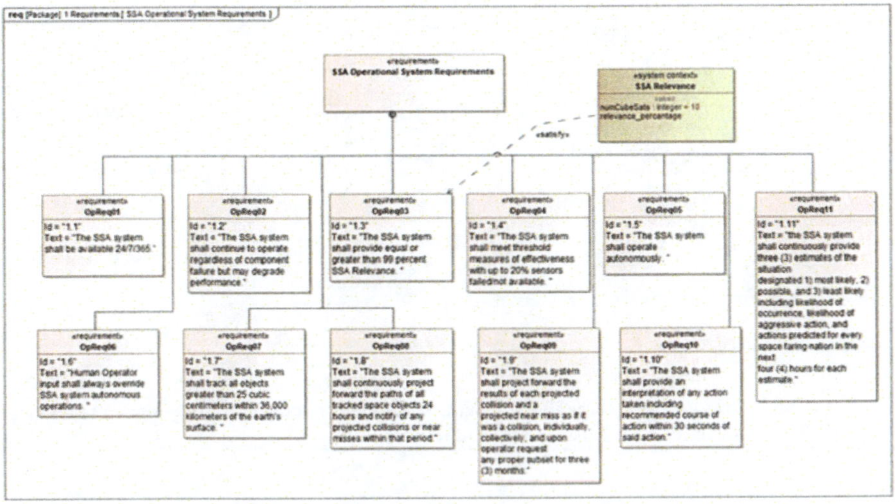

Fig. 3 Operational requirements for CubeSat-SSA system

of SysML Use Case diagram (Fig. 4). SysML State Machine Diagram enabled the creation of Concept of Operations in the model (Fig. 4).

Modeling System Architecture (Physical)/Internal Structure/Subsystem Communication SysML Block Definition Diagram was used to model Physical Architecture of the CubeSat-SSA system. Moreover, the physical architecture leveraged INCOSE's CubeSat Reference Model to represent standard CubeSat components and subsystems (Kaslow et al. 2018). SSA Domain is comprised of Artificial Space

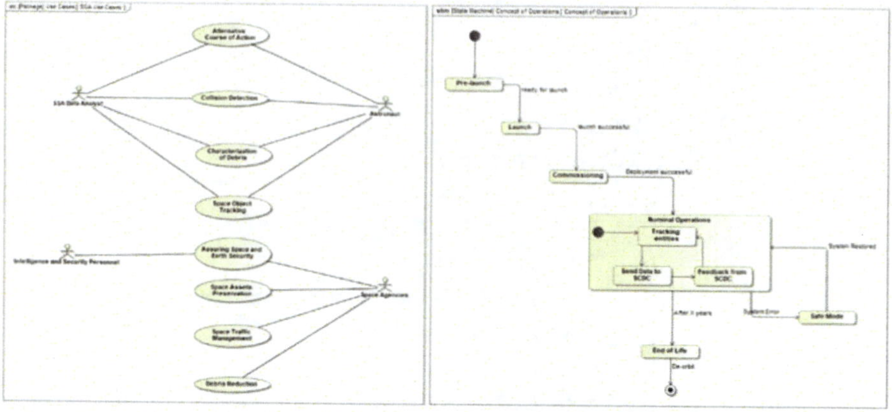

Fig. 4 (**a**) CubeSat-SSA system use cases (left); (**b**) concept of operations (right)

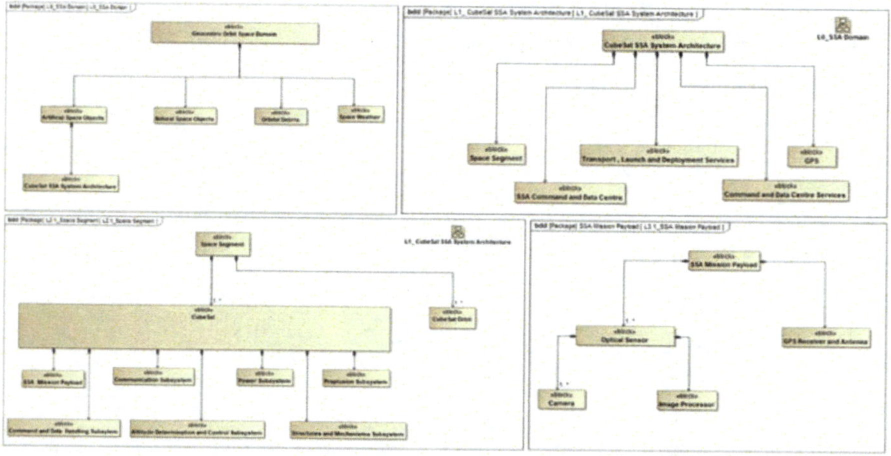

Fig. 5 (**a**) SSA Domain (top left); (**b**) CubeSat-SSA System Architecture (top right); (**c**) Space Segment (bottom left); (**d**) SSA Mission Payload (bottom right)

Objects, Natural Space Objects, Orbital Debris, and Space Weather. CubeSat-SSA System Architecture falls under Artificial Space Objects. CubeSat-SSA System Architecture consists of Space Segment, SSA Command and Data Centre, Command and Data Centre Services, Transport, Launch and Deployment Services, and GPS. Space Segment and SSA Command and Data Centre were further decomposed into subsystems (Fig. 5).

Integration with the COTS Analysis/Simulation Tools After defining the system context and corresponding behaviors for each of the scenario to be simulated, next step was to integrate the COTS tools (MATLAB and STK) with those behaviors. An activity diagram consists of different types of action elements. This research paper used CSM's Opaque Action, Read Structural Feature Action, and Read Self Action to facilitate the integration process. Opaque Action facilitated the integration of MATLAB script into CSM. CSM does not allow integration of Systems Tool Kit (STK), which is a widely used space mission analysis tool. STK was run from CSM via MATLAB scripts. STK and MATLAB interoperate with each other through STK's COM interface. STK_Relevance's classifier behavior used a MATLAB function script, which automated and modified the SSA relevance function defined in a previous research. In that research work, authors used SSA relevance measures of effectiveness for comparing SSA system architectures using a network of CubeSats (Chandra et al. 2018). However, the authors were forced to run the MATLAB script independently due to the non-executable nature of the model they defined.

Illustrative Example Scenarios The first scenario utilizes co-simulation in STK to visualize the scenario and calculate the results. In this scenario, an observation satellite (CubeSat) tracks ten unknown objects. A pointing sensor is attached to the observation satellite, namely, ADSLSat. Random creation of the unknown objects is based on the following assumptions – radius of earth is 6371 km, low earth orbit ranges from 100 to 2000 km, minimum semimajor axis is 6471 km, and maximum semimajor axis is 8371 km. The scenario returns the access values (access start times and access stop times) for each of the unknown object. The scenario can be run/reset/closed solely from SysML model without opening the STK application. SysML state machine diagram perfectly worked to create the scenario in STK by a systematic process with signals and triggers. For example, when the "Access Computation" state is running in CST, STK 3D model sends all the access data to the CST console in real time (Fig. 6). Each state was integrated with activity diagram which defines the MATLAB script.

In the second scenario, SSA Relevance function is simulated in MATLAB taking "number of CubeSats" as input (through read structural feature action in the activity diagram) from the CubeSat-SSA SysML model (Fig. 7). Value properties were added to the system context in order to provide input and accept output from MATLAB tool (Fig. 8). *Relevance_percentage* value was not populated initially because it would be populated by the simulation result of an external MATLAB script. Moreover, for simplicity the number of mesh layers (3), mesh resolution (5 degrees), analysis period (30 days), orbit period (1 day), and time step (1 day) for analysis were kept as constant. Moreover, inertial angle, altitude, camera line of sight, and camera FOV were generated randomly.

After the simulation finishes, a MATLAB plot showing SSA relevance is generated, and SSA relevance percentage is returned to the CubeSat-SSA system model through Opaque Action in the activity diagram. The Operation Requirement-03 states that the SSA system shall provide equal or greater than 99% SSA Relevance. Based on the relevance percentage value property, the OpReq-03 was automatically verified (Fig. 8). Satisfy relationship exists between the OpReq-03 and SSA Relevance block (Fig. 3). When the "*relevance_percantage*" value property automatically populated, it check whether it satisfies the requirement

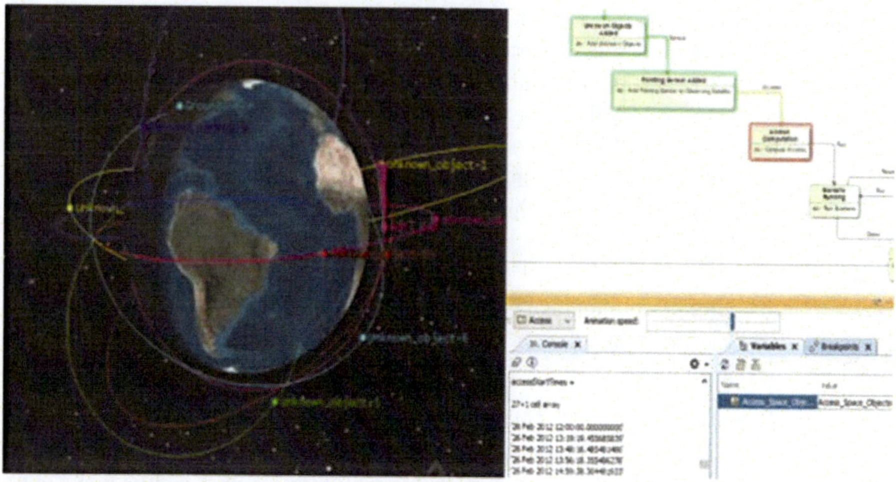

Fig. 6 Synchronization of SysML state machine diagram with STK for Scenario 1

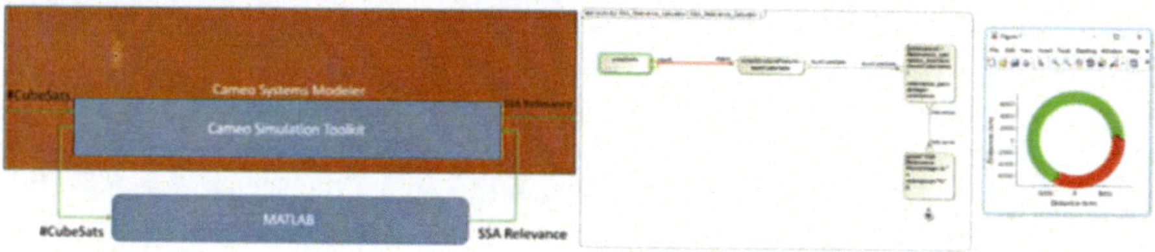

Fig. 7 Interaction between CSM and MATLAB for Scenario 2

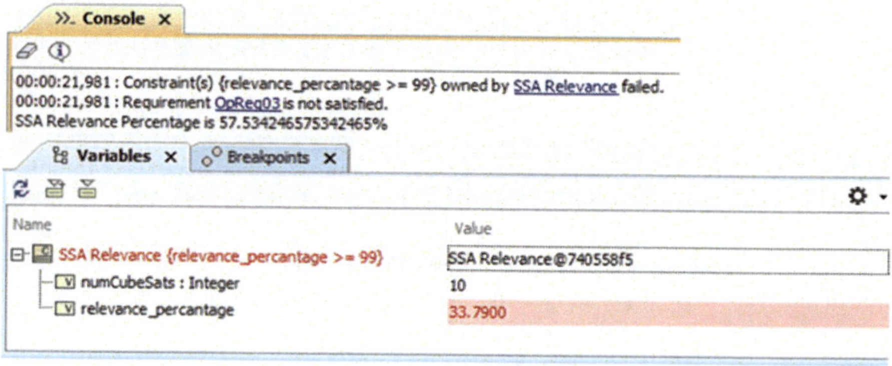

Fig. 8 (**a**) Console panel showing requirements not satisfied (top), (**b**) red color indicating requirement was not satisfied

specified value or not (Fig. 8). The following table summarizes the key Scenario 1 and 2 parameters (Table 1).

Table 1 Data exchange between integrated simulation tool and SysML model

Scenario no.	Integrated simulation tool	Simulation input	Simulation results to SysML model
1	STK	Space object properties	Access between objects
2	MATLAB	Number of CubeSats	Relevance percentage

4 Discussion

The above section demonstrates the benefits of executable modeling instead of modeling for communication purpose. Scenario 1 demonstrated how to create/run/resent a STK scenario inside the SysML CubeSat-SSA system model. Moreover, access values from a known satellite to a number of unknown objects were returned to the system model automatically. In Scenario 2, a number of CubeSats were fed into MATLAB simulation from the SysML CubeSat-SSA system model and simulation result, i.e., relevance percentage was returned to the model in order to verify a requirement. So, different CubeSat-SSA Architecture based on the number of CubeSats network can be evaluated from the SysML CubeSat-SSA system model. It was apparent that some of the SysML diagrams were not executable in nature that may change when SysML v2 will be implemented. There were some lag between the CST simulation and STK/MATLAB simulation scenario. Moreover, CSM does not support all data types as input.

5 Conclusion

The purpose of the SysML model was to demonstrate how executable modeling could incorporate the systems engineering artifacts, namely, system design (System Architecture, ConOps, Use Case, and Requirements), analysis/simulation, and requirements verification into the model itself. Hence, the user does not need to create separate simulation parameters into a separate software as that software can be run from SysML model with the capability of defining/exchanging all the simulation parameters. Hence, SysML models alone will be sufficient for system behavior verifications. Further work needs to be done to make the integration of different COTS tools from a variety of domain with the SysML models to achieve the true purpose of MBSE.

References

Balestrini-Robinson, S., D.F. Freeman, and D.C. Browne. 2015. An Object-oriented and Executable SysML Framework for Rapid Model Development. *Procedia Computer Science* 44: 423–432. https://doi.org/10.1016/j.procs.2015.03.062.

Cawasji, K.A., and J.S. Baras. 2018. SysML Executable Model of an Energy-Efficient House and Trade-Off Analysis. *IEEE International Systems Engineering Symposium (ISSE)* 2018: 1–8. https://doi.org/10.1109/SysEng.2018.8544402.

Chabibi, B., A. Anwar, and M. Nassar. 2015. Towards an alignment of SysML and simulation tools. In *2015 IEEE/ACS 12th International Conference of Computer Systems and Applications (AICCSA)*, 1–6. https://doi.org/10.1109/AICCSA.2015.7507216.

Chabibi, B., A. Douche, A. Anwar, and M. Nassar. 2016. Integrating SysML with Simulation Environments (Simulink) by Model Transformation Approach. In *2016 IEEE 25th International Conference on Enabling Technologies: Infrastructure for Collaborative Enterprises (WETICE)*, 148–150. https://doi.org/10.1109/WETICE.2016.39.

Chandra, A., Lutfi, M., & Gross, D. C. (2018). Leveraging the Emerging CubeSat Reference Model for Space Situational Awareness.

Delligatti, L. (2014). SysML Distilled: A Brief Guide to the Systems Modeling Language. Pearson Education.

esa. n.d. *Space debris by the numbers*. European Space Agency. Retrieved August 3, 2019, from https://www.esa.int/Our_Activities/Space_Safety/Space_Debris/Space_debris_by_the_numbers

Gasparini, G., and V. Miranda. 2010. Space situational awareness: An overview. In *The Fair and Responsible Use of Space: An International Perspective*, ed. W. Rathgeber, K.-U. Schrogl, and R.A. Williamson, 73–87. Vienna: Springer. https://doi.org/10.1007/978-3-211-99653-9_7.

Gauthier, J.-M., F. Bouquet, A. Hammad, and F. Peureux. 2015. Tooled Process for Early Validation of SysML Models Using Modelica Simulation. *FSEN*. https://doi.org/10.1007/978-3-319-24644-4_16.

Hart, L. 2015. *Introduction to Model-Based System Engineering (MBSE) and SysML*. 43.

Inc, A. G. 2013, July 24. *Bringing in External Data to Model Space Objects in STK*. https://vimeo.com/70964608

Inc, N. M. n.d.. *Cameo Systems Modeler*. Retrieved June 19, 2020, from https://www.nomagic.com/products/cameo-systems-modeler

Karban, R., N. Jankevičius, and M. Elaasar. 2016. ESEM: Automated Systems Analysis using Executable SysML Modeling Patterns. *INCOSE International Symposium* 26 (1): 1–24. https://doi.org/10.1002/j.2334-5837.2016.00142.x.

Kaslow, D., B. Ayres, P.T. Cahill, L. Hart, A.G. Levi, and C. Croney. 2018, September 17. *Developing an MBSE CubeSat Reference Model – Interim Status #4. 2018 AIAA SPACE and Astronautics Forum and Exposition*. 2018 AIAA SPACE and Astronautics Forum and Exposition, Orlando, FL. https://doi.org/10.2514/6.2018-5328.

Kennewell, J.A., and B. Vo. 2013. An overview of space situational awareness. In *Proceedings of the 16th International Conference on Information Fusion*, 1029–1036.

Kotronis, C., A. Tsadimas, G.-D. Kapos, V. Dalakas, M. Nikolaidou, and D. Anagnostopoulos. 2016. Simulating SysML transportation models. In *2016 IEEE International Conference on Systems, Man, and Cybernetics (SMC)*, 001674–001679. https://doi.org/10.1109/SMC.2016.7844478.

Nikolaidou, M., G.-D. Kapos, A. Tsadimas, V. Dalakas, and D. Anagnostopoulos 2016. Challenges in SysML Model Simulation.

Seidewitz, E., and J. Tatibouet 2015. Tool Paper: Combining Alf and UML in Modeling Tools - An Example with Papyrus.

Space Situational Awareness. n.d. Retrieved August 4, 2019, from https://www.spaceacademy.net.au/intell/ssa.htm

Tsadimas, A., G.-D. Kapos, V. Dalakas, M. Nikolaidou, and D. Anagnostopoulos. 2014. Integrating simulation capabilities into SysML for enterprise information system design. In *2014 9th International Conference on System of Systems Engineering (SOSE)*, 272–277.

Comparing Weighting Strategies for SME-Based Manufacturability Assessment Scoring

Emily S. Wall, Christina H. Rinaudo, and R. Cody Salter

Abstract Manufacturability involves many different influence factors from product design and geometry to supply chain and ergonomics. Many software packages are available for product assessments; however a gap in the available software packages was discovered. The lack of a software that looked at both product design and the design of the process, along with other variables that affect the total manufacturability, was identified. From this research, the MAKE tool was created; however how to score and weigh each assessed part of a product was still under investigation. This paper outlines a comparison between two proposed weighting methods for a manufacturability assessment. The first was the topic of a 2019 CSER paper describing a weighting method conducted by SME inputs and counting risk concerns and high scores. The second method, which is introduced in this paper, uses a value curve method to assign weights to each aspect of manufacturability. A case study is used to illustrate each method, and the results are compared and graphically displayed to magnify similarities and differences between methods along with conclusions and future research areas.

Keywords Weighting · Manufacturability · Scoring · Value curves

1 Introduction

Previous research has defined manufacturability as the ease in which a product can be produced, delivered, and effectively utilized in its intended environment (McCall et al. 2018). Manufacturability, as defined in this research, goes beyond the focus of product geometry design and includes manufacturing process design

E. S. Wall (✉)
Mississippi State University, Starkville, MS, USA
e-mail: ewall@cavse.msstate.edu

C. H. Rinaudo · R. C. Salter
U.S. Army Engineer Research and Development Center, Vicksburg, MS, USA

A. M. Madni et al. (eds.), *Recent Trends and Advances in Model Based Systems Engineering*, https://doi.org/10.1007/978-3-030-82083-1_41

Fig. 1 Manufacturability interaction matrix (MIM)

Aspect of Design (AD) → / Aspects of Mfg (AM) ↓	Material	Product and Manufacturing Information (PMI)	Design Geometry
Process	X	X	X
Quality	X	X	X
Supply Chain	X	X	X
Capital Tooling & Equipment	X	X	X
Labor	X	X	X
EHS & Ergonomics	X	X	X
Capacity and Scalability	X	X	X

issues, supply chain concerns, and personnel issues such as labor and ergonomic concerns. The research team identified a gap in the current software programs that analyzed products for manufacturing and determined that a tool was needed that not only investigated cost and design features but other factors that directly affect the manufacturability (Shi et al. 2018; Shukor and Axinte 2009; Andersson et al. 2014). When assessing manufacturability, two primary categories emerge as areas of focus: aspects of design (AD) and aspects of manufacturing (AM). Aspects of design include material, product, manufacturing information, and geometry, while aspects of manufacturing consider process, supply chain, quality, capital tooling and equipment, labor, environment/health/safety/ergonomics, and capacity/scalability (McCall et al. 2018). Previous research developed a manufacturability interaction matrix (MIM) (Fig. 1). The three aspects of design and seven aspects of manufacturing create a scoring matrix that is the criteria used to assign rating scores to product components (Fig. 1).

In the previous research, subject matter experts (SMEs) provided an individual manufacturability assessment of the product under review. The assessment includes assigning a risk score (between 1 and 10) from Fig. 2 for each of the interaction's categories in Fig. 1.

This research compares two weighting strategies for manufacturability assessment score generation using SME input. The first method weighs components by tallying poor scores and risk concerns (Wall et al. 2019), while the second method develops component weights using the swing weight matrix (Parnell and Trainor 2009).

SCORING GUIDELINES FOR MAKE

Impact of Interaction on Manufacturability - As determined by Subject Matter Expert (SME)	Rating	Color Scale
Minor: - No significant impact identified by the SME. - Concerns are minor and unnoticeable to the customer. - No impact to product yields.	1-2	
Low: - Slightly significant impact predicted by SME. - Concern may cause slight customer annoyance. - Slight deterioration of the product or service. minimal impact to the next process or minor rework. - Moderate facilitative efforts can be expected to maintain required production levels.	3-4	
Moderate: - Moderately significant impact predicted by SME. - Concern identified may result in customer discomfort and dissasfaction. - May cause the need for unscheduled repairs or rework of the product. - Production yields may be significantly less than required due to one or more quality, delivery, or efficiency performance issues.	5-6	
High: - High degree of impact predicted by the SME. - Concern will result in high customer dissatisfaction due to the nature of the failure such as an inoperable product or process. - Does not involve safety issues or government regulations but will cause the need for unscheduled repairs or rework of the product. - May cause disruptions to subsequent processes with high impact to production yields.	7-8	
Very High: - Very high impact to manufacturability predicted by the SME. - Concern involves loss of primary functionality with very high impact to the customer. - Concern may involve safety or regulatory issues. - Production yields are significantly impacted with significant repairs, rework and retesting required.	9-10	

Fig. 2 Subject matter expert scoring guidelines

1.1 Value Modeling and Weighting Background

Value models could be used to aid with making informed decisions and allow for transparency in the decision-making process. Value modeling is a method for the determination of "how well candidate solutions attain stakeholder value" (Trainor and Parnell 2011). Quantitative value models represent a set of "functions, weights, and mathematical equations that are used to evaluate candidate solutions" (Trainor and Parnell 2011). Value model development includes determining scaling factors which help to "quantify the trade-offs between value measures" (Parnell and Trainor 2009).

Scaling factors indicate the importance of each value measure relative to all other value measures. Although many techniques exist for determining scaling factors, this research focuses on the use of the swing weight matrix methodology. The swing weight matrix is a technique for understanding the importance of each value measure and how impact changes as the value measure swings through its acceptable range (Parnell and Trainor 2009; Trainor and Parnell 2011). The swing weight matrix can consist of a 3 × 3 grid with categories of "Variation in measure range" (High, Medium, and Low) and "Level of importance of the value measure" (Very Important, Important, and Less Important) (Trainor and Parnell 2011).

2 Weighting Strategy Background

2.1 SME-Based Normalized Weighting Strategy

The previous SME-based weighting strategy process included the assumption that SMEs will give red scores (High Impact of Interaction on Manufacturability) and

Impact of Interaction on Manufacturability - As determined by Subject Matter Expert (SME)	Rating	Color Scale	AIAG Rule of thumb for effectiveness
Minor:	1-2		0-5, 6-10
Low:	3-4		11-13, 14-15
Moderate:	5-6		16-18, 19-20
High:	7-8		21-40, 41-60
Very High:	9-10		61-80, 81-100

Fig. 3 AIAG Rule of Thumb Scale with Scoring Scale ("Measurement Systems Analysis," 2010)

more concerns to high-risk interactions and thus drive the weight given to each AM category. The research team realizes that this assumption may not always be true and that SME response will vary. For each AM, the tally of concerns and red scores for an AM category is evaluated against the sum of all other AM tallies to arrive at a percentage of the total number of concerns and red scores that fall in each AM category. This percentage, which must correspond back to the rating scale (Figs. 1 and 2) to determine a weight, is then fitted to the rating scale (Fig. 3). These weights are then multiplied by their corresponding averages of the AM for both assemblies and components. Those final AM-weighted averages are then combined to result in a final score. This score is normalized using a minmax normalization equation of the new weighted scores to linearly transform the minimum and maximum values to map to 0 and 1 score, respectively. By accomplishing the weighting programmatically instead of manually, additional involvement of the SME is avoided, and SME subjectivity is minimized.

2.2 SME-Based Swing Weight Matrix: SME Weighting Methodology

While the previous research method utilized SME information to develop a normalized weighting, the researchers investigated using the swing weight matrix methodology in order to implement a weighting strategy accepted throughout the systems engineering community. The swing weight matrix applies a weight to the aspects of manufacturability average across all parts and assemblies assessed.

In order to transfer the previously generated SME ratings into the swing weight matrix methodology, the seven aspects of manufacturability are evaluated against the level of importance of the value measure and the variation in the measure range. The researchers first analyzed the range of SME raw scores across each aspect of manufacturability. This information was used to determine the variation

		Level of importance of the value measure		
		Very Important	Important	Less Important
Variation in measure range	High	Quality, Process $f_1 = 9$	Capability/ Scalability $f_3 = 6$	$f_6 = 5$
	Medium	$f_2 = 8$	Capital Tooling & Equipment $f_5 = 4$	EHS & Ergonomics $f_8 = 2$
	Low	$f_4 = 7$	$f_7 = 3$	Labor, Supply Chain $f_9 = 1$

Fig. 4 Swing weight matrix for aspects of manufacturing (AM)

in measure range category (high, medium, low) for the AM swing weight matrix assignments. For example, the AM aspects that displayed the highest range value from their given scores were categorized as "high." In order to determine the level of importance of the value measure, the researchers used a short survey to gather from the SMEs which AMs they felt were most important or carried the highest risk in the manufacturing of the product. The survey results allowed the SMEs to rank the aspects of manufacturability by importance to their individual industry and product. Using the two rankings, the seven AMs are applied to the swing weighting matrix that has already had a non-normalized swing weight, between 1 and 9, assigned to each section (Fig. 4).

These weights are then applied to the corresponding aspect of manufacturability raw/average score and multiplied to result in a non-normalized weighted score. The seven AM scores are then averaged and normalized in order to develop a final product score rating score between 0 and 10.

3 Case Study Results

The results of using the case study data from the previous 2019 conference paper "Development of a weighting strategy for a manufacturability assessment" (Wall et al. 2019) research reveal that the swing weight matrix-based weighting strategy generated a higher final score overall than the normalized weighting strategy (Figs. 5 and 6). A higher final score indicates a higher level of concern for the overall manufacturability of the product. By using the range of scores given across all parts assessed and the aspect of manufacturing importance ranking survey, the weights are applied to the scores to amplify risk areas that would otherwise not be as visible using a non-weighting system. The research team has been aware of the need for a logical application of weights to the results and felt that using the unweighted final score did not adequately tell the whole story. A positive finding of the final case study results shows that the weights given on both methods were applied in generally the same areas and with the same amount of intensity. This indicated to

Comparison Case Study: Product A								
Swing Weight method	Assemblies	Components	**Normalized SME-based method**	Assemblies	Components	**Non-Weighted method**	Assemblies	Components
Process	10	8.7	Process	10	5.4	Process	3.2	1.4
Quality	7.6	10	Quality	6.2	10	Quality	2.4	1.6
Supply Chain	0.1	0.2	Supply Chain	0.1	1.4	Supply Chain	1	1.3
CT&E	3.8	2.9	CT&E	1.9	1.2	CT&E	2.8	1.2
Labor	0.1	0.2	Labor	0.1	0.1	Labor	1	1.2
EHS & Ergo.	1.3	1.1	EHS & Ergo.	1.5	0.1	EHS & Ergo	2.3	1.2
Cap. Scal.	5.5	7.2	Cap. Scal.	1.7	9.3	Cap. Scal.	2.7	1.8
Final Score		4.19	Final Score		3.50	Final Score		1.80

Fig. 5 Case study results

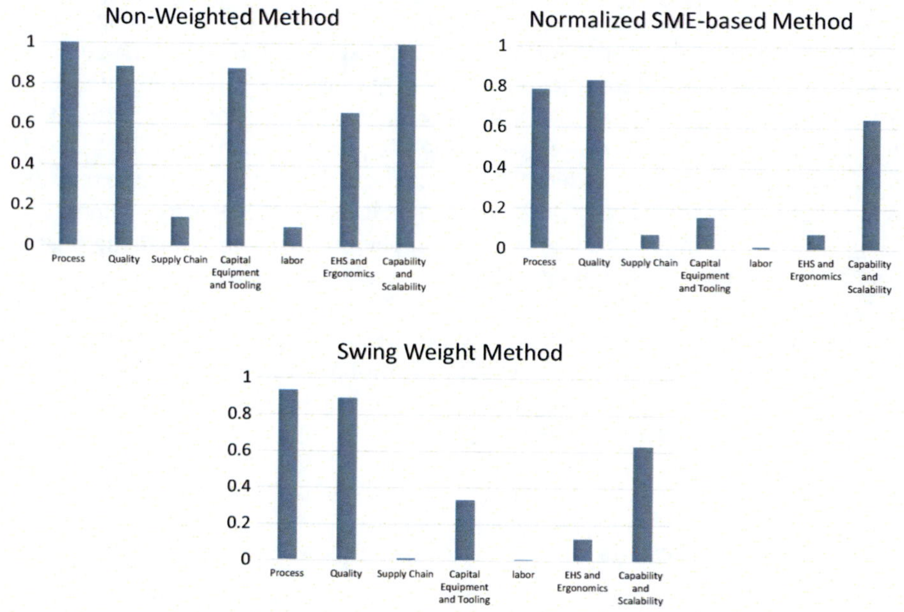

Fig. 6 Case study results – AM scores by method

the research team that the use of the swing weight method accurately placed weights on the same aspects of manufacturability as dictated by the assessment results, both scores and SME given concerns. Even though the final results between the swing weight method and the normalized SME-based method were very similar, it is unexpected that the swing weight method resulted in a higher final score. Initial expected results from the research team suggested that, because of SME biases, the SME-based method would most likely assign harsher weights, thus resulting in higher final scores. The results show the opposite and indicate to the research team that using SME-based judgments for weighting is not a reliable method and that the swing weight method adequately assigns weights where emphasis is needed but still is based indirectly on SME inputs within the assessment process by a calculated and assigned, well-known process using a swing weight matrix. Figure 6 breaks down the normalized final scores for each aspect of manufacturability and displays them via bar graphs. For supply chain and labor specifically, the application of weights affects the overall averaged final score by almost negligible values on the chart. This visualization is not meant to indicate that when weights were applied, supply chain and labor scores were deemed unimportant; rather this shows that overall these two aspects of manufacturability did not affect the overall final score of manufacturability and had less variation in scores and concerns when compared to other areas such as process and quality. The weights are meant to allow for problem areas of a product to appropriately influence the final manufacturability score and not allow other "high-scoring" areas to drown out key issues when calculating final scores.

4 Conclusions and Future Work

Using the swing weight matrix method for SME-based manufacturability assessments provides the researchers with the ability to implement a systems engineering accepted weighting strategy methodology and compare the effects of the weights with other published methods. Although the final score generated using the swing weight matrix resulted in a higher final score than in previous research, additional testing and case study analysis are needed to further investigate the feasibility of using this methodology in future manufacturability assessments. Overall, the swing weight method achieves two of the research goals: (1) reduce the SME interaction in applying weights and (2) highlight risk areas not displayed in the non-weighted method. Previous concerns related to the possibility of a zero or one range factor resulting from common red scores in an interaction negatively affecting the weighting score were addressed. These concerns are part of future research to be investigated, and if the case of the range resulting in the AM ranking being lower, the consistent red scores would already indicate risks, and thus a weighting to indicate risks would not be needed due to the high-risk scores already assigned. Future research could investigate additional case studies to provide the research team with further insight into the most appropriate weighting methodology for the manufacturability assessment.

Acknowledgments This material is based upon work performed under Contract No. W912HZ-17-C-0018, with Mississippi State University. Any opinions, findings, and conclusions or recommendations expressed in this material are those of the authors and do not necessarily reflect the views of the US Army Engineer Research and Development Center. Disclaimer: Reference herein to any specific commercial company, product, process, or service by trade name, trademark, manufacturer, or otherwise does not necessarily constitute or imply its endorsement, recommendation, or favoring by the US government or the Department of the Army (DOA). The opinions of the authors expressed herein do not necessarily state or reflect those of the US government or the DOA and shall not be used for advertising or product endorsement purposes.

References

Andersson, F., A. Hagqvist, E. Sundin, and M. Björkman. 2014. Design for manufacturing of composite structures for commercial aircraft-The development of a DFM strategy at SAAB aerostructures. *Procedia CIRP*. https://doi.org/10.1016/j.procir.2014.02.053.

McCall, T., L. Dalton, S. Fuller, N. Watson, E. Wall, B. Smith, et al. 2018. *Developments of the Manufacturability Assessment Knowledge-based Evaluation (MAKE) and Application of Case Studies*. Vicksburg: Technical Report for Engineering Research and Development Center.

Measurement Systems Analysis. 2010. *Automotive Industry Action Group (AIAG), 4th*, 78.

Parnell, G.S., and T.E. Trainor. 2009. Using the swing weight matrix to weight multiple objectives. *19th Annual International Symposium of the International Council on Systems Engineering, INCOSE 2009* 1 (July 2018): 283–298. https://doi.org/10.1002/j.2334-5837.2009.tb00949.x.

Shi, Y., Y. Zhang, S. Baek, W. De Backer, and R. Harik. 2018. Manufacturability analysis for additive manufacturing using a novel feature recognition technique. *Computer-Aided Design and Applications* 15 (6): 941–952. https://doi.org/10.1080/16864360.2018.1462574.

Shukor, S.A., and D.A. Axinte. 2009. Manufacturability analysis system: Issues and future trends. *International Journal of Production Research* 47 (5): 1369–1390. https://doi.org/10.1080/00207540701589398.

Trainor, T., and G.S. Parnell. 2011. Problem Definition. In *Decision making in systems engineering and management*, ed. G.S. Parnell, P.J. Driscoll, and D.L. Henderson, 2nd ed., 297–352. Wiley.

Wall, E., B. Smith, S. Vick, N. Watson, and T. McCall. 2019. Development of a weighting strategy for a manufacturability assessment. *Procedia Computer Science* 153: 309–316. https://doi.org/10.1016/j.procs.2019.05.084.

A Framework for Using the MAKE Methodology and Tool for Objective Manufacturability Decision Analysis

Sara C. Fuller, Tonya G. McCall, Emily S. Wall, Terril C. Falls, Christina H. Rinaudo, and Randy K. Buchanan

Abstract The objective of the proposed research involves the challenge of developing a methodology and tool to assess the manufacturability of conceptual designs at Milestone A, where minimal system design information is available. From a practical standpoint, the idea of utilizing a subject matter expert SME) as a basis for judgment on a design's manufacturability early in the design process lacks feasibility due to the inability to efficiently and effectively evaluate a large tradespace of unique design alternatives. The practice of Design for Manufacturing (DFM) analysis typically involves having access to design geometry and specifications with some consideration of the manufacturing processes. However, in the conceptual stage, the challenge involves assessing manufacturability based on a significant number of unknown parameters and doing so in a manner which is nonsubjective. Furthermore, evaluation of early stage product designs has significant influence on program cost. So, how can programs realize the impact of design options that may influence manufacturability and relate this back to a common frame of reference (i.e., cost, schedule, risk)? A research challenge includes determining how to harness the knowledge that is used to determine manufacturability from both factual and heuristic-based approaches, which requires some knowledge of the design parameters and the decision-making involved with assessing manufacturability. There are different ways to explore this area of research, but one possible approach rests in the exploration of artificial intelligence and how it can be applied in the area of manufacturability assessments. There are various subsets of artificial intelligence; some involve areas such as rule-based engines and systems, knowledge graphs, and expert systems, while others explore more complex areas such as machine learning and neural networks. The choice on which path to take requires some exploration into these possibilities and an understanding of the design data available in pre-milestone A and how feature-based information can be used to create an objective-

S. C. Fuller (✉) · T. G. McCall · E. S. Wall · T. C. Falls
Mississippi State University, Starkville, MS, USA
e-mail: sfuller@cavse.msstate.edu

C. H. Rinaudo · R. K. Buchanan
U.S. Army Engineer Research and Development Center, Vicksburg, MS, USA

A. M. Madni et al. (eds.), *Recent Trends and Advances in Model Based Systems Engineering*, https://doi.org/10.1007/978-3-030-82083-1_42

based manufacturability assessment. This paper serves to explore the options for incorporating artificial intelligence within the MAKE assessment methodology and related software tool. Based upon feedback from the user community, one or more of these options could be incorporated in future efforts.

Keywords Manufacturability · Analysis of alternatives · Tradespace

1 Manufacturability Assessment Knowledge-Based Evaluation (MAKE) Background

According to McCall and Fuller (2018), the Manufacturability Assessment Knowledge-based Evaluation (MAKE) tool draws upon a taxonomy of manufacturability concerns (i.e., life cycle cost drivers), based on functional areas of a manufacturing system (quality, EHS, supply chain, etc.). It exists to identify concerns within areas of each manufacturing system that are impacted by characteristics of the design. These concerns drive the determination of a manufacturability metric, which can be used to compare alternatives of a design at levels from the individual components up to the final assembly.

The Department of Defense (DoD) science and technology communities support the analysis of model-based engineering early into the design process to support decision-making for analysis of alternatives (AoA). Analysis of alternatives is a DoD requirement of military acquisition policy to ensure that multiple design alternatives have been analyzed prior to making costly investment decisions (US Office of Management and Budget 2008). The objective of this research involves the development of a methodology, more specifically a metric, intended to reflect the manufacturability of a product design. The metric may reflect the manufacturability of a total product design and subcomponents or subassemblies of that design. Ultimately, the metric is intended to provide some guidance during AoA or trade-off studies in order to understand the cost drivers or risk inherent to a particular design. Through the evaluation of different design options, users can arrive at design solutions that best meet the mission goals.

McCall et al. detailed the approach to the research which began with development of a methodology at life cycle Milestone C, where fidelity of the design is at a stage where relevant design and manufacturing parameters exist on which to base the development of the architecture for the manufacturability assessment. As the research progressed, more effort has been spent to understand how far to the "left" in the product life cycle a particular design can be assessed. That is, "what is the earliest point in the life cycle timeline at which a useful assessment can be performed?" In addition, there is also the driving question of what is required of the methodology to allow for the assessment of such designs in the early phases where design fidelity is minimal and multiple alternatives are being considered. Figure 1 depicts the strategy of the manufacturability development.

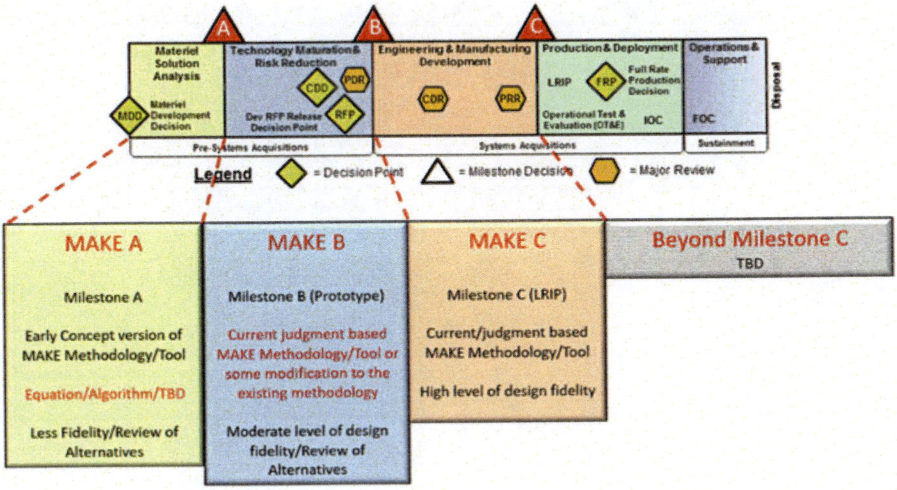

Fig. 1 Strategy of MAKE

Previous case study assessments described in prior work by McCall et al. (2016, 2017) focused on analysis of systems near Milestone C and utilized the methodology referenced in Fig. 1 as MAKE C. Significant research effort has supported understanding this scope of work and the extent to which the current methodology can be applied at Milestones A and B. Understanding the limitations of the current methodology established a research framework for determining the architecture of the manufacturability methodology necessary for design fidelities inherent to Milestone A.

2 MAKE Current Capabilities to Support Tradespace Analysis

2.1 Existing Methodology and Tool Features

The existing structure of the MAKE tool allows for a user to perform an analysis of alternatives in support of the decision-making process during the product design. As part of the assessment process, the user creates a parts list and subsequently a hierarchical bill of materials (BOM). By asking the question, "What is the impact of a particular aspect of design on a particular aspect of manufacturing?", the user is able to review each part of the BOM, documenting design concerns and recommendations for each of 21 different interactions. The "Scores" area of Fig. 2 shows the manufacturability interaction matrix (MIM) that guides the assessment.

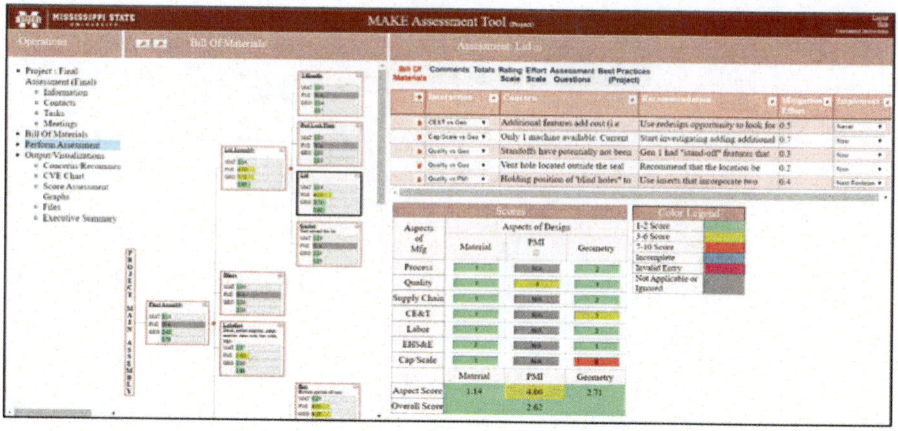

Fig. 2 MAKE tool showing BOM and part assessment

After documenting the concerns for a particular part of the BOM, the assessor then assigns a score to each of the interactions. The tool allows the assessor to quickly and easily add, remove, and substitute parts and subassemblies in the BOM. The design's manufacturability metric is automatically "rolled up" from the BOM giving instant feedback to the assessor as modifications are made. Once the parts have been evaluated, the assessor can evaluate several variations of the design by modifying the BOM as needed and documenting the resulting metric for that variation.

In addition to assigning a score and documenting the concerns for each part, the assessor can upload additional information such as part drawings, comments, and photos to further document and justify the score for a specific part. Maintaining the score and auxiliary information at the part level allows the assessor to:

- Assemble information needed to define the variation.
- List the concerns and recommendation for that variation.
- Assemble data that highlight issues within a variation.
- Collect, develop, and store other information pertinent to the variation.

2.2 *Output Includes List of Prescriptive Measures for Decision-Making*

In addition to documenting the concerns identified during the assessment, documenting recommendations for prescriptive measures to mitigate the concern is an important element of the manufacturability assessment. The tool allows recommendations for a concern to be documented while the assessor is evaluating the part and captures the immediate response to remedy the concern (e.g., "using screws in design rather than glue"). This, however, does not prevent assessors or other experts from later revisiting the concerns and modifying or adding recommendations.

The tool allows the assessor to rate the level of concern and effort to implement a recommendation and provides a listing of concerns for a specific part along with the associated recommendations for mitigation. This list can be produced for an individual part or "rolled up" for a part and its subassemblies. The tool also provides a concern versus effort graph, which is a graph of the score of a concern versus effort to implement recommendations to mitigate that concern.

2.3 Reliance upon Subject Matter Experts

The MAKE methodology relies heavily on subject matter experts (SMEs) to evaluate the design areas in need of assessment. Furthermore, the entire manufacturability product assessment is based on the existence or creation of a reliable BOM and a thorough review of each part and assembly within that BOM. The assessor's knowledge of the best practices within the design area that he/she is evaluating is paramount to the accuracy of identifying the concerns and providing viable prescriptive measures for mitigation of manufacturing risk. The knowledge basis of the SME is essential to the accurate portrayal of a product's manufacturability risk. While the reliance on SMEs may be acceptable for a MAKE C evaluation, it is a significant concern as one looks toward an assessment for early life cycle designs.

2.4 Challenges of Applying MAKE to Early Life Cycle Assessments

The challenge of developing a methodology (MAKE A, Fig. 1) to assess the manufacturability of conceptual designs at Milestone A, where minimal system design information is available, involves both the use of SMEs and the fidelity of the design at this stage. A primary goal for early life cycle manufacturability assessment is to support understanding the impact of these early design decisions on manufacturing cost, time, and quality. From a practical standpoint, using SME judgment on a design's manufacturability at this stage lacks feasibility. The most prominent issue relates to the inability to evaluate the large number of unique alternatives in the tradespace environment.

The practice of DFM analysis typically involves having access to design geometry and specifications with some consideration of the manufacturing processes. However, in the conceptual stage, the challenge involves assessing manufacturability based on a significant number of unknown parameters. Pre-Milestone A design analysis primarily focuses on performance, cost, and other metrics for which the methodology to evaluate at this stage has been developed and optimized through years of experience and research.

Another challenge exists in the foundation for the tradespace evaluations. Previous analysis of point-based design processes (using an existing design as a foundation) has demonstrated that later iterations to refine that design solution can be time-consuming and costly and lead to a suboptimal design (Iansiti 1995; Kalyanaram and Krishnan 1997). The ability to examine many designs is made possible by relying on past products and extrapolating information from those past designs. This extrapolation process relies heavily on years of experience and research. The attempt to similarly provide the manufacturability metric for each of these designs is extremely problematic.

Tradespace exploration research by investigated various methods to integrate cost models with tradespace analysis while requiring minimal user interaction. This research used predictive modeling using artificial neural networks to develop a surrogate model. Furthermore, various expert system methodologies could be investigated for implementation with the MAKE A methodology (Viral and Bhushan 2014). Using similar methods could provide the ability to reduce the reliance on SME input and further automate the process for efficient analysis.

3 MAKE 2.0

3.1 The Connection with Tradespace Exploration

By using tradespace exploration in the AoA process, program analysts and decision makers are provided with early system design and development analysis to support understanding of potential system capabilities, gaps, and potential compromises and implications. It informs decision makers regarding opposing system options and the significance of decisions across various missions and objectives.

Tradespaces are essentially a matrix of information which contains design parameters of a variation of a product's design and the associated results of various analyses for that design. During the product design, in addition to defining various design variations, researchers investigate system attributes in order to derive other parameters for that variation, such as suitability, performance, cost, maintainability, etc. The design parameters and analysis results are collected into the tradespace. The tradespace is analyzed by the teams using common data analytic techniques and system engineering tools (e.g., multi-objective decision analysis) to determine an optimal design.

Generating the manufacturability metric for each design variation of a large tradespace could fully integrate the MAKE methodology into tradespace exploration. However, attempting this with the current tool methodology is impractical due to the requirements for intensive user input for each design. Attempting to use the tool as previously described could require a cost prohibitive amount of effort. The possibility of an objective-based assessment provides the means in which to make the connection between manufacturability and life cycle cost.

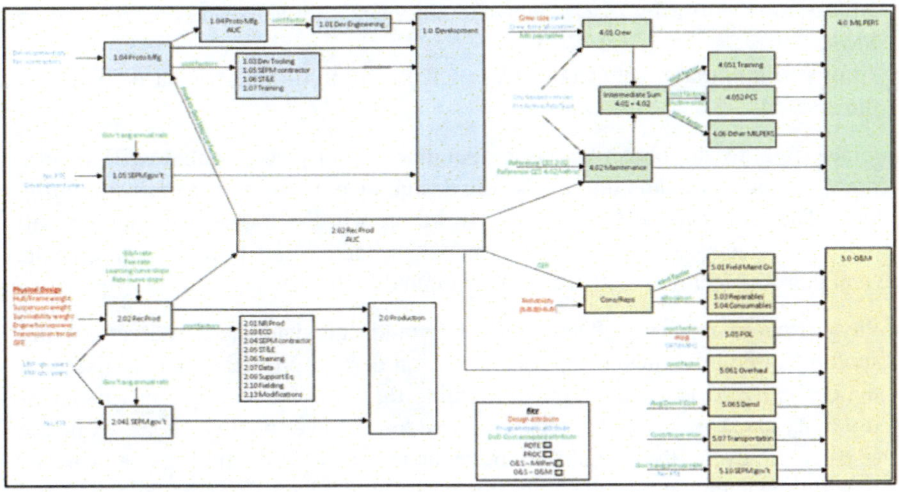

Fig. 3 Cost methodology

Previous research by Buchanan et al. (2018) investigated analyzing set-based design and incorporation of cost estimation using a notional cost model for ground vehicles. The cost methodology in Fig. 3 from Cherwonik (2017) illustrates the flow from physical design parameters (red) combined with programmatic cost drivers (blue) and the cost factors generated from historic data (green) prior to integration. A potential connection point for manufacturability assessment to support early Milestone A tradespace analysis could be made by interfacing with the cost methodology through the "step-up" factors. For example, the manufacturability assessment rating could provide a multiplication factor between the prototype manufacturing (1.04) and production development (2.02) costs. Linking the manufacturability assessment rating to the cost model estimation process could provide decision makers with additional cost and manufacturing insights while making system decisions for a program.

3.2 *Transitioning from Subjective to Objective Analysis*

As stated previously, the current MAKE methodology is highly subjective due to the extensive use of SMEs' knowledge of the manufacturing process. The desire is to remove as much subjectivity as possible and to create an objective system. The advantages of such a system are apparent, such as:

- Remove biases that are the result of isolated events in the SME's career.
- Available tradespaces with manufacturability metric, if the designs have the fidelity required to analyze.

- Ability to perform the assessment without years and years of experience of an SME.
- Ability to mix manufacturing processes without multiple SMEs participating in the assessment.

Converting to an objective-based manufacturability assessment will require creating computer models that assess the design against a set of manufacturability criteria. The fact that the assessment will be used in the tradespace analysis will also drive the decision as to the type of models needed to satisfy both requirements. Several possible techniques that could be applied include:

- Physics-based models – These models would calculate the manufacturability score of alternative design options based on certain physical characteristics of the design (e.g., geometry), thus providing the basis for an objective manufacturability assessment model. However, due to the lower fidelity of data available in early life cycle, they would not be realistic for use in the tradespace analysis. In addition, even the basis of a physics-based model would require some level of heuristic knowledge in order to truly identify the manufacturing impact.
- Expert system (rule-based) models – These models would utilize expert systems, where the SMEs' knowledge and experience for an interaction are coded into rules and then passed through an inference engine to obtain the final metric. Unfortunately, expert systems are not appropriate for use where much of the input is unknown or vague due to the rigid structure of the system.
- Artificial intelligence/machine learning (AI/ML) models – These models use modern AI techniques to build models that can handle the "fuzziness" of the input. Bayesian networks, a type of probabilistic graphical model, appear to be very promising. The networks can better deal with missing or unknown information. The Bayesian network also assigns a probability to the computed value providing a measure of confidence to value. Other techniques such as neural networks not only provide the model but also are able to "learn" when unknown conditions arise.
- Combination/ensemble models – These models use combinations of various models, attempting to use the best part of each technique used. One such possibility is the expert system combined with neural networks, under the assumption that the neural network would learn new rules to handle the fuzziness of the input.

While all the above techniques could possibly meet the requirements, the pure AI/ML techniques may be the most promising in order to meet the requirements of an objective, early life cycle manufacturability analysis resulting in a manufacturability metric that could support for tradespace analysis.

4 Conclusions

While the current version of the MAKE methodology is suitable for a Milestone C assessment, there is a desire to perform early life cycle assessments at Milestone A or pre-Milestone A timing. A secondary goal is to transition the methodology from a very subjective process to an objective assessment, which will integrate the methodology into tradespace environments. This will serve to optimize the assessment process by removing biases from SMEs and reduce the time and expertise needed to perform the assessment.

It is possible to achieve these goals, by using various computer models such as artificial intelligence, machine learning, expert systems, or physics-based models. Each of these options has positives and negatives associated with them. Another option is to use a combination of the aforementioned models. That transition would, theoretically, allow the framework to provide the manufacturability metric for any number of design variations in a tradespace.

Since most tradespaces are created to support the acquisition of new products, solving the early life cycle challenge of low design fidelity is paramount. More research will need to be done in the area of early life cycle assessments.

Acknowledgments This material is based upon work supported by the US Army Engineer Research and Development Center (ERDC) under Contract No. W912HZ-17-C-00218. Any opinions, findings, and conclusions or recommendations expressed in this material are those of the authors and do not necessarily reflect the views of the ERDC.

References

Buchanan, R.K., J.E. Richards, C.H. Rinaudo, and S.R. Goerger. May 2018. Integrating set-based design into cost analysis. In *2018 Conference on Systems Engineering Research (CSER18)*, 8–9. Charlottesville.

Cherwonik, J. 2017. *Engineered Resilient Systems (ERS) Lifecycle Cost Analysis for Trade-space Generation*. Internal ERDC: Vicksburg.

Iansiti, M. 1995. Shooting the Rapids: Managing Product Development in Turbulent Environments. *California Management Review* 38: 37–58.

Kalyanaram, G., and V. Krishnan. 1997. Deliberate Product Definition: Customizing the Product Definition Process. *Journal of Marketing Research* 34 (2): 276–285.

McCall, T., and S. Fuller. 2018. Manufacturability and the Product Design: Case Study Exploration of Manufacturability Assessments across a Product Life Cycle. In *Proceedings from the 2018 American Society for Engineering Management International Conference*. Coeur d'Alene.

McCall, T., C. Walden, L. Dalton, et al. 2016. *Manufacturability Assessment Methodology for Reducing Life Cycle Costs*. Vicksburg: Technical Report for Engineering Research and Development Center.

———. 2017. *Enhancements to the Manufacturability Assessment Knowledge-based Evaluation and a Pilot Case Study*. Vicksburg: Technical Report for Engineering Research and Development Center.

U.S. Office of Management and Budget. 2008. *Circular No. A–11, Preparation, Submission, and Execution of the Budget*. Washington, DC: Executive Office of the President.

Viral, Nagori, and Bhushan Trivedi. April 2014. Types of Expert System: Comparative Study. *Asian Journal of Computer and Information Systems* 02 (02) ISSN: 2321–5658.

A Bioinspired Framework for Analyzing and Predicting the Trade-off Between System of Systems Attributes

Abheek Chatterjee, Richard Malak, and Astrid Layton

Abstract This research investigates a bioinspired framework for analyzing and predicting trade-offs between system of systems' (SoS) performance, affordability, and resilience early in the design process – without the need for highly detailed simulations or disruption models. This framework builds on ecological research that has found a unique balance between redundancy and efficiency in biological ecosystems. This balance implies that highly efficient ecosystems tend to be inflexible and vulnerable to perturbations, while highly redundant ecosystems fail to utilize resources effectively for survival. Twenty architectures for a notional hostiles' surveillance SoS are investigated, showing that highly efficient SoS architectures fail catastrophically in the face of disruptions, while highly redundant architectures are unnecessarily expensive: indicating that engineered SoS architectures follow a *fitness* trend akin to complex ecological networks. The results suggest that SoS may benefit from mimicking a balance of redundancy and efficiency similar to that found in ecological networks.

Keywords SoS architecture · Resilience · System design · Ecological network analysis · Bioinspired design

1 Introduction

Successfully achieving mission objectives requires the intelligent utilization of all participating systems. The term "system of systems" (SoS) is used to describe a collection of independently operating systems, which interact with one another to accomplish mission requirements that cannot be achieved by the individual systems alone (Owens 1996; White 2006). These systems may provide essential capabilities such as surveillance, detection, information processing and exploitation,

A. Chatterjee · R. Malak · A. Layton (✉)
J. Mike Walker '66 Department of Mechanical Engineering, Texas A&M University, College Station, TX, USA
e-mail: alayton@tamu.edu

A. M. Madni et al. (eds.), *Recent Trends and Advances in Model Based Systems Engineering*, https://doi.org/10.1007/978-3-030-82083-1_43

neutralization of hostiles, and the supply of aid, all within the larger SoS context. SoS can have evolutionary tendencies (addition of new systems over time), a range of geographic distributions, and operational independence of participating systems (Maier 1998). Additionally, the functional interdependence of the systems within the SoS creates challenges for their analysis and development. The attributes of performance, affordability, and resilience (Pape et al. 2013; Uday & Marais 2015; Dagli et al. 2013) are most commonly used to analyze and track SoS development. Performance, as used here, quantifies the extent to which a SoS meets mission objectives. Affordability is taken here to account for the capital and operational costs of a SoS. Resilience is defined as the ability of a SoS to survive and recover from disruptions (Uday & Marais 2013, 2015) and is of particular importance when operations are in high-risk environments.

Assessment of SoS resilience requires highly detailed simulations and disruption models, but such information is available only *after* the SoS design is complete. Approaches based on graph/network theory have been used to investigate possible architecture-based metrics that can be utilized to guide SoS designs toward desired objectives (Yang 2014; Raz & DeLaurentis 2017; Dekker 2005). This study attempts to take another step in this direction by proposing and investigating a bioinspired architecture-based metric that can be utilized to guide SoS designs with favorable trade-offs between important SoS attributes.

Ecological principles have the potential to provide design inspirations for engineering resilience (Raz & Kenley 2019). Ecosystems are biological SoS that are resilient to a wide variety of perturbations (Holling 1973). They are made up of independent actors (species or functional groups) that perform essential roles in the network (like producers, predators, and decomposers) and have an evolutionary nature. Ecological network analysis (ENA) is a mathematical analysis method for quantitatively linking ecosystem structure and functions (Ulanowicz 2004). The ENA metric *degree of system order*, developed from Information and Graph Theory, indicates the balance between efficient and redundant interactions in a network (Ulanowicz et al. 2009). This metric ranges between 0 and 1, representing the extremes of pathway redundant and pathway efficient networks, respectively.

Ecologists have found that ecological networks (which had survived for an extended period of time) avoided both these extremes and maintained a unique range of the degree of system order dubbed the "window of vitality [15]." This balanced structure is believed to allow ecosystems to be both productive under normal circumstances and have the reserve capacity and flexibility to survive and recover from perturbations (Ulanowicz et al. 2009), presenting an interesting source of SoS design inspiration.

The analysis of the degree of system order only requires knowledge of network architecture: the participating actors and their interactions (the mathematical framework is discussed in Section 2). As a tool for system engineers, degree of system order could enable selection between SoS architectures for those with the best attributes (such as performance, affordability, and resilience) without the need for highly detailed simulations or disruption models. This paper models and analyzes notional hostile's surveillance SoS architectures as a network of interacting systems to gather evidence about whether the ecologically inspired concept of the "window of vitality" applies to engineered SoS. Do SoS architectures with bioinspired balances of efficient and redundant interactions present better performance, affordability, and response to disruptions? Factors that govern an engineering SoS "window of vitality" are expected to be clarified when compared to highly efficient or redundant architectures.

2 ENA, the Degree of System Order, and the Ecological Fitness Function

Ecologists use ecological network analysis (ENA) to study the complex interactions among the species within a food web (an ecosystem modeled in terms of its predator-prey-based interactions). A directional graph or digraph is created for the food web, where the nodes represent the actors or species and the directed arcs represent the transfer of energy or nutrients between them and their immediate environment. Flow magnitude information between the nodes within the system boundaries, as well as those with the surrounding environment (system inputs, outputs, and dissipations), are all stored in the square (N+3) x (N+3) flow matrix **T** (where N is the number of actors within the system or SoS boundaries). The matrix elements T_{ij} represent the magnitude of flow from node i (producers/prey) to node j (consumers/predators). The nodes 1 to N in the flow matrix represent the actors within the specified system boundary. The nodes 0, N+1, and N+2 are the system imports (row "zero" in the **T** matrix) and system exports and dissipations (columns N+1 and N+2, respectively). Figure 1a illustrates this process, with a hypothetical food web (top-left) modeled as a digraph (bottom-left) and then quantified in its flow matrix **T** (bottom-center). Readers interested in a more detailed description may refer to *Fath* et al. (Fath et al. 2007).

ENA metrics are calculated using the **T** matrix. The sum of all the flows through the network is the *Total System Throughput* (*TSTp*, Eq. 1). *Ascendancy* (*ASC*) measures the network's organizational development, or the ability of the network to efficiently transport the medium of interest from one point to another. When normalized by *TSTp*, *ASC* becomes *Average Mutual Information (AMI*, Eq. 2) *(Ulanowicz* 1986*)*. Nothing can grow in nature without bounds, including network organization. The upper limit on *ASC* is the *Development Capacity* (*DC*). *Shannon Index* (*H*, Eq. 3) is *DC* normalized by *TSTp*. These metrics are rigorously derived using the concepts of information theory in scholarly works by Ulanowicz (Ulanowicz et al. 2009; Ulanowicz 1986).

$$TSTp = \sum_{i=0}^{N+2} \sum_{j=0}^{N+2} T_{ij} \qquad (1)$$

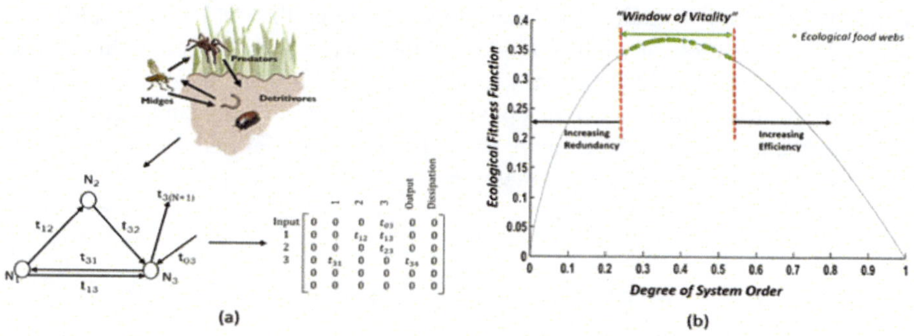

Fig. 1 (**a**) A schematic of the modeling procedure used in ENA, based on *Layton et al.* (Layton et al. 2016), and (**b**) the ecological fitness (R_{ECO}, Eq. 9) curve with food webs (green dots) from the dataset of Borrett et al. (Borrett & Salas 2010) illustrating the "window of vitality" (Ulanowicz et al. 2009)

$$AMI = \sum_{i=0}^{N+2} \sum_{j=0}^{N+2} \frac{T_{ij}}{TST_p} \log_2 \left[\frac{T_{ij} \bullet TST_p}{T_{i.} \bullet T_{.j}} \right] \tag{2}$$

$$H = -\sum_{i=0}^{N+2} \sum_{j=0}^{N+2} \frac{T_{ij}}{TST_p} \log_2 \left[\frac{T_{ij}}{TST_p} \right] \tag{3}$$

where

$$T_{i.} = \sum_{j=0}^{N+2} T_{ij} \tag{4}$$

$$T_{.j} = \sum_{i=0}^{N+2} T_{ij} \tag{5}$$

These metrics are used for networks with only one medium or unit of flow. Human SoS consists of *multiple* distinct and interdependent flows, however, and the translation of these metrics requires their modification to allow for multiple flows: *AMI* and *H* in Eqs. 6 and 7 have been reformulated to accommodate multiple flows (Chatterjee & Layton 2019). The symbols retain their original meaning, and the additional subscript l signifies the different types (mediums) of flows where $l \in [1, M]$ for a network composed of M distinct flow types.

$$AMI = \sum_{i=0}^{N+2} \sum_{j=0}^{N+2} \left\{ \left(\prod_{l=1}^{M} \frac{T_{ijl}}{TST_{pl}} \right) \log_2 \left[\frac{\prod_{l=1}^{M} T_{ijl} \bullet TST_{pl}}{\prod_{l=1}^{M} T_{i.l} \bullet T_{.jl}} \right] \right\} \tag{6}$$

$$H = -\sum_{i=0}^{N+2} \sum_{j=0}^{N+2} \left\{ \left(\prod_{l=1}^{M} \frac{T_{ijl}}{TST_{pl}} \right) \log_2 \left[\prod_{l=1}^{M} \frac{T_{ijl}}{TST_{pl}} \right] \right\} \tag{7}$$

The ratio of *AMI* to *H* is the *degree of system order* (a) (Ulanowicz et al. 2009) and has values ranging from zero to one. A value of a close to zero indicates that a large number of redundant pathways exist in the network, and a value close to one indicates that the network has a minimum number of highly constrained (or efficient) flow pathways. These extreme cases correspond to networks where either all actors are connected to each other (most redundant, $a = 0$) or all actors are connected in a linear series (highest efficiency, $a = 1$). Neither of these extreme cases results in a "fit" network. An excessively redundant network is not effective at utilizing the available resources, and an excessively efficient network will be vulnerable to disruptions. *Fit*, or optimal, network architectures can safely be assumed to lie between these two extremes. This is mathematically reflected in Ulanowicz's formulation of the *fitness function* of a flow network (F), which is as a function of the *degree of system order* and the *Boltzmann measure of its disorder* ($-k \bullet \ln [a]$) (Ulanowicz et al. 2009; Ulanowicz 2009). Equation 8 is the general form of this fitness function. The range of a where mature ecosystems reside, known as the "window of vitality," indicates that complex systems in nature have evolved to a selection of $a \sim 1/e$ (). Equation 9 is the resulting *ecological fitness function*, plotted in Fig. 1b with 48 different ecosystems residing near $a \sim 1/e$.

$$F = -\left(a^{\beta}\right) \bullet \ln \left(a^{\beta}\right) \tag{8}$$

$$R_{eco} = -(a) \bullet \ln(a) \tag{9}$$

3 Investigating the *Fitness* Trends in a Hypothetical Hostiles' Surveillance SoS

Twenty feasible architectures of a hypothetical hostiles' surveillance SoS are investigated here, consisting of a mission command center, the Continental United States (CONUS) headquarters, and on-site surveillance and data exploitation systems. The objective of this SoS is to monitor a 9000 sq. miles area continuously for any sign of hostile activity and then select and implement appropriate response measures. The available on-site surveillance systems are Joint Surveillance and Target Attack Radar (JSTAR) aircrafts, unmanned aerial vehicles (UAV), and a military satellite.

Table 1 Performance characteristics for the available systems within the hostiles' surveillance SoS based on descriptions found in (Dagli et al. 2013)

System	Surveillance quality	Max. surveillance area (sq. miles)	Exploitation capability	Operational cost (10^3 $/hour-unit)
JSTAR	1	>9000	1 JSTAR	18
UAV	0.9	2250	–	2
Military satellite	0.7	>9000	–	1
Theater	–	–	2 UAVs	10
CONUS	–	–	Unlimited	–

Every two UAVs require a local, on-ground control and exploitation unit (theater), while JSTAR has onboard crewmembers to perform data exploitation (Technology AF 2019). CONUS handles the control and data exploitation of the satellite and provides the mission command with the exploited data. CONUS also exchanges commands and reports with the mission command center and acts as the mediator between government agencies and the SoS. The mission command center receives exploited surveillance data from all participating systems, monitors their status, and commands their operations.

Table 1 shows the performance characteristics and operational costs of the systems within this hypothetical SoS based on descriptions found in (Dagli et al. 2013). The satellite is assumed to already be in orbit and operational, only needing to be commissioned for the mission, resulting in a low operational cost (Dagli et al. 2013). All surveillance is assigned a quality value between zero and one for the precision and clarity of the raw data collected. As this was an initial proof of concept investigation, a deterministic approach was used for the performance, disruption, and recovery assessment, and related uncertainties were considered to be outside the scope of this analysis.

3.1 SoS Performance, Cost, and Response to Disruptions

The total operational cost of the SoS is calculated as the sum of the operational costs of each participating system. The performance level (*PL*) of the SoS is formulated as shown in Eq. 10 and is based on (Uday and Marais 2013). A is the area covered by a surveillance system (in sq. miles), q indicates the quality of surveillance by that system ($q \in [0,1]$), s indicates the state of the exploitation system associated with the surveillance unit ($s \in \{0,1\}$), and the subscript i indicates the systems under consideration ($i \in \{1,2,\ldots,N\}$). The maximum surveillance performance level is 9000, for coverage of the entire area of interest.

$$PL = \max\left[\left(\sum_{1}^{N} A_i \bullet q_i \bullet s_i\right), 9000\right] \tag{10}$$

The loss of one, two, and three systems in the SoS (N-1, N-2, and N-3 contingencies) was randomly simulated to investigate the response of the SoS architectures to external disruptions. The worst-case performance level of the SoS right after disruption as well as after the mission command reallocates the remaining systems that were then assessed. These contingency analyses consider any on-site units as candidates for attack by hostiles. The satellite is assumed to have partial immunity to hostiles, resulting in a loss of only communications. The mission command center and CONUS are assumed to be safe from external disruptions.

3.2 *Degree of System Order of SoS Architectures*

The participating systems have two kinds of interactions: (a) the flow of surveillance data and (b) the exchange of commands and reports. The magnitude of surveillance data collected by surveillance units is assumed to be the product of the area they cover and the surveillance quality of the system. Ten percent of the raw surveillance data collected is assumed useful after exploitation. Fifty units of command and report data are assumed to be exchanged between the mission command and each surveillance/exploitation system. Three hundred units of commands and report data are assumed to be exchanged between the mission command, CONUS, and governing authorities.

Figure 2 shows the case study SoS architecture modeled as a flow network and the associated flow matrices. The JSTAR imports surveillance data through imaging (entry T_{01} in Fig. 2b), exploits the raw data, and provides useful data to the mission command center (entry T_{12} in (b)). The remaining (non-useful) raw surveillance data is modeled as dissipation (entry T_{15} in Fig. 2b). Mission command utilizes the exploited data for decision-making, which is modeled as a useful export (entry T_{24} in Fig. 2b). There is also a bidirectional flow of commands and reports between mission command and the JSTAR (entries T_{12} and T_{21} in Fig. 2c), as well as CONUS and mission command (entries T_{32} and T_{23} in Fig. 2c). CONUS receives orders from the governing authorities outside the SoS boundary (import, entry T_{03} in Fig. 2c) and reports back to them (useful export, entry T_{34} in Fig. 2c). *AMI* and *H* were calculated using Eqs. 6 and 7 with the information from the two matrices in Fig. 2c and d. The degree of system order was then calculated using the *AMI* and *H* values.

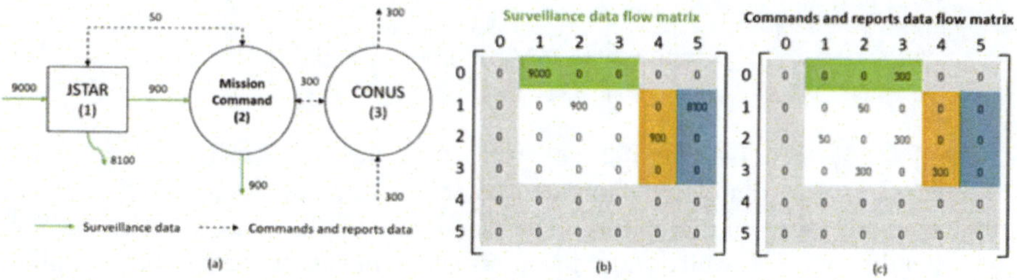

Fig. 2 (**a**) A schematic of the surveillance SoS represented as a data flow network (the numbers on the directed arcs represent the magnitude of data flows) and the associated (b) surveillance data flow matrix and (**b**) commands and reports data flow matrix

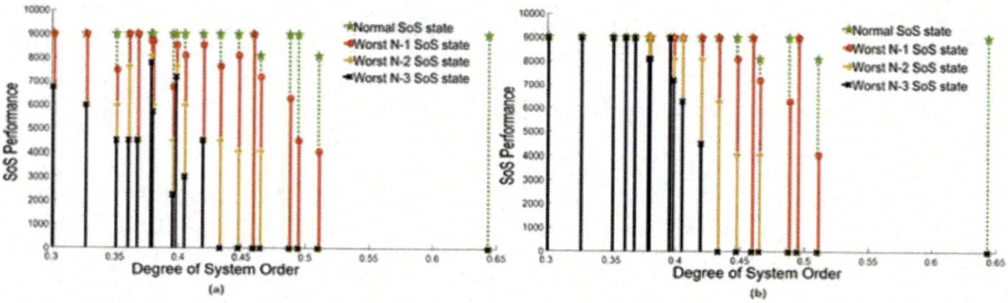

Fig. 3 The worst-credible SoS performance levels (**a**) immediately after the disruption and (**b**) after reallocation of surviving systems

4 Results and Discussion

The twenty SoS architectures investigated utilize the techniques of physical redundancy (e.g., multiple JSTARs), functional redundancy (e.g., surveillance using both JSTAR and the satellite), and localized capacity (e.g., sharing the surveillance load between multiple drones). Degree of system order values of the investigated architectures ranged from 0.3 to 0.65. The worst-case performance levels of the SoS immediately after the disruption and after reallocation of surviving systems are presented in Figs. 3.a and 3.b, respectively.

Networks with a values closest to 1 (highly pathway efficient) suffered complete failures and could not regain any functionality. Their lack of redundancy created an inability to recover the lost functionality. The SoS architectures with lower a values (higher pathway redundancy) were found to have the flexibility needed to survive and recover from external disruptions. However, a tipping point was observed (Fig. 3) beyond which higher pathway redundancy did not equate to higher recoverability in the context of this SoS. This point is closer to one (efficient architecture) for the N-1 scenarios and shifts toward $a = 0$ (redundant pathways) for the more severe N-2 and N-3 scenarios.

Standard techniques for improving resilience in complex systems, like physical and functional redundancy and localized capacity (among others) (Jackson and

Ferris 2013), still beg the question of *how much* redundancy or distribution of capacity is "enough" and how much is "too much?" An excessively redundant or distributed SoS may be indifferent to the loss of a single system but would be extremely expensive (Uday and Marais 2013) or could cause organizational interoperability issues (Dekker 2005); both can reduce overall performance. A highly efficient SoS may have an acceptable performance level at low operational costs but would be vulnerable to catastrophic failure if even one of the participating systems was disturbed.

The behavior of ecosystems offers a potential route to answer this question. The case study here is inadequate to analyze interoperability issues, but it is possible to analyze the trade-off between recoverability and operational cost of the various SoS architectures. Recoverability was defined as the fraction of the undisrupted state performance level that the SoS can regain after the surviving systems are reallocated (Eq. 11). To investigate the trade-off between recoverability and operational cost for the SoS architectures, recoverability to cost ratio (*RCR*) of the SoS architectures was calculated as shown in Eq. 12.

$$\text{Recoverability} = PL_{\text{after re-allocation}}/PL_{\text{undisrupted state}} \tag{11}$$

$$\text{RCR} = \text{Recoverability}/\text{Operational Cost} \tag{12}$$

The recoverability to cost ratios in the worst-case N-1, N-2, and N-3 scenarios for the SoS architectures are plotted against their a values in Fig. 4. This analysis shows that the recoverability to cost ratio of SoS architectures first improves with more redundancy (lower a values) until it reaches a peak, after which there is a decreasing trend. This overall behavior imitates the fitness of complex systems observed in ecology, evidence that SoS architectures can learn to balance efficient and redundant interactions from ecology to better manage their performance, affordability, and response to disruptions.

Figure 4 also shows that each N-X case has a select number of designs covering a slightly different a range that have significantly better recoverability-cost ratios (indicating a better trade-off between affordability and response to disruptions). The range of the favorable a values seems stricter than the ecosystems' *window of vitality*. This may be due to the ecosystem data covering biological SoS that experience a range of disturbances, from mild to extreme, while the hypothetical SoS investigated here has been grouped by specific N-X disturbance levels. The *fittest* SoS architectures can be seen to favor values of a closer to 0 as the severity of disruptions increases (with increasing X values in N-X). This observation indicates that higher threat levels cause the SoS fitness trends to move toward the *ecological fitness function*, suggesting that the level of threats in a SoS's operating environments may provide a governing factor in deciding its specific *window of vitality*.

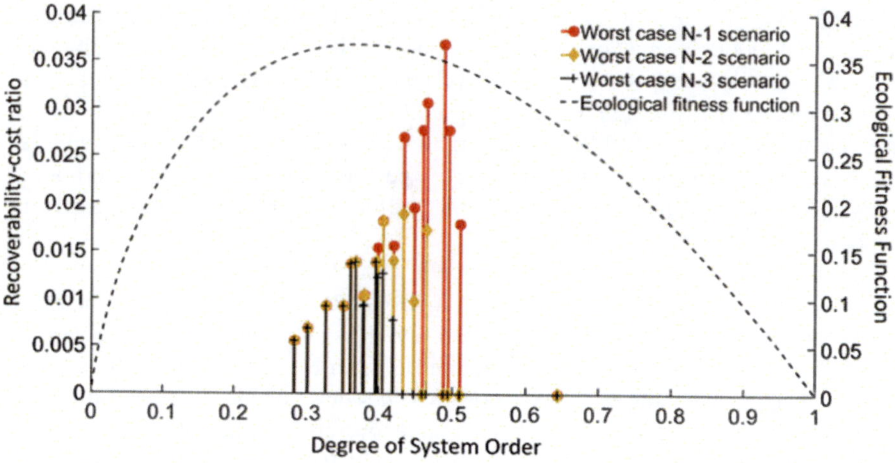

Fig. 4 The recoverability to cost ratios for all SoS architectures investigated, at the worst-credible state after N-1, N-2, and N-3 contingencies. The best SoS designs reside near the peak of the ecological fitness function for the N-3 disruptions

5 Concluding Remarks

This study suggests that the ecologically inspired concept of the *window of vitality* may be useful for the design and management of engineered SoS. While the range of degree of system order values presenting favorable SoS attribute trade-offs seems stricter than for the ecosystems, this presents an opportunity to customize the window of vitality for a SoS based on its specific needs and environment. Future work will analyze SoS response to more complex threat models and seek to develop a mathematical framework to predict the peak fitness (β parameter, Eq. 9) and the range of favorable degree of system order values given SoS operating conditions. More complex SoS case studies are also needed to further test for evidence of organizational interoperability with excessive redundant interactions. Success may be able to provide system engineers with a decision-making and design tool to select SoS architectures without the need for costly simulations or detailed disruption models.

Acknowledgments This manuscript is based on work supported, in whole or in part, by the Systems Engineering Research Centre (SERC) under contract WRT-1011.

References

Borrett, S.R., and A.K. Salas. 2010. *Evidence for resource homogenization in 50 trophic ecosystem networks*. *Ecological Modelling* 221 (13-14): 1710–1716.

Chatterjee, A. and A. Layton. 2019. *Bio-Inspired Human Network Design: A Multi-Currency Robustness Metric Inspired by Ecological Network Analysis*. in *Proceedings of the 2019 International Design Engineering Technical Conferences and Computers and Information in Engineering Conference (IDETC/CIE)*. Anaheim, CA: ASME.

Dagli, C.H., et al., *An advanced computational approach to system of systems analysis & architecting using agent-based behavioral model.* 2013.

Dekker, A.H. 2005. *C4ISR, the FINC methodology, and operations in urban terrain*. *Journal of Battlefield Technology* 8 (1): 25.

Fath, B.D. 2014. *Quantifying economic and ecological sustainability*. *Ocean & Coastal Management* 108: 13–19.

Fath, B.D., et al. 2007. *Ecological network analysis: network construction*. *Ecological Modelling* 208 (1): 49–55.

Holling, C.S. 1973. *Resilience and Stability of Ecological Systems*. *Annual Review of Ecology and Systematics* 4: 1–23.

Jackson, S. and T.L.J. Ferris, *Resilience principles for engineered systems.* 2013. 16(2): p. 152-164.

Layton, A., B. Bras, and M. Weissburg. 2016. *Ecological Principles and Metrics for Improving Material Cycling Structures in Manufacturing Networks*. *Journal of Manufacturing Science and Engineering* 138 (10): 101002-1–101002-12.

Maier, M.W., *Architecting principles for systems-of-systems.* 1998. 1(4): p. 267-284.

Owens, W.A. 1996. *The emerging US system-of-systems*. Washington DC: Institute for National Strategic Studies, National Defense University.

Pape, L., et al. 2013. *A Fuzzy Evaluation method for System of Systems Meta-architectures*. *Procedia Computer Science* 16: 245–254.

Raz, A.K. and D.A. DeLaurentis. 2017. *System-of-Systems Architecture Metrics for Information Fusion: A Network Theoretic Formulation*, in *AIAA Information Systems-AIAA Infotech@ Aerospace*. p. 1292.

Raz, A.K. and C.R. Kenley. 2019. *Multi-Disciplinary Perspectives for Engineering Resilience in Systems*. in *2019 IEEE International Conference on Systems, Man and Cybernetics (SMC)*.

Technology, A.F. 2019. *JSTARS – Joint Surveillance and Target Attack Radar System*. 10/13/2019]; Available from: https://www.airforce-technology.com/projects/jstars/.

Uday, P., and K. Marais. 2013. *Exploiting Stand-in Redundancy to Improve Resilience in a System-of-Systems (SoS)*. *Procedia Computer Science* 16: 532–541.

———. 2015. *Designing Resilient Systems-of-Systems: A Survey of Metrics, Methods, and Challenges*. *Systems Engineering* 18 (5): 491–510.

Ulanowicz, R.E. 1986. *Growth and Development: Ecosystems Phenomenology*. iUniverse.

———. 2004. *Quantitative methods for ecological network analysis*. *Computational Biology and Chemistry* 28 (5–6): 321–339.

———. 2009. *The dual nature of ecosystem dynamics*. *Ecological Modelling* 220 (16): 1886–1892.

Ulanowicz, R., et al., *Quantifying sustainability: Resilience, efficiency and the return of information theory*. Vol. 6. 2009. 27-36.

White, B.E. 2006. *Fostering intra-organizational communication of enterprise systems engineering practices*. in *National Defense Industrial Association (NDIA), 9th Annual Systems Engineering Conference, San Diego CA*.

Yang, G., et al., *Key potential-oriented criticality analysis for complex military organization based on FINC-E model.* 2014. 20(3): p. 278-301.

Model-Based Systems-of-Systems Healthcare: Coordinating the Coordinators

Bernard P. Zeigler, Mark Redding, Pamela J. Boyers, and Ernest L. Carter

Abstract Achieving value-based healthcare – increasing quality, reducing cost, and spreading access – has proven to be extremely challenging. In recent years, a large variety of care coordination organizations have emerged at regional and national scales. Unfortunately, each such health entity lives in its own definition (silo) of care coordination leaving large gaps in care as well as duplicative or inconsistent interventions where care domains overlap. This situation leads to the need for higher-level coordination of the coordinators with well-defined population health metrics and means for sharing of information and control of patient-centered interventions. In "Value-based Learning Healthcare Systems: Integrative modeling and simulation" (Zeigler et al. 2018), we presented a modeling and simulation (M&S) approach to value-based healthcare within a system-of-systems framework that enables designing, testing, and implementing care coordination based on identifying and addressing risks at the individual and family level and tracking progress though health information technologies (HITs). In this paper, we discuss how a model-based system-of-systems design for HIT infrastructure can support innovative "coordination of the coordinators" assuring that critical modifiable risks spanning health and social issues are identified and addressed resulting in better health and social outcomes. We describe existing foundations for implementing such a design such as digital platforms, pathways-based community coordinator organizations, risk factor registries, as well as comprehensive simulation facilities where the design and its components can be tested. Research required to enable integrating such foundations into a working whole is also described.

B. P. Zeigler (✉)
RTSync, Corp, Chandler, AZ, USA
e-mail: zeigler@rtsync.com

M. Redding
Research and Quality Director, Pathways Community HUB Institute, Akron, OH, USA

P. J. Boyers
University of Nebraska Medical Center, Omaha, NE, USA

E. L. Carter
Health Department, Prince Georges County, MD, USA

A. M. Madni et al. (eds.), *Recent Trends and Advances in Model Based Systems Engineering*, https://doi.org/10.1007/978-3-030-82083-1_44

Keywords Value-based healthcare · Care coordination · Health information technology · Pathways · Simulation · Model-based system-of-systems design

1 Introduction

In response to the need for coordination of health and social services to identify and address modifiable risks, a large variety of care coordination organizations have emerged at regional and national scales. Unfortunately, each such entity lives in its own definition (silo) of care coordination leaving large gaps in care as well as duplicative or inconsistent interventions where care domains overlap. This situation leads to the need for higher-level coordination of the coordinators with well-defined population health metrics and means for sharing of information and control of patient-centered interventions. In "Value-based Learning Healthcare Systems: Integrative modeling and simulation" (Zeigler et al. 2018), the objectives of value-based healthcare were broadly stated by the following equation:

$$\text{Objectives} = \text{low Cost} + \text{high Quality} + \text{wide Accessibility}$$

The exact meaning of the attributes, cost, quality, and accessibility can vary, as can their priority, or even applicability, in different contexts. Nevertheless, when we refer to measuring value, we mean some concrete formulation of increase in quality while reducing cost and increasing access. The importance of the equation becomes evident when we recognize that a healthcare service system is composed of a large number of distributed components that are interrelated by complex processes. Understanding the behavior of the overall system is becoming a major concern among health and social service system managers and decision-makers intent on increasing value for their systems.

An optimal health and social service delivery system requires methods to model large-scale distributed complex systems (Dahmann 2018), a challenge that has been identified under the rubric of model-based systems of systems (SoS) engineering (Wymore 1993, Jamshidi 2008, Zeigler et al. 2018) in that the optimization cannot be based on sub-optimization of the component systems but must be directed at the entire system itself. People with multiple health and social needs are high consumers of healthcare services and are thus drivers of high healthcare costs. The ability to provide the right information to the right people in real time requires a system-level model that identifies the various community partners involved and rigorously lays out how their interactions might be effectively coordinated to improve the effectiveness of the system in identifying and addressing modifiable risk in a whole person approach for whose care costs the most.

Modeling and simulation (M&S) brings the latest methods and technologies being adopted in SoS problems, ranging from missile defense systems to population management systems. M&S is fast becoming the core knowledge generator for

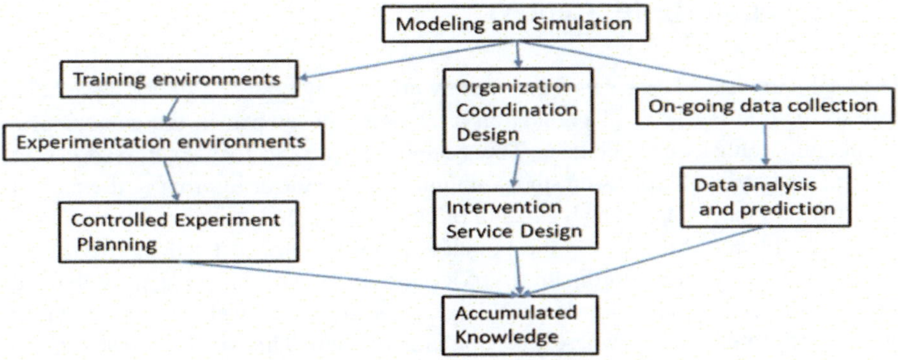

Fig. 1 Modeling and simulation healthcare development paradigm

complex systems engineering. As illustrated in Fig. 1, the new holistic paradigm places M&S at the top of ongoing knowledge accumulation in a learning health and social service system in which it is at the center of multiple related core activities involving a variety of health and social service providers. M&S spawns new ambitious training and experimentation treatment simulation environments enabling controlled experiments, design and engineering of interventions and their coordination across multiple scales, as well as ongoing data collection from multitudes of sensors with associated data analysis and prediction. We return to this discussion in the "Research and Development" section. A modeling and simulation methodology and framework model the entire health and social service system as a loosely coupled distributed system (Zeigler 2016, Madni, 2018). The criteria for such a model were laid out by Zeigler et al. (2018). Such a model systematically represents the behaviors of patients who require coordinated care interventions and the providers of such coordination services to render such behaviors amenable to health and social service system design and engineering. Unfortunately, most of the work concerning health and social service system modeling and simulation (M&S) in the literature is unit or facility specific rather than taking a consistent global whole person and family view. In contrast, Zeigler et al. (2018) present a framework that encompasses common perspectives taken in the research literature but also goes beyond them toward their integration with additional perspectives that are becoming critical in today's environment. It proposes a stratification of the levels of abstraction into multiple perspectives. In each of these perspectives, models of different components of healthcare systems can be developed and coupled together. Concerns from other perspectives can be abstracted as parameters in such models. The resulting global model can be coupled with a holistic experimental frame to derive results that cannot be accurately addressed in any of the perspectives if taken alone.

2 Population Health Context

The health and social service system framework of Zeigler et al. (2018) is extended here to include the level of population as is being developed in the context of a population wellness management. The system design objectives expressed earlier are replaced by the Triple Aim: the simultaneous pursuit of improving the patient experience of care, improving the health of populations, and reducing the per capita cost of healthcare. Introducing the population level is necessitated by the proliferation of various care coordination service providers with independently defined domains of care coordination leaving large gaps in care as well as duplicative or inconsistent interventions where care domains overlap. This situation leads to the need for higher-level coordination of the coordinators which can only be addressed at a population level.

Figure 2 sketches in broad strokes, a UML system design that extends the framework of Zeigler et al. (2018) by employing risk management and pathways of care to provide SoS-level coordination of specialized modifiable risk identification and mitigation interventions and coordination services. The process of induction of a patient into coordinated care starts with a full-scale screening and assessment of medical, social, and behavioral health risks and assignment to one of a small set of categories of risk, each with its own distinct portfolio of interventions and care coordination services. A primary distinction with current practice is that a *patient is assigned to single primary care provider* (*PCP*, *doctor*) and connected care coordination team who are responsible for all subsequent patient interactions with the system. The PCP employs the output of risk screening and patient category of risk to assign pathways that lay out steps of interventions and services toward mitigation of modifiable risks. Interventions for identified risks span medical, social, and behavioral health service interventions. The same PCP team is continually updated with results of the patient's encounters with such interventions and mitigations of risk through system-provided tracking, thus enabling monitoring of progress in addressing modifiable risk.

A Central Referral System (CRS) (Fig.3) provides the underlying digital infrastructure to initiate induction into the system as well as a host of necessary services. The Consent2Share consent management tool is the key application which not only controls the information that a patient allows to be shared but, most importantly, serves as the patient registration portal that guarantees the unique coupling of patient and PCP. Sharing of electronic, medical, and other patient health records is mediated by a regional health information exchange (HIE) which allows health information to move electronically among disparate health information systems employed by hospitals and other providers. The goal of the HIE is to deliver the right health information to the right place at the right time – providing safer, timelier, efficient, effective, equitable, patient-centered care. Further this infrastructure supports monitoring and tracking of a holistic registry of risk mitigation efforts tracked within pathways and includes patient interactions with providers and agencies, progress evaluation, health record keeping, and information sharing.

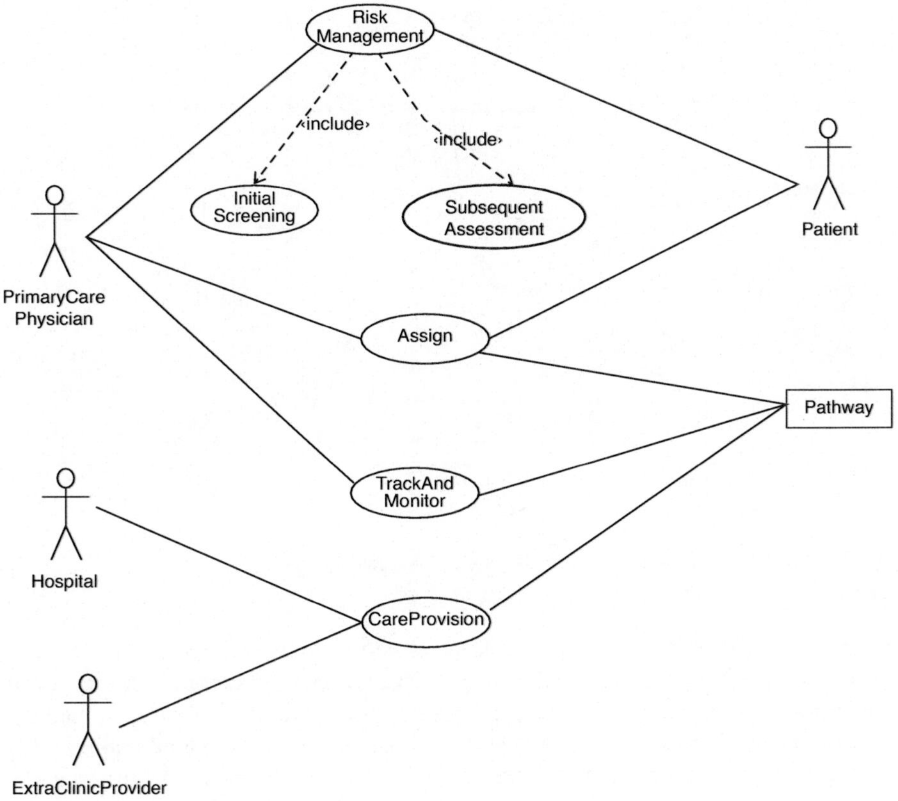

Fig. 2 Sketch of UML system design

3 Risk Factor Registry

The health, social, and behavioral health assessment of modifiable risks and the related risk stratification in the patient induction screening is an ultracritical component of the system design. This sets up requirements for the management of a comprehensive approach to identifying and addressing risk factors, including their individual and combinatorial effects on outcomes. Risk factors spanning medical, social, and behavioral health are interlocking and interconnected in their impact across medical social and behavioral health domains. Current research is starting to develop complete registries of risk factors (Redding et al., 2018) with quantitative information including relative weight of impact of these factors on dependent variable outcomes such as hospitalization, emergency room use, total cost of care, school performance, and employment. Such research is needed to discover the signature risk combinations, groupings of specific factors, across health and social domains that exponentially amplify impact. Such a registry is critical

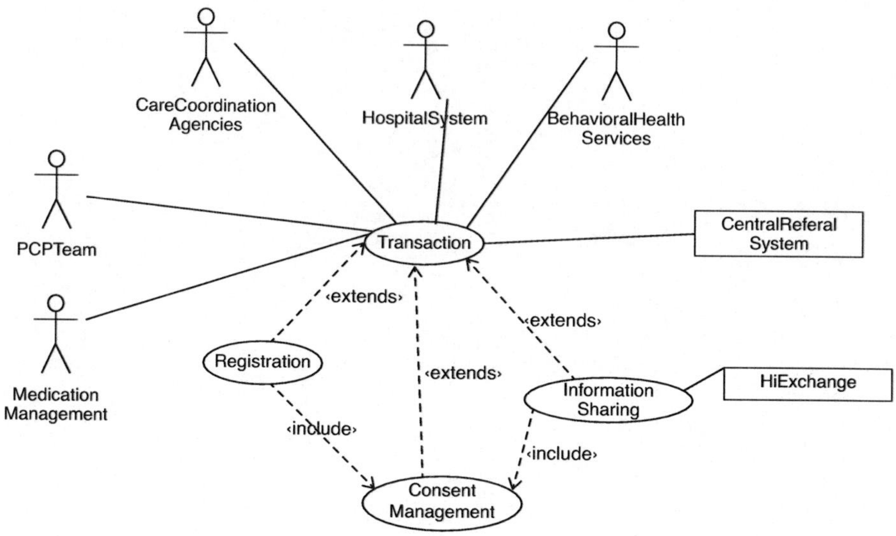

Fig. 3 UML sketch of transactions guarded by consent management

to the ability to intervene, enabling laser targeting of preventive actions toward individuals likely to experience catastrophic and expensive outcomes. Moreover, it is necessary to distinguish *individually modifiable factors of risk* spanning medical social and behavioral health (reference) from those that are not so modifiable so that intervention and coordination services can be targeted in the most effective manner.

4 Comprehensive Healthcare Simulation

For testing of SoS designs such as Fig. 1 prior to fielding, an approach involving live, virtual, and constructive (LVC) simulation is advisable as has been developed in defense system contexts (Wikipedia LVC). Figure 4 illustrates the functionality and enabling capacities of such a comprehensive facility that is designed to replicate the various healthcare settings. The generic description of Fig 4. is based on the Dr. Edwin G. & Dorothy Balbach Davis Global Center (Davis Global Center) at University of Nebraska Medical Center as detailed in the section "Davis Global Simulation Center" (see below). Domains of simulation include both clinical (hospital), social, and behavioral health individually and in combination. Types of simulation include surgical and interventional, as well as simulated communicated care and prehospital preparation. Live simulation includes human acting patients. Virtual simulation includes virtual reality portrayal of internal anatomical and physiological systems and clinical settings. Constructive model-based simulation includes computerized representations such as simulated responses, manikins, robot surrogates, etc. Supporting infrastructure includes networking including middleware

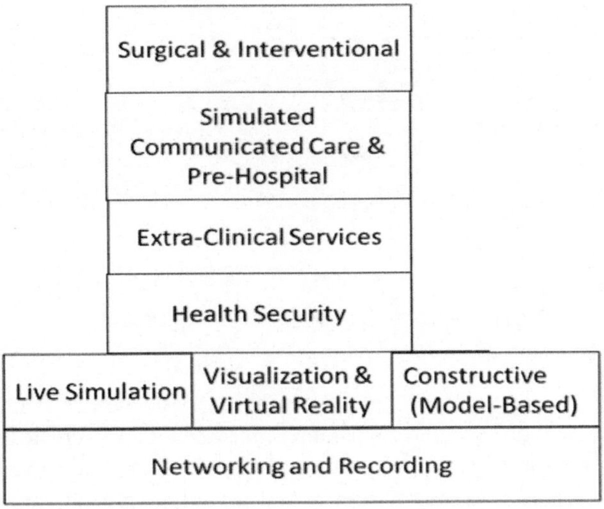

Fig. 4 UNMC Davis Global Center Design: A Comprehensive Healthcare Simulation Facility

supporting real-time and abstract simulation (Wikipedia High-Level Architecture). Furthermore, pervasive recording of all events occurring with simulation experiences enables post-event analysis and rerunning of protocols later for training.

5 Current Research and Development

The system design outlined earlier is forward looking, but several existing technological developments can serve as foundational for its realization. Here we briefly review such technology components as well as the continued research needed to bring them readiness for deployment in the near future.

5.1 Population Health Management

Maryland is the only state with Medicare waiver that affects all patients treated in Maryland hospitals. Under its rules, every payer pays the same charge for the same care. In return, Maryland must slow the rate at which total hospital costs are increasing. The goal of the new waiver is to simultaneously improve health, quality, and affordability. Prevention Link is a program being developed under a 5-year cooperative agreement between the Centers for Disease Control and Prevention (CDC) and the Prince George's County Health Department to lead the collaborative development of regional infrastructure for chronic disease prevention

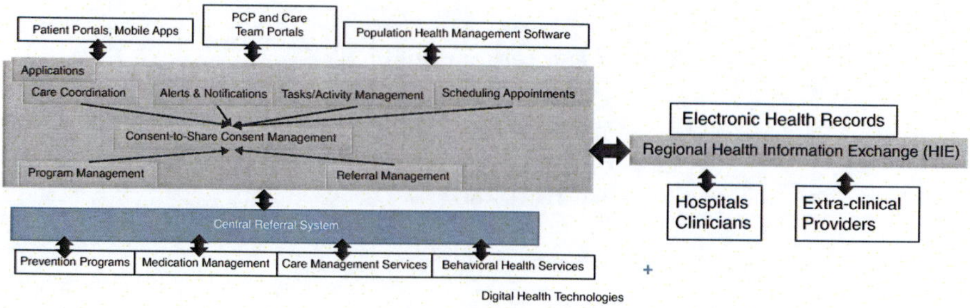

Fig. 5 Central Referral System Platform for Coordination of Coordinators

and care across Southern Maryland. It combines evidence-based prevention and care approaches, technology, and communications strategies to create a multifaceted integrated system for improving health and wellness related to chronic disease in Southern Maryland. The hypothesis is that a well-developed care management program is the key to better outcomes and cost savings, especially in populations with chronic disease. The ultimate goal is an effective, sustainable, and replicable model to demonstrate how the All Payer Maryland model can be adopted by other states. A system design, which is abstracted in Fig. 5, is being developed to implement such a management program. The CRS hosts a suite of tools aimed at improving the facilitation of care for linked care providers. Such software tools include tools for Population Health Management, Care Coordination, Alerts and Notifications, Tasks/Activity Management, Scheduling Appointments, Program Management, Referral Management, and Prevention Programs. Maryland's statewide HIE (CRISP) allows providers to track the current location of their patients and provides a variety of alerts and notifications about their statuses. External services such as Medication Management, Care Management Services, and Behavioral Health Services are integrated into the system via the HIE and CRS.

As described below, pathways developed through the Pathways Community HUB Institute (PCHI) can be utilized to confirm identification and mitigation or failure of mitigation of modifiable risks spanning medical, social, and behavioral health domains. These granular measurements of meaningful work product completions in mitigated risk are critical, quality, payment, and tracking components to implement effective coordination of the coordinators and related reporting required at the population level. As they represent meaningful events with future impact on health and the ability to live normal productive lives, each confirmed risk mitigation within the Pathways Community HUB model approach represents an intermediate or final patient outcome. These risk mitigation outcomes work individually and in combinations to impact larger outcomes of disease control, birth outcomes, cost of care, future school and job performance, etc. Experience with physicians working closely with community health workers in the Pathways HUB located in Alaska should be exploited in the development of PCP-based coordination of coordinators.

5.2 Prerequisites for Healthcare Learning System

A learning health and social service system must provide the right data and models to support human selection of alternatives likely to improve the quality of its services. It follows that a) there must be working definitions of quality of service, for example, Porter's Healthcare Value, defined as outcome divided by cost; b) there must be systems implemented to measure, in an ongoing manner, the elements of clinical, social, and behavioral health interventions that can be aggregated to compute quality of service as defined; c) there must be implemented systems that allow alternative component configurations (protocols, processes, procedures) to be continually tested; and there must be systems to correlate measured quality with component configurations to provide evaluations that humans can employ to help select the most promising options. A prerequisite for such conditions to prevail in such a system is that sufficient organization and infrastructure exist to support their implementation. The CRS infrastructure of Fig. 5 must be extended to satisfy these requirements in order to become a true learning healthcare system.

5.3 Pathways Community HUB

The Pathways Community HUB model is a delivery system for care coordination services provided in a community setting (28). The model is designed to identify the most at-risk individuals in a community, assess modifiable risks, connect them to evidence-based interventions to mitigate risks, and measure the results (17). Specific nationally standardized pathways spanning medical social and behavioral health are utilized to track each specific risk identification including mitigation success, failure, and time to completion. Community care coordination works to coordinate care of individuals in the community to help address health disparities including the social barriers to health.

The Pathways Community HUB model is a construct that enforces threaded distributed tracking of individual clients experiencing certain pathways of intervention, thereby supporting coordination of care and fee-for-performance based on end-to-end outcomes (Redding et al. 2018). As an essential by-product, the pathway concept also opens up possibilities for system-level metrics that enable more coherent transparency of behavior than previously possible, therefore greater process control and improvement reengineering.

Zeigler (2016) developed a *Coordination Model* that abstracts essential features of the Pathways Community HUB model so that the kind of coordination it offers can be understood and employed, in a general SoS context. This allows development of a M&S framework to design, test, and implement such coordination models in a variety of SoS settings, exemplified by healthcare, that present the issues that such coordination models address. Formalization provides a firm basis for capitalizing on the transparency that is afforded by the Pathways Community HUB model

(Redding et al., 2018). Such pathways were represented as DEVS atomic models with implementation in the form of an active calendar that combines event-based control, time management, and data architecture capabilities (Zeigler et al., 2016). Further, such DEVS Pathways can become components of coupled models, thereby enabling activation of successors and sharing of information. Such pathway models represent steps in a pathway as states that can constrain steps to follow each other in proper succession with limited branching as required; external input can represent the effect of a transition from one step to next due to data entry. Moreover, temporal aspects of the pathways, including allowable duration of steps, can be directly represented by the DEVS atomic model's assignment of residence times in states.

The Pathways Community HUB model is being implemented in 40 community networks in eight states. The span of care coordination services extends from infancy through adults and elders. Health, social, and behavioral health factors are all required components of risk identification and mitigation. Initial publications have described and demonstrated improved outcomes and reduced cost of care (Alley et al., 2016; Redding et al., 2015). The Pathways Community HUB Certification Program (PCHCP) within the PCHI holds the national standards and certification of HUBs and has committed to modification of the standards based on the available scientific evidence. The model is currently deployed in programs focusing on maternal and child health, adults with chronic disease, the justice involved, behavioral health, substance abuse disorder, high-risk elementary students, and others.

5.4 *Risk Registry*

Effective modeling and simulation outcomes research is clearly needed to produce the guiding registries to inform the risk mitigation interventions and payment approaches to improve health and economic success, starting with the most at-risk individuals and families. Such research is needed to identify the signature combinations to screen for supporting laser targeting of individuals and families with specific, relationship empowered risk mitigation interventions with proven predictive improvements in health and cost. The same modeling can inform hospital systems, social service, childhood education, strategies for parenting intervention, and related learning and behavior change approaches. The breadth of US health and social services fits well within the strategic analysis of identified and addressed risk. There is no current tracking or accountability for risk identification and mitigation. Consistent with Toyota production methodology, LEAN production strategies, and value chain analysis (Porter and Teisberg, 2006), this new strategic taxonomy, work item tracking, and evaluation structure can help health and social services produce outcomes using the driving and quality-focused approaches in American business. Research is needed to help in finding the granular meaningful chunks of outcome which can then be linked together in value chain analysis.

Multiple national research centers, policy centers, and a network of community initiatives are engaged in research to develop the initial and ongoing registry of trajectory intelligence for modifiable risks. The intent is to inform policy and payment approaches to address risk mitigation. Research lines include:

- Continued growth of the risk registry and validation of the registry contents, as well as the development of educational and training representational forms of the registry.
- Modifiable risks that must be mitigated with learning-based (behavior change) focused interventions represent more than one third of the total registry of risks. These risks that include critical factors such as safe sleep, nutrition, parenting, childhood education support, medication compliance, and many others represent a substantial contribution to the outcomes the health and social service system seeks to improve. The positive effect of supportive professional relationships that serve to personally engage and substantially modify the learning-based risks is a critical component that has been built into successful care coordination approaches (Agency for Healthcare Research and Quality, 2016). It is known that community engagement and involvement of community health workers in these transformative efforts play essential roles in the culturally competent, interpersonal relationships that empower and support behavior change and the related risk reductions achieved as part of coordination of care (Redding et al., 2018).
- Characterization of individually modifiable risks (need of a medical home, medication adherence, clothing) and household risks (lack of housing, utilities, food access) in contrast to population risks (such as lack of housing supply, racism, neighborhood safety). Clarifying this taxonomy of risks further in the published literature will substantially improve our ability to share strategic information and improve the effectiveness of our efforts to focus programming and payment on the specific components of the system where there are gaps. For example, high recordings of incomplete pathways in Ohio for homeless expectant mothers are helping to provide specific numeric data to inform the need for the population-level interventions needed to increase the overall supply of housing in communities.
- Continued development of appropriate metrics for pathways and the comparison of their effectiveness in pay-for-performance in contrast to Healthcare Effectiveness Data and Information Set (HEDIS), shared savings, and others.

6 Davis Global Simulation Center, Omaha, Nebraska

A simulation center and state-spanning network is well under development centered at the University of Nebraska Medical Center in Omaha. The Davis Global Center is a highly advanced clinical simulation facility purposefully designed to foster the practice of patient care in highly functioning and effective interprofessional teams. The Center provides realistic replicated healthcare settings in which teams can practice and experiment safely. Its mission (iEXCEL) is improving human performance and effectiveness in healthcare by providing “Next-Gen” training that is early and throughout a lifetime of training. The aim is to develop a truly interprofessional training model (nurses, doctors, pharmacists, allied health, public health, dentistry, etc.) and work together, using simulation to do hands-on (experiential) procedures, team training, complex medical scenarios, etc. Simulated care is patient centered with special focus on “hand-offs” from one provider or team to a second. Training is to be outcomes-oriented, i.e., competency-based (including communication, professionalism, value-based care, etc.) using XReality (Wikipedia XReality) including holographic technology to foster innovation, new ways of learning, patient care, etc. A high capacity/high speed network that connects across the state (and globally) provides state-wide outreach using digital and visual technologies (remote and distributed learning.) The simulation facility attempts to replicate the entire healthcare system (home to hospital and back!) which includes the following capabilities:

- Advanced simulation – including in situ live, virtual, constructive exercises
- Surgical and procedural labs
- XReality Labs – including immersive environments (CAD walls, interactive digital walls, 5-sided CAVE, etc.)
- Clinical test bed for research and development using simulation and visualization
- Disaster preparedness training and emergency management skills
- Biocontainment and infectious disease readiness training
- Data capture capabilities (enabling computation of metrics for human performance)

7 Conclusion

A design for a model-based system-of-systems design for HIT infrastructure can support innovative high-level coordination of the coordinators. We described existing foundations for implementing such a design such as digital platforms, pathways-based community coordinator organizations, risk factor registries, as well as comprehensive simulation facilities where the design and its components can be tested. Challenging research and development based on modeling and simulation are required to enable integrating such foundations into a working whole. Research in population management coordination of coordinators, risk registry, and comprehensive simulation facility will be particularly motivated in applications that require integration of all three. For example, in learning-based risks, development of appropriate learning pathway resources in which new parents are supported to achieve parenting strategies is proven to reduce adverse childhood events, decrease protective service involvement, and improve school performance. This intervention combined with identifying and addressing interlocking risks of access to medical care, housing, food, and others can be coordinated through interaction with all agencies related to baby care, care team providers, and health workers. Team-based providers of care as well as the individuals and families served can be educated though virtual reality simulation critical to achieve the behavior change of parents.

References

Agency for Healthcare Research and Quality. 2016. *Connecting Those at Risk to Care: The Quick Start Guide to Developing Community Care Coordination Pathways*. https://innovations.ahrq.gov/sites/default/files/Guides/CommHub_QuickStart.pdf

Alley, D.E., C. Asomugha, P. Conway, and D. Sanghavi. 2016. Accountable Health Communities — Addressing Social Needs through Medicare and Medicaid. *The New England Journal of Medicine* 374: 8–11.

CRISP (Chesapeake Regional Information System for our Patients) https://www.crisphealth.org/

Dahmann, J. 2018. *Systems Engineering Guide*., MITRE Publication, McLean VA. https://www.mitre.org/publications/systems-engineering-guide/about-the-seg. Accessed 11 Nov 2018.

iEXCEL https://www.unmc.edu/iexcel/

Jamshidi, M., 2008. (editor) Systems of Systems – Innovations for the 21st Century, Wiley, New York.

Madni, A.M. 2018. *Transdisciplinary Systems Engineering: Exploiting Convergence in a Hyper-Connected World*. Heidelberg: Springer.

ONC, Consent2Share, ttps://www.healthit.gov/topic/health-it-health-care-settings/behavioral-health-consent-management

Porter, M. E. and Teisberg, E. O. 2006. *Redefining Health Care: Creating Value-based Competition on Results*. Harvard Business Review Press; 1 edition

Redding, S., E. Conrey, K. Porter, J. Paulson, K. Hughes, and M. Redding. 2015. Pathways community care coordination in low birth weight prevention. *Journal of Maternal and Child Health* 19 (3): 643–650. https://doi.org/10.1007/s10995-014-1554-4.

Redding, R., Hoornbeek, J., Zeigler, B., Kelly, M., Redding, S., Falletta, L., Minyard, K., Chiyaka, E., Bruckman, D., 2018. *Risk Reduction Research Initiative: A National Community–Academic Framework to Improve Health and Social Outcomes*. Popul Health Manag, Published Online:13 Aug 2018, https://doi.org/10.1089/pop.2018.0099

Wikipedia High Level Architecture, https://en.wikipedia.org/wiki/High_Level_Architecture

Wikipedia LVC, https://en.wikipedia.org/wiki/Live,_virtual,_and_constructive

Wikipedia XReality (XR) https://en.wikipedia.org/wiki/X_Reality_(XR)

Wymore, A.W. 1993. *Model-Based Systems Engineering*. Boca Raton: CRC Press.

Zeigler, B.P. 2016. Discrete event system specification framework for self-improving healthcare service systems. *IEEE Systems Journal*. https://doi.org/10.1109/JSYST.2016.2514414.

Zeigler, B.P., Marvin, J.W.; Cadigan, J.J. 2018. Systems Engineering and Simulation: Converging Toward Noble Causes. In *Proceedings of the 2018 Winter Simulation Conference*. Gothenburg, Sweden.

Zeigler, B.P., S. Redding, B.A. Leath, E.L. Carter, and C. Russell. 2016. Guiding Principles for Data Architecture to Support the Pathways Community HU Model. *eGEMs* 4 (1). https://doi.org/10.13063/2327-9214.1182.

Zeigler, B.P., M. Traoré, and G. Zachariwicz. 2018. *Value-based Learning Healthcare Systems: Integrative Modeling and Simulation*. London: IET Publication.

Model-Based Systems Engineering for CubeSat FMECA

Evelyn Honoré-Livermore and Cecilia Haskins

Abstract The CubeSat standard has given universities, small companies, developing countries, and others a new gateway to space exploration and space knowledge. Combined with shorter development time and Commercial Off-The-Shelf components, the cost has been lowered considerably. However, the combination of use of low-maturity components and inexperienced development teams results in a short lifetime and poor reliability for most CubeSats. The growth of model-based systems engineering (MBSE) supports reuse of design architectures in many industries and has lowered the costs of development and is gaining popularity in CubeSat teams. This paper demonstrates the application of reliability methods for implementing dependability analysis in MBSE and shows how this can benefit CubeSat teams struggling with limited personnel resources and low experience with space systems.

Keywords CubeSat · MBSE · Model · FMEA · FMECA · Risk assessment · Systems engineering

1 Introduction

An increasing number of small satellite projects are conducted at universities, many of which do not have previous space hardware or software experience or relevant curriculum to support the development. There are several reasons for this, such as better access to Commercial Off-The-Shelf (COTS) components, cheaper launch opportunities, introductory courses from the National Aeronautics and Space Administration (NASA) and the European Space Agency (ESA), multiple papers and guidelines on how to develop and build small satellites, and an increasing demand from students for projects with hands-on experience (Langer et al. 2015; Langer and Bouwmeester 2016; Holtstiege and Bridges 2018; Luther

E. Honoré-Livermore (✉) · C. Haskins
Norwegian University of Science and Technology, Trondheim, Norway
e-mail: evelyn.livermore@ntnu.no

A. M. Madni et al. (eds.), *Recent Trends and Advances in Model Based Systems Engineering*, https://doi.org/10.1007/978-3-030-82083-1_45

2016; Berthoud and Schenk 2016; Larsen and Nielsen 2011). Approximately 500 CubeSats are launched each year, where 40% of them are launched by universities (Kulu 2019). However, research shows that the lifetime of a CubeSat mission is short, with over 50% of the satellites DOA (dead on arrival) (Swartout 2019b).

Efforts to increase the reliability of CubeSats should increase the success rate of missions. Having a more systematic approach to reliability and verification and validation (V&V) of the satellite in early phases is strongly recommended by the literature, as well as learning from other CubeSat or small satellite projects through lessons learned databases or reuse of design. Universities without space experience may not be aware of these resources when they start a new project and therefore work in a more ad hoc manner without the rigor and discipline space projects require (Swartout 2019a).

The use of model-based systems engineering (MBSE) has been promoted as an option to provide a development platform to perform different types of analyses to support the design of a small satellite. The International Council on Systems Engineering (INCOSE) Space Systems Working Group (SSWG) is developing a CubeSat Reference Model (CRM) for this purpose (Kaslow and Madni 2017; Kaslow et al. 2018). By using the CRM as a starting point of a CubeSat project, it is possible to take a more systematic approach to reliability and V&V, as the reference model also includes guidelines on how to develop the CubeSat mission itself and pointers on how to work with V&V.

This paper describes the use of MBSE within a RAM-SE (reliability availability maintainability - systems engineering) framework for improved reliability analyses through the application of failure mode, effects, and criticality analysis (FMECA) and subsequent risk management. The case study is the HYPer-spectral Smallsat for ocean Observation (HYPSO) satellite developed at the Norwegian University of Science and Technology (NTNU) (Honoré-Livermore 2019). The paper builds on the work submitted by three bachelor students in 2019 (Moen et al. 2019).

2 Background

CubeSats are satellites built on the standard of a 10 cm x 10 cm x 10 cm cube. While both NASA and ESA recommend following strict reliability and quality approaches for their long-term missions, these practices often are costly and not realistic for universities. Both organizations have recommendations for CubeSat builders, such as derating for electrical components, basic failure detection, isolation, and recovery (FDIR) analysis guidelines, applying FMECA to both the functional and physical product tree, and planning a high degree of early testing and verification of subsystems as well as end-to-end functional testing (TEB 2016; CalPoly Sat Program 2017; Capogna and Gupta 2018).

While traditional space projects conducted by NASA/ESA have access to a large knowledge base and experienced people and may use 5–15 years in development, CubeSat projects approach dependability analysis pragmatically. Access to resources determines the level of analysis performed under conditions of limited knowledge, short schedules (2–3 years), and constrained budgets (Langer and Bouwmeester 2016; Faure et al. 2017). In addition, it is typical for CubeSats to use COTS components that have lower reliability, lower technology readiness level (TRL), and less data available to perform the reliability analysis. Performing FMECA on CubeSats has been discussed (Menchinelli et al. 2018) where a practical approach to managing risk through FMECA is described.

CubeSat university projects face several challenges, most notably in access to knowledge and experienced project management, which often directly influence the success of any mission (Honoré-Livermore 2019; Cho and Mazui 2016; Bouwmeester et al. 2008). University teams struggle with not having enough people, high turnover due to graduation, lack of previous experience with project work and multidisciplinary project teams, and finally, but not the least, the challenge of building a satellite to work in space including the stringent requirements for documentation and evidence of space systems testing results.

NTNU has been working on small satellites and CubeSats for the past decades (Grande et al. 2017) but only recently with a mission based on oceanographic research with funding from the Research Council of Norway for both hardware and launch (Honoré-Livermore 2019). Previously, much of the work was conducted in a student organization or loosely associated with course assignments. Today, there is a team of 20–30 undergraduate and graduate students working on developing the HYPSO satellite. As a part of this activity, there is research in systems engineering and project management on how to improve the processes and methods applied in a university-based CubeSat project to increase the probability of mission success and the satisfaction of students and supervisors for their academic work.

3 MBSE for CubeSat

Systems engineering (SE) emerged from the engineering of complex systems and is today an integral part of product development in many industries and fields of research. Applying model-based systems engineering (MBSE) as a part of the process is gaining popularity as the tools and methods mature and offer greater interoperability between disciplines in complex product development and life cycle management. The MBSE approach proposes a unified model that can provide the different viewpoints necessary to perform analyses in all engineering domains (Piggott et al. 2007). The viewpoints offer a variety of relevant information and hide the nonrelevant data to focus analysis depending on the need (Friedenthal and Oster 2016; Haberfellner et al. 2019). Recent advances in the realm of MBSE show increased interoperability between the system model, mechanical analysis tools, electrical analysis tools, and other design tools (Brower et al. 2019; Madni and Sievers 2018).

Not all SE tools support the modeling elements needed for reliability analysis but most offer customization options. There are also other approaches to add the failure analysis elements to the model, such as using export/import functions between the system model and the safety model (Sango 2018). A failure mode is the absence of a function or a nondelivery of a need (Schindel 2010), and an added benefit of using MBSE is that the failure analysis can be performed at all phases of the development, as it is possible to identify failure modes at the top level of a system even before the system has been designed. Integrating failure analysis in MBSE can lower costs and time spent on dependability analysis through providing a more systematic framework and increase communication on safety analysis and design. Baklouti et al. discuss the state-of-the-art research of FMECA and MBSE in their paper (Baklouti et al. 2019) and describe a use case for analysis of an electromechanical actuator system. The study by Gregory et al. (2020) recommends the use of MBSE in Functional Avionics in spacecraft for Communication and Consistency and for Template Model Framework. NASA discusses the requirements for a framework for safety analysis of space missions in Evans et al. (2018). This is aligned with ESA and NASA's overall strategy to manage complex space missions with MBSE.

4 RAM and SE Framework

Reliability, availability, and maintainability (RAM) analysis aims at using engineering knowledge and techniques to control the risk of experiencing failures and to reduce engineering uncertainties (O'Connor & Kleyner, 2012). The main activities of RAM engineering cover (1) artificial experiments to test out the properties of a given system or parts and (2) analysis and modeling techniques to reveal the cause-effect relationships between failure and specific conditions (Verma et al., 2015).

RAM analysis can be both qualitative and quantitative. Qualitative analysis is used to identify failure modes, mechanisms, and causes (such as FMECA) and determine the possible maintenance and test strategies. As the design matures, these analyses may be iterated and updated via communication and consultation with operators, manufacturers, and designers.

Space system design is a concurrent and collaborative process, where different engineering teams are involved. The RAM issues must be considered as early as possible to support the decision-making about redundancy, modularization, strategies for interventions, and the like. However, the effect of RAM considerations is not easily observed by the whole engineering team, and RAM methods do not have a well-defined interface with other analyses carried out in parallel phases of the design. A similar problem is also identified by Barnard (2008) who points out that the overemphasis on probabilistic modeling frequently leads to misinterpretation of RAM analysis, which can lead to bad design or waste of engineering efforts. A recently proposed RAM-SE framework recommends several activities to integrate both the SE and RAM community as shown in Fig. 1a (Zhang et al. 2018).

Estimation of the reliability of the components and in-house developed software is done by the team when better sources of data are not available. The results of the analysis are thus dependent on the expertise and configuration of the team and may

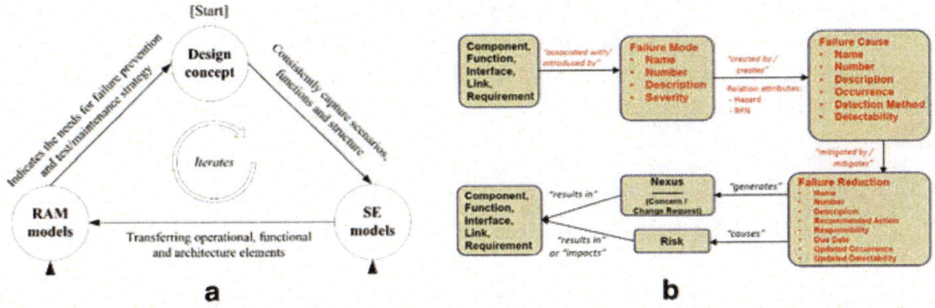

Fig. 1 (**a**) Conceptual RAM-SE (Zhang et al. 2018). (**b**) Classes used for modelling from (Kratzke 2018). The classes in black are the existing ones in the tool, while the classes in red are the ones that needed to be added. Some of the attributes given in red already existed, while others had to be added

Table 1 Advancements for RAM methods in SE context (Zhang et al. 2018)

Methods	Objectives	Advancements of systems thinking
FMECA	Use a basis for detailed RAM analysis and maintenance optimization and planning	Systematically identify all operational modes and functions attached to each potential failure modes
	Document the effect of failure on system	Carry out an extended/revised type of FMECA that is able to involve dynamic aspects of key scenarios; see also the discussion in Issad et al. (2017)

not be valid. However, the exercise of performing the analysis and the ranking of critical failure modes and causes are valuable because they create awareness in the team of potential issues in the design. In addition, the cognitive process of thinking about dependability and reliability can lead to better design processes and decisions in the future.

An FMECA analysis can be categorized broadly into two different types based on the approach: top-down functional FMECA or bottom-up component FMECA – given from MIL-STD-1629A (Department of Defense 1980). Depending on the phase of the project and the maturity of design and other information available, one is more appropriate than the other. Using a model-based approach allows for a unification of these types.

When the system is continuously modeled from the operational, functional, logical, and physical viewpoint, it is possible to combine the different types of analysis and aggregate the data. The operational viewpoint is implemented through the functional design which in turn is allocated onto logical units that are finally realized in the physical design. Depending on the tools used for modeling, real data from testing and verification can also be introduced and give a fifth viewpoint and source of information for the dependability analysis (Schindel 2010; Bürger 2019). Table 1 indicates benefits of integrating RAM and SE models.

Incorporating RAMS aspects as early as possible gives several advantages in the form of engineering efforts and budgets in many industry sectors such as nuclear, satellite, and aviation, where the analysis is further amplified by the complexity of design solutions (Zhang et al. 2018).

5 Modeling and FMECA Implementation

The SE tool used to implement FMECA in this research did not have the necessary modeling elements natively. The classes that had to be added to the modeling tool were based on a presentation given by Kratzke (2018) shown in Fig. 1(b). These classes were added to represent failure mode, failure cause, failure reduction, and the relationships between these and with the existing classes. The modeling and FMECA were performed on the NTNU HYPSO mission. The CubeSat consists of multiple subsystems, shown in Fig. 2. The payload (PLD) subsystem interacts with the subject of interest and is the most critical part of the space segment. The payload consists of several subsystems: HyperSpectral Imager (HSI), RGB camera (RGB), Onboard Processing Unit (OPU), and Break-Out Board (BOB).

System to be analyzed HYPSO CubeSat mission. The mission of the satellite is defined by a mission statement and mission success criteria. These are the most important requirements of the mission, and therefore, the FMECA analysis chose to focus on these. The system was modeled in a MBSE tool, with requirements, component trees, functional trees and chains, and operational user scenarios. Since the HYPSO team is tasked with developing the payload systems and ground segment, while the other subsystems are COTS components with a higher TRL, the analysis was limited to the space segment payload systems and the ground segment.

FMECA workshops FMECA workshops were conducted during which the HYPSO team members were asked to identify the operational modes for each of the subsystems and which functions were necessary for the operational modes

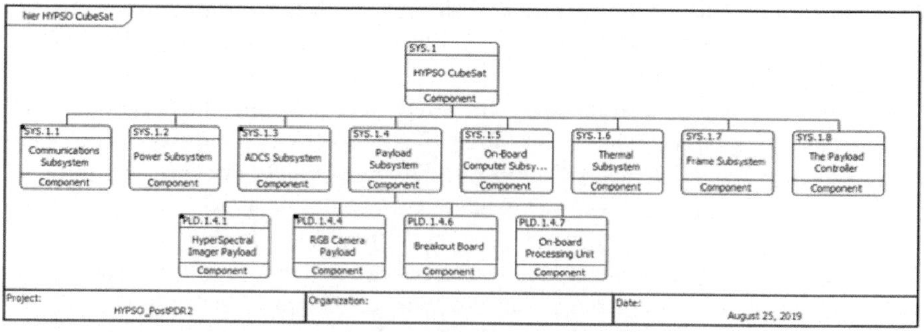

Fig. 2 The HYPSO CubeSat physical hierarchial structure in MBSE

that would then be used as a basis to identify the failure modes. Using the mission success criteria, which are reflected in the concept of operations of the system, the FMECA focused on the critical parts of the performance of the system.

Failure mode assessment The assessment followed a procedure tailored from the aforementioned Mil-Std-1629A (Moen et al. 2019). Each failure mode is assessed with respect to severity (scale 1–5, where 1 is negligible and 5 is absence of function) and occurrence (scale 1–5, where 1 hardly ever occurs and 5 is probable). This analysis provides the criticality number, given by the multiplication of severity and occurrence. Next, the failure modes are assessed according to their detectability (scale 1–10). A high detectability is given index 1, meaning that it is easy to identify the failure, i.e., that only one type of failure can give the effect that is observed. For many CubeSats, having a high level of detectability is more important than avoiding all failure modes, especially if there is a method to reset or remove the failure. Identifying the failure mode can also allow for redesign of the system in the next satellite. The attributes are evaluated and added to the model and used to calculate the Risk Priority Number (RPN), which is given by the multiplication of severity, occurrence, and detectability.

A total of 69 failure modes 65 causes and 45 failure reduction actions were identified and modeled. The PLD subsystem BOB is used here as an example of the actual implementation (see Fig. 3 and Fig. 4). The severity, occurrence, and detectability attributes were determined based on the workshop input, resulting in ranked failure modes. These attributes are assumed to be evaluated for nominal operations, not for ground testing or integration.

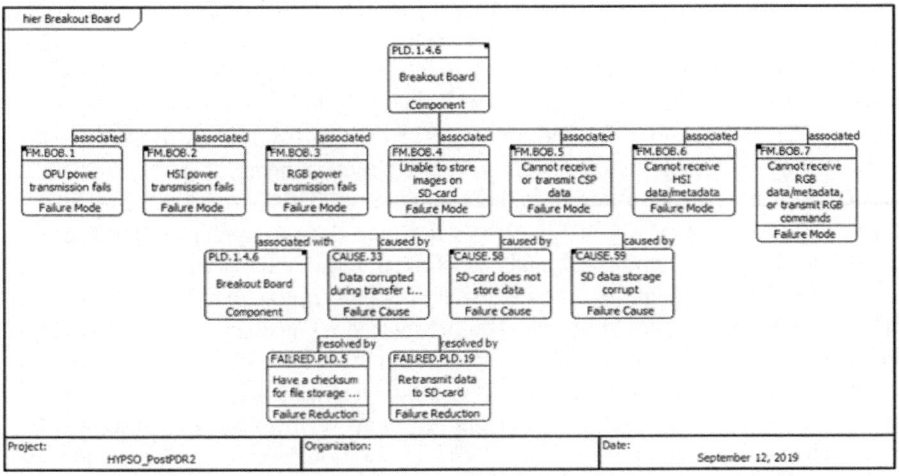

Fig. 3 BOB with its failure modes displayed in a hierarchy diagram. One of the failure modes has been expanded to show the failure causes and reduction methods

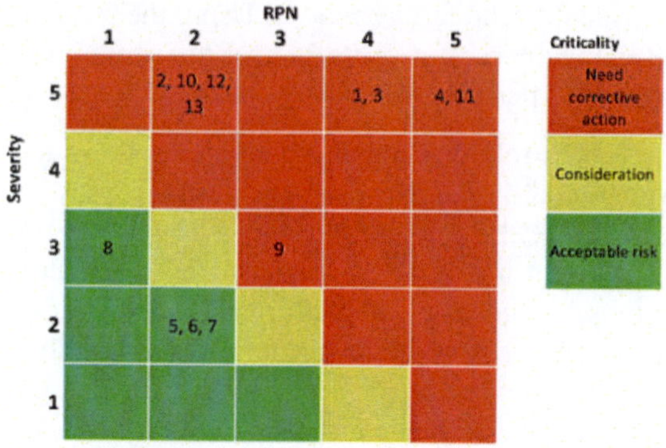

ID	Risk Reference
1	FM.BOB.6-CAUSE.40
2	FM.BOB.6-CAUSE.39
3	FM.BOB.5-CAUSE.31
4	FM.BOB.5-CAUSE.4
5	FM.BOB.4-CAUSE.58
6	FM.BOB.4-CAUSE.59
7	FM.BOB.4-CAUSE.33
8	FM.BOB.3-CAUSE.30
9	FM.BOB.3-CAUSE.27
10	FM.BOB.2-CAUSE.36
11	FM.BOB.1-CAUSE.27
12	FM.BOB.1-CAUSE.30
13	FM.BOB.2-CAUSE.30

Fig. 4 Risk matrix for BOB failure modes. The RPN is referenced to a scale of 100,and then into ranges of 20 to give the resulting 1–5 indexing

The FMECA model and attributes are used to generate a risk matrix composed of RPN and severity. The risk matrix includes severity as a high RPN number may not indicate a true risk to the project. The risk matrices are given per subsystem to make it simpler for the team to use. The critical failure causes and their corresponding reduction methods should then be prioritized in project work.

6 Discussion

The process identified 13 risks for BOB. Of these, eight were of severity rating 5. The risk matrix in Fig.4 was configured such that all causes with severity rating 5 would need corrective action, as this indicates a nonfulfillment of minimum mission success criteria. Prior to the workshop and modeling, only the loss of power transmission had been considered as a risk in design discussions. Based on this analysis, the BOB component has been updated to reduce the number of critical risks through implementation of failure reduction measures. This process agrees with the RAM-SE process proposed in Fig. 1(a) where the design concept is modeled in MBSE and the subsequent RAM analysis is integrated in the model.

The modeling tool supports the inclusion of actors (humans) and their influence on the system through, e.g., use cases, but this was not considered in the modeling. Human error can be a considerable cause of failure modes to the system. The focus on mission success criteria when analyzing the system also left out an important part of the CubeSat development life cycle: test and verification. Use cases can be developed for test activities such as shock or vibration, which carry high risk of damage to the device under test (DUT) and personnel. Including these use cases for future CubeSat modeling would address additional risk areas. Additionally, the choice of perspective when modeling and analyzing did not support evaluation of simultaneous failure. A fault tree analysis (FTA) on a functional level would increase the chance of discovering if two faults would give a severe impact that they individually would not in the FMECA. This is supported in the class relationship implemented according to Fig. 1(b). Limited knowledge about the reliability of low-TRL components lowers the validity of the analysis. It is largely dependent on the people attending the workshop and their ability to properly assess the severity, occurrence, and detectability of individual events. Modeling facilitates visualization of the relationships between the failure modes, the components, the functions, and the overall success criteria.

For CubeSat teams, the starting point of a reference model such as the CRM provided by INCOSE SSWG lessens the burden on the faculty to have prior experience with satellite systems and focuses the teamwork on specific subsystems. The university teams can then populate the model with the known information such as regulations, existing ground systems, communication capabilities, and operational aspects specific to the organization. The reuse of a model with failure analysis built in from the previous mission or other teams' missions will reduce the team resources needed to perform dependability analysis. An added benefit from using MBSE is the potential of automated analyses on the same model. Each element in a model will have attributes that can be analyzed automatically through scripts or functions in the tool. For dependability analysis, it is then a matter of checking what the change of one element will have on the overall dependability of a system.

Swartout (2019a) highlights the need for "streamlined practices, experientially developed" and recommends increasing the time spent on integration and test, also supported in Faure et al. (2017). Designing the system in MBSE through the application of a CRM and incorporated dependability analysis could be a part of the "streamlined practices" to improve the design maturity faster, thereby allowing more time to perform integration and test.

7 Conclusion

Developing and building complex systems is a challenge, and while CubeSats are small, they are still complex systems requiring a systematic engineering approach. Statistics show that there are still many CubeSat missions that fail, mostly because of low dependability. However, performing the full dependability and quality management that NASA and ESA recommend is not feasible for small university teams lacking both people, money, and time. A pragmatic approach to dependability analysis through the usage of MBSE systems that can support the FMECA analysis offers promise to increase dependability for university CubeSat teams.

This paper has shown how the framework suggested by Kratzke (2018) can be applied to an existing CubeSat MBSE model and can support the subsequent risk management from the FMECA workshops. The FMECA workshop and modeling enabled the HYPSO team to understand which failure modes are associated with different subsystems and how these affect the overall success criteria for the mission.

It also established a priority for completing different failure reduction activities and highlighting critical interfaces and components/functions. Using visualization through the operational scenario modeling and focusing the effort on making the mission a success have proven to be of great value to the CubeSat team that did not have previous experience with dependability analysis. It also enables keeping the information connected and ensures traceability for future design decisions.

The results agree with the suggestions of Gregory et al. (2020); MBSE has been used as a tool to improve Communication and Consistency of the RAM (project) information, and it can be reused in future spacecraft building on the same spacecraft subsystem structure with little rework necessary. The framework could be a part of the university MBSE template or used to extend the CRM.

Incorporating the CubeSat Reference Model with FMECA analysis from an operational or functional top-down view or component-based bottom-up view has required less effort than starting from scratch. Furthermore, the analysis could be performed at different phases of the development, as the failure modes are relevant on higher operational levels as well as lower functional or logical levels. Reuse and continuous development of the systems are facilitated using MBSE as suggested by Madni and Sievers (2018).

For this preliminary work, the risk assessment was done in a separate tool from the MBSE tool because the research team uses a project management suite. As future work, this could be automated and linked to improve workflow. Additionally, future work should explore how the lessons learned and failure analysis from multiple teams could be aggregated into the reference model, highlighting typical failure modes that are associated with certain interfaces or types of subsystems.

Acknowledgments This work is supported by the Norwegian Research Council (Grant No. 270959), the Norwegian Space Agency, and the Centre of Autonomous Marine Operations and Systems (NTNU AMOS).

References

Baklouti, A., N. Nguyen, F. Mhenni, J. Choley, and A. Mlika. 2019. Improved Safety Analysis Integration in a Systems Engineering Approach. *Applied Sciences* 9.

Barnard, R. W. A. (2008). *What Is Wrong with Reliability Engineering?* Paper presented at the Proceedings of the 18th Annual INCOSE International Symposium, Utrecht, the Netherlands.

Berthoud, L, Schenk, M., 2016. How to Set Up a CubeSat Project – Preliminary Survey Results. In *30th Annual AIAA/USU Conference on Small Satellites*. Logan, UT.

Bouwmeester, J., G.T. Aalbers, and W.J. Ubbels. 2008. Preliminary Mission Results and Project Evaluation of the Delfi-C3 Nano-Satellite. In *4S Symposium Small Satellites Systems and Services*. Rhodes: Greece.

Brower, E.W., C. Delp, R. Karban, M. Piette, I. Gomes, and E. Wyk. 2019. *OpenCAE Case Study: Europa Lander Concept*. Torrance: In INCOSE International Workshop.

Bürger, E.E., 2019. *A conceptual MBSE framework for satellite AIT planning*. Doctoral thesis. Instituto Nacional de Pesquisas Espaciais.

California Polytechnic's PolySat Program. 2017. CubeSat 101. California Polytechnic State University at San Luis Obispo.

Capogna, F., Gupta, V., 2018. *How to increase CubeSat reliability!* ESA (TEC-QCD).

Cho, M., Mazui, H., 2016. Best practices for successful lean satellite projects. In *4th UNISEC Global Meeting*.

Defense, Department of. 1980. *MIL-STD-1629A Procedures for performing a Failure Mode, Effects and Criticality Analysis*. Washington, DC.

Department of The Air Force. 1988. *MIL-STD-1543B Reliability program requirements for Space and Launch Vehicles*. Washington, DC.

Evans, J., F.J. Groen, L. Wang, S. Okon, R. Austin, A. Witulski, N. Mahadevan, S. Cornford, M. Feather, and N. Lindsey. 2018. Towards a framework for reliability and safety analysis of complex space missions. *International Journal of Human Factors Modelling and Simulation* 6: 203.

Faure, P., A. Tanaka, and M. Cho. 2017. Toward lean satellites reliability improvement using HORYU-IV project as case study. *Acta Astronautica* 133: 33–49.

Friedenthal, S., Oster. C., 2016. Applying SysML and a Model-Based Systems Engineering Approach to a Small Satellite Design. In Richard Curran John Hsu (ed.), *Advances in Systems Engineering* (American Institute of Aeronautics and Astronautics: Reston, Virginia, USA).

Grande, J., Birkeland, R., Gjersvik, A., Mathisen, S., Stausland, C., 2017. Norwegian student satellite program – lessons learned. In *68th IAC*.

Gregory, J., Berthoud, L., Tryfonas, T., Rossignol, A., Faure, L. 2020. The long and winding road: MBSE adoption for functional avionics of spacecraft. *Journal of Systems and Software*, 160, art. no. 110453

Haberfellner, R., O. de Weck, E. Fricke, and S. Vössner. 2019. The Systems Engineering Basics in Our Systems Engineering Concept. In *Systems Engineering: Fundamentals and Applications*. Springer: Cham.

Holtstiege, J., Bridges, C., 2018. Lean satellite design for amateur communications payload in the ESA ESEO mission.

Honoré-Livermore, E. 2019. CubeSats in University: Using Systems Engineering Tools to Improve Reviews and Knowledge Management. *Procedia Computer Science* 153: 63–70.

Issad, M., Kloul, L., & Rauzy, A. (2017). *A scenario-based FMEA method and its evaluation in a railway context*. Paper presented at the Reliability and Maintainability Symposium (RAMS).

Kaslow, D., Ayres, B., Cahill, P., Hart, L., Levi, A., Croney, C., 2018. Developing an MBSE CubeSat Reference Model – Interim Status #4. In *2018 AIAA SPACE and Astronautics Forum and Exposition*. Orlando, FL.

Kaslow, D., Madni, A., 2017. Validation and Verification of MBSE-compliant CubeSat Reference Model. In *Conference on Systems Engineering Research, edited by Barry Boehm Azad M*. Madni. Redondo Beach, CA, USA.

Kratzke, R., 2018. Failure Modes Effects Analysis in MBSE. In *INCOSE Texas Gulf Coast Chapter*. Houston, TX.

Kulu, E., 2019. *NanoSats database*. Accessed June 2019. https://www.nanosats.eu/.

Langer, M., Bouwmeester, J.. 2016. Reliability of CubeSats – Statistical Data, Developers' Beliefs and the Way Forward. In *AIAA/USU Conference on Small Satellites*.

Langer, M., C. Olthoff, J. Harder, C. Fuchs, M. Dziura, A. Hoehn, and U. Walter. 2015. Results and lessons learned from the CubeSat mission First-MOVE. In *Small Satellite Missions for Earth Observation, 10th International Symposium*. Berlin: IAA.

Larsen, J., and J.D. Nielsen. 2011. Development of cubesats in an educational context. In *International Conference on Recent Advances in Space Technologies*, 777–782. Istanbal: IEEE.

Luther, S. T., 2016. *SysML Based CubeSat Model Design and Integration with the Horizon Simulation Framework*. Master of Science, California Polytechnic State University.

Madni, A., Sievers, M., 2018. Model-Based Systems Engineering: Motivation, Current Status, and Needed Advances. In *Disciplinary Convergence in Systems Engineering Research*. Springer International Publishing.

Menchinelli, A., F. Ingiosi, L. Pamphili, P. Marzioli, R. Patriarca, F. Costantino, and F. Piergentili. 2018. A Reliability Engineering Approach for Managing Risks in CubeSats. *Aerospace* 5: 121.

Moen, A. K. A, Sjøvold, E., Jordheim, O.. 2019. *Implementation of FMECA in Small Satellite Development*. Norwegian University of Science and Technology. Bachelor thesis.

O'Connor, P., and A. Kleyner. 2012. *Practical Reliability Engineering*. 5th ed. Hoboken: Wiley.

Piggott, S., P. Melanson, and L. Hartman. 2007. A Vision for Super-Model Driven Systems Engineering. *INCOSE International Symposium* 17: 52–65. https://doi.org/10.1002/j.2334-5837.2007.tb02857.x.

Requirements and Standards Division, ESA-ESTEC. 2009. *Space product assurance: Dependability, FME(C)A*. Noordwijk, NL.

Sango, M., 2018. *Model Based Safety Analysis Thanks to the Bridge Between Capella and Safety Architect*. Accessed August 2019, https://www.youtube.com/watch?v=-NtUKdaUGdc. YouTube.

Schindel, W.D. 2010. Failure Risk Analysis: Insights from MBSE. *INCOSE International Symposium* 20: 1227–1239.

SEMATECH. 1992. *FMEA: A Guide for Continuous Improvement for the Semiconductor Equipment Industry*.

Swartout, M., 2019a. *CubeSat Mission Success: Are We Getting Better?* In 2019 CubeSat Developers' Workshop. San Luis Opisbo, CA.

———. 2019b. *Mission Success for Universities and Professional Programs*. Accessed September 2019. https://sites.google.com/a/slu.edu/swartwout/home/cubesat-database/repeat-success.

TEB. 2016. *Product and Quality Assurance Requirements for IOD CubeSat Projects*. TEC-SY/129/2013/SPD/RW. Noordwijk, NL.

Verma, A.K., S. Ajit, and D.R. Karanki. 2015. *Reliability and Safety Engineering*. United Kingdom: Springer London Ltd.

Zhang, J., Y. Liu, M.A. Lundteigen, and C. Haskins. 2018. The Application of Systems Engineering for Framing Reliability. *Availability and Maintainability in Subsea Design. Systems Engineering*. https://doi.org/10.1002/sys.21462.

Model-Based Systems Engineering for Design of Unmanned Aircraft Traffic Management Systems

Lindsey Martin, Samantha Rawlins, and Leonard Petnga

Abstract As technological advances in material science, computing, electronic, and artificial intelligence are fueling increasingly autonomous unmanned aircraft system (UAS) capabilities, the legal operational framework is progressively catching up. However, for commercial applications involving beyond-visual-line-of-sight (BVLOS) group flights to become ubiquitous, the successful development of safe, integrated UAS traffic management (UTM) systems (UTMSs) is imperative. We propose a generic model-based systems engineering (MBSE) approach for supporting conceptual design and analysis of UTMS. A system-of-systems (SoS) perspective of the UTMS is adopted to drive the design effort. This iterative process is rigorous and technology independent and is powered by the flow-down of requirements driving the specification of each layer of abstraction of the UTMS as a SoS. At each layer (SoS, systems, subsystems, components), functional, structure, and behavior are developed and synthesized and then verified against the requirements. Various analyses, including trade studies, are performed to support system evaluation and guide decision-making throughout the design. The essential features of the framework are highlighted in a simplified UTMS as a directed SoS for a package delivery application developed using the Systems Modeling Language (SysML).

Keywords UTM · Model-based systems engineering · System of systems · SysML

L. Martin · L. Petnga (✉)
ISEEM Department, The University of Alabama, Huntsville, USA
e-mail: lp0053@uah.edu

S. Rawlins
Department of Mechanical and Aerospace Engineering, The University of Alabama, Huntsville, USA

A. M. Madni et al. (eds.), *Recent Trends and Advances in Model Based Systems Engineering*, https://doi.org/10.1007/978-3-030-82083-1_46

1 Introduction

1.1 Problem Statement

By 2022, the Federal Aviation Administration (FAA) predicts that the total unmanned aircraft system (UAS) use will grow by 235%. Within the same time frame, commercial UAS fleets are projected to balloon from 110,604 vehicles to more than 450,000, making them the fastest growing part of the market (Commercial drones are the fastest-growing part of the market - Taking flight 2017; Federal Aviation Administration 2018). While the majority of the small model hobbyist units are both physically and legally restricted in their access to the national or international airspace, unified, comprehensive legislation on how to best incorporate commercial, unmanned aircraft into the preexisting regulation and management system is in the final rule-making stages (Federal Aviation Administration 2018). Failure to implement rules enabling future Unmanned Service Suppliers (USSs) with conflicting priorities to collaboratively and safely manage designated low-altitude airspace for UAS traffic – as a system of systems (SoS) – will lead to intractable operational and legal issues. This will ultimately impede the development of UAS, which, as shown in Fig. 1, is becoming ubiquitous given the increasing number of applications areas.

1.2 Scope and Objectives

Instead of presenting another solution for a UTM architecture, this paper aims to emphasize the benefits of a SoS-enabled model-based systems engineering (MBSE) approach to the design of UTM systems (UTMSs). An SoS perspective is required to be driven by the need to incorporate the priorities of multiple stakeholders, coupled with the one to account for complexity of interactions between components and

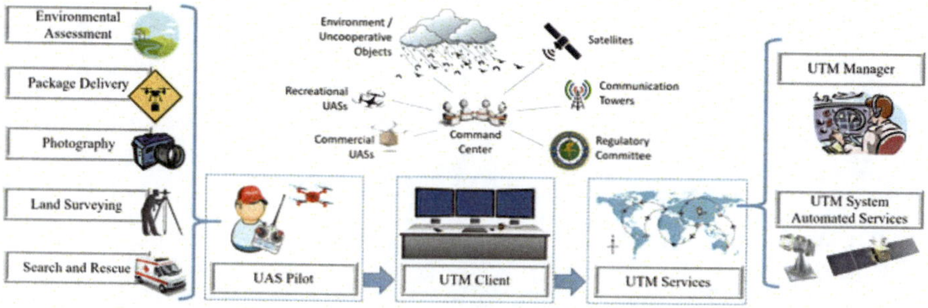

Fig. 1 UTMS Concept of Operations flow chart. (Modified from (National Aeronautics and Space Administration (NASA) 2015))

systems distributed in space and time. We first discuss the current state of the art of design approaches to the UTM and how UTM architecture falls within the SoS framework. Next, the proposed layered MBSE approach is introduced, and we show how it best addresses the challenge of designing an integrated UTMS. Finally, a case study applies the approach to assess how a UAS Package Delivery Service (UPDS) would influence UTM design.

2 Background and Literature Review

2.1 Overview of State of the Art of Unmanned Aircraft Traffic Management Systems

Current UTMS regulations are evolving in a completely different way than the manned civil aviation world. From the experiences gained over the course of decades of trial and error, air navigation services have developed a robust, albeit slow and complex, safety management system that is generally accepted and maintained throughout the world. The major influencers of UTM regulation have begun to split off and create their own UTM architectures. As such, unique frameworks have been proposed and in some cases already implemented, from global, national, state, commercial, and even city perspectives (UTM Current State-of-the-Art 2017; AirCrew 2018; G. Nichols, "Davos develops drone regulation how-to for governments (and the FAA should pay attention)," ZDNet & Nichols, n.d.; P. Butterworth-Hayes, "Unmanned Airspace: The Market for UAS Traffic Management Services – 2018-2021," Unmanned airspace & Butterworth-Hayes, n.d.; Global UTM Association. Global UTM Association, n.d.; F. Schubert. The development of the UAS traffic management (UTM), an air navigation services perspective. p. 9 & Schubert, n.d.).

2.2 System of Systems: Definition, Typology, and Design Challenges

According to the DoD, a system of systems (SoS) is defined as "a set or arrangement of systems that results when independent and useful systems are integrated into a larger system that delivers unique capabilities." Based upon the way that these systems interact with each other, a SoS can be categorized into four different types: (1) *virtual SoS*, which lacks a central management authority and a centrally agreed-upon purpose; (2) *collaborative SoS*, where each entity knows its predefined role and interacts with the other to fulfill agreed-upon central purposes; (3) *acknowledged SoS*, where each constituent system maintains its independent status but a designated manager, and resources, is in charge of ensuring achievement of recognized objectives; and (4) *directed SoS*, where component systems maintain an ability to operate independently, but a central manager is in charge of decision-making to fulfill specific purposes. Thus, the issues of command and control are central to the realization of successful and efficient SoS designs (Department of Defense (DoD) 2008).

2.3 Unmanned Aircraft Traffic Management System as a System of Systems

Successful design and operation of UTMSs whether at local, state, regional, national, or global scales requires the development and integration of multiple evolving technologies and systems beyond the capabilities of single industry suppliers, much like for current ATM systems. No matter the chosen architecture or usage, each successful UTM will have to integrate independently operated components including vehicles (i.e., UASs), pilots (e.g., recreational, commercial), operation centers (e.g., UTM Manager, FAA), and air infrastructure (i.e., radars, launch and landing points, weather). Thus, each individual UTM is an SoS. Plus, the actual USS infrastructure will need to include and integrate shared or proprietary geospatial system (for terrain and weather information), as well as communication, detection, and tracking systems resulting into acknowledged or directed SoS architectures involving UASs, depending on the needs of the applications and design choices. Moreover, USSs operating UTMSs will need to integrate their systems with one another but also with regulator (e.g., FAA), public safety, and air traffic control systems to enable safe and efficient beyond-visual-line-of-sight (BVLOS) flights of each UAS it serves. This organization highlights the need for USSs to participate into a collaborative SoS for effective large-scale operation. Moreover, in either case, a wide range of technologies including airspace design, dynamic geofencing, congestion management, and terrain avoidance will be needed (Jiang et al. 2016).

3 Proposed Model-Based Development of UTM Systems

3.1 Development Approaches and Strategies

The central tenant of the MBSE approach is the use of models (as opposed to documents) as the main artifacts of system development. State-of-the-art research and practice do not identify a unique "one-size-fit-all" winning strategy for MBSE (Estefan 2008). The intrinsic complexity of SoS such as the UTMS makes the challenge of designing this type of system even harder. Previous efforts include (1) an iterative SoS development strategy in which processes captured in "wave models" enable the definition and description of SoS evolution (Bonanne 2014) and

(2) the adoption of an agent-based modeling (ABM) approach that highlights the inherent independence of the systems that comprise the SoS (Achesona et al. 2013). However, experience indicates that good solutions result from a combination of semiformal and formal models. The former capture ideas (i.e., goals and scenarios) and preliminary designs represented in graphical languages such as the Unified Modeling Language (UML) and the SysML (Fridenthal et al. 2012). Given that SoS are mission-oriented, abstraction mechanisms are needed to enable system engineers to efficiently decompose the mission at enough level of detail and identify the appropriate hierarchy of systems, subsystems, and components (decomposition) or, alternatively, systematically assemble systems to be part of the SoS from predefined components (composition). Integrating both the top-down (decomposition) and bottom-up (composition) provides means for the system engineers to adapt to the inherent evolving nature of the SoS while keeping in check the traceability of selected components to the mission. Current approaches lack such capabilities. Thus, we adopt a multilayered development strategy that provides flexibility and modularity to the design and leverages existing, properly vetted, capabilities to support the mission and call for the design of new systems only on the need basis.

3.2 *Elements of MBSE for UTM System Design*

Our approach addresses the UTMS (as a SoS) design challenge through the lens of two integrated dimensions, (1) multiple layers of abstractions organized in a hierarchy, e.g., mission ➔ systems ➔subsystems ➔ to components, and (2) design tasks that can be assigned to either the problem or the solution domain or their interface at each layer of abstraction. For instance, a perspective of the USS UTMS at the mission level would reveal the high-level interactions between the various core systems, e.g., weather system, global positioning system (GPS), communication systems (telemetry), detection and tracking systems (e.g., space-based sensors and radars), traffic control systems, and with UASs as required by the applications to be deployed on the UTMS. The system layer below would focus on individual systems (e.g., the GPS) identified at the layer above. Designing the UTMS will require the translation of various needs into requirements to be captured in the problem domain of the SoS design challenge at various layers of abstractions and then tied to the ones above and below.

Figure 2 shows the design pathway from inputs and outputs and at a representative layer of abstraction. The problem design is formulated in the problem domain through the capture and representation of the USS UTM system requirements and their traceability to the architectural solution and analysis. As a part of this effort, key stakeholders and their interactions with the USS UTMS are mapped out in the concept of operations (CONOPS) along with goals and scenarios. Requirements can be refined with SysML use cases, sequence, and activity diagrams. Key performance indicators are defined to drive formulation of the solution domain but also enable the

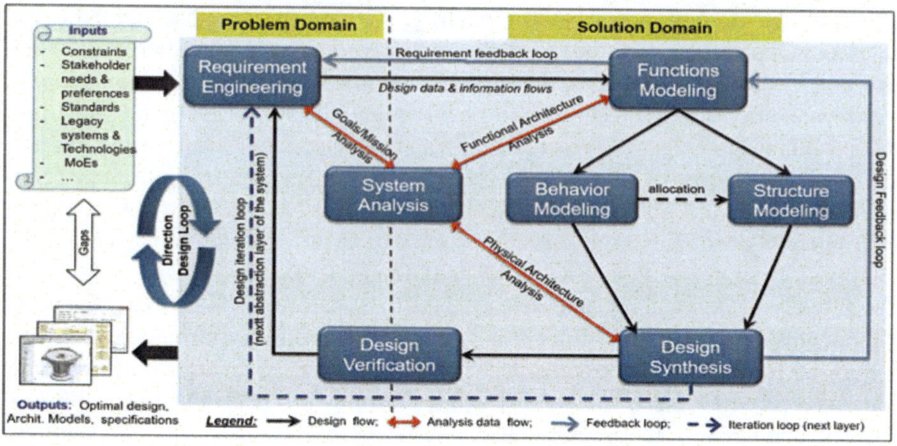

Fig. 2 Proposed MBSE approach for UTM development: pathway from the inputs and outputs for a representative design layer

assessment of how well the USS UTMS under design performs its mission. Functional analysis provides a high-level definition and decomposition of system core function and leads to defining functions, actions, and operations to be performed by individual systems or their core subsystems/components. The specifications of these structural elements are identified and documented during the structure modeling step using SysML block definition diagrams (BDDs). Design synthesis is performed by mapping (allocating) behaviors to structure. For instance, the function of tracking UAS movement in the airspace can be assigned to radars (structures). Throughout the design, analyses are performed to better inform decisions within the system architecture and beyond but also to ensure proper elicitation of requirements as well as functional and physical architectures. Analysis of the dynamic of the system yields physical models at various levels of abstraction that can be captured in SysML parametric diagrams. The latter can be integrated with analysis tools (e.g., MATLAB) for simulation and assessment of the USS UTMS under design. Formal verification of the design is performed to ensure the satisfaction of the requirements. The gap between inputs and outputs (effectively design artifacts) is addressed – with the stakeholders in the loop – until a satisfactory design is formulated for the level of abstraction under consideration. Then, the design can move to the next layer of abstraction below, following the pathway defined by the decomposition of requirements. The specifications are constantly checked against existing solutions for “buy vs. make” decisions and to determine the level of abstraction at which the decomposition is to be stopped.

4 Case Study: MBSE of an UTM System for a Package Delivery Service

4.1 Overview

Multiple USSs have been testing the use of UASs for package delivery. To be successful, UTMs must incorporate all differing stakeholder needs and facilitate coordination across areas of operation controlled by different companies with differing needs and requirements. At this high level of abstraction, the UTMS serves as an acknowledged SoS, where UAS flight is not limited to the airspace of a single USS. In such a model, UASs would cohesively cross from one USS-controlled airspace to another. Considering only one USS in this case study, at a lower level of abstraction, the UTMS is a directive SoS in which the UTMS's operation management relays information between subsystems to direct the flight of UASs. Madison County, Alabama, was used as the intended area of operation for the application of the UTMS in package delivery for this case study. We implement the approach depicted in Fig. 2 and cover two layers of our UTM SoS system as briefly described in the next two sections.

4.2 System-of-Systems Level Design

In addition to input received from a commercial stakeholder, requirements were derived from passive observations of similarities between precedent systems that currently guide package delivery, ground traffic management, and ATM, with the UTMS. To align UTM with current aviation standards, FAA regulations were also referenced during design. From requirements and concept of operation (CONOPS) development, the uses of the UTM were elaborated into a use case diagram using SysML, depicted in Fig. 3. Core use cases identified for the UTM include managing, monitoring, and determining routes for UASs. To perform these services, the UTM relies on multiple subsystems. The UAS GPS system data and the data from the ground radar system provide inputs via telemetry to the central UTMS hub, which centrally controls the UTMS in a directed SoS architecture.

4.3 System-Level Design

After designing and verifying the system level of the UTMS, the systems that comprise the UTMS underwent requirement engineering, as shown in Fig. 2. The UTMS leverages inputs from GPS, radar systems, emergency management systems, weather authorities, FAA and local regulations, as well as UASs. GPS and radar systems monitor the locations of all UASs and communicate with the UTMS's

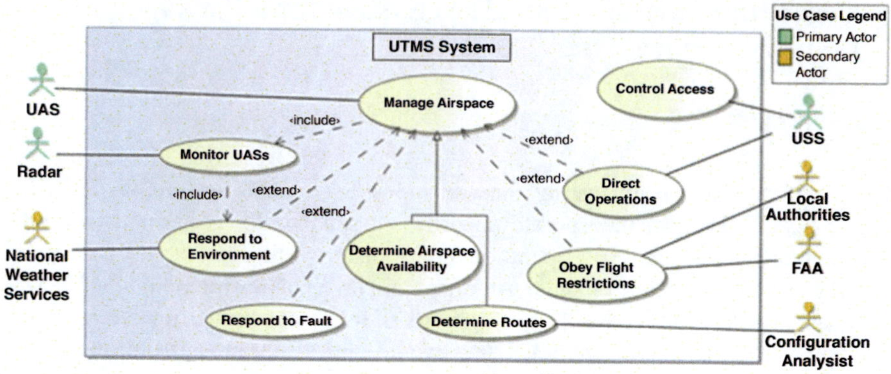

Fig. 3 SysML use case representation of the UTM functionality

operations management system which directs information to UASs. Flying BVLOS, UASs utilize this information to autonomously avoid collision on their flight paths to delivery.

4.4 Physics-Based Modeling and Engineering Analysis

Requirements gathered from the stakeholder placed emphasis on ensuring safety by consistently tracking positions of UASs. Radars were found to be the most effective systems for the task. Each UAS had to be within range of at least one radar to accurately account for location during flight. The position data from radars would be relayed to the UTMS hub, which would determine routes and telemeter, and that data would be sent to flight control onboard the UASs. The parametric diagram in Fig. 4 captures the various constraints for effective radar coverage within the UTMS. Equation 1 shows a simplified formulation of the radar transmission range with parameters shown in Table 1.

$$R = \sqrt[4]{\frac{P_t * G^2 * \lambda^2 * \sigma}{P_{min} * (4\pi)^3}} \tag{1}$$

The radar transmission range is later used to determine the minimum number of radars necessary to fully cover any operational area, defined by the area's average length and width in meters. In this case study, we consider Madison County, Alabama, which has an area of approximately 2,071 square kilometers as shown in Fig 5a. Through integration of the SysML parametric diagram with a MATLAB script, the range of a single radar was compared to the operational area, and the number of radar systems was incremented until the cumulative coverage of the

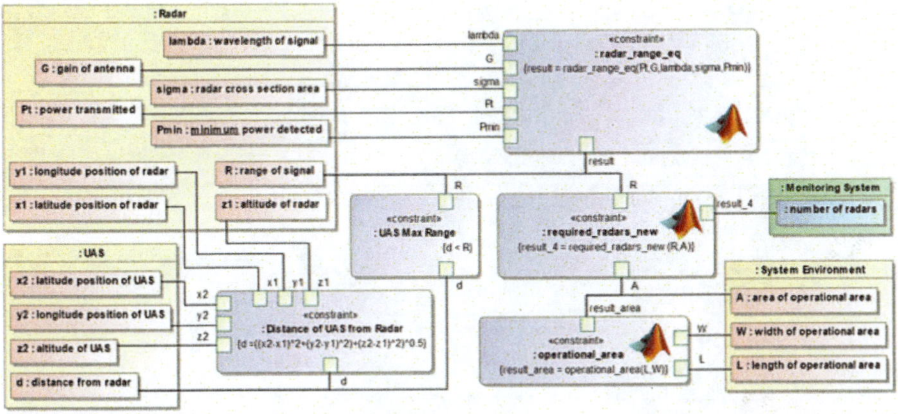

Fig. 4 Parametric diagram capturing system constraints for UTM radar coverage of Madison County, AL

Table 1 Radar equation values

	Peak power transmitted	Minimum signal detected	Frequency/wavelength	Antenna gain	Target cross-sectional area
Variables	P_t	P_{min}	f/λ	G	σ
Values	250,000W	0.00001 W	1.3GHz/0.230609m	38dB/6309 linear	2.5 square meters

radar systems exceeded the operational area. For the given radar parameters, Fig. 5b. depicts a graph of how many radar systems would be necessary to cover operational areas ranging in size from 0 to 2,500 square kilometers. The plot shows that we'll need 81 such radars for our UTMS for Madison County, AL.

5 Conclusion and Future Work

In this work, we look at UTMSs through the lens of SoS. Two important insights are gained from this: (1) the complexity of the UTMS architectures and operations can effectively be captured and managed through the various SoS types, i.e., acknowledged, collaborative, or directed – vantage points – and (2) a multilayered scalable and reusable MBSE approach is effective in assisting USSs in designing and modeling operations of next-generation UTMS. The case study demonstrated the effectiveness of the use of system thinking supported by MBSE techniques in the design of a USS-managed UTMS as a directed SoS. The architecture is developed with the Systems Modeling Language (SysML) as implemented by Cameo Enterprise Architecture toolkit integrated with MATLAB for parametric calculations. This effort has shed light on the benefits and contribution of MBSE

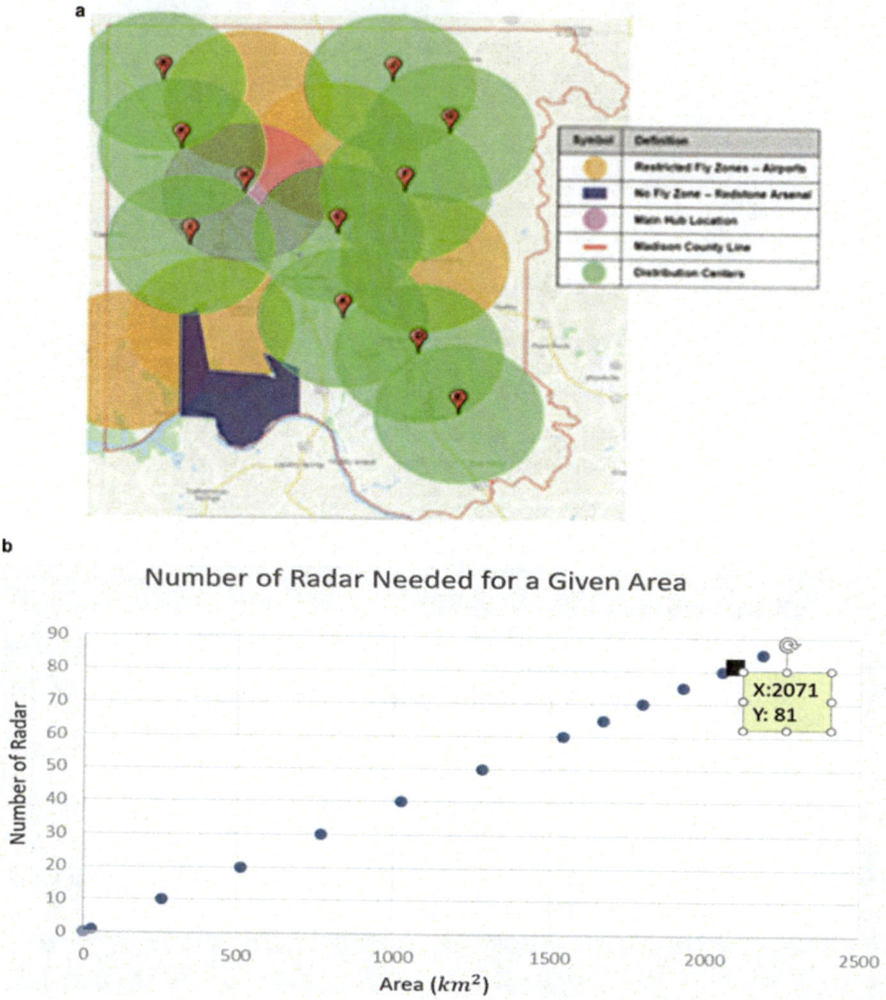

Fig. 5 (**a**) Coverage map of Madison County, AL (**b**) Determination of number of radars for UTMS coverage

approaches to the ongoing effort toward successful design and large-scale safe operation of UTMSs. Going forward, a deeper analysis needs to be conducted on the interactions between systems within the SoS. The potential electromagnetic interference (EMI) between ground radar and onboard telemetry systems needs to be evaluated at with specific operating bandwidths. Benefits of future extensions would lead to considerable savings in design cost and time. This will include detailed selection of technologies as well as trade-off analysis and formal verification of design for various CONOPS.

Acknowledgments This paper is an extension of a student systems engineering class project. Thanks to Christine Aldijali, Laure Campbell, Vanessa Cardwell, Hannah Hallmark, Heidi Miller, Giula Palma, Trent Rich, Hannah Smith, and The Boeing Company Inc.

References

P. Achesona, C. Daglia, N. Kilicay-Erginb 2013. Model-Based Systems Engineering for System of Systems Using Agent-Based Modeling. *Conference on Systems Engineering Research (CSER), In Procedia Computer Science* 16, pp. 11–19.

AirCrew 2018. AirMap Selected as UAS Service Supplier in Six UAS IPP Awards. *AirMap*, 09-May-2018.

K. H. Bonanne 2014. *A Model-Based Approach to System-of-Systems Engineering via the Systems Modeling Language*. Master Thesis, Purdue University, 117 pages

Commercial drones are the fastest-growing part of the market - Taking flight 2017. *The Economist*, 10-Jun-2017.

Department of Defense (DoD) 2008. *Systems Engineering Guide for Systems of Systems*. Office of the Deputy Under Secretary of Defense for Acquisition and Technology.

J. A. Estefan 2008. *A survey of Model-Based System Engineering (MBSE)*, version 8, INCOSE MBSE initiative. INCOSE, 70 pp.

F. Schubert. The development of the UAS traffic management (UTM), an air navigation services perspective. p. 9.

Federal Aviation Administration 2018. FAA Releases Aerospace Forecast. 16-Mar-2018

Fridenthal, S., A. Moore, and R. Steiner. 2012. *A Practical Guide to SysML: The Systems Modeling Language*, 599. The MK/OMG Press.

G. Nichols, "Davos develops drone regulation how-to for governments (and the FAA should pay attention)," ZDNet.

Global UTM Association. Global UTM Association.

Jiang, T., J. Geller, D. Ni, and J. Collura. 2016. Unmanned Aircraft System traffic management: Concept of operation and system architecture. *International Journal of Transportation Science and Technology* 5: 123–135.

National Aeronautics and Space Administration (NASA) 2015, November 15. *UTM: Air Traffic Management for Low-Altitude Drones*. NASA Facts, 2 pages.

P. Butterworth-Hayes, "Unmanned Airspace: The Market for UAS Traffic Management Services – 2018-2021," Unmanned airspace.

UTM Current State-of-the-Art 2017. *Eurocontrol*, 06-Jul-2017.

Exploration of MBSE Methods for Inheritance and Design Reuse in Space Missions

Alejandro E. Trujillo and Azad M. Madni

Abstract Design reuse, a common and valuable practice in complex engineering projects, is particularly important within the space industry. However, design reuse in this discipline is often conducted in an ad hoc fashion that limits the upside while introducing potential pitfalls. Model-based systems engineering (MBSE), while in the early stages of adoption in the industry, may form the foundation of a new strategic and systematic methodology for investigating and implementing design reuse for space missions and campaigns. To that end, this paper explores one aspect of that envisioned methodology, namely, the technical inheritance process which assesses the technical feasibility of adapting design elements from their native missions to new missions of interest. First, we investigate the impact of the level of architectural decomposition of the element to be reused. Then, we synthesize a generalized version of the process from a historical survey of design reuse examples and suggest MBSE methods, specifically using SysML, to implement them. Lastly, we discuss where this inheritance process fits in the larger picture of the systematic reuse strategy.

Keywords Design reuse · Design inheritance · Space systems architecting · Model-based systems engineering (MBSE)

1 Introduction and Motivation

The concept of reuse, and especially design reuse, is common in many engineering disciplines. It is particularly important in space systems engineering as high development and operations costs, long development timelines, and inaccessibility

A. E. Trujillo (✉)
Massachusetts Institute of Technology, Cambridge, MA, USA
e-mail: alextruj@mit.edu

A. M. Madni
Viterbi School of Engineering, University of Southern California, Los Angeles, CA, USA

A. M. Madni et al. (eds.), *Recent Trends and Advances in Model Based Systems Engineering*, https://doi.org/10.1007/978-3-030-82083-1_47

of assets once emplaced make space missions highly risk averse. Thus, any action that may reduce risk – in the case of reuse, by leveraging and redeploying proven concepts – is highly valued. Additional benefits of reuse typically cited include reduced design, development, and testing (DD&T) costs and increased responsiveness (reduced time-to-flight from initial need identification).

One commonly cited issue for reuse in the space industry is the often ad hoc nature with which it is implemented. This leads to efforts which may not realize all the potential benefits of reuse. Indeed, in its Systems Engineering Vision for 2025, INCOSE identifies a key challenge currently facing the systems engineering (SE) discipline, namely, that "Knowledge and investments are lost between projects . . . increasing cost and risk" (INCOSE 2014). A lack of systematic understanding of reuse also leads to mistaken attempts to implement reuse in scenarios where it may in fact be detrimental (i.e., where rework and integration costs exceed costs of designing new elements).

Thus, to fully leverage its potential, a systematic approach that looks at reuse from a strategic perspective is required for implementing reuse (Lange et al. 2018). Developing such an approach within the paradigm of MBSE would help realize an aspect of INCOSE's envisioned solution to the SE challenge mentioned above: that "composable design methods in a virtual environment [will] support rapid, agile and evolvable designs of families of products" (INCOSE 2014).

1.1 Problem Statement

This paper is part of a larger research effort concerned with developing a systematic, MBSE-centric methodology for exploring reuse scenarios for space missions and campaigns. Figure 1 depicts the decision-making process and key factors we have previously identified for a generalized reuse scenario (Trujillo and de Weck 2019). It is a balance of both technical and business factors which combine to yield both costs and benefits to reuse. These must then be compared to the status quo (i.e., the non-reuse approach) to yield important insights for decision makers.

In this paper, we specifically look at the *inheritance process* in the technical factors shown in Fig. 1. Inheritance is the process which explores the technical feasibility of adapting design elements from past missions as solutions for new missions being architected. We define a generalized set of procedures in the inheritance process and discuss how these may be implemented in an MBSE environment (Sect. 3). Then, we investigate how that approach can act as a foundational step in larger reuse efforts including composition of new designs from MBSE libraries/catalogs and reconfiguration of designs and product families to meet new mission needs (Sect. 4). First, we begin with a brief survey of the current state of MBSE in the space industry in Sect. 2.

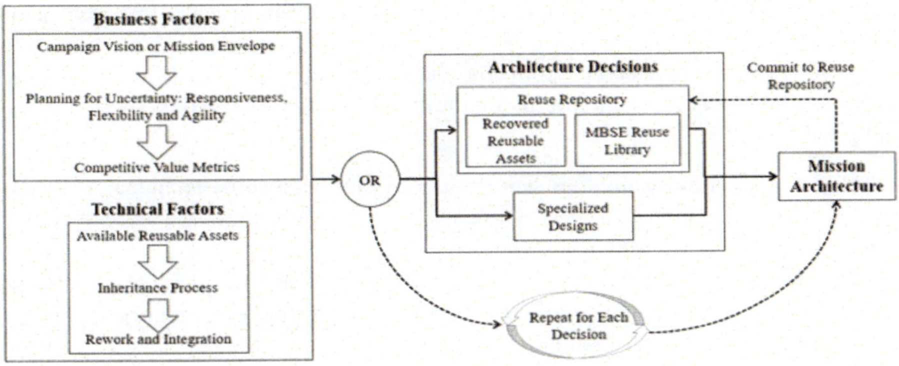

Fig. 1 Key factors and high-level process with which firm may make reuse decisions

2 MBSE, Reuse, and the Space Industry

2.1 MBSE and the Space Industry

Since the mid-2000 timeframe, INCOSE's Space Systems Working Group (SSWG) has been an advocate for demonstrating the potential of MBSE for engineering space systems. The SSWG has collaborated with the space industry, notably with NASA's Jet Propulsion Laboratory (JPL) (Delp 2012), to develop exemplar MBSE capabilities including SysML modeling of hypothetical satellite missions (Kaslow et al. 2015), development of CubeSat reference models (Kaslow and Madni 2017), and application of reference models to real-world CubeSat missions (Spangelo et al. 2012). Partly from these early efforts, the aerospace industry has begun to recognize the value of model-based approaches and increased digitization of system information for complex missions requiring multidisciplinary expertise.

For example, in 2016, NASA conducted a pathfinder effort to develop and advance MBSE capabilities throughout the agency upon discovering that MBSE adoption was still in the nascent stages (Phojanamongkolkij et al. 2017). This led to the application of MBSE methods on various missions with a specific focus on information systems of Orion's Exploration Flight Test 1 (EFT-1) (McVittie et al. 2012), the Asteroid Redirect Mission (ARM) prior to project cancellation, and to a limited extent for the development of the Space Launch System (SLS) (Parrot et al. 2016). Moreover, there is a desire at NASA to demonstrate the capabilities of MBSE at larger scales, including for architecting Moon/Mars campaigns as well as for the development of databases of system models to facilitate future reuse efforts. Beyond NASA, the use of MBSE in space systems development and research has been demonstrated at other national agencies, such as at Germany's DLR for its small satellite multi-mission platform known as S2TEP (Fischer *et al.*, 2017) and for educational efforts such as Pakistan's use of Object-Oriented Systems Engineering

Method (OOSEM) for scalable/repeatable student satellite projects (Waseem and Sadiq 2018).

Upon synthesizing insights from the above examples, we discover that space systems engineering imposes several unique requirements on MBSE methodologies. Among these are:

- System models must integrate a variety of discipline-specific technical analysis tools used by mission designers as a result of the multidisciplinary nature of space systems.
- Interactions within and across all spacecraft systems and mission environments must be modeled to understand performance and capture emergent behaviors – both beneficial and detrimental.
- MBSE methodologies must assess and promote critical value attributes desired of the new generation of space architectures, namely, flexibility, evolvability, reusability, and safety.
- MBSE methodologies must accommodate technological and managerial uncertainties inherent in the long timelines that are typical of space missions and campaigns.

These requirements add to the complexity that MBSE methodologies must address to become value-adding tools for space systems architects. Further, new methodologies must satisfy these requirements in a way that is not mission-specific but rather applicable (i.e., reusable) across diverse missions.

2.2 MBSE and Reuse

We have previously shown in Trujillo and de Weck (2019) that reuse can generally be classified into three broad categories: reuse of *physical artifacts*, reuse of *codified design knowledge*, and reuse of *design effort*. To explore the value of MBSE in enabling reuse, it is important to first define and distinguish the latter two types of reuse. *Codified design knowledge* can be defined as the set of architectural concepts and physical principles (and the relationships between them) that address a particular objective or requirement in a mission of interest. *Design effort* is the set of value-adding work products of the systems architecting process that capture and record design *knowledge. Design efforts* are increasingly being digitized as MBSE system models, metamodels, and patterns. If prudently designed, these are especially amenable to standardization and reuse on missions beyond their original mission.

Metamodels can be defined as the specification (i.e., syntax, structure, relationships) for constructing a system model (Madni and Sievers 2018) .While metamodels can exist at multiple levels of abstraction and for different modeling contexts, a reuse metamodel may be viewed as laying out how the integration of heritage designs may be modeled for a new mission of interest. Conforming to a reuse metamodel is essential for developing the requisite model consistency for building and using model libraries and catalogs. A stretch goal for the usage of such model catalogs is the rapid composition of existing design features or architectural elements into new mission systems. Progress has been made in this direction with the MBSE-based composable design methodology developed at Lockheed Martin and demonstrated on its A2100 line of communications satellites (Kaiser and Oster 2015).

2.3 What Is Missing?

Substantial progress has been made toward broader adoption of MBSE in the space industry. However, usage of MBSE even within companies continues to be a mission-to-mission decision. By building on current work in the area of reference models, metamodels, and pattern-based approaches, it may be possible to further develop MBSE capabilities and foster greater adoption within the space enterprise (Madni and Purohit, 2019). Specifically, in the area of design and model reuse, a productive first step could be the definition and codification of the technical inheritance process.

3 Design Reuse via Inheritance

Most examples in current practice of design reuse in the space industry have not been carried out within an MBSE context. Nevertheless, these examples exhibit commonalities in how they identify reuse candidates, assess their compatibility, and make technical recommendation to decision makers. Thus, it may be possible to synthesize a generalized approach or pattern from these observations which in turn can be used in a systematic MBSE reuse methodology. In what follows, we first develop the steps of the inheritance process and then explore how MBSE techniques can be employed to implement the process, with emphasis on the compatibility assessment step.

3.1 Technical Inheritance Process

Reuse decisions encompass both technical and business considerations. On the technical side, a key analysis supporting that decision-making progress is the determination of the feasibility and technical value of adapting a proven design to satisfy the needs of a new mission. Figure 2 is a SysML activity diagram depicting this generalized technical inheritance process. This diagram was developed from the explored examples of design reuse. In this diagram, solid path lines indicate the flow of information from action to action, while the dashed lines indicate the action-

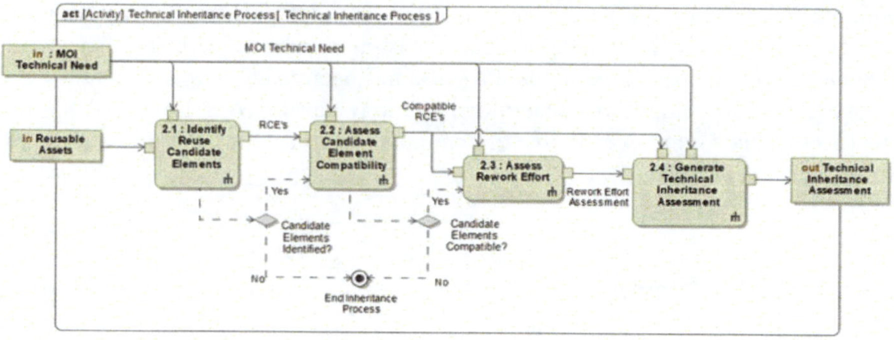

Fig. 2 SysML activity diagram of the technical inheritance process for systematic reuse analyses

token flow which controls whether actions are carried out, depending on conditional rules marked by the diamond-shaped decision nodes.

As shown on the frame of the diagram, the overall inputs to the process are a description of the mission of interest (MOI) and the set of reusable assets available to the developing organization. The MOI description may be sourced from an existing system model or may require digitization of document-sourced information. The MOI contains objectives, system-level requirements, and design decisions made to date. The reusable asset survey may be a compilation of system models for the firm's past and current designs (at an arbitrary level of decomposition). Further development of these inputs to the process is beyond the scope of this paper – it is assumed that these inputs contain sufficient design detail, using a consistent modeling scheme, to enable inheritance analysis. The output of the process is a technical assessment for inheritance of candidate reusable elements with details on required rework and integration (the next technical factor of the reuse process, per Fig. 1). The steps of the process are described below:

Step 1: Identify Reuse Candidate Elements – From the asset survey that acts as input to the process, we first extract those missions whose designs may be applicable to the MOI. This requires mission designers to make a preliminary, subjective judgment on the degree of similarity shared by missions. This judgment includes qualitative comparisons of the major technical and functional overlaps across missions. This acts as a first-pass screening of all of a company's missions recorded in the RAS down to a set of reuse candidate missions (RCMs). Mission designers now step within the identified RCMs to extract specific architectural elements or designs that may be suitable to the MOI. These will likely emerge from the major technical and functional overlaps identified in the previous step and are designated reuse candidate elements (RCEs). Here it is critical to specify the level of decomposition of the element needed to enable inheritance by a new mission. This decompositional level determines the types of interactions that must be considered during adaptation to the new mission and

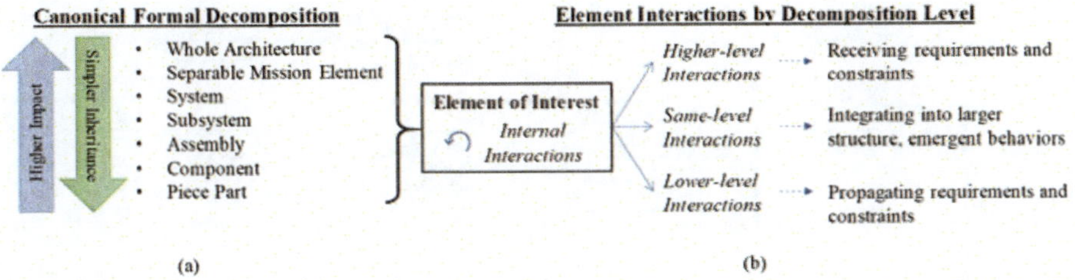

Fig. 3 (**a**) Formal decomposition hierarchy; (**b**) interactions among elements to be investigated/resolved when applying reuse

also defines the potential impacts (in terms of cost and schedule reductions) of reuse decisions.

Figure 3 illustrates (a) a standard architectural breakdown for specifying the RCE's decompositional level within its RCM and (b) a description of interactions that occur between the RCE and the new mission architecture. In (a) we see that while lower-level reuse is generally a simpler practice to implement, greater value-added of reuse is typically expected for higher levels. In (b) we see that these interactions occur at higher, similar, and lower levels of decomposition. Further, the modularity of the reused element determines the balance of internal vs. external interactions. Thus, in order to limit interactions to be considered (a zeroth-order proxy for rework effort), a more modular element would be preferred as a reuse candidate; some interactions cannot be encapsulated in the element of interest, including those dealing with structural, energetic, and informational integration of the RCE into the larger architecture. Additionally, beyond physical modularity, this process must also consider the functional, organizational, and implementation-related aspects of the mission. These also impact the type and degree of interactions that may arise with the reused element.

The output of this step is a listing of RCEs, with decomposition specified, from within the previously identified RCMs along with information about the likely technical overlaps.

Step 2: Assess Candidate Element Compatibility – The preliminary and qualitative judgments made in the previous step are now refined with supporting analyses. A detailed procedure for this compatibility assessment is depicted in the activity diagram shown in Fig. 4. Compatibility, that is, the suitability of an RCE to be redeployed with or without modifications on the MOI, is determined by considering two key indicators: requirements and interfaces. Requirements deal primarily with the functionality and performance of the element considering objectives and design constraints; interfaces capture how the element interacts with other elements in the larger architecture of the RCM or MOI. In both cases, compatibility is assessed by pairing like with like across the native mission and the mission of interest at the appropriate decompositional level.

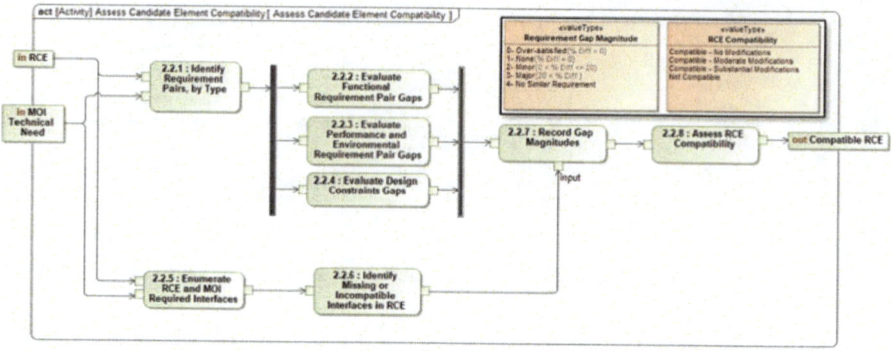

Fig. 4 SysML activity diagram of Step 3: assessment of the RCE's technical compatibility to the new mission context

A pair-wise comparison of relevant requirements recorded for the RCM and MOI yields some combination of (a) similar requirement text with similar metrics, (b) similar requirement texts with dissimilar metrics (i.e., metric gaps), or (c) dissimilar requirement texts. This comparison reveals (a) MOI requirements already satisfied by the RCE, (b) the gap in the RCE's functionality, performance, or design that must be bridged by rework or adaptation efforts, and (c) requirements in the MOI for which no RCE requirement is comparable. In the last case, this indicates where new, from scratch development will be needed to address the unique MOI requirement. RCEs which demonstrate an inability to satisfy the requirements of the MOI (within some rework and adaptation tolerance) are rejected. A similar comparative analysis is conducted for all relevant interfaces – including structural/mounting, data, material, and energy interfaces. Where interfaces are missing or incompatible, additional gaps in compatibility are noted. Gap magnitudes (shown as the inset in Fig. 4) are determined for each requirement and interface pair, and a determination of compatibility is made for each, with qualifiers for estimated rework or modification required for the RCE. The output of this step is an enumeration of "compatible RCEs."

Step 3: Assess Rework Effort – The compatible RCEs that make it to this step are now further evaluated to estimate the amount of rework needed to adapt the RCE to the new MOI context. This analysis begins with the technical gaps (as captured in requirements and interface specifications) identified in the previous step. The reuse engineer then solicits inputs from subject matter experts and discipline engineers about corrective actions necessary to bridge these gaps. For each requirement or interface pair, a set of rework actions is generated. Supporting domain analyses (e.g., thermal, structural, communications, etc.) are integral to this step in order to verify if each rework action is necessary and appropriate. Such analyses ideally would be integrated into the underlying MBSE model which acts as the authoritative data source for necessary parameters and design elements. After determining the feasibility of the rework regime on the RCE and its impact on programmatic aspects such as testing campaigns, a high-level rework work breakdown structure is generated. RCEs for which the rework effort is deemed too extensive or infeasible are rejected. The rest are passed on to the final step of the process.

Step 4: Generate Technical Inheritance Recommendation – The full set of data on an RCE, from preliminary compatibility judgment to an enumeration of necessary adaptation, is now available to the user to develop a technical inheritance summary. These assessments are compiled for the RCEs considered and form on half of the reuse report on an item; the report details those RCEs that are most likely to be adaptable to the needs of the current mission, the rationale for such findings including supporting technical analyses, and estimates (including a preliminary work breakdown structure) for rework efforts required. The other half of the methodology is a valuation and programmatic impact assessment which is ultimately delivered to decision makers. The reuse valuation analysis (which is out of the scope of this paper) relies on the outputs of the technical inheritance process detailed here in order to determine cost and schedule impacts of the reuse scenario.

3.2 MBSE Methods for Compatibility Assessments

While above we depicted the technical inheritance process using SysML formalisms, we now explore how the steps in that process might actually be implemented. Specifically, we explore methods for carrying out Step 3, the RCE compatibility assessment in a SysML environment. Here, it is assumed that a system model of the MOI is being developed and maintained by the hypothetical firm. Additionally, the firm has also developed system models for the RCMs it is exploring (or at the very least, has gathered required information in a traditional, document-centric format). Table 1 describes possible options for each of the sub-steps in the compatibility assessment step as laid out in Fig. 4. It is left to future work to attempt, evaluate, and select a set of MBSE implementation options for each of the sub-steps.

4 Inheritance and MBSE in Larger Systematic Reuse Strategy

The approach developed in Sect. 3 is a building block for a more comprehensive, systematic reuse methodology. The technical inheritance process provides decision makers with sound technical judgments to support prudent reuse efforts. This will be augmented with additional analysis modules for consideration of nontechnical business and programmatic aspects which place an engineering project in a real-

Table 1 Enumeration of MBSE implementation options for sub-steps within the compatibility assessment step

Comparison	Description	MBSE (SysML-specific) implementation options
Functional requirements	Form neutral comparison of delivered vs. desired functionality	• *RD*: compare RCM requirement satisfied by RCE vs. similar requirements in MOI • *BDD*: logical/functional decomposition of architectures + comparison • *UCD*: scenarios for usage of RCE vs. desired element in MOI
Performance and environmental requirements	Does the RCE meet the technical performance required by the MOI objectives? And will it perform in the environmental conditions of the MOI?	• *RD*: compare RCM requirement satisfied by RCE vs. similar requirements in MOI • *PD*: technical analyses of performance metrics and operability of RCE compared with MOI requirements and environmental factors
Physical/design constraints	Is this RCE compatible with the physical and design constraints imposed by the MOI (e.g., SWaP, nonengineering constraints, etc.)?	• *RD*: compare RCM constraint satisfied by RCE vs. similar constraint in MOI • *BDD + Constraints*: impose and check satisfaction of constraints on structure of system
Interfaces	Comparison of material/structural, material, data, or energy interfaces across the RCE and MOI. Are interfaces missing?	• *IBD*: generate/compare interface diagrams from physical decomposition of RCE and MOI models • *PD*: similar to IBD, can "size" interfaces with relevant FOMs

Key: *BDD* block definition diagram, *UCD* use case diagram, *AD* activity diagram, *RD* requirements diagram, *PD* parametric diagram, *IBD* internal block diagram, *SD* sequence diagram

world environment. The inheritance process may be carried out multiple times, in parallel, for various architectural decisions in complex space missions. Thus, future research efforts will take the process laid out here and "zoom out" from the individual element level to the various reuse decisions made throughout the life of a project.

The horizon goal for the framework is to address two major use cases: First, the predecessor-successor use case is a standard heritage reuse scenario where multiple design elements in a current MOI have been adapted from past missions after technical judgments and additional analyses have deemed them compatible and value-adding. Second, the composable design use case is a more agile scenario in which a model library enables rapid configuration of new mission systems from a catalogue of existing and proven capabilities, payloads, infrastructure systems, etc. In either case, the inheritance process is instrumental in rendering the technical judgments about the underlying adaptability of a proven design to a new mission.

5 Limitations, Future Work, and Conclusions

There are several limitations of the generalized inheritance process as developed here. These limitations arise from certain simplifying assumptions that were made. The most important assumption for applying the procedure in an MBSE environment is that consistently represented system models exist for both the MOI and the RCM. These capture sufficient detail to make the necessary comparisons across architectures as well as the supporting technical analyses. The limited level of adoption of MBSE in the industry currently means that this is an unrealistic assumption. It also makes it difficult to test and validate the proposed procedures on real-world examples. Practically, we must accommodate for cases where system information is captured in a variety of locations and inconsistent formats. Secondly, as demonstrated in Table 1, the SysML specification currently allows for multiple ways of accomplishing similar modeling tasks; these alternatives must be explored and down-selected. More immediate future work will seek to further elaborate the other steps in the inheritance process, as was done for the compatibility assessment step.

Codification of the technical inheritance process is an initial step in a larger research effort. Our goal is to enable space mission and campaign architects to fully leverage the potential of reuse of proven designs and concepts. MBSE is a powerful paradigm and SysML is a powerful tool for developing and implementing systematic approaches for realizing reuse benefits while avoiding its pitfalls.

References

Delp, C. 2012. *INCOSE Space Systems Challenge Team and JPL MBSE*. Pasadena.

Fischer, P.M., et al. 2017. Implementing model-based system engineering for the whole lifecycle of a spacecraft. *CEAS Space Journal* 9 (3): 351–365. https://doi.org/10.1007/s12567-017-0166-4.

INCOSE. 2014. *Systems Engineering Vision 2025 – A World in Motion*.

Kaiser, M.J., and C. Oster. 2015. Managing a satellite product line utilizing composable architecture modeling. In *AIAA SPACE 2015 Conference and Exposition*, 1–10. https://doi.org/10.2514/6.2015-4436.

Kaslow, D., and A.M. Madni. 2017. Validation and verification of MBSE compliant CubeSat reference model. *Disciplinary Convergence in Systems Engineering Research*: 381–393. https://doi.org/10.1007/978-3-319-62217-0_27.

Kaslow, D., et al. 2015. Developing and distributing a Cubesat Model-Based Systems Engineering (MBSE) reference model. In *31st Space Symposium, Technical Track*. Colorado Springs.

Lange, C., et al. 2018. Systematic reuse and platforming: Application examples for enhancing reuse with model-based systems engineering methods in space systems development. *Concurrent Engineering Research and Applications* 26 (1): 77–92. https://doi.org/10.1177/1063293X17736358.

Madni, A., and S. Purohit. 2019. Economic analysis of model-based systems engineering. *Systems* 7 (1): 12. https://doi.org/10.3390/systems7010012.

Madni, A.M., and M. Sievers. 2018. Model-based systems engineering: Motivation, current status, and research opportunities. *Systems Engineering* 21 (3): 172–190. https://doi.org/10.1002/sys.21438.

McVittie, T.I., O.V. Sindiy, and K.A. Simpson. 2012. Model-based system engineering of the Orion flight test 1 end-to-end information system. In *2012 IEEE Aerospace Conference*, 1–11. IEEE. https://doi.org/10.1109/AERO.2012.6187440.

Parrot, E., et al. 2016. *NASA GRC MBSE Implementation Status*.

Phojanamongkolkij, N., et al. 2017. Modeling to Mars: a NASA model based systems engineering pathfinder effort. In *AIAA SPACE and Astronautics Forum and Exposition*, 1–15. Orlando, FL. https://doi.org/10.2514/6.2017-5235.

Spangelo, S.C., et al. (2012) Applying model based systems engineering (MBSE) to a standard CubeSat. In *IEEE Aerospace Conference Proceedings*, 1–20. Big Sky: IEEE. https://doi.org/10.1109/AERO.2012.6187339.

Trujillo, A.E., and O. L. de Weck. 2019. Towards a comprehensive reuse strategy for space campaigns Alejandro Trujillo. In *70th International Astronautical Congress*, 21–25. Washington, DC.

Waseem, M., and M.U. Sadiq. 2018. Application of model-based systems engineering in small satellite conceptual design-A SysML approach. *IEEE Aerospace and Electronic Systems Magazine* 33 (4): 24–34. https://doi.org/10.1109/MAES.2017.180230.

Part VII
Future of MBSE

Models in Systems Engineering: From Engineering Artifacts to Source of Competitive Advantage

Azad M. Madni

Abstract Models in systems engineering have existed in various forms dating back to the 1950s. They have been used by engineers to understand various types of phenomena, envision future systems, and generate engineering artifacts. Today the increasing complexity of operational missions and technological advances enabled in part by disciplinary convergence and wide access to data are having a dramatic impact on system modeling. Operational missions are becoming increasingly more complex with multiple sources of uncertainty and subject to a variety of disruptions. Technological advances paced by advances in semantic technologies, machine learning, AI, and applied analytics are transforming model development into a closed-loop process. The advent of Industry 4.0 and digital engineering (including digital twin and digital thread) is causing models to be viewed in an entirely new light. And the convergence of engineering with other disciplines is opening up a whole new way of developing system models. This paper presents a historical perspective on models over several decades and offers a vision of how recent developments are likely to shape the trajectory of system models in the future.

Keywords Engineering models · Deterministic models · Probabilistic models · Learning models · Digital engineering · Industry 4.0

1 Introduction

Models, which have been a mainstay of engineering, are becoming central to systems engineering (SE) with the advent of model-based systems engineering (Madni and Sievers 2018). The questions today are determining where system modeling is headed and what impact it is likely to have on systems and enterprises.

A. M. Madni (✉)
University of Southern California, Los Angeles, CA, USA
e-mail: azad.madni@usc.edu

A. M. Madni et al. (eds.), *Recent Trends and Advances in Model Based Systems Engineering*, https://doi.org/10.1007/978-3-030-82083-1_48

These are some of the questions that the SE community is interested in as SE is being transformed to address the needs and challenges of the twenty-first century.

At the outset, it is worth reminding ourselves of George Box's (1976; Box and Draper 1987) famous refrain, "All models are wrong, but some are useful." The entire quote is: "All models are approximations. Essentially, all models are wrong, but some are useful. However, the approximate nature of the model must always be borne in mind." Box followed up on this cautionary comment with a more insightful and actionable refrain: "Remember that all models are wrong; the practical question is how wrong do they have to be to not be useful." This quote essentially introduces the concept of "model fidelity." In SE, model fidelity is largely driven by the phase in SE life cycle and the intended purpose of the model (i.e., questions the model is expected to help answer).

Against the foregoing backdrop, this paper reviews the chronology of models over the past 60 years – first in engineering, then in SE, and most recently in model-based systems engineering (MBSE). It examines recent business and technology trends that are likely to shape the trajectory of system modeling in the next decade and what they foreshadow for system modeling in the twenty-first century.

This paper is organized as follows. Section 2 discusses the history of models in engineering. Section 3 presents models and modeling advances in SE. Section 4 presents the expanding role of models in MBSE. Section 5 discusses the growing importance of ontologies, knowledge graphs, metamodels, and reference models in MBSE. Section 6 reviews how models have evolved over the last several decades. Section 7 takes a look over the horizon to portend future advances in models. Section 8 summarizes the key points made in this paper.

2 Models in Engineering

Models have been used to envision architectures and engineering artifacts from time immemorial. In the early days, models took the form of sketches, which were a prelude to building physical artifacts. Over the ensuing years, models started to become increasingly more structured. Over the past 50 years, the importance of standardized representation, syntactic correctness, consistent semantic conventions, and the need to enforce semantic consistency in models was gradually recognized. This recognition enabled models to progress beyond drawings and concept maps to computer-based representations with a standard lexicon (vocabulary), syntax (grammar), and semantics (meaning). As important, the use of computer-based models expanded to the understanding of real-world phenomena as models grew in sophistication. Today models are being used to study and build complex sociotechnical systems with the ability to dynamically adapt, learn, and improve (Madni et al. 2018a).

Models are fundamentally abstractions (i.e., a simplified representation of reality) in which the simplifications are achieved through purposeful suppression of irrelevant real-world details (i.e., details that do not contribute to answering questions at hand). Abstractions can take a variety of forms, including generalizations from specific instances, uniform suppression of details not relevant to the purpose of the model, and selective suppression of details (e.g., elimination/simplification of certain perspectives or functions) to reduce complexity of envisioned systems or phenomenon under study (Madni et al. 2018b).

Models can be descriptive, prescriptive, or predictive. *Descriptive* models represent or explain a phenomenon, problem situation, or system to enhance shared human understanding and facilitate collaboration. *Prescriptive* models specify required/desired behaviors or courses of action. *Predictive* models facilitate exploration and help illuminate future outcomes in response to what-if assumptions and decisions/actions. Model purpose (i.e., the questions we want the model to answer at a desired level of detail) determines the scope and fidelity of models needed.

Today models are used in engineering analysis and design to visualize envisioned systems or modifications to existing systems; specify structure and behavior of systems; and understand how parts of a system inter-relate and behave in relation to each other and the external world. Models are also used to guide system development, assemble parts, and identify/generate and evaluate alternatives during design. And, finally, models are used to maintain an audit trail of assumptions and design decisions during system development (Madni et al. 2019).

3 Models in Systems Engineering (SE)

Models in SE have generally followed the evolution of models in traditional engineering disciplines but with a time lag. Over the years, models have appeared in a variety of forms: a "back of the envelope" calculation to ballpark a solution; a sketch on a napkin to communicate a germinating idea or an evolving concept; a computational algorithm to describe a physical law that lends itself to mathematical description; a deterministic representation to describe systems with known cause and effect; a probabilistic representation to capture environmental uncertainties and uncertainties in the knowledge of the system state space, as well as to account for random events; a statistical model to parsimoniously summarize the data collected over space and time; an architectural model to depict system structure and behavior and conduct trade-off analyses among performance and quality attributes associated with a system in its operational context; a logical model to describe how entities relate to each other in implementation-independent form; a data model to represent data in abstract form and organize and standardize how the data elements relate to each other; and a "learning" model that employs supervised, unsupervised, and reinforcement learning to increase model accuracy at system "build-time" and during system "run-time" (i.e., operational use).

Table 1 presents an approximate timeline of modeling methods that have been employed over time to deal with increasing problem and system complexity. Modeling in SE began with the ubiquitous block diagram or black box model,

Table 1 Rough chronology of models in systems engineering

Modeling Construct Name	Year Originated/Discipline	Application within SE (approx.)
Black Box (block diagrams)	1945/electronic circuit theory	1950
Functional Flow Block Diagram	1955/systems engineering	1955
Petri Nets	1962/concurrent hardware communication	1977
Hidden Markov Model/POMDP	1965/operations research, robotics	2015
Reinforcement Learning	1965/psychology, robotics	2014
Structure Analysis and Design Technique	1969/software and systems engineering	1969
Linear Temporal Logic	1977/computer science	1980
Data Flow Diagram	1979/software engineering	1980
N2 Diagram/Design Structure Matrix	1980/software and hardware design	1990
IDEF0	1981/manufacturing	1982
Contract-Based Design	1986/design automation	2000
State M	1987/computation theory	1990
Axiomatic Design	1990/system design	2002
Unified Modeling Language (UML)	1996/software engineering	1997
Digital Twins	2002/manufacturing	2019
Flexible Contract Approach	2010/design automation	2014

in which blocks represented system components and the arcs between the blocks represented the exchange of information, energy, and physical artifacts. Similarly, the N2 diagram and later the design structure matrix (DSM) gained popularity as a parsimonious way to represent a system along with its interactions and dependencies. In systems engineering, the N2 diagram was interpreted from a functional perspective, with the components in the N2 diagram being replaced with major system functions. Thereafter, the functional flow block diagram (FFBD) was developed to capture the dynamics of system behavior in a multitier, time-sequenced flow diagram depicting a system's functional flow. In the meantime, the software engineering community was engaged in developing data flow diagrams (DFDs) to model the data flow aspects of software systems. In the 1969–1973 timeframe, Structured Analysis and Design Technique (SADT) came into being within systems engineering and software engineering methodologies to describe systems as a hierarchy of functions (Ross 1977). It was subsequently formalized and published as Integrated Computer-Aided Manufacturing (ICAM) Definition, or

$IDEF_0$, in 1981 (Marca and McGowan 1987; Davis 1992; Mylopoulos 2004). The $IDEF_0$ representation was primarily championed by the USAF as a viable way to model systems. Not long after, several structured approaches emerged including structured programming, structured design, and structured analysis. It is worth noting that DFDs, N2 charts, and $IDEF_0$ diagrams all capture the same time-lapsed flow of information, energy, and physical artifacts among functions. Then came the recognition of the importance of system states and the advent of state machines (or state transition diagrams), which were adapted by several engineering disciplines to capture dynamic behavior. These methods were rapidly adopted and applied by the SE community to model system modes and states. It soon became evident that state machines suffered from a combinatorial explosion in their state space compromising their scalability. To ameliorate this problem, the SE community turned to heuristics, meta-rules, Petri nets, and Petri net variants (Zisman 1978). This strategy delayed the combinatorial explosion but did not eliminate it.

The past six decades have seen several contributions to systems modeling from a variety of disciplines such as electrical engineering, operations research, design automation, manufacturing, and software engineering. For example, formal modeling approaches for representing, analyzing, and designing systems originated in software engineering and design automation. The early modeling work relevant to SE, which drew on mathematical system representations, includes modeling formalisms (Tarski 1955), homomorphic relational structures (Klir 1991; Lin 1999), axiomatic design (Suh 1998), and structured analysis and design (Yourdon 1989).

It is interesting to note that many of the models being used in systems engineering had their origins in other disciplines such as electronic circuit theory, operations research, software engineering, design automation, robotics, and manufacturing. Also, some modeling approaches from other disciplines were adopted quickly by systems engineers, while others took more than a decade. This time lag was essentially a function of the need expressed by the SE community. For example, increasing system complexity and emphasis on system safety led systems engineers to employ formal and probabilistic methods to address verification and validation needs and uncertainties in knowledge of system states and the environment. Similarly, the advent of machine learning was only recently adopted by the SE community when it became apparent that many complex systems operate in uncertain, partially observable environments in which incoming information from sensor onboard vehicles and the environment help reduce the uncertainty in system models.

4 Models in Model-Based Systems Engineering (MBSE)

The historic use of models in SE is best characterized as "engineering with models." The basic idea is that models from different disciplines can be integrated to provide a solution to a problem that cuts across multiple disciplines (e.g., electrical, mechanical, thermal, optical). However, these models, based on different assumptions, were not designed with integration in mind. Model-based systems engineering (MBSE) is different from engineering with models. In MBSE, models represent an enduring and authoritative source of truth. MBSE replaces the traditional document-centric approach to SE while ensuring that documents can be produced on demand from the unified model and from the perspective of different stakeholders. This is similar to the automatic generation of code on demand in Model-Based Software Engineering. In MBSE, the models have shared assumptions and shared (or compatible) underlying ontologies and representations. Models in MBSE are centralized digital repositories that interconnect information from multiple sources and disciplines such that a change in one part of the model can be traced back to the original/derived requirement or use case. SysML, an extension of a subset of UML developed by the International Council on Systems Engineering (INCOSE) and subsequently advanced by Object Management Group (OMG) and INCOSE, became the popular system modeling language.

Over the past decade and a half, several MBSE methodologies have emerged (Estefan 2008). They include as follows: IBM Rational Unified Process (RUP) supported by IBM Rational Suite; INCOSE Object-Oriented Systems Engineering Method (OOSEM) developed with extensive aerospace involvement, which is supported by commercial SysML tools; Vitech's CORE product suite; JPL State Analysis Methodology; Dori's Object-Process Methodology (OPM); INCOSE MBSE Initiative, OMG's Model-Driven Architecture; and ISO/IEC 42010.

Today MBSE remains an important augmentation of SE as it continues to address additional phases of the system life cycle such as verification, validation, and testing. In this regard, the advent of digital twins (from digital engineering) can be expected to facilitate and accelerate system life cycle coverage (Madni et al. 2019).

5 Growing Importance of Ontologies, Knowledge Graphs, Metamodels, and Reference Models

Two key problems being addressed by the SE community today are to eliminate the miscommunications that frequently occur within SE teams and to assure interoperability among models and between information systems of collaborators. This focus led to the growing importance of ontologies, knowledge graphs, metamodels, and reference models.

Ontologies are thesauri of words representing concepts, the relationships among them, and the rules that help with model correctness checking (Sowa 1996, 2011). The model checking rules help with identification of gaps and semantic inconsistencies in system models. Ontologies have been a subject of study in the systems engineering community as a means to reduce modeling complexity, facilitate model verification, and enhance interoperability. From a data perspective, ontologies are semantic data models that define the types of entities in a particular domain, the relationship among the entities, and the properties that can be used to describe the entities. Ontologies are generic data models in that they only model generic types of entities that share certain properties but do not include information about specific entities in the domain. For example, an ontology might focus on generic vehicles, attempting to capture characteristics that most vehicles might have. By capturing information in this way, the ontology can be used to describe other vehicles in the future. An ontology comprises three main elements: classes, which are distinct types of entities that exist in the domain; relationships, which link any two classes; and attributes, which are properties that describe an individual class. When classes are linked through relationships, the ontology can be visualized as a graph.

A *knowledge graph* acquires and integrates information into an ontology and applies a reasoner to derive new knowledge (Ehrlinger and Wöß 2016). In other words, knowledge graphs are instantiations of ontologies. Using an ontology as an organizing framework, real data about specific entities in the domain can be added to create a knowledge graph. When data about specific entities are added for all entities in the ontology, a knowledge graph emerges. In other words, a knowledge graph is created when an ontology is used as an organizing construct for real-world data. Thus,

$$\boxed{\text{Ontology} + \text{Data} = \text{Knowledge Graph}}$$

Metamodels define the abstract syntax (i.e., grammar) of model description languages (Sprinkle et al. 2014). For example, the Unified Modeling Language (UML) metamodel defines the abstract syntax of various UML diagrams. More generally, metamodels express the logical syntactical structures that domain-specific models need to conform to for scalability, reuse, and extensibility. Metamodels are concerned with defining the symbols and structure for a predefined class of problems, along with rules that operate on the symbols. These properties allow the instantiation of a model from a metamodel. Thus, a metamodel defines the general structure, constraints, and symbols that can be used to model a system. Since metamodels do not specify the semantics of models, they do not have stand-alone use. However, ontologies and metamodels are complementing and synergistic. Specifically, an ontology can represent concepts and relationships formally using the structure provided by the metamodel. While ontologies may not use a metamodel, those that do will have certain desirable properties (e.g., interoperability, reuse, syntactic correctness, semantic consistency).

Reference models are abstract frameworks or domain-specific ontologies consisting of an interlinked set of clearly defined concepts produced by authoritative sources within defined stakeholder communities. A reference model can represent business functions and system components as long as they constitute a complete set. The terms in the reference model can be used to communicate ideas clearly among members of the SE community from vastly different backgrounds. The reference model is distinct from, but can include, related taxonomies, entities, and relationships to reveal hierarchies (e.g., system hierarchy, system architecture) relevant to stakeholders.

6 How Have Models Changed over the Last Several Decades?

After more than 50 years, system modeling has evolved in several important ways shown below:

- The starting point for modeling has changed from choosing a modeling construct to starting with a detailed analysis of needs to derive system modeling requirements which are then used to determine the right combination of models needed to model the system of interest.
- The scope of modeling has expanded – from a single system to networked systems, system of systems, and enterprises.
- Models have grown in sophistication – from deterministic to stochastic, probabilistic, and learning models.
- Engineering models used to be rooted in the engineering discipline. Today they are drawing on other disciplines such as biology, cognitive science, social science, economics, and entertainment arts.
- Earlier models used to be "data hungry." They needed complete information before they could provide value. Today models can cope with partial information and still provide value.
- Modeling used to be an open-loop process. Today modeling is being transformed to a closed-loop process that improves model completeness and accuracy based on data from collection assets, machine learning, and data analytics techniques. For example, virtual system models can now incorporate data from the corresponding physical system and become a digital twin (Madni et al. 2019).
- System representations have expanded from fixed structures to flexible representations which are needed to respond to systemic problems and adapt to external disruptions.
- System models are becoming increasingly more formal and rigorous to enable verification and validation, support simulation-based testing, and facilitate reasoning, interoperability, and reuse.
- Models are beginning to incorporate the capability to explain system behavior, an important characteristic that is key to model acceptance and trust in the engineering community. Explanation capability is needed for black box models, while interpretability is needed for glass box models.
- The SE community is much more cost conscious, with an emphasis on economic value derived from transitioning to MBSE (Madni and Purohit 2019).
- Industry view of models has changed from viewing them solely as engineering artifacts to viewing them as knowledge assets and a source of competitive advantage.

Today models are expanding into the behavioral domain. The human is no longer modeled as a transfer function, optimal controller, or utility maximizer. Rather, the human is modeled with an awareness of strengths (e.g., ability to generate creative options, rapid context awareness) and limitations (e.g., cognitive limitations, biases, tendency to lose focus). The advent of cyber-physical-human systems is a driver in

Table 2 System models: pre-2005 and today

System models	Pre-2005	Today
Comparison factors		
Starting Point	a modeling construct (e.g., IDEF0, SADT)	requirements derived from needs and mapped to appropriate combination of models
Focus	single system	networked systems, SoS, enterprise
Methods	deterministic (mostly)	deterministic, stochastic, probabilistic
Multidisciplinary Emphasis	minimal	significant
Model Requirements	need complete information	can work with partial information (e.g., POMDP)
Process	open loop	closed loop
Representation	fixed	flexible
Correctness Proof	not available	available
Rigor	structured representation; static correctness checking	formal representation; use of ontology and metamodel; support for formal reasoning
Learning	a priori supervised learning	in situ unsupervised and reinforcement learning
Explanatory Capability	none	some; distinguishes between interpretability (glass box models) and explainability (black box models)
Emphasis on ROI	modest	significant
Industry View	engineering artifact	knowledge asset; source of competitive advantage

this regard. As important, the age-old thinking of "humans versus machines" has been replaced by "humans and machines" with a growing emphasis on augmented intelligence (Madni 2020a).

Table 2 provides a comparison of pre-2005 system models and system models today.

7 Looking over the Horizon

With systems continuing to grow in complexity and missions continuing to become increasingly more challenging, the versatility and value of a model depend on its ability to provide useful information despite incomplete information; ability to acquire and reflect valid information pertaining to key system characteristics and behaviors of interest; ability to support simplifications (e.g., assumptions, approximates) while retaining requisite fidelity to provide correct answers; and ability to validate its outputs. Importantly, models require real-world measurements to test the validity of their predictions and explanations and for validation of outputs. It may not be feasible to meet these requirements in circumstances where input conditions cannot be adequately controlled or input and control conditions cannot be replicated.

With the recent surge in interest to transform and underpin SE with formal methods (e.g., linear temporal logic, contract-based design), three necessary characteristics of models surfaced: provably correct representation essential in applications where safety is paramount; flexible representation to support agility and resilience; and evidence-based learning to complete and refine models. Learning ability is crucial when operating in partially observable environments in which information about the system and the environment becomes incrementally available during mission execution. In response to these requirements, probabilistic learning models emerged including the flexible contract approach with the capacity to learn (Sievers and Madni 2016; Madni 2018a). This construct combines traditional contracts, partially observable Markov decision process (POMDP), reinforcement learning, and heuristics to strike an effective balance between model verifiability and flexibility (Sievers and Madni 2017; Madni et al. 2018a, b).

In the light of methodological advances and ongoing integration of MBSE with digital engineering, systems modeling can be expected to evolve in new and exciting directions. We already see evidence of formal methods being introduced within the MBSE rubric. Specifically, the concept of ontologies from computer science is being introduced into MBSE to enhance semantic consistency, enhance interoperability, and formalize scope with respect to the system modeling activity. In particular, ontologies can be expected to play important roles in answering stakeholder/user questions by capturing key concepts and relationships from use cases of interest and supplemented by expert knowledge. The scope of system modeling can be expected to expand to cover probabilistic modeling, formal modeling, modeling with incomplete or partial information, and learning models (i.e., supervised, unsupervised, and reinforcement learning). Models can be expected to have richer semantic foundations to reflect new perspectives made possible by disciplinary convergence. These advances and enhancements will enable more detailed questions to be answered earlier in the system's life cycle. With growing convergence of engineering with entertainment arts, it will be possible to transform system models into stories that can be executed in simulation or in virtual worlds (Madni et al. 2014; Madni 2015). Importantly, enterprises are beginning to increasingly rely on their suppliers, application providers, and tool vendors to create sustainable competitive advantage in their respective markets. This reliance calls for seamless interoperability. The latter can be achieved through models based on domain ontologies with interoperability being enabled by creating a semantic layer between enterprises and their technology/tool providers (Madni 2020b). As a result, system models are no longer being viewed as engineering artifacts but rather as knowledge assets and a source of competitive advantage for organizations.

8 Summary

Models have been a mainstay of systems engineering for several decades. However, the types of models and the value they provide have changed dramatically. The types of models used for system modeling have evolved considerably driven in large part by the increasing complexity of the system and the environment and advances made in formal and probabilistic methods, machine learning, and applied analytics. These advances have transformed modeling from being a one-shot open-loop activity to an iterative closed-loop activity informed by evidence and results of machine learning. Importantly, the historical view of models as engineering artifacts has changed dramatically. Today they are viewed as sources of competitive advantage. The competitive advantage results from the ability to reuse models, completely or in part, to rapidly achieve interoperability in risk-mitigated fashion with third-party applications and tools and enable the use of ontologies and metamodels. The growing importance and adoption of MBSE in major organizations coupled with the advent of digital engineering make digital twin-enabled MBSE especially effective for model-based V&V. This paper has addressed both modeling problems and how far along systems modeling has advanced as a result of problem pull and enabled by advances in system modeling and ongoing convergence of systems modeling with machine learning, data analytics, and entertainment arts (Madni 2018b). This trend can be expected to continue and grow in the future. As a result of these advances, systems models are becoming knowledge assets and a source of competitive advantage in various industries.

References

Box, G.E.P. 1976. Science and Statistics. *Journal of American Statistical Association* 71 (356): 791–799.

Box, G.E.P., and N.R. Draper. 1987. *Empirical Model Building and Response Surfaces*. New York: Wiley.

Davis, W.S. 1992. *Tools and Techniques for Structured Systems Analysis and Design*. Addison-Wesley. ISBN 0-201-10274-9.

Ehrlinger, L., and W. Wöß. 2016. Towards a Definition of Knowledge Graphs. In *SEMANTICS*, September 13–14, Leipzig, Germany.

Estefan, J. 2008. INCOSE Survey of MBSE Methodologies, USA, WA, Seattle: INCOSE TD 2007-003-02.

Klir, G. 1991. *Facets of Systems Science*. New York: Plenum.

Lin, Y. 1999. *General Systems Theory: A Mathematical Approach*. New York: Kluwer Academic/-Plenum.

Madni, A.M. 2015. Expanding Stakeholder Participation in Upfront System Engineering Through Storytelling in Virtual Worlds. *Systems Engineering* 18 (1): 16–27.

———. 2018a. Formal Methods for Intelligent Systems Design and Control. In *AIAA SciTech Forum, 2018 AIAA Information Systems, AIAA InfoTech@Aerospace*, Kissimmee, Florida, January 8–12.

———. 2018b. *Transdisciplinary Systems Engineering: Exploiting Convergence in a Hyperconnected World (forward by Norm Augustine)*. Springer, September.

———. 2020a. Exploiting Augmented Intelligence in Systems Engineering and Engineered Systems. *INSIGHT Special Issue, Systems Engineering and AI*, March.

———. 2020b. Minimum Viable Model to Demonstrate the Value Proposition of Ontologies for Model Based Systems Engineering. In *2020 Conference on Systems Engineering Research (CSER)*, October 8–10.

Madni, A.M., and S. Purohit. 2019. Economic Analysis of Model Based Systems Engineering. In *MDPI Systems, special issue on Model-Based Systems Engineering*, February.

Madni, A.M., and M. Sievers. 2018. Model-Based Systems Engineering: Motivation, Current Status, and Research Opportunities, Systems Engineering. In *Special 20th Anniversary Issue*, Vol. 21, Issue 3.

Madni, A.M., M. Sievers, and D. Erwin. 2019. Formal and Probabilistic Modeling in the Design of Resilient Systems and System-of-Systems. In *AIAA Science and Technology Forum*, San Diego, California, January 7–11.

Madni, A.M., M. Spraragen, and C.C. Madni. 2014. Exploring and Assessing Complex System Behavior Through Model-Driven Storytelling. In *IEEE Systems, Man and Cybernetics International Conference, invited special session Frontiers of Model Based Systems Engineering*, San Diego, CA, October 5–8.

Madni, A.M., M. Sievers, A. Madni, E. Ordoukhanian, and P. Pouya. 2018a. Extending Formal Modeling for Resilient System Design. *Insight* 21 (3): 34–41.

Madni, A.M., M. Sievers, and C.C. Madni. 2018b. Adaptive Cyber-Physical-Human Systems: Exploiting Cognitive Modeling and Machine Learning in the Control Loop. *Insight* 21 (3): 87–93.

Madni, A.M., C.C. Madni, and D.S. Lucero. 2019. Leveraging Digital Twin Technology in Model-Based Systems Engineering. In *MDPI Systems, special issue on Model-Based Systems Engineering*, February.

Marca, D., and C. McGowan. 1987. *Structured Analysis and Design Technique*. McGraw-Hill. ISBN 0-07-040235-3.

Mylopoulos, J. 2004. Conceptual Modelling III. Structured Analysis and Design Technique (SADT).

Ross, D.T. 1977. Structured Analysis (SA): A Language for Communicating Ideas. *IEEE Transactions on Software Engineering* SE-3 (1): 16–34.

Sievers, M., and A.M. Madni. 2016. Agent-Based Flexible Design Contracts for Resilient Spacecraft Swarms. In *AIAA Science and Technology 2016 Forum and Exposition*, San Diego, CA.

———. 2017. Contract-Based Byzantine Resilience for Spacecraft Swarm. In *2016 AIAA Science and Technology Forum and Expo*, Grapevine, Texas, January 9–13.

Sowa, J.F. 1996. Top-Level Ontological Categories. *International Journal of Human-Computer Studies* 43 (5/6): 669–686.

———. 2011. Ontology Metadata and Semiotics. In *Conceptual Structures: Logical, Linguistic, and Computational Issues, LNAI 1867*, ed. B. Ganter and Mineau, 55–81. Berlin: Springer.

Sprinkle, J., B. Rumpe, H. Vangheluwe, and G. Karsai. 2014. Metamodeling: State of the Art and Research Challenges. *arXiv*, September.

Suh, N.P. 1998. *Axiomatic Design Theory for Systems in Research in Engineering Design*. Vol. 10, 189–209. Berlin: Springer.

Tarski, A. 1955. Contributions to the Theory of Models I II II. *Nederlandse Akademie Wetenschappen Proceedings Series A* 57: 572–581.

Yourdon, E. 1989. *Modern Structured Analysis*. Upper Saddle River: Yourdon Press.

Zisman, M. 1978. *Use of Production Systems for Modeling Asynchronous Concurrent Processes, Pattern-Directed Inference Systems*, 53–68. Academic.

Transdisciplinary Systems Engineering Approaches

Bryan Mesmer, Doroth Mckinney, Michael Watson, and Azad M. Madni

Abstract Transdisciplinary systems engineering provides new ways of thinking when developing engineering solutions to complex systems problems. The solution space afforded through the use of transdisciplinary approach tends to be broader than that achievable with traditional approaches. This expanded solution space is a result of new ways of thinking enabled by exploiting disciplinary convergence. A shift to transdisciplinary systems engineering requires a shift in the approaches and tools needed to engineer complex systems. These changes in systems approaches include models that can capture transdisciplinary concepts, inclusion of expansive meta-data in transdisciplinary models, and application of nontraditional engineering methods to explore concepts and outcomes. As engineering takes on the challenges of increasingly more complex systems with different levels of autonomy, we believe that transdisciplinary approaches will enable the realization of more complete solutions for such systems and enable the complexity necessary for these systems. This paper presents specific approaches within the rubric of transdisciplinary systems engineering.

Keywords Elegance · Meta-data · Storytelling · Transdisciplinary

B. Mesmer
University of Alabama in Huntsville, Huntsville, AL, USA

D. Mckinney
Lockheed Martin (retired), Henderson, NV, USA

M. Watson (✉)
NASA Marshall Space Flight Center, Huntsville, AL, USA
e-mail: michael.d.watson@nasa.gov

A. M. Madni
University of Southern California, Los Angeles, CA, USA

A. M. Madni et al. (eds.), *Recent Trends and Advances in Model Based Systems Engineering*, https://doi.org/10.1007/978-3-030-82083-1_49

1 Introduction

A transdisciplinary systems engineering approach enables the exploration of systems needs and generation of solutions through a transdisciplinary lens, where envisioned systems are explored from a holistic perspective. This holistic approach is integral to the transdisciplinary approach. The transdisciplinary space is inherently broader than a disciplinary space because it contains information from multiple engineering, science, and business disciplines. Exploration of this space facilitates the investigation and understanding of the value proposition of convergence.

The concept of transdisciplinarity has been in the social sciences for several decades. The expansion of systems engineering as transdisciplinary (Madni 2007, 2010, 2012, 2019) has cast the discipline into a broader context bringing in new perspectives to think about solutions and broadening the solution space. In addition, transdisciplinary systems engineering has brought a focus on convergence of engineering, science, social, and liberal arts disciplines to enable new system solutions.

The importance of the human in the operation and application of the system is integral to transdisciplinary systems engineering approaches. Transdisciplinary approaches enable solutions which may turn out to be more resilient because they address the problem encountered from multiple perspectives. Specifically, autonomy in many systems demands transdisciplinary thinking. For example, use cases for these systems are also transdisciplinary in nature in that they are highly interconnected in subtle ways leading to unexpected emergent properties. This recognition points to the importance of transdisciplinary model meta-data to capture rich and nuanced information related to these models motivated by the convergence among disciplines (Madni 2018).

A recent research activity has produced results germane to this transformational thinking. The NASA Systems Engineering Research Consortium investigated various aspects of systems engineering. These investigations have led to a set of systems engineering postulates, principles, and hypotheses (Watson et al. 2016). Based on this work, INCOSE adopted a modified set as the systems engineering principles (Watson et al. 2019b). The NASA research focused on the idea of system elegance as defined in a speech given by Robert Frosch (1993), a more recent paper defining four characteristics of system elegance by Michael Griffin (2010), and by an INCOSE journal publication on the subject (Madni 2012). The consortium defined elegance as follows: "A system that is robust in application, fully meeting specified and adumbrated intent, is well structured, and is graceful in operation." To truly explore robustness of a system, a transdisciplinary approach is needed that measures and improves robustness with a holistic perspective. This paper examines multiple aspects of transdisciplinary systems engineering to highlight the unique approaches and findings that it enables (NASA 2019).

System autonomy and systems of systems (SoS) are also important aspects that must be accounted for in systems engineering. The power of liberal arts methods can also be exploited within transdisciplinary perspectives arising from the use of nontraditional engineering disciplines. Meta-data is needed to capture the much broader and nuanced information. This paper examines these transdisciplinary concepts and characteristics.

2 Realizing Elegance Through Transdisciplinary Thinking

Transdisciplinary system elegance involves considering all aspects of the convergent disciplines contributing to the solution space. Considering the NASA definition of system elegance stated above, several transdisciplinary aspects become evident. How well a system meets the intentions of the users of the system, i.e., maintainers and operators, is a key aspect in defining the elegance of the system. When a system is robust in application, the system is meeting the intents of the user community for different purposes. Knowing this in advance has not always been achievable. However, transdisciplinary systems engineering provides a new way of thinking that incorporates more of the social intentions as well as the engineering intentions.

Traditional systems engineering focuses on the specified intentions of the stakeholders as stated in requirements. Requirements often fail to capture all the intentions of users in how they are applying the system and their expectations on how to interact with the systems. In this regard, exploring this space through storytelling is an approach to more thoroughly define these intentions (Madni 2015). Adumbrated intentions, those which are present but not clearly evident, are discovered more thoroughly when the transdisciplinary solution space is explored. This holistic view sheds light on aspects of the system that have not been easy to discern in more traditional systems engineering methods. Being able to identify and understand both the specified and adumbrated intentions of the user community is a key contributor to realizing an elegant system solution.

Elegant systems reflect a structural balance arising from taking into account the disparate intentions for the system. The result is a solution which is well structured in all aspects: social, engineering, political, business, etc. Transdisciplinary thinking enables realization of such a structure. This holistic structure results from a convergent solution space as different engineering, science, and project disciplines contribute knowledge and understanding to the intentions for the system.

A system that is graceful in operation often is simple. "Simplicity" here is in the sense implied by the remark, often attributed to Albert Einstein, that "everything should be made as simple as possible, but not simpler." For example, it may properly be said that while no airplane is actually "simple," some aircraft designs seem to accomplish their intended purposes with considerably less fuss and bother than others and are thus regarded as clearly more elegant. One needs only to compare the Ford Tri-Motor and the Douglas DC-3, which originated at approximately the same time, to appreciate the point.

Simplicity in this sense is not merely the absence of complexity but rather demands a deep understanding of the system's inherent nature. Jony Ive, Chief Designer at Apple, Inc., has said of simplicity, "It's not just minimalism or the absence of clutter. It involves digging through the depth of complexity. To be truly simple, you have to go really deep" (Isaacson 2011).

Context is needed to frame the definition of simplicity for a given design as it relates to the Vitruvian principle of beauty, which must be understood within the transdisciplinary concept of system operation and use. Simplicity is evident in the manner of and the degree to which the integration and unification of design functions are achieved. The quality and style of this integration comprises an essential part of elegant system engineering.

Simplicity is also revealed by the choices which are made as to what the system will not do. An elegant system cannot be all things to all users. An elegant system does what it is intended to do without superfluous functions, where "intention" and "superfluity" are defined by the user mainstream. Apple's Steve Jobs was noted as being a master at maintaining simplicity by keeping "features" out of his designs (Manjoo 2010). This simplicity is achieved by taking into account the transdisciplinary nature of the system, informing the system design and operations by a holistic view.

A system that is graceful in operation invariably exhibits ease of use. Ease of use implies the ability to use the system in its operational environment without superfluous steps or procedures. Ease of use is enabled by the holistic thinking inherent in transdisciplinary systems engineering. Holistic transdisciplinary engineering requires thinking about the system differently, addressing different system viewpoints, and achieving convergence among them leading to seemingly simple, easy to use systems.

3 Transdisciplinary Nature of Autonomy and Systems of Systems

System autonomy and system of systems provide approaches to expand the capabilities of systems in two very different ways. System autonomy adds complexity to the system to provide more responsive interaction with the environment. System of systems provide more of a building block approach to generate new capabilities.

System autonomy implies the need for intelligent decision-making within a system. This can elevate complicated systems to complex adaptive systems (Watson et al. 2019a). These systems responsively interact with their environments: natural, induced (environments caused by system motion), and social. Thus, autonomous systems stand to benefit from transdisciplinary approaches in both system representation and complex behavior realization to properly define the system.

System of systems are a specialized form of transdisciplinary systems. Constructing new capabilities by defining the interaction of independent systems provides a new level of system structure. SoS expand the efficacy of independent systems through interactions among the individual systems themselves. The individual systems can be based on very different engineering characteristics and have very different social interactions. Inherently transdisciplinary, these SoS realize convergence through integration of the independent systems (which themselves can be transdisciplinary). Thus, SoS provide a building block approach to construct new system capabilities.

Smart cities are example of SoS using autonomy to bring about the integration of the separate services of the city. These services are transdisciplinary by nature and involve communication, data, electrical, groceries and retail services, natural gas, transportation, and water. The dependencies of these separate systems are coordinated through autonomous functions to provide a smoother operation of the city. This is viewed from the social, economic, and political aspects forming a tapestry of engineered services and social systems.

Finding the convergence of the different disciplinary contributions to the system solution, whether for autonomous systems, system of systems, or both, is aided by identifying the system integrating perspective. This perspective provides a view of the system which integrates all of the contributing disciplinary views into a holistic perspective. There is still much research to be conducted in finding transdisciplinary integrating perspectives. It appears that these are not found through a single model but perhaps through a set of system models each providing a holistic view of certain system characteristics.

For example, system integrating physics perspectives have been identified which integrate multiple engineering discipline views into a simple relationship (Watson 2018). This provides a physics-based view, but not a social system view. The social system interactions with the system are captured by system dynamics models. System value models also provide an integrated view of stakeholder preferences. System state variable models provide the connection of the systems physical functions to the system preferences in the value model. Statistical models also provide understanding of the uncertainties of the system. Statistics provide a convenient integration approach for convergent disciplinary solutions (NASA 2019). Relationships of processes and information about the system are provided by system relational models (typically referred to as model-based systems engineering models) (NASA 2019). This set of models enables the exploration of the whole transdisciplinary solution space, accounting for all of the convergent discipline solutions. These modeling approaches used as a set aid in the exploration of the transdisciplinary solution space.

4 Storytelling to Understand the Transdisciplinary Nature of System Solutions

Storytelling is another transdisciplinary method to explore the outcome or solution space (Madni et al. 2014, 2016; Madni 2015). Significant work has been performed to integrate engineering and mathematical disciplinary knowledge to improve the exploration of systems. Examples of this work include integrating economics' game theory and design engineering's optimization (Collopy et al. 2012), simulation's agent-based models and systems engineering's approach assessment (Bott and Mesmer 2019), and design engineering's coupling and sensitivity analyses with economics' objective function formation (Kannan et al. 2017). Integration of these disciplines into a trans-discipline has resulted in improvements in design efficiency and effectiveness and better informed approach assessment.

The design and exploration of systems require more than just engineering and mathematical disciplines. A truly trans-discipline to explore systems needs should integrate concepts from a broad swath of disciplines. By doing so, system principles that involve complex behaviors can be fully understood. For example, without the integration of political science and law into the trans-discipline, the social and governmental constraints on a system cannot be fully captured. Without environmental science, the system's impact on society would be incomplete. A broad integration of disciplines is critical to a trans-discipline intended to explore systems.

A critical element in systems that is often overlooked by engineers is the human interaction. If trans-discipline only integrates engineering and mathematics disciplines, then the stakeholders and their behaviors will only be partially represented. The stakeholders are defined here as any person that affects or is affected by the system. Examples of stakeholders are engineers, marketers, company shareholders, the public, etc. System stakeholders are complex, interacting entities, who interact either directly or indirectly through the system. Engineering and mathematics use representations of the stakeholders to improve the system but are not complete representations of stakeholders. Forming stakeholder representations needs to also draw on the social sciences.

The preferences (expressed needs) of the stakeholders are critical to forming a more complete representation of stakeholders that can then be used by engineering and mathematical methods. Traditional elicitation methods are based on interviews and surveys. In recent years, elicitation methods have been explored from the disciplines of communication arts and theatre to integrate into a trans-discipline for exploring systems. Content analysis from communication arts has been used on documents produced by NASA stakeholders to elicit preferences concerning the NextStep-2 Habitat (Palma and Mesmer 2018). This integration of communication arts is important for eliciting preferences from documents that may obscure the true needs; however those needs must be stated in some form in the document to be identified. This leads to a challenge in preference elicitation – what if needs are not stated in interviews, surveys, or documentation?

Hidden truths (i.e., tacit knowledge) are present in many fields. Hidden truths are known by stakeholders but are not explicated without prompting. For example, a hidden truth in the model-based systems engineering (MBSE) community may be the difficulty in programming current MBSE software to interact with third-party software. While this truth may be discussed within the community, outsiders may not know this, and this is not a primary topic of discussion between MBSE practitioners and outsiders, making it a hidden, or difficult to obtain, truth. Recent research has examined disciplines to integrate into a trans-discipline for exploring systems to enable the identification of hidden truths.

Theatre offers methods to elicit hidden truths from communities of stakeholders. Theatre of the oppressed theory poses that theatre can be used to discover oneself. As stated by Augusto Boal in this theory, "We must all do theatre – to find out who we are, and to discover who we could become" (Boal 2006). Specifically, the method of improvisational comedy offers an elicitation technique to integrate into a trans-discipline of exploring systems.

A recent study by the NASA Systems Engineering Research Consortium investigated the use of improvisational comedy to elicit hidden truths in the cost and scheduling community. Improvisational comedy was chosen as "The truth is funny. Honest discovery, observation, and reaction is better than contrived invention" (Halpern et al. 1994). The study's hypothesis is that actors will find that making fun of hidden truths will garner reactions from the audience, leading to them identifying the truths in their comedy. At the 2019 NASA Cost and Schedule Symposium, actors from the community were trained in improvisational techniques and performed a show moderated by an improvisation professional. Cards were distributed prior to the show and asked attendees about issue areas in the community. These cards were used to give the actors starting points.

Hidden truths were identified in the show by the community actors. Such truths identified included stereotypes of positions in the community and different levels of optimism held by those positions concerning projects. Feedback cards gathered from the audience suggest an enjoyment in attending and a benefit from learning about aspects of the community they were not as informed about prior to watching the show. Further work is being conducted to more formally elicit stakeholder preferences for the use of designing systems.

This study on the use of improvisational comedy to elicit hidden truths, along with previous work on the relationships between systems engineering and theatre (Palma et al. 2017, 2019), highlights that the exploration of systems is transdisciplinary in nature. There are questions to be answered in exploring systems that cannot be answered without a transdisciplinary perspective. Nonengineering disciplines should be integrated into the trans-discipline to enable a complete exploration of systems.

5 Importance of Meta-data in Transdisciplinary Systems Engineering

The transdisciplinary systems engineering solution space generates a variety of information from many disparate sources that must be understood holistically. System meta-data contains the holistic data set describing the transdisciplinary system. There are many aspects of this transdisciplinary meta-data and several benefits from capturing this meta-data.

A central responsibility of transdisciplinary systems engineering is to orchestrate interactions between stakeholders. As system architecture and design progress, additional information is developed and needs to be shared. Alternatives considered and choices made, plus the rationale for each choice, are important to capture and share in a transdisciplinary context. Also, consequences and implications of decisions, such as limitations in the functionality of the system which will result from a specific design decision, should be shared. These limitations cross the discipline boundaries, and the capturing of the transdisciplinary constraints is informative to members of the design team who come from varying disciplines. Capturing the rationale and description of this information better informs future systems modifications to update any limitations found in application.

In transdisciplinary systems engineering, as in traditional systems engineering, systematic capture of information in such a way that it is accessible to (and readily understandable by) all stakeholders is critical. Building and maintaining trust among stakeholders with divergent goals and agendas require transparency – and reliable capture and retention of shared information is foundational for transparency.

Creating a consistent and unified body of knowledge is essential in transdisciplinary systems engineering. It is important that the various stakeholder perceptions, languages of expression and capture, modeling techniques, and model purposes all be continuously unified to represent a single shared reality. This means that in addition to the capture of information from the various sources, techniques, tools, models, and practices, the transdisciplinary semantic unification of that information needs to be continuously undertaken to avoid storing contradictory information or constraints about the shared reality. This requires an understanding of transformation techniques from all the various transdisciplinary sources and representations into a common reality. A useful and expressive baseline is the result of a "many-to-one" mapping set of transformations; this "translation" between the different vocabularies of different stakeholders has always been one of the added values of systems engineering. Gathering the information and capturing it are not enough. Unifying that information to ensure domain consistency and uniform understanding about the target of the modeling is as important as the information capture itself. This set of transformation techniques is an essential part of the enhanced model which incorporates meta-data.

Without a systematically organized body of knowledge within a transdisciplinary system context, individual stakeholders do not have any way to discover what they do not know – and so may not even know what questions to ask or what knowledge they may have that others lack. Just looking at the equivalent of the "table of contents" for the project's body of knowledge can help a stakeholders readily see (a) areas of knowledge about which they know nothing (so they can decide whether they need to learn about a topic) and (b) topics they consider important which are not yet addressed in the body of knowledge – so they can see areas where their input is needed by the project community.

Information changes over time during the development of a transdisciplinary system (which may be complex or a large complicated system), as circumstances change, mission needs evolve, existing knowledge is shared, and new knowledge is co-created. Since not all stakeholders will participate in each piece of knowledge discovery or co-creation, each stakeholder needs a way to learn of knowledge newly added to the common understanding.

It is quite common to have changing personnel in many of the stakeholders groups, such as rotations in assignments of military personnel. Systematic capture and retention of information makes it possible for people new to the endeavor to discover past decisions and history so they can understand how the current situation came about. This need to make history available to stakeholders also means that it is very important to "curate" the information so that links between related items of information are maintained. This is actually a very powerful motivation to enhance our models so all relevant meta-data can actually be included in the models; then modeling tool sets can prompt for the inclusion of the data and links between items of data and other model elements and automatically prompt the model user for updates of all linked data when a data item or linked model element is changed.

Modeling is one method of capturing information so it can be used and shared in the transdisciplinary context. The most effective models in use today can show stakeholders critical aspects of the environment and the system under development – and experience has proven that showing combined with telling is much more effective in human communication than describing alone. One of the challenges of transdisciplinary systems engineering is first identifying all of the information/knowledge needed to successfully instantiate a new (or revised) system, then discovering what subset of this information/knowledge is not available, and finally orchestrating interactions to co-create the missing but needed new knowledge/information. Some of this needed knowledge/information can be embodied in the kinds of models available today. The needed information which cannot be embodied in any of the kinds of models available today is referred to in this paper as "meta-data." Several types of information are needed for understanding of transdisciplinary systems as shown in Table 1.

In the past, a few key individuals (e.g., chief engineers of projects) kept much of the critical "meta-data" in their heads. In fact, the ability of a chief engineer to keep all of the critical information in mind was often a major determinant of project success. However, with the increasingly complex transdisciplinary systems being developed, and the increasing complexity in the human and SoS environments in which these new transdisciplinary systems are expected to operate, the involvement of so many stakeholders is needed for success that no one person, however brilliant, can keep all of the critical information in their mind. So, as we transition into the use of transdisciplinary systems engineering, we will need to find more effective ways to capture and share knowledge. Enhancing models so that they include all of the relevant information offers a promising way forward.

There are several benefits of enhancing models to include all of the knowledge and information needed to successfully develop a complex system. Increasing the thoroughness and systematicity of the identification, collection, and co-creation of needed knowledge is essential in transdisciplinary systems engineering. Embedding the transdisciplinary constraints in the model Improves the quality of the system architecture and design. This allows the constraints which currently have to be

Table 1 Meta-data information types

Model meta-data information (e.g., environment, system, and its parts/elements)	Process-related meta-data information	Product-related meta-data information
Stand-alone models	Communications between stakeholders	Operational constraints (such as those necessitated by limitations in system power or by limitations in the capabilities of human operators)
Interconnected models	Information capture	Options considered and the rationale for selection and elimination of various options
	Architecting the system	System architecture description
	Designing the system(s) and elements of the system	Designs of system parts/elements
	Manufacturing, programming, and configuring the system	Interfaces and interactions between parts/elements of the system and between the system and its environment
	Testing the system and its parts/elements	Data used to test the system
	Training users	Results of system tests, including calibration data
	Making changes to the system after it is deployed	

identified, remembered, and enforced by a chief architect or chief engineer to be electronically linked with all of the architecture and design elements affected by the constraints, so everyone using the model can see them. These constraints can be enforced during design – and even during system operation and later during system revisions – to ensure that the system is and remains viable. Transdisciplinary systems engineering meta-data also provides a vehicle that stakeholders can use to understand more about the system and the requirements and constraints driving its design. Collecting needed information in a way which highlights links and interdependencies between different items of information, makes it much more likely that the information will remain current as the system evolves from an idea to an architecture to a design to an operational system, and beyond through system maintenance and revision. Today, the methods we use often end up with a disconnect between documentation and the as-built system, which often has to be remedied by reverse engineering the as-built system to get the final design description (one part of which often ends up being analyzing software code to deduce the final system requirements, in systems controlled by software). Transdisciplinary systems engineering cannot be implemented when the information about the system is siloed and disconnected. Transdisciplinary systems engineering requires new thinking and the capture of the meta-data to support this thinking.

6 Summary and Conclusion

Transdisciplinary systems engineering provides a holistic approach that exploits convergence among disciplines including engineering, cognitive, and social sciences. Elegance is key system characteristic that results from the use of transdisciplinary methods. Elegant systems reflect the holistic solutions based on addressing the perspectives provided by the contributing disciplines and that account for them in system use.

Transdisciplinary systems engineering provides a much broader view of the solution space, greatly expanding this domain. The transdisciplinary view of this space enables system autonomy to be effectively developed and applied. This view also aids in the construction of system of systems. Autonomy elevates the system complexity and expands capabilities through intelligent responsiveness to the system environment. System of systems are constructed on a building block basis, combining functionality of independent systems to create new capabilities. These expansions of system complexity and capability require a transdisciplinary approach.

Exploration of the transdisciplinary solution space requires the use of the integrating perspective, accommodating the viewpoints of many different disciplines. A set of system models can support the exploration of these solutions. Storytelling is a method which is transdisciplinary by its very nature that facilitates the effective exploration of the solution space.

The information produced and needed to be understood about these different exploration methods and their results are captured as meta-data. Transdisciplinary meta-data incorporates many different information types and yields many benefits to retaining system understanding.

Acknowledgments The discussion on the simplicity found in elegance makes use of comments from Michael Griffin which advance systems engineering.

References

Boal, Augusto. 2006. *The Aesthetics of the Oppressed*. Routledge.

Bott, M., and B. Mesmer. 2019. Agent-Based Simulation of Hardware-Intensive Design Teams Using the Function-Behavior-Structure Framework. *Systems* 7 (3): 37.

Collopy, P., C.L. Bloebaum, and B. Mesmer. 2012. The Distinct and Interrelated Roles of Value-Driven Design, Multidisciplinary Design Optimization, and Decision Analysis. In *14th AIAA/ISSMO Multidisciplinary Analysis and Optimization Conference (MA&O)*, Indianapolis, September, 2012

Frosch, R.A. 1993. A Classic Look at Systems Engineering. In *Readings in Systems Engineering*, ed. F.T. Hoban, and W.M. Lawbaugh.

Griffin, M.D. 2010. How Do We Fix Systems Engineering? In *61st International Astronautical Congress*, Prague, Czech Republic.

Halpern, C., D. Close, and K. Johnson. 1994. *Truth in Comedy: Manual for Improvisation*. Meriwether Publishers.

Isaacson, Walter. 2011. Steve Jobs, Simon and Schuster, 343.

Kannan, H., B. Mesmer, and C.L. Bloebaum. 2017. Increased, System Consistency through Incorporation of Coupling in Value-Based Systems Engineering. *Systems Engineering* 20 (1): 21–44.

Madni, A.M. 2007. Transdisciplinarity: Reaching Beyond Disciplines to Find Connections. *Journal of Integrated Design and Process Science* 11 (1): 1–11.

———. 2010. Transdisciplinary System Science: Implications for Healthcare and Other Problems of Global Significance. *Transdisciplinary Journal of Engineering & Science* 1 (1): 38–54.

———. 2012. Elegant Systems Design: Creative Fusion of Simplicity and Power. *Systems Engineering* 15 (3): 347–354.

———. 2015. Expanding Stakeholder Participation in Upfront System Engineering Through Storytelling in Virtual Worlds. *Systems Engineering* 18 (1): 16–27.

———. 2018. *Transdisciplinary Systems Engineering*. Springer.

———. 2019. Transdisciplinary Systems Engineering: Exploiting Disciplinary Convergence to Address Grand Challenges. *IEEE SMC Magazine* 5 (2): 6–11.

Madni, A.M., M. Spraragen, and C.C. Madni. 2014. Exploring and Assessing Complex System Behavior through Model-Driven Storytelling. In *IEEE Systems, Man and Cybernetics International Conference,* invited special session "Frontiers of Model Based Systems Engineering", San Diego, CA, October 5–8, 2014.

Madni, A.M., M. Richey, E. Ordoukhanian, J. Venkatesh, F. Zender, K. Chang, and M. Nance. 2016. Exploiting Storytelling in Collaborative Systems Engineering: Towards a Smart Experiential Dashboard. In *Conference on Systems Engineering Research 2016*, Huntsville, AL.

Manjoo, Rarhad. 2010, July/August. *Apple Nation*, 74–75. Fast Company.

NASA Technical Publication, NASA/TP-TBD-2019. 2019. *Engineering Elegant Systems: Theory of Systems Engineering*. NASA.

Palma, G., and B. Mesmer. 2018. A Preliminary Content Analysis of NASA's Nextstep-2 Habitat Documentation for Preference Representation. In *AIAA SciTech 2018*, Orlando, FL, January, 2018.

Palma, G., B. Mesmer, and A. Guerin. 2017. Similarities of Milestones in Theatre Productions and Systems Engineering. In *ASEM 2017 International Annual Conference,* Huntsville, AL, October, 2017.

———. 2019. Relating Theatre and Systems Engineering: Experiences of a Systems Engineer in Theatre Courses. In *ASEE Annual Conference & Exposition 2019,* Tampa, June, 2019.

Watson, M.D. 2018. System Exergy: System Integrating Physics of Launch Vehicles and Spacecraft. *AIAA Journal of Spacecraft and Rockets* 55 (2): 451–461.

Watson, M.D., B. Mesmer, and P. Farrington. 2016. *Engineering Elegant Systems: Postulates, Principles, and Hypotheses of Systems Engineering*. Charlottesville: CSER.

Watson, M.D., D. McKinney, R. Anway, L. Rosser, and J. MacCarthy. 2019a. Appreciative Methods Applied to the Assessment of Complex Systems. In *INCOSE International Symposium 2019*, Orlando, FL.

Watson, M.D., B. Mesmer, G. Roedler, D. Rousseau, C. Keating, R. Gold, J. Calvo-Amodio, C. Jones, W.D. Miller, D. Long, S. Lucero, R.W. Russell, A. Sedmak, and D. Verma. 2019b. Systems Engineering Principles and Hypotheses. *INCOSE Insight Magazine* 22 (1).

A Systems Science Basis for Compositionality Reasoning

Swaminathan Natarajan, Subhrojyoti Roy Chaudhuri, and Anand Kumar

Abstract Compositionality reasoning is fundamental to engineering. The problem of compositionality is typically framed as: given a configuration of parts with characteristics and interrelationships, how can we derive the characteristics of the configuration as a whole? This paper uses systems science concepts to address a related scope question: how should parts be characterized, and what are the kinds of compositionality reasoning needed, in order to assert that a given configuration will exhibit desired behaviour and characteristics?

Our model structures compositionality reasoning into categories that take a successively wider view of the system: pattern of organization, variety, dynamics, planes of operation, levels of organization and context impacts. The ultimate objective is to explicate the systems science basis for systems engineering and contribute to tooling and engineering practice by identifying the compositionality assertions to be satisfied and the scope of concerns to be covered in systems modelling. To test the coverage and utility of the proposed model, we examine its relationship to systems engineering practices in a large radio telescope project and the extent to which it covers a collection of systems science concepts identified in the literature.

Keywords Compositionality · Systems engineering · Systems modelling · Systems science basis for systems engineering

1 Introduction

Compositionality – how parts come together to form wholes – is central to engineering. Every engineering design involves tacit or explicit) compositionality reasoning: if parts with particular characteristics are put together in a particular

---

S. Natarajan (✉) · S. R. Chaudhuri · A. Kumar
Tata Consultancy Services Research, Pune, India
e-mail: swami.n@tcs.com

A. M. Madni et al. (eds.), *Recent Trends and Advances in Model Based Systems Engineering*, https://doi.org/10.1007/978-3-030-82083-1_50

configuration, what would be the behaviour and characteristics of the resulting whole?

Naturally, there has been extensive research work on compositionality reasoning, including mathematical formulations, in both systems engineering and other engineering disciplines. Typically, it is assumed that each part is characterized mathematically, and the problem is to compose these to characterize the whole. But what should be included in the characterization of each part? For instance, should we assume that the structure of the system (and hence the characterizations) stay fixed over time? Should we model the context, and if so, how?

The goal of this paper is to re-look at compositionality in such a way as to provide conceptual foundations for systems engineering practice: what must we take into account, and what is the range of analysis assertions that must hold, in order to assert that a particular solution configuration will have desired characteristics? Engineering, particularly architecture, is concerned with the life cycle of delivered solutions, so analysis must include the trajectory of characteristics over time, including changes triggered by its various environments.

This work is guided by a key principle: *In order to achieve completeness, compositionality reasoning should cover the various kinds of phenomena that occur in systems*. For this, we rely on systems science (Mobus and Kalton 2015). Systems science studies patterns of organization and has established key determinants of system behaviour and evolution trajectory, including variety, dynamics and mutual interactions and influences between the system and its environments. It has also identified various phenomena, such as faults, pathologies, threats, operating modes, phase changes, tipping points, emergence, fractal patterns, chaos phenomena, self-organization, autopoiesis, combogenesis (Volk 2017), etc. This paper organizes the phenomena and resulting characterization into layered categories: pattern of organization, variety, dynamics, planes of operation, levels of organization and context impacts. This categorization aims to align with the way we organize domain and systems knowledge and with typical engineering analysis practice.

We do not provide a formal theoretical basis to assert the completeness of the proposed categories; instead, we try to obtain confidence in their validity and completeness by (a) intuitive reasoning, (b) comparison with actual practice and (c) comparing them with 45 systems concepts identified in Katina 2016, and Adams et al. 2014. This gives us preliminary confidence in the validity and utility of the model, but completeness is not assured.

Analysis assertions are based on formal, informal or tacit knowledge, so the validity of compositionality reasoning is limited by the accuracy and completeness of knowledge. This work is limited to mechanistic systems with predesigned patterns of organization and behaviour, avoiding considerations such as worldviews (Rousseau 2018) and motivation that applies to analysis of the behaviour of purposeful systems.

The ultimate goal of this work is to help establish the systems science basis for systems engineering practice. In Natarajan et al. 2019, we proposed a Four Worlds model of knowledge formation and discussed how the systems engineering process can be understood in terms of the activities required to synthesize solutions in the real world using model and knowledge worlds. This work is complementary, proposing a systems science basis for the product side of systems engineering, in terms of a model for compositionality reasoning.

Section 2 briefly discusses a fraction of the extensive literature in this area, limiting the discussion to some key ideas that our work draws upon. Section 3 presents our understanding of relevant systems science concepts. Section 4 describes our proposed categories model. Section 5 explores its relationship to typical current practice. Section 6 examines its relationship to the 45 concepts referred to above. Section 7 concludes the paper with some remarks about the contributions, limitations and value of the work.

2 Background

Wymore 1967, in his work on compositionality reasoning, characterized parts in terms of transfer functions: relationships between inputs and outputs. Willems 2007, pointed out the limitations of this approach, particularly for physical dynamical systems where interconnections constrain the behaviour of devices, and proposed a more comprehensive approach to modelling the behaviour of open, interconnected systems called tearing-linking-zooming. The conceptual model that we present in this paper is compatible with Willems' formulation and, from a mathematical perspective, can be viewed as elaborating on it.

Our approach to modelling the constraints arising from interconnections is based on the assume-guarantee concept discussed in Benveniste 2012. Interrelationships with external entities are modelled in terms of *context roles* (e.g. physical environment, partner systems, infrastructure, controller, life cycle manager), each with associated *role profiles* that capture our assumptions about that entity, including patterns of interaction and anticipated trajectory of state and structural change over time. As a corollary, the validity of reasoning is limited by the extent to which actual deployment contexts conform to these assumptions.

3 Systems Concepts

3.1 Systems as Networks of Processes Enabled by Structural Networks

The Systems Phenomenon (Schindel 2016) indicates that systems behaviour arises from a mutual influence cycle between states and interactions: interactions are influenced by the states of participating entities and in turn may modify the state and/or structure of the participants. This captures the fundamental generative dynamics of systems and leads to the understanding that the organizational pattern of systems is as networks of interacting entities and that cycles of interaction produce dynamical behaviour, while variations in entity state and characteristics in turn produce variations in interactions and outcomes. System behaviour arises from the dynamics and variety of interactions over a pattern of organization.

Simon 1962, pointed out the limitation of near decomposability that applies when we try to understand such a network hierarchically, i.e. when we try to collapse a region of the network (a *configuration* of entities that interact with each other and with their environment) into a single node, i.e. we try to characterize the configuration of parts as a unitary whole. Such composition is possible only when there is denser coupling within regions than between regions and the resulting characterization is an approximation of the actual behaviour. This limitation of near decomposability applies to our compositionality reasoning approach.

3.2 Levels of Organization

The behaviour of a system can be observed, understood, modelled and explained at different scales of perception, e.g. a toaster can be understood at the functional, electromechanical, material and atomic levels. This arises from the phenomenon of combogenesis (Volk 2017): entities at one level of granularity combine to produce entities at a higher level of granularity with new degrees of behavioural freedom (Sillitto 2018). We reason about the new behaviours in terms of a new vocabulary of concepts and body of knowledge that describes the entities, interactions and resulting behaviour at the larger granularity. This leads to the formation of multiple knowledge domains with different vocabularies and knowledge items, each of which capture knowledge about phenomena that occur at that level of description. Of course, the toaster is a single whole, involving all of these phenomena, i.e. all this knowledge must be brought to bear to understand its behaviours.

This has two implications for compositionality reasoning:

- Consistency relationships must hold across the levels of description, i.e. the lower levels of description must produce the entities and interactions present in the higher levels of description.
- Higher levels of description are abstractions that assume integrity of configuration and behaviour at lower levels. A complete approach to compositionality reasoning must take into account behaviours and fault situations that may arise from lower-level phenomena.

In engineering, we typically model the system at the functional, technical and technological levels.

3.3 Planes of Operation

A comprehensive approach to compositionality reasoning, which aims to take into account changes to the state and structure of the system over its lifetime, must take into account not only the behaviour of the system configuration but also all the processes in its ecosystem that act upon the system to change it, including control and life cycle processes. The categorization below of the network of system processes into planes builds on concepts from Mobus 2020, and the viable system model (Beer 1979):

- *Operational plane*: Structures and processes involved in delivering basic functional behaviour
- *Resources and structural facilitation plane*: Processes and structures involved in acquisition and distribution of materials, energy and/or information inputs and resources and waste disposal
- *Operational control plane*: Structures and processes involved in orchestration, monitoring, feedback control and situation handling of system functioning towards desired goals
- *Life cycle management plane*: Structures and processes involved in facilitating the life cycle of the system-of-interest, including adaptation to its context (e.g. upgrades)
- *Identity management and governance plane*: Structures and processes involved in forming and evolving the role of the system with respect to its environment and asserting this identity

Figure 1a identifies some of the typical processes involved in each plane. For mechanistic systems, many of these processes involve entities (including people) in its ecosystem. These ecosystem entities, as shown in Fig. 1b (e.g. controller/life cycle manager, resource provider, system owner), are modelled as *context roles*, with associated *role profiles*. These role profiles characterize context entities in the same way parts are characterized: in terms of functions and characteristics, states, interactions and their effects on state and structure and trajectory of change over time (these concepts and Fig. 1b are discussed further in Sect. 4). By including context roles (including their trajectory of change) in compositionality reasoning, we recursively enable characterization of the (anticipated) trajectory of change of the system over its lifetime.

4 Categories of Phenomena to Be Included in Systems Modelling and Compositionality Reasoning

The above concepts, and other systems science phenomena such as listed earlier, all need to be taken into account when determining the behaviour and trajectory of evolution of system consisting of a configuration of parts. The model below

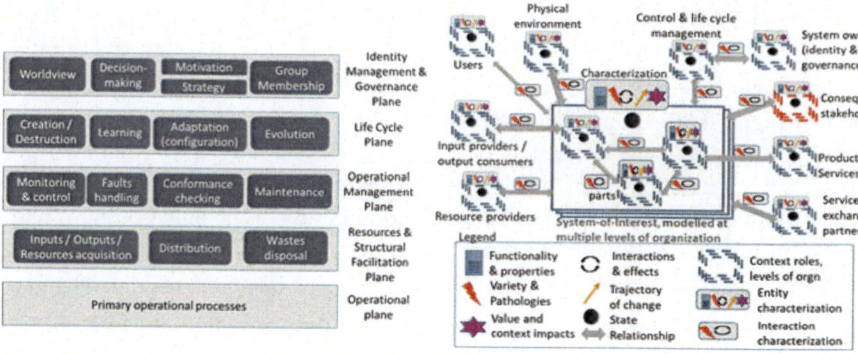

Fig. 1 (**a**) Partial list of processes in each plane of operation. (**b**) Systems modelling concept for compositionality reasoning

is proposed as a categorization of the space of considerations involved, which aligns with the way we organize knowledge and with current typical engineering practice:

- *Pattern of organization*: The nature of patterns knowledge and analysis methods in any domain is that they express mappings between configurations of parts and the resulting functional and quality outcomes. This patterns-based base case reasoning is the foundation for compositionality.
- *Variety, undesired variety and pathologies*: In addition to the base case, the space of possibilities associated with the states, flows, interactions and structural relationships in a system over its lifetime includes boundary conditions, transient situations, spontaneous (non-designed) processes (e.g. rusting, wear-and-tear), undesired/unexpected/non-nominal inputs (e.g. environmental and resource perturbations, security threats, etc.) and pathological situations in which the state and/or structure of the system varies from the defined "healthy" state and structure, resulting in adverse behavioural impacts. Compositionality reasoning must include working through this entire variety space in determining system behaviour.
- *Dynamics*: Variety thinking examines the effect of structure and state variations in the structural network on behaviour. Dynamics analysis complements this by working through the behaviour of the cyclical network of processes. Dynamics analysis can be viewed as including three levels:
  - *Short-term dynamics*: This examines operational behaviour: how state changes resulting from the network of processes produce a trajectory of behaviour over time. Often, the methods are discipline-specific, e.g. fluid flow equations, Nyquist plots, path planning, etc. Functional, fault and security scenarios also involve short-term dynamics.

  - *Medium-term dynamics*: This focuses on structural changes arising during system operations. It includes phenomena such as operating modes, tipping points, emergence, self-organization, homeostasis, autopoiesis, learning, adaptation, combogenesis, etc.
  - *Long-term dynamics*: This focuses on co-evolutionary changes between the system and its environment. For mechanistic systems, this includes the trajectory of changes to stakeholder needs and sources of value, as well as technology and other environments.

  The field of complexity science focuses on identifying phenomena resulting from the network dynamics of systems, building knowledge and developing analysis techniques in this area. Mobus 2020, discusses an approach to modelling and analysing dynamical behaviour.
- *Planes of operation*: Variety and dynamics analysis need to be applied not only to the operational system but to all the planes of operation associated with the system-of-interest. These may involve entities outside the system-of-interest, including people in its ecosystem. This is facilitated by modelling external entities as context roles, as shown in Fig. 1b. The central point there is that compositionality reasoning requires explicating our assumptions about the context as part of the model, including assumptions about its (anticipated possible) trajectories of change over time. These characteristics and trajectories may be specified in terms of possibility spaces, e.g. range or collection of possible values, scenarios, etc. If deployment contexts deviate from these assumptions, compositionality reasoning needs to be updated accordingly.
- *Levels of organization*: As discussed earlier, systems are understood relative to multiple levels of organization (e.g. functional, technical, technological), with associated bodies of knowledge at each level. In order to bring all of that knowledge to bear, we need a family of models corresponding to the different levels, with consistency relationships among them. Consistency relationships enable the propagation of the results of reasoning across levels, so that the complete characterization includes, for example, faults and pathologies identified at each level.
- *Context impacts – influences, value and consequences*: All the above reasoning has focused on establishing/verifying the behaviour and characteristics of the system-of-interest. However, the systems operation may have impacts on its environment, and in turn the environment impacts the system. In particular, the environment may include purposeful entities ("stakeholders"), partner systems and environmental entities who are impacted by the system, and compositionality reasoning should include determining the outcomes for these in terms of value delivered (including experiential and perceived value) and consequences: negative impacts resulting from changes to their state and structure over time. This may require extending the scope of system modelling and reasoning beyond the immediate context roles that interact directly with the system, to include the indirect interactions involving these other affected entities, ensuring that we

capture the essential dynamics that determines the impact. For the same reasons, it is also necessary to include within the scope of modelling and reasoning those context elements that significantly influence the behaviour of the system and its trajectory of change over time.

The focus of the above model is on identifying the *kinds* and *scope* of reasoning needed to understand the behaviour of systems and their trajectory of evolution or to assert that a system will conform to desired behaviours. It does not address *how* to perform such reasoning or indeed whether techniques for such reasoning exist today, e.g. dynamics analysis today may not predict emergent outcomes and self-organization. What it does do is establish based on systems science the responsibilities of systems engineering modelling and reasoning. It should be noted that the concerns identified by the categories need to be considered together during analysis, not independently, e.g. dynamics reasoning needs to include variety considerations.

Figure 1b shows the kind of systems modelling framework required to support such reasoning. It is a family of models at different levels of organization, with context roles covering the various planes of operation. Characterization of both entities and interactions includes variety, dynamics and context impacts.

Performing all the above reasoning on such a model determines/verifies the characteristics of a single "block": one entity at one level of the system hierarchy. All this reasoning must be recursively applied to each entity at each level of the hierarchy, so that the characterization of parts at any level of the hierarchy matches its behaviour determined at the next lower level. It should also be noted that pattern of organization reasoning is based on domain knowledge and typically involves/requires multiple views of the system, aligned with applicable functional and quality concern knowledge domains, based on stakeholder concerns.

5 SKA Radio Telescope Example

The model identifies a wide range of reasoning necessary to assert that a system configuration will deliver desired behaviour. How does this relate to typical current systems engineering practice? We examine a large systems engineering project in which we participated recently the design of the Square Kilometre Array, an international project to create a pair of radio telescopes in Australia and South Africa, involving about 500 engineers and scientists over 3–4 years. The table below maps engineering activities to the reasoning model.

Unsurprisingly, current practice includes design strategies, models and formal/informal reasoning across all the layers – if there were major gaps, such projects would not succeed. Rather, it gives us confidence that our model is in the right ballpark and that there is a match between the systems science theory and the systems engineering practice. Of course, it can also be seen that there are areas where the theory can bring in more systematic thinking, practice and tooling support,

Reasoning type	Engineering modelling and analysis activities in SKA project
Pattern of organization	Functional analysis, activity diagrams, MATLAB and other domain-specific models and analysis, safety/security patterns, quality attributes analysis, interfaces and ICDs
Variety, undesired inputs, pathologies	Observing modes, scenarios analysis, fault scenarios, RFI excision, safety and security threats, equivalence classes, load curtailment, degraded operations
Short-term dynamics Medium-term dynamics Long-term dynamics	Startup/shutdown/end of observations, failover, commissioning, pulsar timing Adaptation dynamics: recalibration, effect of calibration change on data products, observation QA (adapting observations to conditions), add custom devices Provisioning for phase 2, anticipated changes, ability to exploit new capabilities
Planes of operation	Operations and maintenance planning, observation scheduling and management
Levels of organization	Models, prototypes, analysis at functional, technical and technological levels
Context impacts	Radio quiet zone, RFI and aircraft transit, stakeholder value analysis, site impacts

particularly in terms of variety and dynamics analysis, systematic coverage of planes of operations and value and consequences analysis.

6 Coverage of Systems Science Concepts

The previous section gives us some confidence that the proposed compositionality reasoning approach maps well to current systems engineering practice. But since our goal is to link systems science with systems engineering, does the simple proposed categorization achieve coverage of key systems science concepts? To obtain some confidence in this area, the table below examines the mapping between 45 systems concepts identified in Katina 2016, and the proposed categories. It should be noted that the propositions are concerned with systems design guidance, while our focus is on developing a reasoning model that is independent of design prescriptions, ideally applicable even to understanding natural systems. This complicates the mapping.

The "not applicable" entries indicate that the concept relates to design guidance; the text in parenthesis indicates the category of concern addressed. The italicized "not covered" entry notes an important point not explicitly mentioned so far, that reasoning often involves the construction of abstractions (including viewpoints), which deliberately removes detail deemed to be of less relevance to specific reasoning goals, to facilitate application of knowledge and reasoning. This is related to the meta-guidance concerns of system boundary, requisite saliency and (avoiding excessive attention to) events of low probability, which suggest the need to exercise

Principle	Category	Principle	Category	Principle	Category
Communication	Variety	**Control**	Planes of Oper	**Emergence**	Dynamics
Hierarchy	Pattern of org (recursion)	**Complementarity**	Pattern of Org (Viewpoints)	**Darkness**	Pattern of orgn (Interactions)
Holism	Pattern of org (context)	**Minimum crit specification**	Scope of modelling	**Pareto**	*Not covered (abstraction)*
Requisite parsimony	Not applicable (modelling)	**Requisite saliency**	Meta-guidance (modelling)	**Equifinality**	Pattern of Org
Multifinality	Variety, dynamics	**Purposive behaviour**	Planes of Oper (Identity)	**Satisficing**	Not applicable (desired goals)
Viability	Planes of Oper	**Redundancy of potential cmd**	Planes of oper (control plane)	**Information redundancy**	Not applicable (Variety)
Dynamic Equilibrium	Dynamics	**Homeorhesis**	Dynamics, trajectories	**Homeostasis**	Dynamics, planes of oper
Redundancy of resources	Not applicable (Variety)	**Relaxation time**	Short-term dynamics	**Self-organization**	Dynamics
Sub-optimization	*Recursion*	**Circular causality**	Dynamics	**Feedback**	Planes of oper (control)
Recursion	Pattern of org	**Requisite hierarchy**	Planes of oper (control)	**Requisite variety**	Variety
Balance of tensions	Pattern of org, planes of oper	**Basins of stability**	Medium-term dynamics	**Buffering**	Variety, planes of oper (rsrcs)
Resilience	Variety, dynamics	**Eudemony**	Planes of oper (identity, gov)	**Events of low probability**	Meta-guidance (variety)
Least effort	Dynamics	**Omnivory**	Planes of oper (rcrcs), variety	**Transcen-dence**	Identity. limits of knowledge
Incompleteness	Limits of knowledge	**Morphogenesis**	Life cycle, dynamics	**Punctuated equilibrium**	Long-term dynamics
Sociotechnical systems	Planes of oper context impact	**System boundary**	Meta-guidance (modelling)	**System environment**	Context impact

judgment on what to include in modelling and analysis, to keep them effective and tractable. Transcendence and incompleteness point out some fundamental limits of reasoning.

It would be incorrect to conclude from this analysis that the proposed categories achieve full coverage of systems science concepts. Nevertheless, the results improve our confidence in the validity of the proposed model, both in terms of coverage, and that the categories are indeed based on known systems concepts.

7 Conclusion

We have proposed a simple categories model of considerations that must be taken into account in compositionality reasoning: pattern of organization, variety and pathologies, dynamics, the various planes involved in systems operation, the need to combine knowledge relating to different levels of organization and the need to include context impacts in the characterization and reasoning. The proposed model is firmly grounded in systems science concepts and maps well to systems engineering practice. We believe this model can provide a bridge between the disciplines, showing engineers the scientific basis for the product engineering practices that are known to be essential to ensuring that systems will function as desired.

Our work also indicates the information that needs to be included in systems modelling to support such reasoning, as shown in Fig. 1b. This can facilitate more systematic practice, but also help in improving tool support for modelling and reasoning. The work also provides value by pointing out the limitations of reasoning and the need for additional research on reasoning areas not well supported by methods and tools.

Acknowledgements Discussions with colleagues at TCS Research and with members of INCOSE SSWG, SysML v2 submission team and others in the systems engineering and systems science communities have contributed significantly to the ideas.

References

Adams, K.M., P.T. Hester, J.M. Bradley, T.J. Meyers, and C.B. Keating. 2014. Systems Theory as the Foundation for Understanding Systems. *Systems Engineering* 17 (1): 112–123.

Beer, S. 1979. *The Heart of Enterprise*, Vol. 2. Wiley.

Benveniste, Albert, et al. 2012. *Contracts for System Design*. Dissertation Inria.

Katina, P.F. 2016. Systems Theory as a Foundation for Discovery of Pathologies for Complex System Problem Formulation. In *Applications of Systems Thinking and Soft Operations Research in Managing Complexity*, 227–267. Cham: Springer.

Mobus, G.E. 2020. Understanding Complex Systems: Analysis, *Modelling and Design*, forthcoming.

Mobus, G.E., and M.C. Kalton. 2015. *Principles of Systems Science,* Vol. 7, No. 5. New York: Springer.

Natarajan, Swaminathan, et al. 2019. How Do Knowledge Domains Come Together in Systems? In *Systems Engineering in Context*, 137–150. Cham: Springer.

Rousseau, David, et al. 2018. *General Systemology: Transdisciplinarity for Discovery, Insight and Innovation*, Vol. 13. Springer.

Schindel, B. 2016. Got Phenomena? Science-Based Disciplines for Emerging Systems Challenges. In *INCOSE International Symposium*, Vol. 26, no. 1, pp. 2256–2271.

Sillitto, H. 2018. personal communication.

Simon, H. 1962. The Architecture of Complexity. *Proceedings of the American Philosophical Society* 106 (6): 467–482.

Volk, T. 2017. *Quarks to Culture: How We Came to Be*. Columbia University Press.

Willems, J.C. 2007. The Behavioral Approach to Open and Interconnected Systems. *IEEE Control Systems* 27 (6): 46–99.

Wymore, A.W. 1967. *A Mathematical Theory of Systems Engineering*.

Toward the Design of Artificial Swarms Using Network Motifs

Khoinguyen Trinh and Zhenghui Sha

Abstract Many complex systems evolve as a result of interactions among individual entities whose behaviors cannot be directly controlled. This makes the design of such systems inherently challenging. The objective of this research is to develop a new approach in engineering complex swarm systems with desired characteristics based on the theory of network motifs – subgraphs that repeat themselves among various networks. In recent studies, the discovery of network motifs has presented the ability to determine reoccurring similarities between similar functioning networks that were originally believed to have not shared any characteristics. It is therefore hypothesized that manipulating the types of network motifs within a network can help engineer artificial swarms with improved functionality. In this study, artificial swarm systems have been modeled as a dynamic complex network where each node represents an individual foraging entity and links represent as the communication between entities. Additionally, motif-detecting algorithms have been used to extract subgraphs that reoccur in these complex networks. Our research has shown promising results that reveal a statistically significant correlation between network motifs and the performance of simulated swarm networks. This study contributes as a new approach that can potentially be used in the design and engineering of complex swarm systems.

Keywords Network motifs · Complex networks · Artificial swarm · Network analysis · Complex system design

K. Trinh
Department of Mechanical Engineering, University of Arkansas, Fayetteville, AR, USA

Z. Sha (✉)
The Walker Department of Mechanical Engineering, The University of Texas at Austin, Austin, TX, USA
e-mail: zsha@austin.utexas.edu

A. M. Madni et al. (eds.), *Recent Trends and Advances in Model Based Systems Engineering*, https://doi.org/10.1007/978-3-030-82083-1_51

1 Introduction

1.1 Background

The engineering of complex systems has traditionally followed a top-down methodology which creates a framework for the system and adds additional features to meet specific design requirements. This general process is embodied in various existing systematic design methods (e.g., Pahl and Beitz's theory 1996), systems engineering models (e.g., Systems Engineering V (Buede 2000) and Waterfall model (Scacchi 2001)), and system engineering processes adopted by organizations such as NASA (2007). For example, the design of a vehicle system requires the decomposition of high-level requirements (e.g., safety, reliability) into individual units such as an acceleration unit or a braking mechanism which can be further broken down into mechanical components that can be designed and manufactured with pre-existing knowledge and tools.

Over the past decade, there has been a growing interest in developing solutions involving complex systems, where the system-level structure emerges from the behaviors of individual entities and their interactions with each other. Some examples of this approach are the development of swarm robotics to fight forest fires and the use of automated drone system to transport industrial goods. In contrast, the design of these complex multi-agent systems differs from traditional engineering design. Instead, it often draws inspiration from natural swarm, such as ant colonies and bee swarms, where individuals make decisions based on local information. In such systems, there are a large number of individual entities, which are heterogeneous in nature and have private objectives that must be met without compromising system goals. Therefore, engineering such systems towards a desired system-level performance is inherently challenging, and maintaining a direct control of the system is nearly impossible. A bottom-up approach that aims at engineering local interactions must be developed, and a better understanding about the relationships between the local network structures and the system-level performance must be obtained.

1.2 Design Methodology Using Network Motifs

In this paper, we present a new approach based on the theory of network motifs to efficiently gain information of how individuals interact with each other and how those interactions would influence the performance of system. Network motifs are defined as reoccurring subgraphs (or patterns) that repeat themselves among various networks (Milo et al. 2002). Current research on network motifs has been primarily focused on the development of motif-detecting algorithms to identify reoccurring subgraphs within a network to support the analysis of network topologies (Zuba 2009). As a result, many non-commercial algorithms, such as *mfinder* and *MAVisto*,

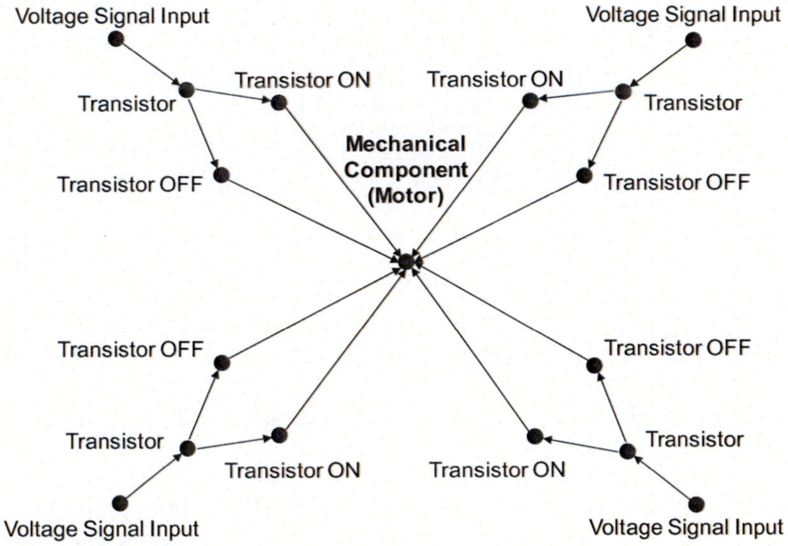

Fig. 1 H-bridge circuit modelled as a network

have been developed to detect these reoccurring subgraphs (Zuba 2009). Other motif research has involved the analysis of biological systems to determine similarities across different organisms (Zabet 2011). Existing research has indicated that network motifs have certain functions to allow for the system to achieve its overall goal. Studies of biological systems, for example, have shown *E. coli* and yeast to share common motif families that make up both of their entire DNA transcription network. Particularly, one motif families, called auto-regulation motifs, allows *E. coli* to repress or accelerate the rate of DNA transcription (Zabet 2011).

One of the main problems currently faced with systems engineering is a lack of holistic methodology (Rousseau 2018). Generally, systems engineering focuses on the design of the system's parts rather than the system as a whole. This results in unexpected behaviors of the system which stem from interactions of different subsystems and their parts. We look to bridge this problem through a methodology that correlates complex system characteristics with network motifs. Inspired by existing studies, we hypothesize that the theory of network motifs can be used in designing complex systems. To give an overview on this approach, we use a simple example – the redesign of a traditional H-bridge circuit – as shown in Fig. 1. The components of this H-bridge circuit can be modeled as various nodes that have connections to other nodes that send various electrical signals. Various subgraphs, such as the one involving the transistor node, can be found repeating themselves within the network (Itzkovitz et al. 2005). The transistor node is directly responsible for receiving signals from one node and sending out new signals to other nodes, making this node crucial in the H-bridge system. The transistor node will receive a signal from voltage signal node that causes the transistor to turn itself "on" or "off"

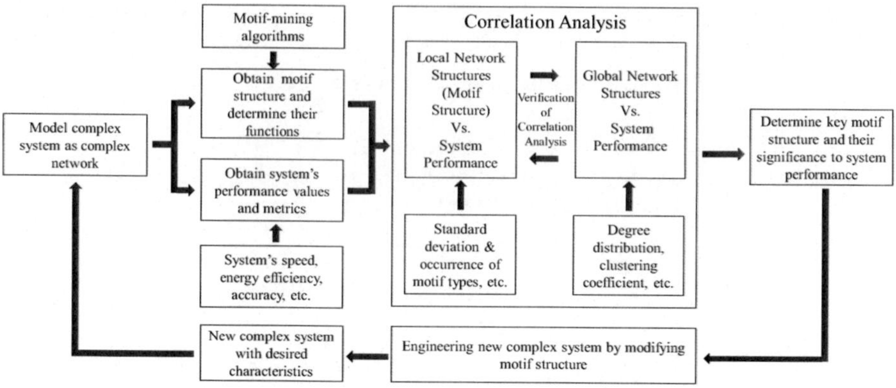

Fig. 2 The proposed approach for engineering complex swarming systems

to allow for electrical current to flow through it. From the signals that the motor node receives, the motor node will either be in its off state, clockwise-rotation state, or counterclockwise-rotation state. This transistor subgraph is not exclusive to the H-bridge only and can be applied to a plethora of other electronic applications. If we have an in-depth understanding about the correlations between this transistor subgraph and its function within the circuit, then a new system could be engineered by promoting the formation of such subgraphs.

In general, the proposed approach can be summarized in Fig. 2. First, a complex system is modeled as a complex network in which nodes and links are defined. With an established network structure, key reoccurring subgraphs can be identified by using motif-detecting algorithms. Their functions must be analyzed and will eventually be used to modify pre-existing systems. To study the relationship between local network structures and system-level performance, correlation analysis has to be performed between the extracted network motif data and the metric values that quantify the system performance. Based on the correlation analysis, the most important network motifs, i.e., the ones with the highest correlations, will become potential solutions in engineering the system's performance level. In order to verify such correlations, we propose to perform additional correlation analyses between the network motifs and network-level properties such as the average degree and the degree distribution. This step is necessary in order to verify if a correlation exists between network motifs and local network structure rather than the global network characteristics. After such verifications, those subgraphs can be confirmed in having a significant influence on system performance. It is, therefore, desired to promote the formation of such subgraphs (local structures) to achieve a higher level of performance. Additionally, these subgraphs can be applied to different complex systems to achieve a similar function as the original system.

In the following sections, this approach is demonstrated through a simulation study of a swarm system foraging for food. This swarm system has been selected for this case study because it represents general functions of many swarm systems in which individual entities communicate locally with nearby entities to collectively accomplish predefined tasks. The rest of this paper is structured as follows: the description of the complex system of interest and the approach used to analyze the system is presented in Sect. 2. The results of the analysis and its discussion are presented in Sect. 3. Finally, we conclude this paper and present our closing insights in Sect. 4.

2 Case Study

2.1 The Swarm Foraging System and Simulation

A case study has been performed on multiple simulations, modeling a swarm of ants foraging for food. In the swarm, there are 100 entities (or ants) which would search in a confined 12-unit by 12-unit arena for food particles over 10,000 timesteps. Each entity has five different states (see Fig. 3): *resting*, *exploring*, *avoidance*, *depositing*, and *homing*.

1. The *resting* state is when entities are in their nest and are inactive.

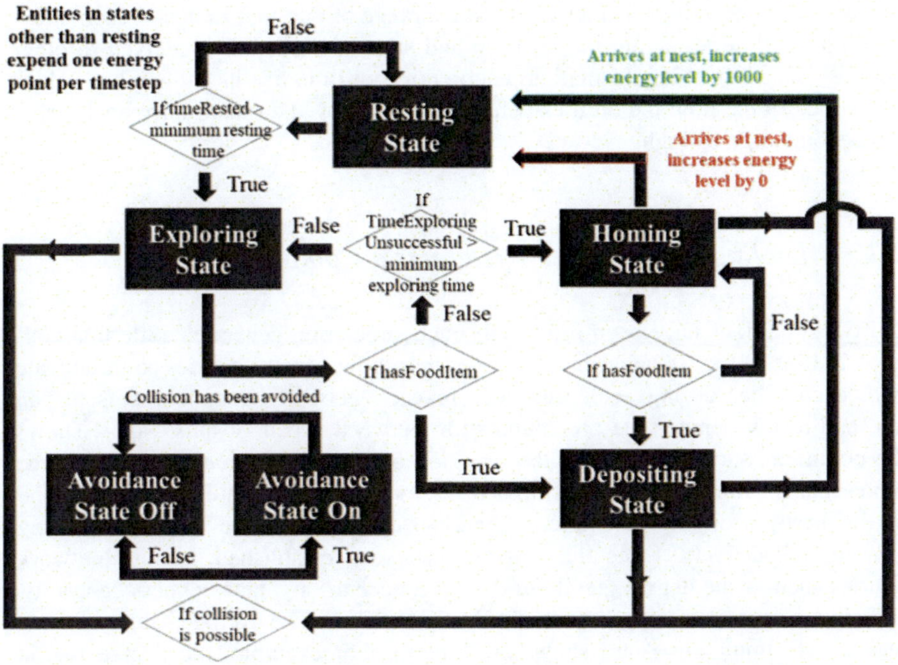

Fig. 3 The model of swarm foraging system

2. The *exploring* state is when entities are searching for food particles.
3. The *avoidance* state results during the exploration state and is caused by an entity nearing collision with another entity or an obstacle (arena wall). The entity will then change its vector to avoid collision.
4. The *depositing* state is when an entity has successfully located a food particle and begins returning to the nest with the food particle.
5. The *homing* state results when the entity has failed to locate any food particles after its resting probability has reached a certain value. This state causes the entity to return to the nest.

At the beginning of the simulation, all entities will begin in the nest area, or the gray area, of the arena. Each simulation begins with an energy level of 0 points. Once the simulation begins, all entities switch to the exploring state and search for food particles in the white area of the arena. For each timestep that an entity is moving, one energy point will be expended from the nest; this rule is implemented to account the energy lost (or food consumed) from active entities searching for food. For example, if 100 entities are active (exploring, avoidance, depositing, or homing) during a timestep, 100 energy points will be taken from the nest resulting in a -100-energy point loss for the nest. For each food particle that is brought back to the nest, 1000 energy points are added to the overall energy level of the nest. As the entities are collecting food, they can communicate with each other by using an RAB (range-and-bearing) sensor. These RAB sensors allow multiple entities to communicate with one another as long as they are within a range of one-unit radius of each other. The information pertaining the position and state of the entity is relayed to other entities. Based on this information, each entity will modify its probability values which determine how the entities behave by changing their state once each value has become high enough.

2.2 *Modeling the Complex System as a Complex Network*

Thirty simulations have been produced. Each simulation generates a file that contains 10,000 100 × 100 adjacency matrices that show how the entities communicate with each other during each timestep. In each network, a node represents an ant, and a link denotes the communication between a pair of ants. Fig. 4 shows the communication networks at the timestep of 10,000 of three swarms in three representative simulations: simulations 8, 18, and 13. These simulations correspond to the best-performing, the intermediate-performing, and the worst-performing swarms, respectively. Once the network has been established, various network metrics such as the degree, geodesic distance, eccentricity, betweenness centrality, and clustering coefficient can be immediately obtained. The average values of these metrics of simulation 8 are shown in Table 1. For example, the degree metric indicates the average number of links an entity has at a particular timestep. This

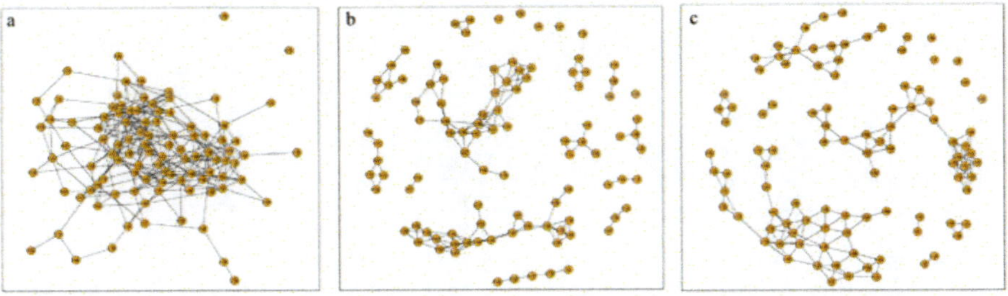

Fig. 4 Three simulations at timestep 10,000 represented by three networks: (**a**) simulation 8, the best performing swarm; (**b**) simulation 18, the intermediate performing swarm; and (**c**) simulation 13, the worst performing swarm

metric gives insight on how many entities are active and how tightly grouped the entities are for communication. The degree metric can help compare whether the overall number of links plays a stronger role in system performance or if various network subgraphs take precedence in system performance.

2.3 Obtaining the Network Motif Data of the Simulations

Among the 10,000 networks in each simulation, 200 networks (1 every 50 timesteps) have been selected for further analysis. This sampling decision has been made due to the computational consideration, and the same approach has been applied to all simulations. The edge list of each network has been used as a data input for a motif-detecting program called *mfinder* (Kashtan et al. 2004). Figure 5 shows the occurrence of an example motif type found in a representative simulation, simulation 8, during the 10,000 timesteps in 10 timeframes.

In this study, 15 representative simulations have been selected for the correlation analysis. To select these simulations, the performance level has to be evaluated. The performance level of each simulation is measured by two aspects: the amount of food collected and the amount of energy expended during the entire foraging period. The performance of these simulations is calculated based on Eq. (1)

$$P_s = \left| \frac{f * 1000}{e_f} \right|, \tag{1}$$

where f is the total number of food particles collected, e_f is the final energy level at timestep 10,000, and P_s is the performance value of the entire simulation. Based on this equation, the following simulations have been selected: the five best-performing simulations (8, 9, 11, 16, 28), the five worst-performing simulations (13, 20, 25, 27, 30), and the five most intermediate-performing simulations (1, 18, 19, 26, 29).

Table 1 Network properties of simulation 8

Timestep	Foraging performance	Degree	Geodesic distance	Eccentricity	Betweenness centrality	Closeness centrality	Edge betweenness centrality	Local clustering coefficient	Global clustering coefficient
1–10000	0.389	3.293	5.261	7.473	0.015	0.017	52.957	0.435	0.534

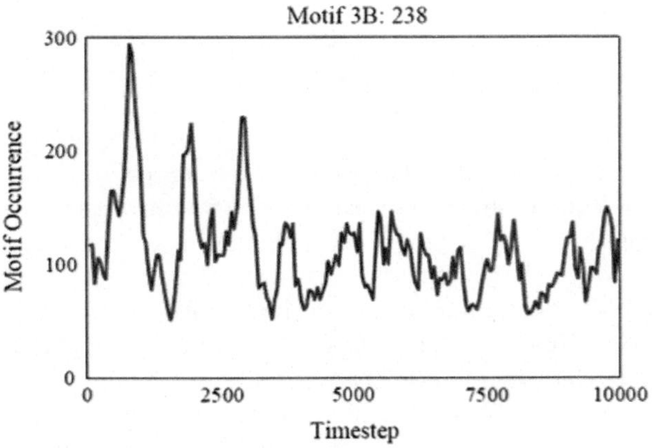

Fig. 5 The occurrence of motif 3B in simulation 8

2.4 Determining Important Network Motif Structures

The average and the standard deviation of network motif occurrence have been calculated across ten different timeframes, i.e., 1–1000, 1001–2000, ... 9001–10,000, for every simulation. The average foraging performance of each simulation has also been calculated across ten different timeframes. This is because there needs to be at least one food particle collected within a selected timeframe in order to evaluate the system performance. Since the swarms in these simulations take over 50 timesteps to collect 1 food particle on average, all simulations have been divided into 10 separate timeframes to ensure that at least 1 food particle is collected within each time interval. The average foraging performance per time interval is calculated by Eq. (2)

$$P_i = \left| \frac{f * 1000}{e_f - e_i} \right|, \tag{2}$$

where e_i is the initial timestep's energy level, e_f is the final timestep's energy level, f is the number of food particles collected within the timestep frame, and P_i is the performance value of the simulation during the interval of interest. In determining which motif structures are significant, correlation analysis has been performed between the average occurrence of each motif structure and the average performance of each swarm (i.e., each simulation). In addition, the standard deviation of each motif structure has also been correlated to the average performance. This is done because observations showed that some simulations yield a better performance with highly fluctuating occurrence values of network motifs in certain timeframes. Both data types have been correlated by using Pearson's correlation coefficient. In this study, a motif is determined to be significant if (a) the correlation coefficients are 0.5 or higher or (b) it appears as one of the simulation's top 10 correlated motif types. These results are discussed in Sect. 3.

3 Results/Data

3.1 Motif Structure Study

The search of important network motifs has been limited to size-3, size-4, and size-5 motif structures (i.e., the motifs that consist of three, four, and five nodes, respectively). The motifs above the size of six nodes have been ignored in this study due to the large computational resources required – the *mfinder* algorithm has to run for about 2 h to analyze a single timestep using a computer configuration of a 64-bit Windows 10 Dell laptop with an Intel Quad Core i7-7700HQ @ 2.80 GHz processor and 8 GB of RAM.

There are 29 different motif types that have been analyzed: 2 of which are size-3 motif structures, 6 of which are size-4 motif structures, and 21 of which are size-5 motif structures. Each motif structure is named by the following format: motif #L, where # represents the motif structure size and L represents a specific motif structure. For example, the two size-3 motif structures have been named as motif 3A and motif 3B, and the motif structures for size-4 motifs have been named motifs 4A, 4B, 4C, 4D, 4E, and 4F. Figure 6a shows the differences between the two types of size-3 motifs. Motif 3A has a centralized node that can freely communicate with two other nodes, while the other two nodes cannot relay any information. This differs from motif 3B where all nodes are able to communicate with each other.

Due to the length of paper, we are unable to present the results of the correlation analysis for all the 15 simulation. Instead, we present the top five swarms (i.e., simulations 8, 9, 11, 16, and 28) as a demonstration of the data. Table 2 shows an abbreviated table of the sorted correlation coefficients between the average motif occurrence and the performance values of the five best-performing simulations. The top of the table shows motifs that have lower correlation coefficients, and the bottom of the table shows higher correlation coefficients. In simulation 8, motif 5K is the least correlated motif type at 0.38, and motif 5N is the highest correlated motif type at 0.76. It has also been observed that certain motifs appear to be highly correlated more often than others. For example, motif 5B (highlighted green) has a correlation coefficient higher than 0.50 for at least three of the five simulations. Motifs 5D and 5H (highlighted blue) appear three times in the top 10 correlated motif types out of the five simulations. This correlation analysis has been applied to all 15 simulations. Based on the criteria of significance aforementioned, the following observations have been acquired:

(a) At least 10 of the 15 simulations studied show 3 motif types that have been valued as significant when correlating the foraging performance to average motif occurrence. In 11 of the simulations (8, 9, 11, 16, 18, 19, 20, 26, 27, 29, and 30), motif 5B has been found to have a correlation coefficient of at least 0.50. The average correlation coefficient of this motif type is 0.525. In 11 of the simulations (9, 11, 13, 16, 18, 20, 26, 27, 28, 29, and 30), motif 5D has been found to be in the top 10 of 29 motif types studied. The average correlation

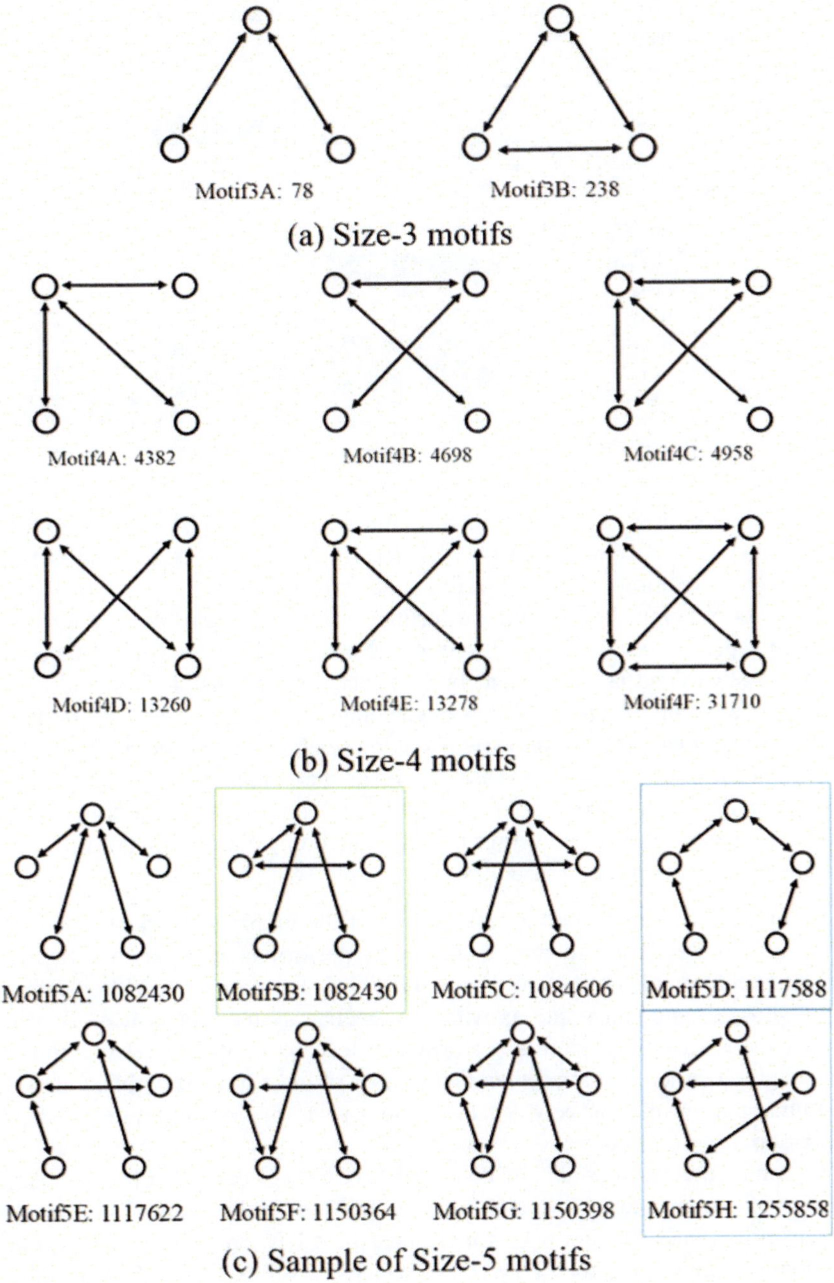

Fig. 6 Different types of network motifs

Table 2 Correlation between motif type occurrence and system performance of the best-performing simulations

Simulation 8		Simulation 9		Simulation 11		Simulation 16		Simulation 28	
Motif5I	0.626	Motif5H	0.484	Motif5T	0.377	Motif5G	0.860	Motif5G	0.280
Motif5L	0.626	Motif5E	0.489	Motif3B	0.396	Motif5B	0.864	Motif5F	0.282
Motif4A	0.627	Motif4B	0.492	Motif5N	0.407	Motif5F	0.869	Motif5B	0.285
Motif5B	0.629	Motif4A	0.497	Motif5M	0.425	Motif3A	0.871	Motif5H	0.288
Motif5F	0.629	Motif5C	0.536	Motif4C	0.434	Motif4C	0.881	Motif4B	0.289
Motif5C	0.636	Motif5D	0.544	Motif4A	0.446	Motif5O	0.882	Motif5J	0.289
Motif5E	0.642	Motif5B	0.549	Motif5H	0.447	Motif5I	0.883	Motif5T	-0.298
Motif5P	0.643	Motif5K	0.554	Motif5E	0.475	Motif5J	0.889	Motif5I	0.300
Motif5J	0.664	Motif5S	-0.524	Motif5A	0.492	Motif4B	0.890	Motif5E	0.306
Motif5O	0.666	Motif5R	-0.583	Motif5D	0.495	Motif5H	0.897	Motif5P	0.307
Motif5G	0.670	Motif5T	-0.583	Motif5B	0.501	Motif5D	0.900	Motif5L	0.310
Motif5M	0.682	Motif5U	-0.602	Motif4B	0.526	Motif5E	0.902	Motif5D	0.323
Motif5N	0.756	Motif5A	0.631	Motif3A	0.549	Motif5M	0.904	Motif5U	-0.345

coefficient of this motif type is 0.542. Similarly, motif 5H is in the top 10 of 29 motif types in 10 simulations (11, 13, 16, 18, 20, 26, 27, 28, 29, and 30). The average correlation coefficient of this motif type is 0.522.

(b) When correlating the foraging performance to the standard deviation of each motif type, at least 10 of the 15 simulations studied shows 2 motif types that have been valued as significant. In 10 of the simulations (1, 8, 9, 11, 16, 18, 20, 26, 27, and 29), motif 5B has been found to have a correlation coefficient of 0.50 or greater. The average correlation coefficient of this motif using the standard deviation metric is 0.544. In 10 of the simulations (1, 9, 11, 13, 16, 19, 20, 26, 28, and 30), motif 5H has been found to be in the top 10 of 29 motif types studied. The average correlation coefficient of this motif using the standard deviation metric is 0.498.

Our initial results indicated a possibility that the global network structure of a swarm system was likely to play a role in its performance rather than the local network structures of the system. Particularly, it was observed that the degree of a node showed potential relations with how well a system performed. To verify the conclusion that the system's performance level is highly correlated with the local subgraph structures rather than the number of links present within a network, we studied the correlation between the system performance and system's degree distribution.

The complementary cumulative distribution (CCD) of each simulation's degree metric has been analyzed. Similarly, each simulation has been divided up into ten different timeframes. Within each timeframe, the CCD curves have been plotted via a logmarithmic scale for every 200 timesteps (see Fig. 8). In characterizing a CCD, the standard deviation has been calculated to determine the variability of the degree metric throughout each timeframe, and the slope has been calculated for the sake of completeness (National Research Council 1996). Therefore, the average and the standard deviation of the slopes of all curves' fitting lines have been calculated within each timeframe and correlated to the system performance values. Based on

It is shown from the results that motifs 5B, 5D, and 5H are significantly influential to the foraging performance of this specific swarm system (see Fig. 7). Therefore, to engineer swarm systems with a higher level of performance, mechanisms can be designed to promote the formation of those motif types. Further validation studies will be required to reevaluate the newly engineered foraging performance and compare it to original simulations. Additionally, the following observation should be noted: the three motifs that have been valued to be significant had two-dangling nodes – two nodes that do not contain outgoing links except with each other. The only other motif structures that contain two-dangling nodes are motif 3A and motif 4B, and they are valued as important in 9 of the 15 simulations.

4 Conclusions

This paper introduces a new approach in engineering complex system based on the theory of network motifs. This theory provides a method to determine the function that is associated with a group of nodes or network motifs that are essential to the system structure and performance. By determining these motifs, complex systems with desired features can be designed. Our approach is network-based, thus is general enough to be applied in other complex system as long as they can be modeled as complex networks.

The complex system used in this study is a swarm simulation modeling a group of ants foraging for food. These simulations have been created in ARGoS – an experimental swarm simulation software. There are 30 simulations that have been created, and 15 of these simulations have been chosen to be used in this study. The results found from this study show a correlation between specific network motifs and swarm foraging performance. The motifs that have been found to have a strong correlation to the foraging performance are motifs 5B, 5D, and 5H; these three motifs all share a two-dangling nodes structure. These motifs have been valued as important in at least 10 of the 15 simulations studied.

Based on these results, new simulations can be generated to study the effect of these motif structures. For example, by controlling how often these motif structures appear, we hypothesize that we can manipulate how well we would want for these systems to perform. These motif structures could then be analyzed on a more microscopic scale to determine why these motifs tend to determine why these motifs tend to determine system performance. This has yet to be achieved due to the complexities required to develop the algorithms to engineer motif structures. We will further investigate it in our future studies. This is also the reason that even if a high correlation has been observed, we are conservative to draw any causations and only provide possible explanations for such phenomenon. Additionally, future studies will be used to determine if the results of this correlation analysis are sensitive to changes in context and architectural parameters such as area dimensions and energy unit assignments.

Acknowledgments We would like to thank Dr. Mengqi Hu and Zishun Yu from the University of Chicago at Illinois for providing the data, images, and simulation code that was necessary to completing this study. We would also like to thank Laxmi Poudel for helping with revisions of the manuscript.

References

Buede, D.M. 2000. *The Engineering Design of Systems: Models and Methods*. New York: Wiley.

Itzkovitz, S., R. Levitt, N. Kashtan, R. Milo, M. Itzkovitz, and U. Alon. 2005. Coarse-Graining and Self-Dissimilarity of Complex Networks. *Physical Review* 71 (1): 016127.

Kashtan, N., S. Itzkovitz, R. Milo, and U. Alon. 2004. Efficient Sampling Algorithm for Estimating Subgraph Concentrations and Detecting Network Motifs. *Bioinformatics* 20 (11): 1746–1758.

Milo, R., S. Shen-Orr, S. Itzkovitz, N. Kashtan, D. Chklovskii, and U. Alon. 2002. Network Motifs: Simple Building Blocks of Complex Networks. *Science* 298 (5594): 824–827.

NASA. 2007. *NASA Systems Engineering Handbook (NASA/SP-2007-6105 Rev1)*. Washington, DC: National Aeronautics and Space Administration.

National Research Council. 1996. *The Waste Isolation Pilot Plant: A Potential Solution for the Disposal of Transuranic Waste*. The National Academies Press.

Pahl, G., and W. Beitz. 1996. *Engineering Design: A Systematic Approach*. 2nd ed. London: Springer.

Rousseau, D. 2018. Three General Systems Principles and Their Derivation: Insights from the Philosophy of Science Applied to Systems Concepts. In *Disciplinary Convergence in Systems Engineering Research*, ed. A. Madni, B. Boehm, R. Ghanem, D. Erwin, and M. Wheaton, 665–681. Cham: Springer.

Scacchi, W. 2001. Process Models in Software Engineering. In *Encyclopedia of Software Engineering*, ed. J.J. Marciniak, 2nd ed. New York: Wiley.

Zabet, N.R. 2011. Negative Feedback and Physical Limits of Genes. *Journal of Theoretical Biology* 284 (1): 82–91.

Zuba, M. 2009. A Comparative Study of Network Motif Detection Tools. *UConn Bio-Grid, REU Summer*.

Enterprise Architecting Applied to Small Unmanned Aircraft System Integration into Low-Altitude Urban Airspace

Raymond T. Vetter and Donna H. Rhodes

Abstract Integrating small unmanned aerial systems (sUAS) into the National Airspace System (NAS) represents a challenging problem set that requires consideration through multiple lenses. Rather than focusing solely on the technological limitations of sUAS operation, this work employs the Architecting Innovative Enterprise Strategy (ARIES) Framework to understand the current and future landscapes for the NAS. The authors use the ARIES elements to holistically describe the current architecture that allows for limited sUAS operations in low-altitude urban airspace. The goal of this work is to develop a level 1 CONOPS as a first step toward a complete enterprise architecture. By identifying current limitations and incorporating emerging concepts and technologies, the authors develop a realistic envisioned future. This future seeks to address externalities that emerge from the increased use of sUAS near the general public. Specific externalities include safety, security, privacy, and transparency concerns. The envisioned future relies on airborne systems to detect and avoid manned aircraft and utilizes an unmanned traffic management system for information sharing and flight coordination. It requires significant investment in developing a shared database to manage unmanned vehicle operations while providing the structure and functions required to make sUAS operations feasible, considering constraints, externalities, and public acceptance.

Keywords System architecture · Small UAS · Systems integration · National airspace · ARIES Framework

R. T. Vetter (✉)
United States Military Academy, West Point, NY, USA
e-mail: raymond.vetter@westpoint.edu

D. H. Rhodes
Massachusetts Institute of Technology, Cambridge, MA, USA

A. M. Madni et al. (eds.), *Recent Trends and Advances in Model Based Systems Engineering*, https://doi.org/10.1007/978-3-030-82083-1_52

1 Introduction and Motivation

This work explores the current environment that has limited the proliferation of sUAS in low-altitude urban airspace and develops an envisioned future that resolves barriers to sUAS operations. For the airspace surrounding major airports (Class B) throughout the United States, current Federal Aviation Administration (FAA) regulations severely limit sUAS operations. Specifically, there are requirements to maintain line of sight with a sUAS and to have adequate detect and avoid capability. These requirements exist for the safety of other aircraft and passengers, as well as ground personnel and property that could be impacted by a sUAS (FAA 2016). However, increasing demand for the use of sUAS, both commercial and public, around major urban areas requires changes to existing regulations. Additional constraints limit the widespread UAS use, from technological limitations to public distrust. The goal of this research is to develop a level 1 CONOPS as a first step toward a complete enterprise architecture.

The addition of sUAS to the National Airspace System (NAS) introduces many positive and negative externalities. Externalities include (among others) safety, privacy, security, and economic benefits and costs. By identifying these emergent characteristics and attempting to understand their implications and interactions, the overall impacts of sUAS operations can be better understood.

This work broadly considers the NAS as an enterprise and uses multiple lenses to understand the current environment and to describe an envisioned future that accounts for the externalities that emerge. Many organizations are working to develop realistic concepts of operations for sUAS within the NAS; however, some of these entities focus on only one aspect (e.g., technical feasibility, regulatory compliance, safety impacts). Considering the NAS as an enterprise and applying the Architecting Innovative Enterprise Strategy (ARIES) Framework allows for holistic sUAS integration solutions in low-altitude urban airspace. With this mindset, the authors develop and propose a future architecture for sUAS operation in low-altitude urban airspace that serves as a possible step to reaching a long-term state of fully integrated sUAS operation within the NAS.

2 ARIES Framework

The ARIES Framework focuses on transforming an enterprise from its current state to a desired future state. Developed by Drs. Deborah Nightingale and Donna Rhodes, the ARIES Framework seeks to develop a thorough understanding of the current enterprise environment based on its defined elements. After conducting this analysis, the architect(s) can establish an envisioned future and select an architecture and implementation plan that moves the enterprise toward that goal. An enterprise "consists of people who generate value for others . . . is a whole system that has a purpose . . . [and] benefits from being part of a larger ecosystem" (Nightingale and

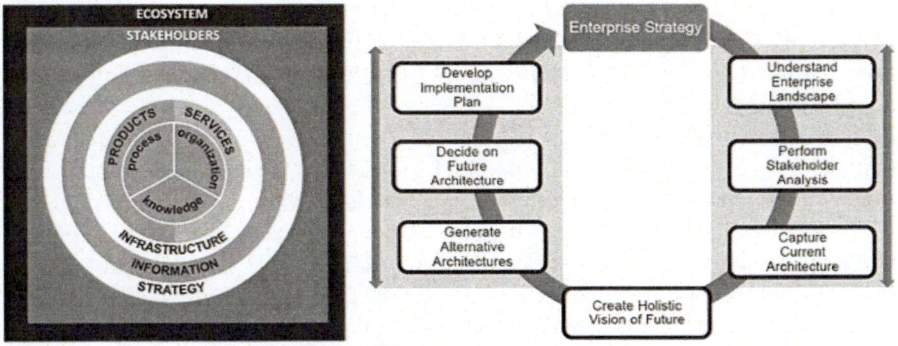

Fig. 1 (**a**) ARIES elements. (**b**) ARIES process (Nightingale and Rhodes 2015, pp. 15–16)

Rhodes 2015, p. 1). Under this definition, it is appropriate to consider the NAS, and specifically low-altitude urban airspace, to be an enterprise.

Using multiple lenses is important when considering an enterprise transformation. Many modern enterprises have failed during transformation initiatives because they focused solely on a new technology and its implementation (Nightingale and Rhodes 2015, p. 5). Figure 1a shows the ARIES elements. The first two elements include entities that may be either internal or external to the enterprise, while the remaining eight elements focus internally on the enterprise. The *ecosystem* is represented as the outermost element, since it includes the external factors that impact the enterprise, specifically, the regulatory, political, economic, and societal factors. *Stakeholders* are the next ARIES element and include the people and organizations that contribute to or are affected by the enterprise (Nightingale and Rhodes 2015, p. 16).

The internal ARIES elements provide different views from which to consider the enterprise. *Strategy* is the element that includes the enterprise core values and overall objectives and sets the direction for the organization. The *information* element entails what information is shared throughout the enterprise and how that material is transmitted. *Infrastructure* is the element relating to the supporting structures (physical facilities, information technology, and communication systems) that allow for continued operations. The *product* element is a physical device or tool related to an enterprise that affects a stakeholder. Similarly, *services* deliver value to stakeholders from enterprise knowledge, skills, and attributes. *Processes* are the established actions and procedures that support the enterprise goals. The *organization* element includes the hierarchical structure and organizational culture for the enterprise and its key stakeholders. The *knowledge* element contains the expertise and intellectual property residing within the enterprise (Nightingale and Rhodes 2015, pp. 16–17).

While the ten ARIES elements are distinct, an enterprise is a complex system where interactions occur. Viewing an enterprise through ARIES elements cannot be done in isolation, as emergent properties from the "entanglement" of elements may develop (Nightingale and Rhodes 2015, p. 20). Thus, after using the elements to decompose and analyze an enterprise, we consider the emergence of unforeseen conditions and externalities.

The ARIES process (see Fig. 1b) consists of seven activities, beginning with understanding the enterprise landscape and progressing through to developing an implementation plan. To *understand the enterprise landscape*, one must consider the effects of external factors (regulatory, political, economic, and societal) that impact the enterprise. *Performing stakeholder analysis* examines the value exchanges and impacts different entities have on the enterprise. During the *capture the current architecture* activity, the current enterprise architecture is described using the ARIES elements. Considering the ecosystem and stakeholder values, a *holistic future vision* establishes the goal of the enterprise transformation. *Generating alternative architectures* involves ideation and creativity to construct feasible architectures, and *deciding on the future architecture* requires scoring and selecting one architecture. Finally, an *implementation plan* establishes the necessary next steps for the enterprise to successfully transform (Nightingale and Rhodes 2015, pp. 22–26). This work focuses on understanding the enterprise landscape, capturing the current architecture, and creating a holistic vision of the future. Stakeholder analysis is ongoing, and the succeeding activities depend upon outcomes from their predecessors.

3 sUAS Enterprise Landscape

Commercial sUAS use has rapidly increased within the United States in recent years, especially with the establishment of 14 CFR Part 107 in June 2016. The FAA's latest forecast indicates sustained commercial sUAS growth for the foreseeable future, with a forecasted average annual increase of at least 33% through 2022 for the United States (Federal Aviation Administration 2018, p. 43). Considering the NAS Landscape using the ARIES ecosystem factors provides context for recommendations.

The regulatory environment for the NAS is multifaceted, but ultimate authority for creating regulations resides with the FAA. Per the FAA's Office of Chief Counsel, "Congress has vested the FAA with authority to regulate the areas of airspace use, management and efficiency, air traffic control, safety, navigational facilities, and aircraft noise . . . " (Federal Aviation Administration, Office of the Chief Counsel 2015, p. 1). Having one regulatory agency establishes a consistent regulatory system to ensure "the highest level of safety for all aviation operations" (Federal Aviation Administration, Office of the Chief Counsel 2015, p. 2). If state or local governments were to regulate aircraft flight or operation, the result would be a "patchwork quilt" of differing restrictions, and ultimately this fractionalizing of airspace could decrease safety and efficiency (National Conference of State

Legislatures 2018). For this analysis, the authors assume compliance with all approved regulations, policies, and procedures.

The political environment related to the NAS and sUAS reflects differing viewpoints from governing bodies and organizations. The White House has emphasized the need to integrate UAS into the NAS to realize economic benefits. One tangible product has been the FAA's Integration Pilot Program, which President Trump signed on October 25, 2017. The pilot program seeks to promote innovation and develop the technologies to have UAS execute potential mission sets (Federal Aviation Administration 2017). However, making UAS use a reality has met several political challenges, mainly from organizations attempting to influence the regulatory process and from conflicts between federal and state/local authorities. The Association for Unmanned Vehicle Systems International (AUVSI) and the Small UAV Coalition and similar entities have been vocal opponents of amendments and legislation that would put the ability to regulate UAS operations at the state and local government level. Pro-UAS groups feel that some state and local governments may impose more restrictive laws and UAS operations may be severely curtailed (Beasley 2018). The Air Line Pilots Association (ALPA) and Aircraft Owners and Pilots Association (AOPA) represent the general aviation and airline industry pilots. ALPA wrote a letter to Congress on February 12, 2018, requesting advancements in UAS anti-collision technology, identification requirements, and tracking requirements before allowing any UAS operations in controlled airspace (Air Line Pilots Association, International 2018).

The potential economic benefits of UAS are significant and represent a driving factor for integration into the NAS. AUVSI estimates that from 2015 to 2025, there will be more than \$82.1 billion of economic impact from UAS integration. In the first 3 years of NAS integration, there may be over 70,000 new jobs created and over 103,000 by 2025. There will be more than \$482 million of tax revenues from 2015 to 2025. More significantly, every year that UAS are not integrated into the NAS, the United States loses more than \$10 billion in potential economic impact (Jenkins and Vasigh 2013). The UAS industry comprises both large firms seeking to incorporate UAS into their operations and small start-ups attempting to develop new technologies. Since 2000, over 300 new firms have entered the UAS space, mainly focusing on hardware, support services, and operations. Operations encompass software and navigational services, unmanned traffic management systems, threat mitigation, and infrastructure construction (Cohn et al. 2017). Industry is making compelling economic arguments advocating for the integration of UAS into the NAS.

Like the economic environment, the societal impacts of UAS integration will be substantial. However, there exists a dichotomy of potential impacts concerning UAS use. The potential economic benefits and convenience that UAS offer are appealing. The proven uses have also been well-documented and include successful search and rescue efforts, drug interdictions, and fugitive investigations. The opportunity for package delivery and safer working conditions also add to the appeal of UAS (U.S. Department of Justice 2015). However, with the impending proliferation of UAS, safety, security, and privacy cause trepidation for the general public. Risks to the public resulting from UAS flights are a legitimate concern. Mid-air collisions with aircraft are an obvious safety issue for passengers aboard aircraft. However, sUAS weighing up to 55 lb can also create a safety hazard to ground personnel if one were to crash into a person, vehicle, or structure. Another concern relates to privacy, with the possibility of sUAS being used to deliver packages or take long-range imagery. Recent events from technology companies (e.g., Facebook) have highlighted the need to properly secure data to maintain some level of privacy. There are also concerns about law enforcement's use of sUAS to collect data or someone flying a sUAS to collect imagery of a person's home (U.S. Department of Justice 2015). The areas of security, safety, and privacy require attention and solutions to increase societal support for sUAS integration.

4 Current Architecture

When using the ARIES elements to consider the current NAS architecture as it relates to sUAS, it becomes clear that, to date, the main emphasis has been placed on policy and products/services. Current operational policy limitations are set forth by Part 107. They are straightforward but restrictive for sUAS operations in the NAS. The Part 107 rule permits daytime operation of sUAS in Class G (uncontrolled) airspace below 400 feet above ground level while remaining within visual line of sight (VLOS) and not flying over people (National Academy of Sciences, Engineering, and Medicine 2018, p. 44). The authorization process has significantly improved with the implementation of the Low Altitude Authorization and Notification Capability (LAANC) system, but UAS Facility Maps (UASFM) represent a significant constraint on sUAS operations in controlled airspace. UASFM limit flight altitudes in urban airspaces near major airports. Processes do allow for Part 107 exceptions, but receiving waivers can take up to 90 days. The FAA's strategy focuses on safety, which is reasonable; however, it fails to consider that sUAS integration may provide economic and societal benefits that may exceed increased level of risk (Center for the Study of the Drone 2018). The FAA's organizational mindset is very safety-centric, and that mentality may prevent some sUAS benefits from being attained. The NAS information, infrastructure, and knowledge relating to sUAS are minimal compared to manned aircraft. While an air traffic controller lacks the capacity to visualize and process every sUAS within controlled airspace while simultaneously monitoring and managing manned air traffic approaches and departures, the limited information and infrastructure prevents almost any situational awareness of sUAS. The knowledge for sUAS is limited partly because sUAS is still a relatively new system. Individuals in government, industry, and academia are working to close the knowledge gap. Even with the current constraints and limitations, sUAS use and implementation continues to evolve.

5 A Holistic Vision of the Future

With an understanding of the current NAS architecture, the next step in the ARIES process is to establish the envisioned future as it relates to sUAS. Describing the envisioned future provides direction for how the enterprise will transform. A clearly defined future needs to consider the appropriate time horizon. This is a critical aspect for sUAS integration in low-altitude urban airspace. Long-term horizons may allow for a future that lacks realism, constraints, or technological limitations. Conversely, if the planned horizon is too short, expectations may fail to be met because of capacity or time limitations (FAA 2016, pp. 71–81). Using the ARIES elements as architectural decisions, considering a 5-year time horizon, and strategically combining concepts elements, we define an attainable envisioned future (see Fig. 2).

The envisioned future is named *Coordinated Flight* to capture the main architectural design theme that we employed in its creation (see Fig. 3). *Coordinated Flight* supports beyond visual line of sight (BVLOS) sUAS operations in urban airspace. Individual sUAS users maintain the responsibility for separation from all obstacles; however, for the *Coordinated Flight Architecture*, there is coordination between other unmanned aircraft. Detect and avoid (DAA) is achieved via an airborne-based DAA system on the unmanned vehicle to enable remote PIC to remain well clear of manned aircraft. UAS Traffic Management (UTM) coordinates sUAS flights, so there is a coordination and scheduling aspect incorporated to sUAS operations. This architecture provides improved coordination and situational awareness and requires additional infrastructure.

Key services for this envisioned future are airborne-based detect and avoid, UTM to coordinate sUAS flights and exchange data, and limited geofencing to prevent unauthorized operations. The envisioned future architecture relies upon an airborne-based DAA system to enable BVLOS. The current FAA requirement for compliance with 14 CFR 91.113 is to detect and avoid manned aircraft, and new airborne sensor systems exist to meet this standard (TechStartups Team 2018).

Coordinated Flight		
Strategy	•Incremental integration •Segregation between sUAS and manned AC	•Focus on noise impacts and societal acceptance
Policies	•Adopt sUAS well-clear recommendation •Maintain requirement for DAA obstacles	•Require remote identification for all users
Organization	•Elevate UAS Integration Office to consolidate knowledge, expertise, and decision-making	•Increased partnerships with industry to promote innovation
Knowledge	•Increased hiring of experts for UAS Integration Office from industry and academia •Development of MOPS for sUAS, DAA, and C2	•Performance-based test for sUAS BVLOS users with not fully autonomous systems
Products/ Services	•UTM to coordinate/schedule individual flights •DAA via airborne sensors using ACAS sXu logic (radar) •UTM provides spatial data and alerts	•UTM provides recommended flight path to user •Remote ID through UTM •Limited geofencing
Processes	•USS approves flights (via UTM) •Deconfliction via airborne DAA and UTM (manned AC and sUAS)	•Expand/repurpose UAS Facility Maps
Information	•UTM coordinates individual flights •UTM provides situational awareness, spatial data and alerts •DAA commands via airborne sensor	•Notifications and UASFM updates via UTM •Remote ID accessible to LE and public via UTM
Infrastructure	•Transmission of data to UTM via cellular/WiFi (adequate service level required)	•UTM network (managed by USS)

Fig. 2 Coordinated Flight architectural decisions

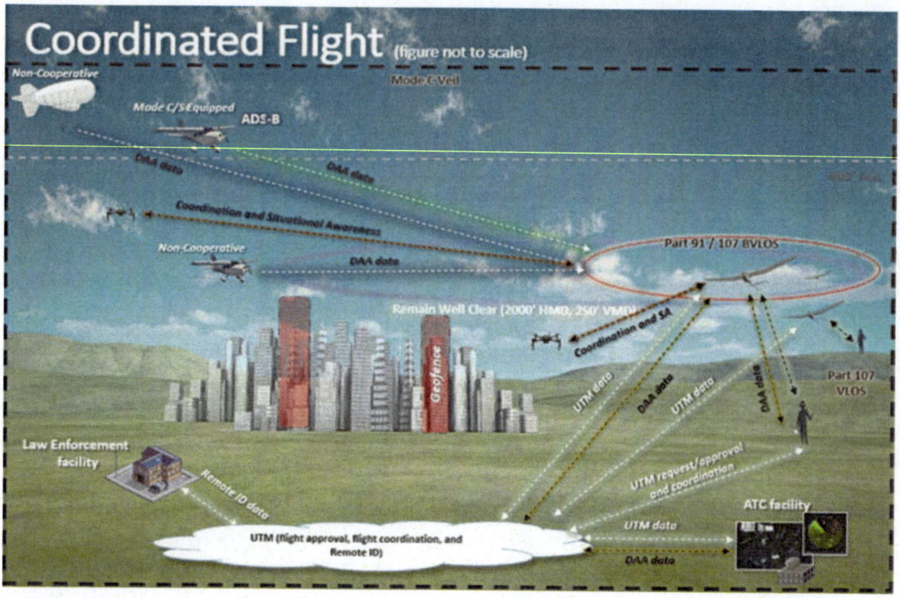

Fig. 3 Coordinated Flight Architecture (adapted) (Intelligent Transportation Society of California 2018, p. 6)

To achieve *Coordinated Flight*, there is an unmanned aerial system traffic management system that provides approval, coordination, and scheduling services. UTM relies on a UAS Service Supplier (USS) to establish a data exchange network that enables automated approval of flight requests and shares NAS information (e.g., temporary flight restrictions, alerts, notifications) to individual sUAS users. After submitting a flight plan, sUAS users receive an approved flight plan from UTM that considers other sUAS operations in the vicinity. The approved flight plan may adjust the proposed flight path or flight times to deconflict with other sUAS operations. This additional level of deconfliction decreases the likelihood of a mid-air collision between sUAS. UTM approval also considers the type of property (residential, commercial, etc.) that the sUAS intends to fly over. To mitigate privacy issues and noise pollution, UTM may approve alternate flight paths that avoid more sensitive properties. With UTM in place, remote identification is possible without a ground sensor network. Upon receiving flight approval via UTM, the ground control station remains connected to UTM. The vehicle transmits the remote identification information back to the GCS, which feeds the information to UTM. UTM stores this data and allow the general public and law enforcement to view relevant information about sUAS operations for transparency and attribution.

Deconfliction between sUAS and other aircraft is a critical process in the *Coordinated Flight Architecture*. While achieving DAA with manned aircraft is a requirement, UTM also provides the additional benefit of deconfliction with other unmanned aerial vehicles. While unable to have transponders that communicate directly between UAVs because of current spectrum limitations, the UTM approval process provides a level of deconfliction between sUAS. When approving flight plans, UTM ensures not only that sUAS operate within established NAS constraints but also that multiple sUAS operations do not conflict with each other. This additional process adds another layer of safety into the enterprise.

Information sharing is primarily accomplished via the UTM network. UTM receives individual flight requests and coordinates those flights. UTM also provides the sUAS users with situational awareness about other sUAS operations in the same airspace, as well as any notifications or alerts. The general public and law enforcement have access to UTM to monitor remote identification information, with law enforcement being able to identify specific users. This remote identification information eliminates anonymity and makes sUAS users more accountable for their actions.

The two infrastructure requirements for the Coordinated Flight Architecture are reliable cellular networks and a robust UTM network. Transmitting data is primarily accomplished through cellular networks, so coverage and reliability must be high. For the urban airspace this architecture considers, the cellular coverage is likely acceptable to enable data transfers between the UAS, ground control station, and UTM network. The UTM network is critical, as it allows for flight coordination, remote identification, and information sharing.

6 Conclusions and Further Research

The ARIES Framework emphasizes that any recommendation for sUAS integration requires changes in multiple domains and that this problem set should be considered through several distinct lenses. While a solution may exist for a specific issue, the ramifications of that decision must be considered for the various stakeholders and across different elements. The goal of this research was to develop a level 1 CONOPS as a first step toward a complete enterprise architecture. The FAA's strategy of incrementally integrating sUAS into the NAS allows for an immediate increase in sUAS operations when compared with awaiting future technologies, processes, and policies to be fully in place. This strategy provides incentive to industry to develop innovative solutions while maintaining safe operations. Organizationally, the FAA should consider elevating the importance and level of the UAS Integration Office. With UAS positioned to increase in both number and capability moving forward, the UAS Integration Office should consolidate the FAA's knowledge, expertise, and decision-making for all UAS-related issues to increase the pace of decision-making and improve communication. Finally, expert knowledge to develop clear standards is essential to providing structure to the rapidly expanding sUAS community and supports both ongoing testing and future operations.

The externalities that emerge from increased sUAS operation and expansion must be fully incorporated in the future architecture design; they cannot merely be acknowledged post hoc. Gaining social acceptance for sUAS operations at low altitudes is challenging because of concerns regarding safety, security, privacy, transparency, and noise pollution; however, gaining public support is essential for long-term success. A recommended architecture needs to adequately address these concerns prior to implementation, as the initial way forward will have a substantial effect on how the NAS integrates sUAS in the future. Future research includes fully developing all key stakeholder perspectives and constructing future architectures that may serve as intermediate steps en route to the proposed holistic vision of the future. Specifically, the authors plan to construct a DoDAF architecture to help capture all relevant viewpoints for sUAS operations in low-altitude urban airspace. After developing future architectures, stakeholders must decide on an architecture and implement it, most likely using a phased approach. Additional safeguards need to be developed to address malicious users in the NAS, as they were considered out of scope for this work. Finally, within the NAS community (government, industry, and academics), continuing to develop performance standards for sUAS is critical for integration.

Acknowledgments This work relied heavily on the support and contributions of many individuals and organizations, including Dr. Wes Olson, MIT Lincoln Laboratory, and the MIT System Design and Management (SDM) program.

References

Air Line Pilots Association, International. 2018, February 12. *Letter for Congress, February 12, 2018*. [Online]. Available: https://www.alpa.org/~/media/ALPA/Files/pdfs/news-events/press-release-content/2018/2018-02-12-alpa-natca-a4a-letter.pdf. Accessed 16 Sept 2019.

Beasley, S. 2018, August 1. Senate FAA Bill Hits More Bumps on the Runway. *Politico*. [Online]. Available: https://www.politico.com/newsletters/morning-transportation/2018/08/01/senate-faa-bill-hits-more-bumps-on-the-runway-302256. Accessed 16 Sept 2019.

Center for the Study of the Drone. 2018, September 25. *Interview: Michael Huerta*. Center for the Study of the Drone at Bard College. [Online]. Available: https://dronecenter.bard.edu/interview-michael-huerta/. Accessed 6 Oct 2019.

Cohn, P., A. Green, M. Langstaff, and M. Roller. 2017, December. *Commercial Drones Are Here: The Future of Unmanned Aerial Systems*. McKinsey & Company. [Online]. Available: https://www.mckinsey.com/industries/capital-projects-and-infrastructure/our-insights/commercial-drones-are-here-the-future-of-unmanned-aerial-systems. Accessed 8 June 2019.

FAA. 2016. *Summary of Small Unmanned Aircraft Rule (Part 107)*. Washington, DC: Federal Aviation Administration.

Federal Aviation Administration. 2017, November 3. *UAS Integration Pilot Program Approval*. U.S. Department of Transportation. [Online]. Available: https://www.faa.gov/news/updates/?newsId=89007&omniRss=news_updatesAoc&cid=101_N_U. Accessed 16 Sept 2019.

———. 2018. *FAA Aerospace Forecast: Fiscal Years 2018–2038*. Washington, DC: Federal Aviation Administration.

Federal Aviation Administration, Office of the Chief Counsel. 2015. *State and Local Regulation of Unmanned Aircraft System (UAS) Fact Sheet*. Washington, DC: U.S. Department of Transportation.

Intelligent Transportation Society of California. 2018. *Urban Air Mobility*. [Online]. Available: https://itscalifornia.org/Content/AnnualMeetings/2018/Presentations/S1P1.pdf. Accessed 4 Sept 2019.
Jenkins, D., and B. Vasigh. 2013. *The Economic Impact of Unmanned Aircraft Systems Integration in the United States*. Arlington: Association for Unmanned Vehicle Systems International.
National Academy of Sciences, Engineering, and Medicine. 2018. *Assessing the Risks of Integrating Unmanned Aircraft Systems into the National Airspace System*. Washington, DC: The National Academies Press.
National Conference of State Legislatures. 2018, September 10. Current Unmanned Aircraft State Law Landscape. *NCSL*. [Online]. Available: http://www.ncsl.org/research/transportation/current-unmanned-aircraft-state-law-landscape.aspx. Accessed 7 Oct 2019.
Nightingale, D.J., and D.H. Rhodes. 2015. *Architecting the Future Enterprise*. Cambridge, MA: The MIT Press.
TechStartups Team. 2018, November 28. *NUAIR Alliance Announces Successful Joint Flight Operations Demonstration of First-of-Its-Kind Detect and Avoid System*. TechStartups. [Online]. Available: https://techstartups.com/2018/11/28/nuair-alliance-announces-successful-joint-flight-operations-demonstration-first-kind-detect-avoid-system/. Accessed 4 Sept 2019.
U.S. Department of Justice. 2015, May 22. *Department of Justice Policy Guidance: Domestic Use of Unmanned Aircraft Systems (UAS)*.

Identification of Elements and Element Relationships for Organizational Architectures for Systems Engineers

Garima Bhatia and Bryan Mesmer

Abstract The lack of and need for theoretical foundations to systems engineering have been recognized by multiple researchers in recent years. The lack of a foundation extends to the positions of systems engineers in organizations. Presently, organizational architectures for systems engineers are based on heuristics. Since systems engineering is required to alleviate certain challenges associated with the development of complex systems, a strong theoretical foundation to the establishment of organizational architectures for systems engineers is imperative. Such a theoretical foundation will ensure that the contribution made by systems engineers to organizational value can be improved. The goal of this paper is to provide a basis for creating a mathematical framework for organizational architectures for systems engineers. A literature review spanning multiple disciplines is conducted to identify elements pertaining to the organizational architectures. These elements are then used in a directed graph to visually represent the relationships between the elements and the mapping between systems engineers and organizational value.

Keywords Organizational architectures · Systems engineers · OASE · Set theory · Functions

1 Introduction

A systems engineering (SE) objective is to facilitate the smooth development of large-scale complex engineered systems (LSCES). A vast number of approaches can be associated with SE. The use of these approaches has changed over time, with certain approaches fading away, new approaches gaining recognition, and other approaches remaining consistently applicable (Bhatia and Mesmer 2019b; Boehm 2006). The instability in the performance of SE approaches over time, and

G. Bhatia (✉) · B. Mesmer
The University of Alabama in Huntsville, Huntsville, AL, USA
e-mail: gb0027@uah.edu

A. M. Madni et al. (eds.), *Recent Trends and Advances in Model Based Systems Engineering*, https://doi.org/10.1007/978-3-030-82083-1_53

the lack of a standard SE methodology may be attributed to a lack of a formal theory for SE. The need for underlying scientific foundations for SE has been recognized at a number of SE and design engineering workshops conducted in the past (DARPA/NSF 2009; Bloebaum et al. 2012; Simpson and Martins 2011; Collopy 2015a) and by researchers (Davendralingam et al. 2016; Collopy 2015b). While researchers are attempting to establish a formalized mathematical theory for SE (Wymore 1993, 1967; Buede 2009), the community still has a long way to go.

The lack of a formal theory for SE also includes the lack of a theoretical foundation to organizational architectures for systems engineers (OASE). Systems engineers interact with and act as the interface between different teams. Although past research has focused on the roles and positions of systems engineers, a mathematical foundation to the establishment of OASE is lacking. Presently, OASE are based on heuristics. The goal of this research is to contribute to the theoretical foundations of SE organizational architectures. A framework for establishing OASE with mathematical underpinnings could be used to guide organizations toward making decisions pertaining to OASE. It is important to focus on OASE because of the following:

- It is impossible for a single person to retain all information pertaining to complex systems; thus, dissemination of information is required, thereby requiring multiple systems engineers (Levchuk et al. 2002a, b).
- Delegation of tasks and decision-making authority is needed in complex organizations in order for them to be effective (Levchuk et al. 2002a).
- As the complexity of projects increases, management of couplings becomes difficult; thus, more specialized personnel are needed (Bloebaum and McGowan 2012).
- Since different activities command different proficiencies, it might be difficult, even impossible at times, for a single person to perform all SE activities. This is true especially in the case of LSCES. However, this statement largely depends on the size and nature of the product (Pyster et al. 2018).
- Overloading personnel with tasks may result in reduced efficiency, which in turn may cause a reduction in the personnel's output (Levchuk et al. 2002a).
- Similarly, redundancy in the organization in terms of SE personnel may lead to wastage of organizational resources, thereby decreasing organizational value (Svejnar and Terrell 1991; Bendor 1985).
- Assignment of roles to the wrong systems engineers may lead to a poor output due to the reduced efficiencies of the systems engineers in completing the tasks.
- An incorrectly assigned systems engineer might require the assistance of other systems engineers to accomplish their tasks, which might cause a reduction in the output of the other systems engineers (Rizzo et al. 1970).

Considering the importance of effective SE, the establishment of effective OASE is key. Thus, a framework, with its foundations grounded in mathematical principles, will provide rigor in establishing OASE. The goal of this paper is to identify key elements pertaining to OASE and the relationships between these elements. A cross-disciplinary literature review is conducted for accomplishing this. A visual mapping between systems engineers and organizational value is then created using the OASE elements. The OASE elements and their relationships will be leveraged in the future to create formal mathematical definitions and theorems, which will together comprise the mathematical framework. Elements of set theory, functions, relations, and graph theory will be leveraged to create the mathematical framework.

2 Cross-Disciplinary Literature Review

For this research, a cross-disciplinary literature review was conducted on OASE elements. An OASE element is any factor that can enable the establishment of an OASE. In addition to studies pertaining to the roles and positions of systems engineers, disciplines including organizational theory, management, operations research, labor economics, and psychology were also reviewed to identify OASE elements.

2.1 Sheard and Helix Study

Twelve roles of systems engineers are examined in an early study conducted by Sheard (1996b). Sheard provides a description of each role and the responsibilities expected to be fulfilled by each role. Heuristics for organizations to determine the appropriate SE role are suggested, and a brief discussion is provided on the interactions between the roles. Following Sheard, a multi-year study called Helix was conducted by the SERC to investigate the effectiveness of systems engineers (Pyster et al. 2013; Hutchinson et al. 2018b). The Helix study interviewed thousands of practicing systems engineers to determine the current state of SE. The Helix study modified the roles defined by Sheard based on the interview data, thereby making them more inclusive. The Helix study also added five new roles. A comparison between the Sheard and Helix role titles can be found in (Bhatia and Mesmer 2019a).

The end products of the Helix study were a self-assessment tool for systems engineers called Atlas (Hutchinson et al. 2018a) and a book summarizing the study (Pyster et al. 2018). In addition to the roles of systems engineers, the book also discussed the proficiencies of systems engineers. The roles defined in these studies will be used in this research to support the mathematical framework for OASE.

2.2 Role Allocation, Role Conflict, and Ambiguity

The assignment of roles and tasks to individuals in organizations has been discussed by multiple researchers. Brandon et al. suggest the use of transactive memory systems for associating tasks, expertise, and people (Brandon and Hollingshead

2004). Acuna and Juristo develop a capability-oriented process model for the assignment of roles in software projects (Acuña and Juristo 2004). The capabilities include intrapersonal, interpersonal, organizational, and management skills, which are behavioral competencies. Hoogendoorn and Treur propose the use of dynamic role allocation to obtain robustness in a multi-agent organization by means of predicate-logic-based temporal logic (Hoogendoorn and Treur 2009). Rizzo et al. discuss how role conflicts and role organizations lead to stress among employees, thereby affecting organizational performance (Rizzo et al. 1970).

2.3 *Optimal Allocation Problem*

The allocation or assignment problem in optimization and its variations are often used for the allocation of roles (Ross and Soland 1975; Campbell and Wu 2011; Nair et al. 2003). Gerkey et al. modified the classic optimal allocation problem (AOP) from operations research to obtain the optimal allocation of roles in RoboCup (Gerkey and Matarić 2003). Farinelli et al. utilize a modified distributed constraint optimization algorithm to account for the dynamic nature of task allocation in extreme teams (Scerri et al. 2005). Dastani et al. use the formalisms of the agent programming language (APL) to determine the allocation of roles in open systems (Dastani et al. 2003).

2.4 *Span of Control (SoC)*

The simple definition of SoC is number of subordinates allotted to a supervisor; however, improvised versions of this definition exist (Ouchi and Dowling 1974). Urwick discusses the relationship between SoC and efficiency from a military perspective (Urwick 1922). Keren and Levhari develop a model to establish the optimum SoC in a pure hierarchical organization with the overall goal of minimizing planning time (Keren and Levhari 1979). Meier and Bohte study the relationships between SoC and organizational performance by conducting a study on 2712 schools (Meier and Bohte 2000). Other studies have focused on control and optimal firm sizes (Williamson 1967).

2.5 *Organizational Context*

Another study of interest to this research is conducted by Ein-Dor and Segev that identified variables related to organizational context that affect the success of management information systems (Ein-Dor and Segev 1978). Examples of the variables include organizational size, structure, and organizational resources.

3 OASE Elements

Based on the cross-disciplinary literature review, a list of elements important to the establishment of OASE displayed in Table 1 is formed. The table provides the names of the elements, the element descriptions, the rationale for including the elements for establishing OASE, and the literature that the element was identified in.

4 Mapping Systems Engineers to Organizational Value

Based on the OASE elements identified in Table 1, a visual representation of the relationships between elements is created using a directed graph. The nodes in the graph are the OASE elements, whereas the relationships between the elements are the edges. The visual representation will aid in understanding how systems engineers impact organizational value by mapping systems engineers to organizational value. The mapping is shown in Fig. 1.

Figure 1 is color-coded in a manner that groups similar elements using colors. For example, elements pertaining to organizations such as capabilities, organization size, etc. are all colored gray. Similarly, elements pertaining to systems engineer roles and elements pertaining to value are grouped together and colored light green and dark green, respectively. Elements that did not fit any groupings were given separate colors. Systems engineers are colored orange, proficiencies are colored beige, and position levels are given the color yellow.

In Fig. 1, it is seen that organizations hire systems engineers based on the requirements of the project and/or organization. The number of systems engineers recruited depends upon the scale of the projects undertaken by the organization and/or the size of the organization. Each system engineer has certain proficiencies. The proficiencies are attributes of a systems engineer, such as education, experience, leadership skills, lifecycle phases worked on, etc. These proficiencies also decide the position of the systems engineer in the organization. The position determines the rank of the systems engineer in the organizational hierarchy. The levels of the organizational hierarchy are influenced by the size of the organization and/or the scope of the project undertaken by the organization. Additionally, the SoC determines the hierarchical levels within organizations. Roles are assigned to systems engineers based on their proficiencies and position levels. Each role is expected to fulfill certain tasks. By completing these tasks, systems engineers create value for the organization. In this case, the value contributed is the impact made on the organizational value ($\Delta OrgVal$).

The performance metrics in Fig. 1 are a quantitative means of measuring $\Delta OrgVal$. A detailed description of the preliminary ideas for measuring the sub-values and $\Delta OrgVal$ are provided in a previous study (Bhatia and Mesmer 2019a). However, the big question that arises is "How are the appropriate roles determined for a project by an organization?". The answer to this question lies in the capabilities

Table 1 List of OASE elements with rationale

Element	Description	Rationale	Literature
Systems engineer	An individual that performs SE activities	The central elements of OASE are the systems engineers	Helix (Pyster et al. 2013; Hutchinson et al. 2018b), Sheard (1996b), SE Handbook (Kapurch 2010; Haskins et al. 2006)
Organizations	Any government or commercial institution developing a product or system or providing a service	The systems engineers are employed by organizations	Org. theory (Jones 2013; Burton et al. (2015), Kinicki and Kreitner (2006)
Systems engineer roles	A role is a collection of related SE activities (Hutchinson et al. 2018b)	The roles are one of the three factors for establishing the mathematical framework for OASE. Correct assignment of roles to systems engineers will lead to a better OASE, leading to an increase in the contribution to organizational value	Sheard (1996b), Helix (Pyster et al. 2013; Hutchinson et al. 2018b)
Number of systems engineers	The total number of systems engineers employed by an organization	This was considered to be an element of OASE, since the mathematical framework will also decide the number of systems engineers required for the architecture	Ein-Dor and Segev (1978), SoC (Ouchi and Dowling 1974; Keren and Levhari 1979; Meier and Bohte 2000)
Positions of systems engineers	A unit of measure for the experience of a systems engineer (Hutchinson et al. 2018b)	The positions decide the size of the hierarchy and the assignment of responsibility to individuals	Helix (Pyster et al. 2018), Ein-Dor and Segev (1978), SoC (Ouchi and Dowling 1974; Keren and Levhari 1979; Meier and Bohte 2000)
$\Delta OrgVal$	The change in an organization's value by virtue of SE activities performed by the organization	This OASE element will be the objective function in this research, with a goal to maximize $\Delta OrgVal$	Helix (Pyster et al. 2013; Hutchinson et al. 2018b), Sheard (1996a), ROI on SE (Honour 2013)
Role tasks	The role tasks are the activities associated with each role	By means of completing these activities, the systems engineers generate value for the organization	Helix (Pyster et al. 2018), Sheard (1996b)
Sub-values	The output generated by each role for performing role tasks	The aggregation of the sub-values is $\Delta OrgVal$	Helix (Hutchinson et al. 2018b; Pyster et al. 2018), Sheard (1996a)

Lifecycle phases	The different segments of a lifecycle	It is assumed in this research that the capabilities of an organization are defined by the work performed by the organization pertaining to different lifecycle phases	Wasson (2015), INCOSE SE Handbook (Robertson 1998), NASA SE Handbook (Kapurch 2010)
Proficiencies	The attributes of a systems engineer, including education, skills, and experience	The proficiencies of a systems engineer are essential in deciding the appropriate roles for the systems engineer	Helix (Pyster et al. 2018), Brandon et al. (Brandon and Hollingshead 2004), Acuna and Juristo (2004), Hoogendoorn and Treur (2009)
Organizational context (org. size, project scale)	Organizational size refers to the number of employees in the organization. Project scale refers to the complexity of the system being developed. A measure of complexity for this research is yet to be determined	The number of systems engineers and the positions will depend on the size of the organization and/or the scope of the project undertaken by the organization	SoC literature (Ouchi and Dowling 1974; Keren and Levhari 1979; Meier and Bohte 2000), Ein-Dor and Segev (1978)
SoC	The number of subordinates allotted to a supervisor	This element is important to determine the optimal number of subordinates that can be assigned to a systems engineer	SoC literature (Keren and Levhari 1979; Ouchi and Dowling 1974; Urwick 1922; Meier and Bohte 2000)
Performance metrics	A measure of effectiveness of systems engineers. Metrics could include efficiency, time to complete tasks, cost of activities, revenue, profit, etc.	These metrics will be used to evaluate the contribution made by systems engineers. This element can be used to determine if a systems engineer is capable enough for a role and/or if the systems engineer is being overburdened with responsibility	Shanteau et al. (2002), Rizzo et al. (1970), Urwick (1922)
Marginal product of labor	The marginal product of labor is defined as the change in output resulting from hiring an additional worker, holding constant the quantities of all other inputs	This element will be useful in assessing an organization's decision to hire an additional systems engineer	Borjas and Van Ours (2010), Cahuc et al. (2014)
Value of marginal product of labor	The value of marginal product of labor is the dollar increase in revenue generated by an additional worker, holding capital constant	This element will be useful in assessing an organization's decision to hire an additional systems engineer	Borjas and Van Ours (2010), Cahuc et al. (2014)

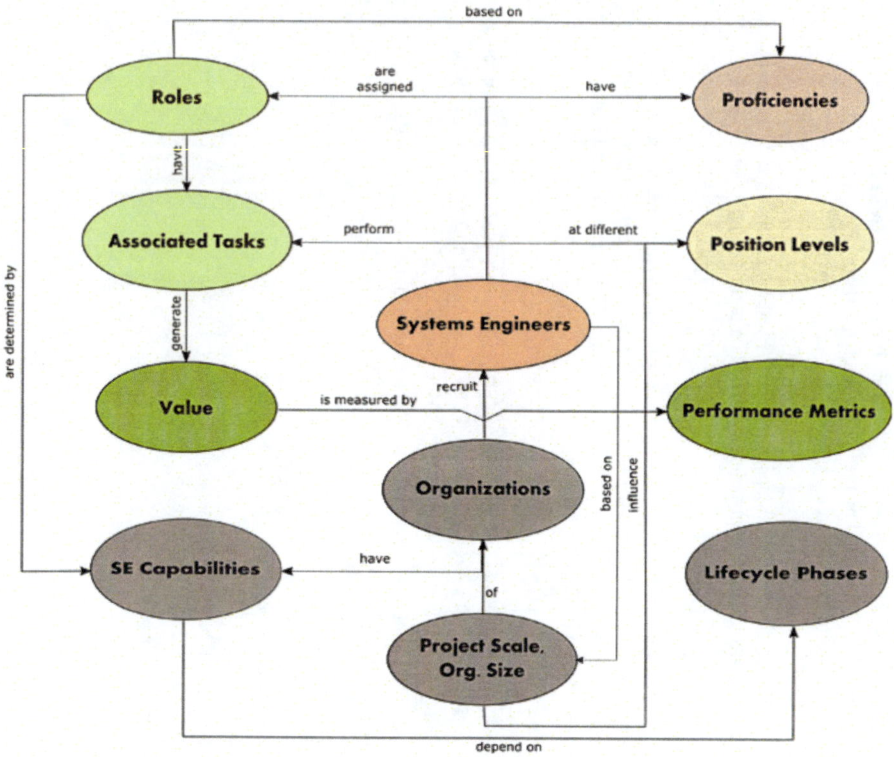

Fig. 1 Mapping between systems engineers and organizational value

offered by the organization. Since the main focus of SE is the development of a system through the different lifecycle phases, lifecycle phase services play a major role in determining the roles for systems engineers required by the organization. For example, if an organization offers conceptual design services only, some of the likely roles required for that organization's capabilities are system architect, requirements owner, and system analyst, among others. The capabilities of the organization might depend on other factors; however, for this research, the lifecycle phase services are considered to be the primary capability determinant of an organization.

The elements and relationships described above will be represented mathematically using set theory, functions, relations, and graph theory. Theorems will then be created to guide organizations toward establishing their OASE. Examples of such definitions are provided below.

1. ***Systems engineer:*** A systems engineer, denoted by *se*, is any person that performs systems engineering activities. The set of all systems engineers is denoted by *SYSENGRS*.

2. ***Systems engineer objective:*** An objective *obj* is a task expected to be completed by a systems engineer. The set of all objectives is denoted by *OBJECTIVE.*
3. ***Satisfaction function:*** The satisfaction function $sat : SYSENGRS \times OBJECTIVE \rightarrow \{FALSE, TRUE\}$ determines the satisfaction of a systems engineer objective by a systems engineer. In this case, we assume that the satisfaction function is binary.

5 Summary and Future Work

There is a lack of foundational theory to SE, which extends to OASE. This paper provides a base for creating a mathematical framework to guide the establishment of OASE. A cross-disciplinary literature review is performed which leads to a list of OASE elements. A directed graph is created which provides a visual representation of the mapping between systems engineers and organizational value by leveraging OASE elements and their relationships. The mapping can be used as a means for understanding the relationships between the elements and to create mathematical definitions and theorems.

Future work will employ set theory and functions to provide formal definitions of OASE elements and their relationships. The definitions will be used to create theorems and corollaries using formal methods of doing proofs. The theorems will be proven using the properties of sets and functions. Together, the mathematical framework comprising the definitions and theorems will provide a means for guiding organizations in establishing their OASE.

References

Acuña, Silvia T., and Natalia Juristo. 2004. Assigning People to Roles in Software Projects. *Software: Practice and Experience* 34 (7): 675–696.

Bendor, Jonathan B. 1985. *Parallel Systems: Redundancy in Government.* University of California Press.

Bhatia, Garima, and Bryan Mesmer. 2019a. Preliminary Analysis of Value Contributed by Systems Engineers to Organizations. In *AIAA Scitech 2019 Forum*, 0766.

———. 2019b. Trends in Occurrences of Systems Engineering Topics in Literature. *Systems* 7 (2): 28.

Bloebaum, Christina L., and A. R. McGowan. 2012. The Design of Large-Scale Complex Engineered Systems: Present Challenges and Future Promise. In *Proceedings of the 14th AIAA/ISSMO Multidisciplinary Analysis and Optimization Conference, Indianapolis, IN, Paper No. AIAA-2012-5571*. https://doi.org/10.2514/6.2012-5571.

Bloebaum, Christina L., Paul D. Collopy, and George A. Hazelrigg. 2012. NSF/NASA Workshop on the Design of Large-Scale Complex Engineered Systems—From Research to Product Realization. In *Proceedings of the 14th AIAA/ISSMO Multidisciplinary Analysis and Optimization Conference, Indianapolis, IN, Paper No. AIAA-2012-5572.*

Boehm, Barry. 2006. Some Future Trends and Implications for Systems and Software Engineering Processes. *Systems Engineering* 9 (1): 1–19.

Borjas, George J., and Jan C. Van Ours. 2010. *Labor Economics*. Irwin Boston: McGraw-Hill.

Brandon, David P., and Andrea B. Hollingshead. 2004. Transactive Memory Systems in Organizations: Matching Tasks, Expertise, and People. *Organization Science* 15 (6): 633–644.

Buede, Dennis M. 2009. *The Engineering Design of Systems: Models and Methods*. Wiley Online Library.

Burton, Richard M., Børge Obel, and Dorthe Døjbak Håkonsson. 2015. *Organizational Design: A Step-by-Step Approach*. Cambridge University Press.

Cahuc, Pierre, Stéphane Carcillo, and André Zylberberg. 2014. *Labor Economics*. MIT press.

Campbell, Adam, and Annie S. Wu. 2011. Multi-Agent Role Allocation: Issues, Approaches, and Multiple Perspectives. *Autonomous Agents and Multi-Agent Systems* 22 (2): 317–355.

Collopy, Paul D. 2015a. Report on the Science of Systems Engineering Workshop. In *53rd AIAA Aerospace Sciences Meeting*, 1865.

Paul Collopy. 2015b. Systems Engineering Theory: What Needs to Be Done. In *2015 Annual IEEE Systems Conference (SysCon) Proceedings*, 536–541. IEEE.

DARPA/NSF. 2009. Systems Engineering and Design of Complex Aerospace Systems. Presented at the DARPA/NSF Systems Engineering and Design of Complex Aerospace Systems Workshop, Arlington, VA.

Dastani, Mehdi, Virginia Dignum, and Frank Dignum. 2003. Role-Assignment in Open Agent Societies. In *Proceedings of the Second International Joint Conference on Autonomous Agents and Multiagent Systems*, 489–496. ACM.

Davendralingam, Navindran, Zhenghui Sha, Kushal Moolchandani, Apoorv Maheshwari, Jitesh H. Panchal, and Daniel A. DeLaurentis. 2016. Scientific Foundations for Systems Engineering: Challenges and Strategies. In *ASME 2015 International Design Engineering Technical Conferences and Computers and Information in Engineering Conference*. American Society of Mechanical Engineers Digital Collection.

Ein-Dor, Phillip, and Eli Segev. 1978. Organizational Context and the Success of Management Information Systems. *Management Science* 24 (10): 1064–1077.

Gerkey, Brian P., and Maja J. Matarić. 2003. On Role Allocation in RoboCup. In *Robot Soccer World Cup*, 43–53. Springer.

Haskins, Cecilia, Kevin Forsberg, Michael Krueger, D. Walden, and D. Hamelin. 2006. Systems Engineering Handbook. In *INCOSE*. Seattle: International Council on Systems Engineering.

Honour, Eric C. 2013. *Systems Engineering Return on Investment*. University of South Australia Australia.

Hoogendoorn, Mark, and Jan Treur. 2009. An Adaptive Multi-Agent Organization Model Based on Dynamic Role Allocation. *International Journal of Knowledge-Based and Intelligent Engineering Systems* 13 (3–4): 119–139.

Hutchinson, Nicole A., Dinesh Verma, Pamela Burke, Megan Clifford, Ralph Giffin, Sergio Luna, and Matthew Partacz. 2018a. *Atlas 1.1: An Update to the Theory of Effective Systems Engineers*. Stevens Institute of Technology Hoboken United States.

Hutchinson, Nicole, Dinesh Verma, Pamela Burke, Megan Clifford, Ralph Giffin, Sergio Luna, and Matthew Partacz. 2018b. *RT-173: Helix, 2017 Helix Technical Report*. Stevens Institute of Technology Hoboken United States.

Jones, Gareth R. 2013. *Organizational Theory, Design, and Change*. Upper Saddle River, NJ: Pearson.

Kapurch, Stephen J. 2010. *NASA Systems Engineering Handbook*. Diane Publishing.

Keren, Michael, and David Levhari. 1979. The Optimum Span of Control in a Pure Hierarchy. *Management Science* 25 (11): 1162–1172.

Kinicki, Angelo, and Robert Kreitner. 2006. *Organizational Behavior: Key Concepts, Skills & Best Practices*. Irwin New York: McGraw-Hill.

Levchuk, Georgiy M., Yuri N. Levchuk, Jie Luo, Krishna R. Pattipati, and David L. Kleinman. 2002a. Normative Design of Organizations. I. Mission Planning. *IEEE Transactions on Systems, Man, and Cybernetics-Part A: Systems and Humans* 32 (3): 346–359.

Levchuk, Georgiy, Yuri Levchuk, Jie Luo, Krishna Pattipati, and David Kleinman. 2002b. Normative Design of Organizations. II. Organizational Structure. *IEEE Transactions on Systems, Man, and Cybernetics-Part A: Systems and Humans* 32 (3): 360–375.

Meier, Kenneth J., and John Bohte. 2000. Ode to Luther Gulick: Span of Control and Organizational Performance. *Administration & Society* 32 (2): 115–137.

Nair, Ranjit, Milind Tambe, and Stacy Marsella. 2003. Role Allocation and Reallocation in Multiagent Teams: Towards a Practical Analysis. In *Proceedings of the Second International Joint Conference on Autonomous Agents and Multiagent Systems*, 552–559. ACM.

Ouchi, William G., and John B. Dowling. 1974. Defining the Span of Control. *Administrative Science Quarterly* 19 (3).

Pyster, Arthur, Deva Henry, Nicole Hutchsion, Kahina Lasfer, and Stan Rifkin. 2013. *The Helix Project*. Hoboken NJ: Systems Engineering Research Center.

Pyster, Arthur, Nicole Hutchison, and Devanandham Henry. 2018. *The Paradoxical Mindset of Systems Engineers: Uncommon Minds, Skills, and Careers*. Wiley.

Rizzo, John R., Robert J. House, and Sidney I. Lirtzman. 1970. Role Conflict and Ambiguity in Complex Organizations. *Administrative Science Quarterly*: 150–163.

Robertson, Tim. 1998. INCOSE Systems Engineering Handbook. *INSIGHT* 1 (2): 20–20.

Ross, G. Terry, and Richard M. Soland. 1975. A Branch and Bound Algorithm for the Generalized Assignment Problem. *Mathematical Programming* 8 (1): 91–103.

Scerri, Paul, Alessandro Farinelli, Steven Okamoto, and Milind Tambe. 2005. Allocating Tasks in Extreme Teams. In *Proceedings of the Fourth International Joint Conference on Autonomous Agents and Multiagent Systems*, 727–734. ACM.

Shanteau, James, David J. Weiss, Rickey P. Thomas, and Julia C. Pounds. 2002. Performance-Based Assessment of Expertise: How to Decide If Someone Is an Expert or Not. *European Journal of Operational Research* 136 (2): 253–263.

Sheard, Sarah A. 1996a. The Value of Twelve Systems Engineering Roles. In *INCOSE International Symposium*, 6:894–902. Wiley Online Library.

Sarah Sheard. 1996b. Twelve Systems Engineering Roles. In *INCOSE International Symposium*, 6:478–485. Wiley Online Library.

Simpson, Timothy W., and Joaquim R.R.A. Martins. 2011. Multidisciplinary Design Optimization for Complex Engineered Systems: Report from a National Science Foundation Workshop. *Journal of Mechanical Design* 133 (10): 101002.

Svejnar, Jan, and Katherine D. Terrell. 1991. *Reducing Labor Redundancy in State-Owned Enterprises*. Vol. 792. World Bank Publications.

Urwick, Lyndall F. 1922. The Manager's Span of Control. *Harvard Business Review* 34 (3).

Wasson, Charles S. 2015. *System Engineering Analysis, Design, and Development: Concepts, Principles, and Practices*. Wiley.

Williamson, Oliver E. 1967. Hierarchical Control and Optimum Firm Size. *Journal of Political Economy* 75 (2): 123–138.

Wymore, Wayne. 1967. *A Mathematical Theory of Systems Engineering: The Elements*. New York: Wiley.

Wymore, A. Wayne. 1993. *Model-Based Systems Engineering*. CRC Press.

Application and Modelling of Systems Engineering Methods to Deployed Enterprise Content Management Systems

Stephan Bren

Abstract This paper presents a proposal for the application and modelling of systems engineering to an already deployed system as a means of resolving challenges frequently associated with deployed enterprise content management systems. It generalizes traditional systems engineering methods as a model that may be applied to already deployed ECM systems. In this paper, the author reviews the traditional systems engineering process followed by a more detailed review of the modified, post-deployment systems engineering process. The author then concludes by discussing the benefits of performing post-deployment systems engineering.

Keyword Systems Engineering

1 Introduction

Enterprise content management systems are frequently deployed as point solutions that are then developed organically into enterprise platforms in response to changing business needs and objectives, without scoping exercises or business planning, on the basis of "let's see what it can do." (Fig. 1)

Such systems may have been originally deployed without a clearly articulated justification as to why it should have been implemented; it may have been deployed in an unplanned manner, without formal business case development prior to deployment or even minimal financial justification. Lack of a clear vision and inadequate planning may contribute to poor user adoption, eventual failure of the system to meaningfully contribute to business needs, and removal of the system at a loss. The lack of adequate business case development and planning hinders management ability to effectively understand and integrate this new tool into organizational IT business processes, such as those defined in ITIL, Internal Standards Organization

S. Bren (✉)
Johns Hopkins University Whiting School of Engineering, Baltimore, MD, USA
e-mail: sbren2@jhu.edu

A. M. Madni et al. (eds.), *Recent Trends and Advances in Model Based Systems Engineering*, https://doi.org/10.1007/978-3-030-82083-1_54

Fig. 1 Enterprise content management system adoption ladder

(ISO), Microsoft Operations Framework (MOF), and others, and this in turn hinders effective organizational adoption. These challenges can be remedied through the application and modelling of post-deployment systems engineering.

2 Traditional Systems Engineering Process

Traditional systems engineering involves an organized and systematic approach to the development of complex systems. Its focus is on the design, development, integration, and testing of a system of interrelated components working together to meet a problem. This approach tends to be a technical one, organized in such a fashion that focuses first on defining the problem in terms of the needs, requirements, and operational concepts as presented by the future users of the solutions and products, before engaging in designing and developing the system to meet the problem. It also engages external factors affecting the development of the system, such as the intended operational environment of the system, its logistical support, and the capabilities of the users of the system. It is iterative in nature, as the lessons learned and emergent discoveries of various tasks in the systems engineering approach are applied to previous tasks and those previous tasks re-engaged so as to improve their output and contribution to the development of the system. This process is engaged with the intent of designing a better product than might otherwise be accomplished.

There are a variety of approaches that may be used for performing systems engineering. The approach that will be engaged here in this effort is best represented using the systems engineering "V" diagram (Kossiakoff et al. 2011). In this approach, the common project tasks of requirements analysis, design, development, testing, and deployment are presented in an orderly fashion that intuitively depicts their sequence and interrelationships.

In traditional systems engineering, the systems engineering process begins with examination of the regional architecture and how that architecture is related to the proposed system; and it involves discovering and collating all artifacts relevant to the project. This effort serves to initially define the system boundary. The feasibility and concept exploration phase examine whether it will be realistically possible to develop a solution to the problem or need, what potential solutions there might be, and what might be the best possible solution. The concept of operations phase presents the understanding of the problem and the proposed solution to that problem in the context of the problem environment. This effort serves to initially define how the system will interact with its environment. During the system requirements phase, the boundaries of the proposed solution with its environment are materialized. In high-level design, also referred to as preliminary design or conceptual design (NASA 2009), effort focuses on developing the overall system architecture in a way that correctly implements system requirements. During the detailed design phase, the system requirements are translated into the specific physical requirements for how the proposed system is to be built.

Actual development of the system proceeds during the software/hardware development and field installation phase. Once the system is developed, it undergoes progressively higher-level testing, from unit to system testing. This first phase of testing serves to verify that the system was built according to its specifications (i.e., "the systems was built right"). The second phase of testing serves to validate that the system meets the intended design needs identified in the concept of operations (i.e., "the right system was built"). On successful completion of verification and validation testing, the system enters into its operational and maintenance phase, where it remains until retirement or replacement. This approach to systems engineering activities is diagrammed on the systems engineering "V" diagram, as presented in Fig. 2. This approach to systems engineering is the context in which this proposal for the application and modelling of systems engineering methods to a deployed enterprise content management system is engaged and understood. In this approach, and, indeed for any systems engineering project, the systems engineering solution development process proceeds from an initial undefined state and advances toward a desired, proposed state, referred to as the "to-be" solution. Each advance in this process incrementally resolves the proposed solution against the continuum of possible solutions, increasingly narrowing the scope of possible solutions until only a few or just one remains as the best possible solution to the original problem.

3 Differences Between Traditional and Post-deployment Systems Engineering

In post-deployment systems engineering, the solution process does not so much advance to a theoretical solution that is increasingly resolved against a continuum of solutions, but to an actual "as-built" solution that is known but more or less

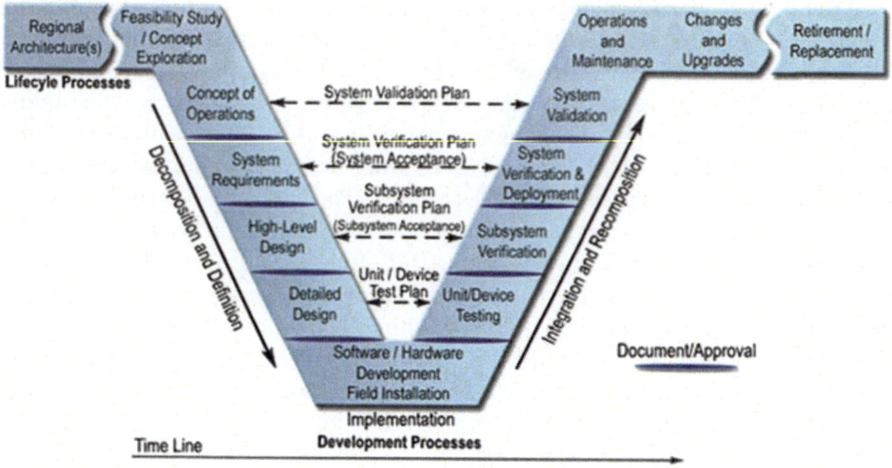

Fig. 2 Traditional systems engineering "V" diagram. (FHWA 2007)

undefined – i.e., the system physically exists and has been deployed, but details of the system may only be partially documented and exist primarily in the minds of its administrators. In effect, the systems engineering approach is generalized as a model that can be applied to any enterprise content management system and indeed to any deployed system.

Post-deployment systems engineering applies the methods of traditional systems engineering to a system that has already been deployed so as to develop the critical system artifacts needed to effectively integrate the system into organizational business processes. The advantage of using systems engineering methods is the structured and systematic approach that it brings to the development of these artifacts and the minimization of stovepipe methods. Just as the application of systems engineering methods to the development of a proposed system helps ensure the highest likelihood of success, so too does the application of systems engineering methods to an already deployed but undefined system helps ensure the highest likelihood of success in accurately defining that system.

In traditional systems engineering, system development proceeds from a large and mostly undefined system solution space to a specific, well-defined "to-be" system solution. This solution is generally not the only possible solution, but is usually the optimal one of a set of possible solutions meeting system requirements. A notional sense of this development is provided in Fig. 3.

In post-deployment systems engineering, the systems engineering process generally follows the same series of steps executed in traditional systems engineering. However, in post-deployment systems engineering, the solution process does not so much advance to a theoretical "to-be" solution that is increasingly resolved against a continuum of solutions, but to an actual "as-built" solution that is known but more or less undefined. These other solutions are not possible solutions in the traditional

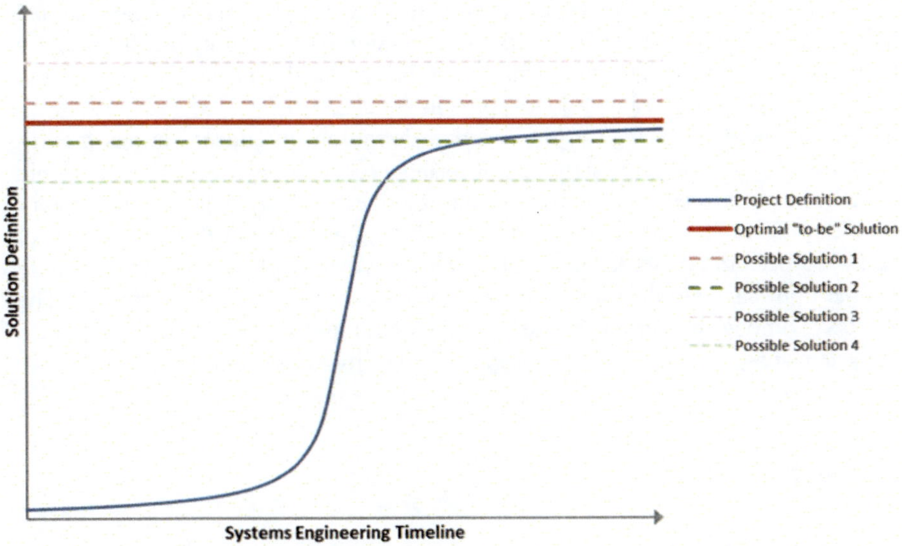

Fig. 3 Traditional systems engineering process timeline

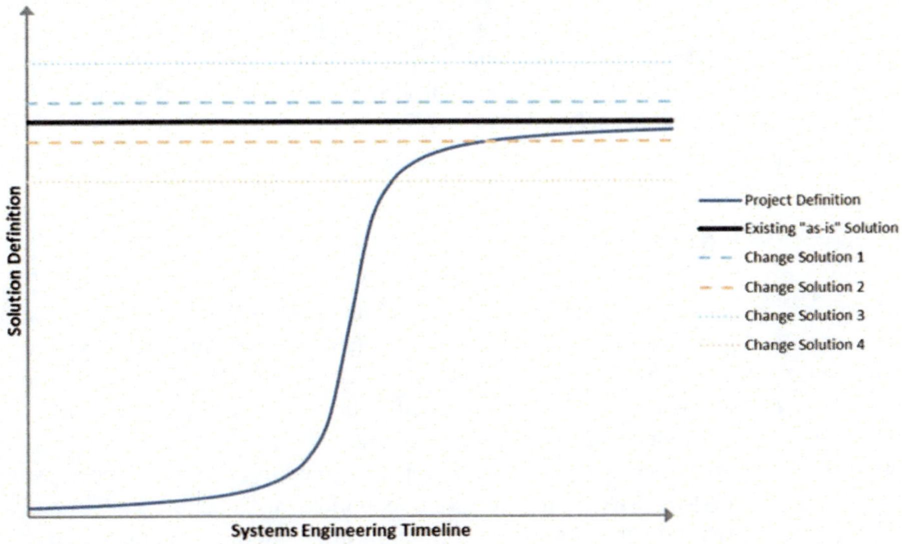

Fig. 4 Post-deployment systems engineering process timeline

systems engineering sense but are solutions in the sense of post-deployment change management. A notional sense of this development is provided in Fig. 4.

Post-deployment systems engineering applies the methods of traditional systems engineering to a system that has already been deployed so as to develop the critical system artifacts needed to effectively integrate the system into organizational business processes. The advantage of using systems engineering methods is the structured and systematic approach that it brings to the development of these artifacts and the minimization of stovepipe methods.

This fundamental change in the solution definition introduces various modifications to the traditional systems engineering processes. Some tasks that would normally be performed in traditional systems engineering may be found to not return significant value in post-deployment systems engineering and therefore may not need to be performed. Some other tasks within a typical systems engineering phase become optional, introducing useful but not necessary value to the process. The core tasks among the various phases remain useful and productively contribute to the systems engineering process as they do for traditional systems engineering.

4 Modified Process

4.1 Overview

Post-deployment systems engineering process follows the traditional "V process" with some modifications. Some phases of the traditional systems engineering process can be skipped, and other phases can be attenuated. Determining the applicability and usefulness of a traditional systems engineering phase to post-deployment engineering is analyzed here on an individual basis.

4.2 Regional Architecture

In this initial systems engineering phase, the modified post-deployment focus is on discovering and gathering relevant artifacts associated with the as-built system and its environment. The general systems engineering plan developed during this phase will not be static but will continually be updated as facts emerge and efforts evolve. The regional architecture phase of the traditional systems engineering approach adds significant value to the post-deployment systems engineering process. This phase defines the boundaries between the more or less undefined "as-built" system and its operational environment, such as network and power interfaces, administrative interfaces, user interfaces, and physical virtual server interfaces. These may be known and understood but insufficiently documented.

4.3 Feasibility Study and Concept Exploration

The feasibility study and concept exploration tasks performed in traditional systems engineering are generally of little value in post-deployment systems engineering. On the other hand, while a concept trade study would not yield value from a technical perspective, it might still return substantive value from a management perspective, such as enabling line managers to engage system policy and budget matters with senior management from a more informed perspective.

4.4 Concept of Operations

The concept of operations is the first significant deliverable that the post-deployment systems engineer will deliver to the system owner/customer. The CONOPS represents a tangible product providing a clear understanding of the system within the organization from the perspective of the system's users. From a post-deployment systems engineering perspective, the CONOPS provides a useful milestone and metric to management in demonstrating that the post-deployment systems engineering task is proceeding productively. It effectively provides management with a key milestone, a Go/No Go decision point.

4.5 System Requirements

The system requirements document (SRD) and the system performance specification will be the two major deliverables resulting from performing the post-deployment systems requirements phase. Specific attention must be given to the clear separation of requirements that apply to the current "as-built" state of the system from its desired, "to-be," state. Capturing both types of requirements adds significant value to the already deployed system.

"As-built" requirements can then be analyzed against known system costs, yielding more accurate and factual cost estimates for system changes and improvements. "To-be" system requirements add value to the post-deployment systems engineering requirements development process as they capture capability gaps that can be addressed through change management.

4.6 System Design

This is the third most significant phase in post-deployment systems engineering of a deployed system. The system specification drafted during this phase effectively defines the baseline of the deployed system and is a necessary component in the development of an Authority to Operate (ATO); building accurate inventory and cost projections; implementing ITIL, ISO, MOF, and other common business processes; and developing reliable continuity of operations procedures.

4.7 Software/Hardware Development and Field Installation

The development stage in traditional systems engineering is of little value in post-deployment systems engineering since the system is already developed and deployed. On the other hand, some of the secondary products of this stage, such as training materials, user manuals, installation, and other documents, may be of value if they have not yet been created.

4.8 Unit Testing

There is no immediate value returned in performing the primary activity of this stage, finding defects. However, some substantive value can still be obtained in performing supporting activities of this stage, such as drafting test cases and drafting a unit verification plan. Test cases and a unit verification plan can be used in developing regression testing procedure for the system, useful for patching and upgrade testing, custom solution deployment testing, and interface modification testing.

4.9 Subsystem Verification

In post-deployment systems engineering, the system has already been integrated from its components and is deployed and operational. Thus, the components are presumed to have met their design objectives and requirements, and subsystem verification against system requirements is not immediately unnecessary. In other words, the subsystems are presumed to have been "built right." Thus, the focus of this phase is less on subsystems meeting requirements and more on whether the requirements accurately reflect them.

4.10 System Verification and Deployment

In post-deployment systems engineering, the system is already deployed and operational. Thus, the system is presumed to have met its operational requirements, and traditional system verification is not immediately necessary. The system is presumed to have been "built right." System verification in post-deployment systems engineering can be used to refine the previously developed as-built and "to-be" requirements against the actual deployed system. Deltas identified from the as-built can be used to optimize the requirements developed in an earlier stage.

Deltas identified from the "to-be" requirements will involve enhancements to the integrated system that can be engaged through change management.

4.11 System Validation

In post-deployment systems engineering, the system has effectively already been validated, since it is fully deployed and operational. The already deployed system is thus deemed to be fully validated and is therefore the right system. Were it otherwise, the post-deployment engineering process would effectively be engaging in traditional systems engineering and working toward a complete solution. Thus, the system validation phase does not yield immediately necessary results. On the other hand, this phase can be used to corroborate the completeness and accuracy of the CONOPS developed previously and identify new capabilities and services for the system to deliver to its users.

4.12 Operations and Maintenance

In post-deployment systems engineering, the system is already accepted, deployed, and deemed to be meeting its objectives. Thus, the traditional activities of this phase yield little immediate value. On the other hand, this phase can be usefully employed in reviewing and optimizing existing performance monitoring and maintenance processes and procedures if they exist or creating new ones if they don't. Optimization of these is facilitated through the application of the rigorously defined and now verified requirements – requirements that may not have existed, or at least been documented, previously. If performance metrics were not previously developed, these can now be created against these requirements.

5 Conclusion

The author has presented a proposal for the application of systems engineering methods to an already deployed system. This proposal employs the known and existing tools commonly and widely used in performing systems engineering in the development of new systems to a system that is already deployed but undefined. The application of systems engineering methods to an already deployed system can help:

- Identify and document the system against its operational environment
- Develop a better understanding of the deployed system in comparison with other systems having similar capabilities

- Identify new capabilities and services for the system to deliver to its users
- Identify and document those requirements that originally motivated deployment of the system and identify obsolete or new requirements
- Establish more accurate and factual cost benchmarks for future changes and enhancements
- Develop accurate system knowledge needed for integrating the system into standardized business processes, such as ITIL, ISO, business continuity planning, and other processes
- Develop the system knowledge needed for improved hiring and training

The value of post-deployment systems engineering are the efficiencies and process improvements made possible through the systematic and integrated application of methods to accomplish common tasks for deployed systems that are otherwise engaged using effective but stovepipe methods.

References

FHWA. 2007, January. Systems Engineering for Intelligent Transportation Systems Guide. Retrieved from United States Department of Transportation - Federal Highway Administration: https://ops.fhwa.dot.gov/publications/seitsguide/.

Kossiakoff, A., W.N. Sweet, S.J. Seymour, and S.M. Biemer. 2011. *Systems Engineering Principles and Practice*. Hoboken: Wiley & Sons.

NASA. 2009, October 19. Preliminary Design Phase. Retrieved Febuary 14, 2017, from Assurance Process for Complex Electronics: http://www.hq.nasa.gov/office/codeq/software/ComplexElectronics/l_prelim_design.htm.

Toward an Enterprise Architecture for a Digital Systems Engineering Ecosystem

Yaniv Mordecai, Olivier L. de Weck, and Edward F. Crawley

Abstract The digital transformation era is upon us. Digital transformation gradually crawls up the value chain from services and manufacturing to product design and systems engineering. In this paper, we envision a cloud-based ecosystem of systems engineering, which is model-based by definition. The ecosystem model we propose is called 2MIDSTARs, which stands for Model, Infrastructure, Data Services, Simulation, Testing, Analysis, and Repositories + Management, Interoperability, Digital Representation, System, Technology, Audit, and Reporting. The first MIDSTAR covers the intrinsic, core MBSE capabilities, while the second MIDSTAR facilitates the integration with the digital enterprise that surrounds the digital systems engineering ecosystem. In this paper, we explain the importance of jointly considering all these elements together and outline the key roles and functionalities of each component.

Keywords Digital transformation · Digital systems engineering · Model-based systems engineering

Nomenclature

DSEE	Digital systems engineering ecosystem/enterprise
IoT	Internet of Things
MBSE	Model-based systems engineering
MIDSTAR(1)	Modeling Services, Infrastructure Services, Data Services, Simulation Services, Testing Services, Analysis Services, Repositories
MIDSTAR(2)	Management Tools, Interoperability Services, Digital Representations, Systems, Things, Auditing, and Reporting Services
OPD	Object-Process Diagram
OPM	Object-Process Methodology

Y. Mordecai (✉) · O. L. de Weck · E. F. Crawley
Department of Aeronautics and Astronautics, Massachusetts Institute of Technology, Cambridge, MA, USA
e-mail: yanivm@mit.edu; deweck@mit.edu; crawley@mit.edu

A. M. Madni et al. (eds.), *Recent Trends and Advances in Model Based Systems Engineering*, https://doi.org/10.1007/978-3-030-82083-1_55

1 Introduction

The digital revolution has been making an impact on everyday lives across various service-oriented ecosystems over the past decade. Coupled with the Fourth Industrial Revolution – Industry 4.0 – and the growing presence of the Internet of Things (IoT), the digital transformation is making its way up the value chain, from advanced manufacturing through product design to systems engineering and business management (Ustundag and Cevikcan 2018).

Model-based systems engineering (MBSE) adoption and utilization have been constantly growing over the past decade as well (Madni and Sievers 2018). Model-based system specifications and design decisions are recorded in conceptual models defined in formal or semi-formal modeling languages with a common database. Until recently, it has been consistently regarded as a possibly better way of conducting systems engineering, as opposed to the document-*based* approach, but more difficult to implement. The debate has been going on for two decades, with the MBSE supporters growing in numbers but still far behind the masses who still use word processors and electronic worksheets or the slightly more scalable but still text-intensive requirement management database tools (Cameron and Adsit 2018). A recent cross-industry survey of MBSE maturity and adoption shows that MBSE is still perceived as immature, on the one hand, but as a critical enabler of digital transformation in research and development into a "Digital Engineering" paradigm, on the other hand (McDermott et al. 2020).

It appears that the scene is set for a major transformation in the way systems engineering is done, communicated, and integrated with other business activities, in many ways a rebirth of the engineering systems paradigm (Crawley et al. 2004). However, while systems engineers are poised to be the leading change agents in socio-technical organizations, there is also a risk to the continuity and viability of systems engineering itself (Peterson 2019). It will no longer be a privilege to use MBSE tools to build and deliver models of complex systems, generate documentation and code, or sync with requirements databases. We believe that systems engineering will need to reinvent itself as a fully digital and integrated business activity. Otherwise, systems engineers will not be able to comply with the digital enterprise strategy (Matt et al. 2015) or the digital enterprise architecture (Goerzig and Bauernhansl 2018). They will fail to catch up with the velocity of the enterprise and gradually become irrelevant or unnecessary.

This paper proposes a systems thinking approach to tackle our concern for the relevance of systems engineering in the digital era. We should begin with understanding what the digital era involves and what it means for organizations to undergo a digital transformation. Market research has shown that 87% of industries are adopting at least one transformative technology, such as IoT, artificial intelligence (AI), blockchain, or 5G (the next generation of cellular communication) (Builta et al. 2019). An extended yet non-exhaustive list of digital transformation technologies is illustrated in Fig. 1.

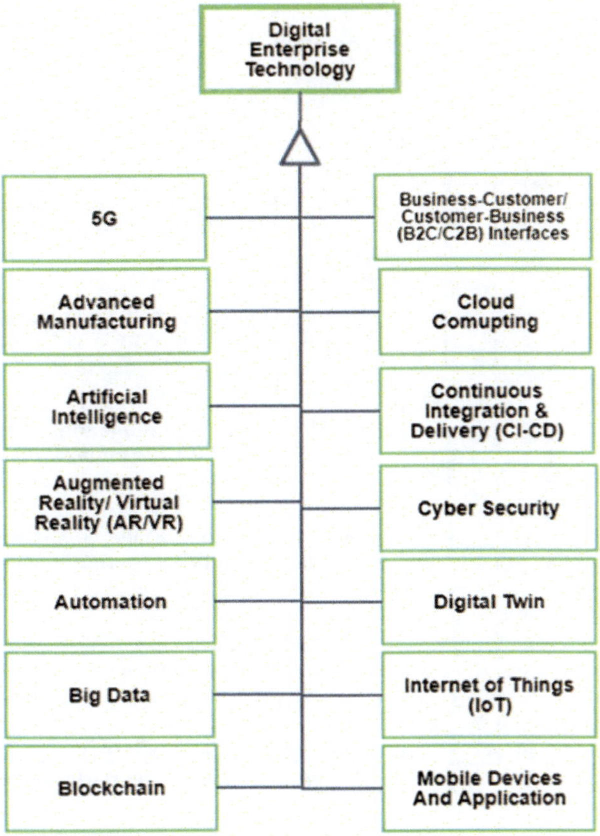

Fig. 1 Digital enterprise technology enabling and driving the digital transformation in socio-technical organizations

Many organizations seeking to undergo a digital transformation – including defense and government agencies, industrial and commercial enterprises, energy facilities, software and hardware manufacturers, and service providers – may be tempted to purchase a few commercial off-the-shelf (COTS) solutions, possibly with the assistance of a consultant to accompany the process. These organizations may not be aware of the need to define the concept of operations for the enterprise that will utilize the capabilities of the digital ecosystem and the operational concept that will characterize the activity of the constituent services.

For instance, imagine a factory that shifts to automated manufacturing based on quickly generated product and part design files that are automatically retrieved and provided to stations along the assembly line or supply chain. If the product and part designers have not been trained with design-to-manufacture techniques and tools, and the factory did not integrate the computer-aided design (CAD), product lifecycle management (PLM), and enterprise resource management (ERM) systems – this digital transformation is bound to fail.

Several approaches to reimagine systems engineering as a digital practice have been suggested, for instance, through a Zachman Framework with various model

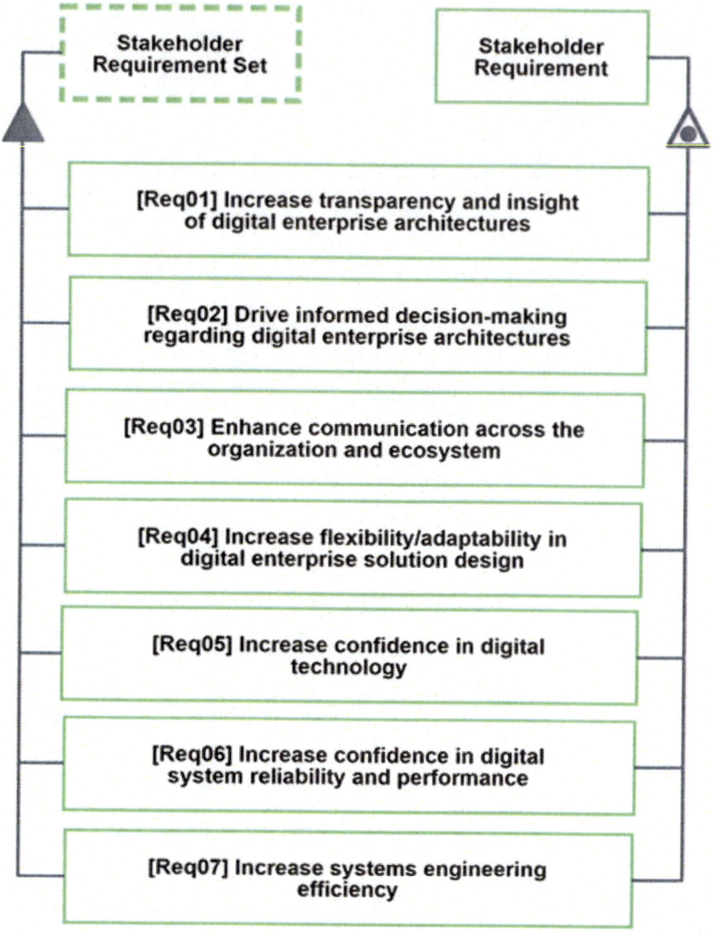

Fig. 2 Stakeholder requirements for the digital systems engineering ecosystem as an enabler of digital transformation

layers, agile management, and novel software development and delivery practices (e.g., microservices) (Bondar et al. 2017).

NASA undertook a digital model-based systems engineering (DMBSE) study to gain better understanding of expectations and challenges associated with such a digital transformation (Hale et al. 2017). NASA's report defined digital model-based engineering (DMbE) as the *use* of digital artifacts, digital environments, and digital tools in the engineering process – as opposed to the traditional documentation-based engineering methods. The NASA team identified several key stakeholder expectations. A set of stakeholder requirements for the digital ecosystem is illustrated in Fig. 2.

NASA's workgroup also identified several challenges: assessing added value, overcoming organizations' culture barriers, regulating the contractual deliverables to meet the standard, building a supporting information technology infrastructure, and ensuring cyber-security.

Additional concerns, mentioned by an anonymous reviewer of this paper, are the setup cost, the challenge of dealing with legacy processes and artifacts, and assuring stakeholders that emerging frameworks will be comprehensive and that they will be viable and deliver return on investment (RoI).

Another ongoing study (Bone et al. 2018; Hagedorn et al. 2020) looked into semantic and ontological integration of models as an enabler for information sharing and collaboration across R&D ecosystems, involving multiple types of models, multiple analysis tools, and multiple data and information consumers.

2 The Digital Systems Engineering Ecosystem

We propose an enterprise architecture for a digital systems engineering ecosystem (DSEE). The architecture has been conceived in order to capture the most relevant aspects of the DSEE: the stakeholders, the core function and purpose of this ecosystem, and the primary constituent systems in this architecture – aligning to the systems architecting process we would have conducted for any system (Crawley et al. 2015).

The proposed enterprise architecture that supports and enables the DSEE is called 2MIDSTARs, a shorthand version of two MIDSTAR acronyms, each consisting of different items. The constituent systems in this architecture are, in the order of appearance in the acronym, Modeling Services, Infrastructure Services, Data Services, Simulation Services, Testing Services, Analysis Services, Repositories ($MIDSTAR_1$) as well as Management Tools, Interoperability Services, Digital Representations, Systems, Things, Auditing, and Reporting Services ($MIDSTAR_2$).

The two MIDSTARs are not grouped together by happenstance, but according to a clear separation of the internal environment ($MIDSTAR_1$) and the external environment ($MIDSTAR_2$). Thus, $MIDSTAR_1$ includes functionalities that are integral and central to a model-based systems engineering discipline. $MIDSTAR_2$ concerns the functionalities that are critical for integrating the digital systems engineering services with the digital enterprise as a whole and includes upstream, downstream, and lateral integration and interaction.

3 Object-Process Methodology

OPM is a conceptual modeling language and model-based systems engineering paradigm for complex and dynamic systems and processes. OPM was standardized as ISO 19450 (Dori 2016; *ISO 19450 Automation systems and integration —*

Object-Process Methodology 2015). OPM relies on the minimal universal ontology principle, whereby stateful objects (things that exist), processes (things that occur), and relations among them constitute a necessary and sufficient ontology for describing any conceivable system in the universe (Dori 2016). OPM's lightweight vocabulary includes ~20 terms.

OPM is visual and textual at the same time. The visual representation is a set of Object-Process Diagrams (OPDs), which are organized hierarchically. OPDs at all levels of the hierarchy retain and allow the same symbol notation, which makes it highly consistent at all decomposition levels. Thus, OPM has only one kind of diagram. Structural, procedural, and functional aspects can reside jointly or exclusively within any OPD. Processes are represented by ellipses, objects by rectangles, and object states by rountangles inside the object rectangle. Objects and processes can be either informatical or physical and either systemic or environmental (external to the boundaries of the system). Links express static and dynamic relations.

OPM's textual representation consists of sentences in Object-Process Language, OPL – a subset of English. Each sentence corresponds to an OPD construct – a set of linked things or states – and vice versa. Each OPD is accompanied by an OPD Specification (OPS) – a set of machine-readable OPL sentences.

There are two software tools for creating OPM models: OPCAT and OPCloud. OPCAT (Dori et al. 2010) is a freely available desktop tool with built-in simulation capabilities, which has been used by thousands of academic and professional users around the world and utilized in hundreds of scientific papers over the last two decades; however, it is based on obsolete desktop software technology, and its development has ended. It can still be downloaded at http://esml.iem.technion.ac.il. OPCloud (Dori et al. 2018) is a relatively new cloud-based modeling studio (accessible online at https://opcloud-trial.firebaseapp.com/), which is under continuous development and evaluation. OPCloud has already been shown to be useful for various domains including medicine (Levi-Soskin et al. 2019), industry (Dori et al. 2020), and enterprise/aerospace architectures (Mordecai et al. 2020). In this paper, we use OPCloud as a modeling tool and framework – which makes perfect sense, since cloud-based capabilities are of utmost importance for such a digital MBSE environment.

4 An Enterprise Architecture of Two MIDSTARs

As explained, the architecture consists of two layers – internal and external. The internal layer consists of all services that make a holistic MBSE environment for the organization. While MBSE focuses on modeling the systems of interest, we extend this scope to cover additional services that we believe are critical for a true MBSE environment, which delivers value to systems engineers and systems engineering stakeholders.

The external layer transforms the MBSE architecture into a digital one and aligns with the digital enterprise as a whole. This layer facilitates interactions with the operational domain that the MBSE focuses on. In a digital world, a model-based design of a critical process can interact with the actual operational enablers or facilitators of that process. The interaction may be possible in both ways: the system of interest and its components are able to consult the model to build machine perception of the process, but also to update configurations and deployments according to revised model structures.

In addition to interacting with the operational domain, the external layer also allows the DSEE to interact with the rest of the digital enterprise for sharing information, dictating solutions, or requesting resources. The architecture should be cloud-based, but this is not mandatory. Utilizing lightweight and easy-to-adjust web services and interfaces that run in or through the cloud will result in significant productivity, streamlining, and synergy. It will also allow for integration with and preservation of legacy assets and reduce transition costs.

Even if the organization is classified or disconnected from the Internet for other reasons, it will be essential for the organization to build a digital laboratory that will allow the enterprise to take advantage of cloud services and adjust them to the needs and challenges of the deployment in question. With commercially available cloud stack packages, this is doable and has been practiced by several classified organizations or sub-organizations in the defense, homeland security, healthcare, finance, and energy domains.

Figure 3 shows the DSEE, the main groups of stakeholders: systems engineers, the systems engineering research community, and the systems engineering software vendors. They all have in common the purpose of generating value in the form of digital systems engineering deliverables: models, tradespace analyses, functional requirements, validation and verification reports, performance assessments, etc. The stakeholder requirements and digital enterprise technologies are both represented as packages that unfold in separate diagrams. The 2MIDSTARs architecture as a collection of services enables the DSEE. The components of MIDSTAR_1 and MIDSTAR_2 are listed in Table 1 and Table 2, respectively.

Several architectural principles are implemented in this architecture:

- **Distributed Data Flow**: all the data is expected to be shared via a central data distribution service, which is part of the infrastructure. This allows for multiple entities of the same type to connect and exchange data with each other; it allows easier virtualization and eliminated interdependency as found in direct interfaces.
- **Expertise**: as opposed to various MBSE platforms which may include a subset of the required capabilities, this architecture advocates isolation and separation of services. These services may still share common user interfaces and other common resources, but the ability to mix and match various software technologies to form an optimal DSEE is essential and more important than a single interface.
- **Focus on Core Competence**: the core MBSE competence includes modeling, simulation, and analysis services, along with supporting data management, access, and storage services. Tools that are available in the software market with expertise in their domain, such as project and task management tools, configuration control and auditing, or dashboards and visualization software –

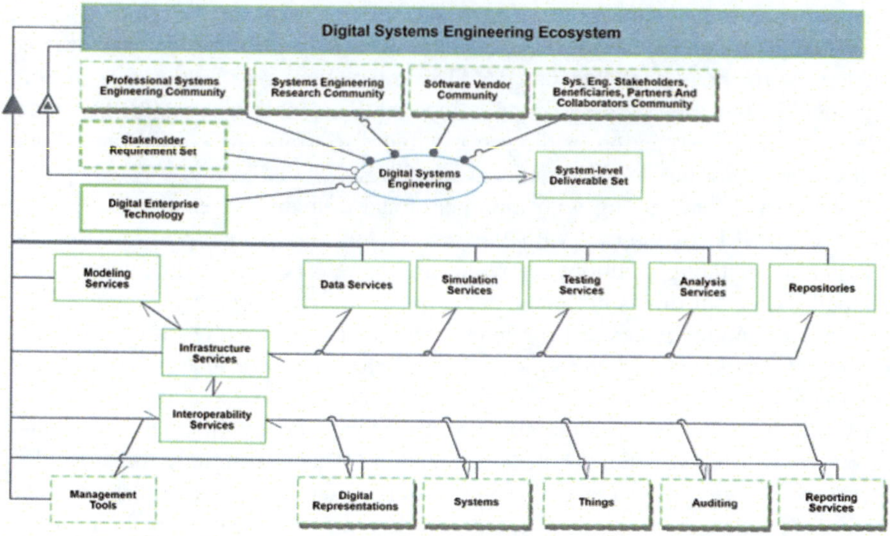

Fig. 3 The 2MIDSTARs enterprise architecture of the digital systems engineering enterprise clearly shows the two layers of services that make up the digital systems engineering ecosystem: the upper, internal layer ($MIDSTAR_1$) and lower, external layer ($MIDSTAR_2$)

Table 1 $MIDSTAR_1$: Internal MBSE layer

Services	Purpose
Modeling services	Build, store, and visualize models in a variety of modeling languages
Infrastructure services	Facilitate interaction among MIDSTAR1 components and, with MIDSTAR2's gateway, provide security and IT governance
Data services	Distribute and retrieve data: Enterprise datasets to inform models, model-generated data, application data, and model metadata
Simulation	Validate and verify system model
Testing services	Connect with test platforms, generate tests in compliance with the models
Analysis services	Analyze, summarize, and validate data, deliver additional value-added capabilities based on the models and simulation results
Repositories	Store and access information of various sorts, including models, analysis results, test plans and results, simulation threads and results, and raw data sources

all do a better job in their area and will better serve the ecosystem than localized developments of similar capabilities.

- **Scalability and Extendibility**: the architecture is built to allow further extension and enhancement for upscaling and broader digital scopes. While this concern is currently beyond the scope of this study, it remains important to ensure this degree of freedom for future enhancements.

Table 2 $MIDSTAR_2$: External DSEE enabler

Services	Purpose
Management tool	Integrate with standard organization management tools to control DSEE activity
Interoperability services	Interact with $MIDSTAR_1$ through its gateway and among $MIDSTAR_2$ members
Digital representation	Build or use digital representations, including engineering designs, software code, digital twins, and virtual environments
System	Deployed realization of a model; interacts with the model that represents it
Thing	Connected entity that models can interact with sensors, actuators, controllers, energy/signal emitters, etc.
Audit	Organizational services that audit activity and ensure viability, quality, transparency, legality, regulation compliance, governance
Reporting	Generate textual, tabular, graphical, visual, and multimedia representations of model information; communicate MBSE outputs and deliverables across the ecosystem

5 Discussion

This paper presents a high-level enterprise architecture for a digital systems engineering ecosystem. By using OPM as a modeling language and the new OPM modeling tool OPCloud as a modeling vehicle, we were able to make the first step of modeling the DSEE using cloud-based tools. Although this is a preliminary model, it serves as a good starting point, capturing core aspects, drawing a clear separation of MBSE core activities from digital interfaces, and clarifying the expertise of each service. We set out with seven stakeholder requirements that the DSEE should tackle. In Table 3, we reflect on the framework's fulfilling (or advances toward fulfilling) of the requirements. This reflection must be fully validated through stakeholder assessment, but it provides a good initial validation for stakeholder focus.

Future research will focus on three directions. First, we plan to extend the architecture to get a better understanding of the microservices required for each service, e.g., what kinds of analysis methods should be included in a model analyzer. This direction will address essential questions that may have naturally arisen on the implementation of proposed constructs, but were beyond the scope of the present paper. In addition, we currently define the data transformation protocols that will allow this transformation to take place. This includes the adoption of mathematical concepts from category theory, which has been mentioned as a potential candidate for a foundational theory of systems engineering, and for a holistic systems engineering platform, of the kind or essence proposed in this paper (Breiner et al. 2017). Finally, we have begun planting the seeds for such a platform for early-adopter government, industry, and research enterprises. The way such organizations can work in a holistic, cloud-based ecosystem must also be explored.

Table 3 Fulfilling of stakeholder requirements using the DSEE – 2MIDSTARs architecture

Stakeholder requirement	Fulfilled by ...
1. Increase transparency and insight of digital enterprise architectures.	Formulation of this reference framework, which informs stakeholders, decision-makers, professionals, and researchers and serves as common ground
2. Drive informed decision-making regarding digital enterprise architectures.	Formulation of this reference framework as the basis for framing decisions in all levels (strategic, tactical, operational) in the context of the critical enablers
3. Enhance communication across the organization and ecosystem.	Facilitation of mechanisms for enterprise interoperability
4. Increase flexibility/adaptability in digital enterprise solution design.	Definition of robust entities and services that can be adapted and shaped gradually, according to evolving needs
5. Increase confidence in digital technology.	Referencing of digital enterprise technology agents as enablers of digital transformation at both the enterprise level and the systems engineering level
6. Increase confidence in digital system reliability and performance.	Inclusion of internal mechanisms for simulation, testing, and analysis, as well as external mechanisms for auditing and reporting
7. Increase systems engineering efficiency.	Formulation of this reference framework which saves time and effort figuring out the issues and allows for prioritization and road-mapping

Acknowledgments We thank the MIT–Technion Post-Doctoral Fellowship Program for funding this research. We also thank the anonymous reviewers for their useful comments and suggestions.

References

Bondar, S., J.C. Hsu, A. Pfouga, and J. Stjepandić. 2017. Agile digital transformation of System-of-Systems architecture models using Zachman framework. *Journal of Industrial Information Integration* 7: 33–43. https://doi.org/10.1016/j.jii.2017.03.001.

Bone, M., M. Blackburn, B. Kruse, J. Dzielski, T. Hagedorn, and I. Grosse. 2018. Toward an Interoperability and Integration Framework to Enable Digital Thread. *Systems* 6: 46. https://doi.org/10.3390/systems6040046.

Breiner, S., Subrahmanian, E., Jones, A., 2017. Categorical foundations for system engineering. In: *15th Annual Conference on Systems Engineering Research*. https://doi.org/10.1007/978-3-319-62217-0

Builta, J., Howell, J., Ambroggi, L. De, Short, M., Grossner, C., Morelli, B., Tait, D., Hall, T., 2019. Digital Orbit - Tracking the development, impact, and disruption caused by transformative technologies across key industries.

Cameron, B., and D.M. Adsit. 2018. Model-Based Systems Engineering Uptake in Engineering Practice. *IEEE Transactions on Engineering Management* 67: 152–162. https://doi.org/10.1109/TEM.2018.2863041.

Crawley, E., B. Cameron, and D. Selva. 2015. *Systems Architecture: Strategy and Product Development for Complex Systems*. Prentice Hall.

Crawley, E., De-Weck, O., Eppinger, S., Magee, C., Moses, J., Seering, W., Schindall, J., Wallace, D., Whitney, D., 2004. Engineering Systems Monograph – The Influence of Architecture in Engineering Systems.

Dori, D. 2016. *Model-Based Systems Engineering with OPM and SysML*. Springer. https://doi.org/10.1007/978-1-4939-3295-5.

Dori, D., A. Jbara, N. Levi, and N. Wengrowicz. 2018. Object-Process Methodology, OPM ISO 19450 – OPCloud and the Evolution of OPM Modeling Tools. *System Engineering Newsl (PPI SyEN)* 61: 6–17.

Dori, D., H. Kohen, A. Jbara, N. Wengrowicz, R. Lavi, Natali Levi Soskin, K. Bernstein, and U. Shani. 2020. OPCloud: An OPM Integrated Conceptual-Executable Modeling Environment for Industry 4.0. In *Systems Engineering in the Fourth Industrial Revolution: Big Data, Novel Technologies, and Modern Systems Engineering*, ed. R.S. Kenett, R.S. Swarz, and A. Zonnenshain. Wiley.

Dori, D., C. Linchevski, and R. Manor. 2010. OPCAT – An Object-Process CASE Tool for OPM-Based Conceptual Modelling. In *1st International Conference on Modelling and Management of Engineering Processes*, ed. P. Heisig, J. Clarkson, and S. Vajna, 1–30. Cambridge, UK: University of Cambridge.

Goerzig, D., and T. Bauernhansl. 2018. Enterprise Architectures for the Digital Transformation in Small and Medium-sized Enterprises. *Procedia CIRP* 67: 540–545. https://doi.org/10.1016/j.procir.2017.12.257.

Hagedorn, T., M. Bone, B. Kruse, I. Grosse, and M. Blackburn. 2020. Knowledge Representation with Ontologies and Semantic Web Technologies to Promote Augmented and Artificial Intelligence in Systems Engineering. *Insight* 23: 15–20. https://doi.org/10.1002/inst.12279.

Hale, J.P., Zimmerman, P., Kukkala, G., Guerrero, J., Kobryn, P., Puchek, B., Bisconti, M., Baldwin, C., Mulpuri, M., 2017. Digital Model-based Engineering: Expectations, Prerequisites , and Challenges of Infusion.

ISO 19450 Automation systems and integration — Object-Process Methodology. 2015. *International Organization for Standardization (ISO)*, Geneva, Switzerland.

Levi-Soskin, N., R. Shaoul, H. Kohen, A. Jbara, and D. Dori. 2019. Model-Based Diagnosis with FTTell: Assessing the Potential for Pediatric Failure to Thrive (FTT) During the Perinatal Stage. In *SIGSAND/PLAIS, LNBIP 359*, ed. S. Wrycza and J. Maślankowski, 37–47. Cham: Springer. https://doi.org/10.1007/978-3-030-29608-7_4.

Madni, A.M., and M. Sievers. 2018. Model-based systems engineering: Motivation, current status, and research opportunities. *Systems Engineering* 21: 172–190. https://doi.org/10.1002/sys.21438.

Matt, C., T. Hess, and A. Benlian. 2015. Digital Transformation Strategies. *Business and Information Systems Engineering* 57: 339–343. https://doi.org/10.1007/s12599-015-0401-5.

McDermott, T.A., Hutchinson, N., Clifford, M., Van Aken, E., Slado, A., Henderson, K., 2020. Benchmarking the Benefits and Current Maturity of Model-Based Systems Engineering across the Enterprise.

Mordecai, Y., James, N.K., Crawley, E.F., 2020. An Object-Process Model-Based Operational Viewpoint Specification Framework for Aerospace Architectures. In: *IEEE Aerospace Conference*.

Peterson, T.A. 2019. Systems Engineering: Transforming Digital Transformation. *INCOSE International Symposium* 29: 434–447. https://doi.org/10.1002/j.2334-5837.2019.00613.x.

Ustundag, A., and E. Cevikcan. 2018. Industry 4.0: Managing The Digital Transformation, Springer Series in Advanced Manufacturing. https://doi.org/10.1007/978-3-319-57870-5.

Collaborative Management of Research Projects in SysML

Benjamin Kruse, Thomas Hagedorn, Mary A. Bone, and Mark Blackburn

Abstract Engineering projects that implement model-based systems engineering encounter the issue of connecting the managing functions of a project with the system engineering model. This research uses a pilot case study to demonstrate how the management of systems engineering can be done in SysML models within the Open Model-Based Engineering Environment (OpenMBEE). The management ability for this pilot includes continuous updates, web-based collaboration, model-based report generation, and enabled semantic reasoning. The semantic reasoning is seen as a key enabler and accomplished using a SysML profile that is aligned to an underlying project ontology. This not only results in utilizing the advantages of a model-based engineering environment for managing the project but also demonstrates the benefit of semantic enabled reasoning that is a focus of research of the research project.

Keywords Model-based systems engineering · SysML · Project management · Semantic web technology · Semantic reasoning

1 Introduction

Within the context of the Department of Defense's (2018) digital engineering strategy and the accelerating evolution and adoption of Model-Centric Engineering (MCE), the Systems Engineering Research Center (SERC) addresses research needs regarding the enabling of practices of model-based systems engineering (MBSE). Following Blackburn et al. (2019), an overall research objective is to understand MBSE practices and measures to enable a progressive transformation toward digital

B. Kruse (✉) · M. A. Bone · M. Blackburn
Stevens Institute of Technology, Hoboken, NJ, USA
e-mail: bkruse@stevens.edu

T. Hagedorn
University of Massachusetts Amherst, Amherst, MA, USA

A. M. Madni et al. (eds.), *Recent Trends and Advances in Model Based Systems Engineering*, https://doi.org/10.1007/978-3-030-82083-1_56

and model-based research and development, including leveraging ontology-based knowledge representation to enable semantic technologies for systems engineering.

As part of this research, web technologies are investigated to provide a solution for easy and consistent capture of information for systems engineering plans, technical data, and reviews through models that formalize traditional document-based artifacts and provide dynamic document generation. OpenMBEE is the open-source Open Model-Based Engineering Environment developed by NASA/JPL (2019). It is used together with the Systems Modeling Language SysML by OMG (2015). OpenMBEE consists of three components: The Model Management System (MMS) for storing the model data in an open way, the Model Development Kit (MDK) plugin for the generation of documents based on views and viewpoints and the synchronization of models between the SysML modeling tool and MMS, and the View Editor, which offers lightweight, web-based, and live access to view-based model data in MMS. Its utilization for project management is explained in Sect. 2, starting with a brief introduction to OpenMBEE in Sect. 2.1.

Using developed SysML profiles with formalized stereotypes aims to enable semantic technologies for systems engineering as part of the Interoperability and Integration Framework (IoIF) by Bone et al. (2018). The SysML-profiled models enable mapping to a suite of underlying ontologies that are based on the Basic Formal Ontology (BFO). Data from the SysML models can be accessed via MMS to be converted into Resource Description Framework (RDF) triples, with the profiles serving as a guide to allow mapping of model constructs to a tool and domain agnostic representation in compliance with the ontologies. Other data sources may then be linked to the SysML model representation using the same taxonomy. The graph data in the triple store can then be analyzed using the ontologies with automated reasoning and semantically enhanced queries, opening the door to applications in model verification, metric assessment, and inferences of complex relations in data arising from multiple sources. Sections 2.2 and 2.3 show further details about the project ontology and the derived SysML profile. They are applied for research projects in Sect. 3. Section 4 finally looks at the enabled semantic representation and reasoning.

Why should an engineering approach be utilized to manage research projects in SysML? The first part of the answer is to demonstrate the engineering approach under investigation by applying it for an additional system, the research project itself. Based on the promising results of prior work by Kruse and Blackburn (2019) or Kruse and Blackburn (2020), the second part of the answer is to use the environment based on SysML and OpenMBEE for utilizing its advantages to manage the systems that are distributed research projects.

Results include the formal capturing of project management details and their interrelations, e.g., assignments, roles, individual responsibilities, and achieved accomplishments. Utilizing the View Editor provides web-based, live, and consistent data in model-derived, yet document-like, views. This can improve communication within the research team, including updates by researchers not familiar with SysML, as well as toward the research sponsor in the form of model-derived project status reports. Finally, and yet to be fully utilized, there is the unambiguous information representation and retrieval that enables sematic reasoning and inferencing about project data. A more detailed discussion of results, lessons learned, and derived conclusions follows in Sect. 5. Despite such advantages, the authors are aware that the presented work is not intended to replace specialized commercial project management tools. For example, it is not possible to include the Gantt chart capabilities of MagicDraw, due to incompatibilities of the required model elements with the data structure in MMS. Certain commercial software products are identified in this material. These products were used only for demonstration purposes. The use does not imply approval or endorsement by Stevens, UMass Amherst, or SERC.

2 Project Modeling Approach

The approach used for project management is presented in three parts: First, a brief overview of the OpenMBEE environment; second, the developed project ontology within its ontology ecosystem; and third, the derived SysML profile.

2.1 OpenMBEE Environment

The tools used for OpenMBEE are MagicDraw and Teamwork Cloud 19.0 SP2, MMS 3.4.2, View Editor 3.5.1, and MDK 4.1.3. The MMS server provides the authoritative source of truth, as explained by Blackburn et al. (2019). The MMS captures all model data in an open and accessible way. It uses JavaScript Object Notation (JSON) format with a Representational State Transfer (REST) interface, to provide versioning, workflow management, and access control for multi-tool integration across engineering and management disciplines, according to NASA/JPL (2019). MMS stores all relevant SysML model elements with their change history, e.g., of blocks, relations, properties, values, and instances, including the view instances for the View Editor. Providing such data also makes OpenMBEE a valuable integration for the IoIF, to, e.g., semantically track relations between heterogeneous data sources.

The MDK plugin is used to synchronize the SysML model data in MMS with the originating modeling tool. It also includes the DocGen language (Delp et al. 2013) for model-based document creation according to the view and viewpoint paradigm which is defined in ISO-42010 (ISO/IEC/IEEE 2011). The view and viewpoint paradigm allows to automatically expose live and stakeholder-specific model content to progress from static documents toward model-based approaches, as shown in Kruse and Blackburn (2019). Views represent exposed system elements from the perspective of a viewpoint, which specifies how to construct the views. Using different viewpoints while exposing the same model elements allows generating different views for different stakeholders. To support the quick creation of consistent documents in which each view represents a chapter, a library of generic and reusable viewpoints was created, including viewpoints for the project model that utilize stereotypes from the profile in Sect. 2.3. By using the viewpoints, project reports are automatically derived from the SysML model that can exist not only as static Portable Document Format (PDF) files but also as editable views in View Editor, i.e., outside their original modeling tool.

The View Editor displays instances of the views in the web browser, showing the formatted live model data in MMS for agile virtual reviews and real-time collaboration. All elements in documents shown in the View Editor are exposed model elements, i.e., objects with unique object IDs, created in the original modeling tool but mostly editable in View Editor. Providing model-based data in a more traditional format, the View Editor supports the communication with stakeholders not familiar with SysML. The editing capabilities of the View Editor do not include the creation of new SysML model elements, but the addition of presentation elements, e.g., new text fields. It is possible to edit the name, value, and documentation of existing model elements that are exposed in View Editor. Changes in the View Editor are directly saved in MMS where they can be synchronized back into the original SysML model.

2.2 *Project Ontology*

Developed in Protégé 5.5 by Rubin et al. (2007), the project ontology is an OWL 2 ontology, as explained by Hitzler et al. (2009). Following Arp et al. (2015), BFO is selected as a top-level ontology to support integration of multiple disparate domain ontologies into a single, interoperable ecosystem. Per BFO style guidelines, terms are refactored from existing ontologies to support development of the project ontology and promote reusability across this research and other projects. These include the Common Core Ontologies, a suite of ontologies providing BFO conformant formalizations of high-level entity types such as information, events, and agents, as well as prior related ontologies of mathematical information, technical models, and decisions. The ontology itself is scoped based on the requirements of the project itself. It focuses mainly on the unambiguous representation of ongoing assignments, the agents responsible (via a role), and the deliverables or accomplishments produced by said assignments. A second focus of the ontology is in representing types of information relating to the project. These rely upon the Common Core's basic treatment of "Information Content Entities," which express information about other entities via a suite of "aboutness" relations, following Ceusters and Smith (2015). These "aboutness" relations allow the project ontology to describe things like roles of responsibility and specifications of system attributes.

The decision ontology includes terminology to support representation of requirements, objectives, needs, and preferences among others. Notably, the terminology may be related to domain-specific information represented using other BFO conformant ontologies. This capability permits, for example, deliverables to be tied to specific model types or requirements to be linked to specifications for how they

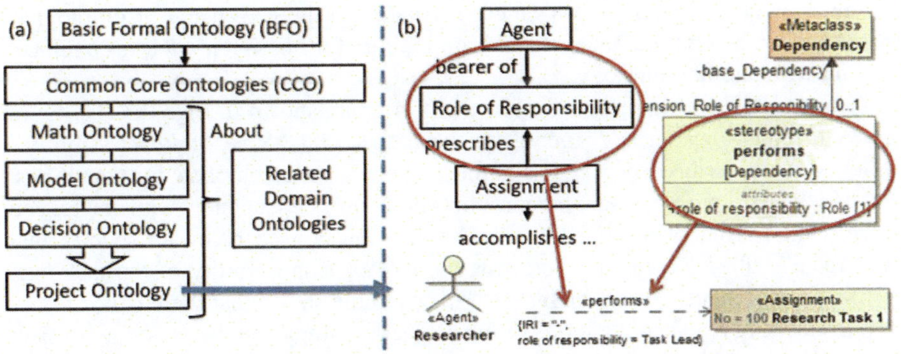

Fig. 1 (**a**) Project ontology ecosystem under BFO. (**b**) Excerpt of project ontology elements with a corresponding SysML profile element and its application (bottom)

might be assessed via measurements of specific phenomena related to a system of interest. The project ontology is thus limited in its scope, but powerful in the types of information it can express once included in a larger ontological ecosystem, as illustrated in Fig. 1a.

Central to this information treatment and integration into a larger ecosystem are logical axioms to assess the relatedness and relevance of information. Using semantic rules and other term-based restrictions, the project ontology introduces several rules to infer consequences of assertions that data are related to one another, allowing it to be paired with reasoning software to make inferences. When this is done, the axioms within the ontology can, for example, assess whether observations violate specifications, infer that two pieces of data have certain types of relatedness to one another, and make explicit aspects of system representation based on a graph-based project description. When combined with the larger ecosystem, these basic reasoning inferences and the ontology's ability to unambiguously represent data may facilitate things like verification of a system across disparate modeling data and promote interoperability with otherwise un-linked tools, as discussed in Sect. 4 and implemented by the IoIF.

2.3 *Derived SysML Profile*

To use the semantically unambiguous context provided by the ontology, matching SysML stereotypes as part of a project profile are defined. They allow to properly map the SysML model elements to the terms of the ontology and enable semantic reasoning. Directly implementing the ontology with its defined interrelations, as is, was not possible due to the necessity to use existing SysML and UML stereotypes and metatypes. The research used a process of creating ontology and SysML profile iteratively, seeking a compromise between parsimonious and correct ontology and profile as well as modeling convenience. Analog to the ecosystem of ontologies, there can also be multiple interrelated profiles, e.g., for mission models, as a related domain with another ontology.

Excerpts of the ontology as well as the profile are shown in Fig. 1b. The figure shows the ontology terms "Agent," as the bearer of a "Role of Responsibility," which gets prescribed by an "Assignment" that is to accomplish further things. On the SysML profile side, there is the "perform" dependency with its tagged value, called "role of responsibility." This relation is used below to specify that an "Agent" called Researcher performs the role of responsibility of the task lead for the "Assignment" Research Task 1. This example shows that there is not a one-to-one mapping between the terms of the project ontology and the matching project profile. For example, the term "Role of Responsibility" gets realized in the form of a subsidiary property, and the relations "bearer of" and "prescribes" are only realized implicitly through the "perform" dependency.

3 Research Project Model in SysML

The presented project profile is used within the modeling environment for two projects, one of which is the continuation of the Surrogate Pilot project by Blackburn et al. (2019), investigating the implementation of the Naval Air Systems Command's (NAVAIR) Systems Engineering Transformation (SET) framework. It utilizes the View Editor not only for continuous updates and comments regarding the various domain models but also for managing itself with the project model. Within the project model, there are two separate documents defined by using viewpoints from the viewpoint library: one auto-generated document for shorter bi-monthly reports and a second, more extensive document that is generated with additional exposed information specifically for editing in the View Editor. This additional information includes placeholder elements, e.g., for accomplishments as seen on Fig. 3, that exist to be adapted by the researchers in the View Editor, to overcome its inability to create new elements from scratch. The latest state of the model and its generated documents under development are captured in a development branch of the baseline master branch in MMS. Read-only tags are used to store earlier versions, i.e., of submitted reports.

The content of the project SysML model includes a hierarchy of assignment elements. Each assignment has a property for its status and can use its documentation for a textual description that also becomes part of the documents. Linked to the assignments are researchers and other stakeholders as "agents" that perform certain roles of responsibility, as shown in Fig. 1b. The interrelations between the different assignments as well as their required inputs and outputs, i.e., the deliverables, are modeled using internal block diagrams, as shown Fig. 2. Figure 2 shows an assignment to align and refactor the "Skyzer Mission Model (IM20)" and "Skyzer System Model (IM30)" according to the "NAVSEM Starter" process model and the "ASRM Framework" while investigating their use and applicability as well as

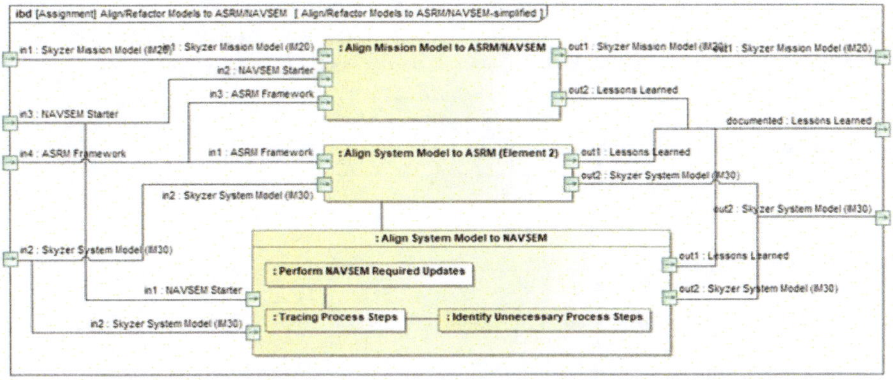

Fig. 2 Simplified internal block diagram of assignments with their interrelations and deliverables

documenting any lessons learned, e.g., about identified unnecessary process steps, as shown as a subsidiary assignment. NAVSEM stands hereby for the NAVAIR Systems Engineering Method, i.e., the used MBSE process that guides creating the designated content of the Acquisition System Reference Model (ASRM). By specifying the assignments as specialized class elements in the profile, they can be modeled as shown in Fig. 2 with their interrelations and deliverables, in contrast to, e.g., extended requirements or activity elements.

The accomplishments of the project are modeled as shown on top of Fig. 3, by using a stereotyped dependency with comment, date, and status properties. The dependency relates the accomplished entity with the achieving assignment. Having a project usage relation in the modeling tool gives direct access to all other used SysML models of the project. This allows to directly refer to the used models, their content, or their documents when capturing accomplishments. Examples are given on the bottom of Fig. 3 with an excerpt from the View Editor showing an accomplished addition to the mission model in the form of an added diagram for the ongoing alignment to ASRM and the completed change of the mission model document's view hierarchy. The shown representation in the View Editor allows researchers to edit the date, status, comment, as well as the names of the accomplished entity and the assignment in the table in a web browser, without a SysML modeling tool. It is also possible to adapt generic placeholder elements, as seen in Fig. 3, into new accomplishments. Similar placeholder elements also exist for assignment elements in the project backlog. Yet, to properly integrate the renamed placeholder elements, additional work in the SysML modeling tool is required.

Finally, the project model contains additional resources about the project goals; the investigated case study in the form of an unmanned aircraft system (UAS), called Skyzer; and a glossary with a list of used acronyms. Based on the model data, several metrics are calculated and exposed within the documents, for example, the number and status of accomplishments.

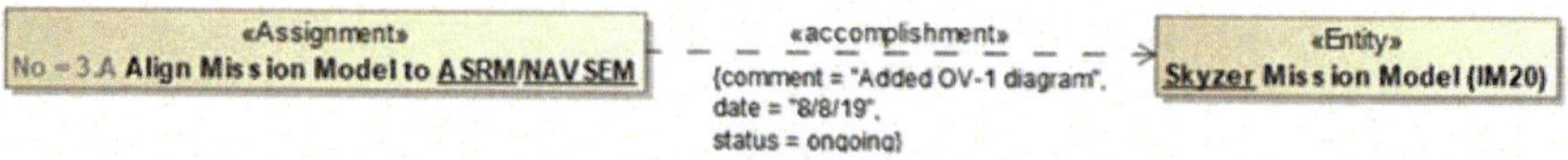

Accomplishment	Part of Assignment	Date Accomplished	Status	Comment
Placeholder01	Align Mission Model to ASRM/NAVSEM	-	unknown	-
Skyzer Mission Model (IM20)	Align Mission Model to ASRM/NAVSEM	8/8/19	ongoing	Added OV-1 diagram
Skyzer Mission Model Document (IM90-20)	Align Mission Model to ASRM/NAVSEM	8/1/19	complete	View hierarchy updated to reflect ASRM structure, while still exposing original Mission Model content

Fig. 3 Example accomplishment in SysML (top) and derived View Editor table (bottom) with placeholder for adaption

4 Semantic Representation and Reasoning

Enabled by the ontology-aligned profile and the supported information retrieval from MMS, it is possible to create a tool-agnostic data representation in RDF graph pattern triples. This is done by parsing and mapping in the IoIF, as explained by Bone et al. (2018). The terms in the SysML profile serve hereby as a guide for the mapping of model constructs to tool and domain agnostic representations that are aligned to the ontologies. This alignment allows semantic queries of the project data while relating it to data from other sources like relating analysis data to project tasks. Semantically conducted weight breakdown calculations followed by analyzing impacts on the formal model-based signoff, as explained by Kruse and Blackburn (2020), have been successfully demonstrated. Ongoing investigations are about relating computational fluid dynamics (CFD) results and other high-fidelity analysis models with SysML models. Such or similar automated and semi-automated inferencing is also planned for project data to (1) assess its completeness and, e.g., determine if there is a task lead for each assignment; (2) identify new relations between assignments, e.g., by searching for implicitly related assignments with related accomplishments; or (3) generally verify and validate the project model data.

5 Discussion and Conclusion

Looking back at the motivation to manage these research projects in the presented way using SysML, it is noticed that the formal documentation of assignments with explicit interrelations, interfaces, and deliverables as well as individual responsibilities within SysML provided benefits for the research project team in the form of a better understanding of what is needed to be achieved. The project model helps to centrally collect information about the case study under investigation in the form of its various models with their interrelations and purposes in respect to the surrogate development process as well as the encompassing research project. This benefit goes in hand with documenting accomplishments, by directly referencing newly added elements from other used SysML models, and tracking the development of the accomplishments or other parts of the project through suitable metrics.

Other advantages for the research project team come with the View Editor that allows real-time collaboration on virtual model-based documents, to enhance communication within the research team. Providing this model information outside of the SysML modeling tool allows direct input through the View Editor's more basic text editing capabilities like adapting provided placeholder elements. The View Editor also allows commenting and captures the full element history across the baseline and branches, e.g., to capture past documents as read-only tags. Reusing standardized viewpoints from a library helps to reduce the required modeling effort when creating suitable view hierarchies and results in more consistent layout of documents. Utilizing the up-to-date model data in document form for communication, reporting, and documentation toward the research sponsors resulted in perceived improvements regarding the reports' level of detail and a more evenly distributed workload, by having the whole team contributing to a consistent and authoritative source of data instead of one person writing the whole report.

Looking at identified issues and disadvantages, it must be said that the whole approach is not meant to or ready to replace specialized commercial tools. While there are some ways to include information about the project schedule, e.g., date properties of assignments or SysML sequence diagrams, it is not possible to include Gantt Charts, even though they are available in the used SysML tool. The structure of the Gantt Charts-specific model elements from the required commercial plugin is not compatible with the general data structure used in MMS, preventing a synchronization through MDK. Additionally, users still require some knowledge about SysML to read the diagrams like Fig. 2. To improve the View Editor, it should be clearer that added presentation elements un-intuitively do not synchronize with the model, while exposed elements, e.g., showing their documentation text, do. This goes in hand with the View Editor's general inability to create new model elements from scratch, resulting in the workaround using placeholders described in Sect. 3. For a distributed collaboration, it is identified to be crucial to make inputs and keep the model up-to-date without necessarily requiring the SysML modeling tool.

Finally, looking at the presented work as an application and test of ongoing research from their own projects, there is the enabled but not yet utilized potential to use the project data for semantic data retrieval and reasoning based on the developed profile and ontology. This future work, as described in Sect. 4, is expected to have the potential to improve model quality and modeling augmentation, for example, by integrating an external project management tool as non-SysML data thought the IoIF or by including accomplishments made outside of SysML. For the presented creation of the project ontology with the derived SysML profile, the best practice found in this research is to find compromises between the ontology and the profile for a lean, parsimonious, and correct setup that still encompasses modeling convenience.

To conclude, applying the here presented approach that utilizes OpenMBEE with an ontology-aligned SysML profile for managing research projects as SysML models shows the general applicability of such an approach that is developed as part of the very research projects managed. It also supports the research goal to develop every document as a model. The project model provides quick and consistent model-based document generation based on views and viewpoints, to be utilized in the View Editor. The View Editor offers live, web-based access for an improved communication within the whole team, without requiring everyone on the team to have knowledge of SysML. Instead, it provides a basis for a more familiar document editing platform, used to generate project reports based on required collaboration and inputs from the team members.

References

Arp, R., B. Smith, and A.D. Spear. 2015. *Building Ontologies with Basic Formal Ontology*. MIT Press. ISBN: 978-0-262-52781-1.

Blackburn, M., R.S. Peak, S. Cimtalay, A. Baker, M. Ballard, D.H. Rhodes, M. Bone, J. Dzielski, R. Giffin, B. Kruse, B. Smith, M. Austin, and M. Coelho. 2019. *Transforming Systems Engineering through Model-Centric Engineering*, SERC-2019-TR-005. Hoboken: SERC.

Bone, M., M. Blackburn, B. Kruse, J. Dzielski, T. Hagedorn, and I. Grosse. 2018. Toward an Interoperability and Integration Framework to Enable Digital Thread. *Systems* 6 (4). https://doi.org/10.3390/systems6040046.

Ceusters, W., and B. Smith. 2015. *Aboutness: Towards Foundations for the Information Artifact Ontology*.

Delp, C., D. Lam, E. Fosse, and C.-Y. Lee. 2013. Model Based Document and Report Generation for Systems Engineering. Aerospace Conference, Big Sky, MT, USA

Department of Defense. 2018. *Digital Engineering Strategy*. Washington, DC, USA.

Hitzler, P., M. Krötzsch, B. Parsia, P.F. Patel-Schneider, and S. Rudolph. 2009. OWL 2 Web Ontology Language Primer. *W3C Recommendation* 27(1).

ISO/IEC/IEEE. 2011. Systems and Software Engineering – Architecture Description. ISO/IEC/IEEE 42010:2011(E).

Kruse, B., and M. Blackburn. 2019. Collaborating with OpenMBEE as an Authoritative Source of Truth Environment. *Procedia Computer Science* 153 (C): 277–284. https://doi.org/10.1016/j.procs.2019.05.080.

Kruse, B., and M. Blackburn. 2020. View and Viewpoint based Digital Signoff using OpenMBEE as an Authoritative Source of Truth. *Journal of Cyber Security and Information Systems (CSIAC)* 7(3): 23–30.

NASA/JPL. Open Model Based Engineering Environment. Accessed 10 25 2019. http://www.openmbee.org/.

OMG. 2015. Systems Modeling Language (OMG SysML) v1.4. formal/2015-06-03.

Rubin, D.L., N.F. Noy, and M.A. Musen. 2007. Protégé: A Tool for Managing and Using Terminology in Radiology Applications. *Journal of Digital Imaging* 20 (1): 34–46.

Supporting the Application of Dynamic Risk Analysis to Real-World Situations Using Systems Engineering: A Focus on the Norwegian Power Grid Management

Michael Pacevicius, Cecilia Haskins, and Nicola Paltrinieri

Abstract Dynamic Risk Analysis (DRA) approaches are virtuous processes enabling the improvement of state-of-the-art techniques for risk calculation in industrial infrastructures. However, they require the existence of an appropriate architecture enabling end-to-end processing of information, which has not yet been defined in practice. This paper aims at discussing the possibilities and the advantages of combining DRA with Systems Engineering (SE) approaches to reach this objective. For that, we define a framework based on SE principles, apply it for the assessment of the role of vegetation on the global risk for power grids, and discuss the benefits it provides.

Keywords Dynamic Risk Analysis · Systems Engineering · Internet of Things · Power grids · Vegetation · Critical infrastructures

1 Introduction

System managers aim, among other objectives, to reduce uncertainties related to process monitoring and to maximize the control efficiency. Those dimensions are, for the most part, defined by the capacity of the stakeholders involved to properly analyze, evaluate, and reduce the risk levels characterizing the system under review. Numerous guidelines and standards have been developed to support these activities.

M. Pacevicius (✉)
eSmart Systems, Halden, Norway

Department of Mechanical and Industrial Engineering, Norwegian University of Science and Technology NTNU, Trondheim, Norway
e-mail: michael.pacevicius@esmartsystems.com

C. Haskins · N. Paltrinieri
Department of Mechanical and Industrial Engineering, Norwegian University of Science and Technology NTNU, Trondheim, Norway

A. M. Madni et al. (eds.), *Recent Trends and Advances in Model Based Systems Engineering*, https://doi.org/10.1007/978-3-030-82083-1_57

However, the way such guidelines are used and the way they affect the quality of monitoring activities show a considerable heterogeneity across fields of application. In addition, the nature and the specificities of the supporting tools also influence the performance level of risk analyses.

Although most risk assessment methods advocate for feedback loops and cyclic processes, the quality of the outputs of these procedures tends to deteriorate over time. This is especially due to the inability for currently implemented tools to integrate in a continuous, reliable, and thus dynamic way information related to the system under review. Researchers aiming to tackle this dilemma have suggested frameworks and approaches leading to more cyclically effective, and thus dynamic, risk analyses and evaluations (Villa et al. 2016). The application of such approaches is still in an embryonic phase in real-world scenarios. This is mainly due to the relative recency of the proposed tools, which still require recommendations for applicability. On the other hand, although such techniques advocate for a better and more intense use of information, they do not describe technical solutions able to properly integrate all the data that is theoretically available and that is suggested to be analyzed in the recommended approaches. In fact, the applicability of such approaches is highly dependent on the existence of an architecture to support this integration in a timely manner. This is especially challenging, considering the general complexity of the systems under review and thus the plurality of influencing factors and data sources to consider.

Systems Engineering (SE) in general offers interesting perspectives for addressing this challenge. By fully considering local specificities while keeping a global understanding of the needs for which a system is developed and the constraints it is and will be subject to, SE offers development perspectives enabling designers to properly attack the resolution of the problem. The present paper discusses the advantages systems engineering can provide to support the application of Dynamic Risk Analysis (DRA). These advantages are demonstrated by an application in risk analysis done for power grids.

2 Risk Analysis: Original Concepts and Requirements for Dynamic Evolutions

Risk is a variable concept across domains due to the different ways people and industries value situations and the consequences of events. The way risk is defined and integrated into models/tools depends on the type of business (Aven 2012), which has led multiple standards to be developed within the diverse fields of application where risk understanding and control is required (ISO – International standardization organization 2007, 2016, 2018; NORSOK 2010).

Generally, the first step for risk analyses consists in properly defining the context for the entire life cycle of the system under review. This includes qualitatively and quantitatively defining the measures of effectiveness relative to risk acceptance, as well as the way performance measures will be verified/validated all along and at the end of the process. This step aims to eventually obtain a risk picture that accurately depicts the true exposure level of the system to specified threats. The risk picture is characterized by the contribution of all scenarios leading to an unwanted event or situation, the likelihood for these scenarios to occur, and the severity of the negative consequences resulting from these scenarios (Kaplan and Garrick 1981).

The tools used for risk analysis in industrial facilities have been observed to have the following limitations:

- Recurrent use of outdated data for frequency evaluation, despite the acquisition of new knowledge based on experiences (Creedy 2011)
- No capture of interactions and dynamic aspect of risk variations (Yang and Haugen 2015)
- Inappropriate consideration of uncertainties related to risk (Aven 2012; Villa et al. 2016)

Furthermore, despite the theoretical inclusion of a cyclic feature in most risk assessment approaches, experience shows that the proper reassessment of establishing the context suffers from some latency as the number of cycles increases. This initial step is often done once and for all at the beginning of the study, which means that further steps of the analyses are not able to react correctly as internal or environmental conditions of the system evolve over time. In addition, accurately characterizing the evolution of the hazards that can impact the system becomes challenging. Thus, although the awareness of requirements for flexibility and adaptability is generally present in the conception phase, the implementations usually suffer from an inability to update and integrate new information. Consequently, the estimated level of risk may, over time, diverge from the true level of risk, which in turns may lead to new potential accidents or catastrophes.

The appropriate processing of correct information – and more especially the capability to consider new variables or be resilient to disturbances (e.g., loss of data source) – represents an area that needs more research. Acknowledging this situation, more dynamic methods have been developed (Villa et al. 2016). In general, these methods require:

- Real-time acknowledgment of information updates for initially considered variables
- The capacity to integrate (or discard) variables and thus restructure the method when new (or old) information is considered relevant (or irrelevant) for the estimation of the current risk image

Thus, dynamic methods require the consideration and integration of data-driven updates rather than only considering long-term calendar-based protocols, which is a common practice across industries today. This means that optimal data management is at the heart of Dynamic Risk Analysis. However, in order to make existing tools structurally updatable and adaptable, there is a need to offer a standardized approach that technically enables automatic integration or suppression of relevant data, thereby enabling evolution from isolated data sources toward informed risk depiction.

3 Key Dimensions of System Engineering and Contributing Potential to DRA

Properly controlling and monitoring a system over time requires an appropriate understanding of the composition of the system and the existence of an appropriate user interface. Furthermore, the behavior of the system needs to be understandable from both a local and a global perspective, in order to identify and assess interactions between subsystems and their respective characterizing variables, thus allowing systems engineers to understand not only the sum but also the product of the subsystems' respective behaviors. Pooling a team of specialists from a diversity of areas supports such actions, as it provides the analysis of multiple dimensions of the system and breaks barriers between fields presenting synergies.

SE advocates for cyclically reviewing and adapting, if necessary, the different phases of the process in order to show evidences that the system fulfills its functions as expected by the customers and stakeholders. Periodically carrying out context evaluation ensures a good understanding, over time, of the system's properties and interactions between its subsystems and helps to avoid the probability of misunderstood or wrongly quantified hazard effects.

The understanding of interrelations between variables and the cyclic requirement in the development of a system represent key dimensions for both efficient SE approaches and DRA tools. Table 1 details further the numerous correspondences which exist between DRA requirements and SE specificities.

4 Approach Description

The approach suggested in the present paper is based on the steps of regular data mining flow processes (Chapman et al. 2000) and is structured using a Systems Engineering mindset in order to provide an efficient analysis of risk over time. It is developed as follows:

(1) Identification of information requirements (business understanding)

Within the context, the different needs of the system stakeholders are gathered. This enables understanding which type of information needs to be available to provide an appropriate solution, as well as to direct the first steps of the research.

(2) Identification of potentially accessible data sources (data understanding)

Based on the requirements formulated by the stakeholders, establish a benchmark of the existing data sources (or of data sources that can be created to reach the defined objectives). The type of data sources corresponds to data sources that, somehow, by their nature, enable a better understanding of the analyzed items. Only those datasets that are accessible for the project are retained.

(3) Filtering of data sources (data understanding, data preparation)

Table 1 Correspondences between DRA requirements and SE specificities

Requirements for efficient DRA tools	Specificities of Systems Engineering
(1) The structure and the architecture of the tools enabling optimal data management in complex environments need to be defined	(1) SE approaches support the construction of complex systems in an efficient and durable way
(2) DRA tools need to be updatable in terms of architecture	(2) Proper system design enables flexibility in the architecture of the system and ability to consider new variables as required
(3) DRA tools need to be resilient to degradation or loss of data sources (quality and quantity of data)	(3) Proper system design enables to make efficient use of correlations between variables and thus provides "as good as knowable" analyses
(4) Cyclic assessments and reconsiderations of context are required to maintain process understanding and thus to keep a realistic risk picture over time	(4) Integration of a cyclic dimension is a pillar of SE which reduces the non-detections of emerging hazards or the apparition of black boxes within the process
(5) Interactions between factors need to be captured to show accurate risk pictures	(5) Implementation of interdisciplinary approaches is a pillar of SE
(6) Frequently updated data need to be considered to show accurate risk pictures	(6) The cyclic dimension of SE supports the frequent updating of information
(7) Use of case-/plant-specific data	(7) Correctly designed systems integrating appropriate data sources enable "as good as knowable" analyses

The informative potential of the retained information sources is initially assessed and used to create a maximum number of scenarios to consider in terms of risk. Discussions between the heterogeneous panels of experts involved enable understanding the importance of the physics for each observed variable, but also – and maybe more importantly – the interrelations and dependencies that can exist between considered variables/phenomena. Data sources providing quantitatively usable information (e.g., databases with numerical values) or convertible (e.g., by some weighting conversion process) are then selected. The informative potential of each of the related retained variables is then assessed in terms of contribution to the calculation of the targeted risk dimension. More especially, each of the variables is evaluated to assess if they provide information relative to the frequency or the consequences of the identified scenarios. Based on the importance of the identified scenarios, requirements for the data acquisition of the related variables are estimated and reported for future performance evaluation of the defined system.

(4) Clustering of data sources (data preparation)

The data sources that are considered are then clustered, based on their resolution and on the reported physical interrelations existing between the observed variables.

(5) Choice of potential environments, frameworks, and algorithms (modelling)

The choice of the environment to work in, as well as the frameworks and algorithms, is based on both the objectives to achieve and the characteristics of the retained data sources.

(6) Structuring of the pipeline (modelling)

The clusters suggested in phase (4) are integrated into layers in which analyses will be done. Layers with lowest resolution are placed in an initial position of the pipeline for optimization of the workload management. The selected frameworks and algorithms are adapted for each layer, depending on the fixed objectives.

(7) Progressive use of outputs and assessment of pipeline final results (evaluation, deployment)

Once the pipeline is prepared, each layer is successively run through in order to eventually reach the final risk picture. Estimations originating from the output of the successive layers are also sent via feedback loops in order to improve the algorithms exploited via approval or rejection of the first results.

A critical requirement of the approach suggested is a good understanding of the techniques used. This will be the only way to properly convert the information they provide into usable inputs for the improved calculation of risk levels.

5 Application: Pipeline Construction for Improved Risk Analyses in Power Grid Management – Focus on Vegetation

5.1 *Situation Overview*

The power grids used daily are exposed to a plurality of hazards (e.g., hurricanes, earthquakes, ice storms, floods), which occurrences can have heavy consequences (DeCorla-Souza 2013; Kenward and Raja 2014). In addition, dimensions such as the size of the grid, the accidental terrain it can be installed in, and the slow, local, and complicated processes used to gather information for inspections and maintenance mean that exercises related to risk analyses often are executed in a suboptimal way. By suggesting a Systems Engineering-based approach, we aim to show how the general level of risk in power grids can be reduced in a continuous way, giving thus evidence that more dynamic approaches can be implemented. For this purpose, we focus here on the impact of vegetation on the power grid. Vegetation was the number one cause of outage in Norway in 2018 (Eggum 2019) and is a main contributing factor for outages in power grids in general (Hansen 2016, 2017, 2018). The most common way for vegetation to affect the power grid is generally by a branch or an entire tree falling directly on a power line. Alternatively, vegetation can also generate outages by simply growing under a line until it makes contact and creates an outage. In the best-case scenario, consequences of such events can be relatively low, with only a few power customers affected. However, such events can also lead to wildfires (Kumagai et al. 2004) or contribute to large blackouts (Alhelou et al. 2019).

Multiple parameters are involved in the occurrence of an outage generated by a tree falling on a power line. The first, obvious ones, are the size of the trees and their physical proximity to the power line. Additional factors are wind or precipitations, variations in temperature, the topography, and the species, health, and shape of the trees – to name just a few.

Grid operators require the following information for decision-making concerning vegetation management:

- Locations of areas that are more likely to face outages involving vegetation to send teams clear-cutting the region before there is a problem
- Level of consequences of such an outage when it happens (particularly in terms of impacted customers)
- Location of areas that are more likely to face outages involving vegetation to know where to look first when those occur and thereby shorten reaction time and eventual power restoration

5.2 *Proposed Architecture*

The architecture proposed to tackle the problem of vegetation is divided into two main phases, as described in Fig. 1.

In a first phase, the first layer, essentially a top-down approach, is initially a "remote-based" information capture. It integrates diverse sources of information, such as optical satellite images, wind exposure, global vegetation characteristics [i.e., dominant species presence, canopy height], human population density, grid topology, temperatures, precipitations, and topography. Information relative to each data source is first collected and stored in a database in such a way that it can be

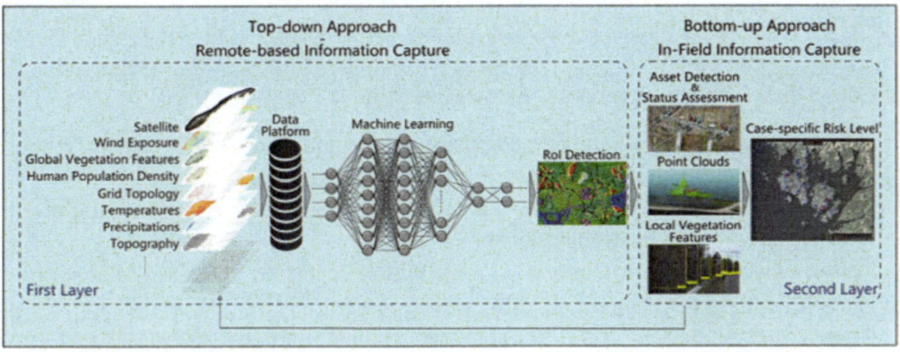

Fig. 1 Architecture for vegetation-focused Dynamic Risk Analyses in Norwegian power grids

used for calculations. Machine learning-based methods enable the combination of the different variables to assess and report the contributing factor of each variable into the first layer risk calculation. This first layer risk calculation aims to provide, by combining both the probability and the consequences of a potential disturbance, estimations of regions where the levels of vegetation-based risk estimation is the highest (Regions of Interest – RoI).

Although calculations are case-specific, results of risk estimation remain characterized by a medium degree of uncertainty because of the resolution of the information used. In order to accept or reject the first-level estimation and thus reduce the level of uncertainty around the risk estimations, a second layer (which represents a bottom-up approach) is introduced in a second phase. This "in-field-based" information capture is here again the fruit of several fields of expertise and enables specific assessment of assets and detection of faulty components using computer vision; centimeter-level distance estimation from trees to power grid components using lidar-based or photogrammetry-based point clouds; and tree-specific characteristics estimation [i.e., species, height, health conditions] using computer vision. Computer vision-based asset information suggests if the probability of outage might be increased by the types and condition of the asset present in the specified RoI. Point clouds enable case-specific high-resolution distance measurements from vegetation to the power line. And computer vision-based tree characterization enables assessment of how the originally estimated level of risk may be affected by the properties of the trees present in the area. These three additional data sources contribute by providing case-specific local information, offering an improvement of both the consequence estimations and the probability estimations of an outage in the RoI, thus enabling a refinement of the final risk image.

The output of the second layer is feedback to the first layer as the re-assessment cycles unfold and helps improve the quality of some first estimations of the local area (e.g., trees height and species), which enables the inclusion of high-resolution time series into the calculations and improved estimations provided by the retrained algorithms.

5.3 Results

The described architecture enables the operator to:

- Discover previously unconsidered risky areas and thus better quantify the consequences of disturbances caused by potential outages.
- Multiply the number of scenarios leading to an unwanted event by highlighting relevant interactions between relevant variables.
- Refine existing scenarios and risk contribution levels of specific variables by enabling a higher-resolution situation understanding.
- Refine estimation of the contribution level of each factor to the global risk picture.

- Increase resilience to loss of information by increasing the number of data sources. This increases the probability of correlation detections among used data sources, which can thus be used as proxies when one/some of them fail or would be removed.
- Make case-specific risk estimations/improvements by gathering local data, avoiding use of averaged values, and reducing thus uncertainty around risk estimation.
- Increase the frequency of risk estimations by benefiting from regular updates (e.g., weather) of the data sources used for the risk estimation.

Based on this dynamic risk estimation, power grid companies can improve the assignments of woodcutting teams by efficiently prioritizing missions based on potential risk, thereby reducing the occurrence of vegetation-influenced outages. Furthermore, they can reduce the time to repair if an outage happens since a product of this architecture increases the probability of spotting the correct areas causing the outage.

The key dimension of the described architecture is to make "as-good-as-knowable" estimations, optimizing the contributing potential of the accessible data sources, increasing the probability of detection of early signals, and reducing the probability of occurrence of events that can be avoided with timely use of information.

6 Discussion and Conclusion

Although the demand and justifications for the development of DRA tools is obvious across industries, it remains a challenging task and a relatively new research area. A theoretical broad access to a large number of data sources and an easy access to powerful IT infrastructures suggest that the main entities that could support the development of DRA tools are already available in practices that support the emerging Internet of Things. Structuring the combination of those different entities and transforming this combination into a useful risk image for an infrastructure under review remains a challenging task requiring competences in a multitude of disciplines, a local understanding of the interrelations, as well as a holistic overview of the constructed system, considered within a specific environment. SE is a particularly supportive field with this regard, as it provides the right framework to develop white box-based systems for which understanding and control can be kept over time.

We illustrated the benefits that can be provided by SE for the development of DRA tools by focusing on the assessment of the impact of vegetation on the true risk level existing in power grids. For that, we showed how relevant data sources should be combined in such a way that decision-makers can optimize their judgments and the management of their resources, as well as implement a pre-event resilience plan and effective post-event restorations. How the risk reduction actions are executed and the way the resulting information is integrated into the new cycle of risk assessment is an additional dimension that needs to be carefully considered to ensure optimal risk management of the infrastructure.

The expected convergence between requirements for efficient DRA tools and solutions provided by the intrinsic properties of Systems Engineering is confirmed in the proposed approach and described in the case study. By enabling the creation of systems that favor cyclic approaches, SE enables a flexible process and offers possibilities for both optimized information management and more resilience. This makes the process more reliable, sustainable, and thus suitable in the long run for the application of Dynamic Risk Analysis.

This work is part of an ongoing project “Dynamic risk management for Smart Grids in large-scale interconnected power systems” funded by eSmart Systems and the Norwegian Research Council. Future steps include the final choice of the best algorithms for the machine learning processes capturing those variation rates of the related data sources. Those elements will be the basis for complementary performance metrics (cf. phase (3) of the suggested approach) and will enable quantification of the level of uncertainty of the calculated risk level.

The generic dimension of the proposed approach offers already a flexibility that enables it to be used for other industries, under the condition that informative data sources are properly identified and accessible. The approach also requires the validation and verification to be continuously ensured by close collaboration with the customer, a condition that has been respected in the present study. Continuously exchanges with the main stakeholders enable appropriate feedback with regard to the performance, the process design, and the context evolution. In this way, the best tradeoff options can be continuously chosen, and divergence between the suggested risk representation and the real risk level over time can be reduced.

References

Alhelou, H.H., M.E. Hamedani-golshan, T.C. Njenda, and P. Siano. 2019. A Survey on Power System Blackout and Cascading Events Research: Motivations and Challenges. *Energies* 12: 1–28.

Aven, T. 2012. The Risk Concept – Historical and Recent Development Trends. *Reliability Engineering and System Safety* 99: 33–44.

Chapman, P., J. Clinton, R. Kerber, T. Khabaza, T. Reinartz, C. Shearer, and R. Wirth. 2000. CRISP-DM 1.0 – Step-by-step data mining guide. p. 76.

Creedy, G.D. 2011. Quantitative Risk Assessment: How Realistic Are Those Frequency Assumptions? *Journal of Loss Prevention in the Process Industries* 24 (3): 203–207.

DeCorla-Souza, K., 2013. Comparing the Impacts of Northeast Hurricanes on Energy Infrastructure, p. 50.

Eggum, E. 2019. Rapport Nr. 29-2019 – Avbrotsstatistikk 2018. Oslo, p. 101.

Hansen, H. 2016. Rapport Nr. 78-2016 – Avbrotsstatistikk 2015. Oslo, p. 99.

Hansen, H. 2017. Rapport Nr. 43-2017 – Avbrotsstatistikk 2016. Oslo, p. 108.

Hansen, H. 2018. Rapport Nr. 64-2018 – Avbrotsstatistikk 2017. Oslo, p. 103.

ISO – International standardization organization. ISO 14971:2007 – Medical Devices – Application of Risk Management to Medical Devices, 2007. Geneva, Switzerland.

ISO – International standardization organization. ISO 17666:2016 – Space Systems – Risk Management, 2016. Geneva, Switzerland.
ISO – International standardization organization. ISO 31000:2018 – Risk Management: Principles and Guidelines, 2018. Geneva, Switzerland.
Kaplan, S., and B.J. Garrick. 1981. On The Quantitative Definition of Risk. *Risk Analysis* 1 (1): 11–27.
Kenward, A., and U. Raja. 2014. *Blackout: Extreme Weather, Climate Change and Power Outages*, 23. Climate Central.
Kumagai, Y., J.C. Bliss, S.E. Daniels, and M.S. Carroll. 2004. Research on causal attribution of wildfire: An exploratory multiple-methods approach. *Society and Natural Resources* 17 (2): 113–127.
NORSOK. Standard Z-013 – Risk and emergency preparedness assessment, 2010. Lysaker, Norway.
Villa, V., N. Paltrinieri, F. Khan, and V. Cozzani. 2016. Towards Dynamic Risk Analysis: A Review of the Risk Assessment Approach and Its Limitations in the Chemical Process Industry. *Safety Science* 89: 77–93.
Yang, X., and S. Haugen. 2015. Classification of Risk to Support Decision-Making in Hazardous Processes. *Safety Science* 80: 115–126.

Toward a Reliability Approach Decision Support Tool for Early System Design: Physics of Failure vs. Historical Failure Data

John Kosempel, Bryan M. O'Halloran, and Douglas L. Van Bossuyt

Abstract The historical failure data reliability prediction method commonly used by systems engineering practitioners has several limitations. Recent literature promotes the physics of failure reliability prediction approach and has seen limited adoption. However, there is limited guidance available to practitioners to determine when the historical failure data reliability approach is appropriate to use and when the physics of failure approach is best applied. This paper presents a decision support framework for practitioners to choose between historical failure data and physics of failure reliability approaches and is specifically meant to be used in early system design. The Reliability Decision Framework RDF) identifies key factors in system design that aid practitioners in the selection of an appropriate reliability prediction approach for systems of interest. The major factors in the decision are (1) relevant historical data, (2) level of complexity, (3) operational life requirement, and (4) criticality of the system.

Keywords Reliability · System design · Physics of failure · Reliability prediction · System architecture

1 Introduction

System reliability estimations are often performed during the system architecture phase of systems engineering to aid in the evaluation of a candidate design with respect to system requirements and to provide a basis for further reliability improvements (Blanchard and Fabrycky 2011). Accurately predicting a system's reliability during early system design is a challenging task due to a lack of significant operational experience of a proposed system. Further, systems engineering practitioners have limited resources to pull data from to formulate how reliable a system of

J. Kosempel · B. M. O'Halloran · D. L. Van Bossuyt (✉)
Naval Postgraduate School, Monterey, CA, USA
e-mail: douglas.vanbossuyt@nps.edu

A. M. Madni et al. (eds.), *Recent Trends and Advances in Model Based Systems Engineering*, https://doi.org/10.1007/978-3-030-82083-1_58

interest is expected to be once the system is fielded. Reliability predictions typically rely on the use of (1) failure data, (2) a statistical model applied to the failure data, and (3) a model of the system's reliability logic. The limiting factor in reliability predictions in many cases is the availability of failure data. Often during the system architecture phase of the systems design process, a constant failure rate is assumed in order to use an exponential statistical model to represent how system failures occur over time. However, a goodness of fit test shows with remarkable consistency that the exponential model is not valid on many existing systems (Leemis 2009). Using an exponential statistical model with historical reliability data may lead to incorrect modeling of the system of interest and misleading reliability predictions (Jones and Hayes 1999). The physics of failure (POF) reliability prediction approach was developed partially in order to not rely on potentially flawed or non-existent historical failure data by developing an understanding of the underlying physical failure mechanisms to evaluate useful life (Schueller 2013).

The systems reliability community currently uses either approach to reliability prediction during the system architecture phase of system design. However, little guidance in the literature is provided on how to determine which approach to use. A significant amount of research exists on the benefits and limitations of each reliability approach; however, the does not address when it is appropriate for a practitioner to use the historical data or the POF approach or if a combination of the two approaches should be performed.

1.1 Specific Contributions

This paper presents the Reliability Decision Framework (RDF) that aids practitioners in determining when it is appropriate to use the historical failure data approach, the POF reliability approach, or a mix of the two approaches during the system architecture phase of system design.

2 Background and Related Work

Several comparisons have been made of both the historical failure data and the POF reliability approaches. (Jones and Hayes 1999; Matic and Sruk 2008; McLeish 2010; Pecht 1996; Aughenbaugh and Herrmann 2009) present the theory behind a POF approach in reliability predictions in relationship to the historical failure data approach and focus in particular on how a POF-based approach can improve current historical failure data approaches (e.g., MIL-HDBK-217F methods). Other research has been done on reliability predictions (Schueller 2013; Varde 2010; Pecht and Gu 2009; Natarajan 2015; Barlow et al. 1993). However, relatively little has been written on decision support methods to aid practitioners in selecting an appropriate reliability approach during early system design. Reliability predictions rely on three critical areas: (1) failure data, (2) statistical modeling of the failure data, and (3) the system's reliability logic model. Failure data can be categorized into three types: (1) field reliability data (system-specific), (2) test reliability data (system-specific), and (3) external data sources (not system-specific).

2.1 Historical Failure Data Approach to Reliability

Due to the limited information provided to the practitioner in the early design stage, the historical failure data reliability approach is often constrained to using external data sources such as MIL-STD-217F although often there is at least some limited historical component data available. The historical failure data reliability approach is commonly used where MIL-STD-217F is the most widely used source for predicting reliability of components (Varde 2010). The historical failure data reliability approach can be broken down into two methods: (1) parts count and (2) parts stress analysis. Both methods are defined in MIL-STD-217F (U.S. Air Force, MIL-HDBK-217F 1995). Multiple publications list the limitations of the historical failure data reliability approach. A brief summary of these limitations is presented in (McLeish 2010; Pecht 1996).

2.2 Physics of Failure Approach

POF is a science-based approach to determining the life of a product through an analysis of the failures. POF emphasizes the root cause of a failure, the identification of failure mechanisms, and a focused analysis of the failures. The POF approach provides the practitioner with a thorough understanding of the cause and effect of failures as well as the strength tolerance of materials and components that lead to a system failure. The strength of a component is measured by the amount of stress it can endure before failing (Pecht 1996; Thaduri 2013). The primary limitation of the POF approach is that it often requires the use of accelerated life testing to get sufficient data.

In the context of PoF, failure mechanisms describe the failure that has occurred and the cause of the failure (O'Halloran et al. 2012). Failure mechanisms are dependent on the system design and the types of components used. Collins provides a mostly complete failure mechanism taxonomy (Collins 1993). Uder, Stone, and Tumer provide an extension of Collin's taxonomy for electrical failure mechanisms (Uder et al. 2004). Failure mechanisms can be categorized into three different types: manufacturing variation, overstress, and wear-out. Each category reflects a stage in the system's lifecycle. Manufacturing variations are the minor changes in production that yield early failures and represent infant mortality. Overstress failure mechanisms are the result of the stress exceeding the strength of the device (Natarajan 2015). Wear-out failure mechanisms are due to the accumulation of stress over time such as fatigue. The majority of mechanical failure mechanisms can be classified as wear-out. A variety of other failure mechanisms and models have been cataloged in the POF literature (Leemis 2009; Varde 2010; ZVEI Robustness Validation Working Group 2013; Dhillon 2015; Safety and Reliability Society 2012; Anderson et al. 2004; Lall 1996; HBM Prenscia 2018; Nelson 1990).

3 Methodology

This section develops the basis for the Reliability Decision Framework (RDF) which is intended to be used during the system architecture phase of the system engineering process and specifically during the functional analysis step (Blanchard and Fabrycky 2011). The RDF aids the system engineering practitioner in choosing the appropriate reliability approach for their system to maximize useful information to support decision-making and tradeoff studies.

The RDF identifies factors that a practitioner should consider before choosing a reliability prediction approach appropriate to the system of interest which include complexity, usable life or operational life, criticality to achieving mission goals, and reliability requirements. It is important to note that the POF approach is more time- and resource-intensive when compared to the historical failure data approach. The RDF is shown in Fig. 1.

3.1 Starting Point: Functional Analysis

The input to the RDF is the system-level functional analysis generated during the system architecture phase of the system design process. In particular, the functional block diagram (FBD) baseline system architecture and the top-level reliability requirements are needed. With this information, the practitioner can generate subsystem designs and allocate performance factors based on a flow down of the reliability requirements.

3.2 Decision Factor: Relevant Historical Data

A major factor in generating a subsystem design is the relevance of a previous similar design. If the system of interest is based on a similar system design or an older configuration, the practitioner will have historical failure rate data available. The relevant historical data can be either operational field data or previously obtained relevant accelerated life test data. The relevancy of the data is dependent on the similarity of the historical system and the system of interest in terms of (1)

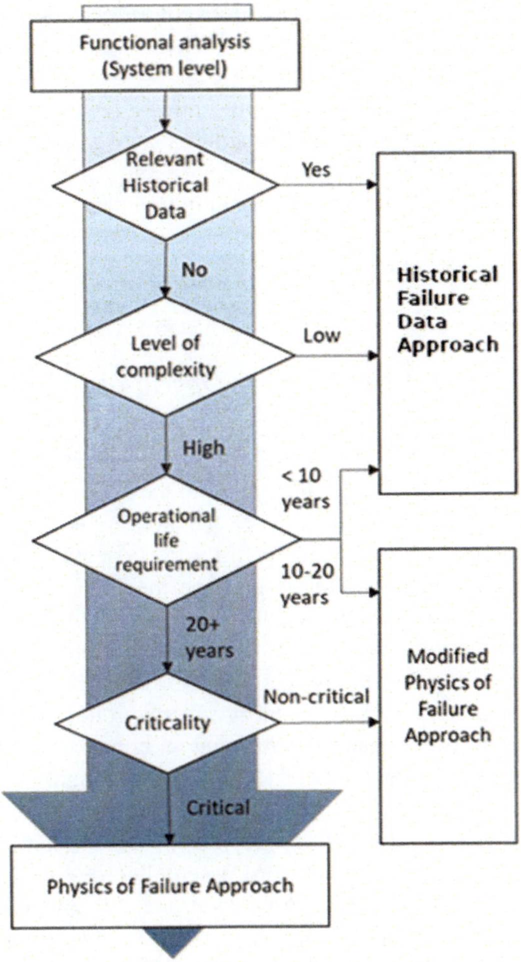

Fig. 1 A decision flowchart of reliability predictions for the Reliability Decision Framework

functionality, (2) architecture, and (3) operational environment. If all three criteria are deemed satisfied, then the historical failure data approach is likely appropriate.

3.3 *Decision Factor: Level of Complexity*

A high level of complexity of the system strongly suggests the rigor of a POF-based reliability assessment may be worthwhile. A system's complexity is difficult to quantify and is based on multiple factors including the number of components, subsystems, emergent behaviors and properties, and nonlinear relationships between components (Body of Knowledge and Curriculum to Advance Systems Engineering (BKCASE) 2017). While the topic of complexity would benefit from a detailed quantification, the limitations of information available while developing FBDs limits the ability to do so. Therefore, the authors propose to rate the level of complexity on a qualitative scale of 1–10, with 10 representing tens of thousands of system components, 5 representing a few thousand components, and 1 representing tens of components. The authors suggest that a qualitative score of 5 be the break point between a low-complexity and a high-complexity system. Low-complexity systems can use the historical failure data approach, while high-complexity systems have additional considerations that must be taken into account before a reliability method can be selected.

3.4 Decision Factor: Operational Life Requirement

The expected operational life of the system is driven by the requirements analysis. This factor dictates how reliable the system needs to be to last through its intended lifecycle. If, for example, the system is expected to last roughly 50 years (a common requirement for military systems), the use of a POF approach becomes more effective. This is because the POF approach analyzes multiple time-based failure mechanisms which can improve understanding of the reliability of the system design. In contrast, a system prone to rapid technology refreshes (2–5-year cycles) requiring regular system redesign lends itself the historical failure data reliability approach which is the most effective reliability prediction method because there is not a strong need to understand all failure mechanisms associated with the system.

For the purposes of RDF, the operational life requirement is divided into three lengths of time. A low operational life is represented as a system expecting to last 10 years or less. Within 10 years, such a system has a high probability of requiring a redesign of circuit card assemblies due to component obsolescence (Torresen and Lovland 2007). An operational life of 10–20 years will generally require a partial redesign of the system and a complete redesign of the subsystems due to the obsolescence of technology (Singh and Sandborn 2006). Systems expected to operate 20 years or greater will require a complete redesign due to diminishing manufacturing sources and material shortages, technology updates, performance increases, and component obsolescence (Singh and Sandborn 2006).

If the practitioner does not have relevant historical data, the level of complexity of the system is high, and the operational life requirement is greater than 10 years, then the reliability prediction becomes increasingly more important. At this point of the RDF, a POF-based approach becomes more effective than the historical failure data reliability approach.

A POF approach can also be modified using principles from both the historical failure data and POF methods. The modified POF approach is a customized approach to suit the practitioner's needs based on the information available. Some publications exist on describing various modified reliability approaches (Aughenbaugh and Herrmann 2009; Thaduri 2013; Thaduri et al. 2015; Yadav et al. 2003). These modified approaches take aspects of the historical failure data and POF approach and specify the reliability assessment based on two primary factors: (1) the type of failure data available to the practitioner in terms of both quantity and quality and (2) the physical architecture of the system. The modified approach becomes relevant to RDF users when the system has a medium operational life.

3.5 *Decision Factor: Criticality*

For a critical system application, the POF approach becomes crucial to increasing the system's survivability under varying operational stresses. This is particularly important in the aerospace, nuclear power, oil and gas, and healthcare industries where system failures may lead to catastrophic outcomes. The evaluation criteria used in the RDF for criticality bins systems into three categories: non-critical (NC), critical application item (CAI), and critical safety item (CSI), respectively (Office of the Under Secretary of Defense for Acquisition, Technology, and Logistics 2016; Bozzano and Villafiorita 2010).

A CSI system in the RDF results in the use of a POF reliability approach. This is because the impact of a failure is significant and will require a thorough analysis of failure mechanisms to design a very robust system to mitigate catastrophic failures. A CAI system results in the use of a modified POF approach. A modified POF approach does not contain as thorough of an analysis on various failure mechanisms as the strict POF approach and uses principles of the historical failure data approach to utilize resources effectively. A NC system may use a historical failure data approach.

A practitioner using the RDF framework is advised to weigh the information gathered at each decision factor above and decide if a historical failure data approach, a POF approach, or a modified POF approach is most appropriate for the system. As the system design evolves, RDF can be re-run to continue to verify that the system meets requirements and to update the reliability models. Through understanding which reliability method is most appropriate to use, a practitioner may achieve a more accurate picture of system reliability.

4 Case Study

This section presents a brief case study to demonstrate how to apply the RDF and to articulate an example of the expected results. While this case study is presented for a real system, the results are only valid for better understanding the RDF framework. Therefore, the results should not be used outside of this paper.

The system being analyzed in this case study is a gas turbine auxiliary power unit (APU) on a military aircraft. As previously mentioned, an input to the method is a system functional analysis which is not shown here due to space limitations. Next the practitioner extends the functional analysis to the subsystems. The reliability requirement then flows down to the design of the subsystems. During this stage, the practitioner has flexibility in allocating reliability requirements to elements of the subsystems and designing the subsystems to optimally meet or exceed the allocated reliability requirements.

A review of previously developed APU designs for commercial applications show similarities in system functionality and architecture. The relevant historical failure data collected by the commercial system is dependent on the environment, and, as the environment for the military application introduces different stressors, the historical data for the commercial system becomes less relevant to the military application. In the RDF, the historical data does not contain all three criteria of relevancy, and therefore the practitioner does not have sufficient relevant historical data.

The level of complexity of the system is analyzed based on the number of subsystems, interfaces, and an estimation of components required for each subsystem. Applying an estimation factor of a hundred components per subsystem gives the system an estimated 100 components. This is equivalent to a complexity level of 5 in the RDF, constituting the system as having a medium complexity level.

The expected operational life requirement for the APU is 35 years. At this point in the RDF, the historical failure data reliability approach is no longer a feasible option for the APU.

The criticality of the system is determined based on the functional requirement, interactions with external systems, safety requirements, and the end application of the APU. The APU provides electrical and hydraulic power to support aircraft systems. The safety requirements are fire prevention, protection for over-speed, rotor containment, and mid-flight engine start. Based on these factors, the APU system is determined to be a mission critical system.

As a result of analyzing the decision factors, the RDF recommends the practitioner to perform a reliability assessment of the APU using a POF reliability approach.

5 Discussion

The RDF presented above provides the practitioner a decision support tool to aid in choosing between a historical failure data approach, a POF approach, and a modified POF approach for analyzing system reliability during conceptual system design. Previously, practitioners may have decided to use one of the three approaches without a formal and repeatable process. As was identified in the case study, if a historical failure data approach for the APU was used, it would produce a system that may not meet reliability requirements in operation due to the lack of reliability enhancements in system design because of inadequate reliability analysis.

The RDF is applicable at the earliest stages of the system design process. This provides significant value to designers by allowing them to make well-informed and impactful decisions. For example, the identification that a system will not meet a reliability requirement can become a catalyst for a substantial design change. If instead a designer were to use an alternative method to RDF, and specifically one employed later in the design process, the flexibility and allowance to make substantial design changes decreases significantly. Further, significant design changes become altogether unrealistic at a certain point in the design process, and therefore an early design approach to choosing and assessing reliability is necessary for fostering a reliable design.

The results of the system-level functional analysis generated in the conceptual design phase provide the practitioner with the necessary information to make a thoughtful decision on an appropriate reliability approach for the system. The RDF highlights the key factors in system design that contribute to an appropriate reliability approach. The reliability approach resulting from the RDF can be used in the refinement of the system and subsystem design. This further enhances the system's reliability throughout the rest of the system design process.

6 Conclusion

In the system architecture phase of system design, relevant system failure data is the limiting factor in reliability predictions. Often practitioners are limited in collected historical failure data and data derived from accelerated life tests. The failure data generally provided by external data sources are very limiting and outdated. Historical failure data reliability prediction methods often rely on the use of external data sources in accurately predicting the reliability of a system. Many reliability predictions do not match experienced operational failures. The POF approach reduces the inaccuracy of reliability predictions by exploring the root causes of failures and defining failure rates for different failure mechanisms. The POF approach results in a more extensive reliability prediction, but often requires failure data derived from accelerated life tests to determine the life-stress profile and properly model the failure mechanism over time. It is important for a practitioner to accurately assess and predict a system's reliability.

The RDF presented in this paper identifies the key factors a practitioner should consider when selecting a reliability approach. Although reliability is an iterative process throughout system design, the RDF is best applied in the system architecture phase of system design when a system-level functional analysis has been performed. In addition to assisting the selection of a reliability prediction method, the results of the RDF may further enhance the system design and the allocation of system requirements in the preliminary design phase.

Acknowledgments This work was supported in part by Naval Research Program NRP-19-085 and the Naval Postgraduate School. All opinions contained herein are those of the authors and do not necessarily reflect the sponsors. No warranty of accuracy or completeness of work is given or implied.

References

Anderson, P., H.J. Jensen, L. Oliveira, and P. Sibani. 2004. Evolution in Complex Systems. *Complexity at Large* 10 (1): 49–56.

Aughenbaugh, J., and J. Herrmann. 2009. Reliability-Based Decision Making: A Comparison of Statistical Approaches. *Journal of Statistical Theory and Practice* 3 (1): 289–303.

Barlow, R.E., C. Claroti, and F. Spizzichino. 1993. *Reliability and Decision Making 1st Ed.* London: Chapman and Hall/CRC.

Blanchard, B.S., and W.J. Fabrycky. 2011. *Systems Engineering and Analysis*. 5th ed. Saddle River: Pearson Education Inc.

Body of Knowledge and Curriculum to Advance Systems Engineering (BKCASE). *Complexity*. Guide to the Systems Engineering Body of Knowledge (SEBoK), 17 November 2017. [Online]. Available: http://www.sebokwiki.org/wiki/Complexity. Accessed Aug 2018.

Bozzano, M., and A. Villafiorita. 2010. *Design and Safety Assessment of Critical Systems*. Boca Raton, FL: CRC Press.

Collins, J.A. 1993. *Failure of Materials in Mechanical Design: Analysis, Prediction, Prevention*. 2nd ed. Wiley-Interscience.

Dhillon, B. 2015. Reliability in the Mechanical Design Process. In *Mechanical Engineers' Handbook*, 1–27. Ottawa, Ontario, Canada: Wiley.

HBM Prenscia. 2018. *ReliaWiki*. ReliaSoft Corporation. [Online]. Available: http://reliawiki.org/index.php/Introduction_to_Accelerated_Life_Testing#Select_a_Life-Stress_Relationship. Accessed July 2018.

Jones, J.A., and J.A. Hayes. 1999. A Comparison of Electronic Reliability Prediction Methodologies. *IEEE Transactions on Reliability* 48 (2): 127–134.

Lall, P. 1996. Tutorial: Temperature As An Input to Microelectronics-Reliability Models. *IEEE Transactions on Reliability* 45 (1): 3–9.

Leemis, L.M. 2009. *Reliability: Probabilistic Models and Statistical Methods*. 2nd ed. Lawrencw M. Leemis.

Matic, Z., and V. Sruk. 2008. The Physics-of-Failure Approach in Reliability Engineering. In *International Conference on Information Technology Interfaces (ITI 2008)*, Cavtat, Croatia.

McLeish, J.G. 2010. Transitioning to Physics of Failure Reliability Assessments for Electronics. In *DFR Solutions*.

Natarajan, D. 2015. *Reliable Design of Electronic Equipment: An Engineering Guide*. Bangalore: Springer.

Nelson, W. 1990. *Accelerated Testing: Statistical Models, Test Plans, and Data Analyses*. New York: Wiley.

Office of the Under Secretary of Defense for Acquisition, Technology, and Logistics. 2016. DOD INSTRUCTION 4140.69. Executive Services Directorate.

O'Halloran, B.M., R.B. Stone, and I.Y. Tumer. 2012. A Failure Modes and Mechanisms Naming Taxonomy. In *Proceedings Annual Reliability and Maintainability Symposium*. Reno.

Pecht, M. 1996. Why the Historical Failure Data Reliability Prediction Models Do Not Work – Is There an Alternative? *Electronics Cooling* 2: 10–12.

Pecht, M., and J. Gu. 2009. Physics-of-failure-based prognostics for electronic products. *Transactions of the Institute of Measurement and Control* 31 (3-4): 309–322.

Safety and Reliability Society. 2012. Applied R&M Manual, for Defence Systems (GR-77 Issue 2012). In *Part C – R&M Related Techniques: Derating*, 1–22. Oldham: Safety and Reliability Society.

Schenkelberg, F. *Norris-Landzberg Solder Joint Fatigue*, Accendo Reliability, [Online]. Available: https://accendoreliability.com/norris-landzberg-solder-joint-fatigue. Accessed July 2018.

Schueller, R. *Introduction to Physics of Failure Reliability Methods*. DfR Solutions, 27 March 2013. [Online]. Available: https://www.dfrsolutions.com/resources/introduction-to-physics-of-failure-reliability-methods-video. Accessed May 2018.

Singh, P., and P. Sandborn. 2006. Obsolescence Driven Design Refresh Planning for Sustainment-Dominated Systems. *The Engineering Economist* 51 (2): 115–139.
Thaduri, A. 2013. *Doctorical Thesis: Physics-of-Failure Based Performance Modeling of Critical Electronic Components*. Luleå, Luleå, Sweden: Universitetstryckeriet.
Thaduri, A., A.K. Verma, and U. Kumar. 2015. Comparison of failure characteristics of different electronic technologies by using modified physics-of-failure approach. *International Journal of System Assurance Engineering and Management* 6 (2): 198–205.
Torresen, J., and T.A. Lovland. 2007. *Parts Obsolescence Challenges for the Electronics Industry. In IEEE Design and Diagnostics of Electronic Circuits and Systems*. Poland: Krakow.
S. J. Uder, R. B. Stone and I. Y. Tumer. 2004. Failure Analysis in Subsystem Design for Space Missions. In *ASME Design Engineering Technical Conferences and Computers and Information in Engineering Conference*. Salt Lake City, Utah.
U.S. Air Force, MIL-HDBK-217F. 1995. *Reliability Prediction of Electronic Equipment*. Griffiss AFB, NY: Department of Defense.
Varde, P.V. 2010. Physics-of-Failure Based Approach for Predicting Life and Reliability of Electronics Components. *BARC Newsletter*, Vols. Mar.–Apr., no. 313, pp. 38–46.
Yadav, O.P., N. Singh, R.B. Chinnam, and P.S. Goel. 2003. A fuzzy logic based approach to reliability improvement estimation during product development. *Reliability Engineering & System Safety* 80 (1): 63–74.
ZVEI Robustness Validation Working Group. 2013. *Handbook for Robustness Validation of Automotive Electrical/Electronic Modules*. Frankfurt am Main: ZVEI – Zentralverband Elektrotechnik- und Elektronikindustrie e. V.

An Approach to Improve Hurricane Disaster Logistics Using System Dynamics and Information Systems

Jeanne-Marie Lawrence, Niamat Ullah Ibne Hossain, Christina H. Rinaudo, Randy K. Buchanan, and Raed Jaradat

Abstract The annual threat of increasingly severe Atlantic hurricanes has raised concerns about the management of logistics in the immediate aftermath and rebuilding phases of catastrophic storms. Logistics challenges include timely delivery of inbound relief supplies, synchronization of supply and demand in the reconstruction phase, and management of the reverse flows of empty containers. Due to the low probability of occurrence, the high level of impact, and the scale and complexity involved, it is often difficult to devise a comprehensive logistics plan in advance of a catastrophe. To ensure satisfactory performance at all stages of recovery, the problem must be conceptualized in systemic terms using information from past experiences to identify causal relationships and opportunities for optimization. This paper highlights the challenges faced at Caribbean ports in the immediate aftermath of a hurricane disaster and in the reconstruction materials supply chain during the rebuilding period and offers a conceptual approach to improve planning by combining holistic thinking with simulation and information technologies.

Keywords Disaster logistics · Hurricanes · Supply chain · System dynamics · Caribbean

1 Introduction

Hurricanes originating in the Atlantic pose an annual recurring threat to the United States and the Caribbean and have serious implications for disaster logistics management in the aftermath. One of the most disastrous seasons on record is the

J.-M. Lawrence · N. U. I. Hossain · R. Jaradat (✉)
Department of Industrial and Systems Engineering, Mississippi State University, Starkville, MS, USA
e-mail: jaradat@ise.msstate.edu

C. H. Rinaudo · R. K. Buchanan
Instiitute of Systems Engineering Research, U.S. Army Engineer Research and Development Centre, Vicksburg, MS, USA

A. M. Madni et al. (eds.), *Recent Trends and Advances in Model Based Systems Engineering*, https://doi.org/10.1007/978-3-030-82083-1_59

2017 hurricane season (Palin et al. 2018), during which 18 named storms were recorded, 10 of which developed into hurricanes, and 6 recorded as major hurricanes in the Category 3, 4, and 5 designations on the Saffir-Simpson scale (National Hurricane Center 2017). Hurricane Harvey made landfall in Texas as a Category 4 hurricane, while Hurricanes Irma and Maria hit Florida, Puerto Rico, Antigua and Barbuda, the British Virgin Islands, the Commonwealth of Dominica, and St. Maarten as Category 4 or 5 storms. Hurricane Maria intensified from a Category 2 to a Category 5 storm in less than 12 hours and tore through the small Caribbean island nation of the Commonwealth of Dominica as a Category 5 storm with little warning (AON 2018) before continuing northward to hit Puerto Rico as a Category 4 storm. Of the five most costly hurricanes on record, Harvey, Maria, and Irma ranked second, third, and fifth, respectively (National Hurricane Center 2017), and occurred within a period of less than 4 weeks. In 2018, the scenario was again repeated with Hurricanes Florence and Michael, which developed into Category 4 storms during the 2018 season (National Hurricane Center 2017), causing severe damage in the Carolinas, Florida, Georgia, and Alabama. Collectively, hundreds of thousands of homes, businesses, and other infrastructure were severely damaged in the affected regions, leaving residents without housing, power, water, food, and medical supplies (Palin et al. 2018). Given these examples and the ongoing predictions of climate change, understanding how post-disaster logistics can be improved is timely, particularly for regions that are isolated due to geography and depend on an existing supply lifeline of food and consumer items routinely imported from overseas.

This paper addresses some of the challenges faced in post-hurricane disaster supply chains in the immediate aftermath and longer-term rebuilding phases. Using the case of Hurricane Maria, the combined experiences of the Commonwealth of Dominica and Puerto Rico are discussed. A systemic approach is taken to identify points of weakness in port operations and the reconstruction materials supply chain. The application of information and simulation tools to plan for improved efficiency and agility in logistics operations are proposed. The paper contributes to the disaster logistics literature by expanding the discourse on the management of hurricane disaster logistics.

2 Background

A disaster is an irregular negative event with serious consequences that results in loss of capacity to operate at pre-disaster levels. Disasters can be locally contained, meaning that the system is able to provide the required supplies to those in need with little external assistance, or it can be catastrophic, resulting in widespread damage and destruction at a scale that requires external intervention (Kunz et al. 2014; Holguín-Veras et al. 2012). In the case of a locally contained disaster, there is some capacity, albeit limited, to continue to meet demand from internal inventory sources (Holguín-Veras et al. 2012). However, in catastrophic situations, commercial supply chains are usually out of commission, and supplies of local inventory are virtually non-existent, requiring external assistance to meet the needs of survivors. In both cases, effective logistics management is critical to ensure rapid recovery.

The Council of Supply Chain Management Professionals defines logistics management as "that part of supply chain management that plans, implements, and controls the efficient, effective, forward, and reverse flow and storage of goods, services, and related information between the point of origin and the point of consumption in order to meet customer requirements." Supply chain flows are subject to both motion or transit time and storage or waiting time in forward and reverse directions (Council of Supply Chain Management 2018). Logistics activities encompass demand forecasting, procurement, production scheduling, material handling, inventory control, transportation, warehousing, packaging, order fulfillment, facility location, reverse logistics, service, and support (Coyle et al. 2017). Another view of logistics is in terms of the creation of utility. Of the five types of utility that add value to a product, place utility (where), time utility (when), and quantity utility (how much) have been identified as the predominant value-creating processes of the logistics function (Coyle et al. 2017). Place utility is created when inventory is moved from points of production or distribution to points of demand. Time utility is produced when supplies are available at the exact time required by customers. Quantity utility is achieved when the amount delivered neither exceeds nor falls short of immediate customer requirements. The management of place, time, and quantity utility are critical logistics functions in post-disaster logistics situations because of the sudden surge in demand, the urgency to deliver products to save lives, and the limitations of infrastructure, particularly transportation and storage infrastructure. Another type of utility, form utility, typically associated with manufacturing processes, may also become important in post-disaster logistics situations. Form utility involves conversion of raw materials into value-added products but can also include breakdown and assembly functions at distribution points (Coyle et al. 2017).

Supply chain management has evolved over time from merely managing the flow of materials between entities to developing tightly integrated and seamless inter- and intra-organizational processes and relationships that emphasize the concept of holism (Fayezi and Zomorrodi 2016). In today's supply chains, integration is necessary to achieve the efficiency, visibility, and agility expected of the logistics function. To create the right supply chain value, logistics professionals must be cognizant of the type of goods supplied and the ultimate purpose of the supply chain. For example, the same type of inventory may require substantially different supply chain configurations and strategies under different circumstances. Efficiency, agility, and combinations of the two have been proposed by various scholars to align with product characteristics and customer demands (Fisher 1997; Simchi-Levi et al. 2008). Efficiency focuses on minimizing the use of input resources without compromising output requirements, while agility addresses resilience – the rapid reconfiguration and execution of the supply chain network in response to customer demands (Christopher 2016).

3 Review of Literature

Supply chain risk and resilience is an emerging body of knowledge in the academic literature (Hohenstein et al. 2015) with much of the focus oriented toward business continuity and firm competitiveness (Chopra and Sodhi 2014) rather than disaster management. Resilience is defined as the ability of the supply chain to adapt, respond, rebound, and grow following unforeseen risk (Ponomarov and Holcomb 2009; Hohenstein et al. 2015). Two proactive approaches to manage supply chain risk and resilience are identified. The traditional view attempts to identify and quantify anticipated or known risks and develop mitigation strategies to enable supply chain continuity. A Failure Mode and Effects Analysis (FMEA) methodology is typically used to quantify the probability of occurrence, severity of impact, and likelihood of detection. The alternative view recognizes that risks related to supply chains are often unknown and unforeseen and, therefore, require a qualitative and systemic approach to develop the adaptive capabilities needed to respond to, and recover from, disruptive events.

There is a paucity of research in the area of post-disaster logistics management despite the fact that rapid recovery is heavily dependent on the effectiveness of systems (Van Wassenhove 2006). This situation is attributed to the small number of practitioners worldwide engaged in humanitarian logistics on a full-time basis (Holguín-Veras et al. 2012). Another reason is the evolution of disaster management practices independent of supply chain coordination and risk management strategies applied to commercial supply chains (Scholten et al. 2013). Studies on disaster management attempt to understand approaches used from a practitioner rather than a theoretical perspective, with little integration of supply chain resilience theory (Scholten et al. 2013). This lack of integration of supply chain principles with disaster management approaches has limited the development of post-disaster humanitarian supply chain logistics management. There is a need to understand the nature of, and environments within which, supply chains operate following a major hurricane disaster to develop strategies to improve performance of post-disaster supply chains.

Various approaches have been recommended for pre-planning to ensure supply chain effectiveness in post-disaster situations. These include inventory management, talent development, and use of information systems. Pre-positioning of inventory in the supply chain is one of the most commonly used strategies. This approach, however, is costly because of the uncertainty regarding where and when a disaster will strike, the quantity of inventory that will be required, and the subsequent deterioration of inventory that could result due to low turnover in periods when there is no disaster (Kunz et al. 2014; Balcik et al. 2010). Another inventory management approach is holding safety stock to minimizing disruptions in critical supply chains (Holguín-Veras et al. 2012; Ozguven and Ozbay 2013). The development of disaster management capabilities has also been suggested as an alternative to the pre-positioning of inventory. The development of human resource capabilities to conduct pre-negotiation of supply contracts, harmonize import procedures, and negotiate customs agreements can be executed in advance of a disaster to increase supply chain agility. In a study using system dynamics modeling to simulate inventory levels in a system with improved disaster response capabilities but no pre-positioning of inventory, Kunz et al. (Kunz et al. 2014) showed that development of disaster response capabilities can significantly reduce lead times and increase supply chain responsiveness. In another study, rapid development of trust, public-private sector partnerships, and quality information sharing were found to be critical enablers of resilience in supply chain networks (Papadopoulos et al. 2016). With the increasing use of technology, other methods have been recommended for increasing visibility and planning. Ozguven and Ozbay (Ozguven and Ozbay 2013) proposed an approach for managing inventory in disaster situations by combining an online system for managing emergency supplies with an offline planning system. The lack of coordination among humanitarian actors is recognized as a major weakness in disaster supply chains as there is no one organization responsible for coordinating the supply chain (Rey 2001).

In the aftermath of severe hurricanes, effective management of logistics is critical for survival and rapid recovery. Two distinct post-disaster phases can be defined: (i) the immediate aftermath of the storm, during which critical relief supplies are required within hours of the disaster and for a period ranging from 30 to 180 days thereafter, and (ii) the re-building phase, which can extend for several months and even years. Consequently, two generic types of post-disaster supply chains can be identified: (i) supply chains in the immediate aftermath that deliver food, water, medical supplies, tarpaulins, fuel, power generators, telecommunications supplies, and equipment for debris removal to save lives (Ozguven and Ozbay 2013) and (ii) supply chains that deliver infrastructural materials to support longer-term rebuilding and reconstruction (Holguín-Veras et al. 2012; Neuman 2017). Each phase is associated with specific performance expectations. While both agility and efficiency are important, early post-disaster phases are expected to prioritize agility over efficiency, while efficiency is expected in longer-term logistics management (Holguín-Veras et al. 2012).

The scope of post-disaster supply chain logistics is broad, and failures can occur at numerous points along the end-to-end supply chain: inbound, operations, distribution, reverse logistics, and disposal stages of supply networks. The extent of these failures is dependent on several factors, including the scale and scope of the disaster, the location of the affected country or region, the degree of development of the affected country or region, the nature and extent of prior preparations, the availability and type of physical and information systems infrastructure, and the capability of the human resource pool to rapidly reconfigure supply chains. Understanding the vulnerable points in post-disaster supply chains is key to determining how to route flows around points of failure (Ozguven and Ozbay 2013), elevate bottlenecks, improve processes, and train people.

To identify risk factors that sub-optimize post-hurricane supply chain performance, the starting point is to develop a systemic view of the logistics operation. A systemic view recognizes that all systems are composed of interrelated subsystems that cannot be reduced to individual components but must be perceived in totality.

Such a view resists thinking in linear terms that attempt to define structured cause and effect relationships. Instead, the entire system of subsystems is conceptualized as a single system and managed holistically to achieve the desired end results (Koskinen 2013). By utilizing past experience and data to understand and model the interrelationships, those involved in planning for post-disasters logistics can be better prepared to address and mitigate the risks involved.

4 Methodology

This paper develops a conceptual approach for pre-planning disaster logistics in the event of a catastrophic disaster. Using the experiences of two Caribbean islands, the Commonwealth of Dominica and Puerto Rico, the challenges faced in the aftermath of Hurricane Maria were investigated. Information was gathered from reports, newspapers, social media, and personal accounts to understand points of weakness. Vensim System Dynamics simulation software was subsequently used to convert these mental models into a system dynamics model to show causal factors and interrelationships. Bottlenecks and inefficiencies are identified, and recommendations are proposed on the use of information technologies to improve agility and efficiency.

4.1 Factors Impacting Post-hurricane Disaster Logistics and Supply Chain Management

Challenges faced in managing supply chains following a large-scale hurricane disaster include (i) adapting to the sudden onset of complexity and (ii) developing the capabilities to respond to survivors rapidly, seamlessly, and completely. These expectations have become more important in a world that is connected by technology and which affords real-time communication globally. Prior to a disaster, there is a baseline understanding of the volume, velocity, and size of material flows through the supply chain. Processes such as order preparation for shipment, vessel loading/unloading, and customer clearance are designed to accommodate specific operational configurations. Following a disaster, these characteristics are altered drastically, requiring rapid adaptation to manage the increase in complexity. Challenges include damaged air and seaport infrastructure that restrict points of entry, reduced storage capacity due to destruction of warehouse facilities, process inefficiency resulting from non-functioning information and communications systems caused by damaged landline and cell phone towers (Kunz et al. 2014), and inadequate capacity at all levels. These factors create bottlenecks that impede the flow of information and inventory. Other challenges result from the origin and volume of inbound flows, demand requirements, and supporting systems and

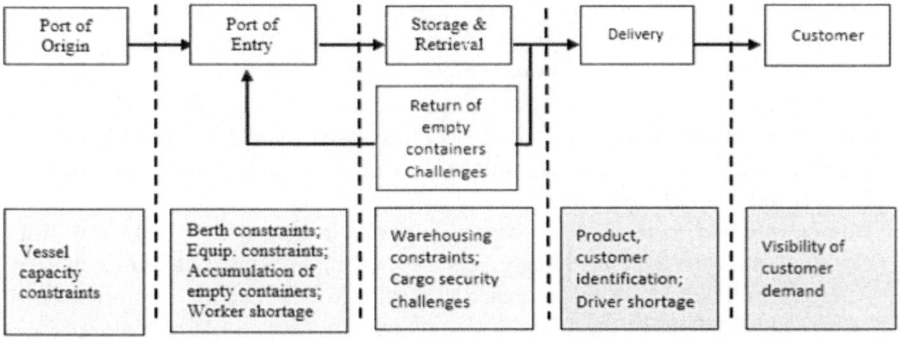

Fig. 1 A generic representation of points of failure in inbound supply chains in the immediate aftermath of Hurricane Maria

networks (Hohenstein et al. 2015). Based on data gathered, major challenges faced in ports in disaster zones were identified as follows:

- **Limited vessel space** for accepting commercial inventory at points of origin in the United States. In the case of Puerto Rico, vessel shipping space from the United States mainland to Puerto Rico was prioritized for the Federal Emergency Management Agency's (FEMA) relief supplies, limiting the available capacity for shipment of commercial goods (Palin et al. 2018; Goentzel 2017).
- **Vessel turnaround delays**. Due to inadequate berthing capacity and lack of systems and procedures to manage receipt of relief items, vessel turnaround time increased. One example is the case of a barge carrying relief items to Dominica that had to wait over 2 weeks to be offloaded in the destination port (Cnc3.news 2018) (Fig. 1).
- **Insufficient material handling equipment capacity** for use in unloading large project-size cargo. An example in the case of Dominica is the arrival of large multipurpose vehicles for use in power restoration. Due to inadequate crane capacity at the main port, the vehicles had to be disassembled to be offloaded from the ship (Gomez 2017) and then reassembled, consuming valuable time, space, and manpower.
- **Capacity constraints at entry ports** due to unusually high volumes of freight from individual donors and humanitarian organizations. In Dominica, freight volumes increased by 60% in the 3 months following Hurricane Maria compared to previous years (Bardouille 2017). This created a backlog, resulting in delays in un-stuffing containers and delivering goods to consignees (Bardouille 2017; (Unstructured Data from Social Media – Facebook posts 2017).
- **Shift in the mix of inbound cargo**. The post-disaster cargo mix shifted from large amounts of containerized merchandise consigned to a few large buyers in pre-disaster situations to a preponderance of small packages from numerous donors and humanitarian organizations consigned to individuals.

- **Storage constraints** arising from damaged and destroyed warehousing infrastructure resulting in insufficient space to **accommodate** the inbound cargo. In the case of Dominica, this was further compounded by the tardiness of consignees in clearing goods (especially vehicles) which had arrived before the hurricane.
- **Security challenges** due to open, uncovered storage, which pose threats of theft, became a major problem in Dominica (Unstructured Data from Social Media – Facebook posts 2017).
- **Inadequate or incomplete labeling of supplies in containers**, making it difficult to identify the contents of containers. An example is the case of Puerto Rico where several containers were labeled "Disaster Relief" making it impossible to identify the contents of the containers until actually opened (Palin et al. 2018).
- **Challenges in moving supplies from the port** due to transportation and fuel shortages. In Puerto Rico, manpower was inadequate as port workers and truckers failed to show up for work.
- **Limited retail inventory** due to government and taxation regulations. Retailers in Puerto Rico held approximately 1 month's supply of inventory under normal situations, reducing the country's ability to respond to survivors' needs following the disaster (Palin et al. 2018).
- **Reverse logistics of empty containers** delivering supplies to affected areas. Due to vessel delays in scheduling the pickup of empty containers, the limited storage capacity was further constrained, requiring alternative storage locations to be found as empty containers accumulated on the port. For example, in Dominica, several months after Hurricane Maria, hundreds of containers awaited pickup from shipping lines. Dominica (Dominica News Online 2017; Dominica Vibes News 2017).
- **Lack of synchronization between supply and demand** for critical items such as food, water, medical supplies, and fuel. In Puerto Rico, food and water aid continued to be requested from FEMA even after grocery stores resumed operations and were able to process transactions using electronic government cards (Ozguven and Ozbay 2013).
- **Lack of synchronization between supply and demand during the rebuilding and reconstruction** phase. In Dominica, ongoing inventory stockouts in the reconstruction supply chain continued to be a problem, hampering the recovery efforts.
- **Parallel imports of reconstruction supplies** by individual residents. Parallel imports by citizens increased the challenge of forecasting and scheduling material requirements as evidenced by the numerous stockouts of building materials.

After Hurricane Maria hit Dominica, the first container ship was not able to berth at the main port until 5 days after the storm due to rough seas. Following the initial receipt of cargo, freight volumes in the ensuing 3 months increased by 60% (Bardouille 2017) compared to the same time in previous years, posing capacity constraints for berthing, unloading, and storage of the cargo. Within the first 6 weeks, 1163 20-foot equivalent units (TEUs) were received at the port, but only 110 empty containers were shipped out as reverse flows in the same time period

(Dominica News Online 2017). As port storage space became constrained, empty containers had to be moved to alternative public open spaces. Eight months after the storm, only 627 empty containers had been shipped out, the vast majority still awaiting pick up by shipping lines (Dominica Vibes News 2017; The Chronicle 2018). Another problem that emerged was the security of cargo due to a lack of secure warehousing due to damage to storage sheds. Most of the freight originating in North America arrived in sealed containers, making it easier to handle and secure the cargo in open yard storage, whereas some of the freight originating at other points of origin as break bulk could not be stored as securely and in some cases was offloaded and stored at ports in other islands until practical to be received in Dominica. In Puerto Rico, the challenge to logistics stemmed primarily from clearing freight from the ports due to shortage of truckers. Based on this information, system dynamics models were developed for the inbound relief supply chain in the immediate aftermath (see Fig. 2) and for the re-construction materials supply chain in the longer-term rebuilding phase (see Fig. 3).

5 Discussion and Recommendation

Post-disaster supply chains are open systems with numerous interactions and relationships between players within and external to the supply chain. Opportunities to improve post-disaster logistics operations can be focused along three dimensions: reduction of complexity, increase in agility, and improvement in efficiency. To address points of weakness, effective information systems are required. Data analytics, centralized systems, and tracking and tracing technologies can be useful in this respect.

Data Analytics At the heart of better planning is access to more accurate data that can be used to model logistics operations and identify points of failure. To improve performance, data must be collected, stored, and used to optimize post-disaster logistics. Devising means to accumulate this data, both during normal operations and in disaster periods, is critical so that over time, an understanding of the dynamics of the system can be gained and fed into system dynamics models. To model the logistics operations using a system dynamics approach, the data required include the following: (i) mix of cargo by type of tertiary packaging (full container, less than full container, break bulk – palletized or not palletized), (ii) mix of cargo based on supplier and consignee (business, individual, humanitarian organization), (iii) cargo volumes for each type of cargo, (iv) points of origin, (v) arrival dates and times, (vi) off-loading dates and times, and (vii) time for each stage of the logistics operation including moving to storage, completing paperwork, and delivery. Big data analytics can be applied to summarize data about past events, make predictions about probable future trends, and evaluate different scenarios to prescribe possible future outcomes. While data may not exist or even be recoverable from a single data source, it is important that it is recoverable as soon as possible

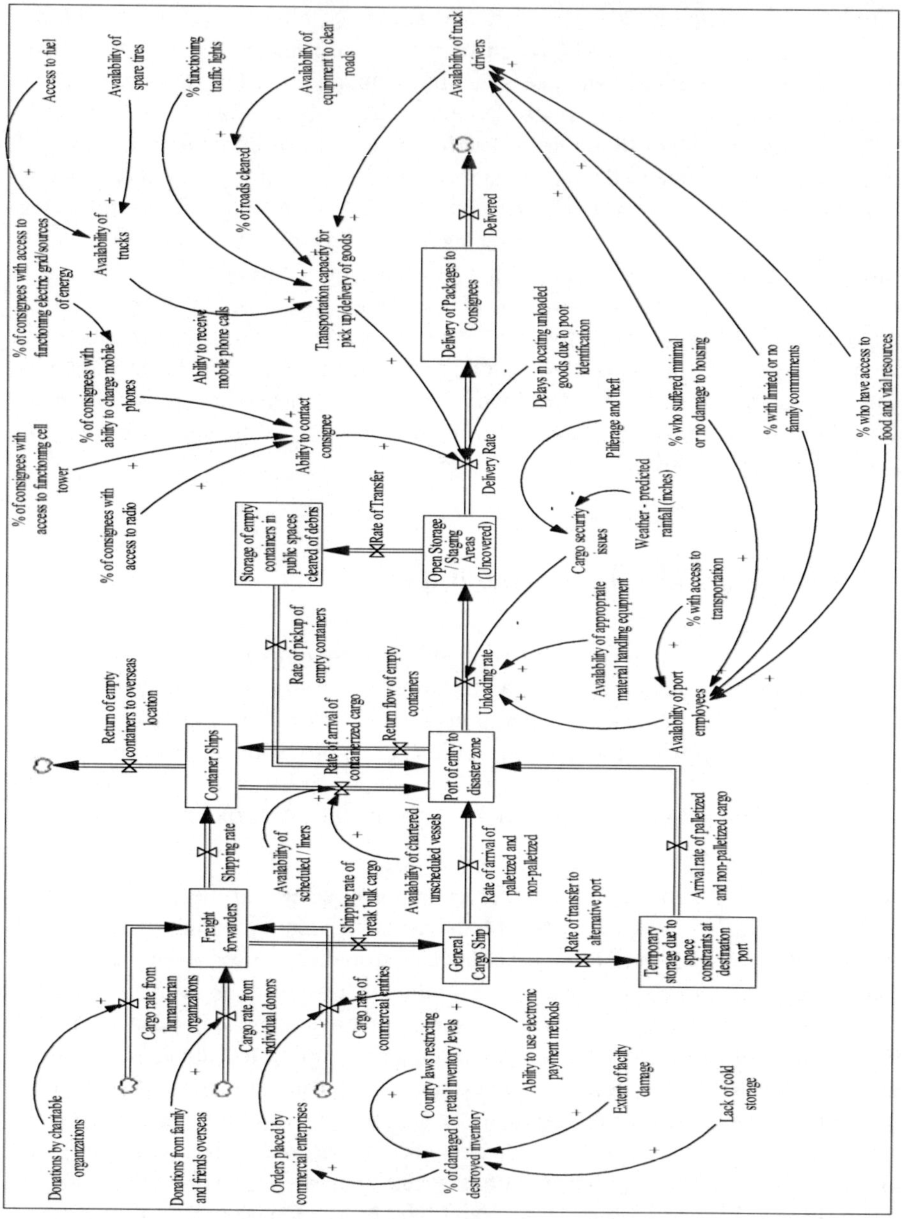

Fig. 2 A generic model of factors impacting port logistics operations and main points of failure in the inbound supply chain in the immediate aftermath of Hurricane Maria based on challenges faced in Dominica

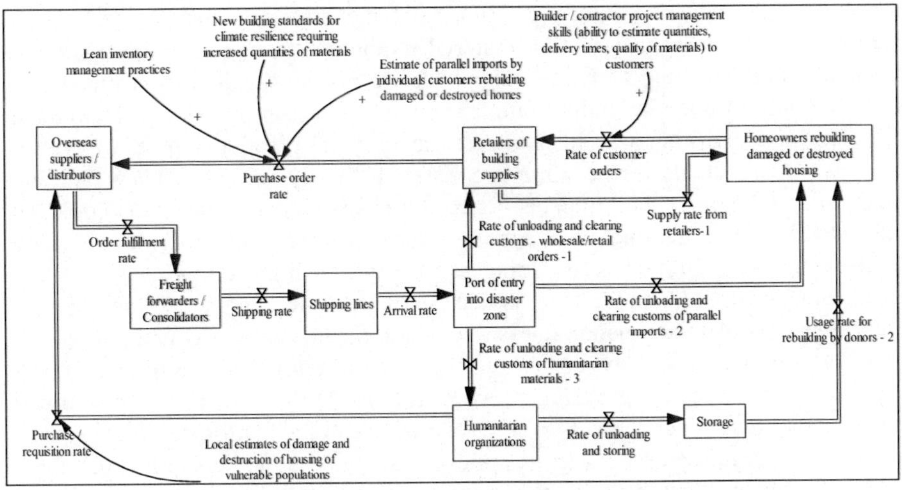

Fig. 3 A generic model identifying main points of failure in the reconstruction materials supply chain following Hurricane Maria based on challenges experienced in Dominica

after the disaster from various sources, such as ship manifests, shipping agents' paperwork, port and customs offices data, surveys of individuals, and social media information. Both structured and unstructured data can be very useful in pre-planning of disaster logistics for better performance. Understanding how the freight data following a disaster compares to pre-disaster situations is fundamental to planning the operational needs to facilitate an agile logistics system.

Tracking and Tracing Tracking and tracing of relief items in post-hurricane disaster supply chains is essential to accelerate delivery to consignees and reduce congestion at entry points. Quick Response (QR) codes have the advantage of being used offline in disaster situations to expedite logistics operations. A pre-designed form provided on the Internet that is accessible to donors following a would keep the data in a consistent format. By scanning documents in pdf format, information can be rapidly retrieved and a copy of the code saved to circumvent formal processes (e.g., customs clearance) without losing pertinent information. QR codes are easy to use with a smartphone and can provide a standard format for providing information such as the complete name, address, and contact telephone numbers of the sender and the consignees. QR codes have also been suggested for use in tracking information on high-value assets commissioned for use in disaster zones.

Centralized Information Systems for Use in Synchronizing Supply and Demand One of the problems in the reconstruction phase is ensuring an uninterrupted supply of building materials to repair destroyed buildings and infrastructure. These problems appear to stem from the inability of businesses to forecast and schedule material quantities to match customer demand. In Dominica, one of the authors witnessed a shortage of 8-inch expanding bolts at every retail store 8 months after Hurricane Maria as reconstruction material requirements surged. To mitigate the stockouts, builders tried to help homeowners to obtain supplies from customers who had unused materials after completing the renovation process. In Dominica, difficulty in forecasting was likely exacerbated by (i) parallel imports of private citizens who travelled overseas to secure materials themselves and (ii) new building guidelines to improve building resilience. One way that this situation could be alleviated is by maintaining a real-time centralized database that provides visibility of inventory stocks of the basic building supplies used in high volumes, e.g., lumber, steel, cement, roofing materials, bolts, hardware, windows, doors, and locks across all retail stores. A recommendation is to develop a database that is updated in real time based on insurance payouts, rebuilding material estimates, and incoming port supplies. Allowances for government aid could also be factored in at a predefined rate. Another recommendation is to create a crowdsourcing platform to allow private citizens to post unused purchased supplies for sale to the public so that customers requiring the inventory have an alternative opportunity to access the materials in timely and less frustrating ways.

6 Conclusion

This paper addressed some of the logistics-related challenges of supply chains following a large-scale hurricane disaster in the Caribbean, with specific focus on port logistics and the re-construction supply chain. The paper took a systemic view of post-disaster logistics in the aftermath of Hurricane Maria and proposed opportunities for utilizing information and simulation technologies to better manage the flow of inventory. There are, however, several other areas in which future research can be conducted. This paper presents the beginning of further study on the development of systemic models to improve efficiency, reduce risk, improve synchronization, and apply systems thinking to manage logistics operations in disaster situations. Future research opportunities include the application of new technologies to integrate the supply chain to improve visibility and agility.

References

AON. 2018. *Hurricane Maria Recap Report*, AON Benfield. Retrieved from: http://thoughtleadership.aonbenfield.com/Documents/20180328-ab-if-hurricane-maria-recap.pdf.

Balcik, B., B.M. Beamon, C.C. Krejci, K.M. Muramatsu, and M. Ramirez. 2010. Coordination in humanitarian relief chains: Practices, challenges and opportunities. *International Journal of Production Economics* 126: 22–34.

Bardouille, B. 2017. *DASPA Calls for Telephone Numbers on Relief Barrels*. Retrieved from: https://dominicanewsonline.com/news/homepage/news/general/daspa-calls-for-telephone-numbers-on-relief-barrels/.

Chopra, S., and M. Sodhi. 2014. Reducing the risks of supply chain disruptions. *MIT Sloan Management Review* 55 (3): 73–80.
Christopher, M. 2016. *Logistics and Supply Chain Management*. 5th ed. Prentice Hall.
Cnc3.news. 2018. *Confusion in Dominica Over Relief Supplies*. Retrieved from:https://www.cnc3.co.tt/news/confusion-dominica-over-relief-supplies.
Council of Supply Chain Management. 2018. Retrieved from: https://cscmp.org.
Coyle, J. J., C.J. Langley Jr., R.A. Novack, and B.J. Gibson. 2017. Supply Chain Management. *Cengage Learning*.
Dominica News Online. 2017. Normalcy Returning to Woodbridge Bay Port Says Bardouille. *DNO*. Retrieved from: https://dominicanewsonline.com/news/homepage/news/general/normalcy-returning-to-woodbridge-bay-port-says-bardouille/.
Dominica Vibes News. 2017. DASPA CEO needs empty containers removed from the port. *DV*.
Fayezi, S., and M. Zomorrodi. 2016. Supply Chain Management: Developments, Theories and Models. In *Handbook of Research on Global Supply Chain Management*, 313–340. IGI Global.
Fisher, M.L. 1997. *What is the Right Supply Chain for Your Product*. Harvard Business Review.
Goentzel, J. (Moderator). 2017. *Supply Chain Resilience: Restoring Business Operations After a Hurricane*. Summary Report, MIT Center for Transportation & Logistics.
Gomez, S.A. 2017. *Dominica Working to Restore Power After Maria Granma*. Retrieved from http://en.granma.cu/mundo/2017-11-07/dominica-working-to-restore-power-after-maria.
Hohenstein, N.-O., E. Feisel, E. Hartmann, and L. Giunipero. 2015. Research on the Phenomenon of Supply Chain Resilience: A Systematic Review and Paths for Further Investigation. *International Journal of Physical Distribution and Logistics Management* 45 (12): 90–117.
Holguín-Veras, J., M. Jaller, L.N. Van Wassenhove, N. Perez, and T. Wachtendorf. 2012. On the Unique Features of Post-Disaster Humanitarian Logistics. *Journal of Operations Management* 30: 494–506.
Koskinen, K.U. 2013. *Systemic View and Systems Thinking, Knowledge Production in Organizations*, 13–30. Springer.
Kunz, N., G. Reiner, and S. Gold. 2014. Investing in Disaster Management Capabilities Versus Pre-Positioning Inventory: A New Approach to Disaster Preparedness. *International Journal of Production Economics* 157: 261–272.
National Hurricane Center. 2017. www.nhc.noaa.gov/data/tcr/index.php?season=2017&basin=atl.
Neuman, S. 2017. *In Devastated Dominica, "Hams" Become Vital Communications Link*. NPR. Retrieve from https://www.npr.org/sections/thetwo-way/2017/09/21/552649149/in-devastated-dominica-hams-become-vital-communications-link.
Ozguven, E.E., and K. Ozbay. 2013. A Secure and Efficient Inventory Management System for Disasters. *Transportation Research Part C* 29: 171–196.
Palin, P.J., L.S. Hanson, D. Barton, and A. Frohwein. 2018. *Supply Chain Resilience and the 2017 Hurricane Season*. CNA Analysis & Solutions in collaboration with the National Academy of Sciences, Engineering, and Medicine.
Papadopoulos, T., A. Gunasekaran, R. Dubey, N. Altay, S. Childe, and S. Fosso-Wamba. 2016. The role of Big Data in explaining disaster resilience in supply chains for sustainability. *Journal of Cleaner Production* 142: 1108–1118.
Ponomarov, Y.S., and M.C. Holcomb. 2009. Understanding the concept of supply chain resilience. *The International Journal of Logistics Management* 20 (1): 124–143.
Rey, F. 2001. The complex nature of actors in humanitarian action and the challenge of coordination. In *Reflections on Humanitarian Action: Principles, Ethics and Contradictions*, ed. Humanitarian Studies Unit. London: TNI/Pluto Press with humanitarian Studies Unit and ECHO (European Commission Humanitarian Office).
Scholten, K., P.S. Scott, and B. Fynes. 2013. Mitigation Processes –Antecedents for Building Supply Chain Resilience, Supply Chain Management. *An International Journal* 19 (2): 211–228.

Simchi-Levi, D., P. Kaminsky, and E. Simchi-Levi. 2008. *Designing and Managing the Supply Chain, Concepts, Strategies, and Case Studies*. 3rd ed. Irwin: McGraw Hill.
The Chronicle. 2018. "Containers Decreased".
Unstructured Data from Social Media – Facebook posts (2017)
Van Wassenhove, L.N. 2006. Humanitarian Aid Logistics: Supply Chain Management in High Gear. *Journal of Operational Research Society* 57 (5): 475–489.

Index

A. M. Madni et al. (eds.), *Recent Trends and Advances in Model Based Systems Engineering*, https://doi.org/10.1007/978-3-030-82083-1

D

E

F

G

I

J

K

L

M

N

O

P